T0190323

Lecture Notes in Artificial Intelligence 12323

Subseries of Lecture Notes in Computer Science

More information about this series at http://www.springer.com/series/1244

Annalisa Appice · Grigorios Tsoumakas ·
Yannis Manolopoulos · Stan Matwin (Eds.)

Discovery Science

23rd International Conference, DS 2020
Thessaloniki, Greece, October 19–21, 2020
Proceedings

Editors
Annalisa Appice
University of Bari Aldo Moro
Bari, Italy

Grigorios Tsoumakas
Aristotle University of Thessaloniki
Thessaloniki, Greece

Yannis Manolopoulos
Open University of Cyprus
Nicosia, Cyprus

Stan Matwin
Dalhousie University
Halifax, NS, Canada

ISSN 0302-9743 ISSN 1611-3349 (electronic)
Lecture Notes in Artificial Intelligence
ISBN 978-3-030-61526-0 ISBN 978-3-030-61527-7 (eBook)
https://doi.org/10.1007/978-3-030-61527-7

LNCS Sublibrary: SL7 – Artificial Intelligence

This Springer imprint is published by the registered company Springer Nature Switzerland AG
The registered company address is: Gewerbestrasse 11, 6330 Cham, Switzerland

Preface

This volume contains the papers selected for presentation at the 23rd International Conference on Discovery Science (DS 2020), which was organized to be held in Thessaloniki, Greece, during October 19–21, 2020. Due to the outbreak of the COVID-19 pandemic, the conference was moved online and held virtually over the same time period. The conference was organized by the Aristotle University of Thessaloniki, Greece, in cooperation with the Open University of Cyprus, Cyprus, Dalhousie University, Canada, and University of Bari Aldo Moro, Italy.

DS is a conference series that started in 1986. Held every year, DS continues its tradition as the unique venue for the latest advances in the development and analysis of methods for discovering scientific knowledge, coming from machine learning, data mining, and intelligent data analysis, with their application in various scientific domains. In particular, major areas selected for DS 2020 include: artificial intelligence applied to science; machine learning; knowledge discovery and data mining; causal modeling; AutoML, meta-learning, and planning to learn; machine learning and high-performance computing; grid and cloud computing; literature-based discovery; ontologies for science, including the representation and annotation of datasets and domain knowledge; explainable AI, interpretability of machine learning, and deep learning models; process discovery and analysis; computational creativity; anomaly detection and outlier detection; data streams, evolving data, change detection, concept drift, and model maintenance; network analysis; time-series analysis; learning from complex data; data and knowledge visualization; human-machine interaction for knowledge discovery and management; evaluation of models and predictions in discovery setting; machine learning and cybersecurity; as well as applications of the above techniques in scientific domains.

DS 2020 received 76 international submissions that were carefully reviewed by three or more Program Committee (PC) members or external reviewers. After a rigorous reviewing process, 26 regular papers and 19 short papers were accepted for presentation at the conference and publication in the DS 2020 volume. Short papers were just allotted a smaller presentation time compared to regular ones.

The conference program included three invited keynotes. Prof. Myra Spiliopoulou from Otto von Guericke University Magdeburg, Germany contributed a talk titled "Knowledge Discovery in mHealth – dealing with few noisy data." Prof. Peter A. Flach from University of Bristol, UK, contributed a talk titled "The highs and lows of performance evaluation: Towards a measurement theory for machine learning." Prof. Gustau Camps-Valls from Universitat de València, Spain, gave a presentation titled "Machine learning for Modelling and Understanding in Earth Sciences." Abstracts of the invited talks with short biographies of the invited speakers are included in this volume.

We would like to sincerely thank all people who helped this volume come into being and made DS 2020 a successful and exciting event. In particular, we would like to express our appreciation for the work of the DS 2020 PC members and external reviewers who helped assure the high standard of accepted papers. We would like to thank all authors of DS 2020, without whose high-quality contributions it would not have been possible to organize the conference.

We are grateful to the Steering Committee chair, Sašo Džeroski, and the whole Steering Committee for their extraordinary support in critical decisions concerning the event plan. We wish to express our thanks to local organization chairs, Anastasios Gounaris and Apostolos Papadopoulos, and the whole organization team for their support and incredible work. We would also thank the treasurer, Richard Chbeir, for his professional work. We would like to express our deepest gratitude to all those who served as organizers, session chairs, and hosts, who made great efforts to meet the online challenge to make the virtual conference a real success. Finally, our thanks are due to Alfred Hofmann and Anna Kramer of Springer for their continuous support and work on the proceedings. We are grateful to Springer for a special issue on Discovery Science to be published in the *Machine Learning* journal. All authors were given the possibility to extend and rework versions of their papers presented at DS 2020 for a chance to be published in this prestigious journal. For DS 2020, Springer also supported a Best Paper Award to Riku Laine, Antti Hyttinen, and Michael Mathioudakis for their paper "Evaluating Decision Makers over Selectively Labelled Data: A Causal Modeling Approach." We would also like to honorary mention the runner-up paper "Explaining Sentiment Classification with Synthetic Exemplars and Counter-Exemplars" by Orestis Lampridis, Riccardo Guidotti, and Salvatore Ruggieri.

September 2020

Annalisa Appice
Grigorios Tsoumakas
Yannis Manolopoulos
Stan Matwin

Organization

General Chairs

Yannis Manolopoulos	Open University of Cyprus, Cyprus
Stan Matwin	Dalhousie University, Canada

Program Committee Chairs

Annalisa Appice	University of Bari Aldo Moro, Italy
Grigorios Tsoumakas	Aristotle University of Thessaloniki, Greece

Program Committee

Martin Atzmüller	Tilburg University, The Netherlands
Viktor Bengs	Paderborn University, Germany
Concha Bielza Lozoya	Universidad Politécnica de Madrid, Spain
Konstantinos Blekas	University of Ioannina, Greece
Alberto Cano	Virginia Commonwealth University, USA
Michelangelo Ceci	University of Bari Aldo Moro, Italy
Paolo Ceravolo	Politecnico di Milano, Italy
Bruno Cremilleux	University of Caen Normandy, France
Claudia d'Amato	University of Bari Aldo Moro, Italy
Nicola Di Mauro	University of Bari Aldo Moro, Italy
Ivica Dimitrovski	Ss. Cyril and Methodius University, North Macedonia
Wouter Duivesteijn	Eindhoven University of Technology, The Netherlands
Sašo Džeroski	Jožef Stefan Institute, Slovenia
Hadi Fanaee-T	University of Oslo, Norway
Nicola Fanizzi	University of Bari Aldo Moro, Italy
Stefano Ferilli	University of Bari Aldo Moro, Italy
Johannes Fürnkranz	Johannes Kepler University Linz, Austria
Mohamed Gaber	Birmingham City University, UK
Dragan Gamberger	Rudjer Bošković Institute, Croatia
Dimitris Gunopoulos	University of Athens, Greece
Makoto Haraguchi	Hokkaido University, Japan
Kouichi Hirata	Kyushu Institute of Technology, Japan
Jaakko Hollmén	Aalto University, Finland
Eyke Hüllermeier	Paderborn University, Germany
Dino Ienco	Irstea, France
Alípio Jorge	University of Porto, Portugal
Ioannis Katakis	University of Nicosia, Cyprus
Masahiro Kimura	Ryukoku University, Japan
Dragi Kocev	Jožef Stefan Institute, Slovenia

Petra Kralj Novak	Jožef Stefan Institute, Slovenia
Stefan Kramer	Johannes Gutenberg University Mainz, Germany
Vincenzo Lagani	Ilia State University, USA
Pedro Larranaga	University of Madrid, Spain
Nada Lavrač	Jožef Stefan Institute, Slovenia
Jurica Levatić	Institute for Research in Biomedicine, Spain
Tomislav Lipic	Rudjer Bošković Institute, Croatia
Francesca A. Lisi	University of Bari Aldo Moro, Italy
Gjorgji Madjarov	Ss. Cyril and Methodius University, North Macedonia
Giuseppe Manco	Institute for High Performance Computing and Networking, Italy
Elio Masciari	Institute for High Performance Computing and Networking, Italy
Anna Monreale	University of Pisa, Italy
Rita P. Ribeiro	University of Porto, Portugal
Panče Panov	Jožef Stefan Institute, Slovenia
George Papakostas	International Hellenic University, Greece
Ruggero G. Pensa	University of Torino, Italy
Bernhard Pfahringer	University of Waikato, New Zealand
Gianvito Pio	University of Bari Aldo Moro, Italy
Pascal Poncelet	LIRMM Montpellier, France
Chedy Raïssi	French Research Institute for Digital Sciences, France
Jan Ramon	French Research Institute for Digital Sciences, France
Kazumi Saito	University of Shizuoka, Japan
Tomislav Šmuc	Rudjer Bošković Institute, Croatia
Jerzy Stefanowski	Poznan University of Technology, Poland
Ljupčo Todorovski	University of Ljubljana, Slovenia
Luis Torgo	Dalhousie University, Canada
Ioannis Tsamardinos	University of Crete, Greece
Herna Viktor	University of Ottawa, Canada
Michalis Vlachos	Université de Lausanne, Switzerland
Albrecht Zimmermann	University of Caen Normandy, France
Blaž Zupan	University of Ljubljana, Slovenia

Additional Reviewers

Besher Alhalabi
Dridi Amna
Giuseppina Andresini
Emanuele Pio Barracchia
Martin Breskvar
Lorraine Chambers
Graziella De Martino
Massimo Guarascio
Julian Hatwell
Theofanis Kalampokas
Ana Kostovska
Ilona Kulikosvkikh
Masahito Kumano
Vladimir Kuzmanovski
Athanasios Lagopoulos
Paolo Mignone
John Mollas
Igor Mozetič
Vu-Linh Nguyen
Luca Oneto
Vincenzo Pasquadibisceglie
Francesco Scicchitano
Tomaž Stepišnik
Bozhidar Stevanoski
Alexander Tornede
Aleš Žagar

Abstracts of Keynote Talks

Knowledge Discovery in mHealth – Dealing with Few Noisy Data

Myra Spiliopoulou

Research Group on Knowledge Management and Discovery (KMD), Faculty of Computer Science, Otto von Guericke University Magdeburg, PO Box 4120, 39016 Magdeburg, Germany
myra@iti.cs.uni-magdeburg.de

Abstract. Patients with chronic diseases can greatly benefit from mHealth technology. There are solutions assisting them in measuring signals (e.g., blood pressure, sugar level, etc.), in keeping a diary with Ecological Momentary Assessments (EMA), such as physical exercise, onset of symptoms, and subjective perception of health condition. Machine learning can deliver useful insights from data thus collected. While sensor signals can be collected without interruption, EMA recording depends on patients' self-discipline and compliance.

The talk starts with an overview of the role of mHealth applications in diagnostics and treatment support. Then, we focus on EMA for chronic conditions. We discuss challenges of learning from few and noisy recordings, and methods for prediction and risk factor identification on these data.

Keywords: mHealth · Multidimensional sequences · Gaps · Time series prediction · Adherence

The Highs and Lows of Performance Evaluation: Towards a Measurement Theory for Machine Learning

Peter A. Flach

Intelligent Systems Laboratory, Department of Computer Science, University of Bristol, Merchant Venturers Building, Woodland Road, Bristol BS8 1UB, UK
Peter.Flach@bristol.ac.uk

Abstract. Our understanding of performance evaluation measures for machine-learned classifiers has improved considerably over the last decades. However, there is a range of areas where this understanding is still lacking, leading to ill-advised practices in classifier evaluation. This is clearly problematic, since if machine learning researchers are unclear about what exactly their experiments are telling them about their machine learning algorithms, then how can end-users trust systems deploying those algorithms?

I suggest that in order to make further progress we need to develop a proper measurement theory of machine learning. Measurement theory studies the concepts of measurement and scale. If you have a way to measure, say, the length of individual rods or planks, this should also allow you to then calculate the combined length of concatenated rods or planks. What relevant concatenation operations are there in data science and AI, and what does that mean for the underlying measurement scale?

I discuss by example what such a measurement theory might look like and what kinds of new results it would entail. I furthermore argue that key properties such as classification ability and data set difficulty are unlikely to be directly observable, suggesting the need for latent-variable models. Ultimately, machine learning experiments need to go beyond simple correlations and aim to make causal inferences of the form 'Algorithm A outperformed algorithm B because two classes were highly imbalanced,' or counterfactually, 'if the classes were rebalanced, the observed performance difference between A and B would disappear.'

Keywords: Machine learning experiments · Classification performance · Psychometrics · latent variables · Levels of measurement · Causal inference

Machine Learning for Modelling and Understanding in Earth Sciences

Gustau Camps-Valls

Image Processing Lab, Universitat de València, Spain
https://isp.uv.es
@isp_uv_es

Abstract. The Earth is a complex dynamic network system. Modelling and understanding the system is at the core of scientific endeavour. We approach these problems with machine learning (ML) algorithms. I will review several ML approaches we have developed in the last years: 1) advanced Gaussian processes models for bio-geo-physical parameter estimation, which can incorporate physical laws, blend multisensor data while providing credible confidence intervals for the estimates, and improved interpretability, 2) nonlinear dimensionality reduction methods to decompose Earth data cubes in spatially-explicit and temporally-resolved modes of variability that summarize the information content of the data and allow for identifying relations with physical processes, and 3) advances in causal inference that can uncover cause and effect relations from purely observational data.

Contents

Classification

Clustering

Data and Knowledge Representation

Data Streams

Dimensionality Reduction and Feature Selection

Distributed Processing

Multi-target Models

Neural Networks and Deep Learning

Spatial, Temporal and Spatiotemporal Data

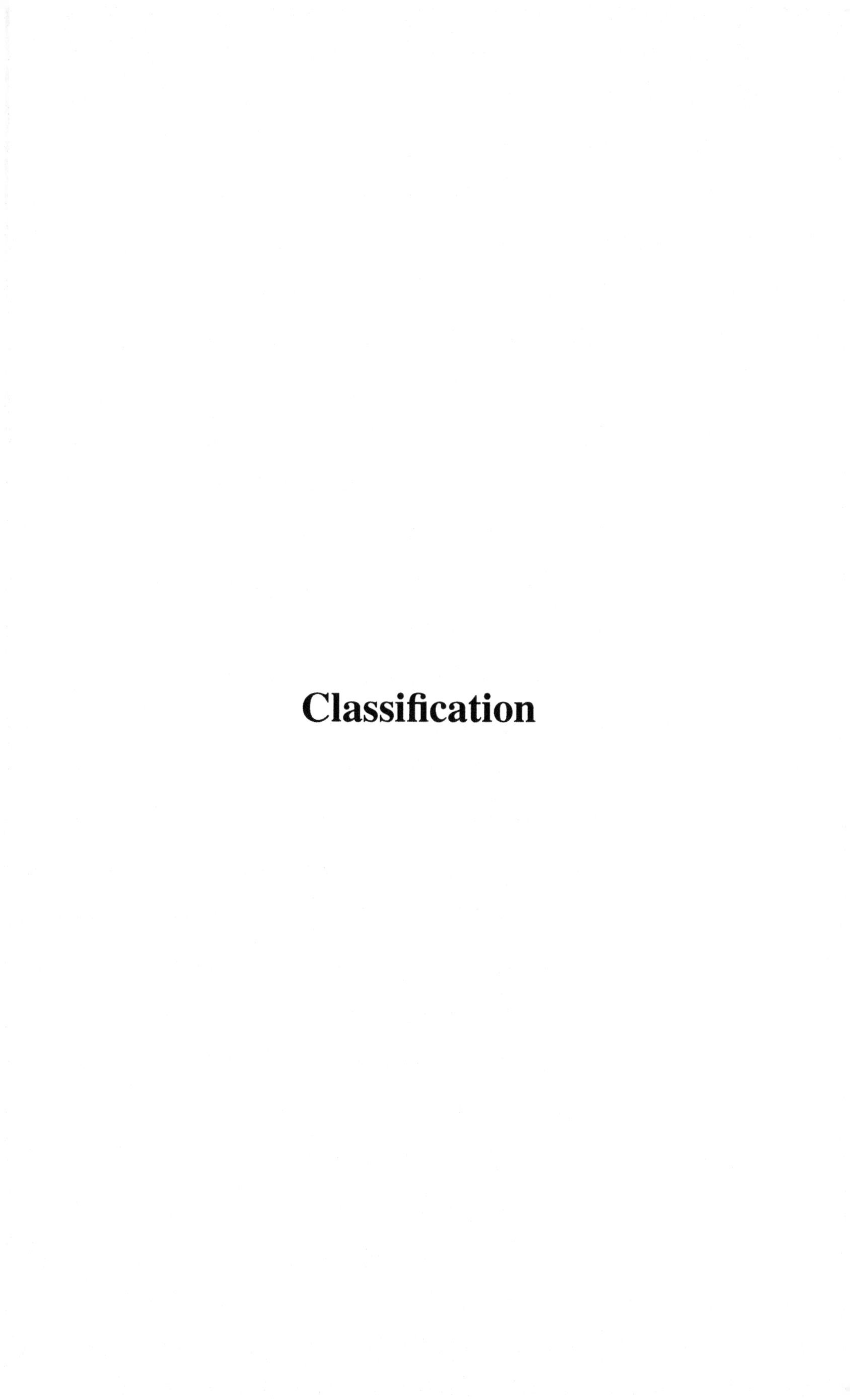

Classification

Evaluating Decision Makers over Selectively Labelled Data: A Causal Modelling Approach

Riku Laine[1](✉), Antti Hyttinen[2], and Michael Mathioudakis[2]

[1] University of Helsinki, Helsinki, Finland
riku.laine@helsinki.fi
[2] HIIT, Department of Computer Science, University of Helsinki, Helsinki, Finland
{antti.hyttinen,michael.mathioudakis}@helsinki.fi

Abstract. We present a Bayesian approach to evaluate AI decision systems using data from past decisions. Our approach addresses two challenges that are typically encountered in such settings and prevent a direct evaluation. First, the data may not have included all factors that affected past decisions. And second, past decisions may have led to unobserved outcomes. This is the case, for example, when a bank decides whether a customer should be granted a loan, and the outcome of interest is whether the customer will repay the loan. In this case, the data includes the outcome (if loan was repaid or not) only for customers who were granted the loan, but not for those who were not. To address these challenges, we formalize the decision making process with a causal model, considering also unobserved features. Based on this model, we compute counterfactuals to impute missing outcomes, which in turn allows us to produce accurate evaluations. As we demonstrate over real and synthetic data, our approach estimates the quality of decisions more accurately and robustly compared to previous methods.

Keywords: Selective labels · Selection bias · Causal modelling · Bayesian inference · Model evaluation

1 Introduction

Today, more and more decisions are made by algorithms based on statistical models learned from data [7,10]. This is quite prevalent on the Web, where algorithms decide search engine results or product recommendations in online stores. But automated decision systems are used also in other situations – for credit scoring, insurance pricing, and also judicial decisions (e.g., COMPAS [2] and RisCanvi [20] are algorithmic tools used to evaluate the risk for recidivism in the US and Catalan prison systems, respectively).

In this work, we present a Bayesian approach to evaluate decision algorithms—a task that often must be performed *before* algorithms are actually deployed. In practice, this is done by simulating the deployment of the algorithm

A. Appice et al. (Eds.): DS 2020, LNAI 12323, pp. 3–18, 2020.
https://doi.org/10.1007/978-3-030-61527-7_1

Judge knew defendant's...				Judge decided ...	Did the defendant violate bail?
crime	age	gender	court behavior		
bad gesture	25	female	calm	**bail**	**yes**
bad words	63	male	bad	**bail**	**no**
bad gesture	44	male	good	**jail**	**no**
bad words	90	female	anxious	**jail**	**no**

Machine knows defendant's...				Machine would have decided ...	Would the defendant have violated bail?
crime	age	gender	court behavior		
bad gesture	25	female	N/A	**bail**	**yes**
bad words	63	male	N/A	**jail**	**no**
bad gesture	44	male	N/A	**jail**	**no**
bad words	90	female	N/A	**bail**	**?**

Fig. 1. Bail-or-jail: a machine decides for the *same defendants* considered by a judge. When the machine decides to allow bail but the judge had denied it, we cannot evaluate directly the machine's decision.

over a log of past cases and measuring how well it would have performed, if it had been used to replace the human decision makers or other decision system currently in place. Herein lies a challenge: previously-made decisions affect the data on which the evaluation is performed, in a way that prevents straightforward evaluation. Let us explain this with an example, also illustrated in Fig. 1.

Example. In some judicial systems, a person who is arrested may stay out of jail if the judge allows them to post bail (i.e., deposit money as a promise to attend the trial and honour other conditions). The decision (bail or jail) is successful if bail is not allowed for defendants who violate its conditions. Now consider a machine-based system with the potential to replace the judge as decision maker. Before the machine is actually deployed, we wish to evaluate its decisions for past cases decided by judges. However, we are only able to directly evaluate the machine's decisions for cases where bail was allowed. Why? Because if bail was not allowed, we do not know whether the defendant would have violated it. In such cases, one approach would be to infer the defendant's behaviour in the hypothetical case they had been allowed bail. Here lie some challenges. First, the inference should take into account the bias in the observed data. And second, the judge might have made their decision based on more information than is available to the machine – e.g., if the judge witnessed aggressive behaviour by the defendant during the ruling.

General Cases. The above exemplifies a general class of cases: a machine is asked to make a binary *decision* (e.g., whether to allow bail, grant a loan, etc.) for a specific case, based on a set of recorded features; the decision leads to an *outcome* that is successful or unsuccessful; and some decisions prevent us from directly evaluating alternative decisions. Our task, then, is to evaluate the quality of machine decisions against a log of past cases. For accurate evaluation, we should account for the bias in the observations and the possibility that non-recorded information influenced past decisions.

Recent Related Work. There is a rich literature on problems that arise in similar settings (see related work in Sect. 5). Recently, Lakkaraju et al. [12] referred to the evaluation in such settings as the '*selective labels problem*' (also considered by [5,10]). They presented *contraction*, a method to evaluate decision makers in a setting where subjects are randomly assigned to decision makers with varying leniency levels. Contraction takes advantage of the assumed random assignment and variance in leniency: essentially it measures the performance of the evaluated system using the cases of the most lenient judge – and so it works well when lenient decision makers decided a large number of cases.

Our Contributions. We build upon the setting of [12] and present a novel approach to evaluate decision makers over selectively labelled data, using causal modelling to represent assumptions about the process that generated the data and counterfactual reasoning to impute unobserved outcomes. We experiment with synthetic data to highlight various properties of our approach and study a case of real recidivism data [2]. Our results indicate that our method achieves more accurate results with considerably less variation than the state-of-the-art, and unlike the contraction approach that is tailored to this setting [12], it does not depend only on the most lenient decision makers in the data.

2 Setting and Problem Statement

We consider data recorded from a decision making process with the following characteristics [12]. Each case is decided by a decision maker and let J index the decision maker the case is assigned to. For each case, described by a set of features F, the assigned decision maker H_j (where j is a particular value for J) makes a binary decision $\mathsf{T} \in \{0, 1\}$, nominally referred to as *positive* ($\mathsf{T} = 1$) or *negative* ($\mathsf{T} = 0$). In our bail-or-jail example of Sect. 1, H_j corresponds to the judge deciding whether to grant bail or not (positive or negative decision, respectively). The decision is followed with a binary outcome Y, which is nominally referred to as *successful* ($\mathsf{Y} = 1$) or *unsuccessful* ($\mathsf{Y} = 0$). An outcome can be *unsuccessful* only if the decision that preceded it was positive. If the decision was negative, then the outcome is considered by default successful. Back in our example, the decision of the judge is unsuccessful only if the judge grants bail and the defendant violates its terms. Otherwise, if the decision of the judge was to keep the defendant in jail, the outcome is by default successful since there can be no bail violation.

For each case, a record (j, x, t, y) is produced that contains only a subset $\mathsf{X} \subseteq \mathsf{F}$ of the features of the case, the decision T of the judge and the outcome Y – but leaves no trace for a subset $\mathsf{Z} = \mathsf{F} \setminus \mathsf{X}$ of the features. In our example, X corresponds to publicly recorded information about the bail-or-jail case (e.g., the charged crime) and Z corresponds to features that are observed by the judge but do not appear on record (e.g., exact verbal response of the defendant in court). The set of records $\mathsf{D} = \{(j, x, t, y)\}$ comprises what we refer to as the **dataset**. Figure 2 shows the causal diagram of this decision making process.

We wish to evaluate a decision maker M that considers a case from the dataset – and makes its own binary decision T based on the recorded features X. In our example, M corresponds to a machine-based system that is considered for bail-or-jail decisions. For M, the definition and semantics of decision T and outcome Y are the same as for decision makers $\mathbf{H}$, described above.

The quality of a decision maker M is measured in terms of its **failure rate** FR – i.e., the fraction of unsuccessful outcomes out of all the cases for which a decision is made. A good decision maker achieves as low failure rate FR as possible. Note, however, that a decision maker that always makes a negative decision $\mathsf{T} = 0$, has failure rate $\mathsf{FR} = 0$, by definition. Thus, for the evaluation to be meaningful, we evaluate decision makers at given leniency levels R, defined as the fraction of cases with positive decisions.

Problem 1 (Evaluation). Given a dataset $\{(j, x, t, y)\}$, and a decision maker M, provide an estimate of the failure rate FR at a given leniency level $R = r$.

3 Counterfactual-Based Imputation for Selective Labels

Problem 1 is challenging because the dataset does not directly provide a way to evaluate FR. If decision maker M makes a positive decision for a case for which the dataset has negative decision by a decision maker H_j, how can we infer the outcome Y in the hypothetical case where M's decision had been followed? Such questions fall straight into the realm of causal analysis and particularly the evaluation of counterfactuals [4] – an approach that we follow in this paper.

A first thought is to simply *predict* the outcomes based on the features of the case. In the bail-or-jail example, we could investigate whether certain features of the defendant (e.g., their age and marital status) are good predictors of whether they comply to the bail conditions – and use them if they do. However, not all features that are available to H_j are available to M in the setting we consider, which forms our second major challenge. These complications mean that making direct predictions based on the available features can be suboptimal and even biased. However, important information regarding the unobserved features Z can often be recovered via careful consideration of the decisions in the data, which our counterfactual approach achieves [8,16].

For illustration, let us consider a defendant who received a negative decision by a human judge H_j. Suppose also that, among defendants with similar recorded features X who were released, none violated the bail conditions – and therefore, judging from observations alone, the defendant should be considered safe to release based on X. However, if the judge was both lenient and precise – i.e., was able to make those positive decisions that lead to successful outcome – then it is very possible that the negative decision is attributed to unfavourable non-recorded features Z. And therefore, if a positive decision were made, *the above reasoning suggests that a negative outcome is more likely than what would have been predicted based alone on the recorded features* X *of released defendants.*

Our approach for evaluating M on cases where negative decision by H_j is recorded in the data, unfolds over three steps: first, we learn a causal model over the dataset; then, we compute counterfactuals to predict unobserved outcomes; and finally, we use these predictions to evaluate a set of decisions by M.

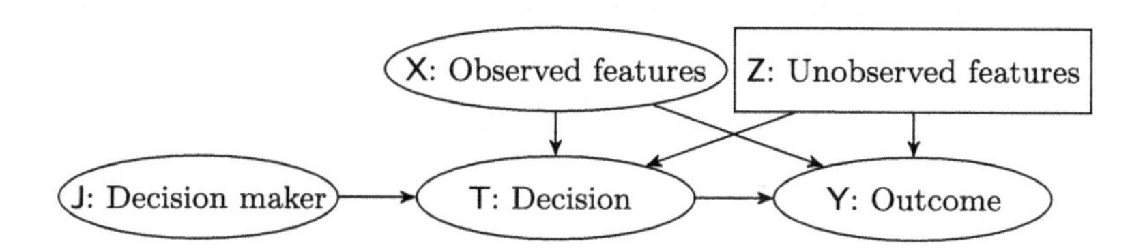

Fig. 2. Causal diagram for the selective labels setting

3.1 The Causal Model

Recall from Sect. 2 that Fig. 2 provides the general structure of causal relationships for quantities of interest. We use the following causal model over this structure, building on what is used by Lakkaraju et al. [12] and others [8,16]. First, we model the unobserved features Z as a (continuous) one-dimensional risk factor. This is motivated by the standard use of propensity scores [1,18] to account for confounding, as well as the COMPAS risk score in the context of recidivism [2]. Motivated by the central limit theorem, we use a Gaussian distribution for it, and since Z is unobserved we can assume without loss of generality that $\mathsf{Z} \sim N(0,1)$. For presentation, we use here a single observed feature X – it is straightforward to extend the model to multiple features X.

In our setting, a negative decision $\mathsf{T} = 0$ leads to a successful outcome $\mathsf{Y} = 1$. When $\mathsf{T} = 1$, the outcome is modelled with a logistic regression model over the features X and Z (σ is the standard logistic function):

$$\mathbf{P}(\mathsf{Y} = 1 \mid \mathsf{T}, x, z) = \begin{cases} 1, & \text{if } \mathsf{T} = 0 \\ \sigma(\alpha_{\mathsf{Y}} + \beta_{\mathsf{X}} x + \beta_{\mathsf{Z}} z), & \text{o/w} \end{cases} \tag{1}$$

We model the decisions in the data similarly with a logistic regression:

$$\mathbf{P}(\mathsf{T} = 1 \mid j, x, z) = \sigma(\alpha_j + \gamma_{\mathsf{X}} x + \gamma_{\mathsf{Z}} z) \tag{2}$$

Although we model the decision makers here probabilistically, we do not imply that their decisions are necessarily probabilistic. The probabilistic model arises from the unknown specific details of reasoning employed by each decision maker H_j. Note also that we are making the simplifying assumption that coefficients $\gamma_{\mathsf{X}}, \gamma_{\mathsf{Z}}$ are the same for all H_j, but decision makers are allowed to differ in intercept α_j. Parameter α_j controls the leniency of a decision maker $\mathsf{H}_j \in \mathbf{H}$.

We take a Bayesian approach to learn the model from the dataset. In particular, we consider the full probabilistic model defined in Eqs. 1 and 2 and obtain the posterior distribution of its parameters $\theta = \{\alpha_{\mathsf{Y}}, \beta_{\mathsf{X}}, \beta_{\mathsf{Z}}, \gamma_{\mathsf{X}}, \gamma_{\mathsf{Z}}\} \cup \bigcup_{\mathsf{H}_j \in \mathsf{H}} \{\alpha_j\}$, which includes intercepts α_j for all H_j employed in the data. We use suitable prior distributions to ensure the identifiability of the parameters (Appendix 2).

3.2 Computing Counterfactual Outcomes

We remind that the goal is to provide a solution to Problem 1 – and, to do that, we wish to address those cases where M decides $\mathsf{T} = 1$ while the data has a negative decision $\mathsf{T} = 0$, where evaluation cannot be performed directly. In other words, we wish to answer a 'what-if' question: for each specific case where a decision maker H_j decided $\mathsf{T} = 0$, what if we had intervened to alter the decision to $\mathsf{T} = 1$? In the formalism of causal inference [17], we wish to evaluate the counterfactual expectation

$$\hat{\mathsf{Y}} = \mathbf{E}_{\mathsf{T} \leftarrow 1}[\mathsf{Y} \mid x, j, \mathsf{T} = 0; \mathsf{D}] \tag{3}$$

The expression above concerns a specific entry in the dataset with features $\mathsf{X} = x$, for which decision maker H_j made a decision $\mathsf{T} = 0$. It expresses the probability that the outcome would have been positive ($\mathsf{Y} = 1$) had the decision been positive ($\mathsf{T} = 1$), conditional on what we know from the data entry ($\mathsf{X} = x$, $\mathsf{T} = 0$, $\mathsf{J} = j$) as well as from the entire dataset D. Notice that the presence of D in the conditional part of 3 gives us more information about the data entry compared to the entry-specific quantities and is thus not redundant. In particular, it provides information about the leniency and other parameters of the decision maker H_j, which is important to infer information about the unobserved variables Z.

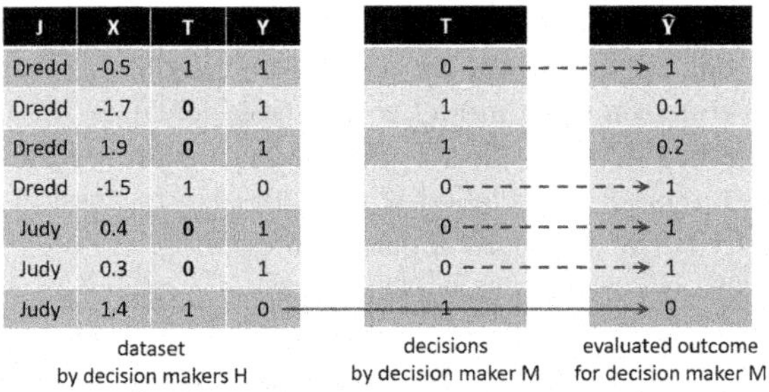

J	X	T	Y	T	$\hat{\mathsf{Y}}$
Dredd	-0.5	1	1	0	1
Dredd	-1.7	0	1	1	0.1
Dredd	1.9	0	1	1	0.2
Dredd	-1.5	1	0	0	1
Judy	0.4	0	1	0	1
Judy	0.3	0	1	0	1
Judy	1.4	1	0	1	0

dataset by decision makers H | decisions by decision maker M | evaluated outcome for decision maker M

Fig. 3. CFBI. Negative decisions ($\mathsf{T} = 0$) by decision maker M are evaluated as successful ($\hat{\mathsf{Y}} = 1$), shown with dashed arrows. Positive decisions ($\mathsf{T} = 1$) by decision maker M for which the decision in the data was also positive ($\mathsf{T} = 1$) are evaluated according to the outcome Y in the data, as marked by the solid arrow. For the other cases (second and third), the evaluated outcomes $\hat{\mathsf{Y}}$ are based on CFBI. The estimated failure rate of the decision maker M is $2.7/7 = 38.6\%$ here.

For the model defined above, the counterfactual $\hat{Y}$ can be computed by the approach of Pearl [17]. For a fully defined model (with fixed parameters) the counterfactual expectation can be determined by the following expression:

$$E_{\mathsf{T} \leftarrow 1}(\mathsf{Y} | j, \mathsf{T} = 0, x) = \int \mathbf{P}(\mathsf{Y} = 1 | \mathsf{T} = 1, x, z) \mathbf{P}(z | j, \mathsf{T} = 0, x) \mathrm{d}z \tag{4}$$

In essence, we determine the distribution of the unobserved features Z using the decision, observed features x, and the leniency of the employed decision maker, and then determine the distribution of Y conditional on all features, integrating over the unobserved features (Appendix 1). Note that the decision maker model in Eq. 2 affects the distribution of the unobserved features $\mathbf{P}(\mathsf{Z}|j, \mathsf{T}=0, x)$.

Having obtained a posterior probability distribution for parameters θ we can estimate the counterfactual outcome value based on the data:

$$\hat{\mathsf{Y}} = \int \mathbf{P}(\mathsf{Y}=1|\mathsf{T}=1, x, z, \theta)\mathbf{P}(z|j, \mathsf{T}=0, x, \theta)\mathrm{d}z\mathbf{P}(\theta|\mathsf{D})\mathrm{d}\theta \tag{5}$$

For all data entries other than the ones with $\mathsf{T}=0$ we have $\hat{\mathsf{Y}} = \mathsf{Y}$ where Y is the outcome recorded in D. The result of Eq. 5 can be computed numerically:

$$\hat{\mathsf{Y}} = \frac{1}{N}\sum_{k=1}^{N} \mathbf{P}(\mathsf{Y}=1|\mathsf{T}=1, x, z_k, \theta_k) \tag{6}$$

where the sums are taken over N samples of θ and Z obtained from their posteriors. In practice, we use the MCMC functionality of Stan to obtain the samples.

3.3 Evaluating Decision Makers

Expression 6 gives us a direct way to evaluate the outcome of a positive decisions for any data entry for which $\mathsf{T}=0$. Note though that, unlike Y that takes integer values $\{0, 1\}$, $\hat{\mathsf{Y}}$ may take also fractional values $\hat{\mathsf{Y}} \in [0, 1]$. Having obtained outcome estimates for all data entries, it is now straightforward to obtain an estimate for the failure rate FR of decision maker M: it can be computed as a simple average over all data entries. Our approach is summarized in Fig. 3. We will refer to it as `CFBI`, for counterfactual-based imputation.

4 Experiments

We test the accuracy and robustness of `CFBI` in evaluating the performance of decision makers of different kinds. Towards this end, we employ both synthetic and real data. We compare `CFBI` especially with `Contraction` [12]. The implementation is available online.[1] Our manuscript contains the specification of the parameters and datasets we used for reproducibility.

4.1 Synthetic Data

We begin our experiments with synthetic data, in order to investigate various properties of our approach. To set up the experimentation, we follow the setting of [12]. Each synthetic dataset we experiment with consists of $n = 5{,}000$

[1] https://version.helsinki.fi/rikulain/CFBI-public.

randomly generated cases. The features X and Z of each case are drawn independently from a standard Gaussian. Each case is assigned randomly to one out of $m = 50$ decision makers, such that each decision maker receives a total of 100 cases. The leniency R of each decision maker is drawn from Uniform(0.1, 0.9). A decision T is made for each case by the assigned decision maker. The exact method of assigning a decision is specified in the next subsection (Sect. 4.2). If the decision is positive, then a binary outcome is sampled from a Bernoulli distribution:

$$\mathbf{P}(\mathsf{Y} = 0 \mid \mathsf{T} = 1, x, z) = \sigma(b_{\mathsf{X}} x + b_{\mathsf{Z}} z + e_{\mathsf{Y}}) \tag{7}$$

with $b_{\mathsf{X}} = b_{\mathsf{Z}} = 1$. Additional noise is added to the outcome of each case via e_{Y}, which was drawn from a zero-mean Gaussian distribution with small variance, $e_{\mathsf{Y}} \sim \mathcal{N}(0, 0.1)$. The data set was split in half to training and tests sets, such that each decision maker appears only in one. The evaluated decision maker M is trained on the training set while the evaluation is based only on the test set.

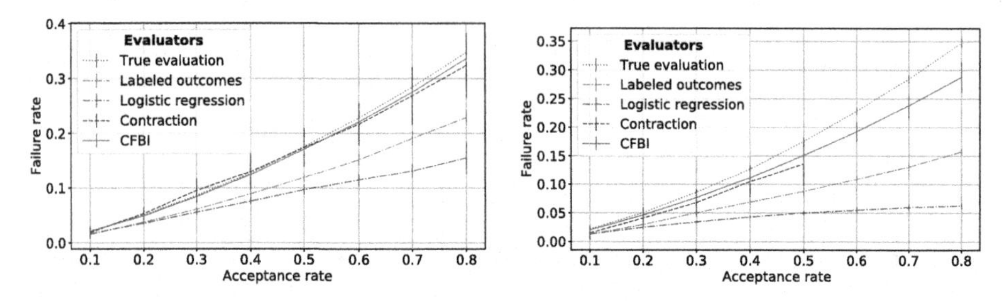

Fig. 4. Left: evaluation of `Batch` decision maker on data with `Independent`. Error bars show std. of the FR estimate across 10 datasets. In this basic setting, both our `CFBI` and contraction follow the true evaluation curve closely but `CFBI` exhibits lower variation. Right: evaluating `Batch` on data employing `Independent` and with leniency at most 0.5. `CFBI` offers sensible estimates of the failure rates for all levels of leniency, whereas `Contraction` only up to leniency 0.5.

4.2 Decision Makers

Our experiments involve two categories of decision makers: (i) the set of decision makers $\mathbf{H}$, the decisions of which are reflected in a dataset, and (ii) the decision maker M, whose performance is to be evaluated on the log of cases decided by $\mathbf{H}$.

Decisions by H. Decision makers $\mathbf{H}$ base their decision on their perception of the dangerousness of a case, to which we refer as the *risk score*. We compute the risk score as

$$\text{risk score} = b_{\mathsf{X}}\mathsf{X} + b_{\mathsf{Z}}\mathsf{Z}. \tag{8}$$

For the *first* type of decision makers we consider, we assume that decisions are rational and well-informed, and that a decision maker with leniency r makes a positive decision only for the r fraction of cases that are most likely to lead to a positive outcome. Specifically, we assume that the decision-makers know the cumulative distribution function F that the risk scores $s = b_{\mathsf{X}}x + b_{\mathsf{Z}}z$ of defendants follow. This is a reasonable assumption to make when decision makers have accurate knowledge of the joint feature distribution. For example, an experienced judge who has tried a large volume and variety of defendants may have a good idea about the various cases that appear at court and which of them pose higher risk. Considering a decision maker with leniency $\mathsf{R} = r$ who decides a case with risk score s, a positive decision is made only if s is in the r portion of the lowest scores according to F. Since in our setting the distribution F is given and fixed, such decisions for different cases happen independently based on their risk score – and we refer to such decision makers as `Independent`.

In addition, we consider a different type of decision makers, namely `Batch`, also used in [12]. Decision makers of this type consider all cases assigned to them at once, as a batch; sort them by the risk score in Eq. 8; and, for leniency $\mathsf{R} = r$, release r portion of the batch with the lowest risk score. Such decision makers still have a good knowledge of the relative risk that the cases assigned to them pose, but they are also short-sighted, as they make decisions for a case *depending* on other cases in their batch. For example, if a decision maker is randomly assigned a batch of cases that are all very likely to lead to a good outcome, a large portion $1 - r$ of them will still be handed a negative decision.

Finally, we consider a third type of decision maker, namely `Random`. It simply makes a positive decision with probability r. We include this to test the evaluation methods also in settings where some of their assumptions may be violated.

Decisions by M**.** For M, we consider the same three types of decision makers as for $\mathbf{H}$ above, with one difference: decision makers $\mathbf{H}$ have access to Z, while M does not. Their definitions are adapted in the obvious way. Risk scores are computed with a logistic regression model which is trained on the training data set. For the `Independent` decision maker M, the cumulative distribution function of the risk scores is constructed using the empirical distribution of risk scores of all the observations in the training data.

4.3 Evaluators

We aim to investigate the performance of different methods in evaluating M on a dataset that records cases decided by $\mathbf{H}$. We call those methods "evaluators". Desired evaluator properties are accuracy (i.e., the evaluator should estimate well the failure rate of M), but also robustness (i.e., consistent performance).

The first evaluator we consider is `CFBI`, the method we propose in Sect. 3. To summarize: `CFBI` uses the dataset to learn a model, i.e., a distribution for the parameters involved in formulas 1 and 2; using this distribution, it predicts the outcome of the cases for which $\mathbf{H}$ made a negative decision and M makes a positive one; and finally it evaluates the failure rate of M on the dataset.

The second evaluator we consider is `Contraction`, proposed in recent work [12]. It is designed specifically to estimate the failure rate of a machine

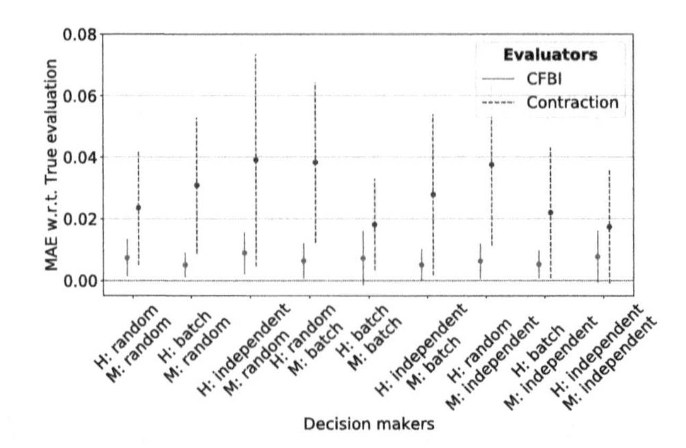

Fig. 5. Mean absolute error (MAE) of estimate w.r.t. true evaluation. Error bars show std. of the absolute error over 10 datasets. `CFBI` offers robust estimates across all decision makers. The error of `Contraction` varies within and across different decision makers.

decision maker in the selective labels setting. Contraction bases its evaluation only on the cases assigned to the most lenient decision maker H_l in the data. Because of the lower leniency of the evaluated decision maker M, the approach assumes M makes a negative decision for all cases for which H_l makes a negative decision. The cases with a positive decision by H_l are sorted according to the lowest leniency level at which they receive positive decisions by M. The sorted list is then *contracted* to match the leniency level r at which M is evaluated. Because all outcomes for cases in this list are available in the data, FR is estimated as the fraction of cases in the contracted list with a negative outcomes among cases assigned to H_l.

In addition, we consider two baselines. As a first baseline, we consider the method that evaluates the failure rate M based only on those cases that received a positive decision by $\mathbf{H}$ in the data. Following [12], it is referred to as `LabeledOutcomes`. As a second baseline, we consider a method that performs straightforward imputation: given a training dataset, it considers only those cases that were accompanied with a positive decision and builds a logistic regression model on them; it then uses the prediction of this logistic regression to impute the outcome in the test data for those cases where M makes a positive decision but $\mathbf{H}$ had made a negative decision. We refer to this evaluator as `LogisticRegression`.

Finally, all evaluators are compared with the optimal evaluator that has access to actual outcomes. While such an evaluator is unavailable in practice, it is available for synthetic data. Following [12], it is referred to as `TrueEvaluation`.

4.4 Results

We show the accuracy of the different evaluators (Sect. 4.3) on different decision makers M over data sets employing different decision makers $\mathbf{H}$ (Sect. 4.2).

The Basic Setting. Figure 4 (left) shows estimated failure rates for each of the evaluators, at different leniency levels, when decisions in the data were made by `Independent` decision maker, while M was of `Batch` type. In interpreting this plot, we should consider an evaluator to be accurate if its curve follows well that of the optimal evaluator `TrueEvaluation`. In this scenario, `CFBI` and `Contraction` are quite accurate, while the naive evaluation of `LabeledOutcomes`, but also the straightforward imputation by `LogisticRegression` perform quite poorly. In addition, `CFBI` exhibits considerably lower variation than `Contraction`.

Figure 5 shows the aggregate absolute error rates of the two evaluators, `CFBI` and `Contraction`. Each error bar is based on all datasets and leniencies from 0.1 to 0.8, for different types of decision makers for $\mathbf{H}$ and for M. The overall result is that `CFBI` evaluates the decision makers accurately and robustly across different decision makers. It is able to learn model parameters that capture the behaviour of decision makers employed in the data and use that model to evaluate any decision maker M. `Contraction` shows consistently poorer performance, and markedly larger variation as shown by the error bars. Again, we postulate this happens because `Contraction` crucially depends on the cases assigned to the most lenient decision makers, while `CFBI` uses all data.

The Effect of Limited Leniency. Figure 4 (right) shows the results when the leniency of decision makers in the data was restricted below 0.5, and not up to 0.9 as for Fig. 4 (left). Here, `Contraction` is only able to estimate the failure rate up to 0.5 – but for higher leniency rates it does not output any results. On the contrary, `CFBI` produces failure rate estimates for all leniencies. We note that, when we compare with `TrueEvaluation`, the accuracy `CFBI` decreases for the largest leniencies – as expected, since such cases do not exist in the data. This observation is important in the sense that decision makers based on elaborate machine learning techniques, may well allow for evaluation at higher leniency rates than those (often human) employed in the data.

The Effect of Unobservables. To explore situations where the importance of unobservables is higher, we now set $b_{\mathsf{X}} = 1$, $b_{\mathsf{Z}} = 5$. The results are shown in Fig. 6. In these settings, the decisions in the data are made mostly based on background factors not observed by the decision maker M being evaluated, thus the performance M is worse than in Fig. 5. Nevertheless, the proposed method (`CFBI`) is able to evaluate different decision makers M accurately. `Contraction` shows again consistently worse performance in comparison. Furthermore, when compared to the basic case (Fig. 5), the performance of `Contraction` is also worse, indicating some sensitivity to unobservables.

Thus overall, in these synthetic settings `CFBI` achieves more accurate results with considerably lower variation than `Contraction`, allowing for evaluation in cases where the strong assumptions of `Contraction` inhibit evaluation altogether.

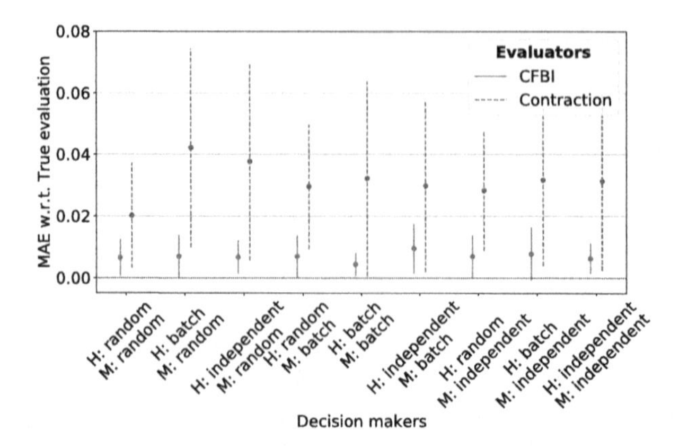

Fig. 6. MAE of estimate w.r.t true evaluation when the effect of the unobserved Z is high ($b_Z = 5$). The decision quality is poorer, but `CFBI` can still evaluate the decisions accurately. `Contraction` shows higher variance and lower accuracy.

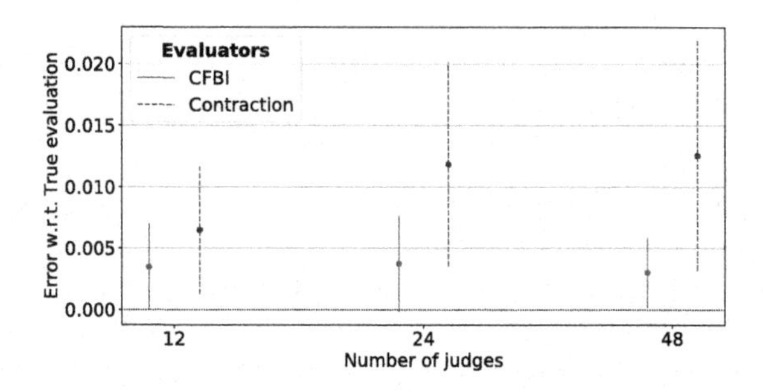

Fig. 7. Results with COMPAS data. Error bars show std. of the absolute FR estimate errors across all levels of leniency w.r.t. true evaluation. `CFBI` gives both more accurate and precise estimates despite of the number of judges used.

4.5 COMPAS Data

COMPAS is a set of tools for assisting decisions in the criminal justice system. It is derived from prior criminal history, socio-economic and personal factors and it predicts recidivism for two years [2]. The COMPAS dataset used in this study is recidivism data from Broward county, California, USA made available by ProPublica. Judges and defendants in the data correspond to decision makers and cases in our setting (Sect. 2), respectively. The original data contained information on 18,610 defendants who were given a COMPAS score during 2013 or 2014. Following ProPublica's data cleaning process, finally the data consisted of $n = 6{,}172$ offenders. Data includes the subjects' demographic information such as gender, age and race together with information on their previous offences.

For the analysis, we deployed $m \in \{12, 24, 48\}$ synthetic judges with fixed leniency levels 0.1, 0.5 and 0.9 so that a third of the decision makers shared a leniency level. The n subjects were distributed to the m judges uniformly at random. In this scenario, the judges based their decisions on the COMPAS score, releasing the fraction of defendants with the lowest score according to their leniency. E.g. if a synthetic judge had leniency 0.5, they would release 50% of defendants with the lowest COMPAS score. Those who were given a negative decision had their outcome label set to positive $\mathsf{Y} = 1$. After assigning the decisions, the data was split 10 times to training and test sets containing the decisions of half of the judges each. A logistic regression model was trained on the training data to predict two-year recidivism from categorised age, race, gender, number of prior crimes and the degree of crime COMPAS screened for (felony or misdemeanour) using only observations with positive decisions. As the COMPAS score is derived from a larger set of predictors than the aforementioned five [2], the unobservable information would then be encoded in the COMPAS score. The built logistic regression model was used in decision maker M in the test data and the same features were given as input for the counterfactual imputation. The deployed machine decision maker was defined to release r fraction of the defendants with the lowest probability for negative outcome.

Figure 7 shows the errors of failure rate of `Batch` as a function of the number of judges in the data (also batch decision makers). The MAE of our `CFBI` at all levels of leniency is consistently lower than that of `Contraction` for each number of judges used in the experiments. Quite notably, the error of `Contraction` gets larger when there are more judges in the data and the variance of the failure rate estimates it produces increases as the most lenient judge is assigned fewer subjects. We attribute this to the fact that `Contraction` crucially depends on the most lenient decision makers, while `CFBI` makes full use of the data.

5 Related Work

We adopted the setting of [12], and showed that causally informed counterfactual imputation can achieve accurate results. In addition, Kleinberg et al. [10] present an in-detail account of employing `Contraction` on real data. In their experiments, they use a decision maker that is set-up similarly to `Independent` decision makers discussed in our work – but that makes decisions determined by utility values.

Unlike our imputation approach (`CFBI`), De-Arteaga et al. [5] directly impute decisions as outcomes and consider learning automatic decision makers from such augmented data. In [10], a multiplicative correction term is used to adjust the bias observed for more conventional imputation. In comparison, `CFBI` uses rigorous causal modelling to account for leniency and unobservables, and gives accurate results even the expert consistency assumption of [5] is violated.

In reinforcement learning, a related scenario is that of offline policy evaluation, where the objective is to determine a quality of a policy from data recorded under some other baseline policy [8,19]. In particular, Jung et al. [8] consider sensitivity analysis in a similar scenario as ours, but without directly modelling decision makers with multiple leniencies. Mc-Candless et al. perform Bayesian sensitivity analysis while taking into account latent confounding [15,16]. Kallus et al. obtain improved policies from data possibly biased by a baseline policy [9].

More generally, our setting exhibits *selection bias* [6], *latent confounding* [17], and *missing data* [13] (depending on how the outcomes for negative decisions are interpreted). In particular, our setting violates *ignorability* and *missing at random (MAR)* in the context of missing data. The effectiveness of causal modelling and counterfactuals is also demonstrated in recent work on e.g. fairness [3,4,11,14,21]. More applied work related to recidivism, can be found e.g. in [2,7,20].

6 Conclusions and Future Work

We considered the task of evaluating (automated) decision makers, which is crucially needed in replacing human decisions with automated ones. We presented CFBI, an approach based on proper causal modelling that makes full use of the available data, and demonstrated that automated decision makers can be evaluated on data that are selectively labelled. Via thorough experimentation we found that CFBI allows for accurate and robust evaluations, also in settings that evaluation was not possible before (i.e., for a leniency level higher than the one present in the data). In future work, we will generalize our setting and modelling assumptions – e.g., to consider more elaborate behaviour for decision makers and additional dependencies between model variables.

Acknowledgments. Authors acknowledge the computer capacity from the Finnish Grid and Cloud Infrastructure (urn:nbn:fi:research-infras-2016072533). RL was supported by HICT; AH by Academy of Finland grants 295673, 316771 and by HIIT; and MM by Research Funds of the University of Helsinki.

Appendix 1 Counterfactual Inference

Here we derive Eq. 4, via Pearl's counterfactual inference protocol involving three steps: abduction, action, and inference [17]. Our model can be represented with the following structural equations over the graph structure in Fig. 2:

$$\mathsf{J} := \epsilon_\mathsf{J}, \quad \mathsf{Z} := \epsilon_\mathsf{Z}, \quad \mathsf{X} := \epsilon_\mathsf{X}, \quad \mathsf{T} := g(\mathsf{H}, \mathsf{X}, \mathsf{Z}, \epsilon_\mathsf{T}), \quad \mathsf{Y} := f(\mathsf{T}, \mathsf{X}, \mathsf{Z}, \epsilon_\mathsf{Y}).$$

For any cases where $\mathsf{T} = 0$ in the data, we calculate the counterfactual value of Y if we had $\mathsf{T} = 1$. We assume here that all these parameters, functions and distributions are known. In the *abduction* step we determine $\mathbf{P}(\epsilon_\mathsf{H}, \epsilon_\mathsf{Z}, \epsilon_\mathsf{X}, \epsilon_\mathsf{T}, \epsilon_\mathsf{Y} | j, x, \mathsf{T} = 0)$, the distribution of the stochastic disturbance terms updated to take into account the observed evidence on the decision maker, observed features and the decision (given the decision $\mathsf{T} = 0$ disturbances are independent of Y). We directly know $\epsilon_\mathsf{X} = x$ and $\epsilon_\mathsf{J} = j$. Due to the special form of f the observed evidence is independent of ϵ_Y when $\mathsf{T} = 0$. We only need to determine $\mathbf{P}(\epsilon_{\mathsf{Z}}, \epsilon_{\mathsf{T}}|h, x, \mathsf{T}=0)$. Next, the *action* step involves intervening on T and setting $\mathsf{T}=1$ by intervention. Finally in the *prediction* step we estimate Y:

$$E_{\mathsf{T}\leftarrow 1}(\mathsf{Y}|j, \mathsf{T}=0, x) = \int f(\mathsf{T}=1, x, \mathsf{Z}=\epsilon_{\mathsf{Z}}, \epsilon_{\mathsf{Y}})\mathbf{P}(\epsilon_{\mathsf{Z}}, \epsilon_{\mathsf{T}}|j, \mathsf{T}=0, x)\mathbf{P}(\epsilon_{\mathsf{Y}})d\epsilon_{\mathsf{Z}}\mathsf{d}\epsilon_{\mathsf{Y}}\mathsf{d}\epsilon_{\mathsf{T}}$$
$$= \int \mathbf{P}(\mathsf{Y}=1|\mathsf{T}=1, x, z)\mathbf{P}(z|j, \mathsf{T}=0, x)\mathsf{d}z$$

where we used $\epsilon_{\mathsf{Z}} = z$ and integrated out ϵ_{T} and ϵ_{Y}. This gives us the counterfactual expectation of Y for a single subject.

Appendix 2 On the Priors of the Bayesian Model

The priors for γ_{X}, β_{X}, γ_{Z} and β_{Z} were defined using the gamma-mixture representation of Student's t-distribution with $\nu = 6$ degrees of freedom. The gamma-mixture is obtained by first sampling a precision parameter from $\Gamma(\nu/2, \nu/2)$ and then drawing the coefficient from zero-mean Gaussian with that precision. This procedure was applied to the scale parameters η_{Z}, $\eta_{\beta_{\mathsf{X}}}$ and $\eta_{\gamma_{\mathsf{X}}}$ as shown below. For vector-valued X, the components of γ_{X} (β_{X}) were sampled independently with a joint precision parameter $\eta_{\gamma_{\mathsf{X}}}$ ($\beta_{\gamma_{\mathsf{X}}}$). The coefficients for the unobserved confounder Z were bounded to the positive values to ensure identifiability.

$$\eta_{\mathsf{Z}}, \eta_{\beta_{\mathsf{X}}}, \eta_{\gamma_{\mathsf{X}}} \sim \Gamma(3,3),\ \gamma_{\mathsf{Z}}, \beta_{\mathsf{Z}} \sim N_{+}(0, \eta_{\mathsf{Z}}^{-1}),\ \gamma_{\mathsf{X}} \sim N(0, \eta_{\gamma_{\mathsf{X}}}^{-1}),\ \beta_{\mathsf{X}} \sim N(0, \eta_{\beta_{\mathsf{X}}}^{-1})$$

The intercepts for the decision makers in the data and outcome Y had hierarchical Gaussian priors with variances σ_{T}^2 and σ_{Y}^2. The decision makers had a joint variance parameter σ_{T}^2.

$$\sigma_{\mathsf{T}}^2,\ \sigma_{\mathsf{Y}}^2 \sim N_{+}(0, \tau^2), \quad \alpha_j \sim N(0, \sigma_{\mathsf{T}}^2), \quad \alpha_{\mathsf{Y}} \sim N(0, \sigma_{\mathsf{Y}}^2)$$

The parameters σ_{T}^2 and σ_{Y}^2 were drawn independently from Gaussian distributions with mean 0 and variance $\tau^2 = 1$, and restricted to the positive real axis.

References

1. Austin, P.C.: An introduction to propensity score methods for reducing the effects of confounding in observational studies. Multivar. Behav. Res. **46**(3), 399–424 (2011)
2. Brennan, T., Dieterich, W., Ehret, B.: Evaluating the predictive validity of the COMPAS risk and needs assessment system. Crim. Justice Behav. **36**(1), 21–40 (2009)
3. Corbett-Davies, S., Pierson, E., Feller, A., Goel, S., Huq, A.: Algorithmic decision making and the cost of fairness. In: Proceedings of the ACM SIGKDD (2017)
4. Coston, A., Mishler, A., Kennedy, E.H., Chouldechova, A.: Counterfactual risk assessments, evaluation, and fairness. In: Proceedings of the FAT, pp. 582–593 (2020)
5. De-Arteaga, M., Dubrawski, A., Chouldechova, A.: Learning under selective labels in the presence of expert consistency. arXiv preprint arXiv:1807.00905 (2018)

6. Hernán, M.A., Hernández-Díaz, S., Robins, J.M.: A structural approach to selection bias. Epidemiology **15**(5), 615–625 (2004)
7. Jung, J., Concannon, C., Shroff, R., Goel, S., Goldstein, D.G.: Simple rules to guide expert classifications. J. Roy. Stat. Soc.: Ser. A **183**, 771–800 (2020)
8. Jung, J., Shroff, R., Feller, A., Goel, S.: Bayesian sensitivity analysis for offline policy evaluation. In: Proceedings of the AIES (2020)
9. Kallus, N., Zhou, A.: Confounding-robust policy improvement. In: Advances in Neural Information Processing Systems, pp. 9269–9279 (2018)
10. Kleinberg, J., Lakkaraju, H., Leskovec, J., Ludwig, J., Mullainathan, S.: Human decisions and machine predictions. Q. J. Econ. **133**(1), 237–293 (2018)
11. Kusner, M.J., Russell, C., Loftus, J.R., Silva, R.: Making decisions that reduce discriminatory impacts. In: Proceedings of the ICML (2019)
12. Lakkaraju, H., Kleinberg, J., Leskovec, J., Ludwig, J., Mullainathan, S.: The selective labels problem: evaluating algorithmic predictions in the presence of unobservables. In: Proceedings of the ACM SIGKDD (2017)
13. Little, R.J., Rubin, D.B.: Statistical Analysis with Missing Data, vol. 793. Wiley, Hoboken (2019)
14. Madras, D., Creager, E., Pitassi, T., Zemel, R.: Fairness through causal awareness: learning causal latent-variable models for biased data. In: Proceedings of the FAT (2019)
15. McCandless, L.C., Gustafson, P.: A comparison of Bayesian and Monte Carlo sensitivity analysis for unmeasured confounding. Stat. Med. **36**(18), 2887–2901 (2017)
16. McCandless, L.C., Gustafson, P., Levy, A.: Bayesian sensitivity analysis for unmeasured confounding in observational studies. Stat. Med. **26**(11), 2331–2347 (2007)
17. Pearl, J.: An introduction to causal inference. Int. J. Biostat. **6**(2) (2010). https://doi.org/10.2202/1557-4679.1203
18. Rosenbaum, P.R., Rubin, D.B.: The central role of the propensity score in observational studies for causal effects. Biometrika **70**(1), 41–55 (1983)
19. Thomas, P.S., Brunskill, E.: Data-efficient off-policy policy evaluation for reinforcement learning. In: Proceedings of the ICML (2016)
20. Tolan, S., Miron, M., Gómez, E., Castillo, C.: Why machine learning may lead to unfairness: evidence from risk assessment for Juvenile justice in Catalonia. In: Proceedings of the Artificial Intelligence and Law (2019)
21. Zhang, J., Bareinboim, E.: Fairness in decision-making - the causal explanation formula. In: Proceedings of the AAAI (2018)

Mitigating Discrimination in Clinical Machine Learning Decision Support Using Algorithmic Processing Techniques

Emma Briggs[1(✉)] and Jaakko Hollmén[2]

[1] Karolinska Institutet, Stockholm, Sweden
emma.briggs@stud.ki.se

[2] Department of Computer and Systems Sciences, Stockholm University, Stockholm, Sweden
jaakko.hollmen@dsv.su.se

Abstract. Discrimination on the basis of protected characteristics - such as race or gender - within Machine Learning (ML) is an insufficiently addressed yet pertinent issue. This line of investigation is particularly lacking within clinical decision-making, for which the consequences can be life-altering. Certain real-world clinical ML decision tools are known to demonstrate significant levels of discrimination. There is currently indication that fairness can be improved during algorithmic processing, but this has not been widely examined for the clinical setting. This paper therefore explores the extent to which novel algorithmic processing techniques may be able to mitigate discrimination against protected groups in clinical resource-allocation ML decision-support algorithms. Specifically, three state-of-the-art discrimination mitigation techniques are compared, one for each stage of algorithmic processing, when applied to a real-world clinical ML decision algorithm which is known to discriminate with regards to racial characteristics. The results are promising, revealing that such techniques could significantly improve the fairness of clinical resource-allocation ML decision tools, particularly during pre- and post-processing. Discrimination is shown to be reduced to arbitrary levels at little to no cost to accuracy. Similar studies are needed to consolidate these results. Other future recommendations include working towards a generalisable framework for ML fairness in healthcare.

Keywords: Fairness · Machine Learning · Clinical decision support · Resource-allocation

1 Introduction

Due to increased attention within data science and medical research, the use of Machine-Learning (ML) techniques to inform clinical decisions is gaining traction which lend themselves to myriad applications within healthcare, such as risk-prediction and resource-allocation [1,2]. Many of these have already successfully been implemented in clinics worldwide [3]. With such techniques so

A. Appice et al. (Eds.): DS 2020, LNAI 12323, pp. 19–33, 2020.
https://doi.org/10.1007/978-3-030-61527-7_2

widely adopted, it is more vital than ever that they are non-discriminatory with regards to protected groups such as ethnic minorities or those originating from disadvantaged socioeconomic backgrounds.

From the data science perspective, fairness in ML is said to occur when the following constraints are satisfied (for direct and indirect non-discrimination respectively):

(1) Those who are similar, as defined by their non-protected characteristics, should receive similar predictions
(2) Any differences in predictions across groups of people must be justifiable by non-protected characteristics [4].

Consequently, failure to meet both these criteria indicates discrimination, which is a violation of both legal and ethical regulations [5,6]. Unfortunately, this easily and often goes undetected, rendering it a particularly pervasive problem [7,8]. Discrimination in clinical resource-allocation decision-support could have severe consequences, such as limiting access to vital resources for those within certain ethnic groups [9]. Alarmingly, even algorithms currently implemented in real-world settings have been found to be highly discriminatory. For instance, one recent study revealed that a US algorithm used to help determine future medical treatment for patients was highly biased against African-Americans [9]. Many such algorithms employ the use of Electronic Health Record (EHR) data, which contains sensitive information. There are known to be three major sources of discrimination on the basis of sensitive attributes when using EHR data to train ML algorithms: missing data, a biased sampling procedure, and misclassification or measurement error [6,10].

Refusing to address issues with discrimination could also lead to the stagnancy of implementation of certain machine-learning decision-support tools in healthcare, on the basis that such tools use algorithms which are simply too unreliable to employ. Unfortunately, it is not explicitly clear how to accommodate for fairness is at any stage of the decision-making process and so the discrimination problem is not trivial [6]. Simple ‘solutions’ such as just removing the sensitive attributes or creating separate models for segregated groups (as organised by sensitive attributes) are considered naïve [5,8].

With regards to the data mining process, there are two parts which the data scientist can control: data collection, and data processing [6]. In terms of data collection, supplying an ML algorithm with biased data has a high potential to reinforce pre-existing discriminatory biases, causing the phenomenon known as *digital redlining* [11] However, unbiased, representative data can be difficult to obtain, and it can be more difficult still to guarantee impartiality. Therefore, taking such measures solely at the collection phase is inadequate and also not always possible, and so action must also be taken at the processing stages of the ML algorithm. There are three major data processing stages at which adjustments may be made: (1) pre-processing, (2) in-processing, and (3) post-processing [12].

Previous research has investigated effective methods of bias mitigation in machine-learning algorithms, and there are already numerous supervised ML models employed in the clinical setting which are used to predict certain risks for patients in hospitals and allocate and distribute medical resources, for example [3,13]. Although limited, and more often performed in other fields outside of healthcare, research into methods to tackle discrimination in such algorithms does exist [1,12]. However, what does not exist is both an assessment of different mitigation methods when applied to the clinical resource-allocation setting, and a direct comparison and appraisal of these methods - which correspond to employment at different stages within the ML approach - within such a setting.

2 Related Research

In this section we briefly review some of the latest related work on mitigating discrimination within ML and the current state of fairness within clinical algorithmic decision-making strategies.

2.1 Measuring Fairness in Machine Learning

It is commonly established that ML fairness can broadly relate to either *group fairness* or *individual fairness* [14]. Although opinions are divided when it comes to emphasising one over the other, much of the current research places a larger focus on ensuring group fairness with a view to avoiding mass discrimination and eliminating prejudice. This is also due to the fact that it can be difficult to concretise individual fairness. Reducing disparate impact is one of the most common investigations in discrimination mitigation and aims to ensure that each protected group does not suffer a disadvantage when compared to the general population. However, solely focusing on group fairness does risk over-generalisation. There is an abundance of existing fairness metrics for ML models and they often correlate with each other [14]. They may broadly be organised into the following four categories [4]:

(1) Statistical tests
(2) Absolute measures
(3) Conditional measures
(4) Situation measures

Ambiguity still exists with regards to concretising the concept of fairness in ML decision tools and an interdisciplinary approach to achieve this, involving policy makers, decision makers and algorithm developers, is strongly recommended [15].

2.2 Algorithmic Processing Techniques

Numerous potential solutions have been proposed with varying success, the majority having been assessed in applications unrelated to the clinical setting, such as credit risk score generation, salary estimation and recidivism rate analysis [14,16,17]. The following table displays an overview of some of the techniques found when scoping the current literature.

Table 1. Summary of variety of pre-, in-, and post- processing techniques for discrimination mitigation in ML

Processing stage	Paper (Mentioned in)	Technique name	Brief description
Pre-processing	[18], Khan S R. et al. 2019	Reweighing	Altering object weights based on the protected attribute value
	[18], Khan S R. et al. 2019	Sampling	Either uniform sampling or preferential sampling using the stratified dampling technique
	[18], Khan S R. et al. 2019	Suppression	Suppressing attributes which have the strongest correlation with the sensitive attribute
	[18], Khan S R. et al. 2019	Massaging	Altering the labels of particular objects according to fairness constraints
	[19], Calmon F P. et al. 2017	Optimised pre-processing	A convex optimisation technique for transforming data with respect to fairness objectives
In-processing	[20], Zhang B H. et al. 2018	Adversarial de-biasing	Introduces an adversary which counteracts fairness constraints
	[21], Slack D. et al. 2019	Fair-MAML	Adjusting the Meta-Agnostic Machine Learning (MAML) algorithm, adding a fairness regularisation term to model learning task losses
	[21], Slack D. et al. 2019	K-Shot fairness	Fair model training with a limited sample size
Post-processing	[17], Kamiran F. et al. 2012	Discrimination-aware ensemble	Examines the disagreement region of theclassifier, making compensations for underprivileged groups
	[17], Kamiran F. et al. 2012	Reject-option - based classification	A cost-based fairness technique which assigns costs to generating false outcomes

Alternative ML Fairness Solutions. This particular study focuses on group fairness as pertains to ensuring distributive fairness in supervised ML models, which aligns with the chosen context of investigating group discrimination in distributive clinical decision-support tools. However, it should be mentioned that there also exists a growing body of work utilising unsupervised learning techniques to improve fairness, in clustering and network representation algorithms, for example [22,23]. Given the wealth of alternative techniques available, Table 1 by no means contains an exhaustive list. Many other techniques exist which seek to satisfy ML fairness beyond the scope of this study, such as manipulation using monotonicity constraints, for example, which especially encourages individual fairness and has been shown to be an effective way to ensure that reasoning behind ML decision-making is fair and justifiable, in contexts ranging from medicine to law [24].

2.3 Discrimination in Clinical Decision-Support

There is plenty of evidence to suggest that discrimination may easily arise in ML algorithms for clinical decision-support, in contexts such as Intensive Care Unit (ICU) monitoring to predict risk of deterioration of patients, length of hospital stay estimation, and admission to specialised healthcare treatment programmes [9,25].

Since most of the current novel techniques for discrimination mitigation have not been assessed in conjunction with clinical datasets, researchers in the clinical field strongly recommend leveraging the discoveries made in other domains and beginning to test the potential of state-of-the-art techniques in healthcare decision-support [10].

3 Empirical Investigation

This paper applies the empirical research methodology of *experiment science* to evaluate current discrimination mitigation techniques when applied to clinical ML decision support.

3.1 Dataset

The dataset selected for the experiments originates from a widely-cited study, [9], investigating the extent to which racial discrimination is embedded in a particular healthcare risk-prediction and resource-allocation ML tool employed in the US to help determine future clinical treatment plans for millions of patients. The researchers worked with a large academic hospital in the US: one of the many hospitals to have adopted the algorithmic decision tool. Together they curated a comprehensive healthcare dataset using a combination of EHR data, containing a rich set of health measures and demographic variables, along with medical claims data. A synthetic dataset which replicates the characteristics and summary statistics of the original dataset (not publicly available on the grounds of patient anonymity) was produced by these researchers with the sole purpose of being conducive to further research in clinical ML fairness.

This dataset is the one employed within this study. It contains 48,784 observations and 150 variables. The protected attribute in this scenario is *race* with *white* being the privileged category and *black* being the underprivileged category. The remaining feature variables may be categorised as follows: *Demographic Variables* (e.g. Age, Gender, Race), *Biomarker Variables* (e.g. Blood Pressure, Creatinine), *Chronic Illness Indicators* (e.g. the presence or absence of conditions such as Hypertension, Diabetes), and *Healthcare Service Use* (e.g. No. of Hospital Days, Emergency Visits).

The label assigned by the original algorithm is a predicted risk score corresponding to estimated future healthcare expenditure for each patient, which helps determine whether or not they are allocated future enrolment in specialised care treatment programs. The algorithm-generated score is said to play a major role in this decision. The authors of [9] who uncovered the discriminatory nature of this algorithmic tool suggest altering the label itself by way of successfully improving model fairness.

However, it is also acknowledged that this is not the only - and not necessarily the optimal - solution to the discrimination problem for clinical resource-allocation algorithms. First of all, many such algorithms necessitate a cost-based design in order to comply with regulations, and using cost predictors can be a very effective way of signalling health requirements [9]. Therefore, label choice is accepted as an ongoing dilemma [9]. The two most popular choices of label for risk-prediction- generating ML algorithms in the clinical setting are *service utilisation* (i.e. a combinatory score for hospital visits, emergency stays etc.) and *medical expenditure forecast* (for the following year, based on the previous year's data) [9]. In addition, past costs and service use to indicate future costs and resource utilisation is a popular approach in real-world clinical resource-allocation tools, especially in the US where this is one of the most widely-adopted techniques [9,26]. Therefore this study adheres to the original label choice, focusing on the ability of algorithmic processing techniques to enhance fairness rather than label adjustment. This line of investigation has a great advantage in that it is not a solution solely applicable to this particular algorithm but a generalisable method of tackling the discrimination problem in clinical ML decision-making. The final algorithmic decision outcome is a binary variable based on the estimated risk score, with 1 denoting acceptance into the care treatment program and 0 denoting non-acceptance.

The dataset is available at [27] and will hereon bc referred to as 'Dissecting-Bias.'

3.2 Algorithmic Setup

The algorithmic processing techniques and model fairness metrics are taken from IBM's comprehensive open-source ML fairness toolkit, Artificial Intelligence Fairness 360 (AIF 360), publicly available on GitHub [28]. This particular toolkit was developed with the aim to consolidate a vast array of current research findings on ML fairness, and therefore seeks to represent pioneering ML bias mitigation techniques from many sources [28]. It currently offers 10 state-of-the-art fairness algorithms and over 70 fairness metrics for predictive Artificial Intelligence (AI) models. It is constantly updated by experts according to new developments and findings. The classification algorithms used for these experiments are the ones prescribed by the examples in the toolkit for each mitigation method.

These experiments are reproducible using the open-source code for the selected algorithmic techniques at [28] and the Dissecting-Bias dataset at [27]. For this study, the dataset and algorithmic techniques were integrated using Jupyter Notebooks.

3.3 Optimising Fairness at the Pre-processing Stage

The selected pre-processing technique for this stage is *Reweighing*. This was chosen primarily because it is a popular method amongst other applications of discrimination mitigation in ML and it is known to work highly successfully for these other settings [18, 29].

Furthermore, this approach works well in optimising group fairness, adhering to the primary goal of the experiments within this study: to reduce discrimination with regards to a specific protected attribute ('race'). This process does also take into account individual fairness constraints; however, these are not guaranteed satisfaction and group fairness takes precedence [18]. The Reweighing technique is mainly recommended if there is pre-existing discrimination in the dataset as it works to combat historical discrimination. Therefore, it is considered highly suitable for the chosen dataset, which is known to contain a high level of pre-existing bias against those from the protected attribute category *black*.

The Reweighing technique was first proposed in paper [30] and works by assigning weights to tuples and then altering these weights in such a way that those from the underprivileged group are more inclined towards the favourable outcome and those from privileged groups are slightly inclined towards the unfavourable outcome. The weights are graded as follows: those from underprivileged groups with a 'positive' decision outcome receive higher weights than those with a 'negative' decision outcome, and those from privileged groups with a 'negative' decision outcome receive higher weights than those with a 'positive' decision outcome. The classifier is then given both the dataset and the weights as decision input. It doesn't require any adjustment in labelling and so doesn't skew the data as much as techniques such as Massaging do. Instead of drastically altering the dataset itself, it rather ensures that underprivileged groups are prioritised in the decision-making process where they would have been discriminated against [29].

3.4 Optimising Fairness at the In-Processing Stage

Adversarial De-biasing is the chosen technique for this stage. It is considered to be more generalisable than comparable in-processing techniques and is therefore often favoured for this reason. It is also model-agnostic and fairly robust [20].

This suits the aim to work towards a more generalisable framework for mitigating discrimination in a variety of clinical decision-making applications.

Adversarial De-biasing, presented in paper [20], learns a classifier to maximise model accuracy whilst introducing another network - known as an adversary – with biased behaviour: it works, for example, to try and detect the protected attribute using only the class label predictions. The adversary's ability to perform such tasks gives an insight into the level of discrimination present in the model. The original learning algorithm is then adjusted to compensate the loss function of the adversary, thereby countering the discrimination.

In this way, the model's predictions are modified based on a singular fairness constraint at a time. Adversarial De-biasing works well with both regression and classification- based models and is a model-agnostic approach [20].

3.5 Optimising Fairness at the Post-processing Stage

The final technique to be adopted is *Reject-Option - based classification* [17]. This is often preferred over other post-processing techniques due to its deterministic nature, as opposed to equalised odds - based post-processing algorithms which contain a randomised component. Other benefits of this technique lie in its ease of use and flexibility.

Reject-Option – based classification manipulates the outcome of the algorithm based on protected groups: it improves the outcomes for underprivileged groups whilst making those for privileged groups slightly less favourable. It achieves this by rejecting traditional decision rules of assigning instances solely based on posterior probabilities, instead establishing a critical region whereby instances are assigned opposing labels to either improve or impair their designated outcome depending on the value of their sensitive attribute.

It is acknowledged that this method works well with any probabilistic classifier and supposedly can greatly improve discrimination-awareness in standard classifiers. Furthermore, it is suggested that this method facilitates interpretability and gives more control to decision-makers, as is often the case for post-processing techniques [17].

3.6 Fairness Metrics

The fairness metrics were selected from the available range in the toolkit, focusing on those which measure group discrimination and were deemed most appropriate for the context of resource-allocation. In light of this, the selected metrics are: *Disparate Impact*, *Statistical Parity*, *Average Odds*, and *Equal Opportunity* [28].

For each metric, the 'ideally fair' value, i.e. the value which denotes no discrimination towards either group, is 0. For Disparate Impact, any value > 0 indicates a decision made in favour of the privileged group, with values < 0 representing a preference for the underprivileged group, and vice versa for all remaining metrics. For an ML model to be recognised as acceptably fair given some fairness metric, it must score beneath a specified 'fairness threshold': a value indicating permissible deviations from the ideally fair value for this metric. The chosen metrics are briefly described below.

Statistical Parity Difference. The Statistical Parity difference is the difference between the rate of desirable outcomes for an underprivileged group and the rate of desirable outcomes for a privileged group. The fairness threshold for Statistical Parity is 0.1. Here, Statistical Parity ensures that any given patient has an equal probability of enrolment in the care treatment program, regardless of race.

Disparate Impact. The Disparate Impact measure is calculated based on the ratio of the rate of a favourable outcome for an underprivileged group against that for a privileged group. It should be noted that the Disparate Impact measure included here is *1 - min(DI, 1/DI)*, where DI is the original Disparate Impact measure with ideally fair value of 1. This transforms the ideally fair value to 0 which is consistent with the remaining fairness metrics. The fairness threshold for Disparate Impact is 0.2. In this scenario, Disparate Impact ensures both black and white patients have an equal probability of enrolment in the care treatment program, but also takes into account proportional representation.

Average Odds Difference. The Average Odds difference is the difference between a privileged group and an underprivileged group in terms of the mean average of the True Positive (TP) rate and False Positive (FP) rate for the classification outcome. The fairness threshold for Average Odds is 0.1. In this scenario, Average Odds ensures that, when making predictions, the classifier correctly assigns patients and incorrectly assigns patients to the care treatment program, on average, at the same rate across both racial categories.

Equal Opportunity Difference. The Equal Opportunity Difference represents the difference of TP rates in the classification outcome when comparing one value of the sensitive attribute with another. The fairness threshold for Equal Opportunity is 0.1. In this scenario, Equal Opportunity ensures that, when making predictions, the classifier correctly assigns patients to the care treatment program at the same rate across both racial categories.

4 Results

Table 2 below presents a summary of the performance of all three techniques and how this relates to that of their equivalent plain model without discrimination mitigation. A 'Mean Fairness Measure' was included to gauge a sense of the overall performance of each model – in terms of fairness – by calculating the mean average of the deviation from the ideally fair value (in this case, 0) across all fairness metrics.

Table 2. Model performance for each technique across a range of performance indicators, including balanced accuracy and fairness metrics

	Balanced accuracy	Statistical parity difference	Disparate impact measure	Average odds difference	Equal opportunity difference	Mean fairness measure: overall	Mean fairness measure: improvement
Reweighing: plain	0.6899	−0.0988	0.2613	−0.0667	−0.0350	0.11545	-
Reweighing: logistic regression	0.6855	−0.0486	0.1312	−0.0364	−0.0236	0.05995	−0.0555
Adversarial de-biasing: plain	0.6048	0.0482	0.1368	−0.00495	−0.0006	0.0475	-
Adversarial de-biasing	0.5987	0.0254	0.1912	−0.0110	−0.0146	0.0723	+0.0248
Reject-option - based classification: plain	0.6853	−0.1324	0.3384	−0.0956	0.0448	0.1530	-
Reject-option - based classification: statistical parity	0.6880	−0.0604	0.1851	−0.0240	0.0260	0.0881	−0.0652

Each technique is represented by the version which outperformed all others for the Dissecting-Bias dataset, other versions here referring to those recommended in the AIF360 toolkit [28]. For example, the Reject-Option - Based Classification experiment included three different versions, each conditioned on a different fairness metric (to construct the critical region for rejection). The version conditioned on Statistical Parity was the one which performed with optimal fairness.

Figure 1 provides an overview for the comparison of the three techniques. Taking the Mean Fairness Measure, it displays the deviation of fairness from the 'ideally fair' value achieved by the adjusted model when compared with the unadjusted model corresponding to the same technique. It is important to note that any reduction in the deviation represents a positive outcome, as the closer to the ideal value, and therefore a deviation value closest to 0 is desired.

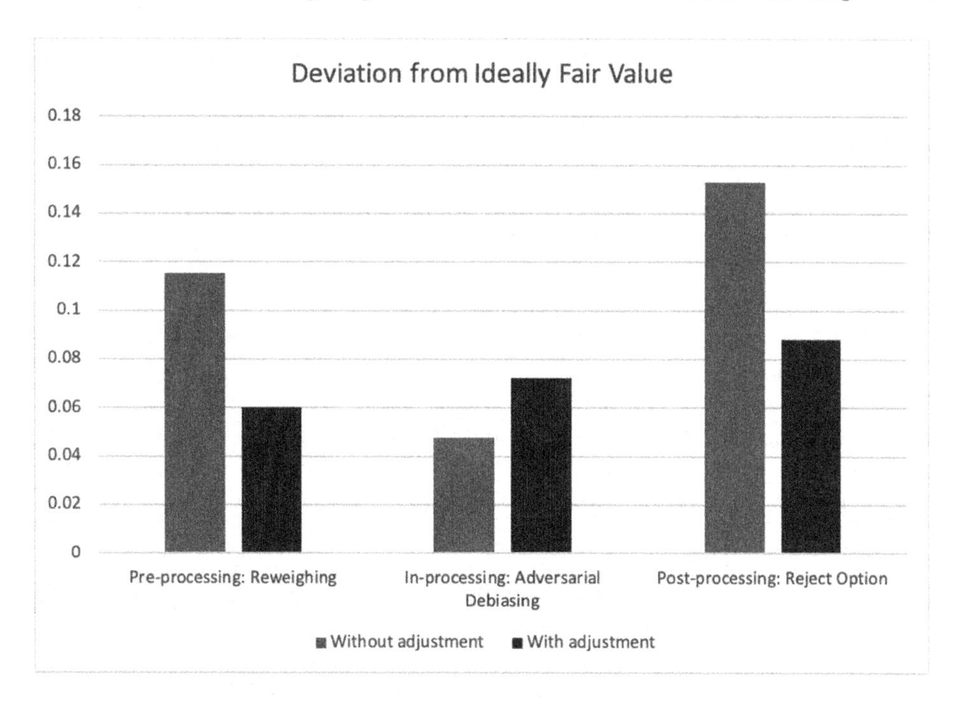

Fig. 1. Bar graph showing overall average deviation from ideally fair value for each technique, both before and after making adjustments for discrimination mitigation

5 Discussion

As is evident from Figure 1, fairness was substantially improved when applying Reweighing and Reject-Option – based classification, but not when applying Adversarial Debiasing. Given the suspiciously low initial fairness value for the model employed when applying Adversarial Debiasing, this increase could be due to the model failing to detect any substantial discrimination when processing the data. Although Reject-Option – based classification saw the greatest reduction in fairness deviation, the final Reweighing model was the most fair of all adjusted models.

In light of this, it became clear that overall performance in this scenario was largely dependent on algorithmic technique. The pre- and post – processing techniques proved to be the most successful, as not only did they improve the fairness measures, they also brought each metric well beneath the threshold for permissible deviations in fairness. This has a twofold benefit: it implies that algorithmic pre- and post- processing techniques have the potential to successfully reduce inexplainable discrimination in clinical resource-allocation algorithms and bring it to an negligible level. Considering the extent of discrimination present in the initial algorithm and the potential implications associated with this, this proves highly promising. An additional merit - as can be observed from Table 2 - was that model accuracy was not demonstrably affected to any great extent when applying mitigation.

Findings from previous research into ML fairness in other disciplines indicate that it is perhaps more beneficial to abate discrimination earliest in the process as possible before it becomes heavily embedded in the model, rendering it difficult to both identify and remove [6,11]. The findings of this study largely support this hypothesis, given the success of the Reweighing technique. This encourages the adoption of fairness-enhancing measures to be implemented during the pre-processing stage of algorithmic processing within clinical ML.

It is important to note that addressing discrimination in ML is a complex, multi-faceted issue, in terms of both measuring and optimising fairness [15]. Therefore, progress may be made towards a more generalisable framework for clinical resource-allocation decision models but no singular method can be said to be best-suited for all contexts especially as it is difficult to even define the 'best' performance in terms of fairness, given the fluidity of the definition of ML fairness [6,15]. In this study, adjustments at the pre- and post- processing stages were shown to have the greatest impact on improving fairness; however, given that only one technique (and its associated variations) was tested at each stage, a repetition of similar experiments using other techniques would be advantageous. Moreover, although the techniques in the AIF360 toolkit are effective in optimising distributive fairness, this particular toolkit was designed exclusively for distributive fairness and so would not be suitable for other types of fairness which may also easily occur in clinical ML-based decision-making, such as fairness in offering equal diagnostic accuracy [2,25,28].

Complications arise when considering the source of discrimination when making decisions guided by ML, as this is often a combination of both computational processes and human error [15]. In clinical decision-making, it is important to be wary of automation complacency and ensure that algorithmic output is used as an aid to physician judgement rather than a replacement [31]. Adjusting algorithmic processes cannot completely eliminate discrimination in clinical decision-making: societal developments to promote fairness are also necessary [15].

6 Conclusion

This work was motivated by the challenge of addressing the current lack of investigation into discrimination in ML in clinical decision-making due to the myriad potential repercussions of this [10,14,25]. The originality of the work arises in the comparison of state-of-the-art fairness-enhancing algorithmic processing techniques across every stage of processing for clinical ML resource allocation.

Specifically, the algorithmic techniques Reweighing, Adversarial De-biasing, and Reject-Option – based Classification were employed and compared by their ability to improve fairness in the context of a known discriminatory clinical ML decision tool used to make decisions on treatment allocation for millions of patients in the US [9]. In this way, this study leverages previous work on algorithmic discrimination mitigation by incorporating scientifically validated techniques available through IBM's AIF 360 toolkit, which offers the latest pioneering techniques substantiated by expert researchers in the bias-mitigation field [28]. In combination with this, it builds upon similar work which aims to combat the known problem of discrimination arising when using ML-based decision support in healthcare [9]. This helps integrate the latest progress of both the data science and healthcare communities.

The results proved to be very promising: two of the three techniques showed a substantial reduction in discrimination against the protected group. Discrimination was reduced to the extent that it was brought under the threshold for acceptable deviation from the 'ideally fair' value. These benefits were reaped without any considerable detriment to model accuracy.

For this particular scenario, mitigating discrimination at the pre-processing stage was found to be the most effective, which is consistent with multiple findings from other applications outside of the medical domain [16,29]. Reduction beneath the threshold for permissible deviations in fairness not only implies that adopting such a technique would make such algorithms fairer, it also suggests that algorithmic pre-processing potentially has the capacity to eradicate all traces of inexplainable discrimination in clinical risk-prediction and resource-allocation models. Considering the extent of discrimination present in the initial algorithm and its possible implications, this is highly encouraging.

It must be noted that discrimination mitigation is a complicated issue in the domain of clinical resource-allocation models, and approaches should also be tailored to individual context. However, given the fact that the methods used in this study were chosen due to their flexibility and generalisability, it can be anticipated that they would also perform successfully given a different dataset. This paper sheds light on the high potential for success when applying discrimination mitigation techniques to algorithmic decision models in the context of clinical resource-allocation (with particular evidence in favour of making adjustments at the pre-processing stage), and therefore further steps should be taken to ensure the inclusion of these or similar methods in similar algorithms.

Finally, it is recommended that future research focuses on consolidating the results found in this paper, expanding the investigation to include wider clinical contexts and other types of fairness, and taking necessary steps to enforce fairness requirements in real-world algorithmic clinical decision models. This topic demands further attention given that one of the key components of high-quality healthcare is equitability [32]. It cannot be ignored that fairness in clinical decision-making of all forms - including algorithmic decision-making - is paramount to providing equitable care.

References

1. Hague, D.C.: Benefits, pitfalls, and potential bias in health care AI. N C Med J. **80**(4), 219–223 (2019). https://doi.org/10.18043/ncm.80.4.219
2. Cai, X., Perez-Concha, O., Martin-Sanchez, F., Day, R., Roffe, D., Gallego, B.: Real-time prediction of mortality, readmission, and length of stay using electronic health record data. J. Am. Med. Inform. Assoc. **23**(3), 553–61 (2016). https://doi.org/10.1093/jamia/ocv110

3. Deo, R.C.: Machine learning in medicine. Circulation **132**(20), 1920–30 (2015). https://doi.org/10.1161/CIRCULATIONAHA.115.001593
4. Žliobaitė, I.: Measuring discrimination in algorithmic decision making. Data Min. Knowl. Discov. **31**(2), 1–30 (2017). https://doi.org/10.1007/s10618-017-0506-1
5. Pedreschi, D., Ruggieri, S., Turini, F.: Discrimination-aware data mining. In: Proceedings of the 14th ACM SIGKDD International Conference on Knowledge Discovery and Data Mining, Las Vegas, USA, August 2008. https://doi.org/10.1145/1401890.1401959
6. Calders, T., Žliobaitė, I.: Why unbiased computational processes can lead to discriminative decision procedures. In: Custers, B., Calders, T., Schermer, B., Zarsky, T., (eds.) Discrimination and Privacy in the Information Society. Studies in Applied Philosophy, Epistemology and Rational Ethics, vol. 3, pp. 43–57. Springer, Berlin, Heidelberg (2013). https://doi.org/10.1007/978-3-642-30487-3
7. Hahn, U., Harris, A.J.L.: What does it mean to be biased: motivated reasoning and rationality. Psychol. Learn. Motiv. **61**, 41–120 (2014). https://doi.org/10.1016/B978-0-12-800283-4.00002-2
8. Haijan, S., Bonchi, F., Castillo, C.: Algorithmic bias: from discrimination discovery to fairness-aware data mining (2016). https://doi.org/10.1145/2939672.2945386
9. Obermeyer, Z., Powers, B., Vogeli, C., Mullainathan, S.: Dissecting racial bias in an algorithm used to manage the health of populations. Science **366**(6464), 447–53 (2019). https://doi.org/10.1126/science.aax2342
10. Gianfrancesco, M.A., Tamang, S., Yazdany, J., Schmajuk, G.: Potential biases in machine learning algorithms using electronic health record data. JAMA Int. Med. **178**(11), 1544–7 (2018)
11. Kamiran, F., Žliobaitė, I.: Explainable and non-explainable discrimination in classification. In: Custers, B., Calders, T., Schermer, B., Zarsky, T., (eds.) Discrimination and Privacy in the Information Society. Studies in Applied Philosophy, Epistemology and Rational Ethics, Chap. 8, vol. 3, pp. 155–170. Springer, Berlin, Heidelberg (2013). https://doi.org/10.1007/978-3-642-30487-3
12. d'Alessandro, B., O'Neil, C., LaGatta, T.: Conscientious classification: a data scientist's guide to discrimination-aware classification. Big Data **5**(2), 120–34 (2017). https://doi.org/10.1089/big.2016.0048
13. Beretta, E., Santangelo, A., Lepri, B., Vetrò, A., De Martin, J.C.: The invisible power of fairness. how machine learning shapes democracy. In: Meurs, M.-J., Rudzicz, F. (eds.) Canadian AI 2019. LNCS (LNAI), vol. 11489, pp. 238–250. Springer, Cham (2019). https://doi.org/10.1007/978-3-030-18305-9_19
14. Friedler, S.A., Scheidegger, C., Venkatasubramanian, S., Choudhary, S., Hamilton, E.P., Roth, D.: A comparative study of fairness-enhancing interventions in machine learning. In: Proceedings of the Conference on Fairness, Accountability, and Transparency (FAT* 1919). Association for Computing Machinery, New York, NY, USA, pp. 329–338 (2019). https://doi.org/10.1145/3287560.3287589
15. Lee, N.T.: Detecting racial bias in algorithms and machine learning. J. Inf. Commun. Ethics Soc. **16**(3) (2018). https://doi.org/10.1108/JICES-06-2018-0056
16. Kamiran, F., Calders, T.: Classifying without discriminating. In: From 2009 2nd International Conference on Computer, Control and Communication, Karachi, pp. 1–6 (2009). https://doi.org/10.1109/IC4.2009.4909197
17. Kamiran, F., Karim, A., Zhang, X.: Decision theory for discrimination-aware classification. In: IEEE 12th International Conference on Data Mining (2012)
18. Khan, S.R., Manialawy, Y., Wheeler, M.B., Cox, B.J.: Unbiased data analytic strategies to improve biomarker discovery in precision medicine. Drug Discov. Today **24**(9), 1735–48 (2019). https://doi.org/10.1016/j.drudis.2019.05.018

19. Calmon, F.P., Wei, D., Vinzamuri, B., Ramamurthy, K.N., Varshney, K.R.: Optimized pre-processing for discrimination prevention. In: 31st Conference on Neural Information Processing Systems (NIPS: Long Beach, CA, USA (2017)
20. Zhang, B.H., Lemoine, B., Mitchell, M.: Mitigating unwanted biases with adversarial learning. In: AAAI/ACM Conference on Artificial Intelligence, Ethics, and Society (2018)
21. Slack, D., Friedler, S., Givental, E.: Fairness warnings and Fair-MAML: learning fairly with minimal data. In: Proceedings of the 2020 Conference on Fairness, Accountability, and Transparency (FAT* 2020). Association for Computing Machinery, New York, NY, USA, pp. 200–209 (2019). https://doi.org/10.1145/3351095.3372839
22. Davidson, I., Ravi, S.S.: Making existing clusterings fairer: algorithms, complexity results and insights. Proc. AAAI Conf. Artif. Intell. **34**(04), 3733–3740 (2020). https://doi.org/10.1609/aaai.v34i04.5783
23. Du, X., Pei, Y., Duivesteijn, W., Pechenizkiy, M.: Fairness in network representation by latent structural heterogeneity in observational data. In: Proceedings of the Thirty-Fourth AAAI Conference on Artificial Intelligence, vol. 34, pp. 3809–3816 (2020)
24. Duivesteijn, W., Feelders, A.: Nearest neighbour classification with monotonicity constraints. In: Daelemans, W., Goethals, B., Morik, K. (eds.) ECML PKDD 2008. LNCS (LNAI), vol. 5211, pp. 301–316. Springer, Heidelberg (2008). https://doi.org/10.1007/978-3-540-87479-9_38
25. Rajkomar, A., Hardt, M., Howell, M.D., Corrado, G., Chin, M.H.: Ensuring fairness in machine learning to advance health equity. Ann. Int. Med. **169**(12), 866–72 (2018). https://doi.org/10.7326/M18-1990
26. Morid, M.A., Kawamoto, K., Ault, T., Dorius, J., Abdelrahman, S.: Supervised learning methods for predicting healthcare costs: systematic literature review and empirical evaluation. AMIA Ann. Symp. Proc. **2017**, 1312–21 (2018)
27. Lin, K., Li, Z.: Dissecting-bias data [Online dataset]. GitHub (2019). https://gitlab.com/labsysmed/dissecting-bias/-/tree/master/data
28. Bellamy, R.K.E., et al.: IBM research. AIF360 (AI Fairness 360) [Online Open Source Toolkit] (2018). https://aif360.mybluemix.net
29. Kamiran, F., Calders, T.: Data preprocessing techniques for classification without discrimination. Knowl. Inf. Syst. **33**, 1–33 (2011). https://doi.org/10.1007/s10115-011-0463-8
30. Calders, T., Kamiran, F., Pechenizkiy, M.: Building classifiers with independency constraints. In: IEEE ICDM Workshop on Domain Driven Data Mining. IEEE press (2009)
31. Mittelstadt, B.D., Allo, P., Taddeo, M., Wachter, S., Floridi, L.: The ethics of algorithms: mapping the debate. Big Data Soc, 1–21 (2016). https://doi.org/10.1177/2053951716679679
32. Williams, J.S., Walker, R.J., Egede, L.E.: Achieving equity in an evolving healthcare system: opportunities and challenges. Am. J. Med. Sci. **351**(1), 33–43 (2016)

WeakAL: Combining Active Learning and Weak Supervision

Julius Gonsior(✉), Maik Thiele, and Wolfgang Lehner

Technische Universität Dresden, Dresden, Germany
{julius.gonsior,maik.thiele,wolfgang.lehner}@tu-dresden.de

Abstract. Supervised Learning requires a huge amount of labeled data, making efficient labeling one of the most critical components for the success of Machine Learning (ML). One well-known method to gain labeled data efficiently is Active Learning (AL), where the learner interactively asks human experts to label the most informative data point. Nevertheless, even by applying AL in labeling tasks the amount of human effort is still too high and should be minimized further.

In this paper therefore we propose WeakAL, which incorporates Weak Supervision (WS) techniques directly into the AL cycle. This allows us to reduce the number of annotations by human experts while keeping the same level of ML performance. We investigate different WS strategies as well as different parameter combinations for a wide range of real-world datasets. Our evaluation shows that for example in the context of Web table classification, 55% of otherwise manually retrieved labels can be generated by WS techniques with a negligible loss of test accuracy by 0.31% only. To further prove the general applicability of our approach we applied it to six datasets from the AL challenge from Guyon et al., where over 90% of the labels could be computed by the WS techniques, while still achieving competitive competition results.

Keywords: Information extraction · Active Learning · Semi-supervised · Machine Learning · Weak Supervision · Classification

1 Introduction

Acquiring training data for supervised learning, such as classification, requires substantial human effort, which already led to many research activities with the goal to increase data efficiency and to minimize the need for manual annotation. The first one is *Active Learning* (AL) that deals with the problem of selecting samples from an unlabeled pool for labeling, e.g. by a human annotator, such that the performance of the model to be learned is maximized. The second one is *Weak Supervision* (WS) that uses a labeled ground-truth to compute labels for the unlabeled data, to improve the quality of the classifier.
Traditionally WS is applied after a small high-quality dataset has been obtained, e.g. through AL. In an optimal setting, AL would query only a few representative samples for each class and the other labels would be derived using WS

A. Appice et al. (Eds.): DS 2020, LNAI 12323, pp. 34–49, 2020.
https://doi.org/10.1007/978-3-030-61527-7_3

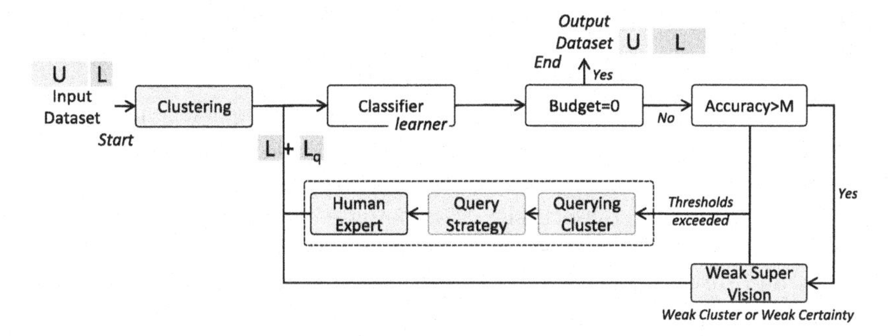

Fig. 1. WEAKAL Overview

techniques. However, in practice this is often not the case: Either too many redundant labels from the AL cycle were obtained, which could also have been generated by WS, or the obtained labels don't work well in combination with WS and produce a lot of false labels.
Therefore, in this paper, we propose WEAKAL, which extends the AL cycle by different WS techniques (see Fig. 1). Given a small initial labeled dataset $\mathcal{L}$ and a large unlabeled pool $\mathcal{U}$, we first cluster the combined samples of $\mathcal{L}$ and $\mathcal{U}$. Then the classifier is trained on $\mathcal{L}$, and as long as the human labor budget is not exhausted, WEAKAL augments the labeled dataset by additional samples. At this point, in the traditional AL cycle, only human experts would be queried. In WEAKAL however, also WS techniques are directly incorporated to obtain labels. If a minimum amount of labeled data is available, which is ensured by an accuracy threshold M, the WS strategies are queried. We propose to use two WS techniques: WEAKCLUST and WEAKCERT. The first one propagates the majority label in a cluster to the unlabeled samples of the cluster, whereas the second one uses the predicted label by the classifier. Based on the parameters for the respective WS strategies, they either return so-called *weak* labels or nothing, indicating that they are not confident enough. Depending on the present labeled data $\mathcal{L}$ and the parameters, the weak labels add more or less label noise. However, by using well-tweaked parameters this can be kept to a minimum. If the WS strategies are not confident enough human experts are consulted, where first a *cluster query strategy* (CQS) identifies a cluster, from which thereafter the *query selection strategy* (QS) selects the samples for the query. The generated labels are be added to the labeled set $\mathcal{L}$ and the cycle starts again.

Contribution. In this paper, we introduce WEAKAL that extends the AL cycle by different WS techniques. In a comprehensive experimental study, we show that combining AL and WS provides very good results in terms of human effort and classification accuracy for many real-world datasets. Our experiments show that the classification models trained on the data determined by AL and WS can safely reduce the amount of human-retrieved annotations by 50%–90% while maintaining the same level of accuracy, and even improving it by a few percentage points.

Outline. The remainder of this paper is organized as follows: In Sect. 2, we present the typical methods used within an AL cycle, which is extended by WS strategies in Sect. 3. Section 4 describes the setup of the experiments we conducted to prove our hypothesis. We compare different evaluation metrics and combinations of WS strategies on multiple real-world datasets. The results are shown and discussed in Sect. 5. Finally, we present related work in Sect. 6 and conclude in Sect. 7.

2 Active Learning Foundations

WEAKAL makes use of typical AL techniques, such as a *cluster query strategy*, a *query strategy* as well as *batching* of samples. Therefore, in Subsect. 2.1, we give an overview of some popular query strategies, which are used in our experiments and emphasize the importance of the right batch size in Subsect. 2.2.

2.1 Active Learning Query Strategies

In this section, we shortly introduce the different strategies for choosing the most informative queries out of a set of given unlabeled samples. Each strategy approximates the contained informativeness of unlabeled data for a potential classifier.

Random Sampling is a common AL query strategy and found application in [1]. Unlike the other methods, random sampling chooses queries at random and fully independently of their informativeness. However, even with this strategy, a rise in prediction accuracy is possible, since the amount of training data is steadily increased. We use random sampling as a baseline to compare the other strategies.

Uncertainty Sampling chooses queries that are the most uncertain to predict. Hence, learning these queries should result in more certain predictions of the classifier. We compare three uncertainty metrics: least confident, margin sampling, and entropy [2]. Least confidence [3] tries to capture the probability, that the classifier is mislabeling the data using the posterior probability P where $\hat{y}$ is the most likely prediction:

$$QS_{x,LC} = \underset{x}{\operatorname{argmax}}\, 1 - P(\hat{y}|x), x \in \mathcal{U} \tag{1}$$

Information about other classes next to the most probable one is not taken into account by this strategy. Margin sampling [4] in contrast uses the posteriors for the first $\hat{y}_1$ and second most probable classes $\hat{y}_2$ and samples the instances with the smallest margin between those two:

$$QS_{x,SM} = \underset{x}{\operatorname{argmin}}\, P(\hat{y}_1|x) - P(\hat{y}_2|x) \tag{2}$$

Entropy uncertainty [5,6] uses all possible classes and captures the entropy of a given distribution. It should, therefore, work well on classification problems with many classes:

$$QS_{x,E} = \underset{x}{\operatorname{argmax}} - \sum_i P(y_i|x) \log P(y_i|x) \tag{3}$$

2.2 Batch Sizes

It is common practice in machine learning to train a model on batches of samples instead of single data points. As the retraining of the classifier cannot be done in real-time it is also easier for human experts to label a batch of data points at once. Batches also allow parallelization of the human annotation process. Early experiments suggested that the batch size has no real impact on the efficiency of the AL process. We use therefore a reasonably small batch size of 10, which is small enough to show changes during the AL cycle in detail, but also large enough to keep the experiment runtime under control.

3 Weak Supervision Enhanced Active Learning Cycle

In this section, we propose WEAKAL, combining the strengths of AL and WS. We claim, that it is beneficial, to prioritize during AL the retrieval of those unlabeled data points, which do not only directly increase the classifier's performance but also lead to more weakly labeled data. We propose an active learning cycle that incorporates WS, resulting in significantly less human interaction, whereas the accuracy achieved is kept on the same level.
Algorithm 1 shows the overall WEAKAL cycle. The AL process starts with two datasets: the unlabeled sample set $\mathcal{U}$ and the already labeled dataset $\mathcal{L}$, where $|\mathcal{L}| << |\mathcal{U}|$. WEAKAL requires that both $\mathcal{L}$ and $\mathcal{U}$ consist of clusters, $\mathcal{L}_c$, and $\mathcal{U}_c$. Each cluster is defined as a tuple consisting of the feature vector x and, in case of the labeled set, the corresponding label y:

$$\begin{aligned} \mathcal{L}_c &= \{(x_{l1}, y_{l1}), (x_{l2}, y_{l2}), \ldots\} \\ \mathcal{U}_c &= \{x_{u1}, x_{u2}, \ldots\} \end{aligned} \tag{4}$$

The main task of the AL cycle is to iteratively increase the set of labeled data $\mathcal{L}$ by identifying the most promising cells in $\mathcal{U}$. The cycle stops when a predefined *budget* B (line 1 in Algorithm 1) of available user interaction is exhausted. At the beginning of each cycle, the classifier f is retrained on the labeled set $\mathcal{L}$.
If a minimum training accuracy M is reached (line 3), WEAKAL utilizes WEAKCLUST and WEAKCERT (Subsect. 3.2) instead of asking the human experts. The budget remains untouched for WS labels, as these queries come for free without human interaction. Both WS strategies have threshold parameters, α, β, and γ. If the thresholds are not met, human experts are used instead. For that, first, a cluster $\mathcal{U}_c$ of the unlabeled data is selected based on the *cluster query strategy CQS* (line 10, (Subsect.. 3.1). Then the utilized *query strategy* (line 14, (Subsect. 2.1) selects as much, as per the *batch size BS* defined, unlabeled

samples q from the selected cluster $\mathcal{U}_c$. The human experts are then asked for the label, and the budget is reduced accordingly. At the end of each cycle the newly labeled data $\mathcal{L}_q$ is added to $\mathcal{L}$ (line 18) and q removed from $\mathcal{U}$ (line 19), and the process starts again by retraining the classifier on the extended dataset.

Algorithm 1 WEAKAL

Input: small clustered labeled start set $\mathcal{L}$, large unlabeled clustered dataset $\mathcal{U}$, query strategy QS, batch size BS, human user interaction budget B, minimum training accuracy before WS M, minimum certainty threshold α, minimum cluster homogeneity β, minimum labeled cluster size γ and a cluster query strategy CQS

Output: labels for $\mathcal{U}$

```
1: while B > 0 do
2:     f ← TRAIN(L)
3:     if ACC(f, L) > M then                                 ▷ Weak Supervision
4:         q, y_q ← WEAKCLUST(L, U, β, γ)
5:         if y_q = ∅ then
6:             q, y_q ← WEAKCERT(U, f, α)
7:         end if
8:     end if
9:     if y_q = ∅ then                                       ▷ Traditional Active Learning
10:        U_c ← CQS(U)
11:        INIT(q)
12:        for 1, ..., MIN(BS, COUNT(U_c)) do
13:            B ← B − 1
14:            APPEND(q, QS(U_c, f))
15:        end for
16:        y_q ← ASKHUMANEXPERTS(q)
17:    end if
18:    MERGE(L, (q, y_q))
19:    REMOVE(U, q)
20: end while
21: return L
```

3.1 Cluster Query Strategies to Support WeakClust

The basic idea of the clustering approach is to save human effort by labeling the entire cluster instead of individual data points. This strategy requires a minimum amount of labels per cluster. We investigate the following three clustering strategies:

Single Cluster Strategy. To compare the approach of limiting the human experts' queries to a single cluster $\mathcal{U}_c$ per AL cycle to the typical approach of using the entire set of unlabeled points $\mathcal{U}$, the *single cluster strategy* puts all unlabeled data into a single cluster, simulating thereby the absence of a cluster strategy.

Random Cluster Strategy. This strategy selects a cluster at random and acts as a second baseline.

Most Uncertain Cluster Strategy. The *most uncertain cluster strategy* can be used in three different flavors, depending on the used uncertainty query strategy: least confidence, smallest margin, and entropy (see Subsect. 2.1). By obtaining the labels for the most uncertain points per cluster only those remain unlabeled that are more likely part of the class-homogeneous core of the cluster. For each cluster $\mathcal{U}_c \in \mathcal{U}$, the query selection strategy is used first to calculate $QS(x)$ for each sample x. After that, the cluster samples are sorted in descending order based on the value of the query selection strategy. The highest most uncertain data points within the batch size BS are stored in $\overline{\mathcal{U}_c}$. The cluster with the highest sum of query selection certainties is selected accordingly:

$$\mathcal{U}_c = \underset{\mathcal{U}_c}{\operatorname{argmax}} \sum_{x \in \overline{\mathcal{U}_c}} QS(x), \text{ for } \mathcal{U}_c \in \mathcal{U} \tag{5}$$

3.2 Weak Supervision Techniques

We selected two WS techniques, WEAKCLUST and WEAKCERT, which we believe work best alongside the AL process, and can easily be incorporated into it. WEAKCLUST propagates the labels of a partially labeled cluster to the entire cluster, and WEAKCERT returns the predicted labels of the trained classifier. Especially the WEAKCLUST technique is optimal for AL, as in an ideal scenario first one sample gets queried per cluster, and then using more most uncertain samples from the cluster, the hypothesis of the first sample gets confirmed or dismissed. Each WS strategy has thresholds that have to be met to confidently add the weak labels to the labeled dataset. A minimum amount of labeled data M needs to be made available first for both WS techniques to justify applying WS. Otherwise, the risk of many false labels from a severely overtrained classifier is increasing. All parameters have to be chosen carefully, as WS automatically computes the annotations and with a suboptimal starting point, many wrong labels can be produced.

Weak Certainty uses the probability of the trained classifier to decide for the unlabeled samples. The pseudocode is given in Algorithm 2. Contrary to the uncertainty AL query strategies, the most certain data points are labeled by this WS strategy. For each unlabeled sample x the predicted label y and the probability σ of the classifier f are calculated (line 4 in Algorithm 2). If the probability is higher than the threshold α (line 5), the predicted label gets assigned. All found labels and samples are stored in the lists ys and q (line 6 and 7). WEAKCERT is therefore basically the application of a single iteration of *self-training* [7].

Algorithm 2 WEAKCERT

Input: unlabeled data points $\mathcal{U}$, trained classificator f, minimum certainty threshold γ

Output: labels y for a set of unlabeled data points q

```
1: INIT(q, ys)
2: for U_c ∈ U do
3:     for x ∈ U_c do
4:         y, σ ← CLASSWITHPROB(f, x)
5:         if σ > α then
6:             APPEND(ys, y)
7:             APPEND(q, x)
8:         end if
9:     end for
10: end for
11: return q, ys
```

Algorithm 3 WEAKCLUST

Input: labeled data $\mathcal{L}$, unlabeled data $\mathcal{U}$, minimum cluster homogeneity size β, minimum ratio labeled-unlabeled samples γ

Output: labels $\hat{y}$ for the cluster of unlabeled data $\mathcal{U}_c$

```
1: for L_c, U_c ∈ L, U do
2:     if COUNT(L_c)/COUNT(U_c) > γ then
3:         ŷ ← MOSTFREQUENTLABEL(L_c)
4:         if COUNT(ŷ)/COUNT(L_c) > β then
5:             return U_c, ŷ
6:         end if
7:     end if
8: end for
9: return ∅, ∅
```

Table 1. Datasets used in experiments

Name	Domain	#Classes	#Features	#Samples	Majority class
DWTC [9]	Table classification	4	227	5,777	39.84%
HIVA [8]	Chemoinformatics	2	1,617	42,678	96.48%
IBN_SINA [8]	Handwriting recognition	2	92	20,722	62.16%
ORANGE [8]	Marketing	2	230	50,000	98.22%
SYLVA [8]	Ecology	2	216	145,252	93.85%
ZEBRA [8]	Embryology	2	154	61,488	95.42%

Weak Cluster identifies clusters that contain i) a lot of labeled data and ii) almost only samples of with the same label (Algorithm 2). To achieve i) the ratio between labeled and unlabeled samples of the cluster is computed. Only clusters where the ratio is above the threshold γ are considered further (line 2). The second criteria, ensuring ii), is checked by calculating the ratio between the most common class $\hat{y}$ and the size of the cluster (line 4). The first cluster, with a ratio above a threshold β, is returned with $\hat{y}$ as the label for the unlabeled portion. The quality of the underlying clusters has a high impact on the quality of this WS technique. Desirable are many smaller clusters containing only samples of the same class. Note that the propagation of labels from the cluster only applies to unlabeled samples. Possible noise in the clusters should have already been removed by the most uncertainty query strategies (see Subsect. 2.1) before the thresholds for WeakClust are met.

4 Experimental Setup

We first introduce the datasets used in our evaluation in Subsect. 4.1. In Subsect. 4.2, we discuss the parametrization of the clustering approaches. To evaluate the performance of WeakAL we conduct a large hyperparameter search on different real-world datasets, which is described in Subsect. 4.3. Finally, in Subsect. 4.4 we present the evaluation metrics used for our experiments. The code for all experiments is publicly available[1] under the AGPL-3.0 license.

4.1 Datasets

We perform our experiments using six real-world datasets described in Table 1. All datasets are used to train classification models and most of them contain noisy data, have missing values, sparse feature representation, and unbalanced class distributions. Except for DWTC, all datasets come from the *Active Learning Challenge* performed by Guyon et al. in 2010 [8]. In our experiments, 50 % of the data was withheld as a test set.

[1] https://github.com/jgonsior/weakal.

Table 2. Hyperparameter search space

Hyperparameter	Search Range
Query Selection	random, uncertainty least confidence, uncertainty max margin, uncertainty entropy
Cluster Selection	dummy, random, most uncertain least confidence, most uncertain max margin, most uncertain entropy
WEAKCLUST?	Yes/No
WEAKCERT?	Yes/No
M, α, β, γ	$[0.5, 1.0]$

4.2 Performed Clustering Strategies

As stated in Sect. 3, WEAKAL expects the input data to be clustered. Since the underlying data characteristics for a dataset to be labeled are often not known, we decided for generally applicable clustering algorithms: For the large datasets, SYLVA and HIVA, we used Mini-batch k-Means [10] and Agglomerative Clustering [11] for the smaller ones. The parameter k, representing the number of clusters, is set to $n\_samples/8$ and the batch size to $min(n\_samples/100, n\_features)$. These parameters ensure an average number of 8 data points per cluster, which proved to work best in our experiments. For our use case, the high number of clusters is not a problem as long as their homogeneity is high. Note, that WEAKAL does not depend on a specific cluster strategy, i.e. others can be used as well.

4.3 Hyperparameter Search

The quality of the used WS technique depends highly on the correct selection of the parameter values. We chose therefore an extensive random hyperparameter search to find optimal values and obtain an understanding of the sensitivity of the WS techniques regarding their parameters. Table 2 lists all the relevant hyperparameters. In total, 37,290 hyperparameters for the DWTC dataset have been tested, which was possible due to its smaller size and 4,922 hyperparameters for all other datasets.

We used a random forest [12] classifier with standard parameters in all experiments since it showed good results for every dataset and is comparatively fast. In addition to that, it has been reported that random forest classifiers are good at dealing with potentially noisy, weak labels [13].

4.4 Evaluation Metrics

To compare the results of an AL run we need to measure its effectiveness in achieving the overall goal of AL, to learn an accurate model with a minimum amount of labeling cost. A desirable metric for WEAKAL takes into account a) amount of user-retrieved labels, b) classifier evaluation metrics, such as accuracy,

F1-Score or AUC, and c) an average of the classifier evaluation metrics throughout all AL iterations. The last two options are quite similar but have different objectives. The average is desirable, to not only compare AL runs where only the final iteration resulted in a high-quality run but also those, where no measurable quality drop occurred. As one normally does not know a priori when to optimally stop the AL process, one has to look at the average to not stop before the final "good" queries. As a direct result of this, the final accuracy is needed, as the average loses the information if the quality is good in the end or just in the beginning.

We determine two basic metrics that should be analyzed in conjunction for a meaningful evaluation: the ratio of weakly labeled data *% saved human effort* hu and the final *test accuracy* acc_end. For the saved human effort 0.0 equals zero savings and 1.0 is the optimal case where no human experts were needed for labeling at all. Besides, two compound metrics are calculated: The first one is called *combined score*, which is the harmonic mean of the two basic metrics:

$$combined\_score = \frac{2 * acc\_end * hu}{acc\_end + hu} \tag{6}$$

It captures the tradeoff between a desired low amount of saved human effort and high test accuracy.

To compare ourselves to the results of the Active Learning Challenge described in Subsect. 4.1, we further report the *global score*, which was used in the challenge [8]. Note, that we compute the AUC values for the global score only for human experts' queries, where WS queries are considered "free".

5 Evaluation

In this section, we want to investigate the feasibility of WeakAL and show whether the integration of weak supervision techniques in the AL cycle has the potential to reduce the human labeling effort. We start in Subsect. 5.1 by analyzing the impact of the human experts' query budget. In Subsect. 5.2, we compare the effect of no WS, WeakCert, and WeakClust individually. Further, we combine both strategies, WeakCert and WeakClust, and report the results for the best working parameter combinations for the DWTC dataset. In Subsect. 5.3, we show that the combination of AL and WS even can achieve higher accuracies than AL alone. In Subsect. 5.4 we show on an example how the two WS strategies are applied in practice. The results on the datasets from the AL challenge are given in Subsect. 5.5. In Subsect. 5.6, we provide some rules of thumb for good hyperparameter values.

5.1 Budget Size Matters

As stated in Subsect. 4.4 the used budget size has a direct impact on the evaluation metrics. Figure 2 plots the best-achieved accuracy for the DWTC dataset for budgets between 0 and 3,000. It can seen that the accuracies for smaller budgets fluctuate a lot. We focus the following analysis therefore on larger budgets, due to stable and more reproducible results. The best result for the *combined score*, representing the balance of the tradeoff between a low amount of saved human effort and high test accuracy, is achieved for a budget of 260 human experts' queries, with an accuracy of 79.20%.

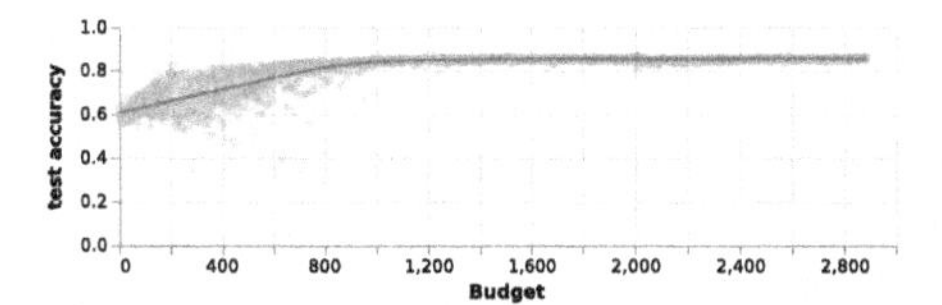

Fig. 2. Comparison of best-achieved test accuracy for different budgets for the DWTC dataset

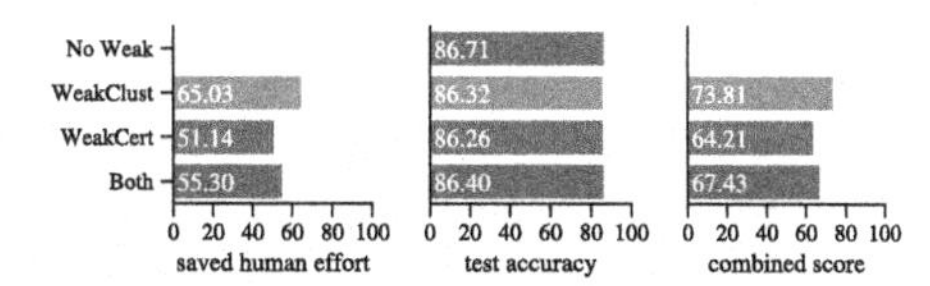

Fig. 3. Comparison of the best result for all possible WS combinations with a budget of 1,500, selected after test accuracy

5.2 Comparison of Best-in-class AL + WS

In this analysis, we compare the results for an AL cycle without WS, with each of WEAKCLUST and WEAKCERT on their own, and with a combination of both. Again we used the DWTC dataset for this experiment. As the saved human effort cannot be calculated, when no WS is being applied, the best results are selected based on the test accuracy, whereas the budget was kept to a fix value of 1,500. Figure 3 shows that the accuracy of WEAKAL using a combination of both WS techniques is only 0.31%, WEAKCERT 0.45%, and WEAKCLUST 1.39% lower than the AL cycle without WS. Hence, it can be concluded that application of WS techniques in WEAKAL provides a significant saving of human effort, with a negligible reduction of the test accuracies. WEAKCERT and WEAKCLUST in combination only achieve a slightly better accuracy than the individual techniques since both often label the same samples. While the savings of human effort are higher for WEAKCLUST compared to WEAKCERT in this example, this is not true in general but highly depends on the budget.

5.3 General Improvement Using WS

So far we only compared the results for selected examples of good parameter combinations. In the following, we investigate the overall distribution of *all possible parameter combinations* for a fairly small budget of 200, due too limitations in compute time, for the DWTC dataset. Figure 4 shows three distributions: in blue all parameter combinations without using WS strategies, in orange all parameter combinations showing an accuracy improvement due to the WS-labels, and in green all parameter combinations using WS and showing a performance decline. The improvement was measured by comparing the test accuracy of a classifier trained on the human experts' queries alone, to a classifier trained on the human expert queries and the automatically generated WS-labels. In addition

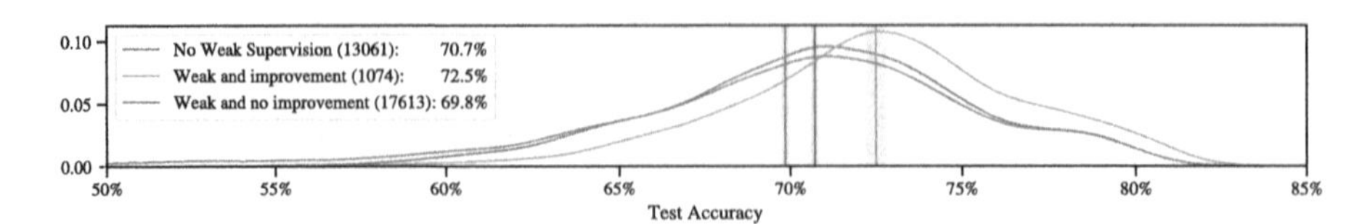

Fig. 4. Kernel density estimation and mean including 95% confidence interval given a budget of 200 samples for the DWTC dataset

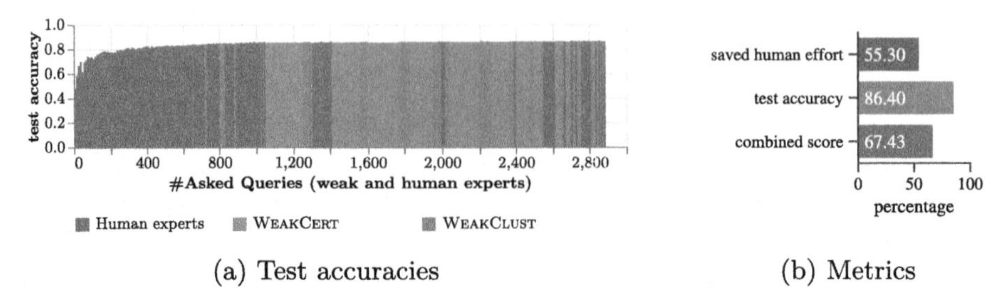

Fig. 5. Highest achieved accuracy result for the DWTC with a budget of 1,500(Color figure online)

to the kernel density estimations of the distribution, the mean value is shown including the 95% confidence interval. It can been seen that incorporating WS into AL even can improve the average test accuracy by 1.81%. There also exists, a large subset of parameters, which consistently achieve a lower accuracy using WS. Nevertheless, using WS directly within the AL cycle, with the right parameters, has the potential to not only lower the human effort drastically but also to even increase the accuracy.

5.4 Detailed Results for DWTC Dataset

This section gives some deeper insights into how WEAKCERT and WEAKCLUST work together in detail, shown exemplarily for the DWTC dataset. Figure 5 shows the best-achieved accuracy result for the DWTC dataset with a budget of 1,500 human experts' queries. Fig. 5.a made up of colored rectangles, one for each iteration of the AL cycle. The width of a rectangle is the number of retrieved labels during the iteration, the height the achieved test accuracy. In the beginning, a lot of human experts' queries (blue) are requested, until WEAKAL is confident enough to apply the WS techniques. From then on, WEAKCLUST (orange), WEAKCERT (green), and the human expert queries alternate constantly. Most of the labels can be generated automatically by WS, without negatively influencing the accuracy. The alternation between WS and the human experts' queries shows, that it is indeed beneficial to apply WS during the AL cycle, and not after a gold standard is obtained.

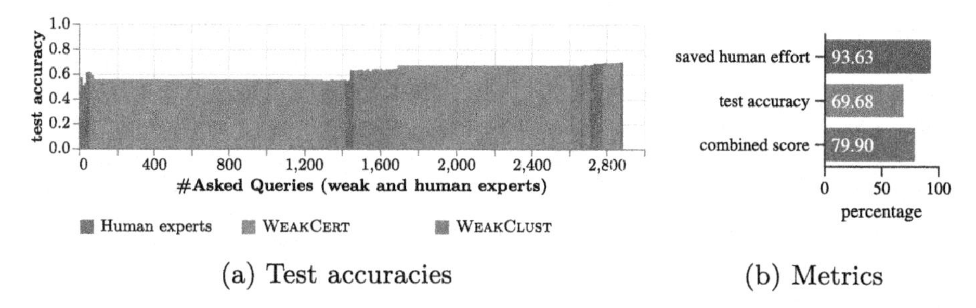

(a) Test accuracies (b) Metrics

Fig. 6. Best-achieved test accuracy result for DWTC with a budget of 200

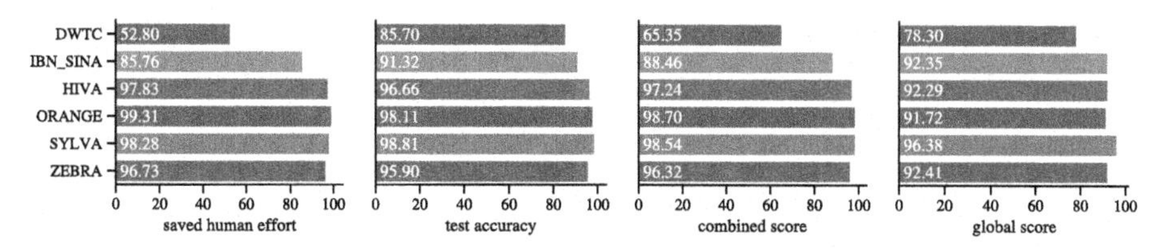

Fig. 7. Comparison of all analyzed datasets with a budget of 1,000 and the combination based on the global score

In contrast, Fig. 6.a displays the test accuracies for a significantly smaller budget of only 200 human experts' queries. Here, most of the labels are generated by WEAKCERT. Interestingly the accuracy is dropping after the first big block of WEAKCERT labels, but rises quickly again after a few oracle queries. After the second smaller block of human expert queries at around 1,450, the accuracy goes even up purely based on WS labels. So without human interaction, the accuracy of the classifier can be improved, which shows that the effectiveness of WS goes further than just producing redundant labels.

The bar charts in Fig. 5.b and Fig. 6.b illustrate the results for different evaluation metrics. It is obvious, that smaller budget results in more saved human effort, accepting a loss of the test accuracy. The *combined score* metric shows, that the tradeoffs between the saved human effort and the test accuracy are worse for the bigger budget. This is not surprising, as it always takes much more data to further improve an already good accuracy than a poor one.

5.5 Active Learning Challenge Datasets

To compare our results to other common AL strategies we selected the training datasets from the AL challenge [8]. The respective best results are shown in Fig. 7. We used a budget of 1,000 and the global score of the AL challenge as the evaluation metric to select the best results. A budget of 500 was too small for most of the datasets and resulted in highly overfitted classifiers with reported test accuracies of under 1%. The figures show, that all AL challenge datasets have high values for all metrics. As the datasets all are highly imbalanced binary decision problems, a vast amount of labels can be generated automatically by WS, as most samples are of the same label anyway. Since we have been not able to determine the budgets used in the AL challenge, a comparison of the results is only partly fair. Nevertheless, under the assumption, that a budget of 1,000 is close to the budget used in the competition, our achieved results are competitive to the winners of the AL challenge. Not all datasets are suited for WEAKCLUST as the underlying data could not be clustered well. Clustering worked good for HIVA, IBN_SINA, and ZEBRA.

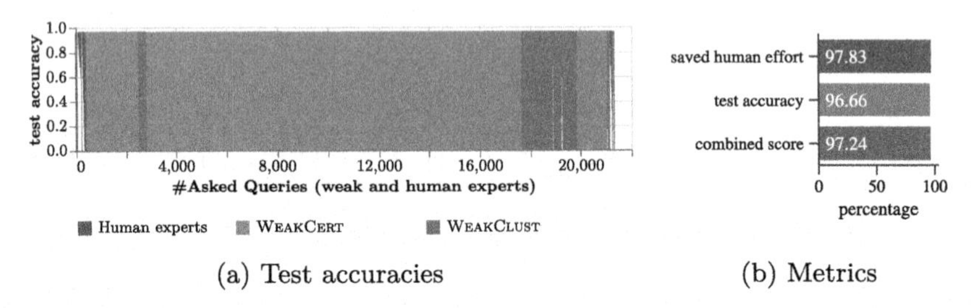

(a) Test accuracies (b) Metrics

Fig. 8. Best-achieved global score result for IBN_SINA with a budget of 1,000(Color figure online)

Figure 8 shows the results for the IBN_SINA dataset in detail. In the beginning, human experts' queries are being collected (blue bars). After that, almost all labels can be generated using the WEAKCERT (orange) and WEAKCLUST (green). Both WS techniques alternate between each other, with few human experts' queries in between. Again, this is an argument for directly embedding WS into the AL process in WEAKAL. The plots for the other datasets from the AL competition looked quite similar. We therefore based our evaluation primarily on the more interesting results for the DWTC dataset.

5.6 Recommended Parameters

Based on the investigation in Subsect. 5.3, we would like to make recommendations which parameter combinations work well in practice: First, the best parameters depend a lot on the desired test accuracy. The higher the test accuracy, the more data is needed, and the higher the thresholds should be set. The minimum training accuracy M should be approximately 10% less than the desired test accuracy. For the query sampling strategies, *uncertainty max margin* performed best, closely followed by *uncertainty least confident*. The selected cluster strategy depends heavily on the quality of the underlying clusters and the amount of available data. For the several datasets, such as ZEBRA, which could be clustered well, and, therefore, WEAKCLUST is applied often, *most uncertain least confident* works best, whereas for those where no meaningful clusters could be found, the dummy cluster strategy is leading. Good values for the threshold α are values between the desired test accuracy up to 1.0. The parameters for WEAKCLUST, β, and γ, should be considered jointly. The lower the cluster homogeneity ratio β, the higher the minimum labeled cluster size γ should be. Good values for both are between 0.75 and 0.95, keeping in mind the reverse dependency between both.

6 Related Work

Semi-Supervised Learning. There exist various techniques to combine the abundance of unlabeled data with labeled data in a classification setting, and the terminology about that is not always clear according to our experience. The most common term is semi-supervised learning, which uses unlabeled data to verify assumptions based on labeled data [14]. Semi-supervised learning also has been incorporated with AL, e.g. using Expectation-Maximization [15] or using multi-view co-training [16]. Adjacent to semi-supervised learning, weak supervision assumes that high-quality ground truth labels exist, and many noisy labels for the rest of the data. In our case, we produce high-quality data when querying the human experts, and noisy labels when using the WS. Following the terminology introduced in [17], we use the term weak supervision when talking about *inaccurate supervision*. We focus on the aspect of generating labels of weak-supervised learning, intending to reduce the amount of needed ground-truth labels.

Clustering. [18] proposes to query only the cluster centers in different feature spaces, and to use a majority vote afterward for the unlabeled data to determine their labels. In [19] graph-based clustering was directly incorporated into an AL setting. Other techniques, such as *label propagation* [20] iteratively propagate labels based on a small labeled ground truth set using a combination of random walk and clamping. Another approach is to use a small set of ground truth labels and program synthesis techniques to automatically generate labeling functions [21].

7 Conclusions

Annotating training data for supervised learning, such as classification, requires substantial human effort. While utilizing Active Learning during the annotation process already decreases the amount of human labor, we argue that AL should be combined with WS to further reduce the number of annotations made by human experts. Therefore, we proposed WEAKAL, a WS extension to a typical AL cycle employing different cluster query strategies to query those samples, which further supports the WS strategies. In a comprehensive study, we selected and compared the proposed strategies as well as multiple parameter combinations. For a Web table classification task the results show that 55.30% of human labeling effort can be saved using automatic WS labels, with only a negligible loss of test accuracy by 0.31%. We showed, that with optimal parameters, a test accuracy improvement by 1.81% can be attributed solely to WS. We further applied WEAKAL on datasets from the AL challenge from Guyon et al., where over 90% of the labels could be generated automatically, while still achieving competitive results, thus proving the general applicability of our proposed approach.

Acknowledgements. This research and development project is funded by the German Federal Ministry of Education and Research (BMBF) and the European Social Funds (ESF) within the "Innovations for Tomorrow's Production, Services, and Work" Program (funding number 02L18B561) and implemented by the Project Management Agency Karlsruhe (PTKA). The author is responsible for the content of this publication.

References

1. Cohn, D., Atlas, L., Ladner, R.: Improving generalization with active learning. Mach. Learn. **15**(2), 201–221 (1994). https://doi.org/10.1007/BF00993277
2. Settles, B.: Active learning literature survey. Computer Sciences. Technical Report 1648 (2010)
3. Lewis, D.D.: A sequential algorithm for training text classifiers: corrigendum and additional data. SIGIR Forum **29**(2), 13–19 (1995)
4. Scheffer, T., Decomain, C., Wrobel, S.: Active hidden Markov models for information extraction. In: Hoffmann, F., Hand, D.J., Adams, N., Fisher, D., Guimaraes, G. (eds.) IDA 2001. LNCS, vol. 2189, pp. 309–318. Springer, Heidelberg (2001). https://doi.org/10.1007/3-540-44816-0_31
5. Shannon, C.E.: A mathematical theory of communication. SIGMOBILE Mob. Comput. Commun. Rev. **5**(1), 3–55 (2001)
6. Baram, Y., El-Yaniv, R., Luz, K.: Online choice of active learning algorithms. JMLR **5**, 255–291 (2004)
7. Scudder, H.J.: Probability of error of some adaptive pattern-recognition machines. IEEE Trans. Inf. Theory **11**, 363–371 (1965)
8. Guyon, I., Cawley, G., Dror, G., Lemaire, V.: Results of the active learning challenge. JMLR **16**, 19–45 (2011)
9. Eberius, J., Braunschweig, K., Hentsch, M., Thiele, M., Ahmadov, A., Lehner, W.: Building the dresden web table corpus: a classification approach. In: BDC, pp. 41–50. IEEE (2015)
10. Sculley, D.: Web-scale k-means clustering. In: WWW, pp. 1177–1178 (2010)
11. Ward Jr., J.H.: Hierarchical grouping to optimize an objective function. J. Am. Stat. Assoc. **58**(301), 236–244 (1963)
12. Breiman, L.: Random forests. Mach. Learn. **45**(1), 5–32 (2001)
13. Folleco, A., Khoshgoftaar, T., Van Hulse, J., Napolitano, A.: Identifying learners robust to low quality data. Informatica (Slovenia) **33**, 245–259 (2009)
14. Zhu, X.: Semi-supervised learning literature survey. Computer Science, University of Wisconsin-Madison 2 (2008)
15. McCallum, A., Nigam, K.: Employing EM and pool-based active learning for text classification. In: ICML, pp. 350–358 (1998)
16. Muslea, I., Minton, S.N., Knoblock, C.A.: Active + semi-supervised learning = robust multi-view learning. In: ICML (2002)
17. Zhou, Z.H.: A brief introduction to weakly supervised learning. Natl. Sci. Rev. **5**(1), 44–53 (2017)
18. Dara, R., Kremer, S., Stacey, D.: Clustering unlabeled data with SOMs improves classification of labeled real-world data. In: Proceedings of the 2002 International Joint Conference on Neural Networks, vol. 3, pp. 2237–2242 (2002)

19. Bodó, Z., Minier, Z., Csató, L.: Active learning with clustering. In: Active Learning and Experimental Design workshop@AISTATS, vol. 16, pp. 127–139 (2011)
20. Zhu, X., Ghahramani, Z.: Learning from labeled and unlabeled data with label propagation. Technical report (2002)
21. Varma, P., Ré, C.: Snuba: automating weak supervision to label training data. VLDB **12**(3), 223–236 (2018)

Clustering

Constrained Clustering via Post-processing

Nguyen-Viet-Dung Nghiem[1](✉), Christel Vrain[1], Thi-Bich-Hanh Dao[1], and Ian Davidson[2]

[1] Univ. Orléans, INSA Centre Val de Loire, LIFO EA 4022, 45067 Orléans, France
nguyen-viet-dung.nghiem@etu.univ-orleans.fr,
{christel.vrain,thi-bich-hanh.dao}@univ-orleans.fr
[2] Department of Computer Science, University of California, Davis, USA
davidson@cs.ucdavis.edu

Abstract. Constrained clustering has received much attention since its inception as the ability to add weak supervision into clustering has many uses. Most existing work is algorithm-specific, limited to simple together and apart constraints and does not attempt to satisfy all constraints. This limits applications including where satisfying all constraints is required such as fairness. In this work, we take the novel direction of post-processing the results of a clustering algorithm (constrained or unconstrained) as a combinatorial optimization problem to find the best allocation of instances to clusters whilst enforcing constraints. Experiments show that when evaluated on a ground truth, our method is competitive in terms of clustering quality with the more recent approaches while being more computationally efficient. Finally, since all constraints are satisfied, our work can be applied to areas such as fairness including both group level and individual level fairness.

Keywords: Constrained clustering · Combinatorial optimization · Fairness constraints

1 Introduction

The area of constrained clustering has been well studied since the seminal papers nearly twenty years ago. The addition of constraints allows domain knowledge or partial supervision from labels to be added to clustering. Constrained versions of a variety of different algorithms exist such as k-means [1], EM [2], spectral [3]. This has allowed application of clustering in domains it has previously been problematic to apply due to strong domain knowledge such as neuroscience [4], intelligent tutoring systems [5] and GPS data [6].

Table 1 surveys classic well known constrained clustering algorithms with respect to the constraint types they can integrate and whether all the constraints are satisfied. As can be seen, most existing work is limited: a) to simple together and apart constraints and b) only partially satisfying them. This is due to a number of reasons including that satisfying constraints is typically intractable [7,8], so approaches for satisfying all constraints will be time-consuming. Finally, a limitation with the entire field is that constrained clustering is typically for a given algorithm (i.e. not algorithm agnostic).

A. Appice et al. (Eds.): DS 2020, LNAI 12323, pp. 53–67, 2020.
https://doi.org/10.1007/978-3-030-61527-7_4

Table 1. A brief survey of constrained clustering algorithms.

Algorithm	Constraint types (Satisfaction of all constraints)
k-Means [1]	Together/Apart (Yes)
EM [2]	Together/Apart (No)
Spectral clustering [3]	Together/Apart (No)
Deep clustering [9]	Together/Apart, Cardinality, Triplet (No)

In this work we take the novel direction of taking the results of an unconstrained or constrained clustering algorithm and post-processing them to satisfy the constraints. The process considers as a combinatorial optimization problem with the objective of assigning instances to the most likely clusters while satisfying all the constraints. This has the benefit of allowing algorithms for which there are no constrained versions to enjoy the benefit of being constrained. Algorithms for which there are no constrained versions are not uncommon and quite popular such as the Louvain method [10] and DBScan [11]. The only requirement of our method is that the result of the clustering algorithm can be represented as a degree of membership to each and every cluster. This is represented by a matrix that we call cluster fractional allocation matrix (CFAM). For probabilistic algorithms such as EM this is directly produced by the algorithm but for other methods it can also be calculated. For example, this can be calculated using the distance to the cluster centroid for k-means. We show in Table 2 for several clustering algorithms how to obtain a CFAM. Our method therefore can be used with a wide variety of clustering algorithms including distance based, probabilistic and deep learning.

Table 2. An overview of how to obtain a cluster fractional allocation matrix from several algorithm outputs.

Algorithm	Method to obtain a CFAM
Centroid based (i.e. K-means)	Normalized distance to centroids
Exemplar based (i.e. K-Medoids)	Normalized distance to exemplar
Graph based (i.e. Spectral Clustering)	Distance in embedded space
Deep clustering (i.e. DEC)	Renormalized embedding vector
Modularity based (i.e. Louvain Method)	Normalized geodesic distance between node and centroid node
Density based (i.e. DB-Scan)	Normalized distance to core points

Our method takes as input a cluster fractional allocation matrix and formulates a combinatorial optimization problem to assign the instances based on these whilst satisfying the constraints. We take the novel direction of modeling this as an affectation problem which can be extended to a general Integer Linear Program (ILP) formulation. This allows us to model a variety of new styles of constraints not previously explored. These new types of constraints and their use are shown in Table 3. Individually though these constraints are useful, together they can be used for innovative applications such as satisfying both individual and group level fairness.

Table 3. An overview of the new constraint types that our approach can integrate.

Constraint	Use	Example
Cluster-overlap	Constrain the amount of overlap between clusters	Each instance can belong to at least r and at most s clusters
Neighborhood	Link the instances to their neighbors	Each person should be in the same group with at least 50% of persons having the same level (individual level fairness)
Property-cardinality	Constrain the number of a specific type of instances in each cluster	The ratio of males/females in the clusters are approximately equivalent (group level fairness)
Attribute level	Constrain number of possible clusters for instances with specific property	Distribute young people across at least r and at most s clusters

Our work makes several contributions:

- We formulate a novel direction of post-processing a clustering to satisfy constraints and we model it by an Integer Linear Program (ILP). We show that it can integrate several existent and new types of constraints.
- We experimentally verify our method can post-process the results of unconstrained and constrained algorithms improving the results of both (Table 4).
- We explore the direction of using our framework to ensure both group level and individual level fairness (Table 10).

We begin by covering related work in Sect. 2, review and introduce constraint types in Sect. 3 then overview preliminaries before covering our formulation in Sect. 4. We experimentally verify our work in the semi-supervised setting in Sect. 5 and in the fairness domain in Sect. 6.

2 Related Work

Constrained clustering is a central area of AI. It enables the integration of prior knowledge, in the form of constraints, to guide a clustering algorithm. To our knowledge all existing work attempts to simultaneously find a clustering which satisfies prior knowledge in the form of constraints.

Prior knowledge is typically from the labels of some objects, which generate pairwise must-link/cannot-link constraints (two objects must be/cannot be in the same cluster) [2]. These kinds of constraints are the most popular and lots of work are developed to handle them [1,3,6,7,12]. Domain expert can also provide guidance beyond pairwise constraints, which can be requirements at cluster level. Several methods have been developed to integrate these constraints: minimal cluster size constraint [13], balanced clustering [14,15], or bounds on the density of the clusters [7]. However, these types of constraints cannot be *combined* as they are formulated in different paradigms.

More generic methods that allow integrating several different types of constraints are declarative approaches, which are developed using a general optimization tool, such as ILP [16–18], SAT [8,19] or constraint programming [20,21]. These methods find a *global* optimal solution that satisfy all the constraints, they suffer however from a lack of efficiency, which prevent them to handle large datasets. Deep clustering methods [22] have recently been proposed that simultaneously optimize the representation and a clustering objective. Based on this approach deep constrained clustering methods such as [9] integrate several types of constraints by adding a satisfaction loss in the objective function. They can handle large datasets but do not guarantee satisfying all the constraints.

In contrast in our work constraints are enforced a posteriori. We use a matrix that presents the degree of membership of each instance to each cluster, which can be computed by a clustering method, or a constrained clustering method, for it does not satisfy all the constraints. Our method finds the best assignment that satisfies all the constraints. It can integrate different types of constraints without relearning the model and can be used with a wide variety of clustering algorithms including distance based, probabilistic and deep learning.

3 Existing and New Constraint Types

Consider the problem of clustering a dataset of N instances into K clusters. Simple pairwise constraints are typically generated from labeled data: must-link (if labels agree) or cannot-link (if labels disagree). An extension of this idea to continuous side information was recently explored [9] to generate triplet constraints: $ab|c$ which means a is more similar to b than to c. As a consequence if a and c belong to the same cluster, or so do b and c, then a, b, c must all belong to the same cluster [23]. Our framework can model both these popular existing constraints and in addition the following constraints. Though some of these have been studied before, our framework allows using all these constraint types simultaneously.

- Geometric constraint gives an upper/lower bound on the diameter of each cluster or on the split between the clusters [7].
- Cluster size constraint requires an upper/lower bound on the number of instances in each cluster. Balanced clustering constraint [13] is a special case, where the clusters are of approximately equal size.
- Cluster-overlap constraint limits the amount of overlapping between clusters by constraining the number of clusters each instance can belong to.
- Property-cardinality constraint gives a lower/upper bound on the number of instances satisfying a *property* in each cluster. Group-level fairness can be expressed, which requires that in each cluster, the ratio of a type of instances (e.g. females) is approximately the same as the ratio in the whole dataset [24].
- Neighborhood constraint links instances with one or more of their neighbors. An example is individual fairness [25] which requires that individuals who are close together should be grouped together.
- Attribute level constraint limits the number of possible clusters for all the instances with a specific property.

Satisfying all these types of constraints is a NP-Hard problem [7] which makes them difficult to satisfy after each iteration of an algorithm. Instead we attempt to satisfy them only once after the clustering algorithm has converged.

4 Constraint Post-processing Formulation

Consider a cluster fractional allocation matrix (CFAM), which is a $N \times K$ matrix P of real numbers, where P_{ik} indicates the allocation score of assigning instance i into cluster k (the larger P_{ik} the more likely instance i is to belong to cluster k). This matrix can be directly the result of any probabilistic (constrained or not) clustering method, such as EM. It can also be computed from the clustering result of other (constrained) clustering algorithms such as k-means, as stated in Table 2. We consider a set of constraints $\mathcal{C}$ that must be satisfied. As it is stated in Sect. 3, satisfying all the types of constraints is typically a NP-Hard problem, we propose a formulation based on Integer Linear Programming for this problem.

Objective. We aim at finding an assignment of instances to the most likely clusters while satisfying all the constraints. The problem is therefore finding a $N \times K$ matrix Z of $\{0, 1\}$ with the objective function

$$\arg\max \sum_{i=1}^{N} \sum_{k=1}^{K} P_{ik} Z_{ik}$$

such that the constraints in $\mathcal{C}$ are satisfied. The matrix Z represents a hard assignment: $Z_{ik} = 1$ means instance i is assigned to cluster k. The constraints in $\mathcal{C}$ can be of different types (see Sect. 3). They can be formulated as follows.

Pairwise Constraint. A must-link (or a cannot-link) constraint on two instances i, j is formulated by $\forall k = 1, \ldots, K$, $Z_{ik} = Z_{jk}$ (or $Z_{ik} + Z_{jk} \leq 1$, for a cannot-link constraint).

Triplet Constraint. A triplet constraint $ab|c$ is formulated by:

$$\forall k = 1, .., K,\ Z_{bk} \geq Z_{ak} + Z_{ck} - 1 \ \text{ and } \ Z_{ak} \geq Z_{bk} + Z_{ck} - 1$$

This formulation yields $Z_{bk} = 1$ if $Z_{ak} = Z_{ck} = 1$ and $Z_{ak} = 1$ if $Z_{bk} = Z_{ck} = 1$.

Cluster-Overlap Constraint. Each instance belonging to at least α and at most β clusters can be expressed by:

$$\forall i = 1, \ldots, N,\ \alpha \leq \sum_{k=1..K} Z_{ik} \leq \beta$$

To enforce hard clustering, i.e. each instance in one cluster, α and β is set to 1.

Cluster Size Constraint. Each cluster must contain at least α and at most β instances can be expressed by:

$$\forall k = 1, .., K,\ \alpha \leq \sum_{i=1..N} Z_{ik} \leq \beta$$

Property-Cardinality Constraint. Let $p(i)$ be 1 when an instance i has a property p and 0 otherwise. The fact that each cluster must have at least α and at most β instances having the property p can be enforced by:

$$\forall k = 1,..,K,\ \alpha \leq \sum_{i=1..N} p(i)Z_{ik} \leq \beta$$

The constraint such that in each cluster, the ratio of the instances having p over the size of the cluster is bounded by $[\alpha, \beta]$ is expressed by:

$$\forall k = 1,..,K,\ \alpha.\sum_{i=1..N} Z_{ik} \leq \sum_{i=1..N} p(i)Z_{ik} \leq \beta.\sum_{i=1..N} Z_{ik}$$

Neighborhood Constraint. A neighborhood constraint requires that each instance i must be in the same cluster with at least a ratio α of its neighborhood N_i. Let $N_i(j) = 1$ if instance j is in the neighborhood of i and 0 otherwise. This constraint is expressed by: $\forall i = 1, \ldots, N, \forall k = 1, \ldots, K,$

$$\sum_{j=1..N} N_i(j)Z_{jk} \geq \alpha(\sum_{j=1..N} N_i(j))Z_{ik}$$

Attribute Level Constraint. An attribute-level constraint limits the number of possible clusters for the instances having a specific property. Let $p(i)$ be 1 when the instance i has the property and 0 otherwise. We introduce a variable $t_k \in \{0, 1\}$ for each cluster k, $t_k = 1$ if and only if the cluster k contains some instances having p. This is expressed by:

$$t_k \leq \sum_i p(i)Z_{ik} \text{ and } \forall i = 1, \ldots, N,\ t_k \geq p(i)Z_{ik}$$

The first constraint ensures $t_k = 0$ if there is no instance having p in cluster k. The second sets $t_k = 1$ as soon as an instance having p is in cluster k. Bounds $[\alpha, \beta]$ on the number of clusters containing the instances having the property p are given by: $\alpha \leq \sum_{k=1..K} t_k \leq \beta$.

5 Experiments

Our experiments attempt to address several core questions to understand how our work can be used in conjunctions with existing algorithms and its limitations and benefits. We attempt to address the following questions.

- Does our method improve the *result/output* of unconstrained and constrained clustering algorithms? (See Table 4)
- Is our method comparable with baseline constrained clustering methods? (See Tables 4 and 6)
- How useful are the new types of constraints and their use in combination? (See Tables 7 and 8)
- How does our method scale to large datasets? (See Tables 6, 7 and 8)

5.1 Experiment Setting

Datasets. We use 3 datasets, which are challenging and also used in a recent deep constrained clustering method [9].
MNIST: The dataset is composed of 60,000 handwritten single-digits, with a size of 28-by-28 pixels.
FASHION-MNIST: The dataset contains 60,000 images associated to a label from 10 classes.
REUTERS-10K: Reuters contains around 810,000 English news stories labeled with a category tree [26]. Following DEC [22], we consider only the single label documents belonging to the `corporate/industrial`, `government/social`, `markets` and `economics` categories. A subset of 10,000 examples is randomly sampled and the TF-IDF measure is computed on the 2000 most frequent words.
Baseline Algorithms. The following systems are used in our experiments.
IDEC: (Improved Deep Embedded Clustering) [27] a popular deep clustering method based on auto-encoder.
Kmeans: the classic algorithm but run on the deep embedding representation learned by IDEC.
COP-Kmeans: the classic constrained clustering algorithm for pairwise constraints but again run on the embedded space [6].

MSE-Kmeans: a modified K-means relying on minimum-cost flow algorithm to satisfy cluster size constraints [15], which is run on the embedded space.
DCC: Deep Constrained Clustering [9] which handles pairwise, triplet and balanced-clustering constraints during the clustering process.
Kmeans-Post, IDEC-Post, DCC-Post: our constraint post-processing method applied on the results of K-means, IDEC and DCC, respectively. IDEC and DCC output a cluster fractional allocation matrix (CFAM) which is then used in our method. For K-means, we generate the CFAM P as follows. Let the centroids be μ_k $(1 \leq k \leq K)$, the matrix P is computed by the t-distribution:

$$P_{ik} = -\log \frac{(1+ \parallel x_i - \mu_k \parallel^2)^{-1}}{\sum_{k'}(1+ \parallel x_i - \mu_{k'} \parallel^2)^{-1}}$$

For fairness, since IDEC is a probabilistic algorithm, we run it once and we gave the learned embedded representation to all the systems. Moreover, DCC is initialized with the network and the parameters learned by IDEC. All algorithms are implemented in Python. We use the ILP solver Gurobi version 8.0[1]. Experiments are run on a 2.8 GHz Intel Core i7 processor with 16GB of RAM. The source code is made available and easy to replicate[2].
Evaluation Metric. In all the datasets, the true class of objects is available and we use it as the ground truth to evaluate the accuracy of the clustering. We consider two measures: Normalized Mutual Information (NMI) and clustering accuracy (ACC), with a one-to-one mapping between clusters and labels, computed by The Hungarian algorithm [28]. In both cases the higher the better.

[1] https://www.gurobi.com/.
[2] https://github.com/dung321046/ConstrainedClusteringViaPostProcessing.

5.2 Baseline Comparisons

Pairwise Constraints. We compare our method with baseline systems on MNIST and Fashion datasets with 3,600, 30,000 and 60,000 pairwise constraints. To measure performance, for each number of constraints we average performance over five sets of constraints and report the average and standard deviation over the five trials. Table 4 reports the results on MNIST. Our post-processing method with pairwise constraints always improves the input in terms of NMI and accuracy, with the benefit of satisfying all the constraints. Moreover, it always obtains better results compared to COP-Kmeans and comparable results to DCC. Similar results are also observed on the Fashion dataset see Table 5.

Table 4. Results with pairwise constraints on MNIST

#Pw	Method	NMI	ACC
0	Kmeans	0.8644 ± 0.0000	0.8838 ± 0.0000
	IDEC	0.8539	0.8799
3600	COP-Kmeans	0.8237 ± 0.0324	0.7372 ± 0.0630
	DCC	0.8637 ± 0.0012	0.8938 ± 0.0075
	Kmeans-Post	0.8649 ± 0.0001	0.8843 ± 0.0001
	IDEC-Post	0.8547 ± 0.0002	0.8804 ± 0.0001
	DCC-Post	0.8640 ± 0.0013	0.8940 ± 0.0077
30000	COP-Kmeans	0.8477 ± 0.0195	0.8302 ± 0.0314
	DCC	0.9407 ± 0.0032	0.9786 ± 0.0013
	Kmeans-Post	0.8689 ± 0.0003	0.8876 ± 0.0003
	IDEC-Post	0.8602 ± 0.0007	0.8839 ± 0.0005
	DCC-Post	0.9429 ± 0.0026	0.9796 ± 0.0011
60000	COP-Kmeans	0.8146 ± 0.0319	0.8039 ± 0.0644
	DCC	0.9549 ± 0.0029	0.9847 ± 0.0012
	Kmeans-Post	0.8739 ± 0.0004	0.8917 ± 0.0003
	IDEC-Post	0.8668 ± 0.0005	0.8887 ± 0.0004
	DCC-Post	0.9581 ± 0.0021	0.9860 ± 0.0009

Cluster Size Constraints. Here we compare our method on MNIST and Fashion, with MSE-Kmeans [15], which is developed specifically for cluster size constraints. We use the minimum and the maximum of the true class sizes as a lower bound and a upper bound on the cluster sizes for all the clusters. The results are shown in Table 6. The results show that our general method of incorporating this constraint is competitive compared to a method which is developed specifically for this type of constraints.

5.3 New Constraint Types and Constraint Combinations

Attribute Level Constraints. This new type of constraints requires that the instances having a specific property cannot be widespread over a large number of clusters.

Table 5. Results with pairwise constraints on Fashion

#Pw	Method	NMI	ACC
0	Kmeans	0.6319 ± 0.0000	0.5877 ± 0.0000
	IDEC	0.6320	0.5879
3600	COP-Kmeans	0.6222 ± 0.0152	0.5808 ± 0.0092
	DCC	0.6403 ± 0.0192	0.6378 ± 0.0277
	Kmeans-Post	0.6306 ± 0.0003	0.5877 ± 0.0002
	IDEC-Post	0.6315 ± 0.0003	0.5880 ± 0.0002
	DCC-Post	0.6402 ± 0.0191	0.6377 ± 0.0276
30000	COP-Kmeans	0.6175 ± 0.0043	0.5974 ± 0.0199
	DCC	0.7421 ± 0.0158	0.7989 ± 0.0279
	Kmeans-Post	0.6253 ± 0.0005	0.5901 ± 0.0005
	IDEC-Post	0.6293 ± 0.0005	0.5905 ± 0.0003
	DCC-Post	0.7446 ± 0.0159	0.8019 ± 0.0282
60000	COP-Kmeans	0.6023 ± 0.0059	0.5853 ± 0.0175
	DCC	0.6430 ± 0.2882	0.7180 ± 0.2891
	Kmeans-Post	0.6219 ± 0.0007	0.5923 ± 0.0008
	IDEC-Post	0.6276 ± 0.0008	0.5940 ± 0.0008
	DCC-Post	0.6624 ± 0.2705	0.7409 ± 0.2671

Table 6. Clustering accuracy and NMI for clustering with cluster size constraints

Data - Method	NMI	ACC	Data - Method	NMI	ACC
MNIST - IDEC	0.8539	0.8799	Fashion - IDEC	0.6320	0.5879
MNIST - MSE	0.8536	0.8816	Fashion - MSE	0.5363	0.5387
MNIST - Post	0.8520	0.8796	Fashion - Post	0.5301	0.5425

We consider Reuters-10K and we require that documents that contain some given words should be covered by at most s clusters. We consider two cases with 5 (resp. 10) constraints by randomly selecting 5 (resp. 10) sets of three words, among those whose documents widespread on at most 2 clusters ($s = 2$). We post-process the initial clustering given by IDEC.

Table 7 reports the quality in NMI and accuracy (ACC), the number of instances involved in all the constraints, the number of constraints that are not satisfied by the clustering given by IDEC (it is the input clustering of our system, note that our system produces a clustering satisfying all the constraints), the number of documents that have been assigned to a different cluster after post-processing and the runtime in seconds. The impact of attribute level constraints is quite high. While a pairwise constraint only affects two instances, the average number of instances that are concerned in an attribute level constraint is around 200. The number of instances that have been reassigned is therefore also high. That could explain the slight decrease of NMI and the

Table 7. Impacts of attribute level constraints on Reuters-10K

#Constraints	0	5	10
NMI	0.5279	0.5253 ± 0.0039	0.5219 ± 0.0055
ACC	0.7452	0.7474 ± 0.0070	0.7499 ± 0.0088
#Involved Inst	-	1168.0000 ± 50.5332	2128.0000 ± 143.6760
#Unsat Constr	-	4.8000 ± 0.4000	9.8000 ± 0.4000
#Changed Inst	-	96.6000 ± 57.4895	212.4000 ± 42.9679
Runtime (s)	-	0.0831 ± 0.0147	0.3052 ± 0.3010

slight increase of accuracy, the random constraints without specific domain knowledge could be too strong.

Combinations of Constraints. One benefit of our framework is that it can integrate and satisfy several types of constraints. Among existing constrained clustering methods, only declarative methods can integrate several types of constraints [8,21] while satisfy them all. However they suffer from scalability and cannot handle datasets as big as MNIST and Fashion. In this part, we consider both pairwise (PW) and cluster size (CS) constraints simultaneously. The number of pairwise constraints is set to 30,000 for both runs.

Table 8. Runtime (in seconds) with/without pairwise (PW) and cluster size (CS) constraints on MNIST and Fashion.

Cases	NMI	ACC	# Changes	# Positive changes	Runtime (s)
MNIST					
No constraint	0.85	0.88	-	-	-
CS	0.85	0.88	445.00	18.00	3.95 ± 0.12
PW	0.86 ± 0.00	0.88 ± 0.00	878.80 ± 21.50	507.00 ± 25.27	3.86 ± 0.32
PW + CS	0.86 ± 0.00	0.88 ± 0.00	1067.00 ± 18.77	596.20 ± 23.28	3.86 ± 0.32
Fashion					
No constraint	0.63	0.59	-	-	-
CS	0.53	0.54	8748.00	977.00	4.06 ± 0.21
PW	0.63 ± 0.00	0.59 ± 0.00	2747.40 ± 43.51	977.40 ± 13.84	3.51 ± 0.02
PW + CS	0.54 ± 0.00	0.55 ± 0.00	9600.80 ± 38.28	1580.20 ± 28.08	24.78 ± 2.76

Table 8 reports results for the MNIST and Fashion datasets. It reports the number of instances that have been assigned to a different cluster by the post-process and the number of changes that have lead to the right cluster.

We notice a difference in behavior between MNIST and Fashion. For the first dataset, adding cluster size constraints improves the results while it is not true for Fashion. It can perhaps be explained by a tighter constraint (upper bound minus lower bound is smaller) for Fashion than for MNIST.

For the pairwise constraint, adding constraints has been shown in our experiments to consistently improve the quality of cluster, while the use of cluster size constraints needs more careful consideration. It is worth to study further for a way to relax this constraints when its impact is too high.

Runtime. We report the runtime in seconds of COP-Kmeans, DCC and our method with pairwise constraints in Table 9. To have a fair comparison, we focus only on the computational time for integrating constraints to the initial clustering provided by IDEC without any constraints.

Table 9. Runtime (in seconds) with pairwise constraints using COP-Kmeans, DCC and postprocess.

Data	#Pairwise	IDEC-Post	COP-Kmeans	DCC
MNIST	3600	1.00 ± 0.04	132.38 ± 18.72	1013.76 ± 790.91
	30000	3.86 ± 0.32	103.77 ± 56.39	1381.73 ± 1067.07
	60000	6.81 ± 0.39	70.95 ± 38.41	3277.37 ± 1555.83
Fashion	3600	0.99 ± 0.02	71.00 ± 57.77	5579.30 ± 2761.33
	30000	3.51 ± 0.02	103.71 ± 35.26	7359.35 ± 3927.79
	60000	6.55 ± 0.43	95.28 ± 37.90	3207.00 ± 1057.71

The runtime of each test mainly depends on the number of pairwise constraints. Our method usually gives comparable results for quality but it is substantially faster. On average, our method is more than 10 times faster than COP-Kmeans and 500 times faster than DCC. Indeed, COP-Kmeans has to compute the distance matrix after updating the cluster centers, whereas DCC has to apply back propagation to update all the model parameters. Moreover, post-processing time has a smaller variance between each test than the other methods.

Concerning other constraints, as given in Tables 7 and 8 our method performs in a very reasonable time.

6 Application: Improving the Fairness of Clustering Algorithms

The previous section performed standard comparisons to show our method was comparable in accuracy to existing methods. Here we show the real worth of our approach as it allows combining multiple constraints to address the challenging problem of fairness in clustering. The area of fair clustering has drawn much recent attention. Fairness in clustering can be classified into group fairness and individual fairness. Group-level fairness usually represents statistical fairness notions based on a protected status variable (PSV). Group fairness typically requires that in each cluster, the ratio of each PSV value is approximately equal to the ratio of this type in the whole dataset [24]. Individual-level fairness corresponds to requirements made to individuals. An example of individual fairness requires individuals who are close together to be treated in the same way [25].

Existing work usually ensures just one of these types of fairness, either group fairness [24, 29] or individual fairness [30]. In our best knowledge, no work has considered both individual fairness and group fairness. Taking advantage that our constraint post-processing method can integrate different types of constraints, both types of fairness can be ensured.

We consider the classic fairness dataset Adult with 48,842 instances. Data to cluster on is described by continuous attributes such as age or working hours, and PSV such as gender, education or marital-status.

Group and Individual Fairness. Group fairness is expressed by the requirement that in each cluster, the ratio of each type of instances is approximately the same as this ratio when computed on the whole dataset. The ratio of females in the dataset is about 33.15%. To ensure group fairness, we require that in each cluster, the ratio of females is between $0.3315 - \epsilon$ and $0.3315 + \epsilon$ with $\epsilon = 0.01$. This is ensured by property-cardinality constraints, as defined in Sect. 4.

To ensure individual fairness, we require that each instance i must be in the same cluster as at least 50% of the elements in its neighborhood N_i. For each instance i, the neighborhood N_i is defined by the set of instances having exactly the same education and occupation, and a difference in age less than or equal to 2. This requirement is ensured by neighborhood constraints, as defined in Sect. 4, with $\alpha = 0.5$. We prove that with $\alpha \geq 0.5$, if x and y have exactly the same value on the attributes used to define the neighborhood, then to satisfy the fairness constraints x and y must be in the same cluster.

Baseline Individual Fairness - Most Votes Greedy Method. In order to ensure individual fairness, as a baseline method, we have implemented a greedy method as below. We iterate t times the following procedure: for each unfair instance x, find the cluster k that contains the most instances in the neighborhood of x and change the assignment of x to the cluster k. In the experiment $t = 10$.

Baseline Group Fairness - Fairlet. We use the code produced by [29] which is an improvement of the method ensuring group fairness using fairlets [24]. Fairlets are subsets of objects that respect the given ratio between the two values of a binary attribute. They are computed first then clustering is achieved on them to ensure group fairness. For the dataset Adult, we require the minimum ratio of females over males is higher than 49.37% so that the lower bound for the percentage of females in each cluster is 33.05%.

Our Post-processing K-means Results. Let μ_k $(1 \leq k \leq K)$ be the cluster centers obtained from K-means with the input $X = \{x_i : i \in [1, N]\}$. Then, the allocation matrix P is computed using the t-distribution.

$$P_{ik} = -\log \frac{(1+ \parallel x_i - \mu_k \parallel^2)^{-1}}{\sum_{k'} (1+ \parallel x_i - \mu_{k'} \parallel^2)^{-1}}$$

Finally, we optimize P under three scenarios: only with individual constraints (Post-Ind.), only with group constraints (Post-Group), and both (Post-Combine).

Results and Analysis. Table 10 reports the result obtained on 10 runs with $K = 5$. It reports the clustering quality in terms of the within cluster sum WCS, the number of instances that are unfairly grouped according to individual fairness, the number of clusters that are unfair according to group fairness and the runtime in seconds. The within cluster sum WCS is defined by the sum of squared distances from each instance to the centroid of its cluster.

K-means gives a clustering that is unfair with respect to both group fairness and individual fairness. As expected, ensuring fairness decreases the clustering quality measured by WCS. However, post-processing achieves better quality than the greedy method. We can observe that group fairness and individual fairness are not relevant, ensuring one type of fairness does not ensure the other type, but even worsen. For instance, Fig. 1 shows that clusters computed while ensuring only individual fairness (Post-Ind.) are group unfair (see for instance Cluster 4 with a very low rate of females).

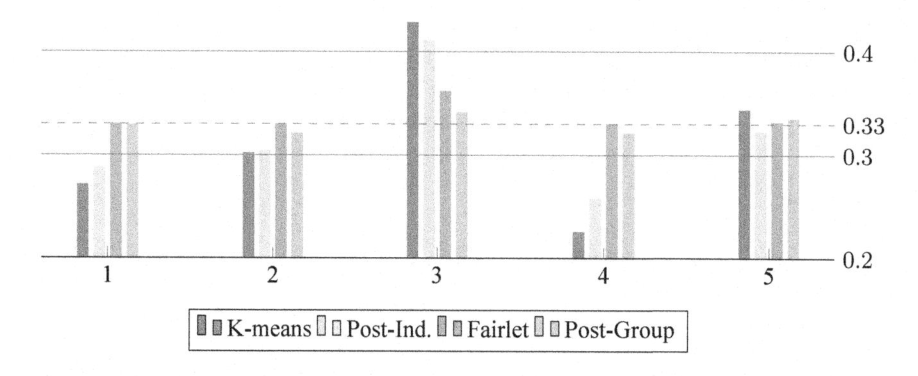

Fig. 1. The ratio of females in each of the five clusters by different methods.

On the other hand, Table 10 shows that Fairlet and Post-Group methods that ensure group fairness do not ensure individual fairness, even the numbers of unfair individuals are higher than the traditional K-means (around 7812 and 5809 respectively, compared to 5686). For group fairness, without upper-bound constraint, Fairlet sometimes produces unfair groups (averaging 0.6 group per test case) while Post-Group ensures both bounds.

In terms of efficiency, the post-processing performs with a very reasonable runtime.

Table 10. Runtime and constraint satisfaction with individual fairness or/and group fairness.

Method	WCS	# Indiv. unfair inst.	# Group unfair clust	Runtime (s)
K-means	8477.14 ± 1.89	5685.60 ± 21.12	5.00 ± 0.00	2.62 ± 0.30
Individual-level fairness				
Most-vote	9071.63 ± 1.84	113.10 ± 5.68	4.70 ± 0.46	18.06 ± 0.71
Post-Ind	9064.86 ± 1.83	0	4.00 ± 0.00	3.93 ± 0.11
Group-level fairness				
Fairlet	9587.47 ± 113.95	7812.30 ± 1102.79	0.60 ± 0.49	36.90 ± 0.84
Post-Group	8581.52 ± 1.30	5809.40 ± 41.81	0	3.16 ± 0.17
Both individual-level and group-level fairness				
Post-Combine	9175.91 ± 1.50	0	0	6.66 ± 1.23

7 Conclusion

Constrained clustering methods can integrate prior knowledge in term of constraints, but they are usually limited on the type of constraints. Moreover, they do not guarantee the satisfaction of the constraints. Declarative methods can handle several types of constraints and satisfy all of them, but they suffer from a lack of efficiency, which prevent them to handle large datasets. In our work, we propose the novel direction of post-processing the results of an unconstrained or constrained clustering algorithm to enforce the constraints a posteriori. Given a matrix that presents the cluster fractional allocation of instances to clusters, our method assigns instances to the most likely clusters while satisfying all the constraints. Our method can handle large datasets, it can integrate all types of popular constraints as well as a variety of new styles of constraints. It can be used with a wide variety of clustering algorithms, including deep learning and we demonstrated its use in the complex setting of ensuring group and individual level fairness using multiple constraints which to our knowledge has not been attempted.

References

1. Wagstaff, K., Cardie, C.: Clustering with instance-level constraints. In: ICML 2000, pp. 1103–1110 (2000)
2. Basu, S., Davidson, I., Wagstaff, K.: Constrained Clustering: Advances in Algorithms, Theory, and Applications, 1 edn. Chapman & Hall/CRC, London (2008)
3. Wang, X., Davidson, I.: Flexible constrained spectral clustering. In: KDD 2010, pp. 563–572 (2010)
4. Walker, P.B., Davidson, I.N.: Exploring new methodologies for the analysis of functional magnetic resonance imaging (fMRI) following closed-head injuries. In: Schmorrow, D.D., Fidopiastis, C.M. (eds.) FAC 2011. LNCS (LNAI), vol. 6780, pp. 120–128. Springer, Heidelberg (2011). https://doi.org/10.1007/978-3-642-21852-1_15
5. Harley, J.M., et al.: Clustering and profiling students according to their interactions with an intelligent tutoring system fostering self-regulated learning. JEDM— J. Edu. Data Min. **5**(1), 104–146 (2013)
6. Wagstaff, K., Cardie, C., Rogers, S., Schrödl, S.: Constrained K-means clustering with background knowledge. In: ICML 2001, pp. 577–584 (2001)
7. Davidson, I., Ravi, S.S.: Clustering with constraints: feasibility issues and the k-means algorithm. In: SDM 2005, pp. 138–149 (2005)
8. Davidson, I., Ravi, S.S., Shamis, L.: A SAT-based framework for efficient constrained clustering. In: ICDM 2010, pp. 94–105 (2010)
9. Zhang, H., Basu, S., Davidson, I.: Deep constrained clustering - algorithms and advances. In: ECML 2019 (2019)
10. Blondel, V.D., Guillaume, J.-L., Lambiotte, R., Lefebvre, E.: Fast unfolding of communities in large networks. J. Stat. Mech. Theor. Exp. **2008**(10), P10008 (2008)
11. Schubert, E., Sander, J., Ester, M., Kriegel, H.P., Xu, X.: DDSCAN revisited, revisited: why and how you should (still) use DBSCAN. ACM Trans. Database Syst. (TODS) **42**(3), 19 (2017)
12. Bilenko, M., Basu, S., Mooney, R.J.: Integrating constraints and metric learning in semi-supervised clustering. In: ICML 2004, pp. 11–18 (2004)
13. Bradley, P., Bennett, K., Demiriz, A.: Constrained k-means clustering, Technical report. MSR-TR-2000-65, Microsoft Research (2000)

14. Ge, R., Ester, M., Jin, W., Davidson, I.: Constraint-driven clustering. In: KDD 2007, pp. 320–329 (2007)
15. Tang, W., Yang, Y., Zeng, L., Zhan, Y.: Optimizing MSE for clustering with balanced size constraints. Symmetry **11**(3) (2019)
16. Babaki, B., Guns, T., Nijssen, S.: Constrained clustering using column generation. In: Simonis, H. (ed.) CPAIOR 2014. LNCS, vol. 8451, pp. 438–454. Springer, Cham (2014). https://doi.org/10.1007/978-3-319-07046-9_31
17. Mueller, M., Kramer, S.: Integer linear programming models for constrained clustering. In: DS 2010, pp. 159–173 (2010)
18. Ouali, A., Loudni, S., Lebbah, Y., Boizumault, P., Zimmermann, A., Loukil, L.: Efficiently finding conceptual clustering models with integer linear programming. In: IJCAI 2016, pp. 647–654 (2016)
19. Métivier, J.-P., Boizumault, P., Crémilleux, B., Khiari, M., Loudni, S.: Constrained clustering using SAT. In: Hollmén, J., Klawonn, F., Tucker, A. (eds.) IDA 2012. LNCS, vol. 7619, pp. 207–218. Springer, Heidelberg (2012). https://doi.org/10.1007/978-3-642-34156-4_20
20. Dao, T.-B.-H., Vrain, C., Duong, K.-C., Davidson, I.: A Framework for actionable clustering using constraint programming. In: ECAI 2016, pp. 453–461 (2016)
21. Dao, T.-B.-H., Duong, K.-C., Vrain, C.: Constrained clustering by constraint programming. Artif. Intell. **244**, 70–94 (2017)
22. Xie, J., Girshick, R., Farhadi, A.: Unsupervised deep embedding for clustering analysis. In: ICML 2016, pp. 478–487 (2016)
23. Liu, E.Y., Zhang, Z., Wang, W.: Clustering with relative constraints. In: KDD 2011, pp. 947–955 (2011)
24. Chierichetti, F., Kumar, R., Lattanzi, S., Vassilvitskii, S.: Fair clustering through fairlets. In: Advances in Neural Information Processing Systems, pp. 5029–5037 (2017)
25. Dwork, C., Hardt, M., Pitassi, T., Reingold, O., Zemel, R.S.: Fairness through awareness. In: Innovations in Theoretical Computer Science 2012, pp. 214–226 (2012)
26. Lewis, D.D., Yang, Y., Rose, T.G., Li, F.: RCV1: a new benchmark collection for text categorization research. J. Mach. Learn. Res. **5**, 361–397 (2004)
27. Guo, X., Gao, L., Liu, X., Yin, J.: Improved deep embedded clustering with local structure preservation. In: IJCAI 2017, pp. 1753–1759 (2017)
28. Kuhn, H.W.: The Hungarian method for the assignment problem. Naval Res. Logistics Quart. **2**(1–2), 83–97 (1955)
29. Backurs, A., Indyk, P., Onak, K., Schieber, B., Vakilian, A., Wagner, T.: Scalable fair clustering. In: ICML 2019, pp. 405–413 (2019)
30. Kearns, M.J., Roth, A., Sharifi-Malvajerdi, S.: Average individual fairness: algorithms, generalization and experiments, CoRR, vol. abs/1905.10607 (2019)

Deep Convolutional Embedding for Painting Clustering: Case Study on Picasso's Artworks

Giovanna Castellano and Gennaro Vessio(✉)

Department of Computer Science, University of Bari "Aldo Moro", Bari, Italy
{giovanna.castellano,gennaro.vessio}@uniba.it

Abstract. Clustering artworks is a very difficult task. Recognizing meaningful patterns in accordance with domain expertise and visual perception, in fact, can be extremely hard. On the other hand, applying traditional clustering and feature reduction techniques to the highly dimensional raw pixel space can be ineffective. To overcome these problems, we propose to use a deep convolutional embedding clustering framework. The model simultaneously optimizes the task of mapping the input pixel data to a latent feature space and the task of finding cluster centroids in this latent space. A quantitative and qualitative preliminary study on a collection of artworks made by Pablo Picasso shows the effectiveness of the model. The proposed method may assist in art-related tasks, in particular visual link retrieval and historical knowledge discovery in painting datasets.

Keywords: Deep clustering · Autoencoders · Cultural heritage

1 Introduction

Cultural heritage, in particular visual arts, are of paramount importance for the cultural, historic and economic growth of the human society. In recent years, due to technology improvements and dramatically declining costs, a large scale digitization effort has been made leading to a growing availability of large digitized art collections. Remarkable examples include WikiArt[1] and the MET collection.[2] This availability, along with the recent advancements in Pattern Recognition and Computer Vision, has opened new opportunities to computer science researchers to assist the art community with intelligent tools to analyze and further understand visual arts. Among the other benefits, a deeper understanding of visual arts has the potential to make them more accessible to a wider population, both in terms of fruition and creation, thus supporting the spread of culture.

The ability to recognize meaningful patterns in visual artworks inherently falls within the domain of human perception [11]. Distinguishing stylistic and semantic attributes of a painting, in fact, originates from the composition of the

[1] https://www.wikiart.org.
[2] https://www.metmuseum.org/art/collection.

A. Appice et al. (Eds.): DS 2020, LNAI 12323, pp. 68–78, 2020.
https://doi.org/10.1007/978-3-030-61527-7_5

colour, texture and shape features visually perceived by the human eye. Unfortunately, this human perception can be extremely hard to conceptualize and verbalize. However, visual-related features, such as those Convolutional Neural Networks (CNNs) are able to automatically learn (e.g., [3,4]), can be effective to tackle the problem of extracting useful patterns from the low-level colour and texture features. These patterns may be beneficial to various art-related tasks, ranging from object detection in paintings [9] to artistic style categorization [20].

While a large body of literature deals with the application of Pattern Recognition and Computer Vision strategies to art-related supervised tasks, e.g. [7,10,12,17], little work has been done in the clustering setting [1,6,13,19]. Having a model capable of clustering artworks in accordance with their visual appearance can be useful for many tasks. The model can be used to support art experts in finding trends and influences among painting schools, i.e. in performing historical knowledge discovery. Analogously, it can help discover different periods in the production of the same artist. The model may discover which artworks mostly influenced the work of current artists. Moreover, it may support interactive navigation on online art galleries by finding visually linked artworks.

In this paper, by taking inspiration from the deep convolutional embedding clustering (DCEC) model recently introduced in [14], we propose DCEC-Paint as a method for grouping digitized paintings in an unsupervised fashion. To develop DCEC-Paint, we added some implementation changes to the original DCEC definition, making the model better suited to the specific image domain. Experimentally, we report the results of a preliminary case study, aimed at evaluating the effectiveness of the method in finding meaningful clusters in a dataset of works made by the same artist, namely Pablo Picasso.

In the rest of this paper: Sect. 2 describes the proposed method; Sect. 3 reports the experimental results; Sect. 4 concludes the work.

2 Method

Clustering is one of the historical tasks of Machine Learning. It is notoriously difficult, mainly because of the absence of supervision in the model generation and evaluation process. In particular, since its appearance, k-means has been extensively used for its ease of implementation and effectiveness [15]. Nevertheless, especially in complex image domains, such as the artistic one, its application may be useless. On one hand, it is widely acknowledged that clustering based on traditional distance measures in the highly dimensional raw pixel space is ineffective. Beside, extracting meaningful features in accordance with domain-specific expertise can be extremely difficult. On the other hand, applying well-known dimensionality reduction techniques, such as PCA, either to the original pixel space or to a manually engineered feature space, can ignore possible nonlinear relationships between the original input and the latent feature space.

The last few years have been witnessing the emergence of a deep clustering paradigm, in which the capability of deep neural networks of finding complex nonlinear relationships among data is exploited for clustering purposes [14,21,22].

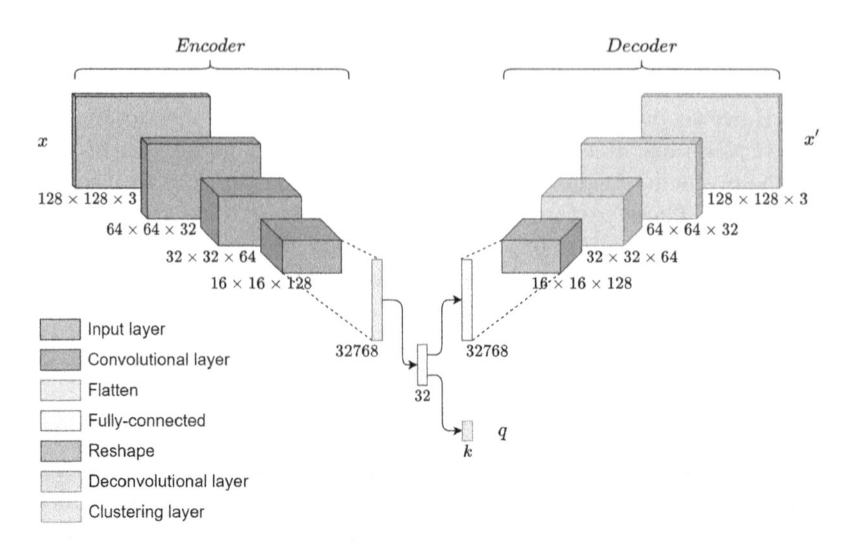

Fig. 1. Architecture of DCEC-Paint.

The idea is to simultaneously optimize the task of mapping the input data to a lower dimensional space and the task of finding a set of centroids in this latent feature space.

Inspired by the deep convolutional embedding clustering (DCEC) framework recently proposed by Guo et al. in [14], we propose DCEC-Paint, as a neural network model for clustering images of digitized paintings. The proposed method is depicted in Fig. 1. The network is based on a convolutional autoencoder and on a clustering layer attached to the embedded layer of the autoencoder. Autoencoders are neural networks that are trained to reconstruct their input. More precisely, an autoencoder is made up of two modules: an encoder ϕ, which learns a nonlinear mapping between the input data and a smaller hidden latent space, and a decoder ψ, which learns to reconstruct the original input by using these latent features. The parameters of the model are updated by minimizing a classic mean squared reconstruction loss:

$$\mathcal{L}_r = \frac{1}{n}\sum_{i=1}^{n}(x'_i - x_i)^2 = \frac{1}{n}\sum_{i=1}^{n}(\psi(\phi(x_i)) - x_i)^2,$$

where n is the number of samples, x_i is the i-th input sample and x'_i its reconstruction. The network receives an input consisting of 128 × 128 RGB images, scaled in the range $[0, 1]$. This input is then propagated through a stack of convolutional layers which learn to extract hierarchical visual features. The first convolutional layer is characterized by 32 filters, with kernel size 5 × 5. The second convolutional layer by 64 filters, with kernel size 5 × 5. The third convolutional layer by 128 filters, with kernel size 3 × 3. All convolutional layers adopt strides 2 and zero-padding, and they are followed by an exponential linear unit (ELU) nonlinearity [8]. We preferred this activation function to the originally proposed ReLU, as ELU tries to make the mean activations closer to zero, thus speeding up learning. All units in the last convolutional layer are flattened and given as an input to a fully-connected layer with 32 units, which represent the latent embedding space. In the original formulation [14], the number of units in this layer is 10. However, we found that this low dimensionality is too constraining, making the reconstruction of complex artistic images slower. The embedding features are then reshaped and propagated through deconvolutional layers, which mirror, in a reverse layer-wise fashion, the hyper-parameters of the encoder and restore the embedding features back to the original input.

As in [14], the formulation of the clustering layer is based on the Deep Embedded Clustering (DEC) proposed in [21]. This layer is connected to the embedding layer of the autoencoder and its task is to assign the latent features of each sample to a cluster. Given an initial estimate of the nonlinear mapping $\phi : X \rightarrow Z$ and initial cluster centroids $\{\mu_j\}_{j=1}^{k}$, the clustering layer maps each latent point, z_i, to a cluster centroid, μ_j, by using Student's t distribution:

$$q_{ij} = \frac{\left(1 + \|z_i - \mu_j\|^2\right)^{-1}}{\sum_j \left(1 + \|z_i - \mu_j\|^2\right)^{-1}},$$

where q_{ij} represents the membership probability of z_i of belonging to cluster j (as if it were a soft assignment). The membership probabilities are then used to compute an auxiliary target distribution P:

$$p_{ij} = \frac{q_{ij}^2 / \sum_i q_{ij}}{\sum_j \left(q_{ij}^2 / \sum_i q_{ij}\right)},$$

where $\sum_i q_{ij}$ are soft cluster frequencies. Clustering is finally performed by minimizing the Kullback-Leibler (KL) divergence between P and Q:

$$\mathcal{L}_c = KL(P \parallel Q) = \sum_i \sum_j p_{ij} \log\left(\frac{p_{ij}}{q_{ij}}\right).$$

In practice, each q_{ij} provides a measure of the similarity between a data point and the different k centroids. Higher values for q_{ij} indicate a higher confidence in assigning a data point to a particular cluster. The auxiliary target distribution is conceived to put more emphasis on the data points assigned with higher confidence while normalizing the loss contribution of each centroid. Hence, by minimizing the divergence between the membership probabilities and the target distribution, the network improves upon the initial estimate by learning from previous high confidence predictions.

In [21], the network discards the decoder and fine-tunes the encoder by only minimizing the clustering loss $\mathcal{L}_c$. However, this approach could distort the embedded space, harming the clustering performance. Instead, as in [14], we propose to keep the decoder attached to the encoder during training. This can help

DCEC-Paint preserve the data structure of the latent feature space. Overall, the network tries to minimize the following composite loss function:

$$\mathcal{L} = \lambda \mathcal{L}_r + (1 - \lambda) \mathcal{L}_c,$$

where $\lambda \in [0, 1]$ is a hyper-parameter that balances the contribution of $\mathcal{L}_r$ and $\mathcal{L}_c$. In the original formulation [14], λ is set to $= 0.1$ and weights are inverted, thus giving more importance to the reconstruction loss rather than the clustering loss. However, since the accuracy of the reconstruction is not the primary task of the model, we found that putting more emphasis on the clustering term improves cluster assignment.

The overall training works in two steps. In a first pre-training step, the convolutional autoencoder learns an initial set of embedding features, by minimizing $\mathcal{L}_r$ and keeping $\lambda = 1$. After this pre-training, the learned features are used to initialize the cluster centroids μ_j by applying traditional k-means. Finally, embedding feature learning and cluster assignment are jointly optimized by setting $\lambda = 0.1$. The overall weights are updated by backpropagation. It is worth noting that, to avoid instability, the auxiliary distribution P is not updated at each iteration using only a batch of data, but by using all embedded points every t iterations. The training procedure stops when a termination criteria is met, that is a change in cluster assignment between two consecutive updates less than a given threshold δ.

3 Experiment

To evaluate the effectiveness of the proposed DCEC-Paint method, we employed a database collecting 439 artworks of a very popular artist, i.e. Pablo Picasso. More precisely, we used a subset of the data provided by the Kaggle platform,[3] scraped from an art challenge website.[4] This was done in order to evaluate the effectiveness of the proposed method in finding meaningful clusters within the production of the same artist.

3.1 Setting

Experiments were carried out on an Intel Core i5 equipped with the NVIDIA GeForce MX110 (dedicated memory of 2 GB). As deep learning framework, we used TensorFlow 2.0 and the Keras API. To speed up calculations, each image was scaled to 128×128 pixels; moreover, to improve the network's performance, images were normalized in the range $[0, 1]$ before training.

It is worth noting that the convolutional autoencoder integrated within the framework was pre-trained end-to-end for 200 epochs using the AdaMax optimizer [16] and mini-batches of size 128. To initialize cluster centroids, we run

[3] https://www.kaggle.com/ikarus777/best-artworks-of-all-time.

[4] http://artchallenge.ru.

k-means with 20 restarts, picking the best solution. Finally, the convergence threshold δ was set to 0.001 and the update interval t to 140.

Since clustering is unsupervised, it is hard to know *a priori* which grouping of paintings is the best. Moreover, since two artworks made by the same artist could have been produced in different stylistic periods, it is difficult to assign a precise label to a given painting, thus providing a form of supervision. Hence, for clustering evaluation, we used two classic internal metrics: the silhouette score [18] and the Calinski-Harabasz index [2], which are based on the model itself. The silhouette score is defined for each sample and is calculated as follows:

$$s = \frac{b - a}{max(a, b)},$$

where a is the mean distance between a data point and all other points in the same cluster, and b is the mean distance between a data point and all other points in the nearest cluster. The final score is obtained by averaging over all data points. The silhouette score ranges between -1 and 1, which respectively represent the worst and best possible value. Values nearby 0 indicate overlapping clusters. The Calinski-Harabasz index is the ratio of the sum of between-cluster dispersion and of inter-cluster dispersion for all clusters. More precisely, for a dataset D of size n_D, which has been partitioned into k clusters, the index is defined as:

$$i = \frac{\text{tr}(B_k)}{\text{tr}(W_k)} \times \frac{n_D - k}{k - 1},$$

where $\text{tr}(B_k)$ is the trace of the between group dispersion matrix and $\text{tr}(W_k)$ is the trace of the within-cluster dispersion matrix. These matrices are defined as follows:

$$W_k = \sum_{q=1}^{k} \sum_{x \in C_q} (x - c_q)(x - c_q)^T,$$

$$B_k = \sum_{q=1}^{k} n_q (c_q - c_D)(c_q - c_D)^T,$$

where C_q is the set of points in cluster q, c_q the center of cluster q, c_D the center of D, and n_q the size of cluster q. It is worth noting that the Calinski-Harabasz index is not bounded within a given interval, but its value tends to grow. For this reason, in the following, we report only normalized values, obtained by dividing the original score by the maximum score. Finally, we also drew some qualitative observations on the cluster assignments provided by the method.

3.2 Results

Table 1 reports the clustering performance of the proposed DCEC-Paint over the whole dataset, by varying the number of clusters k. By looking at the silhouette score, it can be seen that well-defined clusters are obtained in all cases, with

Table 1. Clustering performance.

# clusters	Silhouette score	Calinski-Harabasz index
2	0.933	0.737
3	0.936	0.771
4	0.951	0.768
5	0.965	1.000
6	0.962	0.812

the highest value at $k = 5$. The values for the Calinski-Harabasz index tend to increase or decrease accordingly.

From a qualitative point of view, Figs. 2, 3, 4, 5 and 6 show sample images from the clusters obtained with DCEC-Paint when $k = 5$. The cluster assignment suggests that the model is able to separate artworks in accordance with their stylistic and semantic features. The first two clusters are clearly related to the "cubist" period, with the first cluster containing works mostly depicting people, while the second cluster concerns with objects of the daily life. The third cluster is made up of paintings of the typical "rose" period of the author. Analogously, the fourth cluster contains works from the so-called "blue" period. Finally, the last cluster contains some drawings the dataset we used is composed of.

It is worth remarking that we quantitatively compared the proposed method to other traditional and deep clustering approaches, particularly k-means and DEC. DEC compared favorably with DCEC-Paint, obtaining slightly lower performance. Instead, k-means turned out to be ineffective. We also compared the proposed variant to the originally proposed DCEC, finding out that giving gradually more importance to the clustering loss rather than the reconstruction loss improves prediction performance, while reducing computational effort. The interested reader may refer to [5], for comparative evaluations.

Fig. 2. Sample artworks from the first cluster.

Fig. 3. Sample artworks from the second cluster.

Fig. 4. Sample artworks from the third cluster.

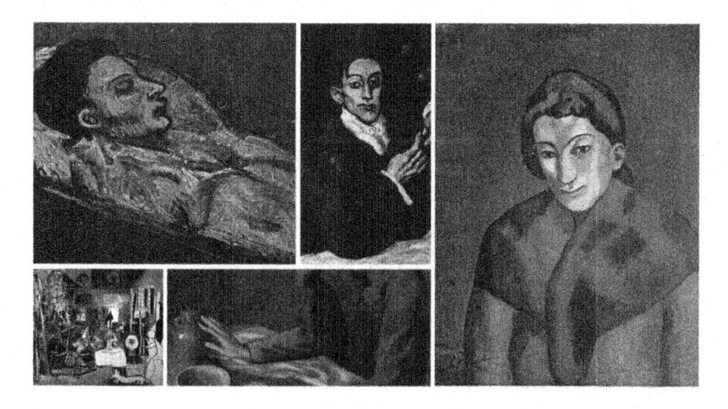

Fig. 5. Sample artworks from the fourth cluster.

On the overall, we can conclude that the proposed method is able to group together works whose distinctive features are not in contrast with the human perception. The clusters discovered by the method are sufficiently justifiable by a human observer and in most cases resemble the intrinsic criteria humans adopt to group artworks together. These criteria combine visual elements, such as colors, and conceptual elements, such as subject matter and visual style.

Fig. 6. Sample artworks from the fifth cluster.

4 Conclusion

We addressed the problem of grouping together digitized paintings in a fully unsupervised way. To this end, we proposed to use a deep convolutional embedding clustering framework which relies only on visual features to automatically group paintings (and drawings). The model was able to find well-separated clusters when focusing on the works produced by the same artist. Quantitative and qualitative results, in fact, confirmed the effectiveness of the method. In particular, from a qualitative point of view, it seems that the model is able to recognize stylistic or semantic attributes of paintings to group them. The proposed method may assist in several art-related tasks, particularly historical knowledge discovery and visual link retrieval. More in general, the experimental results here reported are encouraging, as they confirm the effectiveness of the deep clustering approach for tackling highly complex image domains, such as the artistic one.

Future work will use much of the existing literature on Picasso to try to label the paintings in the dataset we used in order to perform a much more systematic evaluation, in accordance with not only internal but also external clustering criteria. Finally, it is worth noting that the first convolutional layer of the encoder could be analyzed to find out what are the distinctive objects in the paintings that have led to their clustering.

Acknowledgement. Gennaro Vessio acknowledges funding support from the Italian Ministry of Education, University and Research through the PON AIM 1852414 project.

References

1. Barnard, K., Duygulu, P., Forsyth, D.: Clustering art. In: Proceedings of the 2001 IEEE Computer Society Conference on Computer Vision and Pattern Recognition, CVPR 2001, vol. 2, pp. II-II. IEEE (2001)
2. Caliński, T., Harabasz, J.: A dendrite method for cluster analysis. Commun. Stat. Theor. Methods **3**(1), 1–27 (1974)

3. Castellano, G., Castiello, C., Mencar, C., Vessio, G.: Crowd detection for drone safe landing through fully-convolutional neural networks. In: Chatzigeorgiou, A., et al. (eds.) SOFSEM 2020. LNCS, vol. 12011, pp. 301–312. Springer, Cham (2020). https://doi.org/10.1007/978-3-030-38919-2_25
4. Castellano, G., Castiello, C., Mencar, C., Vessio, G.: Crowd detection in aerial images using spatial graphs and fully-convolutional neural networks. IEEE Access **8**, 64534–64544 (2020)
5. Castellano, G., Vessio, G.: Deep convolutional embedding for digitized painting clustering. arXiv preprint arXiv:2003.08597 (2020)
6. Castellano, G., Vessio, G.: Towards a tool for visual link retrieval and knowledge discovery in painting datasets. In: Ceci, M., Ferilli, S., Poggi, A. (eds.) IRCDL 2020. CCIS, vol. 1177, pp. 105–110. Springer, Cham (2020). https://doi.org/10.1007/978-3-030-39905-4_11
7. Cetinic, E., Lipic, T., Grgic, S.: Fine-tuning convolutional neural networks for fine art classification. Expert Syst. Appl. **114**, 107–118 (2018)
8. Clevert, D.A., Unterthiner, T., Hochreiter, S.: Fast and accurate deep network learning by exponential linear units (ELUs). arXiv preprint arXiv:1511.07289 (2015)
9. Crowley, E.J., Zisserman, A.: In search of art. In: Agapito, L., Bronstein, M.M., Rother, C. (eds.) ECCV 2014. LNCS, vol. 8925, pp. 54–70. Springer, Cham (2015). https://doi.org/10.1007/978-3-319-16178-5_4
10. Crowley, E.J., Zisserman, A.: The art of detection. In: Hua, G., Jégou, H. (eds.) ECCV 2016. LNCS, vol. 9913, pp. 721–737. Springer, Cham (2016). https://doi.org/10.1007/978-3-319-46604-0_50
11. Cupchik, G.C., Vartanian, O., Crawley, A., Mikulis, D.J.: Viewing artworks: contributions of cognitive control and perceptual facilitation to aesthetic experience. Brain Cogn. **70**(1), 84–91 (2009)
12. Garcia, N., Renoust, B., Nakashima, Y.: Context-aware embeddings for automatic art analysis. In: Proceedings of the 2019 International Conference on Multimedia Retrieval, pp. 25–33. ACM (2019)
13. Gultepe, E., Conturo, T.E., Makrehchi, M.: Predicting and grouping digitized paintings by style using unsupervised feature learning. J. Cult. Heritage **31**, 13–23 (2018)
14. Guo, X., Liu, X., Zhu, E., Yin, J.: Deep clustering with convolutional autoencoders. In: Liu, D., Xie, S., Li, Y., Zhao, D., El-Alfy, E.S. (eds.) International Conference on Neural Information Processing. LNCS, vol. 10635, pp. 373–382. Springer, Cham (2017). https://doi.org/10.1007/978-3-319-70096-0_39
15. Jain, A.K.: Data clustering: 50 years beyond K-means. Pattern Recogn. Lett. **31**(8), 651–666 (2010)
16. Kingma, D.P., Ba, J.: Adam: a method for stochastic optimization. arXiv preprint arXiv:1412.6980 (2014)
17. Mao, H., Cheung, M., She, J.: Deepart: learning joint representations of visual arts. In: Proceedings of the 25th ACM International Conference on Multimedia, pp. 1183–1191. ACM (2017)
18. Rousseeuw, P.J.: Silhouettes: a graphical aid to the interpretation and validation of cluster analysis. J. Comput. Appl. Math. **20**, 53–65 (1987)
19. Spehr, M., Wallraven, C., Fleming, R.W.: Image statistics for clustering paintings according to their visual appearance. In: Computational Aesthetics 2009: Eurographics Workshop on Computational Aesthetics in Graphics, Visualization and Imaging, pp. 57–64. Eurographics (2009)

20. Van Noord, N., Hendriks, E., Postma, E.: Toward discovery of the artist's style: learning to recognize artists by their artworks. IEEE Signal Process. Mag. **32**(4), 46–54 (2015)
21. Xie, J., Girshick, R., Farhadi, A.: Unsupervised deep embedding for clustering analysis. In: International Conference on Machine Learning, pp. 478–487 (2016)
22. Yang, B., Fu, X., Sidiropoulos, N.D., Hong, M.: Towards k-means-friendly spaces: simultaneous deep learning and clustering. In: Proceedings of the 34th International Conference on Machine Learning, vol. 70, pp. 3861–3870 (2017)

Dynamic Incremental Semi-supervised Fuzzy Clustering for Bipolar Disorder Episode Prediction

Gabriella Casalino[1(✉)], Giovanna Castellano[1], Francesco Galetta[1], and Katarzyna Kaczmarek-Majer[2]

[1] Department of Computer Science, University of Bari, Bari, Italy
gabriella.casalino@uniba.it
[2] Systems Research Institute, Polish Academy of Sciences, Warsaw, Poland

Abstract. Bipolar Disorder (BD) is a chronic mental illness characterized by changing episodes from euthymia (healthy state) through depression and mania to the mixed states. In this context, data collected through the interaction of patients with smartphones enable the creation of predictive models to support the early prediction of a starting episode. Previous research on predicting a new BD episode use mostly supervised learning methods that require labeled data and hence force a filtering of the available data to retain only those data that have valid labels (from the psychiatric assessment). To avoid limitations of supervised learning, in this paper we investigate the use of a semi-supervised learning approach that combines both labeled and unlabeled data to derive a model for BD episode prediction. Specifically we apply the DISSFCM (Dynamic Incremental Semi-Supervised Fuzzy C-Means) algorithm which offers the possibility to process in an incremental fashion the data stream of the voice signal captured by the smartphone, thus exploiting the evolving time structure of data which is ignored by static learning methods. DISSFCM processes data in form of chunks and creates a dynamic collection of clusters thanks to a splitting mechanism that generates new clusters to better capture the hidden geometrical structure of data. This gives DISSFCM the ability to detect changes in data and dynamically adapt the model to them, thus improving the prediction accuracy. Preliminary results on real-world data collected at the Department of Affective Disorders, Institute of Psychiatry and Neurology in Warsaw (Poland) show that DISSFCM is able to predict some of healthy episodes (euthymia) and disease episodes even when only 25% of labeled data are available. Moreover DISSFM performs better than its previous version without split (ISSFCM) and it also overcomes the batch algorithm (SSFCM) that uses the whole dataset to create the model.

Keywords: Semi-supervision · Fuzzy C-Means · Clustering · Incremental learning · Bipolar Disorder episode prediction · Smartphone data · Acoustic features · Pervasive computing · mHealth · Digital health

A. Appice et al. (Eds.): DS 2020, LNAI 12323, pp. 79–93, 2020.
https://doi.org/10.1007/978-3-030-61527-7_6

1 Introduction

In the last decades, with the increase of technologies and the spreading of intelligent objects, the way medicine and healthcare are practiced is rapidly changing. In 2019, the World Health Organization officially proposed *Digital health* as new discipline that combines Artificial Intelligence, Internet of Things, Big Data and Data Analytics techniques for the health sector[1]. Mobile Health and Remote Monitoring, Sensors and Ambient Assisted Living together with Artificial Intelligence, Robotics and Data Analytics are just some examples of this new research branch [8].

From one hand, automatic techniques have became necessary to manage and analyse the huge amount of daily produced medical data. In this context learning and predictive algorithms are critical to support the medical decisions [28,32,34]. On the other hand, everyday objects have became more and more *smart* by embedding computational capabilities. Pervasive computing has grown interest in the medical field since it minimizes the need of interaction between patients and physicians, by collecting daily information that will be analyzed by the medical experts. Wearable objects and smartphones are widely used to easily acquire the most varied daily users' information such as kinematic and physiological data [9,31], visual scenes and geolocation information to assist visually-impaired and blind people [22] or to monitor vital parameters [12,25] such as oxygen saturation to prevent COVID-19 infection [3,7], just to mention few examples. This has led to a new frontier of tele-medicine known as mHealth (mobile health) [33].

A promising mHealth application is represented by smartphone-based monitoring of patients affected by mental disorders [20,21,26]. In this work, we focus on Bipolar Disorder (BD) which is a serious mental disorder characterized by manic episodes (states) of elevated mood and overactivity, interspersed with periods of depression. Since BD is a chronic and recurrent disease, effective monitoring of changing state is of particular importance. Typically, the psychiatric assessment of patient's state is carried out by a psychiatrist during routine check-up visits. However, the frequency of control visits is usually insufficient to provide early intervention at the start of the episode. The management of BD could be significantly improved by real-time monitoring of illness activity via smartphone [1] and early detection of changes of patient's phase. In this context, the time structure of data should be taken into account to model the interepisodic mood instability [2].

Previous studies on BD episode prediction concentrated primarily on the phase detection formulated as a supervised classification task [13,23]. However, the use of supervised learning methods in this context requires labeled data and hence forces a filtering of the available data to retain only those data that have valid labels. Labels describing patients state are assigned by a doctor during an interview, the so called the psychiatric assessment. Following [13], data collected 7 days before and 2 days after the assessment can be labeled with the outcome of this psychiatric assessment. In [11], label validity is considered even less - only to 3 days before the

[1] Global Strategy on Digital Health 2020–2024 https://extranet.who.int/.

assessment and on the day of it as the depression (HDRS-17) and mania (YMRS) rating scales address symptoms over the last four days. Furthermore, the frequency of interviews is usually insufficient to assign labels to all data.

Contrary to the supervised approaches, there are works that apply completely unsupervised approaches to monitor changes in the severity of the depressive and manic symptoms [24] or to analyze behavioural data about smartphone usage [15,16]. However, unsupervised learning approaches insufficiently benefit of the a-priori knowledge given by labeled data of the psychiatric assessments [27].

Alternatively to state-of-art methods that use supervised or unsupervised learning approaches, in [5] we initially explored the use of a semi-supervised learning approach that combines both labeled and unlabeled data to derive a model for BD episode prediction. Specifically, the ISSFCM (Incremental Semi-Supervised Fuzzy C-Means) [6] and its batch version SSFCM [30] were compared with standard classification algorithms to predict bipolar disorder. ISSFCM offers the possibility to process in an incremental fashion the data stream of the voice signal obtained from the interaction of the patients with their smartphone, thus exploiting the time structure of data which is ignored by static learning methods [18]. Moreover, the results in [5] showed that ISSFCM achieves good results thanks to fuzzy clustering that enhances the detection of patterns in data and provides interpretable results in terms of labeled prototypes that represent data in a synthetic manner.

However, one limitation of ISSFCM when applied to data streams for BD prediction is the static nature of the created model which is based on a fixed number of cluster prototypes that is given in advance according to the known number of classes. In an attempt to better cope with the evolving nature of acoustic data produced by patients interacting with their smartphone, in this paper we explore the use of DISSFCM (Dynamic Incremental Semi-Supervised FCM) algorithm [4] that improves ISSFCM by adding the possibility to dynamically adapt the number of clusters.

The structure of the paper is as follows. Section 2 describes materials and methods applied in this research, starting from the observational study on bipolar disorder that led to data collection, up to the description of the DISSFCM algorithm. Then, experimental results are presented in Sect. 3 where a comparison among DISSFCM, ISSFCM and SSFCM is also reported. In Sect. 4, main conclusions are discussed and future work is outlined.

2 Materials and Methods

2.1 Observational Study on Bipolar Disorder Patients

Motivation for this research comes from analyzing real-world data collected in a recent observational study[2] that was conducted in the Department of Affective

[2] Data considered in this paper come from CHAD project – entitled "Smartphone-based diagnostics of phase changes in the course of bipolar disorder" (RPMA.01.02.00-14-5706/16-00) that was financed from EU funds (Regional Operational Program for Mazovia) in 2017–2018.

Disorders, Institute of Psychiatry and Neurology in Warsaw, Poland. The study included patients diagnosed with bipolar disorder (F31 according to ICD-10 classification). In total, 33 patients were enrolled and used a dedicated smartphone application in everyday life for up to 15 months. This dedicated mobile application, called *BDMon* was able to collect acoustic features about phone calls. Within this paper, preliminary results for data of two exemplary patients are presented.

During the observational study, each patient was associated to a psychiatrist who evaluated his mental state. Psychiatrists were using standardized measures of depressive and manic symptoms, that are Hamilton Depression Rating Scale and Young Mania Rating Scale to assess the psychiatric state of a patient (BD episode). The interviews were performed with various frequency depending on the need identified by the doctor or a patient. The outcomes of psychiatric assessments are used as labels for classification, namely depression, mania, mixed state and euthymia (the only healthy state). Following [13], labels from the psychiatric assessments are assigned to smartphone data collected 7 days before the psychiatric assessment and 2 days after the day of the visit. Within this research, we concentrate on predicting healthy from unhealthy episodes and therefore, annotate data with either *healthy* or *sick* label.

2.2 Acoustic Feature Extraction and Selection

The sound signal was obtained directly from the smartphone's microphone to avoid recording of the interlocutor's speech. The voice signal was processed in real time to extract its physical descriptors, which were transferred to a secure server and stored in a database for analyses. Each phone call was divided into short 10–20 ms frames, in which it is approximately stationary. An adopted version of openSMILE library [10] was used to collect a rich dataset of 86 acoustic features, such as time-domain descriptors (zero crossing rate, amplitude statistics, signal energy), spectral features (distribution of energy, mel-cepstral coefficients, fundamental frequency and its harmonics), voice quality (jitter, shimmer, harmonics to noise ratio) and prosodic features (voicing probability, normalized loudness).

To reduce the number of attributes, in [13] a manual selection of the features for classification is performed. In [17] an advanced approach to feature selection using fuzzy clustering, self organizing maps and psychiatric assessments is presented. In this study, to obtain most significant voice features, Recursive Feature Elimination (RFE) [14] is applied and as a result, the following features are considered in the semi-supervised approaches: (1) envelope of the fundamental frequency contour (2) spectrum slope in the range 0–500 Hz (3) spectrum slope in the range 500–1500 Hz (4) the Zero-Crossing rate (5) the first MFCC mel-cepstral coefficient. Data representing each feature from each phone call are summarized with their average values and standard deviation. Thus, 10 variables are used as input data to describe patient state.

2.3 Dynamic Incremental Semi-supervised Fuzzy C-Means

In this work we investigate the use of DISSFCM (Dynamic Incremental Semi-Supervised Fuzzy C-Means) [4] to analyze the acoustic data that have been collected during the clinical study. The basic idea of the DISSFCM algorithm is to incrementally apply the semi-supervised fuzzy clustering algorithm (SSFCM) [30] to subsequent portions (*chunks*) of a data stream. SSFCM is an alternate optimization algorithm that iteratively updates both cluster centers and the partition matrix by minimizing the objective function in (1). Whilst the first part of the equation is the classical objective function of FCM (Fuzzy C-Means), the second part adds partial supervision into the clustering process.

$$J = \sum_{k=1}^{K} \sum_{j=1}^{N_t} u_{jk}^2 d_{jk}^2 + \alpha \sum_{k=1}^{K} \sum_{j=1}^{N_t} \left(u_{jk} - b_j f_{jk}\right)^2 d_{jk}^2 \tag{1}$$

where $K \geq C$ is the number of clusters, $N_t = |X_t|$ is the cardinality of the t-th chunk in the data stream, $u_{jk} \in [0, 1]$ is the membership degree of a sample $\mathbf{x}_j$ in the k-th cluster, d_{jk} is the Euclidean distance between jth sample and center $\mathbf{c}_k$ of the k-th cluster, $\alpha \geq 0$ is a regularization parameter for the second part of the objective function that exploits class information, $b_j = b(\mathbf{x}_j)$ and $f_{jk} = 1$ iff the j-th sample belongs has the same class label of the k-th cluster.

The DISSFCM process starts when the first chunk of data is available. At the beginning, one cluster prototype is assigned to each class by randomly selecting a labeled sample from the chunk. SSFCM is then applied to all data in the chunk, and cluster prototypes together with membership matrix are returned. Cluster prototypes are defined as medoids, namely the prototype of each cluster is the data sample that is closest to the cluster center $\mathbf{c}_k$. Each prototype is tagged with the class label of the corresponding cluster. When a new chunk arrives, the process is repeated starting from the prototypes generated on the previous chunk. In this way information about old data are injected in the new chunk in a synthetic form (the prototypes) thus preserving history of data without saving the old data.

After each chunk is processed, the quality of the resulting clusters is evaluated in terms of *Reconstruction Error* (*RE*) [29] and a split mechanism is applied to the worst-quality cluster, i.e. the cluster corresponding to the highest value of RE (indicated by HRE). Specifically, while the value of HRE in the current chunk is higher than the HRE in the previous chunk, a splitting step is performed to divide the worst cluster in two novel clusters. The maximum number of splits is set to 10 to avoid an excessive growth of the number of clusters. In DISSFCM the splitting is performed by means of the CFCM (Conditional FCM) algorithm [19] which is applied to data belonging to the worst cluster in order to derive two new cluster prototypes. After splitting the same data are re-assigned to the new clusters according to the membership degree computed by CFCM.

Cluster prototypes generated by DISSFCM provide an evolving classification model than can be used to classify new unlabeled data leveraging a simple matching mechanism. Indeed, any new data sample is matched against all prototypes and the label of the closest prototype is assigned to the sample.

The incremental nature of DISSFCM, as well as its semi-supervised learning mechanism, make it a suitable candidate to process evolving data streams including labeled and unlabeled data. The acoustic data considered in this study may be actually characterized by a partial labeling, since not all data coming from interaction of BD patients using their smartphone app may be annotated by the psychiatrist. Moreover these data may contain both abrupt concept drifts as well as smoother changes in the data distribution. The aim of this empirical study is to investigate potential applicability of DISSFCM to deal with these problems that are common in real-world data streams.

3 Experimental Results

Experiments have been conducted in order to prove the effectiveness of DISSFCM in classifying bipolar disorder data, available at different time intervals (chunks). In particular the aim was to assess the extent to which DISSFCM is capable to create classification models that can adapt to the data by coping with abrupt concept drifts. Moreover we evaluated the influence of the labeling percentage of data on the performance of DISSFCM.

3.1 Experimental Settings

In this preliminary study, the BD event prediction problem was reduced to a binary classification task by considering only the *sick* class including depression, mania and mixed states of BD, and the *healthy* class corresponding to the euthymic state. Data from two patients were considered. The number of total labeled data extracted from phone calls for each patient is summarized in Table 1. As it can be observed, data coming from Patient 1 are imbalanced. Indeed, Patient 1 was in a sick state almost as much as in healthy episode. Differently, there are almost the same number of healthy and sick episodes for Patient 2. Due to the very different behavior of the two patients, data streams of each patient were treated separately to create a classification model for each patient.

To better evaluate how the presence of unlabeled data may influence the classification results of DISSFCM, for each patient we started with a set of labeled data and then randomly removed labels to simulate different labeling percentages, i.e. 25%, 50%, 75% and 100% of labeled data. To reduce the random factor in the simulation of unlabeled data, five fold cross validation was used. For each fold and for each percentage of labeling, five different sets of labeled and unlabeled data have been used. Furthermore, in order to investigate the DISSFCM capabilities in dealing with changes in data distributions, the available data stream of each patient was temporally split in two different sets of chunks, named *set(a)* and *set(b)*. The first set is very challenging since it contains abrupt concept drifts from one chunk to the subsequent, whilst the second set presents more smooth changes. Figures 2 and 3 show the number of chunks in each set and the class distributions for Patient 1 and Patient 2, respectively. For each chunk, 70% of data were used as training set (i.e. to create the prototype-based

Table 1. Class distribution of smartphone-based acoustic data considered in this study.

Condition	Class	Patient 1	Patient 2
Euthymia	Healthy	142	148
Depression, mania, mixed	Sick	261	144
	Tot	403	292

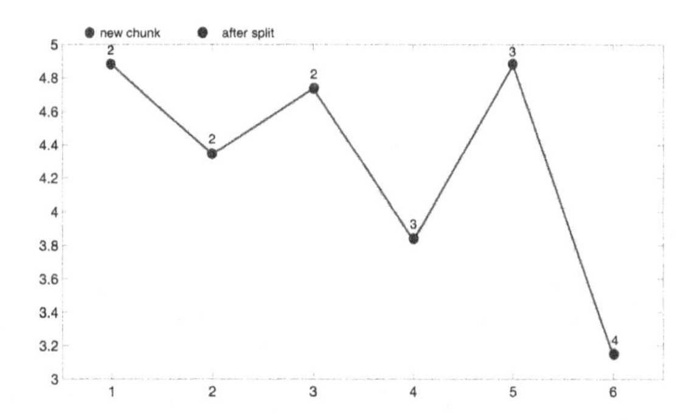

Fig. 1. An example of the HRE trend. (Color figure online)

classification model) and 30% as test set. It should be noted that we created a reduced number of chunks so as to assure the presence of both classes in each chunk. Indeed, due to the sequential nature of data, the creation of a higher number of chunks would lead to smaller chunks where each chunk would contain only samples from one of the two classes.

Standard classification evaluation measures were considered, namely accuracy, precision and recall. After the processing of each chunk, the average values of such measures were computed on the corresponding test sets.

First of all, to show how the quality of clusters created by DISSFCM improves during the processing of chunks, an example of the trend of the HRE value is reported in Fig. 1 for a run on patient 2. Green dots indicate the HRE computed on clusters generated when a new chunk arrives, whilst blue dots indicate the HRE computed on clusters after a split. It can be observed that when the third chunk arrives, the HRE increases with respect to the previous chunk, so the splitting mechanism is activated and the cluster with the worst RE value is divided in two clusters. The number of clusters is thus increased by one unit. It can be seen that the HRE decreases after splitting, indicating that the splitting actually improves the quality of clusters. When the forth chunk is processed, the HRE raises again, so a new split is necessary. Finally, the DISSFCM ends up with four clusters that adequately represent the two classes.

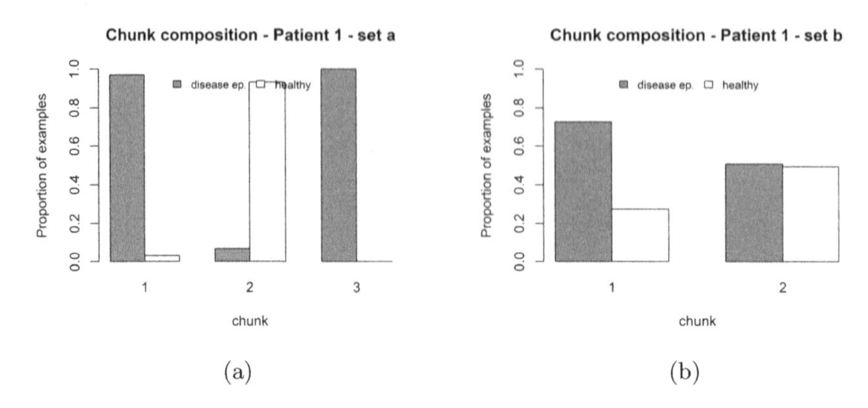

(a) (b)

Fig. 2. Proportion of learning examples from the sick (disease episode) and healthy classes in chunks of Patient 1.

Table 2. Average accuracy through the data chunks for Patient 1.

Chunk	Accuracy							
	Set(a)				Set(b)			
	25%	50%	75%	100%	25%	50%	75%	100%
#1	0.66	0.61	0.47	0.74	0.61	0.65	0.61	0.65
#2	0.31	0.45	0.83	0.83	0.59	0.59	0.59	0.66
#3	0.69	1.0	1.0	1.0	–	–	–	–

3.2 Patient 1

DISSFCM was firstly evaluated on acoustic data of Patient 1. Table 2 shows the average accuracy obtained on the test set, through the different chunks of set(a) and set(b), varying the labeling percentage.

It can be seen that in case of set(a) (which is the more challenging than set(b)) with a low labeling percentage ($\leq$50%) the chunk #1 and the chunk #3 are more easily modeled than the chunk #2, where the class *healthy* appears with more evidence, and the class *sick* almost disappears. Thus the model is not able to cope properly with the drift that occurs with the second chunk, when data are poorly labeled. It is worth noting that, when processing the third chunk, the model evolves and succeeds again to catch the data distribution, even if one class is not represented at all in this chunk. Hence we can say that the high variance and imbalancing of data distribution through the chunks, strongly influence the results of DISSFCM in set(a). A similar behavior is observed from the recall and precision measures in Table 3. These values are very dependent on the chunk composition. Precision strongly depends on the most represented class, and when the percentage of labeled data increases, the algorithm discriminates

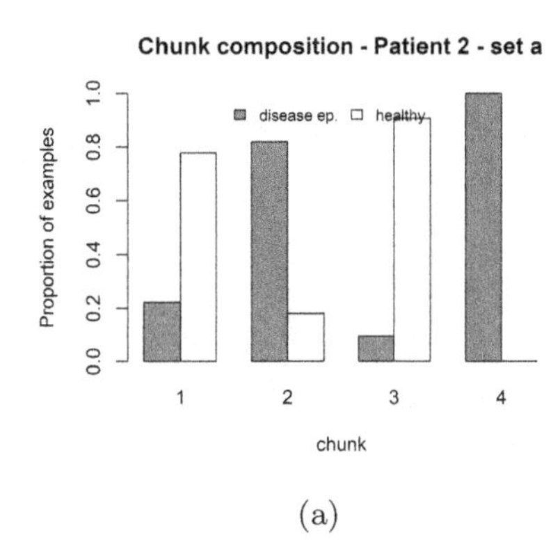

(a)

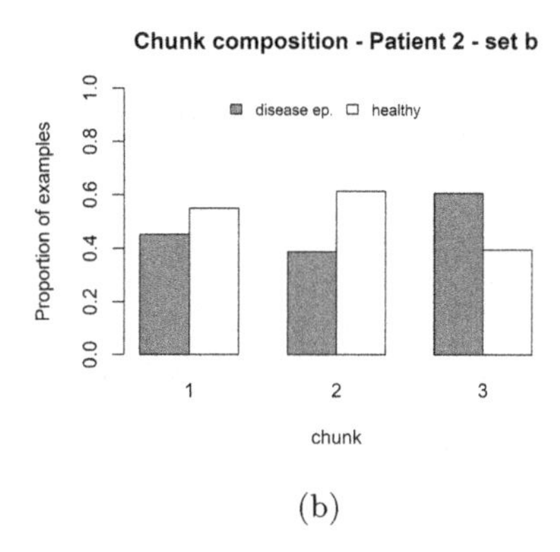

(b)

Fig. 3. Proportion of learning examples from the sick (disease episode) and healthy classes in chunks of Patient 2.

better the two classes. However it is worth noting that DISSFCM is also able to correctly detect at least the 70% of sick episodes with only 25% of data labeling.

Conversely, in set(b) more stable results are observed when varying the labeling percentages. In this case data are divided in two chunks, hence there are more samples per chunks and changes in class distributions are more gradual, if compared to set(a). Thus, the greater the chunks are, the less the influence of the labeling percentage on the classification results of DISSFCM.

On the overall, the classification results achieved by DISSFCM on the two different sets of chunks are quite satisfying if we consider that BD episode prediction is a very difficult task due to several factors involved in the human brain activities. This is confirmed also by the empirical study in [5] where standard classification algorithms such as SVM, Decision Trees and Random Forest return relatively low accuracy values on the same dataset. The best accuracy achieved by Random Forest, with all the data labeled is of 67% on patient 1, and 57% on patient 2. Moreover, these methods do not exploit the evolving nature of the data stream being all batch learning algorithms. Conversely, DISSFCM implements an incremental learning scheme, thus it is more suitable to process streams of data like the acoustic data considered in this study.

3.3 Patient 2

As for Patient 1, the set(a) of Patient 2 is more critical than set(b) as it includes four chunks containing abrupt drifts and highly imbalanced classes, whilst set(b) divides data in three more balanced chunks.

When applied to set(a), DISSFCM generates a classification model that is capable to detect the sick class more easily than the healthy one. This justifies high accuracy values (see Table 4), for those chunks (#2 and #4) where the sick class is the over represented. On the contrary the accuracy decreases on the other chunks (#1 and #3) where the sick class is under represented and the most of

Table 3. Recall and Precision values through the chunks for Patient 1.

Chunk	Recall								Precision							
	25%		50%		75%		100%		25%		50%		75%		100%	
	S	H	S	H	S	H	S	H	S	H	S	H	S	H	S	H
	Set(a)															
#1	0.65	1.0	0.59	1.0	0.46	1.0	0.49	1.0	1.0	0.07	1.0	0.06	1.0	0.05	1.0	0.25
#2	1.0	0.26	1.0	0.41	0.5	0.85	0.5	0.85	0.09	1.0	0.11	1.0	0.2	0.96	0.06	0.92
#3	0.69	–	1.0	–	1.0	–	1.0	–	1.0	–	1.0	–	1.0	–	1.0	–
	Set(b)															
#1	0.57	0.71	0.62	0.71	0.54	0.79	0.62	0.71	0.84	0.38	0.85	0.42	0.87	0.39	0.85	0.42
#2	0.60	0.57	0.60	0.57	0.60	0.57	0.69	0.64	0.60	0.57	0.67	0.55	0.60	0.57	0.78	0.66

the data belong to the healthy class. Of course, when all the data are labeled high accuracy values are returned.

Moreover, the labeling percentage highly influences the recall and precision results. Indeed when the labeling percentage is low, the sick class is more easily recognized (recall values higher than 0.75 with only 25% of labels). On the contrary, when more labels are available (≥50%), healthy and sick classes are both recognized. This result suggests that with highly imbalanced data streams, the labeling percentage highly influences the classification results of DISSFCM. Indeed, when the chunk is imbalanced, the under represented class has a low number of samples, that become lower due to the presence of unlabeled data. Thus, there are not enough labeled data to properly learn that class.

As concerns the set(b), the DISSFCM algorithm classifies better samples in the second chunk, but it needs labels to better perform on the first and third chunk. This behavior is confirmed by the recall and precision values in Table 5. Also in this case, it is worth noting that DISSFCM correctly recognizes almost the 90% of sick episodes using only 25% of labeled data. Instead a higher percentage of labeled data (≥75%) is necessary to correctly detect both classes.

3.4 Empirical Comparison

A further set of simulations was devoted to assess the effectiveness of the split mechanism implemented in DISSFCM which provides an adaptation of the number of clusters during the processing of chunks. To this aim, we compared DISSFCM with its previous version ISSFCM [6] that keeps unchanged the number of clusters and with its batch version SSFCM [30].

For a fair comparison the same experimental setting (as described in the previous section) was used for the three algorithms. However, since SSFCM is not an incremental clustering algorithm, the whole dataset was considered to apply SSFCM, i.e. all chunks have been merged to compose the training ad test set.

Table 4. Average accuracy through the data chunks for Patient 2.

Chunk	Accuracy							
	Set(a)				Set(b)			
	25%	50%	75%	100%	25%	50%	75%	100%
#1	0.43	0.79	0.57	0.79	0.55	0.4	0.55	0.6
#2	0.73	0.87	0.93	0.87	0.93	0.93	0.8	0.93
#3	0.33	0.39	0.55	0.83	0.65	0.61	0.78	0.74
#4	0.92	1.0	0.75	1.0	–	–	–	–

Table 5. Recall and Precision values through the chunks, varying the labeling percentage. Patient 2, set(a) and set(b).

Chunk	Recall								Precision							
	25%		50%		75%		100%		25%		50%		75%		100%	
	S	H	S	H	S	H	S	H	S	H	S	H	S	H	S	H
	Set(a)															
#1	1.0	0.27	0.67	0.82	0.67	0.82	0.33	0.45	0.27	1.0	0.5	0.90	0.20	0.78	0.5	0.90
#2	0.75	0.67	1	0.33	1	0.67	0.92	0.67	0.9	0.4	0.86	1	0.92	1	0.92	0.67
#3	1.0	0.25	1.0	0.31	1.0	0.5	0.5	0.88	0.14	1.0	0.16	1.0	0.2	1.0	0.33	0.93
#4	0.92	–	1.0	–	0.75	–	1.0	–	1.0	–	1.0	–	1.0	–	1.0	–
	Set(b)															
#1	0.89	0.27	0.78	0.09	0.44	0.64	0.78	0.45	0.5	0.75	0.41	0.33	0.5	0.58	0.54	0.71
#2	0.83	1.0	1.0	0.89	1.0	0.67	0.83	1.0	1.0	0.9	0.85	1.0	0.66	1.0	1.0	0.9
#3	0.93	0.22	0.79	0.33	0.79	0.78	1.0	0.33	0.65	0.67	0.65	0.5	0.85	0.7	0.7	1.0

Figures 4 and 5 report for each patient and for each chunk set the average accuracy of the classification model after each chunk processing, varying the percentage labels. We can observe that in all cases DISSFCM overcomes the results obtained with the other algorithms. Particularly, thanks to the splitting mechanism, DISSFCM is able to cope with abrupt changes as well as imbalanced classes that characterize data chunks in set(a). In this case, ISSFCM obtains lower results than the baseline SSFCM. When the class distribution through chunks changes slowly and more samples are present in chunks (as in set(b)), ISSFCM outperforms the batch algorithm, and the dynamic approach of DISSFCM further improves the results. The improved classification results of DISSFCM show that the dynamic mechanism of cluster generation through splitting provides classification models that can better capture the structure of the data stream with respect to more static incremental methods like ISSFCM where the number of clusters is fixed a-priori and kept unchanged.

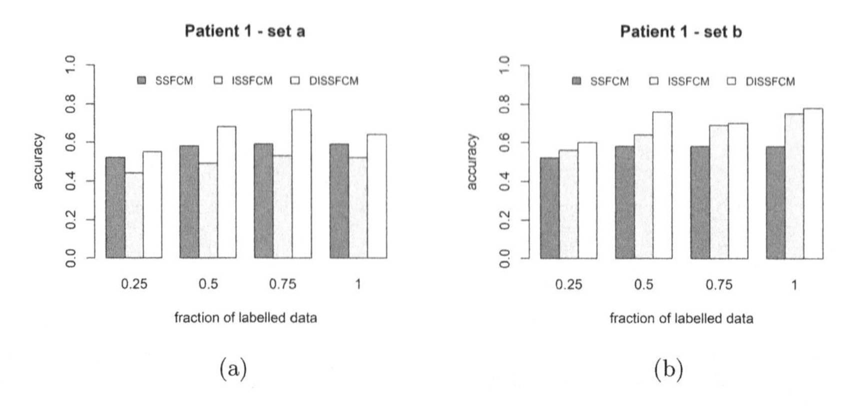

Fig. 4. Comparison in terms of average accuracy through chunks, varying the labeling percentage for set(a) and set(b) of Patient 1.

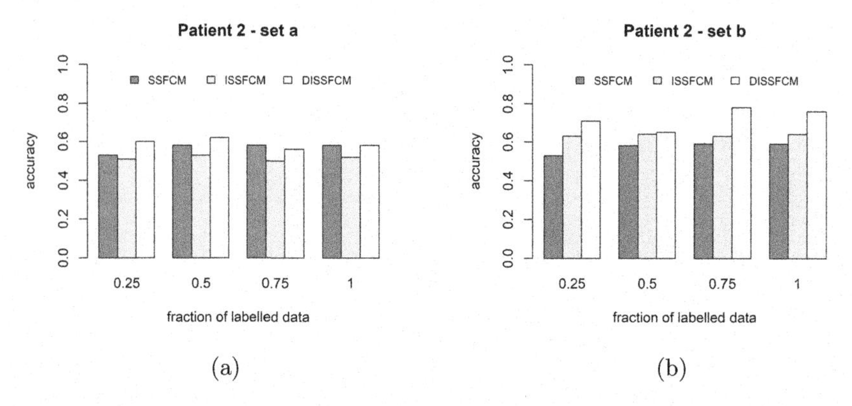

Fig. 5. Comparison in terms of average accuracy through chunks, varying the labeling percentage for set(a) and set(b) of Patient 2.

4 Conclusions

In this work, we have investigated the effectiveness of the DISSFCM data stream classification method to predict Bipolar Disorder episodes on the basis of acoustic data collected during the interaction of patients with a dedicated smartphone application. Preliminary results on data of two different patients showed that DISSFCM provides good results with respect to the baseline provided by fully supervised learning methods applied to the same BD data [5], even when working on partially labeled data. Compared to its previous static versions, DISSFCM shows a better capability in capturing changes in data thanks to a splitting mechanism that adapts the number of clusters.

It should be noted that the obtained results are of preliminary nature since they are limited to only two BD patients. Further work is in progress to derive BD event prediction models for all patients of the considered study. Moreover, a more extensive comparison among different semi-supervised learning algorithms will be carried out to better assess the effectiveness of the semi-supervised approach in the context of BD episode prediction from acoustic data streams.

Acknowledgment. Datasets considered in this paper were collected in the CHAD project – entitled "Smartphone-based diagnostics of phase changes in the course of bipolar disorder" (RPMA.01.02.00-14-5706/16-00) that was financed from EU funds (Regional Operational Program for Mazovia) in 2017–2018. The authors thank psychiatrists and patients that participated in the observational study for their commitment. The authors thank the researchers Olga Kamińska, Karol Opara and Weronika Radziszewska from Systems Research Institute, Polish Academy of Sciences for their support in data preparation and analysis, as well as the researchers Monika Dominiak, Anna Antosik-Wójcińska and Łukasz Święcicki from Institute of Psychiatry and Neurology for their advice and comments. This work has been partially supported by the GNCS-INDAM (Gruppo Nazionale per il Calcolo Scientifico of Istituto Nazionale di Alta Matematica) within the research project "Computational Intelligence methods for Digital Health".

References

1. Antosik-Wójcińska, A.Z., et al.: Smartphone as a monitoring tool for bipolar disorder: a systematic review including data analysis, machine learning algorithms and predictive modelling. Int. J. Med. Inform. **138**, 104131 (2020)
2. Bonsall, M., Swallace-Hadrill, S., Geddes, J., Goodwin, G., Holmes, E.: Nonlinear time-series approaches in characterizing mood stability and mood in stability in bipolar disorder. Proc. R. Soc. Lond. B Biol. Sci. **279**, 916–924 (2012)
3. Casalino, G., Castellano, G., Zaza, G.: A mHealth solution for contact-less self-monitoring of blood oxygen saturation. In: Proceedings of IEEE Symposium on Computers and Communications 2020 (ISCC 2020). IEEE (2020)
4. Casalino, G., Castellano, G., Mencar, C.: Data stream classification by dynamic incremental semi-supervised fuzzy clustering. Int. J. Artif. Intell. Tools **28**(08), 1960009 (2019)
5. Casalino, G., Dominiak, M., Galetta, F., Kaczmarek-Majer, K.: Incremental semi-supervised fuzzy C-Means for bipolar disorder episode prediction. In: Proceedings of the 2020 IEEE Conference on Evolving and Adaptive Intelligent Systems (EAIS 2020) (2020)
6. Castellano, G., Fanelli, A.M.: Classification of data streams by incremental semi-supervised fuzzy clustering. In: Petrosino, A., Loia, V., Pedrycz, W. (eds.) WILF 2016. LNCS (LNAI), vol. 10147, pp. 185–194. Springer, Cham (2017). https://doi.org/10.1007/978-3-319-52962-2_16
7. Chan, C., et al.: A smartphone oximeter with a fingertip probe for use during exercise training: usability, validity and reliability in individuals with chronic lung disease and healthy controls. Physiotherapy **105**(3), 297–306 (2019)
8. Chang, A.: The role of artificial intelligence in digital health. In: Wulfovich, S., Meyers, A. (eds.) Digital Health Entrepreneurship. HI, pp. 71–81. Springer, Cham (2020). https://doi.org/10.1007/978-3-030-12719-0_7

9. Coviello, G., Avitabile, G., Florio, A.: A synchronized multi-unit wireless platform for long-term activity monitoring. Electronics **9**(7), 1118 (2020)
10. Eyben, F., Weninger, F., Gross, F., Schuller, B.: Recent developments in openSMILE, the Munich open-source multimedia feature extractor. In: Proceedings of the 21st ACM International Conference on Multimedia, pp. 835–838. ACM (2013)
11. Faurholt-Jepsen, M., Vinberg, M., Debel, S., Bardram, J.E., Kessing, L.V.: Behavioral activities collected through smartphones and the association with illness activity in bipolar disorder. Int. J. Methods Psychiatr. Res. **25**(4), 309–323 (2016). https://doi.org/10.1002/mpr.1502
12. Ganesh, A., Sahu, P., Nair, S., Chand, P.: A smartphone based e-Consult in addiction medicine: an initiative in COVID lockdown. Asian J. Psychiatry **51**, 102120 (2020)
13. Grünerbl, A., Muaremi, A., Osmani, V.: Smartphone-based recognition of states and state changes in bipolar disorder patients. IEEE J. Biomed. Health Inform. **19**(1), 140–148 (2015)
14. Guyon, I., Weston, J., Barnhill, S., Vapnik, V.: Gene selection for cancer classification using support vector machines. Mach. Learn. **46**(1–3), 389–422 (2002)
15. Iglesias, J.A., Ledezma, A., Sanchis, A., Angelov, P.: Real-time recognition of calling pattern and behaviour of mobile phone users through anomaly detection and dynamically-evolving clustering. Appl. Sci. **7**(8), 798 (2017)
16. Kaczmarek-Majer, K., et al.: Control charts designed using model averaging approach for phase change detection in bipolar disorder. In: Destercke, S., Denoeux, T., Gil, M.Á., Grzegorzewski, P., Hryniewicz, O. (eds.) SMPS 2018. AISC, vol. 832, pp. 115–123. Springer, Cham (2019). https://doi.org/10.1007/978-3-319-97547-4_16
17. Kamińska, O., Kaczmarek-Majer, K., Hryniewicz, O.: Acoustic feature selection with fuzzy clustering, self organizing maps and psychiatric assessments. In: Lesot, M.-J., et al. (eds.) IPMU 2020. CCIS, vol. 1237, pp. 342–355. Springer, Cham (2020). https://doi.org/10.1007/978-3-030-50146-4_26
18. Leite, D., Škrjanc, I., Gomide, F.: An overview on evolving systems and learning from stream data. Evol. Syst. **11**, 1–18 (2020)
19. Li, P., Wu, X., Hu, X., Wang, H.: Learning concept-drifting data streams with random ensemble decision trees. Neurocomputing **166**(C), 68–83 (2015). https://doi.org/10.1016/j.neucom.2015.04.024
20. Linardon, J., Fuller-Tyszkiewicz, M.: Attrition and adherence in smartphone-delivered interventions for mental health problems: a systematic and meta-analytic review. J. Consult. Clin. Psychol. **88**(1), 1 (2020)
21. Luxton, D.D., McCann, R.A., Bush, N.E., Mishkind, M.C., Reger, G.M.: mHealth for mental health: integrating smartphone technology in behavioral healthcare. Prof. Psychol.: Res. Pract. **42**(6), 505 (2011)
22. Mataró, T.V., et al.: An assistive mobile system supporting blind and visual impaired people when are outdoor. In: 2017 IEEE 3rd International Forum on Research and Technologies for Society and Industry (RTSI), pp. 1–6. IEEE (2017)
23. Maxhuni, A., Munoz-Melendez, A., Osmani, V., Perez, H., Mayora, O., Morales, E.: Classification of bipolar disorder episodes based on analysis of voice and motor activity of patients. Perv. Mob. Comput. **31**, 50–66 (2016)
24. Vazquez-Montes, M.D.L.A., Stevens, R., Perera, R., Saunders, K., Geddes, J.R.: Control charts for monitoring mood stability as a predictor of severe episodes in patients with bipolar disorder. Int. J. Bipolar Disord. **6**(1), 1–15 (2018). https://doi.org/10.1186/s40345-017-0116-2

25. Meng, X., Dai, Z., Hang, C., Wang, Y.: Smartphone-enabled wireless otoscope-assisted online telemedicine during the COVID-19 outbreak. Am. J. Otolaryngol. **41**, 102476 (2020)
26. Mohr, D.C., Zhang, M., Schueller, S.M.: Personal sensing: understanding mental health using ubiquitous sensors and machine learning. Annu. Rev. Clin. Psychol. **13**, 23–47 (2017)
27. Kamińska, O., et al.: Self-organizing maps using acoustic features for prediction of state change in bipolar disorder. In: Marcos, M., et al. (eds.) KR4HC/TEAAM 2019. LNCS (LNAI), vol. 11979, pp. 148–160. Springer, Cham (2019). https://doi.org/10.1007/978-3-030-37446-4_12
28. Pazienza, A., et al.: Adaptive critical care intervention in the internet of medical things. In: 2020 IEEE Conference on Evolving and Adaptive Intelligent Systems (EAIS), pp. 1–8. IEEE (2020)
29. Pedrycz, W.: A dynamic data granulation through adjustable fuzzy clustering. Pattern Recogn. Lett. **29**(16), 2059–2066 (2008)
30. Pedrycz, W., Waletzky, J.: Fuzzy clustering with partial supervision. IEEE Trans. Syst. Man Cybern. Part B Cybern. **27**(5), 787–95 (1997)
31. Picerno, P., Pecori, R., Raviolo, P., Ducange, P.: Smartphones and exergame controllers as BYOD solutions for the e-tivities of an online sport and exercise sciences university program. In: Burgos, D., et al. (eds.) HELMeTO 2019. CCIS, vol. 1091, pp. 217–227. Springer, Cham (2019). https://doi.org/10.1007/978-3-030-31284-8_17
32. Rajkomar, A., Dean, J., Kohane, I.: Machine learning in medicine. N. Engl. J. Med. **380**(14), 1347–1358 (2019)
33. Swan, M.: Health 2050: the realization of personalized medicine through crowdsourcing, the quantified self, and the participatory biocitizen. J. Pers. Med. **2**(3), 93–118 (2012)
34. Vessio, G.: Dynamic handwriting analysis for neurodegenerative disease assessment: a literary review. Appl. Sci. **9**(21), 4666 (2019)

Iterative Multi-mode Discretization: Applications to Co-clustering

Hadi Fanaee-T[1](✉) and Magne Thoresen[2]

[1] Center for Applied Intelligent Systems Research, Halmstad University, Halmstad, Sweden
hadi.fanaee@hh.se
[2] Department of Biostatistics, University of Oslo, Oslo, Norway
magne.thoresen@medisin.uio.no

Abstract. We introduce a new concept called "Iterative Multi-Mode Discretization (IMMD)" which is a new type of efficient data sparsification that can scale up many tasks in data mining. In this paper we demonstrate the application of IMMD in co-clustering, i.e. simultaneous clustering of the rows and columns in a matrix. We propose IMMD-CC, a novel co-clustering algorithm, which is developed based on IMMD. IMMD-CC has attractive properties. First, its time complexity is linear, so it can be used in large-scale problems. In addition, IMMD-CC is able to estimate the number of co-clusters automatically, and more accurate than state-of-the-art methods. We demonstrate the performance of IMMD-CC in comparison to several state-of-the-art methods on 100 data sets from a benchmark cohort, as well as 35 real-world datasets. The results show the promising potential of the proposed method.

Keywords: Co-clustering · Bi-clustering · Discretization

1 Introduction

Many real-life datasets are represented by a two-dimensional n by m data matrix, sometimes referred to as two-mode data [5]. The first mode is usually samples (e.g. cases, persons) and the second dimension refers to features (e.g. measurements, genes). It is of great interest in many applications to group rows or columns to meaningful clusters. In the recent years several algorithms are developed for clustering. Recently, motivated by applications in text/web mining, bioinformatics, marketing, and ecology [5] a new version of clustering methods are being developed that simultaneously cluster rows and columns of a data matrix. These methods are usually referred to as co-clustering, biclustering, block clustering, or two-mode clustering. For instance, in analyzing gene expression data, co-clustering can discover functionally related gene sets under different subsets of experimental conditions [17].

The main problem with current co-clustering methods is that many of the them are developed for applications where the volume of data is considered small, so the majority of developed methods are not scalable and are not feasible for very large datasets. Second, the majority of methods, as their clustering counterparts, require the number of clusters as an input, which is much more complicated to set when there are complex

A. Appice et al. (Eds.): DS 2020, LNAI 12323, pp. 94–105, 2020.
https://doi.org/10.1007/978-3-030-61527-7_7

two-way structures in the data. Besides, since some of these methods have no restriction regarding overlapping co-clusters, they produce an extremely large solution space as the output.

In a very large matrix with billions of rows and columns, where we should look for the co-clusters? Can we somehow limit our search space to more *interesting* parts of the matrix? Can we find the approximate locations of co-clusters more efficiently (preferably with a linearly in time) and without requiring the user to set the number of co-clusters? These are the questions we attempt to answer in this research. We demonstrate that they can be addressed via Iterative Multi-Mode Discretization (IMMD), a new general concept that we propose. So, our contributions are as follows.

- We propose "Iterative Multi-Mode Discretization" (IMMD), a general and efficient sparsification method for large matrices with real values.
- We propose a new method based on the IMMD concept to extract the matrix's point of interests and then approximate co-clusters without any requirement for the user to specify the number of co-clusters.
- We provide empirical evidences on the promising performance of our method in spotting the known clusters by exhaustive comparison against state-of-the-art methods on tens of benchmark simulated and real-life data sets.

2 Background

Suppose that A is an $n \times m$ data matrix, where the elements of this matrix are real numbers and n and m represent the number of samples and features, respectively represented by sets of rows X and columns Y. Each element of the matrix a_{ij} represents the relation between row i and column j. The matrix A can be defined with its set of rows $X = \{x_1, .., x_n\}$ and columns $Y = \{y_1, .., y_m\}$, so that we can denote A with (X, Y).

Also let's denote a set of rows $I \subseteq X$ and columns $J \subseteq Y$, that constitute the submatrix $A_{IJ} = \{I, J\}$ that contains only those elements of a_{ij} belonging to the subset of rows $I = \{i_1, .., i_d\}$; $d \leq n$ and columns $J = \{j_1, .., j_r\}$; $r \leq m$, and vice versa.

Co-clustering is defined as a task of finding a set of elements $C_k = (I_k, J_k)$, specifically sets of rows I_k that exhibit similar behavior across the set of columns J_k, and vice versa.

Several methods have been developed for co-clustering, which the majority of them are covered in the recent surveys [1,12,13,19,23].

3 Co-clustering via Iterative Multi-mode Discretization

In this section we introduce the idea of Iterative Multi-Mode Discretization and later show how we extend this idea to co-clustering.

3.1 Multi-mode Discretization

Continuous data are among the most common data types and can be found in all areas. They can be positive or negative with different units and variances. Hence, normalization or transformation is an essential step in any co-clustering algorithm. Normalization is typically performed column-wise using the following formula.

$$A^c_{i,j} = \frac{a_{ij} - min_j}{max_j - min_j} \tag{1}$$

Similarly, a row-wise normalization can be performed as follows.

$$A^r_{i,j} = \frac{a_{ij} - min_i}{max_i - min_i} \tag{2}$$

We can extend Eq. 1 to obtain the discretized matrix Z^c with elements $Z^c_{i,j}$ between 1 and s.

$$Z^c = round((s-1) \times A^c) + 1 \tag{3}$$

Similarly, we can obtain the row-wise discretization as well:

$$Z^r = round((s-1) \times A^r) + 1 \tag{4}$$

Here, s is a scale parameter which should be an odd, natural number ≥ 3. This *multi-mode transformation* converts the continuous data matrix A with elements $\in \mathbb{R}$ to two discrete matrices Z^r and Z^c with elements $\in \{1, \ldots, s\}$.

The benefit of this type of discretization is that similar data points in any row or column will automatically get allocated the same value. So there will be no need to perform the exhaustive search to find groups of similar elements across rows and columns. This is the main and the most important trick. A real-world example of this can be wavelengths in the eye's color perception. Among several lights arriving to our eye we simply focus on one specific wavelength to identify, for example a red object. So, in the co-clustering case, if we want to find the first group of data items, we just need to retrieve the elements whose values are equal to 1 in the transformed matrices.

The reason why we do the transformation across both rows and columns (multi-mode) is to capture the joint information among samples (rows) and features (columns), in a more efficient way without the need for searching.

3.2 Iterative Removal of Non-interesting Elements

The second component of IMMD relies on the removal of non-interesting data elements in an iterative process.

After obtaining Z^r and Z^c with scale parameter s, from these matrices we remove those elements that are not equal to either $\{1, v, s\}$, where v is the median of $\{1, \ldots, s\}$. For instance, if $s = 9$, v will be equal to 5. So all elements except those that are equal to 1, 5, and 9 are removed. The goal of this stage is to reduce the search space as much as possible by keeping only floor, median, and ceiling values at each iteration. Our assumption is that the co-clusters of interest should appear with more signal power in the final sparse matrices. After the removal phase we generate a new copy of the input matrix that contains only the values of non-removed indices, and for the rest we put missing value (denoted by N/A). We repeat this procedure iteratively, for both modes until we obtain a high-quality sparse discrete matrix for both modes.

3.3 Intersection of Hotspots Across the Modes

After removing phase in an iterative process on both modes and after reaching to the stopping criterion (obtaining an empty matrix), we have to the select the iteration that leads to a better sparsification quality (the statistics we use is described in the algorithm details). In the final matrices we replace those elements that are equal to either 1, v, or s with 1 and replace the other elements with zero, which leads to a binary matrix respectively for column-wise and row-wise discretization. We then add these two binary matrices, which leads to the intersection matrix. The elements with value of 2 in the intersection matrix represent the matrix's point of interest that appeared as hotspots in both row-wise and column-wise discretization.

3.4 Finding Co-clusters

Due to the high-level of sparsity in the output matrices, co-clusters normally appear with a reasonable distance from each other and we can identify them by connectivity criteria (a bit similar, but different to algorithms like DBSCAN [4]). We first sort the elements based on the repeated counts of their indices in the remaining items. Then we start from the points whose elements have higher frequent indices. We create a cluster for the first point. If the second point is reachable (either via row or column) from the first point with a reasonable support it goes to the same cluster, otherwise we create a new cluster for the second point. We continue this procedure until we have no further non-allocated points.

3.5 Pruning Co-clusters

In this phase we remove those clusters that have a low number of members, lower than a predefined threshold. This is a controlling parameter to avoid overwhelming number of clusters with low number of members.

3.6 Algorithm IMMD-CC

Our proposed algorithm IMMD-CC is presented in Algorithm 1. The input of the algorithm are as follows. The input matrix $A(n \times m)$ with continuous real numbers; ϵ, co-cluster's connectivity support threshold; q, the minimum number of members for each co-cluster; h, the top-h number of hotspots in the mode intersection phase; s, the scale parameter for discretization. The output of the algorithm is the co-clusters in the form of indices of rows and columns of the matrix grouped in each cluster.

Note that lines 4–13 and 14–23 are identical. The former is the process based on row-wise normalization and the latter is same process with column-wise normalization. Here we describe the process for the former, but the same explanations can be used for the column-wise part.

We begin with multi-mode discretization as described in Sect. 3.1. We first obtain the discretized version of the matrix (Z^r) in line 7 using the parameter s. Then we create a copy of the original matrix A called I_i and remove from I_i those elements where their corresponding values in Z^r is not either 1 (floor), v (median), nor s (ceiling).

This is done in line 9. For instance, suppose a 3×2 matrix which original values are $A = \{0.1, 0.1; 0.2, 0.3; 0.3, 0.9\}$. Its column-wise discretization transformation is $Z^c = \{1, 1; 3, 2; 5, 5\}$ in the scale of $s = 5$. The median of $\{1 : s\} = \{1, 2, 3, 4, 5\}$ is 3. So, from A we remove those elements that are neither 1, 3 nor 5. So the new data becomes $\{0.1, 0.1; 0.2, N/A; 0.3, 0.9\}$. Then this new data gets replaced with the original data (new A) and we repeat this procedure until all values of the matrix becomes N/A.

At each iteration we also keep a statistics $O_{i,k}$ (line 12) that allows us chose the best sparsification. Let us denote the number of elements whose values are equal to 1, v, or s in the matrix Z^r, respectively with n_1, n_v, and n_s. Then our statistics is computed as follows. The sum of absolute values of pairwise differences between n_1, and n_v and n_s divided by number of available elements in the iteration. For instance, in the above example $n_1 = 1$, $n_3 = 2$, $n_5 = 1$. Also the number of available elements is 4 out 5 (only one N/A at the first iteration). So, O_i at iteration $k = 1$ is computed as $(|1 - 2| + |1 - 1| + |2 - 1|)/4 = 2/4 = 0.5$. We keep this statistics at each step, because later at line 24 we want to select the best sparsification based on the lowest obtained value for O over all iterations.

After selecting the best sparsification, the lines after line 26 correspond to the co-clustering phase. Based on the selected iteration ks and ps we pick the data output at the end of these iterations. Then we make two temporary matrices T_i and T_j, that are binary (boolean) copy of the data for respectively row-wise and column-wise scenarios. Following the above example, let us suppose that we conclude the first iteration ($k = 1$) gives the best sparsification. Then we have $T_j = \{1, 1; 1, 0; 1, 1\}$, and $T_i = \{1, 1; 1, 1; 1, 1\}$. Since normalization based on row and column give different discretization transformation the values of these are certainly different.

At line 28 we add T_i to T_j and generate the intersection matrix T. Again following the above example our intersection matrix is $T = \{2, 2; 2, 0; 2, 2\}$. Since both matrices are binary, the values equal to 2 in T demonstrate the intersection of hotspots obtained both row-based and column-based. They are somehow important points of data that depending on the level of sparsification can be exactly the centroid or important members of co-clusters. So, in the next lines (30–31) we obtain the corresponding indices related to these hotspots.

Next, in a procedure that is outlined in lines 32–41 we first generate a new empty matrix G which is the same size as the original data matrix. Then we start from the list of hotspots and add +1 to the corresponding indices of rows and columns of hotspot. If we sort the elements based on their count in the G matrix from higher to lower values we obtain a list of ranked hotspots from the highest important ones to less important items at the end of list.

The first item in the ranked list is with high probability the centroid of the most important cluster. So we create a cluster with this item as its first member (line 42). Then we check the second-ranked item in the list of hotspots. If there is a sufficient connectivity support of this point with the previously allocated element, then it goes to the same cluster, otherwise we create a new cluster and put the new element inside that. The criterion we use for connectivity support is that we count of the number of times that either the row or column index of the candidate has appeared jointly in the pair of hotspots. Then we divide this by the maximum obtained repeating count, which gives a value between 0 and 1. Values close to 0 indicates that the candidate is less connected to the previous member, and values close to 1 means that the candidate is very well connected to the previous element. This procedure continues until all items of hotspots are allocated to a cluster. In order to make the algorithm more efficient we set a parameter h to only do the allocation for top-h hotspots.

Finally, the algorithm ends after a pruning step (line 51) to remove those co-clusters that do not have sufficient number of members. This is controlled with the parameter q.

4 Experimental Evaluation

4.1 Datasets

Simulated Datasets. The simulated datasets are generated by [15]. We have five groups of 20 datasets, each composed of 500 rows and 200 columns with a constant number of k co-clusters in each group. Each co-cluster (sub-matrix of 50×50) contains zero values on its elements and the remaining elements of each dataset that do not belong to that co-cluster, are generated i.i.d. from $N(0, 1)$.

Real-World Datasets. The real datasets are publicly available benchmark datasets [21], which consists of 35 microarray datasets related to cancer for different tissue types (eg., blood, lung, colon, breast, skin, prostate, etc.).

4.2 Compared Methods

We compare our method with 11 state-of-the-art methods from five categories of techniques: (1) Greedy methods: Cheng and Church (CC) [2], Large Average Submatrices (LAS) [20]; (2) Divide-and-conquer algorithms: Binary Inclusion-Maximal Biclustering Algorithm (BiMax) [18], Qualitative BiClustering (QUBIC) [11]; (3) Distribution parameter identification method: Modified Plaid Algorithm (Plaid) [22], Spectral Biclustering (kSpectral), Bipartite Spectral Graph Partitioning (BSGP) [10], Factor Analysis for Bicluster Acquisition (FABIA) [6]; and (4) Information-theoretic algorithms: Information-Theoretic Co-clustering (ITL) [3]; and Fuzzy methods: FuzzyBiClustering [8].

4.3 Evaluation Metric

Accuracy. Horta and Campello [7] benchmarked 14 quality metrics for co-clustering algorithms and suggested that co-clustering error (CE) proposed by [16] is one of the most appropriate metrics for comparison of co-clustering algorithms. So, we exploit CE as our accuracy metric. Let us denote the found co-cluster with C_1 and the reference co-clusters with C_2. Then let us denote the union of two co-clusterings by $U = |C_1 \cup C_2|$. Now we can define CE as:

$$\mathbb{S}_{ce} = 1 - \frac{|U| - D_{max}}{|U|} \tag{5}$$

Where D_{max} is the sum of the diagonal elements of the confusion matrix of C_1 and C_2 whose elements are the intersection of C_1 and C_2, i.e., $m_{ij} = |C_1 \cap C_2|$.

Algorithm 1. IMMD-CC

```
    Input A (n × m), ε, q, h, s                                    ▷ Refer to section 3.6
    Output C
 1: v ← med(1 : s)                                                  ▷ Median, Refer to section 3.2
 2: A_i' ← A
 3: A_j' ← A
 4: while A_i' is not empty do
 5:     k ← k + 1
 6:     A^r ← Row-wise Normalization of A_i'
 7:     Z^r ← round((s − 1) × A^r) + 1                                ▷ Refer to section 3.1
 8:     I_i ← A_i'
 9:     A_i' ← Remove from I_i if their Z^r ≠ {1, v, s}
10:     A_{i,k}' ← A_i'
11:     U_i ← Count of Non-zero elements in A_i'
12:     O_{i,k} ← (|n_v − n_1| + |n_s − n_v| + |n_1 − n_s|) / U_i
13: end while
14: while A_j' is not empty do
15:     p ← p + 1
16:     A^c ← Column-wise Normalization of A_j'
17:     Z^c ← round((s − 1) × A^c) + 1                                ▷ Refer to section 3.1
18:     I_j ← A_j'
19:     A_j' ← Remove from I_j if their Z^c ≠ {1, v, s}
20:     A_{j,p}' ← A_j'
21:     U_j ← Count of Non-zero elements in A_j'
22:     O_{j,p} ← (|n_v − n_1| + |n_s − n_v| + |n_1 − n_s|) / U_j
23: end while
24: ks ← iteration k that have Min(O_{i,k})
25: ps ← iteration p that have Min(O_{j,p})
26: T_i ← A_{i,ks}', T_i = 1 if A_{i,ks} ≠ 0, else T_i = 0
27: T_j ← A_{j,kp}', T_j = 1 if A_{j,ps} ≠ 0, else T_j = 0
28: T ← T_i + T_j
29: W ← T, where T == 2
30: H_1 ← Row indices of W
31: H_2 ← Column indices of W
32: c_max ← length of H_1 or H_2
33: G ← Zeros(n, m)
34: while c ≤ c_max do
35:     c ← c + 1
36:     L_i ← id of nonzero elements H_{1,c}'s column of T
37:     L_j ← id of nonzero elements H_{2,c}'s column of T
38:     G(L_i, L_j) ← G(L_i, L_j) + 1
39:     H_{3,c} ← G(L_i, L_j)
40: end while
41: Descend Sort H based on third column H_3
42: Create cluster C_1 and Add (H_{1,1}, H_{2,1}) to it
43: c ← 0, i ← 1
44: while c ≤ h do
45:     c ← c + 1
46:     if ConnSupport ≤ ε then                                       ▷ Refer to section 3.6
47:         i ← i + 1
48:     end if
49:     Add (H_{1,c}, H_{2,c}) to C_i
50: end while
51: Remove co-clusters that do not have at least q members
```

Estimation of Number of Clusters. Another important factor that needs to be taken into account when choosing a co-clustering method is how accurately the method can identify the true clusters without having prior knowledge about the true number of clusters. We use this as one of our evaluation metrics.

Coverage. Co-clustering is a more difficult problem compared to regular clustering algorithms that do not have any restrictions with regard to overlapping co-clusters. So we might have two co-clusters with a large portion of overlap. In order to evaluate this we introduce a measure called *Coverage* as the following. This metric is relative and can be used only for comparison. It penalizes the methods that produce lots of overlapping co-clusters.

$$Covg = \frac{\sum C_c}{M} \tag{6}$$

Where $\sum C_c$ represents the total area of found co-clusters and M is total area. Values closer to zero demonstrates compact solution space and less overlapping clusters.

4.4 Experimental Configurations

We set the parameters as follows. For our method IMMD-CC we set the following parameters $\epsilon = 0.3$, $q = 2$, $s = 9$, and $h = \infty$. For other methods we use the implementations available in the MTBA toolbox [9] (Available publically at http://www.iitk.ac.in/idea/mtba/) with the default parameters suggested by the toolbox. Two methods (kSpectral and FuzzyBiClustering) were excluded on experiments with real-life datasets, due to the some bugs in the implementation, which made them infeasible for some datasets.

4.5 Results

In this section we report the results from the analysis of both simulated datasets and real-world datasets.

Simulated Datasets. The box plot of co-clustering accuracy (measured by CE) for 100 datasets is presented in Fig. 1. The surprising observation is that among 11 compared methods only five has a reasonable performance on the simulated datasets. The reason might be that our simulated datasets include only non-overlapping co-clusters and some of these methods may be designed for particular types of co-clusters with specific patterns (see examples in Fig. 1 of [14]).

As we can see from the results, among the 11 compared methods, only ITL [3] outperforms IMMD-CC. However, when we look at the number of times where IMMD-CC identifies exactly the correct number of simulated co-clusters with zero error tolerance ($k = \{1, 2, 3, 4, 5\}$) with respect to the ground truth, IMMD-CC stands out by a large difference beyond the other methods, including ITL. In 35 out of 100 datasets, IMMD-CC finds the exact number of simulated co-clusters, while the second-ranked method, CC [2] and third-ranked method, Qubic [11] find the correct number in 20, and 17 respectively of the datasets.

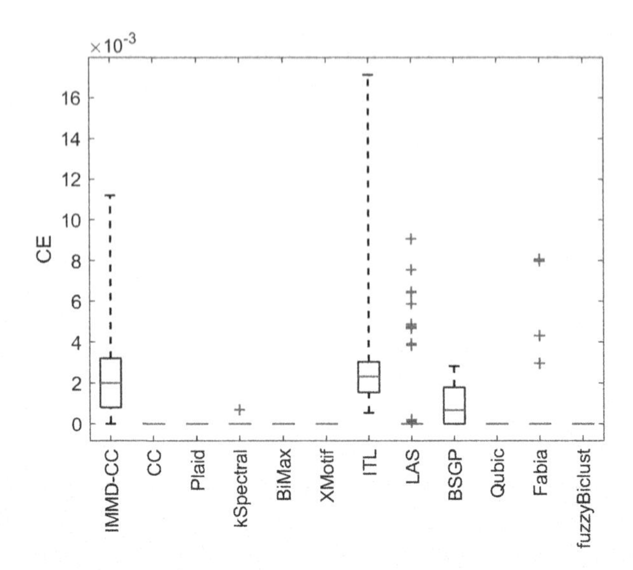

Fig. 1. Co-clustering quality

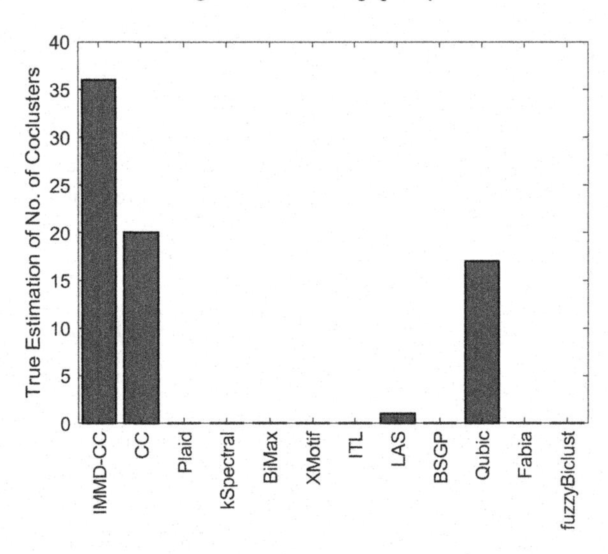

Fig. 2. True estimation of number of co-clusters

Real-Life Datasets. One of the limitations of the studies on co-clustering problems is the lack of availability of datasets with a ground truth. This is already pointed out by [15] in their co-clustering benchmarking study. They used regular clustering datasets to evaluate the performance of co-clustering algorithms. This makes sense because each cluster in a matrix is a special case of a co-cluster, where one mode includes all rows or columns. However, still the real value of co-clustering algorithms that generate sub-matrices cannot be evaluated via this strategy.

Table 1. Coverage and number of detected clusters on real-life datasets.

Dataset	True Clusters	**IMMD-CC** Found Clusters	**IMMD-CC** Coverage	**CC** Found Clusters	**CC** Coverage	**Plaid** Found Clusters	**Plaid** Coverage	**BiMax** Found Clusters	**BiMax** Coverage	**XMotif** Found Clusters	**XMotif** Coverage	**ITL** Found Clusters	**ITL** Coverages	**LAS** Found Clusters	**LAS** Coverage	**BSGP** Found Clusters	**BSGP** Coverage	**Qubic** Found Clusters	**Qubic** Coverages	**Fabia** Found Clusters	**Fabia** Coverages
alizadeh-2000-v1	2	2	0.13	0	100	0	0.35	0	0.11	0	0	0	54.34	0	22.49	0	99.91	0	0.65	0	3.76
alizadeh-2000-v2	3	2	0.02	0	100	0	4.41	0	0.09	0	0	0	33.10	0	29.20	0	33.16	0	0.92	0	6.39
alizadeh-2000-v3	4	2	0.02	0	100	0	5.53	0	0.13	0	0	0	25.98	0	38.74	0	25.75	0	1.19	0	7.55
armstrong-2002-v1	2	2	0.03	0	0.19	0	0	0	100	0	3	0	49.47	0	8.54	0	99.81	0	13.70	0	0
armstrong-2002-v2	3	2	0.01	0	0.07	0	0	0	100	0	2.92	0	34.70	0	22.09	0	55.70	0	3.33	0	0
bhattacharjee-2001	5	0	0	0	0.51	2	1.22	0	100	0	1.08	0	14.80	0	39.31	0	90.54	0	13.91	0	0
bittner-2000	2	2	0.15	0	100	0	0	0	0	0	5.07	0	55.82	0	55.26	0	0	0	1.62	0	2.09
bredel-2005	3	3	0.11	0	100	0	0	0	0.07	0	0	0	30.94	0	9.95	0	18.63	0	0.79	0	5.27
chen-2002	2	0	0	0	100	2	4.50	1	0.42	0	0	1	63.23	1	8.08	2	50.13	2	0.55	2	4.42
chowdary-2006	2	0	0	1	5.45	0	0	0	100	1	1.80	1	51.27	2	39.77	0	98.08	1	11.87	0	0
dyrskjot-2003	3	0	0.01	2	0.04	0	0	0	100	0	0	2	33.11	2	9.31	0	99.17	2	8.16	0	0
garber-2001	4	3	0.02	0	100	0	0	0	0.02	0	0	0	29.70	0	21.23	0	49.85	0	0.90	0	9.07
golub-1999-v1	2	2	0.01	0	0.03	0	0	0	100	0	1.17	0	56.02	0	4.05	0	99.79	0	1.62	0	0
golub-1999-v2	3	2	0.01	2	0.04	0	0.26	0	100	2	1.77	3	41.47	3	42.94	0	99.57	1	2.37	0	0
gordon-2002	2	0	0	0	0.08	2	0.49	0	0	0	0	0	54.30	0	7.85	0	99.75	0	2.35	0	0
khan-2001	4	0	0	0	100	0	0.09	1	0.22	0	1.27	3	27.67	2	42.26	0	97.66	1	10.83	4	8.61
laiho-2007	2	0	0	1	0.03	1	0.08	0	100	0	0	2	48.07	1	4.12	0	99.73	1	8.78	0	0
lapointe-2004-v1	3	2	0.01	0	84.68	0	8.54	0	0.08	0	0	0	38.72	0	43.58	0	99.82	0	2.10	0	6.79
lapointe-2004-v2	4	0	0	0	100	0	7.52	1	0.11	0	0	3	24.46	3	59.82	0	29.37	1	1.76	0	6.49
liang-2005	3	3	0.04	0	100	0	0	0	0.18	0	0	0	37.26	0	22.08	0	33.85	0	2.41	0	5.42
nutt-2003-v1	4	0	0.01	1	0.19	0	0	0	100	0	0	2	21.45	2	106.98	0	99.27	1	15.36	0	0
nutt-2003-v2	2	0	0.02	1	0.30	0	0	0	100	0	0	1	48.68	1	44.08	0	0	1	6.48	0	0
nutt-2003-v3	2	0	0	2	0.49	0	0	0	0	0	0	1	51.48	2	72.73	0	99.83	1	5.19	0	0
pomeroy-2002-v1	2	0	0	0	0.06	1	0.17	0	100	0	0	0	50	0	34.14	0	98.60	0	2.42	0	0
pomeroy-2002-v2	5	0	0.01	2	1.94	0	0	0	100	3	11.18	3	18.16	4	41.75	0	28.86	0	14.15	0	0
ramaswamy-2001	14	0	0	0	1.76	0	0	0	100	6	8.83	6	7.23	6	52.87	0	0	0	218.94	0	0
risinger-2003	4	2	3.24	0	100	0	0	0	0.05	0	0	0	25.93	0	24.43	0	91.19	0	4.28	0	9.09
shipp-2002-v1	2	0	0	1	0.22	1	1.48	0	100	1	2.53	1	41.46	2	11.66	0	98.25	1	5.43	0	0
singh-2002	2	0	0	0	0.03	1	3.77	0	100	0	1.56	0	25.81	0	37.60	0	0	0	12.83	0	0
su-2001	10	0	0	6	0.33	0	1.37	0	100	0	1.13	4	7.91	8	25.43	0	0	4	5.06	0	0
tomlins-2006-v2	4	4	0.03	0	100	0	0	0	0	0	0	0	27.76	0	26.11	0	51.70	0	0.56	0	8.50
tomlins-2006	5	3	0.01	0	100	0	0	1	0.01	0	0	2	24.72	2	22.54	0	99.65	2	0.75	3	8.20
west-2001	2	1	0.01	1	0.03	1	0.25	0	100	2	1.75	2	50.08	1	3.76	0	99	1	1.85	0	0
yeoh-2002-v1	2	1	0	0	0.21	0	0.02	0	100	0	0.49	0	47.88	0	50.79	0	99.92	0	100	0	0
yeoh-2002-v2	6	0	0	2	0.43	0	0.04	0	100	3	1.57	2	15.55	4	51.18	0	0	0	100	0	0

Regarding the simulated datasets, since the ground truth on the exact location of co-clusters is available we could directly measure the accuracy of co-clustering using co-clustering evaluation metrics. However, regarding the real datasets, since we do not have real co-clusters, maybe using the same measures for co-clustering might not be ideal. Besides, neither of the metrics measure the compactibility of solution space. For this reason, we instead measure the performance of algorithms in estimation of clusters with respect to their output coverage (see Sect. 4.3). In Table 1 we present the estimated number of clusters together with the coverage.

Table 1 should be read as follows. The first column is the true number of clusters chosen by domain experts in different cancer studies. Then we have pairs of columns for each method. The first column of each pair illustrates how many of the clusters were found by each method, and the next column is the *Covg* value described in Sect. 4.3.

IMMD-CC has identified the true clusters with the tolerance of 1 error on at least 14 out of 35 datasets. Regarding these 14 datasets the average coverage is 0.04. On the other hand ITL can find the true clusters in 12 datasets with a tolerance of 1 error. However, average coverage for ITL is 43.72 which is almost 1000 times larger than IMMD-CC. This is an evidence on the effectiveness of the IMMD method in high-quality sparsification, so that the clusters of interest can be identified from a much more compact subspace, compared to methods like ITL (Fig. 2).

5 Conclusion and Future Work

We introduce a new family of co-clustering methods based on a new concept called iterative multi-mode discretization. We demonstrate the effectiveness of the method both on simulated and real-life benchmark datasets. Although in terms of accuracy, methods such as ITL [3] present a better average performance compared to our method, if we consider other factors such as true estimation of co-clusters, as well as coverage, our approach is a more competitive technique. Still there are some issues left for future research. As an example, applying IMMD-CC to higher order tensors is one potential direction. Another direction is tuning of the parameters for performance improvement. For instance, we empirically found the scale parameter $s = 9$ good enough for all datasets we experimented. However, still there is no systematic way to choose this parameter for other datasets, which should be investigated further. The more important future research should be focused on other applications of IMMC, and not necessarily co-clustering application per se. The IMMC can be considered a highly potential versatile tool in data mining and machine learning. Preliminary work have shown some good results in applications in other problems such as sorting and regular clustering.

References

1. Charrad, M., Ben Ahmed, M.: Simultaneous clustering: a survey. In: Kuznetsov, S.O., Mandal, D.P., Kundu, M.K., Pal, S.K. (eds.) PReMI 2011. LNCS, vol. 6744, pp. 370–375. Springer, Heidelberg (2011). https://doi.org/10.1007/978-3-642-21786-9_60
2. Cheng, Y., Church, G.M.: Biclustering of expression data. In: ISMB, vol. 8, pp. 93–103 (2000)

3. Dhillon, I.S., Mallela, S., Modha, D.S.: Information-theoretic co-clustering. In: Proceedings of the Ninth ACM SIGKDD International Conference on Knowledge Discovery and Data Mining, pp. 89–98. ACM (2003)
4. Ester, M., Kriegel, H.P., Sander, J., Xu, X.: A density-based algorithm for discovering clusters a density-based algorithm for discovering clusters in large spatial databases with noise. In: Proceedings of the Second International Conference on Knowledge Discovery and Data Mining, KDD 1996, pp. 226–231. AAAI Press (1996)
5. Govaert, G., Nadif, M.: Co-clustering: Models, Algorithms and Applications. Wiley, Hoboken (2013)
6. Hochreiter, S., et al.: FABIA: factor analysis for bicluster acquisition. Bioinformatics **26**(12), 1520–1527 (2010)
7. Horta, D., Campello, R.J.: Similarity measures for comparing biclusterings. IEEE/ACM Trans. Comput. Biol. Bioinform. **11**(5), 942–954 (2014)
8. Huang, S.Y., Sun, H.J., Huang, C.D., Chung, I.F., Su, C.H.: A modified fuzzy co-clustering (MFCC) approach for microarray data analysis. In: 2014 IEEE International Conference on Fuzzy Systems (FUZZ-IEEE), pp. 267–272. IEEE (2014)
9. Gupta, J.K., Singh, S., Verma, N.K.: MTBA: MATLAB toolbox for biclustering analysis, pp. 94–97. IEEE (2013)
10. Kluger, Y., Basri, R., Chang, J.T., Gerstein, M.: Spectral biclustering of microarray data: coclustering genes and conditions. Genome Res. **13**(4), 703–716 (2003)
11. Li, G., Ma, Q., Tang, H., Paterson, A.H., Xu, Y.: QUBIC: a qualitative biclustering algorithm for analyses of gene expression data. Nucleic Acids Res. **37**(15), e101–e101 (2009)
12. Madeira, S.C., Oliveira, A.L.: Biclustering algorithms for biological data analysis: a survey. IEEE/ACM Trans. Comput. Biol. Bioinform. (TCBB) **1**(1), 24–45 (2004)
13. Mounir, M., Hamdy, M.: On biclustering of gene expression data. In: 2015 IEEE Seventh International Conference on Intelligent Computing and Information Systems (ICICIS), pp. 641–648. IEEE (2015)
14. Orzechowski, P., Boryczko, K., Moore, J.H.: Scalable biclustering—The future of big data exploration? GigaScience **8**(7), giz078 (2019)
15. Padilha, V.A., Campello, R.J.: A systematic comparative evaluation of biclustering techniques. BMC Bioinform. **18**(1), 55 (2017)
16. Patrikainen, A., Meila, M.: Comparing subspace clusterings. IEEE Trans. Knowl. Data Eng. **18**(7), 902–916 (2006)
17. Pontes, B., Giráldez, R., Aguilar-Ruiz, J.S.: Biclustering on expression data: a review. J. Biomed. Inform. **57**, 163–180 (2015)
18. Prelić, A., et al.: A systematic comparison and evaluation of biclustering methods for gene expression data. Bioinformatics **22**(9), 1122–1129 (2006)
19. Saber, H.B., Elloumi, M.: DNA microarray data analysis: a new survey on biclustering. Int. J. Comput. Biol. (IJCB) **4**(1), 21–37 (2015)
20. Shabalin, A.A., Weigman, V.J., Perou, C.M., Nobel, A.B., et al.: Finding large average submatrices in high dimensional data. Ann. Appl. Stat. **3**(3), 985–1012 (2009)
21. de Souto, M.C., Costa, I.G., de Araujo, D.S., Ludermir, T.B., Schliep, A.: Clustering cancer gene expression data: a comparative study. BMC Bioinform. **9**(1), 497 (2008)
22. Turner, H., Bailey, T., Krzanowski, W.: Improved biclustering of microarray data demonstrated through systematic performance tests. Comput. Stat. Data Anal. **48**(2), 235–254 (2005)
23. Xie, J., Ma, A., Fennell, A., Ma, Q., Zhao, J.: It is time to apply biclustering: a comprehensive review of biclustering applications in biological and biomedical data. Brief. Bioinform. **1**, 16 (2018)

Data and Knowledge Representation

COVID-19 Therapy Target Discovery with Context-Aware Literature Mining

Matej Martinc[1,2(✉)], Blaž Škrlj[1,2], Sergej Pirkmajer[3], Nada Lavrač[1,2,4], Bojan Cestnik[1,5], Martin Marzidovšek[1,2], and Senja Pollak[2]

[1] Jožef Stefan International Postgraduate School, Ljubljana, Slovenia
matej.martinc@ijs.si
[2] Jožef Stefan Institute, Ljubljana, Slovenia
[3] Institute of Pathophysiology, Faculty of Medicine, University of Ljubljana, Ljubljana, Slovenia
[4] University of Nova Gorica, Vipava, Slovenia
[5] Temida d.o.o, Ljubljana, Slovenia

Abstract. The abundance of literature related to the widespread COVID-19 pandemic is beyond manual inspection of a single expert. Development of systems, capable of automatically processing tens of thousands of scientific publications with the aim to enrich existing empirical evidence with literature-based associations is challenging and relevant. We propose a system for contextualization of empirical expression data by approximating relations between entities, for which representations were learned from one of the largest COVID-19-related literature corpora. In order to exploit a larger scientific context by transfer learning, we propose a novel embedding generation technique that leverages SciBERT language model pretrained on a large multi-domain corpus of scientific publications and fine-tuned for domain adaptation on the CORD-19 dataset. The conducted manual evaluation by the medical expert and the quantitative evaluation based on therapy targets identified in the related work suggest that the proposed method can be successfully employed for COVID-19 therapy target discovery and that it outperforms the baseline FastText method by a large margin.

Keywords: Knowledge discovery · Literature mining · Representation learning · Contextual embeddings · COVID-19

1 Introduction

Scientific knowledge for a specific domain is in most cases given in an unstructured form, as a set of scientific papers covering a variety of findings, experiments and methodologies related to a specific scientific field or problem. The current speed and quantity of scientific research production makes manual inspection of the literature from a specific field virtually impossible. The recent trend of interdisciplinary research complicates things even more, as it would require from a researcher to understand all the aspects, from which a specific research problem can be covered in order to "connect all the dots" and advance the field by the discovery of the so-called latent scientific knowledge.

A. Appice et al. (Eds.): DS 2020, LNAI 12323, pp. 109–123, 2020.
https://doi.org/10.1007/978-3-030-61527-7_8

To solve this problem, several automated strategies for uncovering this knowledge have been proposed. Somewhat older studies proposed literature-based discovery (LBD) [8] focusing especially on cross-domain literature mining, which aims at finding interesting bridging terms (b-terms) or bridging links revealing the potentially new connections between separate domain corpora of interest. On the other hand, more recent approaches to latent knowledge discovery from the scientific literature employ word embeddings [26]. For example, a study by [34] showed that latent knowledge regarding future discoveries is to a large extent embedded in past publications by retrieving information from the scientific literature with the usage of Word2Vec embeddings [26].

The latest development in the natural language processing (NLP) is a new type of embeddings called contextual embeddings. ELMo (Embeddings from Language Models) [29] and BERT (Bidirectional Encoder Representations from Transformers) [12] are the most prominent representatives of this type of contextual embeddings, and have been also adapted to scientific literature [3]. The main difference between these novel contextual embeddings and older "static" embeddings is that in these embeddings a different vector is generated for each context a word appears in, i.e., for each specific word usage in the corpus. These new contextual embeddings solve the problems with word polysemy and other changes in word meaning given different context. On the other hand, it is not entirely clear how to generate a meaningful general word representation from the word usage embeddings. This means that the usage of contextual embeddings for LBD is not entirely straight forward, since they can not be used in the same way as the traditional static embeddings, and have at least to our knowledge not been used for the task at hand.

In this work, we explore how contextual embeddings can be leveraged for the task of discovering latent scientific knowledge in the very topical scientific literature about the COVID-19 disease. More specifically, we are interested in the discovery of new COVID-19 therapy targets from the targets discovered in the past research. The novelty of this work is two-fold:

- The paper contributes a new methodology of generating general word representations from contextual embeddings, proposes an entire workflow for acquisition of novel COVID-19 therapy targets and shows that our method of using contextual embeddings for LBD outperforms the baseline method of using static embeddings by a large margin.
- Medically, the paper contributes to identifying new potential COVID-19 therapy targets, motivated by a recent proof-of-concept study that used a state-of-the-art omics approach to identify new possible targets for existing drugs, such as ribavirin [5].

2 COVID-19 Medical Background and Recent Therapy Targets

In late 2019 a novel coronavirus disease (COVID-19), caused by severe acute respiratory syndrome coronavirus 2 (SARS-CoV-2), emerged in China [38,39].

COVID-19 quickly spread and was declared a pandemic by the World Health Organization.

While new targeted therapies and vaccines against SARS-CoV-2 virus are being actively developed, their potential use in the clinics is not imminent. Therefore, until effective pharmacological therapies and/or vaccines are available, medicine needs to resort to other approaches to treat patients with COVID-19 or prevent transmission of SARS-CoV-2. One approach is to identify which among the antiviral drugs that were developed to treat other viral diseases might be effective against SARS-CoV-2. A preliminary report suggests that remdesivir seems to be the most promising candidate among these drugs [2]. Another approach is to identify drugs that are used for other purposes but also exert antiviral effects. The most prominent example among these is hydroxychloroquine, which is used for chronic treatment of rheumatic diseases but also suppresses SARS-CoV-2 in vitro [22]. Identifying a known drug with well-characterized adverse effects would certainly save time and lives before more specific treatments are developed. However, repurposing of existing drugs is also a challenge as highlighted by a recent controversy with hydroxychloroquine [7,25] and new candidate drugs and/or therapeutic targets are needed.

3 Related Work

The related work is divided into three Sections, namely related work on Literature-based discovery in Sect. 3.1, related work on text representation learning in Sect. 3.2 and selected overview of recent NLP research on COVID-19 in Sect. 3.3.

3.1 Literature-Based Discovery

Literature-based discovery (LBD) aims to generate new knowledge by combining what is already known in the literature. It has been used to (semi-automatically) identify new connections between genes, drugs and diseases, etc. [18]. Traditionally, LBD has been addressed as finding interesting bridging terms revealing the potentially new connections between separate domain corpora of interest [8]. Swanson [33] developed one of the early LBD approaches, the so-called ABC model, to detecting interesting b-terms to uncover the possible cross-domain relations among previously unrelated concepts.

On the other hand, a more recent state-of-the art tool LION LBD [31] enables researchers to navigate published information and supports hypothesis generation and testing. The system is built with a particular focus on the molecular biology of cancer. LBD has led to discovery of potential treatments in other domains, including multiple sclerosis [19], and has been applied successfully in drug development and repurpusing [11]. Recent LBD approaches benefit from word embeddings. One is the study by [34] already mentioned in Sect. 1 and the other is the work by [9], who proposed graph-based, neural network methods to perform open and closed LBD and demonstrated improved performance on existing tasks.

3.2 Text Representation and Embeddings

Recently, the embedding approach became a prevalent way to build representations for many different types of entities, e.g., texts, graphs, electronic health records, images, relations, recommendations, etc. Text embeddings use large corpora of documents to extract vector representations for words, sentences, and documents. The first neural word embeddings like Word2vec [26] produced one vector for each word, irrespective of its polysemy. These so-called static embeddings have been further developed and the most popular static embeddings currently in use besides Word2Vec are GloVe (Global vectors for word representation) [28] and FastText [4]. Recent developments like ELMo [29] and BERT [12] take a context of a sentence into account and produce different word vectors for different contexts of each word. Another novelty of these approaches is the employment of the transfer learning technique, which has recently become a well established procedure in the field of NLP. This procedure relies on a language model pretraining on very large unlabeled textual resources and after that transfer of the knowledge obtained by the language model onto a specific downstream task by further fine-tuning the model.

3.3 Text Mining and NLP Research Related to COVID-19

With regard to biomedical research on COVID-19, time is a central factor as scientists try to design treatments and vaccines amid the pandemic caused by the SARS-CoV-2 virus, therefore leveraging LBD and its potential to reduce scientific discovery time could prove crucial.

Many search platforms emerged for retrieving COVID-19 related papers. For example, Neural Covidex[1] is based on neural ranking architecture and provides information access capabilities to the COVID-19 Open Research Dataset (CORD-19) (see Sect. 4.1). SciSight [17] in contrast to standard targeted search facilitates finding connections between biomedical concepts that are not obvious from reading individual papers. It displays a network of top related terms mined from the corpus, based on the co-appearance in the same sentence.

Studies that can generate new knowledge about COVID-19 by applying embeddings are still scarce but do exist. For example, a recent study has projected Covid-related medical texts in a 3D human atlas space that helps to navigate the literature [14]. The objective was to learn semantically aware groundings of sentences with five different BERT models [12].

4 Background Knowledge and Resources

We describe the CORD-19 corpus (Sect. 4.1) and embeddings technology (Sect. 4.2) used in this study.

[1] https://covidex.ai/.

4.1 CORD19 Database

The scientific literature considered in this work has been recently introduced as the CORD-19 corpus[2]. CORD-19 is a resource of over 135,000 scholarly articles, including over 68,000 with full text, about COVID-19, SARS-CoV-2, and related coronaviruses. This freely available data set is provided to the global research community to apply recent advances in NLP and other AI techniques to generate new insights in support of the ongoing fight against this infectious disease.

We use the corpus version 12, published on May 1st 2020, from which we extract only full text scholarly articles converted into xml from a pdf format. This results in altogether 48,410 papers, which are summarized in Table 1.

Table 1. CORD-19 dataset statistics.

Origin	Number of papers	Number of tokens
Commercial use subset	9,918	46,206,453
Non-commercial use subset	2,584	10,732,608
PMC custom license subset	32,450	156,247,363
bioRxiv (not peers reviewed)	2,670	8,968,183
medRxiv subset (not peer reviewed)	788	3,285,558
All	48,410	225,440,165

4.2 Considered Embeddings

We use FastText [4] embeddings as a baseline in this study. The main advantage of FastText embeddings is its word representation as a sum of n-grams, which allows the model to, in addition to leveraging semantic relations, also leverage morphological information.

One of the most oftenly used models for the generation of contextual embeddings is the BERT model [12] that was originally pretrained on the Google Books Corpus (800 million tokens) and Wikipedia (2,500 million tokens). This pretraining is however not entirely appropriate for the text mining tasks on the scientific literature due to specificities of the scientific language and vocabulary. For this reason, in this research we opted for SciBERT [3], a version of BERT pretrained on a large multi-domain corpus of scientific publications, a random sample of 1.14 M papers from Semantic Scholar. SciBERT model has 12 encoder layers with the attention mechanism and a hidden layer size of 768.

5 Methodology

In this section, we present the methodology of the proposed approach by explaining how we obtain word representations, how we acquire therapy target candidates and how we evaluate the approach.

[2] https://www.kaggle.com/allen-institute-for-ai/CORD-19-research-challenge.

5.1 Word Representations

First, we fine-tune SciBERT as a masked language model for domain adaptation on the lowercased CORD-19 dataset. Next, we generate word representations for each word in the vocabulary. Figure 1 visualizes the process described below. The documents from the corpus are split into sequences of byte-pair encoded tokens [20] of a maximum length of 256 tokens and fed into the fine-tuned SciBERT model. For each of these sequences of length n, we create a sequence embedding by summing the last four encoder output layers. The resulting sequence embedding of size n times *embeddings size* represents a concatenation of contextual embeddings for the n tokens in the input sequence. By chopping it into n pieces, we acquire a representation, i.e. a contextual token embedding, for each word used in the corpus. Note that these representations vary according to the context in which the token appears, meaning that the same word has a different representation in each specific context (sequence).

Finally, the resulting embeddings are aggregated on the token level (i.e. for every token in the corpus vocabulary, we create a list of all their contextual embeddings) and are averaged, in order to get one representation for each token in the vocabulary. We enforce a constraint that a list of contextual embeddings for a specific token should contain at least five elements, otherwise the specific token is discarded. This is done in order to remove tokens that do no appear in the corpus enough times for the model to learn a meaningful representation (e.g., mostly tokens that contain typos or very rare technical terms). Since the byte-pair input encoding scheme [20] employed by the SciBERT model does not necessarily generate tokens that correspond to words but rather generate tokens that correspond to parts of words, we also propose the following *on the fly* reconstruction mechanism that allows us to get word representations from byte pair tokens. If a word is split into more than one byte pair token, we take an embedding for each byte pair token constituting a word and build a word embedding by averaging these byte pair tokens. The resulting average is used as a context specific word representation.

The final result are static embeddings for each word in the vocabulary, capable of leveraging a broader semantic knowledge due to the SciBERT being pretrained on a large corpus of scientific articles. As a baseline, we also train a FastText skip-gram model with an embedding dimension of 100 (which is the default) on the lowercased CORD-19 dataset. Once again we enforce the constraint that a word should appear in the corpus at least five times.

5.2 Synonym Resolution

Once embeddings are generated, we conduct synonym resolution with the help of a list of 19,302 gene names and their most common synonyms [30]. The embedddings belonging to the synonyms of the same gene are averaged in order to combine contextual information of different identifiers referring to the same gene and in order to avoid possible mismatches due to different naming.

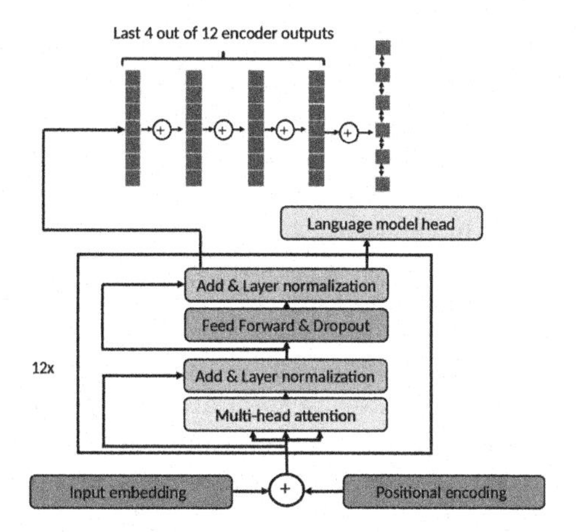

Fig. 1. Extraction of word usage embeddings from BERT. Note that only the last 4 out of 12 BERT encoder layers are used for the embedding generation. This was done in accordance with the previous studies that suggested that the last four layers carry the bulk of the semantic information obtained by the model [24].

5.3 Candidate Acquisition

The main idea of our approach is to leverage semantic similarity in order to derive new scientific knowledge from an already existing one. For this to work, some initial seed concepts need to be acquired and used as a starting point. We explore two possibilities for this:

- **Seed Concepts Recommended by the Expert:** The experts with a medical background were asked to recommend genes and/or proteins with a known and confirmed link to COVID-19. The final consensus was to focus on angiotensin-converting enzyme 2 (ACE2) and transmembrane protease serine 2 (TMPRSS2). ACE2, a receptor for the spike S protein, is important because SARS-CoV-2 uses it to enter the host cell [16]. TMPRSS2 promotes SARS-CoV-2 entry into the cell by priming the spike S protein [16]. Blockage of binding of SARS-CoV-2 to ACE2 or inhibition of TMPRSS2 are therefore two possible approaches to treat COVID-19.
- **Seed Concepts Found in the Literature:** Due to the abundance of recent research on COVID-19 it is also possible to find seed concepts in the related research. We opted for a study by [5] in which a set of COVID-19 therapy targets were identified. The considered list of altogether 2802 potential targets[3] is the result of a large-scale screening for active proteins, and offers a starting set of candidates obtained empirically. The list is ranked according to the increase or decrease of production of a specific protein at a specific

[3] Note that the original list contains 2715 targets (see Supplementary Table 1 in [5]). Some of them are however represented as a set of similar genes/proteins belonging to the same family. On the other hand, we treat each individual gene/protein as a separate target, which results in a set of 2802 targets.

time point. We explore what is the optimal number of seed candidates by exponentially enlarging the size of the seed candidate set. Sampling from the list is conducted according to the ranking of the protein candidates, i.e., we sample 2, 4, 8, 16, 32 and 64 best ranked seed candidates according to the increase in their production 24 h after the infection (column Ratio 24h in Supplementary Table 1 in the study by [5]).

Once seed concepts are acquired, we calculate their embeddings and look for semantically similar concepts by finding the concepts that are the closest to seed concepts according to the cosine distance between the embeddings[4]. More specifically, we find a set of 2802 closest candidate concepts for each gene/protein in each seed candidate set, and the acquired candidates are ranked according to the cosine similarity. Finally, we calculate the average ranking for each candidate (i.e. by averaging ranks for each seed concept in the set) and therefore obtain NumOfCandidatesInSet $*$ 2802 closest candidates for each of the seed concept sets with possible duplicates originating from different seed concepts.

Since the initial experiments showed that many of the most similar concepts are in fact variations of the same base concept (e.g., the closest neighbours to ACE2 being ACE, ACE2M, ACE2S...) and since we are interested in maximizing the variety of the acquired candidates, we conduct an additional filtering according to the normalized Levenshtein distance defined as:

$$\textsc{norm}\text{LD} = 1 - \frac{LD}{\max(\text{len}(w_1), \text{len}(w_2))},$$

where NORMLD stands for normalized Levenshtein distance, LD for Levenshtein distance, w_1 is either a seed concept or a concept already in the list of acquired neighbours and w_2 is the new candidate neighbour. Concepts for which normalized Levenshtein difference is bigger than 0.7 are discarded[5]. The filtering is conducted in order from the top of the list (neighbours with the best average rank) to the bottom.

At the end of the candidate acquisition process, we cut the ranked list of neighbours at 2802 target candidates for each of the distinct seed concept sets used in the evaluation.

5.4 Evaluation

The methods for discovering new therapy targets are evaluated in two evaluation settings, quantitative and qualitative.

[4] Note that these concepts obtained according to semantic similarity are not necessarily proteins/genes but rather any word in the embedding vocabulary.

[5] The normalized Levenshtein difference threshold of 0.7 was chosen empirically.

Quantitative Evaluation. We evaluate if therapy target candidates acquired in the previous step have been confirmed as targets in the study by [5], i.e. how many of them appear in the list of 2802 candidates they identified[6]. Note that in this setting we only evaluate the proposed method on the previously existing knowledge, therefore in the quantitative evaluation we can not asses, if the method has managed to discover some potentially useful and previously undiscovered knowledge.

We are interested in precision at rank k. This means that only the candidates ranked equal to or higher than k are considered and the rest are disregarded. Precision is the ratio of the number of relevant candidates divided by the number of candidates returned by the system, or more formally:

$$\text{precision} = \frac{|\text{relevant candidates@}k|}{|\text{returned candidates}|}$$

Recall@k is the ratio of the number of relevant candidates ranked equal to or higher than k by the system divided by the number of correct ground truth candidates:

$$\text{recall} = \frac{|\text{relevant candidates@}k|}{|\text{correct candidates}|}$$

We measure precision and recall at k = 100 and k = 2802 in order to investigate how different number of retrieved candidates for each seed concept set affects the precision and recall of the methods. More specifically, we are trying to confirm or deny a hypothesis that larger k values degrade the overall precision of the method.

The relevance of the candidate is determined according to two matching criteria. First one is the **exact** match, where the candidate is deemed relevant if it appears in the list of identified targets in the study by [5]. The second is the **fuzzy** match, where we check if the targets belong to the same "family" as a specific confirmed target. This strategy was proposed by the medical experts and checks whether the prefix of the specific gene (characters in the gene name that appear before the first digit in the name) matches a prefix of a specific gene name in the list. We enforce an additional constraint that the matching prefixes need to be at least three characters long for a successful match in order to minimize the false positive rate.

Qualitative Evaluation. We generated two distinct therapy target candidate lists using the proposed SciBERT based embedding method. First one contained 100 closest neighbours to the protein ACE2 according to the cosine distance between embeddings, and the second one contained 100 closest neighbours to the protein TMPRSS2. Both lists were given to the medical expert who inspected the list for possible previously undiscovered candidates.

[6] Note that the study by [5] is not included in the CORD-19 corpus used for training the embeddings, since it was published on May 14th 2020 and we use the CORD-19 version published on May 1st 2020.

6 Results

Here we present the results of the quantitative and qualitative evaluation.

6.1 Results of the Quantitative Evaluation

The results of the quantitative evaluation are presented in Table 2. In column ACE2 + TMPRSS2 we present results when these two proteins are used as seed concepts, and in column UBA2 + NCKAP1 we present results when these two proteins, which were chosen according to the largest value of the Ratio 24h criterion (see Sect. 5.3) are used as seed concepts. Left part of the Table presents results for the proposed approach based on SciBERT and the right part of the Table presents results for the baseline FastText approach in terms of precision and recall at two distinct k values (100 and 2802). EXACT indicates that exact matching is used and FUZZY indicates fuzzy matching (see Sect. 5.4).

SciBERT based method outperforms the FastText baseline by a large margin in both seed therapy target acquisition scenarios and according to all the criteria. Using UBA2 + NCKAP1 works better than using ACE2 + TMPRSS2, achieving the best fuzzy precision@100 of 0.490 and the best exact precision@100 of 0.220. FastText baseline also works fairly well in this scenario, achieving fuzzy precision@100 of 0.380 and the best exact precision@100 of 0.170. When more (2802) candidates are obtained, the recall increases for both methods but at an expense of a significant drop in precision for both methods and for almost all configurations. The only exception is the increase in fuzzy precision by about 2% points when FastText method and ACE2 + TMPRSS2 seed concepts are used. The most likely reason for the drop is that at larger k values some of the target candidates acquired by the method might be semantically too dissimilar to the seed targets, since more candidates per each seed therapy target need to be acquired in order to get the required amount of semantic neighbours (e.g., for k=2802, we get about 1401 semantic neighbours for each of the seed genes).

This raises the question of how many seed terms should be supplied to the system for the best performance when a large number of target candidates is

Table 2. Results (precision@k and recall@k) of the quantitative evaluation for two seeds by the expert and two seeds from the literature. Best result in each row is bolded.

	SciBERT		FastText	
	ACE2 + TMPRSS2	UBA2 + NCKAP1	ACE2 + TMPRSS2	UBA2 + NCKAP1
EXACT P@100	0.110	**0.220**	0.040	0.170
EXACT R@100	0.004	**0.008**	0.001	0.006
EXACT P@2802	0.097	**0.118**	0.025	0.076
EXACT R@2802	0.097	**0.118**	0.025	0.076
FUZZY P@100	0.290	**0.490**	0.070	0.380
FUZZY R@100	0.010	**0.017**	0.002	0.014
FUZZY P@2802	0.222	**0.252**	0.092	0.183
FUZZY R@2802	0.222	**0.252**	0.092	0.183

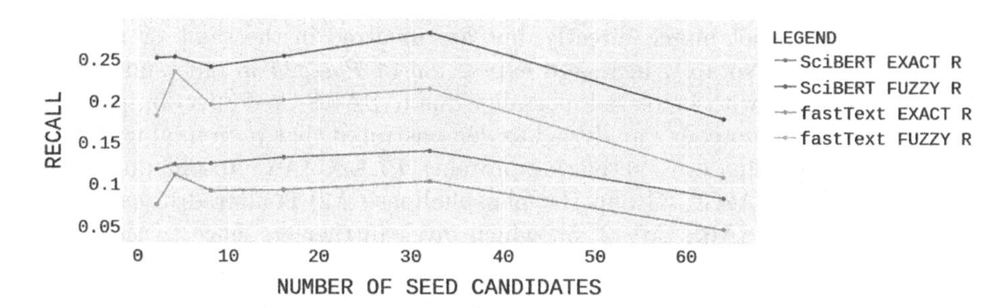

Fig. 2. Relation between recall and the number of seed candidates.

required as output. Figure 2 shows the relation between the achieved recall@2802 (exact and fuzzy) of both methods when we increase the number of seed candidates (see Sect. 5.3 for details about our sampling procedure). For SciBERT based method, the best fuzzy and exact recalls are achieved when 32 seed candidates are used (28.2% and 14.1% respectively). On the other hand, the FastText based method shows a spike in performance when 4 seed concepts are used. This indicates that for some reason the two seed candidates ranked third and fourth (ENO1 and ATP5O, respectively) according to the Ratio 24 criterion have a very positive effect on the FastText model but not SciBERT. While we do not have a clear explanation for this phenomenon, it is hypothesized that it might be connected with morphological similarity between these two genes and other genes in the list of candidates proposed by [5], since FastText can also leverage morphological similarity. Spikes asides, the general trend for both methods and both recalls is quite similar. There is a gradual increase in performance for up to 32 seed candidates and after that the performance decreases.

6.2 Results of the Qualitative Evaluation

Nine genes/proteins were the same in the ACE2 and TMPRSS2 lists, indicating they might be important for pathogenesis of COVID-19. The role of these genes/proteins in pathogenesis of COVID-19 has not been established, but indirect evidence supports this notion at least for some of them. Indeed, most of these genes/proteins have been previously linked to viral diseases, including those caused by SARS-CoV (a virus, which causes SARS, and is related to SARS-CoV-2), and other coronaviruses (Table 3). Furthermore, METAP2 and DPP7, which we identified as potentially relevant for COVID-19, were altered in cells infected with SARS-CoV-2, although the difference for DPP7 did not reach the level of statistical significance [5].

Interestingly, three proteins in Table 3 (PTGS2, CRTH2, and PLA2R1) are linked to infection with coronaviruses as well as metabolism of phospholipids and/or prostaglandin synthesis and action. Furthermore, both the ACE2 and TMPRSS2 lists contain genes/proteins, such as PLA2 (phospholipase A2, PLA2G2D (Group IID secretory phospholipase A2), and SPLA2 (secretory

PLA2), which do not match directly, but are involved in the same or related cellular processes. Notably, increased expression of *Pla2g2d* in older mice was shown to be linked with increased mortality due to SARS-CoV infection [36]. In addition, a recent proteomic analysis has demonstrated that protein abundance of PLAA (phospholipase A2-activating protein), PLA2G4A (cytosolic phospholipase A2), and PLA2G2 (Group IIA phospholipase A2) is altered in cultured cells infected with SARS-CoV-2 [5], which gives further credence to the idea that phospholipid metabolism is important under these conditions. In summary, taken together with published experimental data, our analysis suggests that phospholipases and/or prostaglandins might represent a target for treatment of COVID-19.

Table 3. Genes/proteins (in alphabetical order) which are common to the TMPRSS2 and ACE2 list and their (putative) relevance to COVID-19.

Gene	Protein	Relevance to COVID-19
ATP2B2 (PMCA2)	Plasma membrane Ca^{2+}-transporting ATPase	?
CRTH2 (PTGDR2)	Prostaglandin D2 (PGD2) receptor	PGD2 is important for survival of mice infected with neurotropic coronavirus. Increased production of PGD2 is linked to increased mortality in aged mice. PGD2blockade improves survival in mice infected with SARS-CoV [35,37].
DPP7 (DPP2)	Dipeptidyl peptidase 2	DPP7 is associated with the magnitude of the antibody response to influenza vaccination [15].
MECP2	Methyl-CpG-binding protein 2	MECP2 duplication in humans is associated with IgA/IgG2 antibody deficiency and severe infections. Mice overexpressing MECP2 are hypersensitive to influenza A virus [1,10].
METAP2 (P67EIF2)	Methionine aminopeptidase 2 (Initiation factor 2-associated 67 kDa glycoprotein)	Plays a role in regulation of protein synthesis during vaccinia virus infection [6].
PLA2R1	Secretory phospholipase A2 (PLA2) receptor	Restricted activity of PLA2 is associated with improved survival in mice infected with HCoV-OC43. Inhibition of cytosolic PLA2 suppresses replication of HCoV-229E [13,27].
PTGS2 (COX2)	Prostaglandin G/H synthase 2 (cyclooxygenase-2)	SARS-CoV induces cyclooxygenase 2 [23].
SOX2	Transcription factor SOX-2	SOX2+ cells are important for regeneration of airway epithelium after severe influenza infection in mice [32].
SSTR2 (SST2)	Somatostatin receptor type 2	?

7 Conclusions and Further Work

In this paper we presented a method for discovering new COVID-19 therapy targets by leveraging contextual embeddings, which outperforms the method based on FastText embeddings. We explored the best tactics for acquiring seed targets from the related work if expert knowledge is not available. The results of the manual qualitative evaluation by the expert indicate that at least two groups of novel therapy target candidates have been discovered.

The proposed method outperforms the baseline FastText method by a large margin, which can be explained by the fact that SciBERT is also leveraging knowledge gained during the pretraining on the large corpus of scientific literature, which enables the model to generate vector representations that reflect this wider semantic context. The drawback is however the difference in the amount of computational resources required by the two methods. We also acknowledge that the proposed method, which constructs static embeddings from the SciBERT contextual embeddings is not the only possibility for construction of meaningful semantic representations. Other possibilities and models (e.g., BioBERT [21]) will be explored in the future work. The quantitative evaluation indicates that the precision and recall of the method are still relatively low in most cases. This can on one side indicate that COVID-19 topic is not researched enough to confirm relations between COVID-19 and some candidates found by the proposed method. Another indication of this is the qualitative study, which confirmed that some of the proposed candidates found by the system have research potential but have not yet been explicitly confirmed as being related to COVID-19 in the existing literature.

On the other hand, low precision most likely also indicates that there is still a large amount of proposed candidates, which play no role in the advancement and prevention of the COVID-19 disease. Some of these false positives can be attributed to inadequate synonym resolution since the list used for that task (see Sect. 5.2) most likely covers only a small percentage of genes and their synonyms found in the CORD-19 corpus. Other mistakes can be contributed to the byte pair encoding scheme SciBERT employs. Since the model generates embeddings for subword tokens instead for an entire words (see how we deal with this problem in Sect. 5.1), some words with similar roots or affixes can perhaps appear closer in the semantic space as they should according to their semantic relatedness because of the morphological resemblance. We will address this issues in the future work.

Acknowledgements. This work was funded by the Slovenian Research Agency (ARRS) through core research programme *Knowledge Technologies* (P2-0103), research project *Semantic Data Mining for Linked Open Data* (financed under the ERC Complementary Scheme, N2-0078), and a young researcher grant (BŠ). The work was also supported by EU Horizon 2020 research and innovation programme under grant agreement No 825153, project EMBEDDIA (Cross-Lingual Embeddings for Less-Represented Languages in European News Media). The publication reflects only the authors' views and the EC is not responsible for any use that may be made of the information it contains.

References

1. Bauer, M., et al.: Infectious and immunologic phenotype of MECP2 duplication syndrome. J. Clin. Immun. **35**(2), 168–181 (2015)
2. Beigel, J.H., et al.: Remdesivir for the treatment of Covid-19—preliminary report. New Engl. J. Med. (2020)
3. Beltagy, I., Cohan, A., Lo, K.: Scibert: pretrained contextualized embeddings for scientific text. arXiv preprint arXiv:1903.10676 (2019)
4. Bojanowski, P., Grave, E., Joulin, A., Mikolov, T.: Enriching word vectors with subword information. Trans. Assoc. Comput. Linguist. **5**, 135–146 (2017)
5. Bojkova, D., et al.: Proteomics of SARS-CoV-2-infected host cells reveals therapy targets. Nature **583**, 1–8 (2020). https://doi.org/10.1038/s41586-020-2332-7
6. Bose, A., Saha, D., Gupta, N.K.: Viral infection: I. Regulation of protein synthesis during vaccinia viral infection of animal cells. Arch. Biochem. Biophys. **342**(2), 362–372 (1997)
7. Boulware, D.R., et al.: A randomized trial of hydroxychloroquine as postexposure prophylaxis for covid-19. New Engl. J. Med. (2020)
8. Bruza, P., Weeber, M.: Literature-Based Discovery. Information Science and Knowledge Management. Springer Science & Business Media, Heidelberg (2008). https://doi.org/10.1007/978-3-540-68690-3
9. Crichton, G., Baker, S., Guo, Y., Korhonen, A.: Neural networks for open and closed literature-based discovery. PLOS ONE **15**(5), 1–16 (2020)
10. Cronk, J.C., et al.: Influenza a induces dysfunctional immunity and death in MECP2-overexpressing mice. JCI Insight **2**(2) (2017)
11. Deftereos, S.N., Andronis, C., Friedla, E.J., Persidis, A., Persidis, A.: Drug repurposing and adverse event prediction using high-throughput literature analysis. Wiley Interdisc. Rev. Syst. Biol. Med. **3**(3), 323–334 (2011)
12. Devlin, J., Chang, M.W., Lee, K., Toutanova, K.: Bert: pre-training of deep bidirectional transformers for language understanding. arXiv preprint: 1810.04805 (2018)
13. Do Carmo, S., Jacomy, H., Talbot, P.J., Rassart, E.: Neuroprotective effect of apolipoprotein d against human coronavirus OC43-induced encephalitis in mice. J. Neurosci. **28**(41), 10330–10338 (2008)
14. Grujicic, D., Radevski, G., Tuytelaars, T., Blaschko, M.B.: Self-supervised context-aware Covid-19 document exploration through atlas grounding (2020)
15. HIPC-I Consortium, et al.: Multicohort analysis reveals baseline transcriptional predictors of influenza vaccination responses. Sci. Immunol. **2**(14), eaal4656 (2017)
16. Hoffmann, M., et al.: SARS-COV-2 cell entry depends on ACE2 and TMPRSS2 and is blocked by a clinically proven protease inhibitor. Cell **181**, 271–280 (2020)
17. Hope, T., et al.: SciSight: combining faceted navigation and research group detection for COVID-19 exploratory scientific search. arXiv preprint: 2005.12668 (2020)
18. Korhonen, A., et al.: Improving literature-based discovery with advanced text mining. In: DI Serio, C., Lió, P., Nonis, A., Tagliaferri, R. (eds.) CIBB 2014. LNCS, vol. 8623, pp. 89–98. Springer, Cham (2015). https://doi.org/10.1007/978-3-319-24462-4_8
19. Kostoff, R.N., Briggs, M.B., Lyons, T.J.: Literature-related discovery (LRD): potential treatments for multiple sclerosis. Technol. Forecast. Soc. Change **75**(2), 239–255 (2008)
20. Kudo, T., Richardson, J.: Sentencepiece: a simple and language independent subword tokenizer and detokenizer for neural text processing. arXiv preprint:1808.06226 (2018)

21. Lee, J., et al.: BioBERT: a pre-trained biomedical language representation model for biomedical text mining. Bioinformatics **36**(4), 1234–1240 (2020)
22. Liu, J., et al.: Hydroxychloroquine, a less toxic derivative of chloroquine, is effective in inhibiting SARS-CoV-2 infection in vitro. Cell Discov. **6**(1), 1–4 (2020)
23. Liu, M., Gu, C., Wu, J., Zhu, Y.: Amino acids 1 to 422 of the spike protein of SARS associated coronavirus are required for induction of cyclooxygenase-2. Virus Genes **33**(3), 309–317 (2006)
24. Martinc, M., Novak, P.K., Pollak, S.: Leveraging contextual embeddings for detecting diachronic semantic shift. arXiv preprint arXiv:1912.01072 (2019)
25. Mehra, M.R., Desai, S.S., Kuy, S., Henry, T.D., Patel, A.N.: Retraction: cardiovascular disease, drug therapy, and mortality in Covid-19. New Engl. J. Med. (2020)
26. Mikolov, T., Chen, K., Corrado, G., Dean, J.: Efficient estimation of word representations in vector space. arXiv preprint arXiv:1301.3781 (2013)
27. Müller, C., Hardt, M., Schwudke, D., Neuman, B.W., Pleschka, S., Ziebuhr, J.: Inhibition of cytosolic phospholipase a2α impairs an early step of coronavirus replication in cell culture. J. Virol. **92**(4), JVI.01463-17 (2017)
28. Pennington, J., Socher, R., Manning, C.D.: Glove: global vectors for word representation. In: Proceedings of the 2014 Conference on Empirical Methods in Natural Language Processing (EMNLP), pp. 1532–1543 (2014)
29. Peters, M.E., Neumann, M., Iyyer, M., Gardner, M., Clark, C., Lee, K., Zettlemoyer, L.: Deep contextualized word representations. arXiv preprint:1802.05365 (2018)
30. Povey, S., Lovering, R., Bruford, E., Wright, M., Lush, M., Wain, H.: The HUGO gene nomenclature committee (HGNC). Hum. Genet. **109**(6), 678–680 (2001)
31. Pyysalo, S., et al.: LION LBD: a literature-based discovery system for cancer biology. Bioinformatics **35**(9), 1553–1561 (2018)
32. Ray, S., et al.: Rare SOX2+ airway progenitor cells generate KRT5+ cells that repopulate damaged alveolar parenchyma following influenza virus infection. Stem Cell Rep. **7**(5), 817–825 (2016)
33. Swanson, D.R.: Medical literature as a potential source of new knowledge. Bull. Med. Libr. Assoc. **78**(1), 29 (1990)
34. Tshitoyan, V., et al.: Unsupervised word embeddings capture latent knowledge from materials science literature. Nature **571**(7763), 95–98 (2019)
35. Vijay, R., et al.: Virus-induced inflammasome activation is suppressed by prostaglandin d2/dp1 signaling. Proc. Natl. Acad. Sci. **114**(27), E5444–E5453 (2017)
36. Vijay, R., et al.: Critical role of phospholipase A2 group IID in age-related susceptibility to severe acute respiratory syndrome-CoV infection. J. Exp. Med. **212**(11), 1851–1868 (2015)
37. Zhao, J., Zhao, J., Legge, K., Perlman, S.: Age-related increases in PGD 2 expression impair respiratory DC migration, resulting in diminished T cell responses upon respiratory virus infection in mice. J. Clin. Invest. **121**(12), 4921–4930 (2011)
38. Zhou, P., et al.: A pneumonia outbreak associated with a new coronavirus of probable bat origin. Nature **579**(7798), 270–273 (2020)
39. Zhu, N., et al.: A novel coronavirus from patients with pneumonia in China, 2019. New Engl. J. Med. (2020)

Semantic Annotation of Predictive Modelling Experiments

Ilin Tolovski[1], Sašo Džeroski[2,3], and Panče Panov[2,3(✉)]

[1] Hasso Plattner Institute, Potsdam, Germany
ilin.tolovski@hpi.de
[2] Department of Knowledge Technologies, Jožef Stefan Institute, Ljubljana, Slovenia
{saso.dzeroski,pance.panov}@ijs.si
[3] Jožef Stefan International Postgraduate School, Ljubljana, Slovenia

Abstract. In this paper, we address the task of representation, semantic annotation, storage, and querying of predictive modelling experiments. We introduce OntoExp, an OntoDM module which gives a more granular representation of a predictive modeling experiment and enables annotation of the experiment's provenance, algorithm implementations, parameter settings and output metrics. This module is incorporated in SemanticHub, an online system that allows execution, annotation, storage and querying of predictive modeling experiments. The system offers two different user scenarios. The users can either define their own experiment and execute it, or they can browse the repository of completed experimental workflows across different predictive modelling tasks. Here, we showcase the capabilities of the system with executing multi-target regression experiment on a water quality prediction dataset using the Clus software. The system and created repositories are evaluated based on the FAIR data stewardship guidelines. The evaluation shows that OntoExp and SemanticHub provide the infrastructure needed for semantic annotation, execution, storage, and querying of the experiments.

Keywords: Computational experiments · Semantic annotation · Ontology · Predictive modelling

1 Introduction

Data mining and machine learning experiments are conducted in higher volume than ever before, in various settings and domains. In the case of predictive modelling, the users usually aim to produce a model that will provide the best predictive performance. However, in practice, almost none of the settings regarding the experimental setup are stored. We usually do not keep track of the exact software environment, the exact dataset that was used to train the model, the duration of the experiment, and the hardware specification of the machine the experiments were performed on.

I. Tolovski—Work done while the author was at the Jožef Stefan Institute, Ljubljana, Slovenia.

A. Appice et al. (Eds.): DS 2020, LNAI 12323, pp. 124–139, 2020.
https://doi.org/10.1007/978-3-030-61527-7_9

The same problem arises when it comes to the models produced by the performed experiments. Regarding the algorithm setup, almost no information is stored about the parameter values of the algorithm implementation which produced the models, and the evaluation scenario used to validate the results. These predicaments make the conducted research hard to verify, reproduce and reuse. There have been previous efforts to address this problem such as the ones developed by Vanschoren et al. [18], Google (AI hub)[1], Schelter et al. [14], and others.

Having access to a repository of computational experiments that are represented by a schemata based on logical formalism is beneficial from several perspectives. First, the results can be accessed, easily verified, and predictive models can be retrieved for their further reuse. We can utilize the logic behind the schema to pose queries that will allow searching not only through the explicit axioms that are asserted but also on the implicit axioms that the reasoners have produced. From this, we can derive new information based on results already stored in the knowledge base.

However, producing the experiments and then transforming their outputs into this logical formalism can be a tedious and error-prone task when repeated for each experiment. Therefore, one can assume that having a framework that will execute the experiments and format the output according to the defined logical formalism, i.e., ontology-based annotation schema, store the annotations in a database, which will be open for querying through a query endpoint, will provide an easy access to a vast knowledge base of experimental workflows, benefiting both data mining practitioners and domain experts. In the literature, there had been efforts for development of ontological resources that allow semantic representation of different entities in the domain of data mining and machine learning. Examples of state-of-the-art resources OntoDM [12], DMOP [9], Exposé [17], MEX [7], MLSchema [6] and others.

The paper is organized as follows. In Sect. 2, we introduce an ontology module for semantic representation of predictive modelling experiments named OntoExp. Next, in Sect. 4, we demonstrate the use of OntoExp within SemanticHub, a system for execution, semantic annotation, storage and querying of experiments. In Sect. 5, we showcase the use of SemanticHub in a water quality prediction use case scenario. Finally, in Sect. 6, we evaluate the system and the created repository according to the DANS FAIR questionnaire, and in Sect. 7 we give our concluding remarks.

2 Representation of Experiments with OntoExp

To create a repository of semantically annotated predictive modeling experiments, we need to create a more granular representation of a predictive modeling experiment that will enable annotation of the experiment's provenance, algorithm implementations, and results.

[1] https://cloud.google.com/ai-hub/.

Table 1. Examples of competency questions addressed by OntoExp.

#	Competency question
1	List all experiments that address the multi-target regression task
2	List all experiments that used an ensemble learning algorithm
3	List all experiments that had a specific dataset, given as input
4	Return the best predictive model learned on a specific given dataset
5	List all experiments that used cross-validation

For this purpose, we introduce OntoExp, an extension of OntoDM-core [12] for representation of predictive modeling experiments. OntoExp provides a representation of different types of predictive modeling experiments on the execution level. Each experiment type, as well as all of the involved entities and processes, need to be formally represented and connected to provide an annotation schema that will used to produce a comprehensive metadata for the experiment.

The main focus of OntoExp is on representing different types of experimental data mining workflow executions, including the executions of different algorithm implementations together with their parameter setup for various data mining tasks. A connection is made with the inputs and outputs of the execution process, i.e., the datasets, predictive models, and experimental results as concretizations of the evaluation measure implementations. In Table 1, we outline the competency questions that are addressed by our developed extension.

Ontology Design. OntoDM [12] was developed in a modular fashion, making it suitable for extension. It adheres to the Open Biomedical Ontologies (OBO) Foundry principles [16] for ontology design. These include the use of an upper-level ontology, formal ontology relations, absence of orphan classes, single inheritance, as well as integration and reusing of terms that are already defined in other ontologies. It is based on the Basic Formal Ontology (BFO) [1] as the upper-level ontology, and the relations are reused from the Relations Ontology (RO) [15]. Furthermore, OntoDM reuses classes defined in other ontologies which are relevant to the domain, such as the Information Artefact Ontology (IAO) [3], OntoDT ontology of datatypes [13], Software Ontology (SWO) [11], Ontology of Biomedical Investigations (OBI) [2], and others. All of the reused classes from these ontologies are imported following the Minimum Information to Reference an External Ontology Term (MIREOT) [5] principles.

OntoExp builds on top of the current ontology structure following the same class taxonomy, as well as design and class reuse principles. Supporting the modularity of OntoDM, it is designed as a separate module that can be used by itself or preferably together with OntoDM for a more comprehensive representation of the domain. The ontology module consists of 296 classes in total, 146 of which are novel classes, and 150 are reused from OntoDM. Since the resource is not introduced as a novel standalone ontology, it is licensed under the same license as OntoDM. The ontology module is available at the following PURL https://w3id.org/ontoexp.

Core Classes. OntoExp follows the Algorithm-Implementation-Execution Design Pattern and principles defined by Lawrynowicz et al. [10]. The most important classes are *algorithm*, *algorithm implementation*, and *algorithm execution*.

Data mining algorithms are represented as a subclass of the general *algorithm* class from the IAO ontology [3] and represents a specification of an algorithm. We distinguish between DM algorithms that output a single model, and the ones that output an ensemble of models.

Algorithm implementation is a concretization of a algorithm specification, implemented in some software product, and written in a specific programming language. In the ontology, we also explicitly represent the provenance information for both the software and the programming language.

An *algorithm execution* is a process that represents the training part of a predictive modeling experiment. It `realizes` the *algorithm implementation*, receives a *DM-dataset* with a *train set role* as an input, and outputs a *predictive model*. This process precedes the *predictive model execution* that represents the process which `realizes` the *predictive model* in order to output the *DM-dataset* with the predicted values of the target variables.

Following the *predictive model execution* (see Fig. 1c), there is the *evaluation calculation* process which uses the predicted *DM-dataset* from the *predictive model execution* process as an input to calculate a specific implementation of an *evaluation measure*. Depending on the type of experiment, or task, the calculation process can vary in different ways. If the experiment has a N-fold cross-validation model evaluation, we need to represent each per-fold *evaluation measure calculation*, and calculate the average value across all measurements. Additionally, if we have complex tasks, such as for example multi-target prediction task, we need to calculate the evaluation measures for each target separately. More details and examples are provides further on in Sect. 3.

Workflow Representation. In OntoExp, we represent the *predictive model evaluation train/test workflow execution* and the *N-fold cross-validation workflow execution* processes and their inputs and outputs. The first workflow can use either a separate test set for evaluation or validation of results on the training set, while the second uses N-fold cross-validation as the evaluation method.

N-fold cross-validation workflow execution contains the *predictive model train/test evaluation workflow execution* as one of the sub-process. In a cross-validation scenario (see Fig. 1a), we first perform the sampling of the dataset on N folds, and in each iteration (see Fig. 1b), we build a predictive model on N−1 folds and evaluate it on the one fold that was not used for training. Finally, at the end we calculate the average average value of the evaluation measure from all the folds. These repetitive evaluations are represented by the *per fold evaluation workflow execution* process (see Fig. 1b) which consists of two subprocesses, i.e., *train/test dataset construction*, and *predictive model evaluation workflow execution*. Each per fold evaluation process is a sub-process of the *N-fold cross-validation workflow execution* process connected to it with the `has part` relation.

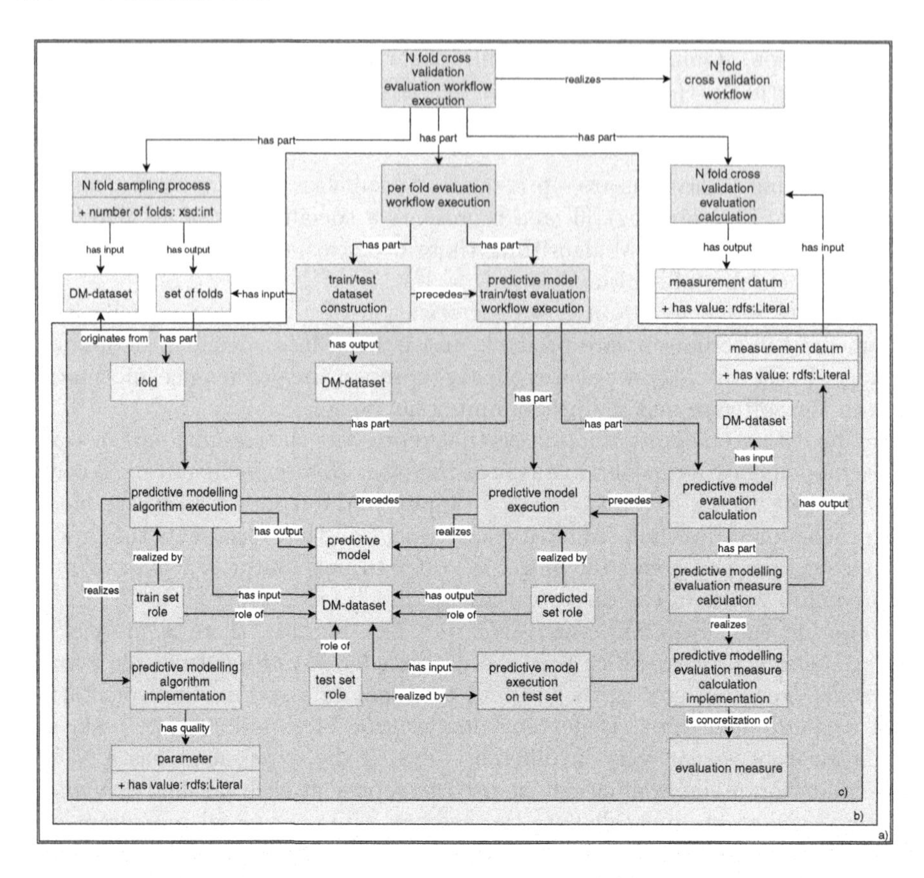

Fig. 1. A representation of a predictive modeling experiment: a) N-fold cross-validation scenario. b) Representation of the evaluation for each fold. c) Representation of the train/test evaluation workflow execution. Red boxes represent processes, blue boxes represent information entities, green boxes represent roles and pink boxes represent realizable entities. (Color figure online)

3 Semantic Annotation of Experiments Using OntoExp

In this section, we describe the complete annotation schema derived from OntoDM and OntoExp on an example of an experiment that involves a cross-validation evaluation for a multi-target regression task using an algorithm that solves that task. In order to represent a cross-validation experimental scenario, we use the *N-fold cross-validation evaluation workflow execution* class (see Fig. 2). This evaluation scenario consists of three consecutive processes: data sampling process, model construction process, and model evaluation process.

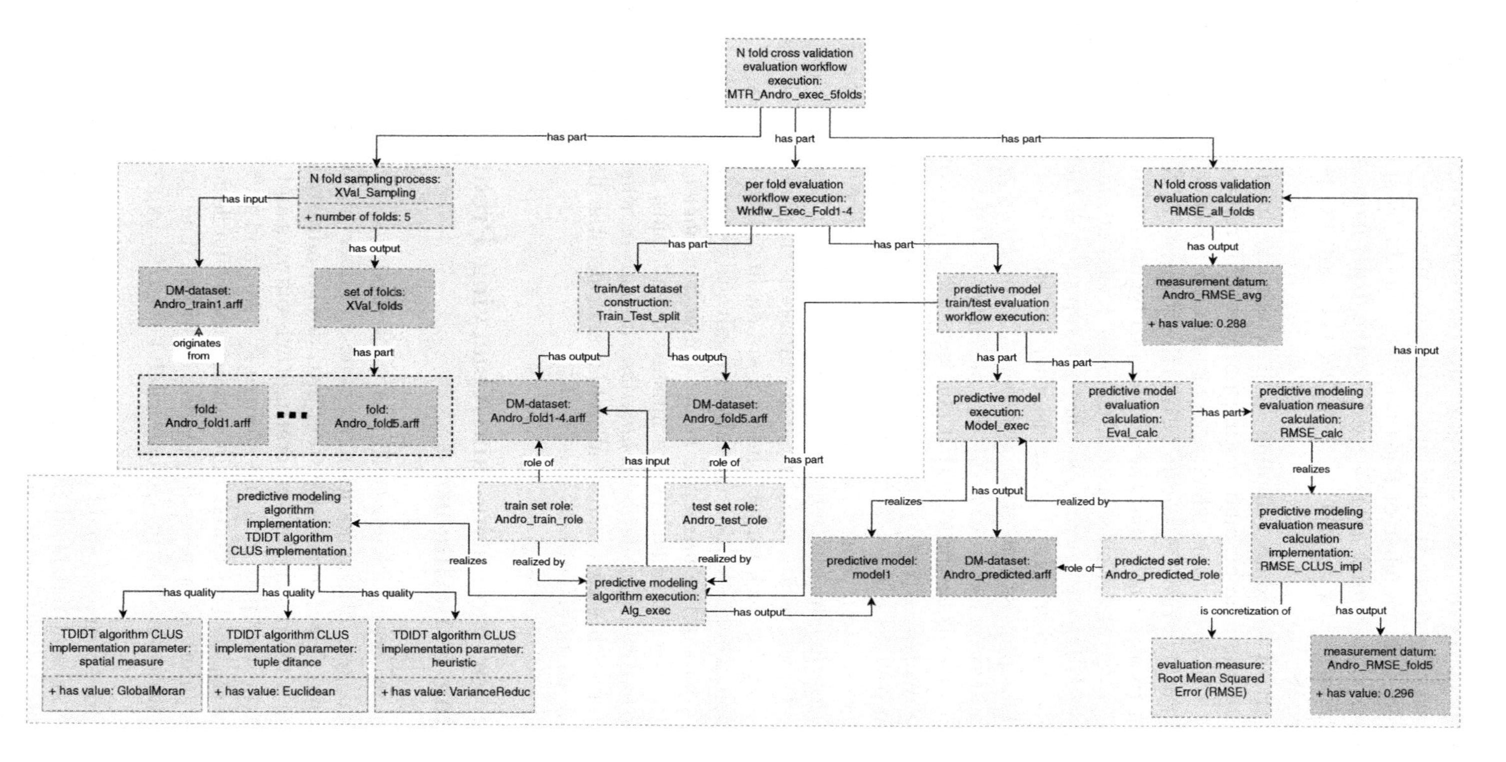

Fig. 2. An example of semantically annotated multi-target regression experiment.

First, we focus on the data sampling process represented with the *N-fold sampling process*. We relate this process with a datatype property that carries information about the number of folds the data should be split in. The input of the process is the original *dataset* used for the experiment, while the output is a *set of folds* which is related to each *fold* that will be used for the model evaluation process.

Another part of the data sampling process are the different combinations of folds used for the training and testing purpose. We represent this with the *train/test dataset construction*, which is a part of the *per-fold evaluation workflow execution* and outputs two *DM-datasets*, one that consists of N−1 folds and will be used with a *train set role*, and the one fold to be used with a *test set role*.

Next, in a cross-validation scenario, there is a separate model creation and evaluation process for each fold. This is then repeated N times, N being the number of folds. To represent this, we use the *per fold evaluation workflow execution* class. This process consists of two parts, *train/test dataset construction*, which we introduced before, and *predictive model train/test evaluation workflow execution*. The latter is the process that connects the model creation and model evaluation process for a given training and test set. The resulting output of the model creation process is a predictive model.

Next is the evaluation process, which starts with a *predictive model execution* process that uses the already built predictive model to produce a *dataset* with the predictions for the target variables. This dataset is then used to calculate the evaluation metrics for each target, since we are dealing with the task of multi-target prediction. The evaluation measures are always dependent on the task at hand. This is a part of the *predictive model evaluation calculation* process.

Finally, once these calculations are finished for all folds, we use them as an input of the *N-fold cross-validation evaluation calculation*, which then calculates the averages for each target across all folds, and also the final average value across the final per-target values.

4 System for Executing and Querying Predictive Modeling Experiments

In this section, we present SemanticHub, a web-based system for remote experiment execution, semantic annotation, storage, and querying of predictive modeling experiments. The presented system provides an infrastructure for running experiments on a remote server, annotating their outputs and experimental settings, storing the raw files in a file storage system, and the annotations in a triple store database. The stored annotations are available for querying either through a user interface or using a querying endpoint. The prototype version of the system is available at http://semantichub.ijs.si/clus/experiment.

System Architecture. SemanticHub is constructed in modular fashion as a synthesis of several independent web services (see Fig. 3). First, the input datasets are sent to our file storage through the a FTP server. The experiment is

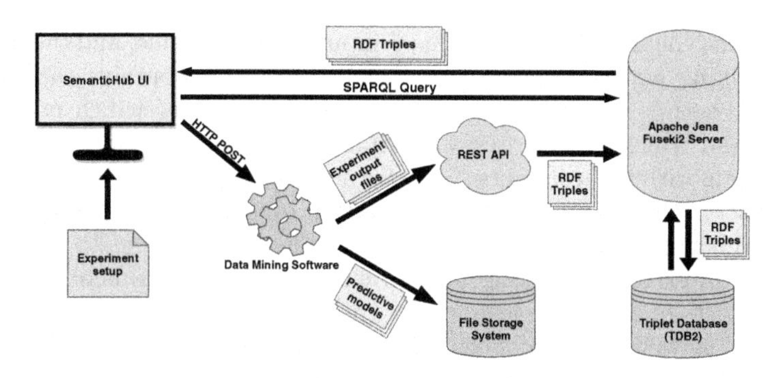

Fig. 3. The architecture of SemanticHub.

defined through interaction with SemanticHub's UI, which sends the parameters and the setup to our server where our data mining software is hosted. Currently, we are using the Clus software[2] for executing the experiments. Clus is a decision tree and rule induction system that implements the predictive clustering framework and has been applied to many different tasks including multi-task learning, structured output learning, multi-label classification, hierarchical classification, and time series prediction.

The whole setup as well as the experimental outputs are annotated with entities and processes defined in OntoExp using a REST API. The annotated experimental setup, metrics, and results are sent to the Fuseki2 server[3] as sets of RDF triples[4]. The resulting predictive models are stored as raw files the file system. The Fuseki2 server hosts the triple-store database which is used for storage and retrieval of the RDF triples. These triples are available to the users through SemanticHub's querying engine, which generates SPARQL queries based on the user's input. The results are shown to the users in SemanticHub's UI.

Running Computational Experiments. Here, we describe the implementation of the framework that allows users to set up and execute computational experiments on our remote servers. One of the two user scenarios for this system allows users to run their own experiments. This is done in two stages. First, the user needs to define the experimental setup, by uploading an experimental specification or setting up the experiment through the user interface. This step includes selecting the datasets, the algorithm for training the models, as well as its parameter values. The dataset is uploaded through a HTTP request to a repository that is open to the users through FTP requests.

For running predictive modeling experiments we use the Clus software for data mining. As examples here, we focus on the tasks of single-target regression and classification, as well as multi-target regression and multi-label classification.

---

[2] http://clus.sourceforge.net/doku.php.
[3] https://jena.apache.org/documentation/fuseki2/.
[4] https://www.w3.org/TR/rdf11-primer/.

For these tasks, the software has two main inputs: a settings file, and the datasets, both for training and testing purposes. The settings file contains the complete experimental setup, i.e., it defines the input data, specifies which part of it is descriptive and which are the target attributes, as well as all of the model constraints and algorithm parameters (for more details please check [5]). Depending on the criteria set in the settings file, Clus can output predictive models, experimental results, as well as predictions for each test example.

The user interface for setting up and executing an experiment in the Clus software consists of a single screen, where the user needs to upload the train and test datasets, as well as the settings file. Additionally, there are two checkboxes, one for the selection of the validation scenario, and one for defining whether a single model or an ensemble algorithm will be run on top of the selected data. Each of these settings/flags runs Clus in a different mode, changing the number of output files, as well as the type of the output files.

Semantic Annotation Workflow. Following the experiment execution, the system utilizes the designed annotation schemata for predictive modeling experiments to create semantic annotations in the form of RDF graphs. The RDF graphs consist of triplets representing the inputs and outputs of the experiment, the algorithm used, its parameter values, as well as the evaluation results (see Fig. 4). Formalized in this way, the RDF graphs are then uploaded to a TDB2 triplet database hosted on a Fuseki2 server. The upload is executed by the SPARQL Graph Store HTTP Protocol.

The CLUS library we use for running predictive modeling experiments provides a comprehensive output, once the experiment is completely finished. Thus, we semantically annotate the experiments after the execution of the experiment. The complete settings file with all default, and user-defined values are contained in the output file for each experiment, enabling us to annotate the experimental setup, runtime provenance information, and results in one step. We should also note that the annotations are solely based on the annotation schemata designed for predictive modeling experiments that use the OntoExp ontology.

Querying the Repository of Predictive Modeling Experiments. The second user scenario of our system is the one where users can query or browse through the database of completed experiments. We use the Graph Store HTTP Protocol for storing and querying the semantic annotations, which are in the form of RDF graphs. The SPARQL endpoint provides the presence on the HTTP network for receiving and handling Graph Store HTTP or SPARQL Protocol requests. The SPARQL querying interface enables users to write raw SPARQL queries directly for each RDF dataset. However, our system provides a simple graphical user interface, where users can define their queries by interacting with the user interface.

For the predictive modeling experiments conducted in the Clus framework, users can query the repository of experiments based on several criteria. These

[5] http://clus.sourceforge.net/doku.php?id=doc:main.

```
Individual: <http://www.ontodm.com/OntoDM-Workflow#Mean_absolute_error(MAE)_Default_test_1>

    Annotations:
        rdfs:label "Mean absolute error (MAE)_Default_test_evaluation"

    Types:
        j.1:evaluation_measure_calculation

    Facts:
     j.0:has_output  <http://www.ontodm.com/OntoDM-Workflow#Mean_absolute_error(MAE)_Default_test_target_1>,
     j.0:has_output  <http://www.ontodm.com/OntoDM-Workflow#Mean_absolute_error(MAE)_Default_test_target_2>,
     j.0:has_output  <http://www.ontodm.com/OntoDM-Workflow#Mean_absolute_error(MAE)_Default_test_target_3>,
     j.0:realizes  j.1:mean_absolute_error_clus_implementation
```

RDF triplets representing an evaluation measure calculation process outputting measurements for 3 target variables.

```
Individual: <http://www.ontodm.com/OntoDM-Workflow#Mean_absolute_error(MAE)_Default_test_target_1>

    Annotations:
        rdfs:label "Mean absolute error (MAE)_Default_test_target_1"

    Types:
        j.0:measurement_datum

    Facts:
     j.1:has_value "2.2689"
```

Measurement datum and its value for the first target.

Fig. 4. Examples of annotations of experimental results.

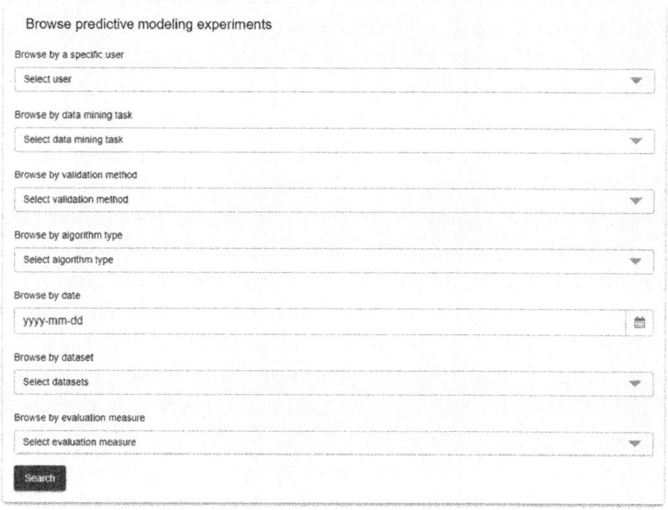

Browse predictive modeling experiments

Browse by a specific user

Select user

Browse by data mining task

Select data mining task

Browse by validation method

Select validation method

Browse by algorithm type

Select algorithm type

Browse by date

yyyy-mm-dd

Browse by dataset

Select datasets

Browse by evaluation measure

Select evaluation measure

Search

Fig. 5. An example of querying interface for CLUS experiments.

include the user, the data mining task that was addressed, the validation method, algorithm type, datasets included, evaluation measure, as well as the date or range of dates when the experiment was conducted. The querying screen from the user interface is shown in Fig. 5. All of the fields allow multiple selections, therefore, the query result can be a set of experiments, not just a single instance.

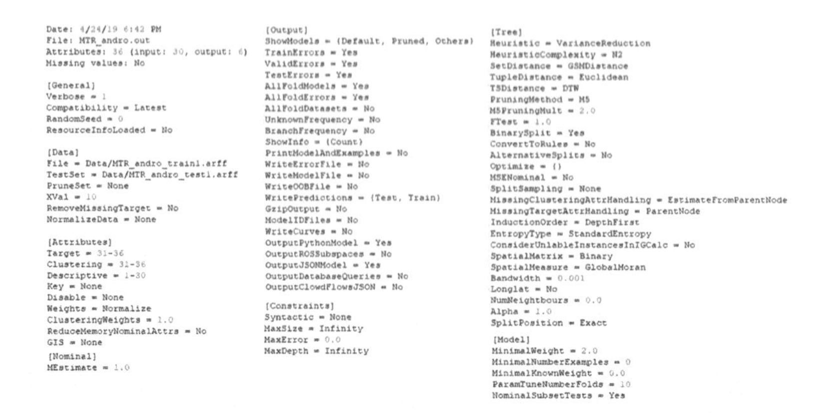

```
Date: 4/24/19 6:42 PM
File: MTR_andro.out
Attributes: 36 (input: 30, output: 6)
Missing values: No

[General]
Verbose = 1
Compatibility = Latest
RandomSeed = 0
ResourceInfoLoaded = No

[Data]
File = Data/MTR_andro_train1.arff
TestSet = Data/MTR_andro_test1.arff
PruneSet = None
XVal = 10
RemoveMissingTarget = No
NormalizeData = None

[Attributes]
Target = 31-36
Clustering = 31-36
Descriptive = 1-30
Key = None
Disable = None
Weights = Normalize
ClusteringWeights = 1.0
ReduceMemoryNominalAttrs = No
GIS = None
[Nominal]
MEstimate = 1.0

[Output]
ShowModels = {Default, Pruned, Others}
TrainErrors = Yes
ValidErrors = Yes
TestErrors = Yes
AllFoldModels = Yes
AllFoldErrors = Yes
AllFoldDatasets = No
UnknownFrequency = No
BranchFrequency = No
ShowInfo = {Count}
PrintModelAndExamples = No
WriteErrorFile = No
WriteModelFile = No
WriteOOBFile = No
WritePredictions = {Test, Train}
GzipOutput = No
ModelIDFiles = No
WriteCurves = No
OutputPythonModel = Yes
OutputROSSubspaces = No
OutputJSONModel = Yes
OutputDatabaseQueries = No
OutputClowdFlowsJSON = No

[Constraints]
Syntactic = None
MaxSize = Infinity
MaxError = 0.0
MaxDepth = Infinity

[Tree]
Heuristic = VarianceReduction
HeuristicComplexity = N2
SetDistance = GSMDistance
TupleDistance = Euclidean
TSDistance = DTW
PruningMethod = M5
M5PruningMult = 2.0
FTest = 1.0
BinarySplit = Yes
ConvertToRules = No
AlternativeSplits = No
Optimize = {}
MSENominal = No
SplitSampling = None
MissingClusteringAttrHandling = EstimateFromParentNode
MissingTargetAttrHandling = ParentNode
InductionOrder = DepthFirst
EntropyType = StandardEntropy
ConsiderUnlableInstancesInIGCalc = No
SpatialMatrix = Binary
SpatialMeasure = GlobalMoran
Bandwidth = 0.001
Longlat = No
NumNeightbours = 0.0
Alpha = 1.0
SplitPosition = Exact

[Model]
MinimalWeight = 2.0
MinimalNumberExamples = 0
MinimalKnownWeight = 0.0
ParamTuneNumberFolds = 10
NominalSubsetTests = Yes
```

Fig. 6. A specification of the experiment (a screenshot of a Clus settings file).

5 Use Case: Water Quality Prediction

In this section, we present a use case scenario of SemanticHub's predictive modeling system integration with the Clus data mining framework. Specifically, we will showcase the data import, construction of the experimental setup through the settings file, as well as the remote execution of an experiment. Finally, we will use SemanticHub's SPARQL endpoint access to formulate a SPARQL query.

We define the experiment specification through the settings file, as shown in Fig. 6. Here, we provide information about the input datasets for this experiment, together with the parameters and constraints for the model that will be trained. Namely, we will use the Andro datasets[6] for water quality prediction [8] for training a single multi-target predictive clustering regression tree. Additionally, we define the descriptive, clustering, key, and target variables. The datasets contain 30 descriptive, as well as 6 target features (temperature, turbidity, oxygen, pH, conductivity, salinity). Regarding the model constraints, we do not limit the tree size in terms of depth, or the minimum number of examples in the leaf (see Fig. 6). We choose the variance reduction heuristic for making the splits, with N^2 complexity. Additionally, we set the rest of the parameters with their default values recommended by the Clus development community.

Finally, we define the output settings, i.e., the verbosity of Clus. Normally, in this section, the user can choose which resources are to be stored for memory optimization. However, in this case, our system overrides the user's preference and selects the settings for maximum verbosity. Doing so, we can successfully annotate not only the experimental setup but the experimental outputs as well.

Once the users have set their experimental setup, the data and settings files are uploaded through SemanticHub's API in our file system. At this point, the experiment is set and the execution has begun. The user is notified when the execution has finished with a server response.

[6] https://www.openml.org/d/41392.

```
1  PREFIX rdfs: <http://www.w3.org/2000/01/rdf-schema#>
2  PREFIX rdf: <http://www.w3.org/1999/02/22-rdf-syntax-ns#>
3  PREFIX OntoExp: <http://www.semanticweb.org/ilint/ontologies/2019/3/untitled-ontology-142#>
4  PREFIX OntoDM: <http://www.ontodm.com/OntoDM-core/>
5
6
7  SELECT ?exp ?expTypeLabel ?trainData ?trainDataLabel ?algTypeLabel
8
9  WHERE {
10   ?exp rdf:type ?expType .
11   ?expType rdfs:label ?expTypeLabel .
12   ?expType rdfs:subClassOf ?expType1 .
13   ?expType1 rdfs:subClassOf OntoDM:OntoDM_U1_059913 .
14   ?exp <http://www.obofoundry.org/ro/ro.owl#has_part> ?alg_exec .
15   ?alg_exec <http://purl.obolibrary.org/obo/OBI_0000308> ?alg_impl .
16   ?alg_impl rdf:type ?algType .
17   ?algType rdfs:label ?algTypeLabel .
18   ?alg_exec <http://purl.obolibrary.org/obo/OBI_0000293> ?trainData .
19   ?trainData rdfs:label ?trainDataLabel .
20   FILTER (?expTypeLabel IN ('predictive modeling train/test evaluation workflow execution'@en)) .
21   FILTER (?algTypeLabel IN ('top down induction of decision trees CLUS implementation'@en)) .
22   FILTER (?trainDataLabel in ('Data/MTR_andro_train1.arff_1'))
23   ?trainData <http://purl.obolibrary.org/obo/OBI_0000316> ?trainRole .
24   ?trainRole rdf:type OntoDM:OntoDM_000269 .
25 }
```

Fig. 7. Generated SPARQL query in the SPARQL endpoint for the experiment ran in Clus.

Once a user has completed the experiments, we can use SemanticHub's SPARQL endpoint to browse through the repository of completed experiments. For example, we can formulate a query that returns details for the experiment executed previously. For this purpose, we search for an experiment that has the Andro dataset as input, addresses the multi-target regression task, has a train-test evaluation scenario, and outputs a single model. The generated SPARQL query, by the UI, is shown in Fig. 7.

6 Evaluation According to the FAIR Guidelines

The FAIR principles are focused on the findability, accessibility, interoperability, and reproducibility of the resources [19]. Here, we evaluate our experiment repositories based on the checklist[7] for evaluation of data FAIRness introduced by the Data Archiving and Networked Services (DANS). We can distinguish between five types of questions, regarding the trustworthiness, findability, accessibility, interoperability, and reproducibility of the repository. In Table 2, we present the assessment questions and discuss the results of the evaluation.

Trustworthiness Assessment. Since we strongly abide by the FAIR principles, we cover the questions of public findability and accessibility of our repository of computational experiments. Additionally, we provide metadata that enables reproducibility of the experimental results, together with raw files that allow reusability of the trained models. However, one additional criteria for the trustworthiness is the CoreTrustSeal [4] certificate that unfortunately we have not obtained yet, hence we obtain two out of four points for this assessment (Q1).

[7] https://tinyurl.com/yyx5uc5k.

Table 2. Evaluation questions from the DANS FAIRness assessment

	Data trustworthiness	Answer	Points
Q1	Is the data repository you have chosen trustworthy?	Yes	2/4
	Data findability	Answer	Points
Q2	Will your dataset have a Persistent Identifier after deposit?	Yes	1/1
Q3	Did you provide enough information (metadata) about your data for others to understand and reuse your data?	Yes	1/1
Q4	Did you provide rich additional documentation?	Yes	1/1
	Data accesibility	Answer	Points
Q5	Is the metadata publicly accessible?	Yes	1/1
	Data interoperability	Answer	Points
Q6	Are the data stored and archived in preferred archival formats?	Yes	1/1
Q7	Did you use standardized vocabulary?	Yes	1/1
	Data reusability	Answer	Points
Q8	Did you give detailed provenance information for the data?	Yes	1/1
Q9	Do you make use of relevant community standards?	Yes	1/1
Q10	Does the data have a usage licence?	Yes	1/1

Findability Assessment. Both our resources and our repository, have persistent URIs (Q2). Next, the annotation schemata introduced in Sect. 2 provides a comprehensive representation of the experiments in the domain of predictive modelling, together with essential provenance information such as creator, date, software environment, hardware capabilities of the machine, etc. (Q3). Finally, we (can) provide additional documentation in the form of a link to the publication where a certain computational algorithm was introduced (Q4).

Accessibility Assessment. All of the metadata stored about the conducted computational experiments, as well as the ontologies describing them, are publicly available through SemanticHub's querying interface, and the SPARQL endpoint hosted on the Fuseki2 server (Q5).

Interoperability Assessment. The metadata that we generate for the computational experiments is stored in the RDF format. This format is preferred in the knowledge representation and semantic web community. Additionally, RDF has several syntax variations and the users can switch between different syntax models to their preference (Q6). For semantic annotation, we used ontologies designed by following the state-of-the-art best practices in ontology engineering. All of the resources, are publicly available and uniquely identified (Q7).

Reusability Assessment. We provide provenance data for each computational experiment regarding the creator, software environment, hardware capabilities, as well as the date of the creation of the experiment (Q8). Since we create and are in full control of the data that enters our repository, we make sure that the generated metadata is in consistent format, which was previously determined to follow community standards regarding the information it contains (Q9). Finally, all of our resources, as well as the metadata in the experiment repository are published under the Creative Commons CC 4.0 usage license, which enables free use for all non-commercial use provided the work is referenced (Q10).

Summary. To evaluate our input for this questionnaire, each of the questions participate with one point in the final score with the exception of the first question regarding the trustworthiness of the repository which has a score of 4 points. For this question, we achieve 2 out of the 4 possible points, since we have not yet obtained the CoreTrustSeal certificate. Regarding the data findability, since we have positive score on all three questions, we achieve 3/3 points. For the accessibility of our repository, we achieve 1/1 point. Storing the data in community-preferred and versatile archival format combined with the use of standardized vocabulary helps us score 2/2 points for the interoperability of our metadata. For the reusability of our metadata for computational experiments we score 3/3 points since we have affirmative answers to the listed questions. In total, we achieve 11/13 points for this assessment.

7 Conclusion

In this paper, we focus on the semantic representation and annotation of predictive modelling experiments. First, we outlined the need and the benefits of creating a semantically annotated repository in the domain. We proposed OntoExp, a resource that provides a semantic representation for each conducted experiment. In addition, we incorporate OntoExp in SemanticHub, a system that can execute, annotate and store the experiments. The conducted experiments in the system are annotated and stored in a TDB2 triplet database hosted on Fuseki2 server. SemanticHub allows for these experiments to be executed through its own infrastructure, meaning that the users can define and run the experiments on our physical servers. In addition, we provided a querying interface from which the users can query the repositories of experiments. Finally, we evaluated the produced experiment repository according to the DANS FAIRness checklist.

In future work, we plan to upgrade this prototype system with more functionalities. These will include the use of different software platforms to execute the experiments and building a user management module. We also plan to use the representational power of ontologies and reasoners to enhance the system's querying engine and capabilities.

Acknowledgements. The authors would like to acknowledge the support of the Slovenian Research Agency through the grant J2-9230.

References

1. Arp, R., Smith, B., Spear, A.D.: Building Ontologies with Basic Formal Ontology. MIT Press, Cambridge (2015)
2. Bandrowski, A.: The ontology for biomedical investigations. PLOS One **11**, e0154556 (2016)
3. Ceusters, W.: An information artifact ontology perspective on data collections and associated representational artifacts. In: MIE, pp. 68–72 (2012)
4. Coretrustseal for data repositories (2019). https://www.coretrustseal.org
5. Courtot, M., et al.: MIREOT: the minimum information to reference an external ontology term. Appl. Ontol. **6**(1), 23–33 (2011)
6. Esteves, D., Lawrynowicz, A., Panov, P., Soldatova, L., Soru, T., Vanschoren, J.: ML schema core specification. W3C (2016). http://www.w3.org/2016/10/mls
7. Esteves, D., et al.: MEX vocabulary: a lightweight interchange format for machine learning experiments. In: Proceedings of the 11th International Conference on Semantic Systems, pp. 169–176 (2015)
8. Hatzikos, E.V., Tsoumakas, G., Tzanis, G., Bassiliades, N., Vlahavas, I.: An empirical study on sea water quality prediction. Knowl. Based Syst. **21**(6), 471–478 (2008)
9. Keet, C.M.: The data mining optimization ontology. J. Web Semant. **32**, 43–53 (2015)
10. Lawrynowicz, A., Esteves, D., Panov, P., Soru, T., Dzeroski, S., Vanschoren, J.: An algorithm, implementation and execution ontology design pattern. Adv. Ontol. Des. Patterns **32**, 55 (2017)
11. Malone, J.: The Software Ontology (SWO): a resource for reproducibility in biomedical data analysis, curation and digital preservation. J. Biomed. Semant. **5**(1), 25 (2014)
12. Panov, P., Soldatova, L., Džeroski, S.: Ontology of core data mining entities. Data Min. Knowl. Discov. **28**(5), 1222–1265 (2014). https://doi.org/10.1007/s10618-014-0363-0
13. Panov, P., Soldatova, L.N., Džeroski, S.: Generic ontology of datatypes. Inf. Sci. **329**, 900–920 (2016)
14. Schelter, S., Böse, J.H., Kirschnick, J., Klein, T., Seufert, S.: Automatically tracking metadata and provenance of machine learning experiments. In: Machine Learning Systems Workshop at NIPS (2017)
15. Smith, B., et al.: Relations in biomedical ontologies. Genome Biol. **6**(5), R46 (2005)
16. Smith, B., et al.: The OBO Foundry: coordinated evolution of ontologies to support biomedical data integration. Nat. Biotechnol. **25**(11), 1251 (2007)
17. Vanschoren, J., Soldatova, L.: Exposé: an ontology for data mining experiments. In: International Workshop on Third Generation Data Mining, SoKD-2010, pp. 31–46 (2010)
18. Vanschoren, J., Van Rijn, J.N., Bischl, B., Torgo, L.: OpenML: networked science in machine learning. ACM SIGKDD Exp. Newslett. **15**, 49–60 (2014)
19. Wilkinson, M.D.: The fair guiding principles for scientific data management and stewardship. Sci. Data **3**(1), 1–9 (2016)

Semantic Description of Data Mining Datasets: An Ontology-Based Annotation Schema

Ana Kostovska[1,2](✉), Sašo Džeroski[1,2], and Panče Panov[1,2]

[1] Department of Knowledge Technologies, Jožef Stefan Institute, Ljubljana, Slovenia
[2] Jožef Stefan International Postgraduate School, Ljubljana, Slovenia
{ana.kostovska,saso.dzeroski,pance.panov}@ijs.si

Abstract. With the pervasiveness of data mining (DM) in many areas of our society, the management of digital data, readily available for analysis, has become increasingly important. Consequently, nearly all community accepted guidelines and principles (e.g. FAIR and TRUST) for publishing such data in the digital ecosystem, stress the importance of semantic data enhancement. Having rich semantic annotation of DM datasets would support the data mining process at various choice points, such as data understanding, automatic identification of the analysis task, and reasoning over the obtained results. In this paper, we report on the developments of an ontology-based annotation schema for semantic description of DM datasets. The annotation schema combines three different aspects of semantic annotation, i.e., annotation of provenance, data mining specific, and domain-specific information. We demonstrate the utility of these annotations in two use cases: semantic annotation of remote sensing data and data about neurodegenerative diseases.

Keywords: Data mining · Datasets · Knowledge representation · Semantic annotation · Ontology

1 Introduction

Recently, the success of Data Mining (DM) and Machine Learning (ML) in a broad range of applications has led to a growing demand for ML systems. However, this success heavily relies on the ML expertise of the practitioners, and on the quality of the analyzed data, both of which are in short supply. One potential solution for overcoming the shortage of expertise is to develop more intelligent data analysis systems, that will assist domain practitioners in the construction of analysis pipelines and the interpretation of results. Such an intelligent DM system would we able to reason over distributed heterogeneous data and knowledge bases, automatically define the learning task, recommend the most suitable algorithms for the task at hand, and correctly interpret the induced predictive models [17,18].

A. Appice et al. (Eds.): DS 2020, LNAI 12323, pp. 140–155, 2020.
https://doi.org/10.1007/978-3-030-61527-7_10

The first step towards the development of such systems is the improvement of data management and data understanding. Research data must be enriched with formal and logical descriptors that capture the characteristics of the data relevant for the task of automation of the data analysis process. Additionally, these descriptors have the potential to significantly improve interdisciplinary research by helping ML practitioners better understand the data originating from the application domains, as well as easily incorporate domain knowledge in the process of analysis. Formal descriptors, when published on the Web, can also improve the accessibility and reusability of scientific data.

Many academic institutions have recognized the importance of effective management of scientific data, making it their central mission. For example, the FAIR (Findable, Accessible, Interoperable, and Reusable) principles [26] are a set of guiding principles that have been introduced to support and promote proper data management and stewardship. In that context, data must be discoverable and it should be semantically annotated with rich metadata. The metadata should always be accessible by standardized communication protocols. The data and the metadata have to be interoperable with external data from the same domain. Finally, both data and metadata should be released with provenance details so that the data can be easily replicated and reused.

Another set of principles that builds upon FAIR data are the TRUST principles [13]. The TRUST principles go a level higher by focusing on data repositories and providing them with guidance to demonstrate Transparency, Responsibility, User focus, Sustainability, and Technology (TRUST).

At the core of both principles lies the semantic enrichment of research data. Semantic annotation of data, as a powerful technique, has attracted attention in many domains. Unfortunately, semantic annotation of DM and ML datasets is still in the early phases of development. To the best of our knowledge, there are no semantic dataset repositories from the general area of data science that completely adhere to the FAIR and TRUST principles.

In this paper, we report on the development of an ontology-based annotation schema for semantic annotation of DM datasets. Our main objective is to provide a rich vocabulary for data annotation, that will serve as a basis for the construction of a dataset repository that closely follows the FAIR and TRUST principles. The annotation schema we proposed includes three different types of information: provenance, DM-specific, and domain-specific. The provenance information improves the transparency and reusability of data. The DM-specific information provides means for reasoning over the analyzed data and helps (in a semi-automatic way) in the construction of the DM workflows (or pipelines). The domain-specific information helps to bridge the gap between ML practitioners and domain experts, as well as to improve cross-domain research. Finally, we demonstrate the utility of domain-specific annotations in two use cases from the domains of neurodegenerative diseases and Earth Observation (EO), respectively.

2 Background and Related Work

In the context of computer science, ontologies are "an explicit formal specifications of the concepts and the relations among them that can exist in a given domain" [9]. In other words, they provide the basis for an unambiguous, logically consistent, and formal representation of knowledge. It is important to note that, the logical component of ontologies allows knowledge to be shared meaningfully both at machine and human level. Also, an immediate consequence of having formal ontologies based on logic is that they can be used in a variety of reasoning tasks, as well as in the inference of new knowledge. The benefits of having ontology-based knowledge representations have been demonstrated in many data- and knowledge-driven applications. The research areas that retained most attention and contributed the most to the technological breakthrough of ontologies are bioinformatics and biomedicine. For example, the Open Biological and Biomedical Ontology (OBO) Foundry [21] is a collective of ontology developers that have developed and maintain over 100 publicly-available ontologies related to the life sciences. When it comes to the process of ontology engineering, the OBO Foundry has played a key role, as they have proposed ontology design principles that promote open, orthogonal, and strictly-scoped ontologies with collaborative development. These principles have further widened the use of ontologies across different fields of science.

In the area of DM and ML, a large body of research has focused on the development of ontologies, vocabularies and schemas that cover different aspects of the domain. Examples of such resources include the Data Mining OPtimization Ontology (DMOP) [11], Exposé [24], MEX vocabulary [8], and the ML schema [7]. DMOP has been designed to support automation at various choice points of the DM process. The Exposé ontology provides the vocabulary needed for a detailed description of machine learning experiments. MEX represents a lightweight interchange format for ML experiments. ML Schema represents an effort to unify the representation of machine learning entities.

The OntoDM suite of ontologies is of particular interest, as this paper extends its line of work. OntoDM includes three different ontologies: OntoDM-core, OntoDM-KDD, and OntoDT. OntoDM-core [17] is an ontology of core data mining entities, such as dataset, DM task, generalizations, DM algorithms, implementations of algorithms, and DM software. OntoDM-KDD [16] is an ontology for representing the process of knowledge discovery following the CRISP-DM methodology [5]. OntoDT [18] is a generic ontology for the representation of knowledge about datatypes.

Another type of information related to DM datasets that is important to be formally represented is the provenance information. Provenance information refers to the kind of information that describes the origin of a resource (in our case a dataset), i.e., who created the resource, when was it published, and what is its usage license. Provenance information is valuable when it comes to deciding whether a specific resource can be trusted. This extra information also helps the users better understand it, easily cite and reuse the resource for their purposes. For the computers to make use of the provenance information, it has to be given explicitly, and it has to be based on common provenance vocabularies, such as the Dublin Core vocabulary [25], the PROV ontology [2], Data Catalog Vocabulary [1], or Schema.org [3].

3 Semantic Description of DM Datasets

To semantically describe a DM dataset, we consider three different types of vocabularies/ontologies: (1) vocabularies for annotation of provenance information, such as title, description, license, and format; (2) ontologies for annotation of datasets with DM-specific characteristics, i.e., data mining task, datatypes, and dataset specification; and (3) ontologies for annotation of domain-specific knowledge that helps to contextualize the data originating from a given domain.

In this section, we discuss the first two aspects of the semantic enrichment of datasets. We describe the Schema.org vocabulary, which we reuse for the purpose of annotation of the dataset's provenance details. Also, we outline the main characteristics of the OntoDT and OntoDM-core ontologies and we further extend their structure with terms essential for semantic description from a DM perspective. In Sect. 4 we discuss the annotation of domain specific knowledge through examples from two different domains.

3.1 Provenance Information Annotation

To annotate DM datasets with provenance information, we have chosen the Schema.org vocabulary, one of the most widely used vocabularies that provides descriptors for provenance information in a structured manner. When annotating the datasets, we usually use a subset of the list of provided descriptors as the complete provenance information is not always available.

Figure 1 depicts an example annotation of provenance information in `JSON-LD` format[1]. For this example, we used a dataset from the domain of Earth Observation (EO), named *Forestry_Kras_LiDAR_Landsat*. The dataset was used in a study that investigates the possibility of predicting forest vegetation height and canopy cover in the Karst region in Slovenia by building predictive models using EO data [23]. For semantic annotation of provenance information for this dataset, we used several terms from Schema.org, such as name, description, URL, keywords, creator, distribution, temporal and spatial coverage, citation, and license.

3.2 Data Mining Specific Annotations

The second type of annotation considers explicit specification of dataset characteristics from a DM perspective, e.g., the format of the data, the type of learning task, and the features' datatypes. Data used in the process of DM can take various forms, but the standard one assumes that there is a set of objects of interest

[1] https://json-ld.org/.

```
{
  "@context":"https://schema.org/",
  "@type":"Dataset",
  "name":"Forestry_Kras_LiDAR_Lansat",
  "description":"This dataset was employed in a study that investigates the possibility of predicting forest
         vegetation height and canopy cover in the Karst region, Slovenia by building predictive models
        using remotely sensed data."
  "url":"http://semantichub.ijs.si/ontodm",
  "keywords":["remote sensing", "Karst region", "LiDAR", "Landsat"],
  "creator":{
     "@type":"Person",
     "url": "https://www.researchgate.net/profile/Daniela_Stojanova2",
     "name":"Daniela Stojanova"
  },
  "distribution":{
        "@type":"DataDownload",
        "encodingFormat":"ARFF",
        "contentUrl":""
  },
  "temporalCoverage":"2001-08-03, 2002-05-18, 2002-11-10, 2003-03-18",
  "spatialCoverage":{
     "@type":"Place",
     "geo": {
        "@type": "GeoCoordinates",
        "latitude": 45.3818,
        "longitude": 13.4815
     }
  },
  "citation": {
     "@type": "ScholarlyArticle",
     "name": "Estimating vegetation height and canopy cover from remotely sensed data with machine learning"
     ,
     "identifier": "https://doi.org/10.1016/j.ecoinf.2010.03.004"
  },
  "license": "https://creativecommons.org/licenses/by/4.0/"
}
```


Fig. 1. An example provenance information annotation for the Forestry Kras Lidar/Landsat dataset [23] using the Schema.org vocabulary.

described with features (or attributes). In that sense, the term data example, or (more commonly) data instance, refers to a tuple of feature values corresponding to an observed object.

The features are formally typed, meaning that each of them has a designated datatype. In general, there are many different datatypes such as boolean, real, discrete datatype, to name a few. Having standardized datatype information at disposal can enable the development of knowledge-based systems that automate parts of data analysis workflows, e.g., assist DM practitioners in choosing a suitable learning algorithm for the data at hand.

Data examples in DM can be described with different characteristics, which can lead to treating the data in radically different ways. We identified four different (orthogonal) characteristics that we believe are important to be represented appropriately. These include (1) the availability of data examples, (2) the existence of missing values, (3) the mode of learning, and (4) the type of target in the case of (semi-)supervised learning tasks.

Extending the OntoDT and OntoDM-core Ontologies. While the OntoDT and OntoDM-core ontologies offer a rich vocabulary for the annotation of DM datasets, they do not cover all of the above aspects. Thus, we extended OntoDT with new DM-specific datatypes and provided an updated datatype taxonomy that allows us to properly describe DM datasets. The proposed taxonomy of datatypes was then used as a basis for the update of the taxonomies of DM tasks and data specification, which are part of the OntoDM-core ontology. The extended OntoDT and OntoDM-core ontologies are available at https://w3id.org/OntoDT-extended and https://w3id.org/OntoDM-core-extended, respectively.

Availability of the Data Examples. Based on the availability of the data examples, we distinguish between two types of data, i.e., batch data (or datasets) and online data (or data streams). The batch setting is the more traditional approach where large volumes of data are collected over a longer period. On the other hand, online data refers to the type of data that is continuously being generated by heterogeneous data sources.

The availability of data examples is the first dimension we considered when we updated the taxonomies of core classes of OntoDT and OntoDM-core. In Fig. 2, we depict the top-level classes of the taxonomies of datatypes, data mining tasks, and data specifications. At the second level, we have the corresponding classes that represent the specifications of the availability of data examples. For instance, the *OntoDT: record(tuple) datatype* and *OntoDT: sequence datatype* classes refer to the datatypes of data examples in batch and online mode, respectively.

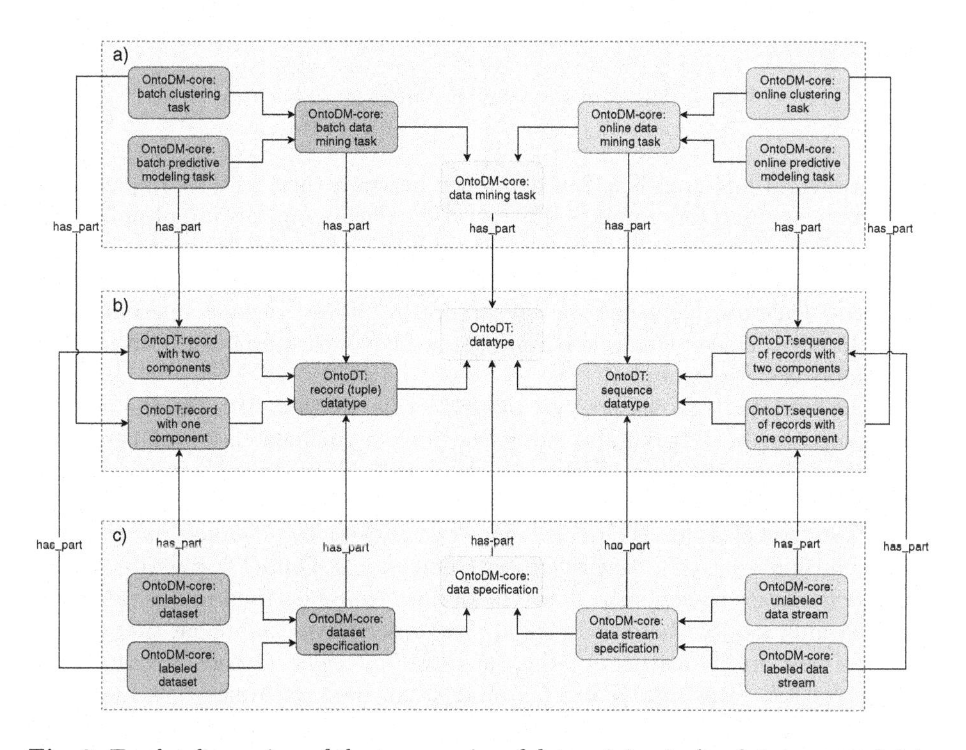

Fig. 2. Top level overview of the taxonomies of data mining tasks, datatypes and data specifications for the batch setting (right-hand side) and online setting (left-hand side).

Type of Learning. According to the type of learning, DM learning methods can be categorized into three groups, i.e., unsupervised, supervised, and

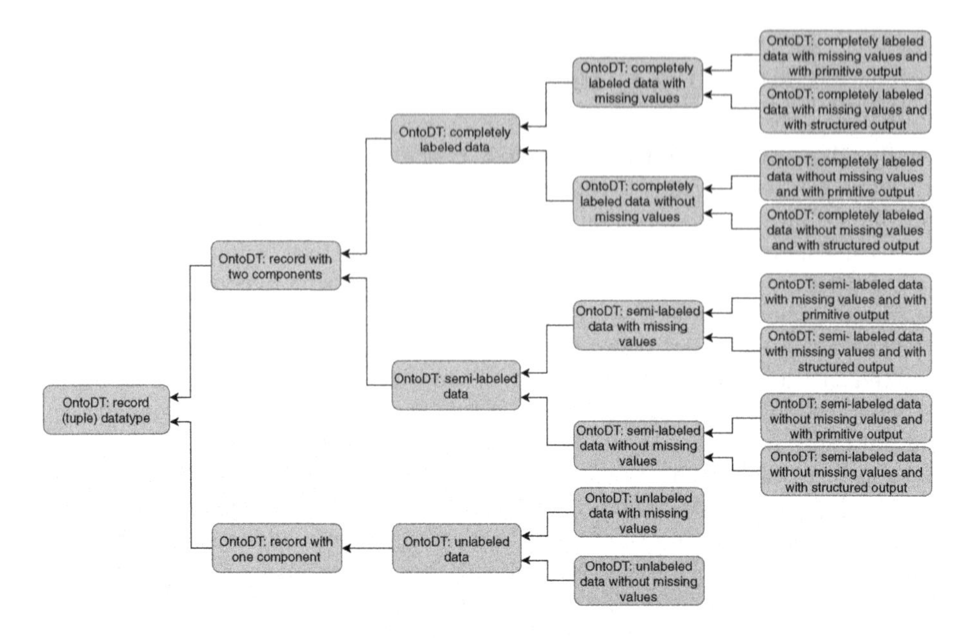

Fig. 3. A part of the OntoDT datatype taxonomy.

semi-supervised learning. The key difference between them is the completeness of the data they use for training. Unsupervised learning makes use of unlabeled data examples that are only composed of descriptive features. Supervised learning, in contrast to unsupervised learning, uses labeled data that, apart from the descriptive features, has some special feature of interest usually referred to as target. Finally, in semi-supervised learning, we have learning from both labeled and unlabeled data examples.

In the updated taxonomies, we modeled this characteristic at the second level. Hence, for both batch and online learning, we defined classes that specify information about the type of learning (see Fig. 2). If we take the taxonomy of data types as an example, in the batch learning scenario the *OntoDT: record (tuple) datatype* class further resolves into two classes: *OntoOT: record with one component* and *OntoDT: record with two components*. *OntoOT: record with one component* class represents the datatype of data examples used in unsupervised batch learning mode, where there is only one descriptive component that aggregates the descriptive features of the data example. The *OntoDT: record with two components* class represents the datatype of data examples that have one descriptive and one target component and are used in either supervised or semi-supervised learning. Figure 3 illustrates in greater detail the taxonomy of data types and the four dimensions that it is based on. Finally, the taxonomies of tasks and dataset specifications are designed similarly following the same principles (see Fig. 2).

Missing Values. Another property we consider when describing data examples, which is important for DM algorithms, is the existence of missing values (see Fig. 3) since some DM algorithms cannot function properly in the presence of missing values. We say that one data example has missing values when there is no recorded value for at least one descriptive feature. This is different from having missing values in the target space, which, as we discussed above, leads to semi-labeled data. Missing values affect the data quality; thus they must be handled accordingly by the DM algorithms.

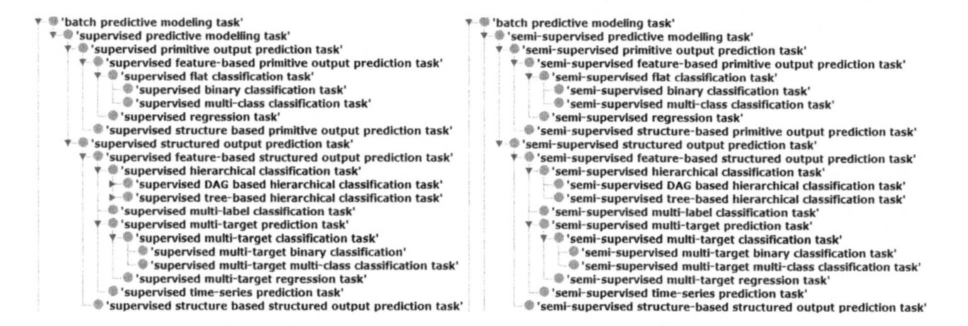

Fig. 4. A Protégé snapshot of the taxonomy of supervised and semi-supervised batch predictive modeling tasks.

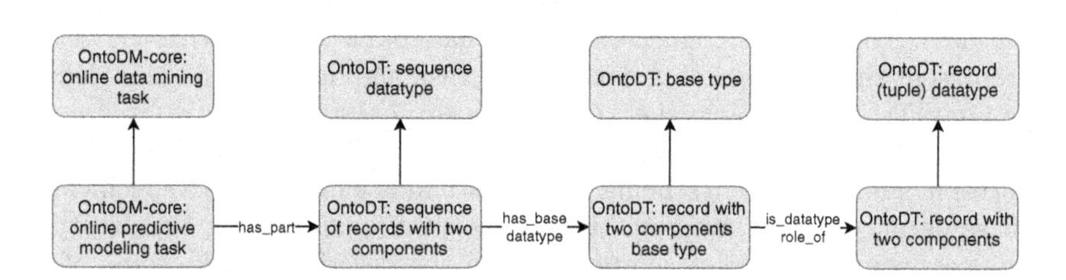

Fig. 5. An example of modeling online data mining tasks with the corresponding datatypes from OntoDT.

Type of Target. In the case of (semi-)supervised learning, data examples can become even more complex as the target/output itself can have a complex structure. Based on the type of target we have primitive and structured output prediction tasks. Primitive output prediction tasks predict a single target, as in classification (a discrete value) and regression (a real value). In the case of structured output prediction tasks, there is more than one target that has to be predicted. Examples of such tasks are multi-target regression, multi-label classification, and hierarchical multi-label classification. Figure 4 presents the complete taxonomy of supervised and semi-supervised predictive modeling tasks.

Concerning the (semi-)supervised online predictive modeling tasks, the base datatypes of the target can be the same as the target datatypes in the batch predictive modeling tasks. Figure 5 illustrates how this is achieved in the OntoDT and OntoDM-core ontologies. For instance, *OntoDM-core: online predictive modeling task* class is related with the *OntoDT: sequence of records with two components* class. Sequence datatypes have a base datatype, in this example, it is the *OntoDT: record with two components base type*, which has the datatype role of *OntoDT:record of two components.* Note that *OntoDT:record of two components* is the same class used for the representation of the data examples' datatype in the batch predictive learning mode.

3.3 Example Annotations of DM Datasets

Using this annotation schema, we have annotated 496 DM datasets in total, all containing data from different application domains. The generated semantic annotations are publicly available in RDF format and can be queried via the Jena Fuseki server[2].

After describing the four characteristics that govern the modeling of the taxonomies of datatypes, data specification, and tasks, we provide an illustrative example that shows how we can combine them in a single annotation schema for the purpose of semantic annotation of DM datasets. Namely, Fig. 6 depicts the classes needed for annotation of a data stream with missing values applicable to the learning task of semi-supervised multi-label classification.

To represent the datatype of the data examples, we use the *OntoDT:feature-based semi-labeled stream data with missing values and with a set of discrete output* class. This class is connected via the `has-part` relation with the classes that represent the corresponding data mining task and data specification defined in the OntoDM-core ontology, i.e., *OntoDM-core: online semi-supervised multi-label classification task* and *OntoDM-core: multi-label semi-labeled classification data stream.* The annotation schema for data streams includes also a specification of a base datatype. Next, we have the classes used for describing the datatypes of the descriptive and target component. On the descriptive side, some of the examples can have missing values, thus, we use a record/tuple of choice (primitive, void) datatypes. For the target component, we have two alternatives, one of which is a discrete datatype used for annotation of labeled examples, and the other is a void datatype used to annotate unlabeled data examples.

4 Domain-Specific Annotations: Use Cases

In this section, we demonstrate the utility of the annotation schema we introduced in the previous section on two use cases, i.e., annotation of datasets for the domains of neurodegenerative diseases and Earth Observation (EO). For the two

[2] Fuseki dataset containing the semantic annotations in RDF format: http://semantichub.ijs.si/fuseki/dataset.html?tab=query&ds=/DMDatasets.

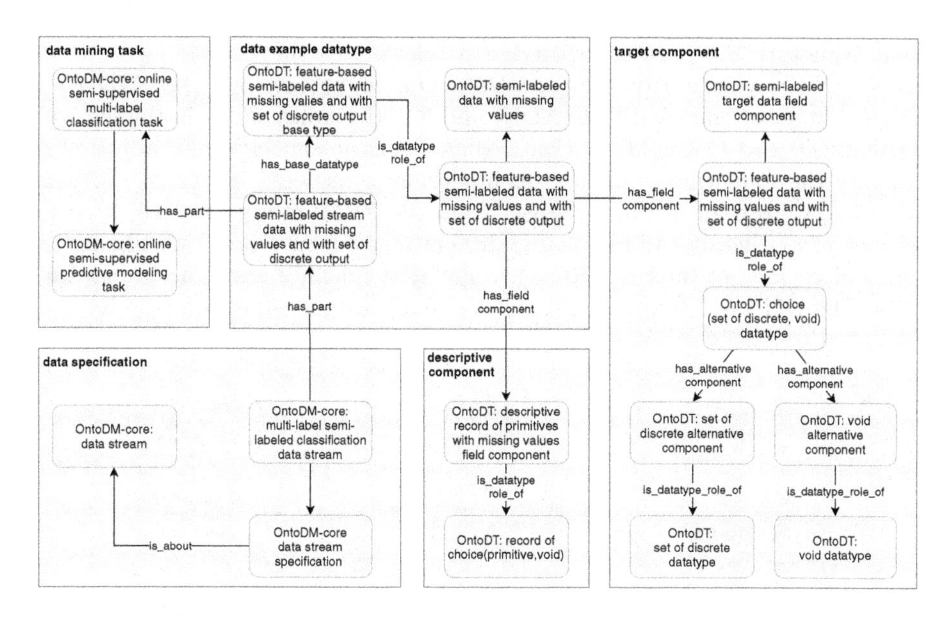

Fig. 6. An example of an annotation schema for data streams applicable to semi-supervised multi-label classification.

use cases, we also enriched the annotation schema with terminology specific to the domain at hand. The inclusion of domain-specific annotations improves the representation of the datasets, making them accessible and reusable, offers the possibility of execution of advanced query scenarios, and enables interoperability with other data from the domain.

On a technical level, the alignment of the DM-specific annotation schema with the annotation schemas designed for the particular domains is straightforward. In that sense, the proposed ontology-based annotation schema enables the direct extension of the datatype classes at any level in the taxonomy with classes that define the semantic meaning of the domain-specific datatypes. The newly introduced datatype classes are then linked to the corresponding entities in the domain ontologies.

4.1 Neurodegenerative Disease Datasets

Neurodegenerative diseases such as Alzheimer's disease (AD), Parkinson's disease (PD), amyotrophic lateral sclerosis (ALS), and Huntington's disease (HD) are a group of diseases caused by a progressive loss of structure or function of neurons. They can lead to irreversible deterioration of cognitive functions like memory loss, cause problems with movement, and spatial orientation. In the past two decades, researchers have been investigating new treatments that can slow or stop the progression of the diseases. There are two widely-known studies concerning neurodegenerative diseases, i.e., Alzheimer's disease Neuroimaging Initiative (ADNI) [19] and Parkinson's Progression Markers Initiative (PPMI) [4].

To annotate the datasets with terms relevant to the domain, we use the NDDO (Neurodegenerative Disease Data Ontology) ontology [12]. NDDO is designed in accordance with the ADNI and PPMI studies and it is aligned with the OntoDT and OntoDM-core ontologies. Thus, it can be easily adjusted to annotate the four aspects of data examples we considered in Sect. 3. To illustrate this, we use an instance dataset from the PPMI study that [15] used for the task of predicting the motor impairment assessment scores by utilizing the values of regions of interest (ROIs) from fMRI imaging assessment and DaT scans. The DM task they were solving was multi-target regression (MTR).

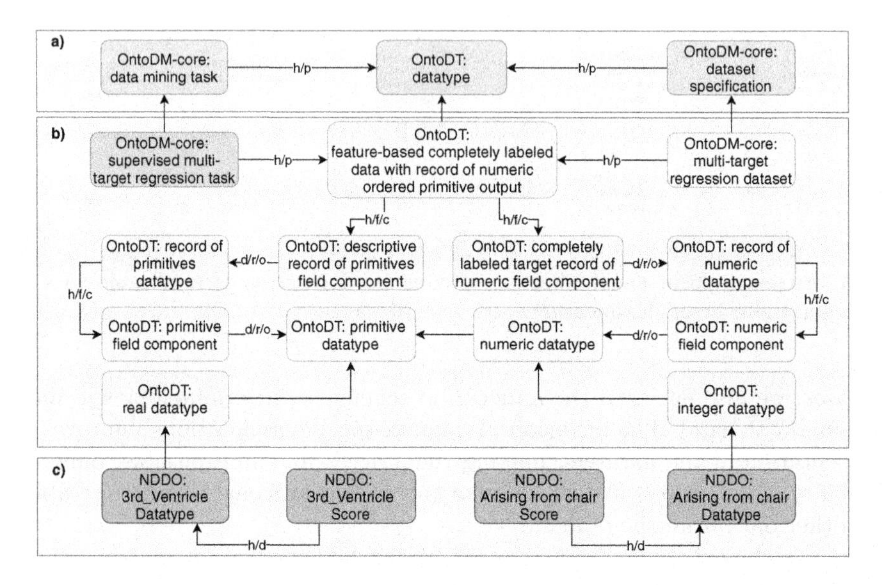

Fig. 7. A semantic annotation schema for the PPMI dataset [15]: a) top level classes from OntoDM and OntoDT; b) specific classes and relations required for annotation of datasets used in cluster analysis and c) specific NDDO datatype classes.

Figure 7, depicts the point of alignment of the domain classes defined in NDDO with classes from the extended versions of the OntoDT and OntoDM-core ontologies. To represent the MTR task and MTR dataset specification, we use the classes defined in OntoDM-core, and connect them with the corresponding datatype class from OntoDT (in our case *OntoDT: feature-based completely labeled data with record of numeric ordered primitive output*) (see Fig. 7 b). This class has two field components. The first one describes the datatypes of the descriptive features, which are of a primitive datatype. The latter describes the datatypes of the features on the target side. In the MTR learning setting each target feature is described with the numeric datatype. The sub-classes of the numeric datatype, real and integer datatype, are positioned at the bottom of the datatype taxonomy, and we link them with the domain datatypes.

For example, *NDDO: 3rd_Ventricle Score* is one of the descriptive features present in the PPMI dataset and it is linked with the *NDDO: 3rd_Ventricle Datatype* class that semantically defines its datatype. Similarly, *NDDO: Arising from the chair Score* is a target and its associated datatype is the *NDDO: Arising from the chair Datatype* class. Other features are connected with the respective datatypes in the same way.

4.2 Earth Observation (EO) Datasets

Remote sensing (RS) is the process of monitoring specific physical characteristics of an area of interest by measuring the reflected and emitted energy at a distance from the target area. Satellite-based remote sensing technologies are commonly used for Earth Observation (EO) to monitor characteristics that change over time, i.e., weather prediction, natural changes of the Earth, and development of the urban area.

Due to the increasing availability of EO data, it is essential to develop an ontological approach to managing this kind of data. However, to the best of our knowledge, a general ontology that systematically describes the EO domain is still lacking. Nonetheless, some ontologies formalize the knowledge of specific parts of the domain, i.e., Semantic Sensor Network (SSN) ontology [6], SOSA (Sensor, Observation, Sample, and Actuator) ontology [10], Semantic Web for Earth and Environment Technology (SWEET) ontology [20], and the Extensible Observation Ontology (OBOE) [14].

For semantic annotation of EO data, we have designed a lightweight ontology that is aligned with the aforementioned EO ontologies. The ontology is available at https://w3id.org/eo-ontology. The ontology was constructed using the bottom-up approach, based on 4 instances of datasets we have available at our side from previous research [22].

The datasets contain two target features (forest vegetation height and canopy cover) whose values are obtained via the LiDAR technology. But since LiDAR can sometimes be inconvenient or expensive, [22] examined the possibility of using remote sensing data generated from satellites, such as Landsat 7, IRS-P6, SPOT, as well as aerial photographs for the construction of descriptive features that can be relevant for the prediction of the two targets. The Landsat 7, IRS-P6, and SPOT satellites use multiple channels for collecting reflected energy, and one channel of emitted energy, that operate on different wavelengths.

In this study, when designing the EO ontology, we took into consideration the process of data collection and data preprocessing described in the study mentioned above. In the preprocessing phase, the raw satellite image is converted into a standard geo-referenced data format, which then undergoes the process of image segmentation (see Fig. 8). A key characteristic of the different image segments is the resolution of the segment size. The image segment size is modeled as a data property of the image segmentation specification class.

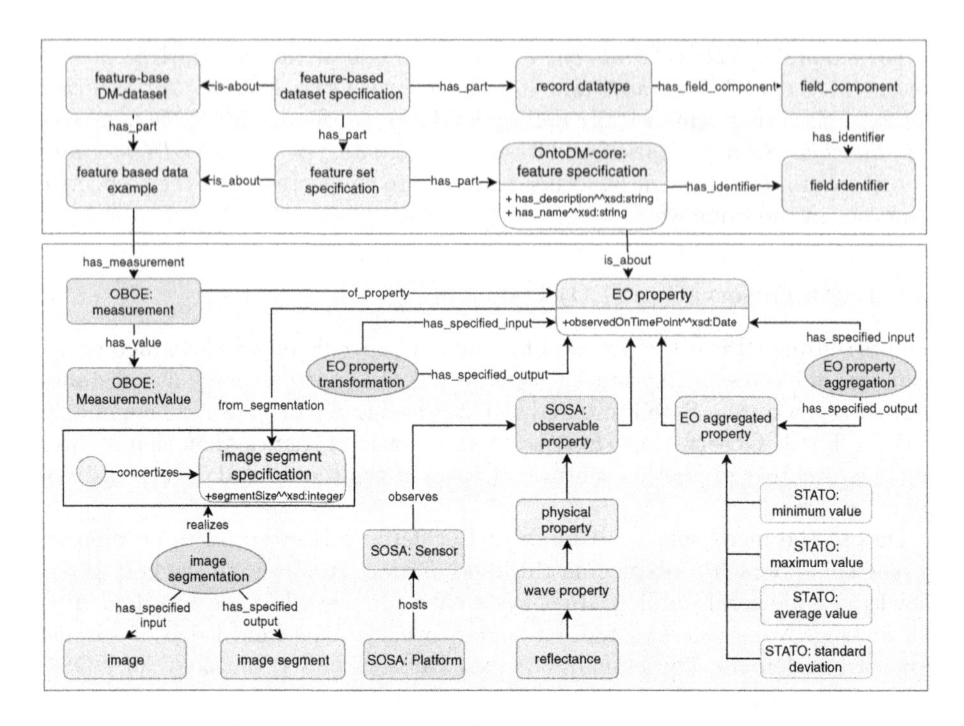

Fig. 8. Core entities of the Earth Observations ontology. Rectangular boxes represent continuant classes, while ellipses represent process classes. The color scheme was chosen for better visual perception.

All features present in the datasets are EO properties observed at a specific point in time, and they are related to a specific image segment. We define two subclasses of the EO property class, i.e., *SOSA: observable property* class and *EO aggregated property* class. The first one refers to the properties observed with a remote sensor (*SOSA: Sensor*) hosted on a given platform/satellite (*SOSA: Platform*). The latter defines the type of properties that are the result of some process of EO property aggregation that transforms the originally observed measurements. The process uses multiple EO properties as input and produces one EO aggregated property. The aggregation can be based on some statistical characteristics, such as *STATO: minimum value*, *STATO: maximum value*, *STATO: average value* and *STATO: standard deviation*, where STATO is an ontology of statistical methods. This was also the case in our observed datasets. Additionally, we define the EO property transformation process that transforms one EO property into another.

Similarly, as in Sect. 4.1, to achieve full interoperability, we integrated the general DM annotations with the domain-specific ones. The integration was perforemed at the level of features appearing in the dataset. Thus, the *OntoDM-core: feature specification* class connects with the datatype of the feature via the `has-identifier` relation, while it also connects with the EO property class via the `is-about` relation. Additionally, the *OntoDM-core: feature-based data example* class is composed of multiple *OBOE: Measurements*. In OBOE, measurement represents a measurable characteristic of an observed property, which in our case is *EO property*.

5 Conclusions and Future Work

We have developed an ontology-based annotation schema for rich semantic annotation of DM datasets that takes into consideration 3 different semantic aspects of the datasets: provenance, DM-specific characteristics of the data, and domain-specific information. The annotation schema is generic enough to support the easy extension of its core classes with information relevant to the application domain. The utility of the designed schema was demonstrated through semantic annotation of data from two different domains: neurodegenerative diseases and Earth observation.

Annotations based on this schema provide means for support of the complete data analysis process, e.g., enable cross-domain interoperability, assist in the definition of the learning task, ensure consistent representation of datatypes, assess the soundness of data, and automatically reason over the obtained results. These annotations also enable the development of applications that require advanced data querying capabilities. They also enable the development of data repositories that adhere to the highest standards of the Open data initiative.

Acknowledgements. The authors would like to acknowledge the support of the Slovenian Research Agency through the project J2-9230 and the young researcher grant to AK.

References

1. The Data Catalog vocabulary (DCAT) vocabulary (2019). https://www.w3.org/TR/vocab-dcat/
2. The PROV Ontology (PROV-O) (2019). https://www.w3.org/TR/prov-o/
3. The Schema.org vocabulary (2019). https://schema.org/
4. PPMI website (2020). http://www.ppmi-info.org/publications-presentations/
5. Chapman, P., et al.: Crisp-DM 1.0 step-by-step data mining guide. Technical report, The CRISP-DM consortium, August 2000
6. Compton, M., et al.: The SSN ontology of the W3C semantic sensor network incubator group. Web Semant. Sci. Serv. Agents World Wide Web **17**, 25–32 (2012)
7. Esteves, D., Lawrynowicz, A., Panov, P., Soldatova, L., Soru, T., Vanschoren, J.: Ml schema core specification. W3C (2016). http://www.w3.org/2016/10/mls
8. Esteves, D., et al.: Mex vocabulary: a lightweight interchange format for machine learning experiments. In: Proceedings of the 11th International Conference on Semantic Systems, pp. 169–176 (2015)
9. Gruber, T.: Toward principles for the design of ontologies used for knowledge sharing? Int. J. Hum. Comput. Stud. **43**(5–6), 907–928 (1995)

10. Janowicz, K., Haller, A., Cox, S., Le Phuoc, D., Lefrançois, M.: SOSA: a lightweight ontology for sensors, observations, samples, and actuators. J. Web Semant. **56**, 1–10 (2019)
11. Keet, M., et al.: The data mining optimization ontology. Web Semant. Sci. Serv. Agents World Wide Web **32**, 43–53 (2015)
12. Kostovska, A., Tolovski, I., Maikore, F., Soldatova, L., Panov, P.: Neurodegenerative disease data ontology. In: Kralj Novak, P., Šmuc, T., Džeroski, S. (eds.) DS 2019. LNCS (LNAI), vol. 11828, pp. 235–245. Springer, Cham (2019). https://doi.org/10.1007/978-3-030-33778-0_19
13. Lin, D., et al.: The trust principles for digital repositories. Sci. Data **7**(1), 1–5 (2020)
14. Madin, J., Bowers, S., Schildhauer, M., Krivov, S., Pennington, D., Villa, F.: An ontology for describing and synthesizing ecological observation data. Ecol. Inf. **2**(3), 279–296 (2007)
15. Mileski, V., Kocev, D., Draganski, B., Džeroski, S.: Multi-dimensional analysis of PPMI data. In: Proceedings of 8th Jožef Stefan International Postgraduate School Students Conference, pp. 175–178. Jožef Stefan International Postgraduate School, Ljubljana, Slovenia (2016)
16. Panov, P., Soldatova, L., Džeroski, S.: OntoDM-KDD: ontology for representing the knowledge discovery process. In: Fürnkranz, J., Hüllermeier, E., Higuchi, T. (eds.) DS 2013. LNCS (LNAI), vol. 8140, pp. 126–140. Springer, Heidelberg (2013). https://doi.org/10.1007/978-3-642-40897-7_9
17. Panov, P., Soldatova, L., Džeroski, S.: Ontology of core data mining entities. Data Min. Knowl. Discov. **28**(5), 1222–1265 (2014). https://doi.org/10.1007/s10618-014-0363-0
18. Panov, P., Soldatova, L., Džeroski, S.: Generic ontology of datatypes. Inf. Sci. **329**, 900–920 (2016)
19. Petersen, R.C., et al.: Alzheimer's disease neuroimaging initiative (ADNI): clinical characterization. Neurology **74**(3), 201–209 (2010)
20. Raskin, R., Pan, M.: Knowledge representation in the semantic web for Earth and environmental terminology (SWEET). Comput. Geosci. **31**(9), 1119–1125 (2005)
21. Smith, B., et al.: The obo foundry: coordinated evolution of ontologies to support biomedical data integration. Nat. Biotechnol. **25**(11), 1251 (2007)
22. Stojanova, D.: Estimating forest properties from remotely sensed data by using machine learning. Master's thesis, Jožef Stefan International Postgraduate School, Ljubljana, Slovenia (2009)
23. Stojanova, D., Panov, P., Gjorgjioski, V., Kobler, A., Džeroski, S.: Estimating vegetation height and canopy cover from remotely sensed data with machine learning. Ecol. Inf. **5**(4), 256–266 (2010)
24. Vanschoren, J., Soldatova, L.: Exposé: an ontology for data mining experiments. In: International Workshop on Third Generation Data Mining: Towards Service-Oriented Knowledge Discovery (SoKD-2010), pp. 31–46 (2010)
25. Weibel, S.: The Dublin Core: a simple content description model for electronic resources. Bull. Assoc. Inf. Sci. Technol. **24**(1), 9–11 (1997)
26. Wilkinson, M., et al.: The fair guiding principles for scientific data management and stewardship. Sci. Data **3**, e0153507 (2016)

Data Streams

FABBOO - Online Fairness-Aware Learning Under Class Imbalance

Vasileios Iosifidis(✉) and Eirini Ntoutsi

L3S Research Center, Leibniz University of Hannover, Hanover, Germany
{iosifidis,ntoutsi}@l3s.de

Abstract. Data-driven algorithms are employed in many applications, in which data become available in a sequential order, forcing the update of the model with new instances. In such dynamic environments, in which the underlying data distributions might evolve with time, fairness-aware learning cannot be considered as a one-off requirement, but rather it should comprise a continual requirement over the stream. Recent fairness-aware stream classifiers ignore the problem of class distribution skewness. As a result, such methods mitigate discrimination by "rejecting" minority instances at large due to their inability to effectively learn all classes. In this work, we propose FABBOO, an online fairness-aware approach that maintains a valid and fair classifier over a stream. FABBOO is an online boosting approach that changes the training distribution in an online fashion based on both stream imbalance and discriminatory behavior of the model evaluated over the historical stream. Our experiments show that such long-term consideration of class-imbalance and fairness are beneficial for maintaining models that exhibit good predictive- and fairness-related performance.

Keywords: Data streams · Fairness-aware classification · Class-imbalance

1 Introduction

Data-driven decision support systems have become a necessity nowadays for many applications where huge amounts of historical data are available for analysis. Their performance in many tasks is comparable or has even surpassed human performance [15] and therefore, for many processes, human decisions are substituted by algorithmic ones. Such a replacement, however, has raised a lot of concerns [4] regarding the fairness, accountability and transparency of such methods in domains of high societal impact such as *risk assessment*, *recidivism*, *predictive policing*, etc. For example, Google's *AdFisher* online recommendation tool showed significantly more highly paid jobs to men than women [10]. Many similar incidents of algorithmic unfairness have been reported in recent years [1,18,26].

As a result of the ever-increasing interest in issues of fairness and responsibility of data-driven systems, a large body of work exists already in fairness-aware learning [17,19–21,23–25,31]. Only a few recent works, however, investigate the problem of fair learning in non-stationary environments [22,30]. Nonetheless, these methods ignore an important aspect of the learning problem, namely that the majority of (streaming) datasets suffer from class-imbalance. Class imbalance refers to the disproportion

A. Appice et al. (Eds.): DS 2020, LNAI 12323, pp. 159–174, 2020.
https://doi.org/10.1007/978-3-030-61527-7_11

among classes i.e., when one class, called *minority* class, has significantly fewer examples than another class, called *majority* class. If the imbalance problem is not tackled, the learner mainly learns the majority class and strongly misclassifies/rejects the minority. Such methods might appear to be fair for certain fairness definitions that rely on parity in the predictions between the protected and non-protected groups. In reality though the low discrimination scores are just an artifact of the low prediction rates for the minority class. This observation has been made in [21] but for the static case. We observe the same issue for the streaming case and propose an imbalance monitoring mechanism based on which we adapt the weighted training distribution.

Moreover, in a stream environment the decisions do not only have a short-term effect, but rather they might incur long-term effects. In case of discrimination, this means that discriminatory model decisions affect not only the immediate outcomes, but they might also affect future outcomes [9]. For example, [9] indicates small wage gaps between college-educated blacks and whites when they are first hired, but the pay gap increased over the years. To this end, we propose to define discrimination cumulatively over the stream rather than based only on the most recent outcomes. This is in contrast to recent stream fairness-aware approaches that focus only on short term outcomes, e.g., [22]. Our experiments verify that when treating for *short-term* discriminatory outcomes, the *cumulative* effects can be substantially higher over time and therefore, a cumulative approach is better.

Our contributions are summarized as follows: i) we propose FABBOO, a fairness and class imbalance-aware boosting method that is able to tackle class-imbalance as well as mitigate different parity-based discriminatory outcomes, ii) we introduce the notion of *cumulative fairness* in streams, which accounts for cumulative discriminatory outcomes, iii) our experiments, in a variety of real-world and synthetic datasets, show that our approach outperforms existing approaches that either do not consider class-imbalance or are based on short-term fairness evaluation.

2 Basic Concepts and Problem Definition

Let X be a sequence of instances $x_1, x_2, \cdots$, arriving over time at timepoints $t_1, t_2, \cdots$, where each instance $x \in \mathbb{R}^d$. Similarly, let y be a sequence of corresponding class labels, such that each instance in X has a corresponding class label in y. Without loss of generality, we assume a binary classification problem, i.e., $y = \{+1, -1\}$, and we denote by y^+ (y^-) the positive (negative, respectively) segments. We denote the classifier by $f : X \rightarrow y$. We follow the *online learning setting*, where new instances from the stream are processed one by one. For each new instance x arriving at t, its class label $f_{t-1}(x)$ is predicted by the current model f_{t-1}. The true class label of the instance is revealed to the learner before the arrival of the next instance, and it is used for model updating, thus resulting into the updated model f_t. This setup is known as first-test-then-train or prequential evaluation [14].

We assume that the underlying stream distribution is *non-stationary*, that is, the characteristics of the stream might change with time leading to *concept drifts*, i.e., changes in the joint distribution so that $P_{t_1}(X, y) \neq P_{t_2}(X, y)$ for two different timepoints t_1 and t_2. We are particularly interested in real concept drifts, that is when

$P_{t_1}(y|X) \neq P_{t_2}(y|X)$, as such changes make the current classifier obsolete and call for model update. Moreover, we consider the scenario where the stream population is *imbalanced*, that is, one of the classes dominates the stream impacting the learning ability of the classifiers that traditionally tend to ignore the minority to foster generalization and avoid overfitting [29]. We, however, do not require that the minority class is predefined and fixed over the course of the stream. Instead, we assume that this role might alternate between the two classes.

We also assume the existence of a sensitive feature *SA*, e.g., gender or race, which is binary with values $SA = \{z, \overline{z}\}$, e.g., gender=$\{$female, male$\}$; we refer to z, $\overline{z}$ as protected, non-protected group respectively[1]. Traditional *fairness-aware classification* aims to learn a mapping $f : X \rightarrow y$ that accurately maps instances x to their correct classes without discriminating between the protected and non-protected groups. The discrimination is assessed in terms of some fairness measure. Formalizing fairness is a hard topic per se, and there has already been a lot of work in this direction. For example, [27] overview more than twenty measures of fairness; however, there is no clear indication which measure is the most appropriate for classification tasks. In this work, we investigate *parity-based* notions of fairness such as the well-known *statistical parity* [23] and *equal opportunity* [16]; however, FABBOO can accommodate various parity-based fairness notions such as *disparate mistreatment* [31], *predictive quality* [27], and so on.

Statistical parity (S.P.) measures the difference in the probability of a random individual drawn from $\overline{z}$ to be predicted as positive and the probability of a random individual drawn from the complement z to be predicted as positive:

$$S.P. = P(f(x) = y^+|\overline{z}) - P(f(x) = y^+|z) \tag{1}$$

The S.P. values lie in the $[-1, 1]$ range, with 0 meaning the decision does not depend on the sensitive value (aka fair), 1 meaning that the protected group is totally discriminated (aka discrimination), and -1 that the non-protected group is discriminated (aka reverse discrimination).

S.P. does not take into account the real class labels, and therefore may allow individuals to be assigned to the positive class, even though they do not satisfy the requirements, thus causing *reverse discrimination.* Equal opportunity (EQ.OP.) resolves this issue by measuring the difference in the True Positive Rates (TPR) between the two groups, i.e.:

$$EQ.OP. = P(f(x) = y^+|\overline{z}, y^+) - P(f(x) = y^+|z, y^+) \tag{2}$$

Similar to S.P., EQ.OP's values lie in the $[-1, 1]$ range.

Our work investigates the problem of fair classification in a stream environment. *Fairness-aware stream learning* refers to the problem of maintaining a valid and fair classifier over the stream. The term *valid* refers to the ability of the model to adapt to the underlying evolving population and deal with concept drifts. At the same time, the classifier should be fair according to the adopted S.P. or EQ.OP. fairness measures. Ensuring fairness is much harder in such an online environment comparing to the traditional batch setting. First, the model should be continuously updated to reflect the underling non-stationary population. The typically accuracy-driven update of the model

[1] SA definition could also be extended to cover feature combinations such as race and gender.

cannot ensure fairness, so even if the initial model was fair, its discriminatory behavior might get affected by the model updates. Second, small amounts of unfairness at each time point might accumulate into significant discrimination as the learner typically acts as an amplifier of whatever biases exist in the data and furthermore, reinforces its errors. So, model update should consider fairness constraints and long term effects of discrimination beyond the point of its evaluation.

3 Related Work

Static Fairness-Aware Learning: Static fairness-aware approaches have received a lot of attention over the recent years. Literature in this area can be categorized in: i) pre-processing methods [5,20,23], where data are processed, transformed, or augmented to reduce discrimination or remove the correlation between various attributes and the sensitive attribute. ii) In-processing methods [21,25,31] focus on facilitating a fairness notion into a model's objective function. iii) Post-processing methods [12,16,19] alter a model's predictions or adjust a model's decision boundary to reduce unfairness.

Stream Fairness-Aware Learning: Stream fairness-aware approaches aim to remove unfair outcomes when data are presented sequentially. In [22], authors present a chunk based stream classification approach in which they apply pre-processing methods, such as *label swapping*, to remove discrimination, which is measured by statistical parity, from data before updating an online classifier; however, this approach accounts for short-term outcomes. In [30], they incorporate the notion of statistical parity into Hoeffding's Tree split criterion so that it accounts for cumulative discriminatory outcomes.

Stream Learning: In stream learning, data arrive sequentially and their distributions can change over time, the so-called concept drifts [14]. Concept drifts can be handled explicitly through *informed adaptation*, where the model adapts only if a change has been detected, or implicitly through *blind adaptation*, where the model is updated constantly to account for changes in the underlying data distributions. In addition, models developed for stream learning are categorized as *incremental* and *online* [28]. Incremental models are trained in batches [13], with the help of a chunk (window), while online models are updated continuously to accommodate newly incoming examples [7].

The goal of this paper is to highlight the importance of class-imbalance problem in fairness-aware stream learning; therefore, we select as competitors fairness-aware stream learners [22,30] and omit class-imbalance stream learners.

4 Online Fairness- and Class Imbalance-Aware Boosting

An overview of FABBOO, standing for online fairness and class imbalance-aware boosting, is shown in Fig. 1. Our method consists of a *class-imbalance monitoring component* that keeps track of the class ratios over the stream and adjusts the weights of the new training instances accordingly to ensure that the learner properly learns both classes (Sect. 4.1), while adapting to concept drifts via *blind model adaptation* [14]. In addition, the *cumulative* discriminatory behavior of the learner is monitored, and when it exceeds a user-defined tolerance threshold ϵ, the decision boundary is adjusted to ensure that the learner does not incur discrimination (Sect. 4.2).

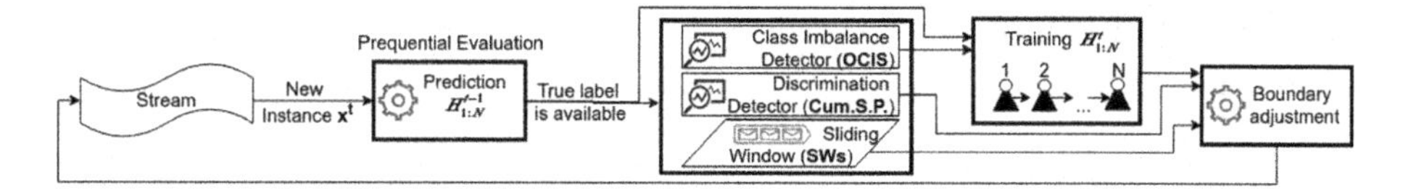

Fig. 1. An overview of FABBOO

4.1 Online Monitoring of Class Imbalance and Model Update

In evolving data streams, the role of minority and majority classes can exchange and what is now considered to be minority might turn later into a majority class or vice versa [28]. Knowing the class ratio over the stream is important for our method as it directly affects the instance weighting during training. Therefore, we keep track of the stream imbalance using the online class imbalance monitor (OCIS) of [28].

$$OCIS_t = W_t^+ - W_t^- \tag{3}$$

where W_t^y is the percentage of class y at timepoint t maintained in an online fashion. In particular, upon the arrival of a new instance x at timepoint t, the percentage of a class y is updated as follows:

$$W_t^y = \lambda \cdot W_{t-1}^y + (1 - \lambda) \cdot \mathbb{I}[(y_t, y)] \tag{4}$$

where $\lambda \in [0, 1]$ is a user-defined decay factor that controls the extent to which old class percentage information should be considered, and $\mathbb{I}[(y_t, y)]$ is an identity function which equals to 1 if the true class label of x_t is y, otherwise 0.

The imbalance index OCIS takes values in the $[-1, 1]$ range, with 0 indicating a perfectly balanced stream and -1 or 1 indicating the total absence of one class.

Model Adaptation: Our basic model is OSBoost [7] that generates smooth distributions over the training instances, and guarantees to achieve small error if the number of weak learners and training instances is large enough. We extend OSBoost to take into account class imbalance by changing the weighted instance distribution so that minority instances become more prominent during the training process.

The pseudocode of the algorithm is shown in Algorithm 1. OSBoost comes with a set of predefined parameters: $\gamma \in [0, 1]$ that is an online analog of the "edge" of the weak learning oracle, and $N \in \mathbb{Z}^+$ that is the number of online weak learners. Upon the arrival of a new instance x at timepoint t, the class imbalance status is updated (line 2) according to Eq. 3. Then, the weak learners are updated sequentially (lines 4–11) so that the predictions of model H_i^t (line 6) affect the training of its successor model H_{i+1}^t by changing the weight/contribution of instance x to the model accordingly. The weight of instance x is tuned per learner H_i^t based on the error of the predecessor model H_{i-1}^t on x, but also based on current class imbalance (lines 8–11).

To summarize, traditional OSBoost performs error-based instance weight tuning but does not adjust for class-imbalance. On the contrary, FABBOO adjusts the instance weights also based on the dynamic class ratio (c.f. Eq. 3) so that minority instances receive extra "boosting" during training. Note that if the stream is balanced, i.e., $W_t^+ - W_t^- \approx 0$, the weights are only slightly affected.

Algorithm 1 FABBOO training procedure

1: **procedure** TRAIN($x_t, y_t, \gamma, H^{t-1}_{1:N}$) ▷ x_t: newly arrived instace, y_t: label of x_t, γ: learning rate, $H^{t-1}_{1:N}$: current ensemble
2: $OCIS_t = W^+_t - W^-_t$ ▷ Update the class imbalance status
3: $w_1 = 1, q_0 = 0$
4: **for** $i = 1$ to N **do**
5: Train H^t_i on x_t with weight w_i
6: $q_i = q_{i-1} + y_t \cdot H^t_i(x_t) - \frac{\gamma}{2+\gamma}$
7: $w_{i+1} = min\{(1-\gamma)^{q_i/2}, 1\}$
8: **if** $x_t \in y_+$ and $OCIS_t < 0$ **then** ▷ y^+ is minority at timepoint t
9: $w_{i+1} = \frac{w_{i+1}}{1+OCIS_t}$
10: **if** $x_t \in y^-$ and $OCIS_t > 0$ **then** ▷ y^- is minority at timepoint t
11: $w_{i+1} = \frac{w_{i+1}}{1-OCIS_t}$
12: **return** updated ensemble $H^t_{1:N}$

4.2 Online Monitoring of Cumulative Fairness and Boundary Adjustment

Methods which restore fairness only on *short-term* (recent) outcomes fail to mitigate discrimination over time as discrimination scores that might be considered negligible when evaluated individually (i.e., at a single time point) might accumulate into significant discrimination in the long run [9]. In this work, we aim to mitigate cumulative discrimination accumulated from the beginning of the stream in order to remove such long term discriminatory effects and adjust the decision boundary not only based on the recent behavior of the model, but rather on its historical performance.

Cumulative fairness monitoring accounts for discriminatory outcomes from the beginning of the stream until time point t. We introduce the cumulative fairness notion for non-stationary environments w.r.t. statistical parity and equal opportunity as follows:

Definition 1. *Cumulative Statistical Parity (Cum.S.P.)*

$$\frac{\sum_{i=1}^{t} 1 \cdot \mathbb{I}[f_i(x_i) = y^+ | x_i \in \bar{z}]}{\sum_{i=1}^{t} 1 \cdot \mathbb{I}[x_i \in \bar{z}] + l} - \frac{\sum_{i=1}^{t} 1 \cdot \mathbb{I}[f_i(x_i) = y^+ | x_i \in z]}{\sum_{i=1}^{t} 1 \cdot \mathbb{I}[x_i \in z] + l}$$

Definition 2. *Cumulative Equal Opportunity (Cum. EQ.OP.)*

$$\frac{\sum_{i=1}^{t} 1 \cdot \mathbb{I}[f_i(x_i) = y^+ | x_i \in \bar{z}, y_i^+]}{\sum_{i=1}^{t} 1 \cdot \mathbb{I}[x_i \in \bar{z}, y_i^+] + l} - \frac{\sum_{i=1}^{t} 1 \cdot \mathbb{I}[f_i(x_i) = y^+ | x_i \in z, y_i^+]}{\sum_{i=1}^{t} 1 \cdot \mathbb{I}[x_i \in z, y_i^+] + l}$$

where parameter l is employed for correction in the early stages of the stream. Cum.S.P. or Cum.EQ.OP. are maintained online using incremental counters updated with the arrival of new instances from the stream, and therefore, it is appropriate for stream applications where typically random access to historical stream instances is not possible. The cumulative fairness notions are employed by FABBOO for discrimination monitoring. When their values exceed a user-defined discrimination tolerance threshold ϵ, the decision boundary should be adjusted i.e., $Cum.S.P. > \epsilon$ or $Cum.EQ.OP. > \epsilon$.

Decision Boundary Adjustment: Post-processing adjustment of the decision boundary for discrimination elimination has been investigated in the literature, e.g., [12, 16]. Closer to our approach is [12], where the authors adjust the decision boundary of an AdaBoost classifier based on the (sorted) confidence scores of misclassified instances of the protected group. However, in contrast to [12], we deal with stream classification, and therefore, we do not have access to historical stream instances in order to adjust the boundary accurately. Except for the access-to-the-data constraint, another reason for not considering the whole history for the adjustment of the boundary is the non-stationary nature of the stream. In such a case, adjusting the boundary based on the whole history of the stream will hinder the ability of the model to adapt to the underlying data and will eventually hurt predictive performance.

To overcome this issue, we use a sliding window model of a pre-defined size M for the adjustment. In particular, we maintain a sliding window of size M for each segment to allow for boundary adjustment for different parity-based notions based on each discriminated segment. In the case of statistical parity or equal opportunity, the only relevant sliding window is the one for the protected positive segment (denoted by SW_z^+). The number of examples (n_t) which are needed in order to mitigate discrimination at timepoint t is given by:

$$n_t = \left\lfloor \sum_{i=1}^{t} 1 \cdot \mathbb{I}[x_i \in z] \cdot \frac{\sum_{i=1}^{t} 1 \cdot \mathbb{I}[f_i(x_i) = y^+ | x_i \in \bar{z}]}{\sum_{i=1}^{t} 1 \cdot \mathbb{I}[x_i \in \bar{z}]} - \sum_{i=1}^{t} 1 \cdot \mathbb{I}[f_i(x_i) = y^+ | x_i \in z] \right\rfloor \quad (5)$$

Similar to statistical parity, to estimate the number of examples (n_t) for equal opportunity, we follow the same logic:

$$n_t = \left\lfloor \sum_{i=1}^{t} 1 \cdot \mathbb{I}[x_i \in z, y_i^+] \cdot \frac{\sum_{i=1}^{t} 1 \cdot \mathbb{I}[f_i(x_i) = y^+ | x_i \in \bar{z}, y^+]}{\sum_{i=1}^{t} 1 \cdot \mathbb{I}[x_i \in \bar{z}, y_i^+]} - \sum_{i=1}^{t} 1 \cdot \mathbb{I}[f_i(x_i) = y^+ | x_i \in z, y^+] \right\rfloor \quad (6)$$

Afterwards, the misclassified instances in SW_z^+ are sorted based on the confidence scores in a descending order. The decision boundary is adjusted according to the n^t-th instance of the sorted window (SW_z^+). In particular, if θ^{t-1} is the decision boundary value (original value θ^0 is 0.5) of the n_{t-1}-th, the fair-boundary is adjusted to θ^t. Note that in the early stage of the stream, where the sliding window does not contain a sufficient number of instances, the boundary is tweaked based on the misclassified instance with the highest confidence within the window.

4.3 FABBOO Classification

FABBOO is an online ensemble of sequential weak learners that tackles class imbalance and cumulative discriminatory outcomes in the stream. Moreover, FABBOO deals with concept drifts, through *blind adaptation*, by employing a base learner that is able to react to concept drifts. In particular, we employ Adaptive Hoeffding Trees (**AHT**) [3] as weak learners; AHT is a decision-tree induction algorithm for streams that ensures DT model adaptation to the underlying data distribution by not only updating the tree with new instances from the stream, but also by replacing sub-trees when their performance decreases.

The classification of a new unseen instance at time point t, i.e., x_t, is based on weighted majority voting and depends on its membership to z. If the instance does not belong to z (i.e., it is a non-protected instance), then the standard boundary of the ensemble is used. Otherwise, the adjusted boundary is used. More formally:

$$f_t(x_t) = \begin{cases} y^+ & if\ x_t \in z\ and\ H^t_{1:N}(x_t) \geq \theta^t \\ H^t_{1:N}(x_t) & \text{otherwise.} \end{cases} \tag{7}$$

where N is the number of weak learners of the ensemble, and θ^t is the fair adjusted boundary at timepoint t. For Cum.S.P. and Cum.EQ.OP., only the boundary of the protected group is tweaked. Other parity-based notions (such as *Disparate Mistreatment* [31]) may also tweak the boundary of the non-protected group. Note that the adjustment of the boundary based on θ^t is applied at the ensemble level and not at each individual weak learner predictions.

5 Evaluation

In this section, we introduce the employed baselines as well as variants of FABBOO[2] that help us to demonstrate the behavior of FABBOO's individual components. The employed datasets as well as the performance measures are given below. For the experimental evaluation, in order to get the best γ, λ and M parameters, we performed a grid-search and selected $\gamma = 0.1, \lambda = 0.9$, $M = 2,000$ that showed an overall good performance across all datasets. We also set $N = 20$ for all the ensemble methods and a very small value $\epsilon = 0.0001$, which means no tolerance to discriminatory outcomes. Finally, for the prequential evaluation of the non-stream datasets, we report on the average of 10 random shuffles (same as in [22,30]).

5.1 Competitors and Performance Measures

We evaluate FABBOO against two recent state-of-the-art fairness-aware stream classifiers [22,30] and the fairness agnostic non-stationary OSBoost [7]. We also employ two variations of FABBOO to show the impact of its different components, namely class-imbalance and cumulative fairness. All methods employ AHTs as weak learners and therefore are able to handle concept drifts. The only exception is FAHT [30] which is an incremental Hoeffding Tree that not tackle concept drifts. An overview follows:

1. **Fairness Aware Hoeffing Tree (FAHT)** [30]**:** FAHT is an extension of the Hoeffding tree that accounts for statistical parity by alternating the node split procedure to facilitate information as well as fairness gain (statistical parity). FAHT grows

[2] https://iosifidisvasileios.github.io/FABBOO.

Table 1. An overview of the datasets.

	#Instances	#Attributes	Sen.Attr.	z	$\bar{z}$	Class ratio (+:-)	Stream	Positive class	Source
Adult Cen	45,175	14	Gender	Female	Male	1:3.03	-	<50K	[2]
Bank	40,004	16	Marit. Status	Married	Single	1:7.57	-	subscription	[2]
Default	30,000	24	Gender	Female	Male	1:3.52	-	default payment	[2]
Kdd Cen	299,285	41	Gender	Female	Male	1:15.11	-	<50K	[2]
Loan	21,443	38	Gender	Female	Male	1:1.26	✓	paid	[8]
NYPD	311,367	16	Gender	Female	Male	1:3.68	✓	felony	[6]
Synthetic	150,236	6	synth.	synth.	synth.	1:3.13	✓	synth.	[22]

according to the joint split of information and fairness gain, thus accounts for cumulative outcomes; however, it does not handle concept drifts nor class-imbalance.

2. **Massaging (MS)** [22]**:** a chunk based model-agnostic approach which minimizes S.P. on recent discriminatory outcomes. It detects and removes discrimination within the chunk by performing label swaps and retrains the model based on the "corrected" chunk. MS is dealing with concept drifts by blind adaptation (using an adaptive learner), but is considering short-term discrimination outcomes and does not account for class imbalance. We use the default chunk size of 1,000 instances.
3. **Online Smooth Boosting (OSBoost)** [7]**:** OSBoost does not consider fairness nor class imbalance.
4. **Online Fair Imbalanced Boosting (OFIB):** A variation of FABBOO that does not account for class imbalance i.e., it does not use OCIS during training. This variation helps to show the importance of tackling class imbalance.
5. **Chunk Fair Balanced Boosting (CFBB):** A variation of FABBOO that tackles short-term, instead of cumulative, discrimination. This variation helps to show the importance of long term fairness assessment. Instead of accounting for discrimination from the beginning of the stream, it monitors the 1,000 most recent instances.

To evaluate the performance of FABBOO and baselines, we employ a set of measures which are able to show the performance in the presence of class-imbalance. Same as in [11], we employ *gmean*, *recall*, and *balanced accuracy* (Bal.Acc.). For measuring discrimination, we report on *cumulative statistical parity* in Sect. 5.3 and *cumulative equal opportunity* in Sect. 5.4.

5.2 Datasets

To evaluate FABBOO, we employ a variety of real-world as well as synthetic datasets which are summarized in Table 1. The datasets vary in terms of class imbalance, dimensionality and volume. Same as in [22,30], we use **Adult census** dataset (Adult) and **Kdd Census** dataset (Kdd Cen.) as well as **Bank** dataset, and **Default** dataset by randomly shuffling them, since they are not streaming datasets. We also employ **Loan**, **NYPD** and a **synthetic** dataset, all of which have temporal characteristics. For synthetic dataset, we follow the authors' initialization process [22], where each attribute corresponds to a different Gaussian distribution, and also inject class-imbalance and concept drifts to the stream. Concept drifts in this scenario are performed by shifting the mean average of

Table 2. Overall predictive and fairness performance for Cum.S.P. (Winner in bold)

	Method	Bal. Acc. (%)	Gmean (%)	Recall (%)	Cum.S.P. (%)
Adult	FAHT	72.14 ± 1.4	68.65 ± 2.2	50.11 ± 3.5	16.51 ± 1.3
	MS	72.31 ± 1.2	69.00 ± 1.8	50.91 ± 2.9	22.93 ± 1.6
	OSBoost	73.90 ± 0.5	71.11 ± 0.8	53.73 ± 1.3	18.05 ± 0.6
	OFIB	74.21 ± 0.3	72.92 ± 0.4	60.01 ± 1.0	0.26 ± 0.1
	CFBB	74.12 ± 0.6	73.74 ± 0.6	66.95 ± 1.1	−5.00 ± 2.0
	FABBOO	**76.58 ± 0.1**	**76.57 ± 0.1**	**73.98 ± 0.7**	**0.21 ± 0.1**
Bank	FAHT	61.92 ± 2.0	50.64 ± 4.4	26.51 ± 4.4	2.58 ± 0.5
	MS	63.21 ± 1.9	53.54 ± 3.6	29.75 ± 4.1	8.10 ± 1.2
	OSBoost	64.41 ± 0.6	55.54 ± 1.1	31.81 ± 1.3	3.37 ± 0.2
	OFIB	67.90 ± 0.7	62.04 ± 1.2	40.21 ± 1.6	**0.22 ± 0.1**
	CFBB	78.37 ± 0.5	78.08 ± 0.6	71.24 ± 1.5	−6.06 ± 1.3
	FABBOO	**83.39 ± 0.4**	**83.38 ± 0.4**	**83.36 ± 1.4**	**0.22 ± 0.1**
Default	FAHT	62.72 ± 0.6	53.48 ± 1.2	29.95 ± 1.4	1.80 ± 0.4
	MS	63.76 ± 0.5	55.53 ± 1.4	32.4 ± 2.0	12.16 ± 1.5
	OSBoost	63.06 ± 0.6	53.87 ± 1.3	30.32 ± 1.7	1.89 ± 0.4
	OFIB	63.79 ± 0.7	55.41 ± 1.6	32.36 ± 2.1	0.29 ± 0.1
	CFBB	65.82 ± 0.6	65.44 ± 0.4	58.58 ± 3.0	−7.74 ± 2.2
	FABBOO	**67.49 ± 0.6**	**66.89 ± 0.5**	**58.66 ± 2.8**	**0.17 ± 0.1**
Kdd Cen.	FAHT	62.80 ± 2.3	51.04 ± 4.6	26.45 ± 4.7	2.82 ± 0.6
	MS	62.02 ± 1.2	49.71 ± 2.3	24.91 ± 2.4	15.8 ± 0.97
	OSBoost	65.55 ± 0.8	56.28 ± 1.3	31.97 ± 1.5	3.62 ± 0.3
	OFIB	67.55 ± 0.9	60.48 ± 1.5	37.59 ± 1.9	0.13 ± 0
	CFBB	78.40 ± 0.5	77.58 ± 0.6	66.60 ± 1.1	1.34 ± 0.5
	FABBOO	**81.48 ± 0.3**	**81.41 ± 0.4**	**77.98 ± 0.6**	**0.04 ± 0**
Loan	FAHT	62.61	60.14	70.21	6.41
	MS	61.44	59.64	69.31	60.13
	OSBoost	**63.84**	**60.31**	76.13	8.14
	OFIB	62.41	58.34	78.63	1.12
	CFBB	63.15	60.05	79.73	−2.72
	FABBOO	63.47	60.22	**79.91**	**0.51**
NYPD	FAHT	50.15	6.13	0.37	0.09
	MS	56.93	41.06	17.47	5.87
	OSBoost	52.24	24.33	6.01	0.75
	OFIB	52.32	24.96	6.36	0.05
	CFBB	62.48	59.48	43.63	−6.46
	FABBOO	**62.96**	**60.78**	**46.83**	**0.03**
synthetic	FAHT	57.12	42.56	18.90	8.31
	MS	62.43	53.81	30.90	15.26
	OSBoost	63.42	54.87	31.61	7.97
	OFIB	64.01	57.54	35.85	**−0.56**
	CFBB	65.93	64.75	53.75	−9.68
	FABBOO	**69.09**	**69.01**	**60.11**	0.66

each Gaussian distribution (5 non-reoccuring concept drifts have been inserted at random points, see Fig. 2 or 3).

5.3 Results on Cumulative Statistical Parity

In this section, we compare our approach against the employed competitors for Cum.S.P., and report the overall results in Table 2. As we see, FABBOO is able to mitigate unfair outcomes and maintain the best performance in terms of balanced accuracy, gmean, and recall for all datasets. E.g., for *Adult Cen.*, the best balanced accuracy is achieved by FABBOO followed by OFIB (2.3%↓), the best gmean is achieved by FABBOO followed by CFBB (2.8%↓), and the best recall is achieved by FABBOO followed by CFBB (7%↓). OFIB is able to reduce discrimination, same as FABBOO, in expense of sacrificing 2.3%↓ balanced accuracy.

Overall, FABBOO achieves the best balanced accuracy, across all datasets, with an average score of 72.01%, followed by CFBB with an average score of 69.73%. In terms of discrimination, FABBOO is the clear winner, across all datasets, with an average score of 0.26%, followed by OFIB with an average score of 0.37%. Although the difference in terms of discrimination is small, OFIB has an average balanced accuracy

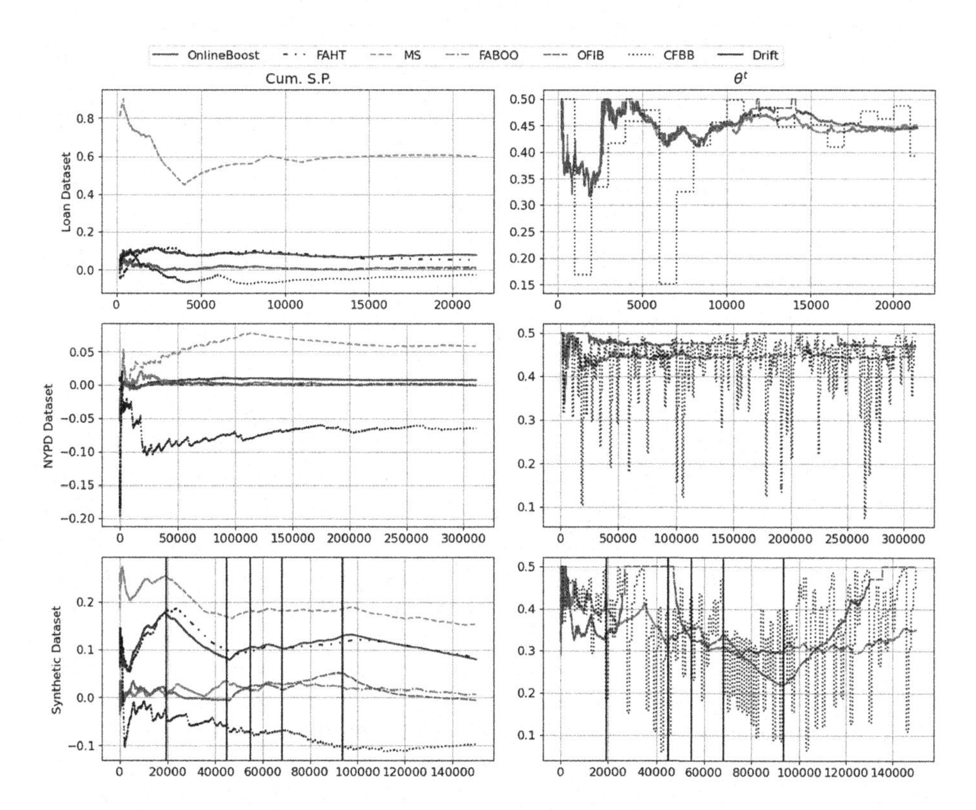

Fig. 2. Cum.S.P. and boundary adjusting for Loan (top), NYPD (middle) and Synthetic (bottom) datasets

score of 64.57%. CFBB achieves an average score of 5.57% in terms of Cum.S.P, while FAHT and MS achieve an average score of 5.49% and 20.02%, respectively.

To get a closer look at the over time performance of the different methods, we show in Fig. 2 the Cum. S.P. (left) and the required decision boundary adjustment (right), i.e., the boundary threshold θ^t, for the datasets with temporal information. Looking at the Cum.S.P.(left), we see that for all datasets, CFBB is not able to mitigate discrimination; instead, it propagates *reverse discrimination* (negative Cum. S.P.) and discriminates the non-protected group. MS falls in the same pitfall; by "correcting" the data based solely on the chuck it is not able to tackle unfair cumulative outcomes. Both CFBB and MS results show that a short-term consideration of fairness is unable to tackle discrimination propagation and reinforcement in the stream. The fairness-agnostic OSBoost is also not able to tackle discrimination. The only exception is the *NYPD* dataset. However, a closer look shows that the achieved low S.P. is only a result of vast rejecting the minority class (c.f., Table 2). On the other hand, FABBOO and OFIB (the FABOO variation that does not tackle class-imbalance) are able to tackle discrimination overtime, and outperform FAHT and MS.

Looking at the required adjustments of the decision boundary (right), we notice that OFIB tends to produce higher boundary values than FABBOO. This is caused due to OFIB's inability to learn the minority class effectively; therefore, it rejects more minority instances from both protected and non-protected groups. For Loan dataset FABOO and OFIB are performing similarly since the dataset is not severely imbalanced. Finally, we observe that CFBB has high fluctuation when adjusting the decision boundary due to its inability to adapt to underlying changes in data distributions w.r.t. fairness.

5.4 Results on Cumulative Equal Opportunity

For Cumul. EQ.OP., we report the results of OSBoost, OFIB, CFBB, and FABOO on Table 3. We exclude FAHT and MS since they are designed to mitigate unfair outcomes based on statistical parity. To the best of our knowledge, there are no fairness-aware stream learning methods that mitigate unfair outcomes based on equal opportunity.

The results indicate that FABBOO performs good in terms of balanced accuracy, gmean, and recall in all datasets except Compass and Loan, which are balanced datasets. E.g., for Adult Cen. dataset, the best balanced accuracy is achieved by FABBOO followed by CFBB (2%↓), the best Gmean is achieved by FABBOO followed by CFBB (2.9%↓), and the best recall is achieved by FABBOO followed by CFBB (7.9%↓). OFIB achieves slightly better Cumul. EQ.OP. than FABBOO (0.01%↓), however OFIB rejects more instances in the positive class. Similar behavior can be observed in all datasets, where FABBOO is able to tackle class imbalance and mitigate unfair outcomes better than the other methods. OSBoost fails to learn the positive (minority) class, thus under-performs in almost all datasets. In some cases, it produces low discriminatory outcomes; however, this is a result of misclassifying huge portions of the positive class.

We also demonstrate how Cumul. EQ.OP. and the decision boundary (FABBOO, OFIB and CFBB) vary over time for the stream datasets in Fig. 3. In all datasets, we observe that CFBB's decision boundary is highly fluctuating in contrast to OFIB and FABBOO. CFBB is also unstable in terms of Cumul. EQ.OP., since it is not mitigating

Table 3. Overall predictive and fairness performance for Cum.EQ.OP. (Winner in bold)

	Method	Bal.Acc. (%)	Gmean (%)	Recall (%)	Cum.EQ.OP. (%)
Adult	OSBoost	73.90 ± 0.5	71.13 ± 0.8	53.73 ± 1.3	18.41 ± 3.2
	OFIB	74.74 ± 0.5	72.36v0.7	56.07 ± 1.1	**3.14 ± 1.4**
	CFBB	76.70 ± 0.4	75.46 ± 0.5	62.96 ± 1.1	9.34 ± 1.6
	FABBOO	**78.71 ± 0.2**	**78.38 ± 0.3**	**70.83 ± 0.9**	3.27 ± 1.6
Bank	OSBoost	64.41 ± 0.6	55.54 ± 1.1	31.81 ± 1.3	5.51 ± 1.1
	OFIB	65.42 ± 0.6	57.46 ± 1.2	34.17 ± 1.4	1.5 ± 0.8
	CFBB	76.74 ± 0.8	75.27 ± 1.1	61.88 ± 2.0	−1.85 ± 1.2
	FABBOO	**82.58 ± 0.5**	**82.44 ± 0.5**	**78.05 ± 1.7**	**0.1 ± 0.6**
Compass	OSBoost	**65.25 ± 0.3**	**64.91 ± 0.4**	58.74 ± 1.4	29.81 ± 1.7
	OFIB	64.58 ± 0.2	64.53 ± 0.2	62.51 ± 1.5	4.84 ± 2.5
	CFBB	64.76 ± 0.4	64.69 ± 0.4	62.07 ± 1.3	14.73 ± 3.3
	FABBOO	64.52 ± 0.3	64.50 ± 0.3	**64.40 ± 1.5**	**4.76 ± 2.9**
Default	OSBoost	63.06 ± 0.6	53.87 ± 1.3	30.32 ± 1.7	79.01 ± 0.9
	OFIB	63.14 ± 0.6	54.06 ± 1.5	30.57 ± 1.8	**0.26 ± 0.6**
	CFBB	66.61 ± 0.3	65.75 ± 0.4	56.31 ± 2.6	−2.21 ± 0.9
	FABBOO	**67.55 ± 0.5**	**66.78 ± 0.5**	**57.79 ± 2.7**	0.93 ± 0.7
Kdd Cen.	OSBoost	65.55 ± 0.8	56.28 ± 1.3	31.97 ± 1.5	15.99 ± 0.3
	OFIB	66.85 ± 0.8	58.88 ± 1.3	35.21 ± 1.6	0.83 ± 0.2
	CFBB	78.52 ± 0.5	77.30 ± 0.7	64.75 ± 1.2	2.72 ± 0.9
	FABBOO	**82.39 ± 0.4**	**82.16 ± 0.4**	**76.26 ± 0.5**	**0.6 ± 0.3**
Loan	OSBoost	**63.84**	**60.31**	76.13	1.25
	OFIB	61.51	58.59	78.31	0.12
	CFBB	62.61	59.84	79.03	12.89
	FABBOO	63.06	60.18	**80.73**	**0.07**
NYPD	OSBoost	52.24	24.33	6.01	1.25
	OFIB	52.31	24.75	6.22	0.12
	CFBB	62.17	58.84	42.08	12.89
	FABBOO	**62.65**	**60.38**	**45.92**	**0.07**
synthetic	OSBoost	63.42	54.87	31.61	5.18
	OFIB	63.75	56.67	34.55	−6.04
	CFBB	66.97	65.02	50.92	−18.10
	FABBOO	**69.13**	**68.17**	**57.68**	**−0.17**

cumulative unfair outcomes. OFIB tweaks the boundary less than FABBOO, while it fails to learn the minority class well enough, thus rejects more positive instances.

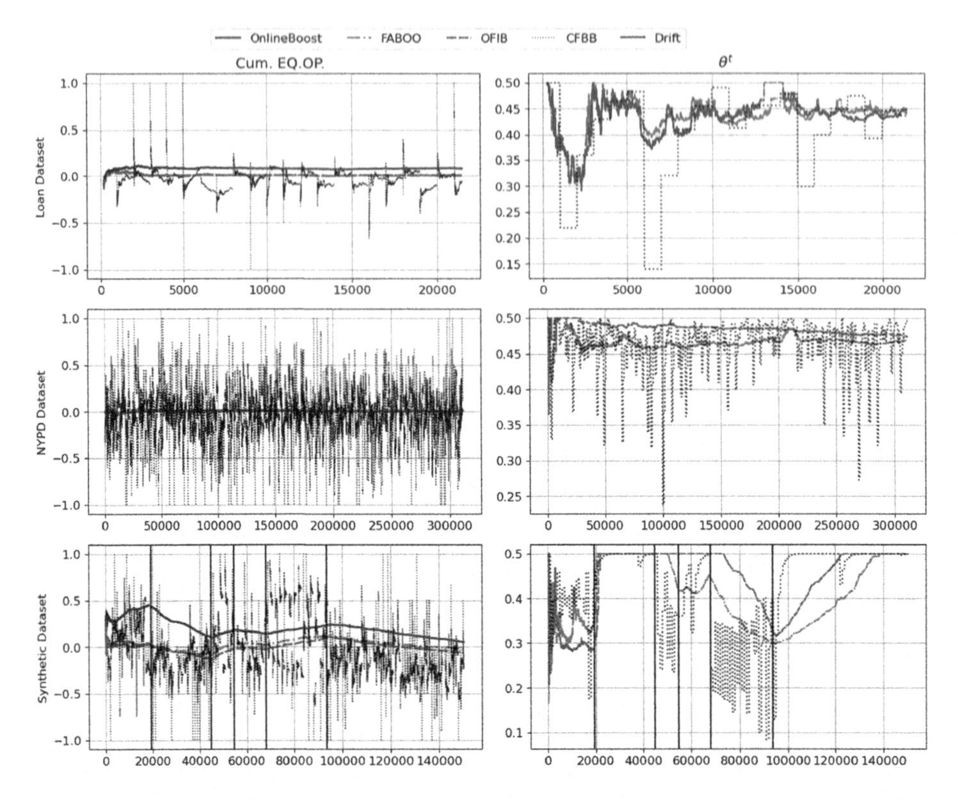

Fig. 3. Cum.EQ.OP. and boundary adjusting for Loan (top), NYPD (middle) and Synthetic (bottom) datasets

6 Conclusion

In this paper, we proposed FABBOO, an online fairness-aware learner for data streams with class imbalance and concept drifts. Our approach changes the training distribution online taking into account class-imbalance. Moreover, our method can facilitate different fairness notions by adjusting the decision boundary on demand. Our experiments show that our approach outperforms other methods in a variety of datasets w.r.t. both predictive- and fairness-performance. In addition, we show that recent fairness-aware methods reject the minority class at large to ensure fair results. On the contrary, our class-imbalance-oriented approach effectively learns both classes and fulfills different fairness criteria while achieving good predictive performance for both classes. Finally, we show that our cumulative definitions enable the model to mitigate long-term discriminatory effects, in contrast to a short-term definition like in CFBB and MS which are unable to deal with discrimination propagation and reinforcement in the stream. As part of our future work, we plan to embed the decision boundary adjustment directly into the training phase by altering the weighted training distribution, as proposed in [21]. Finally, we have assumed that the role of the minority class is not fixed over the stream;

however, we have assumed that the protected group is fixed over the stream. We intend to waive this assumption and extend FABBOO to tackle *reverse discrimination* as well.

References

1. Ali, M., et al.: Discrimination through optimization: how facebook's ad delivery can lead to skewed outcomes. arXiv preprint arXiv:1904.02095 (2019)
2. Bache, K., Lichman, M.: UCI machine learning repository (2013)
3. Bifet, A., Gavaldà, R.: Adaptive learning from evolving data streams. In: Adams, N.M., Robardet, C., Siebes, A., Boulicaut, J.-F. (eds.) IDA 2009. LNCS, vol. 5772, pp. 249–260. Springer, Heidelberg (2009). https://doi.org/10.1007/978-3-642-03915-7_22
4. Calders, T., Žliobaitė, I.: Why unbiased computational processes can lead to discriminative decision procedures. In: Custers, B., Calders, T., Schermer, B., Zarsky, T. (eds.) Studies in Applied Philosophy, Epistemology and Rational Ethics. Discrimination and Privacy in the Information Society, vol. 3, pp. 43–57. Springer, Heidelberg (2013). https://doi.org/10.1007/978-3-642-30487-3_3
5. Calmon, F., Wei, D., Vinzamuri, B., Ramamurthy, K.N., Varshney, K.R.: Optimized pre-processing for discrimination prevention. In: Advances in Neural Information Processing Systems, pp. 3992–4001 (2017)
6. Chapman, D., Ryan, P., Farmer, J.P.: Introducing alpha.data.gov. Office Sci. Technol. Policy (2013). www.whitehouse.gov/blog/2013/01/28/introducing-alphadatagov
7. Chen, S.T., Lin, H.T., Lu, C.J.: An online boosting algorithm with theoretical justifications. arXiv preprint arXiv:1206.6422 (2012)
8. Cortez, V.: Preventing discriminatory outcomes in credit models (2019). https://github.com/valeria-io/bias-in-credit-models
9. Council, N.R., et al.: Measuring Racial Discrimination. National Academies Press, Washington, DC (2004)
10. Datta, A., Tschantz, M.C., Datta, A.: Automated experiments on ad privacy settings. Priv. Enhancing Technol. **2015**(1), 92–112 (2015)
11. Ditzler, G., Polikar, R.: Incremental learning of concept drift from streaming imbalanced data. IEEE Trans. Knowl. Data Eng. **25**(10), 2283–2301 (2012)
12. Fish, B., Kun, J., Lelkes, Á.D.: A confidence-based approach for balancing fairness and accuracy. In: Proceedings of the 2016 SIAM International Conference on Data Mining, pp. 144–152. SIAM (2016)
13. Forman, G.: Tackling concept drift by temporal inductive transfer. In: Proceedings of the 29th Annual International ACM SIGIR Conference on Research and Development in Information Retrieval, pp. 252–259. ACM (2006)
14. Gama, J.: Knowledge Discovery from Data Streams. Chapman and Hall/CRC, New York (2010)
15. Grace, K., Salvatier, J., Dafoe, A., Zhang, B., Evans, O.: When will ai exceed human performance? evidence from ai experts. J. Artif. Intell. Res. **62**, 729–754 (2018)
16. Hardt, M., et al.: Equality of opportunity in supervised learning. In: Advances in Neural Information Processing Systems, pp. 3315–3323 (2016)
17. Hu, H., et al.: Fairnn-conjoint learning of fair representations for fair decisions. arXiv preprint arXiv:2004.02173 (2020)
18. Ingold, D., Soper, S.: Amazon Doesn't Consider the Race of Its Customers. Should It, Bloomberg (2016)
19. Iosifidis, V., Fetahu, B., Ntoutsi, E.: Fae: a fairness-aware ensemble framework. In: 2019 IEEE International Conference on Big Data (Big Data), pp. 1375–1380. IEEE (2019)

20. Iosifidis, V., Ntoutsi, E.: Dealing with bias via data augmentation in supervised learning scenarios.In: Bates, J., Clough, P.D., Jäschke, R., Otterbacher, J.: International Workshop on Bias in Information, Algorithms, and Systems (BIAS). Proceedings of the International Workshop on Bias in Information, Algorithms, and Systems (BIAS). CEUR Workshop Proceedings, pp. 24–29 (2018). http://ceur-ws.org/Vol-2103/#paper_5
21. Iosifidis, V., Ntoutsi, E.: Adafair: cumulative fairness adaptive boosting. In: CIKM (2019)
22. Iosifidis, V., Tran, T.N.H., Ntoutsi, E.: Fairness-Enhancing Interventions in Stream Classification. In: Hartmann, S., Küng, J., Chakravarthy, S., Anderst-Kotsis, G., Tjoa, A.M., Khalil, I. (eds.) DEXA 2019. LNCS, vol. 11706, pp. 261–276. Springer, Cham (2019). https://doi.org/10.1007/978-3-030-27615-7_20
23. Kamiran, F., Calders, T.: Data preprocessing techniques for classification without discrimination. Knowl. Inf. Syst. **33**(1), 1–33 (2012)
24. Kamiran, F., Mansha, S., Karim, A., Zhang, X.: Exploiting reject option in classification for social discrimination control. Inf. Sci. **425**, 18–33 (2018)
25. Krasanakis, E., Xioufis, E.S., Papadopoulos, S., Kompatsiaris, Y.: Adaptive sensitive reweighting to mitigate bias in fairness-aware classification. In: WWW, pp. 853–862. ACM (2018)
26. Vafa, K., Haigh, C., Leung, A., Yonack, N.: Price discrimination in the princeton review's online sat tutoring service. JOTS Technol, Sci (2015)
27. Verma, S., Rubin, J.: Fairness definitions explained. In: 2018 IEEE/ACM International Workshop on Software Fairness (FairWare), pp. 1–7. IEEE (2018)
28. Wang, S., Minku, L.L., Yao, X.: A learning framework for online class imbalance learning. In: 2013 IEEE Symposium on Computational Intelligence and Ensemble Learning (CIEL), pp. 36–45. IEEE (2013)
29. Weiss, G.M.: Mining with rarity: a unifying framework. ACM SIGKDD Explor. Newsl. **6**(1), 7–19 (2004)
30. Wenbin, Z., Ntoutsi, E.: Faht: an adaptive fairness-aware decision tree classifier. arXiv preprint arXiv:1907.07237 (2019)
31. Zafar, M.B., Valera, I., Gomez Rodriguez, M., Gummadi, K.P.: Fairness beyond disparate treatment & disparate impact: learning classification without disparate mistreatment. In: Proceedings of the 26th International Conference on World Wide Web, pp. 1171–1180. WWW (2017)

FEAT: A Fairness-Enhancing and Concept-Adapting Decision Tree Classifier

Wenbin Zhang[1(✉)] and Albert Bifet[2,3]

[1] University of Maryland, Baltimore County, MD 21250, USA
wenbinzhang@umbc.edu
[2] University of Waikato, Hamilton 3216, New Zealand
albert.bifet@waikato.ac.nz
[3] LTCI, Télécom Paris, Institut Polytechnique de Paris, Paris, France

Abstract. Fairness-aware learning is increasingly important in socially-sensitive applications for the sake of achieving optimal and non-discriminative decision-making. Most of the proposed fairness-aware learning algorithms process the data in offline settings and assume that the data is generated by a single concept without drift. Unfortunately, in many real-world applications, data is generated in a streaming fashion and can only be scanned once. In addition, the underlying generation process might also change over time. In this paper, we propose and illustrate an efficient algorithm for mining fair decision trees from discriminatory and continuously evolving data streams. This algorithm, called FEAT (Fairness-Enhancing and concept-Adapting Tree), is based on using the change detector to learn adaptively from non-stationary data streams, that also accounts for fairness. We study FEAT's properties and demonstrate its utility through experiments on a set of discriminated and time-changing data streams.

Keywords: AI ethics · Online fairness · Online classification

1 Introduction

Artificial Intelligence (AI)-based decision making systems are routinely being used in both online as well as offline settings to assist or even completely automate the decision-making. Yet, these automated data-driven tools may, even in the absence of intent, lead to a loss of fairness and accountability in the employed models. A plethora of such kind of AI-based discriminatory incidents have been observed and reported [1,2,7,12]. As a recent example, the AI algorithm behind Amazon Prime has suggested signs of racial discrimination when deciding which areas of a city are eligible for advanced services [13]. Areas densely populated by black people are excluded from services and amenities even though race is blind to the AI algorithm. Such incidents have sparked heated debate on the bias and

A. Appice et al. (Eds.): DS 2020, LNAI 12323, pp. 175–189, 2020.
https://doi.org/10.1007/978-3-030-61527-7_12

discrimination in AI decision systems, pulling in scholars from a diverse of areas such as philosophy, law and public policy.

The growing concern over discriminative behavior of AI models has motivated a number of approaches, ranging from defining discrimination to discrimination discovery and prevention for the development of AI tools that are discrimination-conscious by-design. Up to now, more than twenty notions have been proposed to measure the discriminative behavior of AI models [19]. One of the most widely used measures is the *statistical parity* [19] which examines whether the probability of being assigned a positive target class, for example allocating healthcare resources, is the same for both privileged and unprivileged groups. Formally put:

$$Discrimination(D) = \frac{PP}{PP + PN} - \frac{UP}{UP + UN} \tag{1}$$

where D is the labeled dataset, PP and PN refer to privileged community receiving positive and negative classification, respectively. So are UP and UN for unprivileged community. Here, the attribute that distinguishes privileged groups from unprivileged ones is referred as the *sensitive attribute* with the *sensitive value* defining the unprivileged community. Take "race" as the *sensitive attribute* for example, then the *sensitive value* is "black" and the positive class value as allocating healthcare resources. The four communities PP, PN, UP and UN therefore represent "non-black" being allocated healthcare resources, "non-black" being denied healthcare resources, "black" receives healthcare resources and "black" does not receive healthcare resources, respectively.

The aim of fairness-aware learning is then to train a decision model which provides accurate predictions, yet does not unduly bias against unprivileged groups. That is to say, from *statistical parity* point of view, equally granting a benefit to both privileged and unprivileged groups. While a large number of methods have been proposed to achieve this goal, most of them tackle fairness as a static problem. In many applications, however, data is generated sequentially and its characteristics might also evolve over time. Therefore, fairness-aware learning for such sort of applications should also be able to adapt to non-stationary distribution simultaneously.

Compared with the booming approaches in static settings, fairness-aware learning in data stream is highly under-explored because of its significant challenges [23]. To address this issue, this paper introduces a fairness-enhancing classifier that also equips with drift adaptation capability. The contribution of this paper is three-fold:

- We define the problem of fairness-aware learning in non-stationary data distribution. Then, we propose FEAT, a discrimination-conscious learner with add-on concept drift adaptation ability to handle discriminated and non-stationary data streams.
- We introduce fair-enhancing information gain that also accounts for the local discrimination to maximize the cumulative fairness, thus providing enhanced fairness-awareness learning.

- The conducted experiments verify the capability of the proposed model in online settings. *To the best of our knowledge, this is the first work that jointly addresses fairness and concept drift.*

The rest of the paper is organized as follows. Related studies are first reviewed in Sect. 2. We describe the proposed FEAT in Sect. 3 and discuss the experimental results in detail in Sect. 4. Finally, Sect. 5 concludes the paper.

2 Related Work

The tremendous societal importance of AI fairness has arose growing concern with ever increasing amount of discrimination-conscious models being proposed [1,2,24]. These approaches typically can be categorized into three main families: i) pre-processing approaches, ii) in-processing approaches and ii) post-processing approaches, based on whether they mitigate bias at the data level, the algorithm design or the output of model, respectively.

The first strategy, *pre-processing solutions*, consists of performing different data level operations such as transformation and augmentation to neutralize or eliminate the extent of inherited bias of the data. The rationale for such type of approaches is that classifiers trained on the fairly represented data could make fair predictions. These methods are model-agnostic and can be employed in conjunction with any applicable classifier after the pre-processing step. Representative works include massaging [15] and reweighting [5]. The former directly swaps the class labels of selected instances to change data distribution for the sake of balanced representation. The swapped instances are selected using a ranker based on the potential accuracy deterioration in order to minimize accuracy loss while reducing discrimination. While the latter, instead of intrusively relabeling the instances, assigns different weights to different communities to reduce discrimination. Instances belonging to the protected group will receive higher weighs comparing to instances from the unprotected group. In [14], these two methods have also been extended for online classification. However, methods in this category are typically not quite effective as standalone approaches unless being used in conjunction with other methods with sophisticated design.

In contrast, the second category, *in-processing approaches*, consists of modifying existing algorithms, usually integrating fairness as a part of the objective function through constraints or regularization, to mitigate discrimination, and is therefore algorithm-specific. [16] is one of the seminal in-processing works, in which discrimination, reflected by the entropy w.r.t. sensitive attribute, is incorporated into the splitting criterion for fair tree induction. In [20], the measure of "decision boundary fairness" is leveraged to penalize discrimination in the formulation of a set of convex margin-based classifiers. More recently, [23] improves the splitting strategy of [16] and operates their model in the online setting. However, research efforts in this direction have still been limited. Our work situates in this highly under-explored research direction to provide fair online decision-making.

The last category, *postprocessing techniques*, consists of either adjusting the decision boundary of a model or directly changing the prediction labels. [12] processes with additional prediction thresholds to work against discrimination while the decision boundary of AdaBoost is shifted w.r.t. fairness in [6]. The latter approaches pay attention to the outcome of a classifier. In [16], for example, relabeling is performed on selected leaves of the decision tree to decrease discrimination while minimizing the effect on predictive accuracy. We emphasize that transferring such techniques to online settings is not straightforward as the boundary/prediction could evolve themselves due to the non-stationary distributions in online settings.

Fairness in data streams further requires the addressing of non-stationary distributions, known as concept drift [4,10,22,25]. The learning algorithms therefore should be able to remain stable on previously learned and not outdated concepts while adapting to such drifts. The adaptation is typically enabled by learning incrementally from new instances [11,17] and by forgetting outdated information from the model [4,18]. A significant amount of work has been done with respect to this specific issue. However, the combined approach of addressing both fairness and concept drift has enjoyed relatively little research. Our work situates in this research direction to enable fairness-aware learning in non-stationary data streams.

3 FEAT: Fairness-Enhancing and Concept-Adapting Tree

This section first outlines the vanilla Hoeffding Tree (HT), then the reformulated fair information gain splitting criterion for fairness enhancement is introduced, followed by the adaption of changes in the example-generating process. A number of refinements and modifications that instantiate the fairness enhancement and concept-adapting learning are specified thereafter.

3.1 The Hoeffding Tree (HT) Classifier

Our Fairness-Enhancing and concept-Adapting hoeffding Tree (FEAT) is built on top of the Hoeffding Tree (HT) classifier [9]. To mine high-speed data stream, HT induces a decision tree from the given stream incrementally, briefly scanning each example in the stream only once and storing sufficient information in its leaves in order to grow. The crux decisions needed during the induction of the tree are when to split a node and with which example-discriminating test. To this end, the authors employ the Hoeddding bound [9] to guarantee that the tree learned probably converges to the conventional static tree built by a batch learner, given enough examples. In HT, these two decisions are based on the $informationgain$, which is exclusively accuracy-oriented and does not consider fairness. In addition, the construction of tree assumes the distribution generating examples does not change over time.

In this work, to enable fairness-aware learning and concurrently adapt to non-stationary data distributions, we extend the HT model in two ways: i) by introducing an enhanced fair splitting criterion that enables the fairness-aware learning (c.f., Subsect. 3.2) and ii) by adding the ability to detect and adapt to the evolution of underlying distribution (c.f., Subsect. 3.3).

3.2 The Fair-Enhancing Information Gain

The $informationgain$ (IG) [21] splitting criterion measures the uncertainty reduction due to a split during the tree construction. It is proposed purely from the data encoding perspective without considering fairness of the tree construction. To address the fairness-free issue of IG, previous studies reformulate the IG by incorporating the discrimination gain into the splitting criterion of the decision tree construction [16,23]. Inspired by these ideas, we propose the *fair-enhancing information gain (FEIG)* as follows,

$$FEIG(D,A) = \begin{cases} IG(D,A) & \text{, if } FEG(D,A) = 0 \\ IG(D,A) \times FEG(D,A) & \text{, otherwise} \end{cases} \tag{2}$$

where A is an attribute relative to the collection of instances D that stored in sufficient statistics, $D_v, v \in dom(A)$ are the partitions/subsets induced by A, and FEG refers to *fair-enhancing gain (FEG)* that measures the difference in discrimination due to the split and is formulated as:

$$FEG(D,A) = |Disc(D)| - \sum_{v \in dom(A)} |Disc(D_v)| \tag{3}$$

where each corresponding discrimination value *Disc* is gauged according to Eq. (1).

In *fair-enhancing gain*, different from the previous proposed fair splitting criteria [16,23], the gain in fairness is directly gauged according to the discrimination difference due to the split rather than entropy in regards to the sensitive attribute. In addition, in fairness-aware learning, it is expected that all groups being treated equally regardless of their population sizes. That is to say, discrimination is discrimination regardless the number of population being discriminated. To align with this idea, our splitting evaluation metric also cares for local discrimination to maximize the cumulative fairness by assigning equal weights to different discrimination representations. Specifically, each partition induced by the attribute A contributes equally to the cumulative fairness of A regardless the number and size of branches. In the general case, the higher reduction in discrimination the merrier, the *fair-enhancing gain* therefore would like a larger merit to be assigned when evaluating the fairness suitability of a candidate splitting attribute and ignores the number of its distinct values and of each specific value.

The FEG is then tied with IG through multiplication as the FEIG. Multiplication is favoured, when combining them as a conjunctive objective, over other operations for example addition as the values of these two metrics could be in different scales, and in order to promote fair splitting which results in a reduction in the discrimination after split, i.e., FEG is a positive value. In the end, this conjunctive metric would be used as the alternative fair-enhancing splitting criterion during the construction of the tree to enable discrimination-aware learning while maintaining predictive performance over the course of the stream.

3.3 The FEAT Algorithm

HT learns incrementally from the high-speed data streams by incorporating the incoming data in the stream into the model while simultaneously maintaining the performance of the classifier on the previous information. The tree is adapted, in practice grow, based on the newly available data in the stream and does not forget the obsolete concept that not following the current example-generating process. Therefore, HT assumes the distribution generating examples does not change over time and cannot adapt to the evolving example-generating process.

To overcome this drawback, we further extend HT and propose FEAT which maintains HT's capabilities of processing high speed data stream and data-driven encoding, also with enhanced fairness-aware learning by employing the previous introduced *fair-enhancing information gain* as well as the ability of change detection and concept forgetting.

To detect and react promptly to the evolution of the stream, FEAT keeps its model consistent with the example-generating process of the current stream, creates and replaces alternative decision subtrees when evolving data distribution is detected at a node. FEAT extends HT which is incremental, so the tree is adapted based on new instances. General speaking, the performance of such model, under stationary distribution without drift, improves over the course of the stream as it generalizes better after incorporating more examples into the model. Therefore, performance deterioration is a good indicator of drift. FEAT employs the sliding window size free ADWIN [3] to monitor the error rate of the non-leaf node and declare when branch replacement is necessary. ADWIN recomputes online whether two "large enough" subwindows of the most recent data exhibit "distinct enough" averages, and the older portion of the data is dropped when such distinction is detected. ADWIN therefore eases the burden of selecting a fixed window size that the distribution likely remains to be stationary within this window and adapts to the rate of change observed in the data itself. The use of ADWIN and the sketch of FEAT is shown in Algorithm (1).

FEAT grows similarly to HT (line 1–2 and 6–16). The difference is that HT depends on IG while FEAT employs FEIG to enable accuracy-oriented and fairness-enhanced construction of the tree. What's more, in order to keep the model it is learning in sync with changes in the example-generating process, FEAT continuously monitors the quality of old search decisions with respect to the latest instances from the data stream (line 17). FEAT creates an alternative subtree for each node that change in the underlying distribution is detected by ADWIN (line 19). Under the condition that an alternative subtree already exists, FEAT checks whether the alternative branch performs better than the old branch (line 21). The old branch will be replaced by the alternative one if so

Algorithm 1. The FEAT induction algorithm

Input: a discriminated data stream D,
confidence parameter δ,
tie breaking parameter τ.

FEAT(D, δ, τ)
1: Let $FEAT$ be a tree with a single leaf (the root)
2: Init sufficient statistics at root
3: **for** each instance x in D **do**
4: FEATGrow(x, FEAT, δ, τ)
5: **end for**

FEATGrow(x, $FEAT$, δ, τ)
1: Sort example into leaf l using $FEAT$
2: Update sufficient statistics in l and nodes traversed in the sort
3: **for** traversed node that has an alternate tree T_{alt} **do**
4: FEATGrow(x, T_{alt}, δ, τ)
5: **end for**
6: **if** examples seen at l are not all of the same class **then**
7: Calculate $FEIG_l(A_i)$ for each attribute according to Equation (2)
8: Let A_a be the attribute with highest $FEIG_l$
9: Let A_b be the attribute with second-highest $FEIG_l$
10: Compute Hoeffding bound $\varepsilon = \sqrt{\frac{R^2 \ln(1/\delta)}{2n_l}}$
11: **if** $A_a \neq A_\emptyset$ and $(FEIG_l(A_a) - FEIG_l(A_b) > \epsilon$ or $\epsilon < \tau)$ **then**
12: **for** each branch of the split **do**
13: Start a new leaf and initialize sufficient statistics
14: **end for**
15: **end if**
16: **end if**
17: **for** non-leaf node that its $ADWIN$ detects change **do**
18: **if** T_{alt}== null **then**
19: Create an alternative subtree T_{alt}
20: **else**
21: **if** T_{alt} is more accurate **then**
22: replace current node with its T_{alt}
23: **else**
24: prune its T_{alt}
25: **end if**
26: **end if**
27: **end for**

(line 22), otherwise the alternative branch will be pruned (line 24). Compared to HT, FEAT also maintains sufficient statistics of the nodes traversed in the sort in order to update alternative branches (line 3–5). The learning process is therefore fairness-enhancing and concept-adapting.

3.4 The FEAT System

Our FEAT induction algorithm is built on top of the HT classifier. FEAT therefore still holds HT's theoretical guarantees and theorems can be proven accordingly. Moreover, FEAT aims at enhancing fairness-aware learning while optimizing predictive performance by alleviating the discrimination bias towards the unprivileged group through the proposed fair splitting criterion, the *fair-enhancing information gain* (Subsect. 3.2), and by equipping itself with the ability of change detection and concept forgetting (Subsect. 3.3). The modifications and refinements being included to Algorithm (1) to instantiate the fairness-enhancing and concept-adapting learning over streams are discussed hereafter.

Pre-pruning. HT detects the case of not splitting a node benefits more than splitting by considering the merit of no split, represented by the null attribute $X_\emptyset$ at each node to enable pre-pruning. A node is thus only allowed to split when the candidate attribute is sufficiently better, according to the same Hoeffding bound test that determines differences among other attributes, than $X_\emptyset$. In the implementation of FEAT, the merit to be maximized is the previous introduced FEIG. Thus, the FEIG of the best split found should be sufficiently better than $X_\emptyset$'s. In terms of the FEIG of the null attribute, the current level of class distribution and discrimination are used to represent IG and FEG, respectively.

Sufficient Statistics. HT briefly inspects each instance in the stream only once and store sufficient information in the leaves to enable the calculation of the splitting merit afforded by each possible split. In FEAT, the statistics required for the calculation of FEIG should also be maintained. For the *discrete attributes*, each node in the tree maintains a separate table per attribute, containing the counts of the class labels that apply for each attribute value for the calculation of IG, and the counts of unprivileged group and privileged group as well as receiving positive classification in unprivileged group and privileged group that apply for each attribute value for the calculation of FEG. The learning process updates appropriate entries based on the attribute value, sensitive attribute value and class of the examples over the stream accordingly. As for the *numeric attribute*, FEAT maintains a separate Gaussian distribution per class label that apply for each attribute. So are the four previous mentioned FEG calculation related statistics. The appropriate distribution statistics is updated according to the sensitive attribute value and class of the examples over the stream. The most appropriate binary split point for each distribution is evaluated based on the allowing test and the merit of each allowed threshold candidate is also calculated according to the proposed FEIG. With the selected split points, the weight of values to their either side are approximated for each class and four FEG calculation related statistics, and the FEIG merit of each *numericattribute* candidate is thus computed from these weights.

Memory Management. Efficient storage of the sufficient statistics is crucial in stream environment. In case of the non-leaf node, FEAT prunes the alternative branch if its performance is inferior to the old one. FEAT also reduces the size of the sufficient statistics in each leaf by removing poor attributes when their FEIG is less than the current best attribute by more than the Hoeffding bound. The rationale is that, according to the bound, such attributes are unlikely to be selected in that leaf. In addition, assuming there are d attributes with a maximum number of v values per attribute and c possible classes in total, the required memory of FEAT is $O((d+2)vc)$ compared to the $O(dvc)$ of HT. FEAT therefore incurs negligible extra costs especially when $d \gg 2$.

4 Experimental Evaluation

In this section, we conduct experiments to evaluate the accountability and fairness of the proposed discrimination-aware data stream learner. To this end, we first investigate the enhanced discrimination reduction capability of the proposed fair splitting criterion. We also show a comprehensive quantitative evaluation to verify the concept adaptation capability of our approach.

4.1 Dataset

Contrary to a growing body of discrimination-conscious approaches motivated by the increasing attentive AI fairness, related datasets and benchmarks are still in a shortage [1]. With respect to the highly under-explored online fairness, this challenge is further magnified by the drift and the demanding requirement of the number of instances contained therein. We evaluate our approach on the datasets used in the recent work of this research direction [23], the *Adult* and the *Census* datasets [8] both targeting on identical learning task of determining whether a person earns more than 50K dollars per annum.

There are 48,843 instances in the *Adult* dataset and each instance is described by 14 employment and demographic attributes (attribute "fnlwg" is removed as suggested). We follow the same options in [23] by setting "gender" as the sensitive attribute with sensitive value equals to "female" being the protected group. The positive class is people making an annual income of more than 50K dollars. The *Census* dataset is significantly bigger in size including 299,285 instances and 41 attributes. It has an identical prediction task as the *Adult* dataset. So are the setting of sensitive attribute, sensitive value and positive classification. The intrinsic discrimination levels, according to Eq. (1), of the these two datasets are 19.45% and 7.63%, respectively.

Existing works mostly address these two datasets from the static learning perspective [19,24,27]. In our experiments we randomize the order of the instances then process them in sequence to simulate discriminated data streams, following [23]. The prequential evaluation [10] is employed in which each incoming instance is first being predicted upon arrival then is available for model training.

4.2 Justification of FEIG

The proposed FEIG is designed to enhance the learning idea of all groups being treated equally regardless of their population sizes for fair-enhancing learning. To validate this enhanced fairness-aware learning, we incorporate FEIG into the model proposed in [23] denoted as FAHT+ and FEAT- representing FEAT driven by the splitting criterion proposed in [23] and compare them respectively. We further incorporates the discrimination-aware splitting criterion of [16] into our model in replacing of FEIG, referred as Kamiran's. We do not incorporate FEIG into their model as it is designed for offline setting. Our motivation for using the identical classifiers is that, since our main interest at this stage is to compare the fair-enhancing learning of FEIG with other discrimination-aware splitting criteria, we would like to minimize the influences on the results from the bias of classifiers due to their versatile difference. The obtained results are shown in Table 1.

Table 1. Accuracy-vs-discrimination between FEIG and other discrimination-aware splitting criteria. Percentage in parenthesis is the relative difference over the performance of its corresponding comparing method.

Methods	Metric			
	Adult dataset		Census dataset	
	Discrimination	Accuracy	Discrimination	Accuracy
FAHT	16.29%	81.83%	3.20%	94.28%
FAHT+	15.62% (**−4.11%**)	81.01% (**−1.0%**)	2.61% (**−18.44%**)	92.82% (**−1.55%**)
FEAT-	19.14%	83.76%	2.20%	94.14%
FEAT	15.26% (**−20.27%**)	84.01% (**+0.3%**)	1.25% (**−43.18%**)	95.03% (**+0.95%**)
Kamiran's	22.61%	83.92%	6.59%	94.82%
FEAT	15.26% (**−32.51%**)	84.01% (**+0.11%**)	1.25% (**−81.03%**)	95.03% (**+0.22%**)

As shown in Table 1, it is clear that FEIG consistently enhances the fairness-aware learning by diminishing the discrimination to a lower level while maintaining a high prediction capability. The best discrimination reduction obtained by FEIG is 81.03% on *Census* dataset comparing with the discrimination-aware splitting criterion proposed by Kamiran et al. [16]. FEIG's learning idea of all groups being treated equally therefore indeed pushes the discrimination to a lower level, which is consistent with its theoretical design. This enhanced anti-discrimination ability is also statistically verified, comparing to the more effective fair splitting criterion among the baseline criteria, as shown in Table 2.

Table 2. The McNemar's test on the datasets for two different splitting criteria: FEIG and FAHT, testing whether FEIG worked to enhance the positive classification of the unprivileged group.

FAHT \ FAHT+	Adult dataset[1]		Census dataset[2]	
	Granted	Rejected	Granted	Rejected
Granted	716	110	1,120	263
Rejected	173	15,193	468	153,924

[1] Chi-squared = 13.583, df = 1, p-value = 0.0002282
[2] Chi-squared = 56.93, df = 1, p-value 4.516e-14

FEAT- \ FEAT	Adult dataset[3]		Census dataset[4]	
	Granted	Rejected	Granted	Rejected
Granted	1,127	80	1,331	359
Rejected	153	14,832	658	153,427

[3] Chi-squared = 22.249, df = 1, p-value = 2.395e-06
[4] Chi-squared = 87.32, df = 1, p-value < 2.2e-16

With respect to the attributes being selected for the construction of trees, both FAHT+ and FEAT select "marital status" as their root on *Adult* dataset, while FAHT and FEAT- are rooted on "age". Neither of these two attributes is discrimination-inclined compared to the root attribute "capital gain" of Kamiran's, which encodes the intrinsic discrimination bias of the historic data as members from the unprivileged group, i.e., the sensitive value is female, are less like to receive higher capital-gain than the privileged group's. On the other hand, age, generally speaking, is positively correlated with income per annum and holds that regardless of the sensitive attribute value. However, it is also possible that age could have local discrimination. That is to say, within a small age range, male could more likely to have a higher income than female as they tend to mature at different ages therefore differ in career age which could reflect income. FEIG's learning idea of all groups being treated equally regardless of their population sizes aims to detect and reflect such type of discrimination encoding. Such fair-enhancing attribute selection can also be concluded from the Pearson correlation coefficients between sensitive attribute and decision boundaries as shown in Table 3. As one can see, FEIG based models' predicted boundaries are less correlated with the sensitive attribute than FAHT's due to its fairness-enhancing ability. In addition, different from the tree induction in static setting, the selected attributes are still splitting candidates for the succeeding splitting selection, such fairness-enhancing decisions therefore have impacts on following decisions as well (feedback loops) and could further enhance fairness-aware learning.

Table 3. Pearson analysis on sensitive attribute, predicted decision boundary and actual decision boundary. Comparison values within each cell are formatted by (FAHT: FAHT+|| FEAT-: FEAT) with results on *Adult* and *Census* dataset in the above and below table, respectively.

Entity	Sensitive attribute	Predicted boundary	Actual boundary
Sensitive attribute	1:1 || 1:1	-0.16:-0.14 || -0.19:-0.14	-0.21:-0.21 || -0.21:-0.21
Predicted boundary	-0.16:-0.14 || -0.19:-0.14	1:1 || 1:1	0.44:0.41 || 0.49:0.50
Actual boundary	-0.21:-0.21 || -0.21:-0.21	0.44:0.41 || 0.49:0.50	1:1 || 1:1

Entity	Sensitive attribute	Predicted boundary	Actual boundary
Sensitive attribute	1:1 || 1:1	-0.09:-0.07 || -0.07:-0.05	-0.16:-0.16 || -0.16:-0.16
Predicted boundary	-0.09:-0.07 || -0.07:-0.05	1:1 || 1:1	0.56:0.53 || 0.57:0.57
Actual boundary	-0.16:-0.16 || -0.16:-0.16	0.56:0.53 || 0.57:0.57	1:1 || 1:1

4.3 Drift Adaptation Capability

FEAT is designed for enhanced fairness-aware learning with add-on concept drift adaptation ability to handle non-stationary discriminated data streams. For comparison, we implemented two recently proposed fairness-aware online learners FAHT [23] and FEI [14]. In addition, we compared against two baselines, the Hoeffding Tree (HT) and Kamiran's which incorporates the discrimination-aware splitting criterion of [16] into FEAT in replacing of FEIG. We also trained a concept-adapting learner, denoted HAT [4], as a baseline. All methods are trained in the same way for all datasets and the results are summarized in Table 4.

Table 4. Accuracy-vs-discrimination between FEAT and baseline models. The best performance of the compared baselines is marked in boldface. Percentage in parenthesis is the relative difference over the performance of the best baseline method.

Methods	Metric			
	Adult dataset		Census dataset	
	Discrimination	Accuracy	Discrimination	Accuracy
HT	22.59%	83.91%	6.84%	95.06%
Kamiran's	22.61%	83.92%	6.59%	94.82%
FAHT	**16.29%**	81.83%	**3.20%**	94.28%
FEI	22.16%	75.51%	6.34%	81.26%
HAT	22.3%	**84.7%**	6.54%	**95.64%**
FEAT	**15.26%** (**−6.32%**)	**84.01%** (**−0.7%**)	**1.25%** (**−60.94%**)	**95.03%** (**−0.64%**)

As one can see, FEAT consistently pushes the discrimination to lower values while maintaining fairly comparable predictive performance in all datasets. Compared with the best accuracy results, FEAT has a small drop of 0.7% and 0.64% on *Adult* and *Census* dataset, respectively. This is expected as HAT is exclusively accuracy-driven while FEAT optimizes for data encoding as well as enhanced discrimination reduction. In comparison with the most fair baselines, FEAT achieves 6.32% and 60.94% discrimination reduction on *Adult* and *Census* dataset, respectively. We also observe that FEI performances poorly although it is proposed for online setting. This verifies that online fairness cannot be trivially solved by a simple combination of existing techniques from corresponding communities. We further posit that such theoretical design is fundamental to progress in fairness in evolving data streams and not ad hoc.

5 Conclusions

This paper focuses on the highly under-explored discrimination-conscious learning in evolving data streams. To address this challenge, we propose FEAT with embedded fair-enhancing splitting criterion and further equip it with the ability of change detection and concept forgetting to handle discriminated and non-stationary data streams. The positive results of conducted experiments show the versatility of FEAT in online settings. One immediate future direction is to have an ensemble as random forests based on FEAT. A different avenue is to extend these results in conjunction with our previous work [26] to situations where the class label is not available for fair clustering. Here there are multiple unique challenges including appropriately defining and assessing fairness in the unsupervised scenarios.

References

1. Beutel, A., et al.: Putting fairness principles into practice: challenges, metrics, and improvements. In: AAAI Conference on Artificial Intelligence, Ethics, and Society (AIES) (2019)
2. Beutel, A., Chen, J., Zhao, Z., Chi, E.H.: Data decisions and theoretical implications when adversarially learning fair representations. arXiv preprint arXiv:1707.00075 (2017)
3. Bifet, A.. Gavalda. R.: Learning from time-changing data with adaptive windowing. In: Proceedings of the 2007 SIAM International Conference on Data Mining, pp. 443–448. SIAM (2007)
4. Bifet, A., Gavaldà, R.: Adaptive learning from evolving data streams. In: Adams, N.M., Robardet, C., Siebes, A., Boulicaut, J.-F. (eds.) IDA 2009. LNCS, vol. 5772, pp. 249–260. Springer, Heidelberg (2009). https://doi.org/10.1007/978-3-642-03915-7_22
5. Calders, T., Kamiran, F., Pechenizkiy, M.: Building classifiers with independency constraints. In: 2009 IEEE International Conference on Data Mining Workshops, pp. 13–18. IEEE (2009)

6. Calders, T., Verwer, S.: Three naive Bayes approaches for discrimination-free classification. Data Min. Knowl. Disc. **21**(2), 277–292 (2010)
7. Chen, I.Y., Szolovits, P., Ghassemi, M.: Can AI help reduce disparities in general medical and mental health care? AMA J. Ethics **21**(2), 167–179 (2019)
8. Dheeru, D., Karra Taniskidou, E.: UCI Machine Learning Repository (2017)
9. Domingos, P., Hulten, G.: Mining high-speed data streams. In: Proceedings of the 6th ACM SIGKDD International Conference on Knowledge Discovery and Data Mining, pp. 71–80. ACM (2000)
10. Gama, J.: Knowledge Discovery from Data Streams. CRC Press, Boca Raton (2010)
11. Gomes, H.M., Read, J., Bifet, A.: Streaming random patches for evolving data stream classification. In: 2019 IEEE International Conference on Data Mining (ICDM), pp. 240–249. IEEE (2019)
12. Hardt, M., Price, E., Srebro, N.: Equality of opportunity in supervised learning. In: Advances in Neural Information Processing Systems, pp. 3315–3323 (2016)
13. Ingold, D., Soper, S.: Amazon doesn't consider the race of its customers. Should it? Bloomberg News (2016)
14. Iosifidis, V., Tran, T.N.H., Ntoutsi, E.: Fairness-enhancing interventions in stream classification. In: Hartmann, S., Küng, J., Chakravarthy, S., Anderst-Kotsis, G., Tjoa, A.M., Khalil, I. (eds.) DEXA 2019. LNCS, vol. 11706, pp. 261–276. Springer, Cham (2019). https://doi.org/10.1007/978-3-030-27615-7_20
15. Kamiran, F., Calders, T.: Classifying without discriminating. In: 2nd International Conference on Computer, Control and Communication, pp. 1–6 (2009)
16. Kamiran, F., Calders, T., Pechenizkiy, M.: Discrimination aware decision tree learning. In: 2010 IEEE International Conference on Data Mining, pp. 869–874. IEEE (2010)
17. Krawczyk, B., Minku, L.L., Gama, J., Stefanowski, J., Woźniak, M.: Ensemble learning for data stream analysis: a survey. Inf. Fusion **37**, 132–156 (2017)
18. Read, J., Tziortziotis, N., Vazirgiannis, M.: Error-space representations for multi-dimensional data streams with temporal dependence. Pattern Anal. Appl. **22**(3), 1211–1220 (2018). https://doi.org/10.1007/s10044-018-0739-7
19. Verma, S., Rubin, J.: Fairness definitions explained. In: 2018 IEEE/ACM International Workshop on Software Fairness (FairWare), pp. 1–7. IEEE (2018)
20. Zafar, M.B., Valera, I., Gomez Rodriguez, M., Gummadi, K.P.: Fairness beyond disparate treatment & disparate impact: learning classification without disparate mistreatment. In: World Wide Web, pp. 1171–1180 (2017)
21. Zhang, L., Zhang, W.: A comparison of different pattern recognition methods with entropy based feature reduction in early breast cancer classification. Eur. Sci. J. **10**(7), 304 (2014). COBISS. MK-ID 95468554
22. Zhang, W.: PhD Forum: recognizing human posture from time-changing wearable sensor data streams. In: 2017 IEEE International Conference on Smart Computing (SMARTCOMP), pp. 1–2. IEEE (2017)
23. Zhang, W., Ntoutsi, E.: FAHT: an adaptive fairness-aware decision tree classifier. In: Proceedings of the 28th International Joint Conference on Artificial Intelligence, pp. 1480–1486. AAAI Press (2019)
24. Zhang, W., Tang, X., Wang, J.: On fairness-aware learning for non-discriminative decision-making. In: 2019 International Conference on Data Mining Workshops (ICDMW), pp. 1072–1079. IEEE (2019)
25. Zhang, W., Wang, J.: A hybrid learning framework for imbalanced stream classification. In: 2017 IEEE International Congress on Big Data, BigData Congress, pp. 480–487. IEEE (2017)

26. Zhang, W., Wang, J., Jin, D., Oreopoulos, L., Zhang, Z.: A deterministic self-organizing map approach and its application on satellite data based cloud type classification. In: 2018 IEEE International Conference on Big Data (Big Data), pp. 2027–2034. IEEE (2018)
27. Zliobaite, I.: A survey on measuring indirect discrimination in machine learning. arXiv preprint arXiv:1511.00148 (2015)

Unsupervised Concept Drift Detection Using a Student–Teacher Approach

Vitor Cerqueira[1(✉)], Heitor Murilo Gomes[2], and Albert Bifet[2,3]

[1] LIAAD-INESCTEC, Porto, Portugal
cerqueira.vitormanuel@gmail.com
[2] University of Waikato, Hamilton, New Zealand
[3] Télécom ParisTech, Paris, France

Abstract. Concept drift detection is a crucial task in data stream evolving environments. Most of the state of the art approaches designed to tackle this problem monitor the loss of predictive models. Accordingly, an alarm is launched when the loss increases significantly, which triggers some adaptation mechanism (e.g. retrain the model). However, this *modus operandi* falls short in many real-world scenarios, where the true labels are not readily available to compute the loss. These often take up to several weeks to be available. In this context, there is increasing attention to approaches that perform concept drift detection in an unsupervised manner, i.e., without access to the true labels. We propose a novel approach to unsupervised concept drift detection, which is based on a student-teacher learning paradigm. Essentially, we create an auxiliary model (student) to mimic the behaviour of the main model (teacher). At run-time, our approach is to use the teacher for predicting new instances and monitoring the *mimicking* loss of the student for concept drift detection. In a set of controlled experiments, we discovered that the proposed approach detects concept drift effectively. Relative to the gold standard, in which the labels are immediately available after prediction, our approach is more conservative: it signals less false alarms, but it requires more time to detect changes. We also show the competitiveness of our approach relative to other unsupervised methods.

Keywords: Concept drift detection · Data streams · Unsupervised learning · Model compression

1 Introduction

Learning from data streams is a continuous process. When predictive models are deployed in environments susceptible to changes, they must detect these changes and adapt themselves accordingly. The phenomenon in which the data distribution evolves is referred to as *concept drift*, and a sizeable amount of literature has been devoted to it [7].

Concept drift detection and adaptation are typically achieved by coupling predictive models with a change detection mechanism [10]. The detection algorithm launches an alarm when it identifies a change in the data. Typical concept

A. Appice et al. (Eds.): DS 2020, LNAI 12323, pp. 190–204, 2020.
https://doi.org/10.1007/978-3-030-61527-7_13

drift strategies are based on sequential analysis [18], statistical process control [6], or monitoring of distributions [3]. When change is detected, the predictive model adapts by updating its knowledge with recent information. A simple example of an adaptation mechanism is to discard the current model and train a new one from scratch. Incremental approaches are also widely used [9].

The input data for the majority of the existing drift detection algorithms is the performance of the predictive model over time, such as the error rate. In many of these detection methods, alarms are signalled if the performance decreases significantly. However, in several real-world scenarios, labels are not readily available to estimate the performance of models. Some labels might arrive with a delay or not arrive at all due to labelling costs. This is a major challenge for learning algorithms that rely on concept drift detection as the unavailability of the labels precludes their application [10].

In this context, there is increasing attention toward unsupervised approaches to concept drift detection. These assume that, after an initial fit of the model, no further labels are available during the deployment of this model in a test set. Most works in the literature handle this problem using statistical hypothesis tests, such as the Kolmogorov-Smirnov test. These tests are applied to the output of the models, either the final decision or the predicted probability.

1.1 Contributions and Paper Organisation

Our goal in this paper is to address concept drift detection in an unsupervised manner. To accomplish this, we propose a novel approach to tackle this problem using a student-teacher (ST) learning paradigm. The gist of the idea is as follows. On top of the main predictive model, which we designate as the teacher, we also build a second predictive model, the student. Following the literature on model compression [5] and knowledge distillation [13], the student model is designed to mimic the behaviour of the teacher.

Using the ST framework, our approach to unsupervised concept drift detection is carried out by monitoring the mimicking loss of the student. The mimicking loss is a function of the discrepancy between the prediction of the teacher and the prediction of the student in the same instance. In summary, we use the loss of the student model as a surrogate for the behaviour of the main model. Accordingly, we can apply any state of the art approach in the literature which takes the loss of a model as the main input, for example, ADWIN [3] or the Page-Hinkley test [18].

When concept drift occurs, it causes changes in the prior probabilities of the classes or changes in the class conditional probabilities of the predictor variables. In effect, we hypothesise that these changes disrupt the collective behaviour between the teacher and student models. In turn, this change of behaviour may be captured by monitoring the mimicking loss of the student model.

We validate the proposed method using a set of experiments with an artificial drift process, which we adapt from Žliobaite [23]. The proposed method is publicly available online to support reproducible science[1]. Our implementation is written in Python and is based on the scikit-multiflow framework [17].

2 Background

2.1 Problem Definition

Let $D(X, y) = \{(X_1, y_1), \ldots, (X_t, y_t)\}$ denote a possibly infinite data stream, where each X is a q-dimensional array representing the input predictor variables. Each y represents the corresponding output label. We assume that the values of y are categorical. The goal is to use this data set $\{X_i, y_i\}_1^t$ to create a classification model to approximate the function which maps the input X to the output y. Let $\mathcal{T}$ denote this classifier. The classifier $\mathcal{T}$ can be used to predict the labels of new observations X. We denote the prediction made by the classifier as $\hat{y}_{\mathcal{T}}$.

Many real-world scenarios exhibit a non-stationary nature. Often, the underlying process causing the observations changes in an unpredictable way, which degrades the performance of the classifier $\mathcal{T}$. Let $p(X, y)$ denote the joint distribution of the predictor variables X and the target variable y. According to Gama et al. [7], concept drift occurs if $p(X, y)$ is different in two distinct points in time across the data stream. Changes in the joint probability can be caused by changes in $p(X)$, the distribution of the predictor variables or changes in the class conditional probabilities $p(X|y)$ [8]. These may eventually affect the posterior probabilities of classes $p(y|X)$.

2.2 Label Availability

When concept drift occurs, the changes need to be captured as soon as possible, so the decision rules of $\mathcal{T}$ can be updated. The vast majority of concept drift detection approaches in the literature focus on tracking the predictive performance of the model. If the performance degrades significantly, an alarm is launched and the learning system adapts to these changes.

The problem with these approaches is that they assume that the true labels are readily available after prediction. In reality, this is rarely the case. In many real-world scenarios, labels can take too long to be available, if ever. If labels do eventually become available, often we only have access to a part of them. This is due to, for example, labelling costs. The different potential scenarios when running a predictive model are depicted in Fig. 1.

Precisely, a predictive model is built using an initial batch of training data, whose labels are available. When this model is deployed in a test set, concept drift detection is carried out in an unsupervised or supervised manner.

In unsupervised scenarios, no further labels are available to the predictive model. Concept drift detection must be carried out using a different strategy other than monitoring the loss. For example, one can track the output probability of the models [23] or the unconditional probability distribution $p(X)$ [15].

[1] https://github.com/vcerqueira/unsupervised_concept_drift.

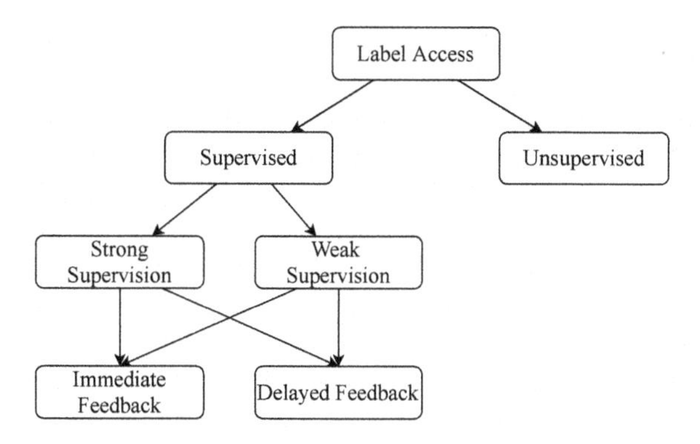

Fig. 1. The distinct potential scenarios regarding label access after the initial fit of the model (adapted from Gomes et al. [9]).

Concept drift detectors have access to labels when the scenario is supervised. On the one hand, the setting may be either strongly supervised or weakly supervised [22]. In the former, all labels become available. In the latter, the learning system only has access to a part of the labels. This is common in applications which data labelling is costly. On the other hand, labels can arrive immediately after prediction, or they can arrive with some delay. In some domains, this delay may be too large, and unsupervised approaches need to be adopted.

In this paper, we address concept drift detection from an unsupervised perspective. In this setting, we are restricted to use $p(X)$ to detect changes, as the probability of the predictor variables is not conditioned on y.

3 Related Research

3.1 Concept Drift Detection

Concept drift can occur in three different manners: suddenly, in which the current concept is abruptly replaced by a new one; gradually, when the current concept slowly fades; and reoccurring, in which different concepts are prevalent in distinct time intervals (for example, due to seasonality).

We split concept drift detection into two dimensions: supervised and unsupervised. The supervised type of approaches assumes that the true labels of observations are available after prediction. Hence, they use the error of the model as the main input to their detection mechanism. On the other hand, unsupervised approaches preclude the use of the labels in their techniques.

Plenty of error-based approaches have been developed for concept drift detection. These usually follow one of three sort of strategies: sequential analysis, such as the Page-Hinkley test (PHT) [18]; statistical process control, for example the Drift Detection Method (DDM) [6] or the Early Drift Detection Method (EDDM) [1]; and distribution monitoring, for example the Adaptive Windowing (ADWIN) approach [3].

Although the literature is scarce, there is an increasing interest in approaches which try to detect drift without access to the true labels. Žliobaite [23] presents a work of this type. She proposed the application of statistical hypothesis testing to the output of the classifier (either the probabilities or the final categorical decision). The idea is to monitor two samples of one of these signals. One sample serves as the reference window, while the other represents the detection window. When there is a statistical difference between these, an alarm is launched. In a set of experiments, Žliobaite shows that concept drift is detectable using this framework. The hypothesis tests used in the experiments are the two-sample Kolmogorov-Smirnov test, the Wilcoxon rank-sum test, and the two-sample t-test.

Reis et al. [20] follow a strategy similar to Žliobaite [23]. They propose an incremental version of the Kolmogorov-Smirnov test and use this method to detect changes. In the same line of research, Yu et al. [21] apply two layers of hypothesis testing hierarchically. Kim et al. [14] also apply a windowing approach. Rather than monitoring the output probability of the classifier, they use a confidence measure as the input to drift detectors.

Pinto et al. [19] present an automatic framework for monitoring the performance of predictive models. Similarly to the above-mentioned works, they perform concept drift detection based on a windowing approach. The signal used to detect drift is computed according to a mutual information metric, namely the Jensen-Shannon Divergence [16]. The window sizes and threshold above which an alarm is launched is analysed, and the approach is validated in real-world data sets. The interesting part of the approach by Pinto et al. [19] is that their method explains the alarms. This explanation is based on an auxiliary binary classification model. The goal of applying this model is to rank the events that occurred in the detection window according to how these relate to the alarm. These explanations may be crucial in sensitive applications which require transparent models.

Gözüaçık et al. [11] also develop an auxiliary predictive model for unsupervised concept drift detection, which is called D3 (for (Discriminative Drift Detector). The difference to the work by Pinto et al. [19] is that they use this model for detecting concept drift rather than explaining the alarms.

3.2 Student–Teacher Learning Approach

Model compression, also known as student-teacher (ST) learning, is a technique presented by Buciluǎ et al. [5]. The goal is to train a model, designated as a student, to mimic the behaviour of a second model (the teacher). The authors use this approach to compress a large ensemble (the teacher) into a compact predictive model (the student).

Hinton et al. [13] developed the idea of model compression further, denoting their compression technique as knowledge distillation. Distillation works by softening the probability distribution over classes in the softmax output layer of a neural network. The authors address an automatic speech recognition problem by distilling an ensemble of deep neural networks into a single and smaller deep neural network.

Both Buciluă et al. [5] and Hinton et al. [13], show that combining the predictions of the ensemble leads to a comparable performance relative to a single compressed model.

While our concerns are not about decreasing the computational costs of a model, we can leverage model compression approaches to tackle the problem of concept drift detection. Particularly, by creating a student model which mimics the behaviour of a classifier, we can perform concept drift detection using the loss of the student model. Since this loss is not conditioned on the target variable y, concept drift detection is carried out in an unsupervised manner.

4 Methodology

In this section, we formalise our approach to concept drift detection. Our method is based on a student-teacher (ST) learning approach. The only information required from the environment is predictor variables of testing instances (X). Since the proposed method is not conditioned on the labels of the target variable (y), we refer to it as unsupervised.

From a high-level perspective, the proposed approach settles on three main steps:

1. Creating the main model $\mathcal{T}$, which is the teacher;
2. Creating the student model $\mathcal{S}$, which mimics the behaviour of $\mathcal{T}$;
3. Deploying the main model $\mathcal{T}$ and performing concept drift detection based on the loss of $\mathcal{S}$;

In the next subsections, we will detail each step in turn.

4.1 Creating the Teacher and Student Models

Main Classifier $\mathcal{T}$. Let $D_{tr}(X, y)$ denote the available training instances. We use $D_{tr}(X, y)$ to train the classifier $\mathcal{T}$, where $D_{tr}(X, y)$ is an initial batch of training instances. This model is used to make predictions on new upcoming instances in the stream D, which we denote as X_{new}. We assume that the model is incremental [9]. $\mathcal{T}$ is updated when new labels become available.

Student–Teacher Approach. We assume that the corresponding labels of X_{new} are not available for a long period after making the prediction. Hence, we cannot rely on approaches that monitor the loss of $\mathcal{T}$ to detect concept drift.

We adopt a student-teacher (ST) learning approach to circumvent this problem. In the ST framework, $\mathcal{T}$ is the teacher model. Then, we create a second predictive model $\mathcal{S}$, the student, which is trained to mimic the behaviour of the teacher, $\mathcal{T}$. This is accomplished as follows. We obtain the predictions $\hat{y}_{\{\mathcal{T},tr\}}$ of $\mathcal{T}$ in the available data D_{tr} used to create it. In effect, we can set up a new data set $D_{tr}(x, \hat{y}_{\mathcal{T}})$, which can be used to train $\mathcal{S}$. In other words, we train the student model using the same observations used to train the teacher. However, the target variable is replaced with the predictions of the teacher.

It might be argued that using the same instances to train both the teacher and the student models leads to overfitting. However, Hinton et al. [13] show that this is not a concern.

In the typical student-teacher approaches, designed for model compression [5] or knowledge distillation [13], the goal is to compress a model with a large number of parameters (usually an ensemble) into a more compact model with comparable predictive performance. In these cases, the student model is deployed in the test set, while the teacher is not used. Conversely, in our methodology, we leverage the student-teacher framework differently. The student model is regarded as a model which can make predictions regarding the behaviour of $\mathcal{T}$, i.e., what the output of $\mathcal{T}$ will be for a given input observation. Both $\mathcal{T}$ and $\mathcal{S}$ models are used in practice, as explained below.

4.2 Concept Drift Detection

Since we assume that the true labels are unavailable, we cannot measure the loss of the main model, $\mathcal{T}$. But we can measure the loss of the student model: the discrepancy between the prediction of $\mathcal{T}$ ($\hat{y}_{\mathcal{T}}$) and the prediction of $\mathcal{S}$ about $\hat{y}_{\mathcal{T}}$ ($\hat{y}_{\mathcal{S}}$). The loss of $\mathcal{S}$ is then defined as $L(\hat{y}_{\mathcal{T}}, \hat{y}_{\mathcal{S}})$, where L is the loss function, for example, the error rate.

In effect, our approach to unsupervised concept drift detection is to monitor the error of the student model. This can be accomplished with any state of the art concept drift detection approach, e.g. ADWIN [3]. Essentially, we use the loss of the student model as a surrogate signal for concept drift detection. While $\mathcal{S}$ is monitored for detecting drift, the model $\mathcal{T}$ is used to make predictions on new instances X_{new}.

Our working hypothesis is the following. When concept drift occurs, it potentially causes changes in the posterior probability of classes $p(y|X)$. Consequently, this change in the behaviour of $\mathcal{T}$ will perturb the mimicking loss of the student model, $L(\hat{y}_{\mathcal{T}}, \hat{y}_{\mathcal{S}})$. Therefore, tracking this signal may enable us to capture changes in the environment without access to any labels.

5 Experiments

In this section, we detail the experiments carried out to validate the proposed approach to unsupervised concept drift detection.

The experiments are designed to address the following research questions:

- **RQ1:** Is the proposed unsupervised ST approach able to detect concept drift in the data?
- **RQ2:** How does the proposed method compare with the gold standard, in which all the labels are immediately available after prediction? Note that even though this scenario is unlikely in real applications, it may still serve as a benchmark of performance for other approaches;

- **RQ3:** How does the proposed approach compare with other unsupervised approaches, namely the statistical hypotheses tests described by Žliobaite [23]?
- **RQ4:** Finally, what is the relative drift detection performance between the different label availability scenarios (see Fig. 1)?

We used two data sets in our experiments: electricity demand [12], and forest cover type [2]. The electricity data set refers to the electricity market in Australian New South Wales. There are a total of 45.312 observations in the data set, which are captured every half-hour. There are eight predictor variables, all of which numeric. The predictive task is binary classification; to predict whether the price will go up or down relative to a moving average of 24 h.

The second data set represents the forest cover type and was obtained by the US Forest Service. In the data set, there is a total of 581.012 observations and 54 predictor variables (10 of which numeric, and the remaining are binary). The data set contains five classes regarding the cover type. Both these data sets have been used in multiple works on data stream mining [2,6,9].

5.1 Synthetic Drift Injection

We perform experiments using artificial drift in order to understand better the relative behaviour of drift detectors in different scenarios. This is accomplished by following a process similar to that described by Žliobaite [23]. We assume that the initial 60% of the observations are labelled and that these observations are used for an initial fit of the models $\mathcal{T}$ and $\mathcal{S}$. Then, we proceed as follows.

1. We randomly select a point between 70% and 90% of the total observations available. After this point, all subsequent observations are *contaminated* with drift (see Fig. 2);
   - Note that we leave a 10% interval on each side (after 60% and before 100%) for securing enough observations to evaluate the behaviour of a concept drift detector; for example, its rate of false alarms or its reactiveness to drift;
2. We randomly select half of the predictor variables from a randomly selected class. The values of these variables are randomly shuffled. Žliobaite [23] handpicks the columns to be swapped. Conversely, we introduce randomness. Essentially, this process injects drift in the conditional probabilities $p(X|y)$.

In order to produce a robust estimate of performance, this process is repeated 50 times in a Monte Carlo approximation manner.

5.2 Methods

We carry out a learning plus testing cycle for each one of the 50 Monte Carlo repetitions. We use an Adaptive Random Forest (ARF) as learning algorithm [9] for training both $\mathcal{T}$ and $\mathcal{S}$. As the name implies, the ARF method extends the

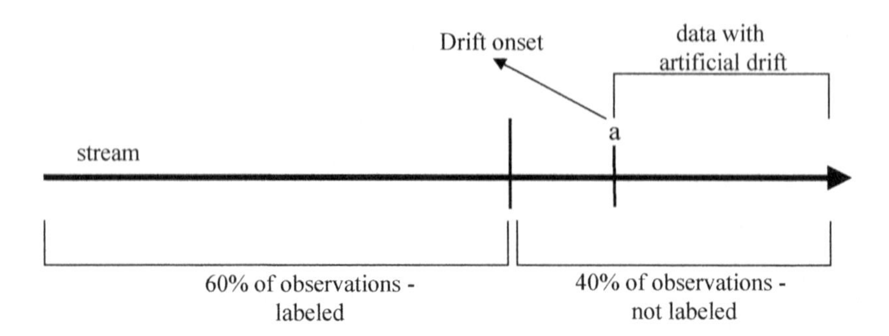

Fig. 2. Workflow for injecting drift. This random process is repeated 50 times following a Monte Carlo approximation approach.

widely used Random Forest approach by Breiman [4] to evolving data stream classification problems.

We focus on performing concept drift detection in an unsupervised manner. Notwithstanding, we also test several variants of supervised scenarios. All of these are detailed below.

Unsupervised Approaches (U). After the initial training with 60% of the observations, our unsupervised setup assumes that no further label is available. For each upcoming instance, we only have access to the predictor variables (X). Accordingly, as we described in Sect. 4, concept drift is performed by monitoring the error of $\mathcal{S}$. In this scenario, we also apply the ADWIN and PHT methods using the error rate of $\mathcal{S}$. These are denoted as `U-ADWIN` and `U-PHT`, respectively. Note that the model $\mathcal{S}$ can be updated online, because its labels are the predictions of model $\mathcal{T}$.

As benchmarks, we also include the following statistical tests suggested by Žliobaite [23]:

- `U-KS`: The two-sample Kolmogorov-Smirnov test, which tests whether two samples come from the same distribution;
- `U-WRS`: The Wilcoxon rank sum test, which tests whether two samples have equal medians;
- `U-TT`: The two-sample t-test, which tests whether two samples have equal means;

Each of these tests are applied using the class output predicted by $\mathcal{T}$ using a sliding window fashion [23]. Specifically, suppose we are at time step i. We create two contiguous samples of the same size (w) up to point i (see Fig. 3). The first sample, which represents the reference window, includes the data in the interval $[i - 2 \times w + 1; i - w]$. The second sample, which denotes the detection window [23], contains the information in the interval $[i - w + 1; i]$. An hypothesis test is carried out using these two samples. An alarm is issued, and $i + 1$ is a change point, if the p-value returned by the test is below α.

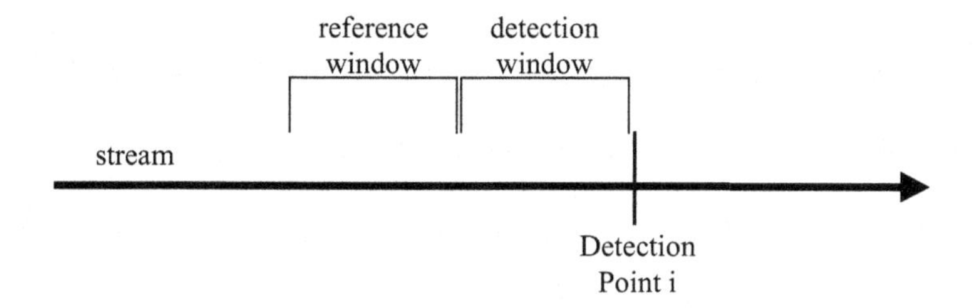

Fig. 3. Detection framework based on windowing.

Supervised Approaches. We also include the following supervised approaches in our experiments.

- **Strongly Supervised (SS):** We apply the typical procedure which assumes that all the true labels are immediately available after making a prediction. This can be regarded as the gold standard. The term *strong* means refers to the fact that all labels are available during testing [22];
- **Weakly Supervised (WS):** In many real-world scenarios, particularly in high-frequency data streams, data labelling is costly. Hence, predictive models can only be updated using a part of the entire data set. This process is commonly referred to as weakly supervised learning [22]. We simulate a weakly supervised scenario in our experiments. Accordingly, predictive models only have access to *l_access*% of the labels. In other words, after a model predicts the label of a given instance, the respective label is immediately available with a *l_access*% chance;
- **Delayed Strongly Supervised (DSS):** Labels can take some time to arrive. We study this aspect by artificially delaying the arrival of the labels by *l_delay* instances. After a label becomes available, the respective predictive model is updated;
- **Delayed Weakly Supervised (DWS):** We combine the two previous scenarios. In the DWS setup, only *l_access*% of the labels are available. Those which are available arrive with a delay of *l_delay* observations.

In the supervised variants, the classifier $\mathcal{T}$ is updated online as soon as each new label is available. In all these variants, concept drift detection is carried out using the error rate of $\mathcal{T}$ with two state of the art approaches: the ADWIN [3] method and the PHT [18] approach. We denote these approaches as *Prefix*-ADWIN and *Prefix*-PHT, respectively. For example, SS-ADWIN refers to the ADWIN method applied using a strongly supervised model.

Parameter Setup. The significance level for the hypothesis tests, also known as p-value, is set to 0.001. The window size for carrying these tests is set to 500 for the electricity data set and 1500 for the cover type data set (see Fig. 3). As Žliobaite [23] points out, these values usually depend on the task at hand. The number of trees in the ARF models is set to 25. ARF comprises an internal concept drift detection mechanism based on ADWIN [3]. Since our goal is to detect drift, we disabled this process within the ARF model. We set *l_access* to 50 for both data sets. This setup means that, in the weakly supervised schemes, only about 50% of observations are available. The value of *l_delay* was set to 1000 on the electricity data set, and 5000 on the cover type data set. These values were set arbitrarily; the difference between data sets is related to their difference is the sample size. The remaining parameters are set to default according to their respective implementation.

5.3 Evaluation Metrics

We apply the following metrics described by Bifet [2] to evaluate the drift detectors:

- Mean Time between False Alarms (MTFA): How often there is a false alarm when there is no change (the higher, the better). MTFA is measured by averaging the distance between consecutive (false) alarms before the change point. Moreover, this score is also averaged across the 50 Monte Carlo repetitions;
- Mean Time to Detection (MTD): After a change occurs, how long it takes for the method to detect it, on average (the lower, the better). In practice, we measure the number of points between the change point and the next alarm launched by the respective method. Similarly to MTFA, the score of MTD is averaged across the 50 repetitions;
- Missed Detection Ratio (MDR): The probability of failing to detect a drift. This is measured by taking the fraction of repetitions (across the 50 simulations) in which the drift method fails to launch an alarm after the onset of the drift. Ideally, this value should be zero, meaning all drifts are captured irrespective of how long it takes to accomplish this;

On top of these, we also include the total number of detections (ND) launched by a model. This metric is also averaged across the 50 repetitions.

5.4 Results

The results of our experiments are reported in Tables 1 and 2, for the electricity and cover type data sets, respectively. For the MTFA, MTD, and ND metrics, we also include the standard deviation of the results across the 50 Monte Carlo repetitions.

The first research question is related to the analysis of the ability of the proposed approach to detect concept drift (RQ1). According to the results obtained, the proposed methods (`U-ADWIN`, `U-PHT`) have an MDR of 12 and 6% (`U-ADWIN`) and 18 and 17% (`U-PHT`) in the electricity and cover type data sets, respectively. This result shows that most of the drifts introduced synthetically were captured by both approaches. Moreover, it also shows that the proposed concept drift detection methodology is not constrained to a single detector: both ADWIN and PHT present a good detection ability. Other approaches could be

Table 1. Results on the electricity dataset

Method	MTFA	MTD	MDR	ND
SS-ADWIN	18372 ± 29514	986 ± 689	0.00	7 ± 2
SS-PHT	1024 ± 284	508 ± 687	0.88	9 ± 4
DSS-ADWIN	677 ± 216	525 ± 572	0.00	19 ± 4
DSS-PHT	357 ± 60	346 ± 290	0.00	26 ± 5
WS-ADWIN	21287 ± 34634	1229 ± 646	0.00	6 ± 2
WS-PHT	1355 ± 451	503 ± 301	0.90	8 ± 5
DWS-ADWIN	6749 ± 1672	1274 ± 1448	0.00	10 ± 3
DWS-PHT	517 ± 158	533 ± 629	0.06	21 ± 7
U-ADWIN	11439 ± 30002	2786 ± 3225	0.12	4 ± 4
U-PHT	11823 ± 28244	3050 ± 4778	0.18	4 ± 2
U-WRS	2040 ± 158	597 ± 246	0.04	17 ± 3
U-TT	611 ± 36	316 ± 284	0.00	27 ± 2
U-KS	2071 ± 200	574 ± 268	0.04	16 ± 3

used for detection, e.g. DDM [6] or EDDM [1]. We focus on only two methods for two main reasons: first, these are representative of the state of the art; second, the detection method is orthogonal to our contributions. Therefore, we designed the experiments to show the effectiveness of the proposed method rather than comparing the performance of multiple detectors.

Relative to the gold standard (RQ2), the proposed approach is more conservative. Both U-ADWIN, and U-PHT present a higher MTFA and MTD scores relative to a strongly supervised approach in most of the cases. The unsupervised approaches take more time to detect changes (higher MTD). However, they also show a larger interval between false alarms (higher MTFA) – except for the SS-ADWIN approach.

We also compare the hypothesis tests suggested by Žliobaite [23] with the proposed approach (RQ3). Overall, the statistical tests are more sensitive as they launch more alarms (column ND). The MDR is close or equal to zero in all variants, which means they capture almost all the drifts injected in the data.

Finally, we analyse the different scenarios in terms of label availability (RQ4). We start by comparing the strongly supervised scenario with their weakly supervised counterpart, i.e., SS with WS and DSS with DWS. Overall, the weakly supervised approaches tend to launch fewer alarms. The results suggest that having less information leads to more conservative behaviour by the ADWIN and PHT detectors.

We now analyse the impact of delaying information. This is done by comparing the SS variants with the DSS variants, and the WS variants with DWS variants. Contrary to weak supervision, delaying the arrival of the labels appears to lead to much more sensitive detectors. Note that these results are constrained on many aspects, for example, data sets, parameter setup, or the drift synthetic process.

Table 2. Results on the cover type dataset

Method	MTFA	MTD	MDR	ND
SS_ADWIN	5806 ± 3354	3070 ± 2919	0.00	12 ± 2
SS_PHT	6317 ± 1117	6012 ± 8017	0.22	14 ± 6
DSS_ADWIN	1532 ± 493	2154 ± 2802	0.00	28 ± 4
DSS_PHT	389 ± 44	406 ± 410	0.00	140 ± 34
WS_ADWIN	14731 ± 6036	4185 ± 3579	0.00	9 ± 2
WS_PHT	7099 ± 2286	10200 ± 13882	0.28	16 ± 7
DWS_ADWIN	4925 ± 1117	2689 ± 3043	0.00	14 ± 3
DWS_PHT	629 ± 477	5441 ± 5902	0.06	42 ± 16
U_ADWIN	15792 ± 3213	5411 ± 6465	0.06	10 ± 4
U_PHT	21864 ± 7720	5877 ± 6857	0.17	6 ± 3
U_WRS	2055 ± 348	1341 ± 1118	0.00	25 ± 2
U_TT	2208 ± 384	1416 ± 1056	0.00	24 ± 3
U_KS	1771 ± 138	850 ± 867	0.00	30 ± 2

5.5 Discussion

In the experiments, we showed that the proposed student-teacher approach can detect concept drift. While it shows a more conservative behaviour relative to the gold standard, the probability of detecting a drift is comparable.

We focused on two state of the art drift detection approaches; ADWIN and PHT. The underlying method applied is orthogonal to our contributions, and we designed the experiments to show the usefulness of the student-teacher approach to unsupervised concept drift detection. In this context, other detectors can be used, such as DDM [6] or EDDM [1].

We controlled the experiments by injecting artificial concept drift in the conditional probabilities $p(X|y)$. This was achieved by randomly swapping the predictors variables in a randomly selected class. The goal of this synthetic process was to enable us to analyse better how the different detection approaches react to drift. In future work, we will develop this analysis from two perspectives:

1. We can analyse the behaviour of the detectors in the presence of drift in the class priors, $p(y)$. To accomplish this, we can follow the strategy by Žliobaite [23], which deletes randomly selected instances from a selected class;
2. We will study the application of the proposed method in a real-world setup without any synthetic process. For example, we can measure its impact by computing the difference in predictive performance. Alternatively, a trade-off between predictive performance and the cost of retrieving a batch on labels to run a supervised approach.

Besides showing the usefulness of the proposed method, we also analysed the behaviour of the two detectors (ADWIN and PHT) under different supervised conditions. Specifically, whether the supervision was strong or weak, in which the latter means that only part of the labels become available. We also analysed the impact of feedback delay, in which the labels take a fixed time to arrive.

We will extend this analysis in future work. For example, we will perform a sensibility analysis to study how these conditions affect not only the performance of drift detectors but also the performance of the predictive models. We will also evaluate the proposed approach on purely real data sets.

6 Final Remarks

The literature for concept drift detection is mostly focused on detecting changes by discovering significant deviations in the loss of the model. In this paper, we follow the hypothesis that it is too optimistic to assume that the labels are readily available for computing the loss [19,23]. Therefore, we tackle the concept drift detection problem in an unsupervised manner.

We develop a novel approach based on an ST learning paradigm. ST approaches are commonly applied to model compression [5] or knowledge distillation [13]. To our knowledge, this is the first work attempting to use an ST approach for concept drift detection.

We validate our proposal with synthetic experiments using two benchmark data sets. The results are promising. The developed method can detect the drifts induced artificially as well as the gold standard, which represents the approach that assumes that labels are immediately available after prediction. Our approach is more conservative relative to the gold standard, and competitive relative to other unsupervised baseline approaches.

Acknowledgements. This work is financed by National Funds through the Portuguese funding agency, FCT - Fundação para a Ciência e a Tecnologia, within project UIDB/50014/2020.

References

1. Baena-Garcıa, M., del Campo-Ávila, J., Fidalgo, R., Bifet, A., Gavalda, R., Morales-Bueno, R.: Early drift detection method. In: 4th International Workshop on Knowledge Discovery from Data Streams, vol. 6, pp. 77–86 (2006)
2. Bifet, A.: Classifier concept drift detection and the illusion of progress. In: Rutkowski, L., Korytkowski, M., Scherer, R., Tadeusiewicz, R., Zadeh, L.A., Zurada, J.M. (eds.) ICAISC 2017. LNCS (LNAI), vol. 10246, pp. 715–725. Springer, Cham (2017). https://doi.org/10.1007/978-3-319-59060-8_64
3. Bifet, A., Gavalda, R.: Learning from time-changing data with adaptive windowing. In: Proceedings of the 2007 SIAM International Conference on Data Mining, pp. 443–448. SIAM (2007)
4. Breiman, L.: Random forests. Mach. Learn. **45**(1), 5–32 (2001)
5. BuciluÇŽ, C., Caruana, R., Niculescu-Mizil, A.: Model compression. In: Proceedings of the 12th ACM SIGKDD International Conference on Knowledge Discovery and Data Mining, pp. 535–541. ACM (2006)

6. Gama, J., Medas, P., Castillo, G., Rodrigues, P.: Learning with drift detection. In: Bazzan, A.L.C., Labidi, S. (eds.) SBIA 2004. LNCS (LNAI), vol. 3171, pp. 286–295. Springer, Heidelberg (2004). https://doi.org/10.1007/978-3-540-28645-5_29
7. Gama, J., Žliobaitė, I., Bifet, A., Pechenizkiy, M., Bouchachia, A.: A survey on concept drift adaptation. ACM Comput. Surv. (CSUR) **46**(4), 1–37 (2014)
8. Gao, J., Fan, W., Han, J., Yu, P.S.: A general framework for mining concept-drifting data streams with skewed distributions. In: Proceedings of the 2007 SIAM International Conference on Data Mining, pp. 3–14. SIAM (2007)
9. Gomes, H.M.: Adaptive random forests for evolving data stream classification. Mach. Learn. **106**(9), 1469–1495 (2017). https://doi.org/10.1007/s10994-017-5642-8
10. Gomes, H.M., Read, J., Bifet, A., Barddal, J.P., Gama, J.: Machine learning for streaming data: state of the art, challenges, and opportunities. ACM SIGKDD Exp. Newslett. **21**(2), 6–22 (2019)
11. Gözüaçık, Ö., Büyükçakır, A., Bonab, H., Can, F.: Unsupervised concept drift detection with a discriminative classifier. In: Proceedings of the 28th ACM International Conference on Information and Knowledge Management, pp. 2365–2368 (2019)
12. Harries, M., Wales, N.S.: Splice-2 comparative evaluation: Electricity pricing (1999)
13. Hinton, G., Vinyals, O., Dean, J.: Distilling the knowledge in a neural network. arXiv preprint arXiv:1503.02531 (2015)
14. Kim, Y., Park, C.H.: An efficient concept drift detection method for streaming data under limited labeling. IEICE Trans. Inf. Syst. **100**(10), 2537–2546 (2017)
15. Kuncheva, L.I.: Classifier ensembles for changing environments. In: Roli, F., Kittler, J., Windeatt, T. (eds.) MCS 2004. LNCS, vol. 3077, pp. 1–15. Springer, Heidelberg (2004). https://doi.org/10.1007/978-3-540-25966-4_1
16. Lin, J.: Divergence measures based on the Shannon entropy. IEEE Trans. Inf. Theor. **37**(1), 145–151 (1991)
17. Montiel, J., Read, J., Bifet, A., Abdessalem, T.: Scikit-multiflow: a multi-output streaming framework. J. Mach. Learn. Res. **19**(1), 145–151 (2018)
18. Page, E.S.: Continuous inspection schemes. Biometrika **41**(1/2), 100–115 (1954)
19. Pinto, F., Sampaio, M.O., Bizarro, P.: Automatic model monitoring for data streams. arXiv preprint arXiv:1908.04240 (2019)
20. dos Reis, D.M., Flach, P., Matwin, S., Batista, G.: Fast unsupervised online drift detection using incremental Kolmogorov-Smirnov test. In: Proceedings of the 22nd ACM SIGKDD International Conference on Knowledge Discovery and Data Mining, pp. 1545–1554 (2016)
21. Yu, S., Wang, X., Principe, J.C.: Request-and-reverify: Hierarchical hypothesis testing for concept drift detection with expensive labels. arXiv preprint arXiv:1806.10131 (2018)
22. Zhou, Z.H.: A brief introduction to weakly supervised learning. Nat. Sci. Rev. **5**(1), 44–53 (2018)
23. Žliobaite, I.: Change with delayed labeling: when is it detectable? In: 2010 IEEE International Conference on Data Mining Workshops, pp. 843–850. IEEE (2010)

Dimensionality Reduction and Feature Selection

Assembled Feature Selection for Credit Scoring in Microfinance with Non-traditional Features

Saulo Ruiz[1](✉), Pedro Gomes[1], Luís Rodrigues[1], and João Gama[2]

[1] Pelican Rhythms, Porto, Portugal
{saulo,pedro,luis}@quickcheck.ng
[2] Laboratory of Artificial Intelligence and Decision Support, University of Porto, Porto, Portugal
jgama@fep.up.pt

Abstract. Since early 2000, Microfinance Institutions (MFI) have been using credit scoring for their risk assessment. However, one of the main problems of credit scoring in microfinance is the lack of structured financial data. To address this problem, MFI have started using non-traditional data which can be extracted from the digital footprint of their users. The non-traditional data can be used to build algorithms that can identify good borrowers as in traditional banking. This paper proposes an assembled method to evaluate the predictive power of the non-traditional method. By using the Weight of Evidence (WoE), a transformation based on the distribution within the feature, as feature transformation method, and then applying extremely randomized trees for feature selection, we were able to improve the accuracy of the credit scoring model by 20.20% when compared to the credit scoring model built with the traditional implementation of WoE. This paper shows how the assembling of WoE with different feature selection criteria can result in more robust credit scoring models in microfinance.

Keywords: Credit scoring · Microfinance · Logistic regression · Weight of Evidence · Emerging markets · Feature selection

1 Introduction

Microfinance in emerging markets is an exciting and growing market with challenges very different from the ones present in developed economies. Even though studies show that 85% of the world population is in emerging markets [1], they still lack a proper finance infrastructure. According to the World Bank, it is estimated that there are 2.5 billion unbanked adults who lack access to financial services. From these financial services, access to credit is the most relevant and requested service. In emerging markets, customers cannot rely on banks to have access to credit as they usually lack a verifiable credit history. Microfinance Institutions (MFI) target these customers, by providing access to basic financial

A. Appice et al. (Eds.): DS 2020, LNAI 12323, pp. 207–216, 2020.
https://doi.org/10.1007/978-3-030-61527-7_14

services. However, due to the high risk involved in this kind of service, MFI's loan process tends to be slow and cumbersome. In most processes, customers are required to provide an identification card, employment letter, utility bills, loan application letter, and guarantors. Although it is a common practice to require this type of information in developed economies, most customers in emerging markets do not have them. Also, it represents too much effort for such a small amount. Furthermore, MFI apply high-interest rates which can directly affect the utility of this service. These factors reduce significantly the number of customers that can repay a loan.

Digital technologies bring a new dynamic to the finance market in emerging markets. Smartphone adoption in these markets is approaching the numbers of developed economies [9] and new fintech solutions for unbanked people are surfacing. On this decade, several companies proposed loan products across emerging markets where a customer can apply for a loan through a mobile application [6]. However, challenges in customer classification and eligibility for a loan arise. Credit scoring has been the way to go in traditional credit institutions and normally rely on a reliable credit history of the customer under evaluation. These new loan pipelines lack access to traditional data. They only have access to data that is input by customer and data collected from the logs of their smartphones, such as Short Message Service (SMS) logs, mobile applications installed and social network relationships.

In [10] we have shown how a credit scoring algorithm built with these non-traditional features can help to improve the risk evaluation process. However, that was only the first step as we aim for an optimal solution. This paper proposes an assembled feature selection method based on the traditional Weight of Evidence (WoE) and replacing the rules of the Information Value (IV) by a second feature selection mechanism as used in [10].

The remainder of this paper is organized as follows. In the next section, we will revise related work on credit scoring and how different approaches have been taken to select the relevant features. Section 3 describes the dataset used in the experimentation phase. Section 4 describes the methodology used in the experiment. Section 5 presents the results of the credit scoring model built with a different subset of features generated with the different techniques. Conclusions and the limitations of this approach are highlighted in the Sect. 6.

2 Related Work

Since the early 2000s Mark Schreiner, one of the main contributors of credit scoring for microfinance, started using structured databases to train credit scoring algorithms. These algorithms were implemented on MFIs located in Latin America. The credit scoring models built by Schreiner were based on scorecards that include details from the customer, loan, and loan officer. The features were selected by Schreiner based on his experience in microfinance. He tried to know as much as he could about the financial capability of the person based on the goods the customer had and the conditions of the accommodation. The scoring model had positive results, however, it was difficult to implement since loan officers had to do the process manually. In some cases, the loan officer had to visit the customer in order to validate the information about household goods. These models can be seen on [13,14] and [12]. Schreiner proved to MFIs that credit scoring could work for their institutions. His approach worked both with non-traditional variables and also with financial system information from the bureaus.

More recently, at the start of the decade of 2010, the usage of credit scoring in microfinance has become a common practice as seen in [4,11,18], among others. However, there is not a consensus on the algorithm to use and even less on which features to use. [4] shows a classical approach to the feature selection by looking at the significant variables using Linear Discriminant Analysis (LDA) and Logistic Regression (LR). In [18] we can observe a more classic approach in finance, the use of WoE. In the second encoding method presented in [18], the features are transformed and filtered by using the WoE. The WoE can also be seen in [16], where is used for feature selection on credit scoring for mortgages and credit cards. However, in our case, we do not have structured financial data as in [4] or [16]. As referred to in the previous section, we have data gathered only by mobile phone. A similar case can be seen in [11]. Where the use of features derived from mobile phone usage is used to build credit scoring algorithms for credit cards. There is no specific method for feature selection shown in [11] and one of the conclusions is to further evaluate the predictive power of their features. In [8] the use of Feature Importance (FI) is applied in microfinance using a Support Vector Machine (SVM) model. The model built in [8] outperformed the previous models without the FI filtering. In our case, we intend to evaluate the predictive power of the mobile phone features with an assembling method. We have shown in [10] that a logistic regression built with a dataset transformed with WoE and filtered with IV performs well even while lacking structured financial data.

3 Dataset

This section describes the dataset used to test the different feature selection methods and their performance. For this study, we granted a dataset of loans from a MFI based on the Sub-Saharan region. The original dataset had ten times more, good users, than bad users. Therefore, we applied random undersampling of good users in order to achieve a balanced dataset. After then undersampling, the dataset contained 3,094 loans and it was balanced by the target. This is 1,547 of loans paid on time (good class) and an equal amount of unpaid or paid after an additional process of recovery (bad class). To be more precise, the bad class contains all the unpaid loans and the loans that were paid 5 or more days after their respective due date. The dataset contains only the first application of each customer. Therefore, no previous financial information is present in this dataset.

The dataset is composed by a list of features gathered from the mobile device, those features can be grouped into three sources, personal information, Mobile Network Operator (MNO) and device information and loan characteristics.

First, we gathered the personal information, this personal information, including demographics, is collected when the customer opens the mobile application for the first time and set up his/her customer profile. Some variables refer to the goods the customers have (e.g., house, car) and go through their employment status (e.g., employed, self-employed). It also collects information about the dependents of the customer thought marital status and number of children. This type of data has been considered since the building of scorecards on [13] to the use of more advanced classification techniques as shown in [4,11,18], among others.

We also gathered device characteristics and MNO related features. These variables capture whenever the customer uses one of the services provided by the MNO, e.g. sending or receiving SMS, buying data, etc. [2] show that having a service of a given MNO can have an impact when building a scoring model. The hypothesis is that customer with higher acquisition level will relate with MNO that provide more complete services hence more expensive. From the mobile phone, we collected the system information, e.g. mobile applications installed, and android system version. With the categories of the applications installed in the device, we can create a complete profile of the customers by grouping the applications by their core function (e.g., financial applications, social media, etc.). The applications installed are a proxy to the *likes* and *shares* presented in [5]. Providing an idea of the real interests of the customers.

Finally, we add loan characteristics and conditions, which are: length of the loan, amount requested, and purpose of the loan.

Notice that the features gathered through the logs are features that come directly from the mobile device log on which the user have no direct way to modify (e.g., number of SMS received, device brand, number of applications installed). While the features from the personal information are filled by the user before the credit application and some of them cannot be verified (e.g., education level, number of children, employment status).

Table 1. Table of features considered on datasets.

Features considered on datasets grouped by source		
Personal information	MNO and device features	Loan characteristics
Age	Airtime	Loan amount
Gender	Airtime top ups	Loan length
Marital status	Number of calls	Loan reason
Education level	Number of SMS	
Number of children	Device Brand	
Employment status	Last mobile update	
Ownership of house	Number of apps installed	
Monthly Income		

Table 1 shows in detail the list of the features used. This list presents only the core variables, we did feature engineering to create the following variables:

- **debt_ratio** = amount requested/monthly income
- **telco_usage** = airtime/monthly income
- **income_split** = monthly income/number of dependants
- **average_top_up** = airtime/number of airtime top ups
- **delta_calls** = number of calls 0–30 days/number of calls 31–60 days
- **delta_sms** = number of SMS 0–30 days/number of SMS 31–60 days.

The notation 0–30 days refers to the number of calls made in the last 30 days prior to applying for a loan, the same logic applies to 31–60 days which refers to the number of calls made from 31 to 60 days prior to applying. Note that for building the **income_split** we have used the number of dependants which is calculated as *1 + number of children* as it considers that the applicant depends on itself. Some previous works also have a similar building of features. Such is the case of the MobiScore [11], which considers the number of SMS and the number of calls as relevant features. Also, [3] conclude that the use of mobile data usage patterns can be a valuable input to build a scoring model even when lacking formal financial history.

4 Assembling of Feature Selection Methods

As presented in the Sect. 3, the target to predict is a binary variable where 1 represents the bad class and 0 represents the good class.

First, we applied the WoE to all the features. The WoE is a method that transforms a feature based on the distribution of good cases and bad cases within the feature. Assume a feature X with domain $\{X_1, \ldots, X_i\}$ The WoE of X_i is computed as:

$$WoE(X_i) = ln\left(\frac{TotalGoods(X_i)}{TotalBads(X_i)}\right) \tag{1}$$

In Eq. 1, TotalGoods(X_i) refers to the number of customers that paid on time for category i in feature X. The same concept applies to TotalBads(X_i) but considering the customers that did not pay or paid late. Note that the value of the WoE is 0 when both the distribution of the good class and the distribution of the bad class are equal. This means that the category evaluated does not allow to differentiate between classes. Using the transformed dataset, we calculated the IV of each feature. The IV for a category i of a feature X with n number of categories is as follow:

$$IV(X_i) = \frac{WoE_{X_i}}{\sum_{i=1}^{n} WoE_{X_i}} \tag{2}$$

The IV for feature X is the sum of the IV of each category i of X. It can be calculated as *IV(X)*:

$$IV(X) = \sum_{i=1}^{i}\left[(TotalGoods_i - TotalBads_i) * ln\left(\frac{TotalGoods(X_i)}{TotalBads(X_i)}\right)\right] \tag{3}$$

The IV is the main indicator used by [15] to select the features that will be used in the modelling phase. [15] proposes intervals to relate the IV to the strength of the relationship with the target to predict. These intervals can relate the feature under analysis in order to determine a weak, medium, strong or non-existent relationship between a feature and the target variable. The relation can be set as shown in the conditions below.

When IV:

- Less than 0.02, the variable does not differentiate the Goods/Bads odds ratio.
- Between 0.02 to 0.1, the variable has only a weak relationship to the Goods/Bads odds ratio.
- Between 0.1 to 0.3, the predictor has a medium-strength relationship to the Goods/Bads odds ratio.
- Equal 0.3 or higher, the predictor has a strong relationship to the Goods/Bads odds ratio.

In Table 2 we present an example of the calculation of IV and the corresponding WoE for each category of X. In the example, the values of each customer on the feature X will be replaced by the corresponding WoE for that category. A customer with X_1 on feature X will now have -0.2337 on feature X. The IV for variable X is **0.0257** meaning it has a weak relationship with the Goods/Bads odds ratio.

We can say that the methodology proposed by [15] has two phases, a transformation based on the distribution within the feature (WoE) and filtering of features based on rules using the IV. On this experiment, we challenged the rules presented by [15] when introducing the WoE and the IV. We propose to replace the filtering part with a feature selection method. Therefore, we will transform the initial dataset following the WoE but instead of selecting the variables by their IV we will apply FI with Extra-trees Feature Selection (ETFS), Recursive Feature Elimination (RFE) and Lasso-Based Feature Selection (L1). The results of these methods will then be compared to the filtering by rules using the IV.

The FI is one of the measures obtained while training a Random Forest (RF). The importance of a feature is obtained as the loss of accuracy in the classification caused by the random permutation of feature values between objects. The accuracy loss is calculated for all trees in the forest which use the feature for classification. Then the average and standard deviation of the accuracy loss are computed. Then the feature importance is obtaining by dividing the average accuracy loss over its standard deviation [7].

For the ETFS we use a similar approach as in FI. The main difference between ETFS and FI lies in the fact that, instead of computing the locally optimal feature/split combination as done for RF, for each feature under consideration, a random value is selected for the splitting. The goal of this mechanism is to obtain more diversified trees and fewer split points to evaluate when training the extra-tress.

The RFE is a simple yet powerful method. Given an external estimator that assigns weights to features, in our case the weights assigned to the features by the LR, the RFE will select features by recursively considering smaller and smaller sets of features. First, the estimator is trained on the initial set of features and the weight of each feature is obtained through the coefficient attribute. Then, the least important features are pruned from the current set of features. That procedure is recursively repeated on the pruned set until the desired number of features to select is eventually reached (in our case ten) or the performance gain in between iteration is smaller than a given threshold. This technique has been proven to provide good results when paired with LR as seen in [19].

The Least Absolute Shrinkage and Selection Operator (LASSO), also known as L1 is a common practice for feature selection introduced in [17]. The L1 technique has is able to shrink some of the coefficients to zero. The L1 method puts an upper limit to the sum of the absolute values of the weights of the features. In order to avoid passing the limit, the method applies a shrinking process where it penalizes the coefficients of the regression features. In this process some of the features can end with a coefficient of zero, therefore, eliminated from the training set.

Table 2. Example of IV calculation and WoE for variable X.

Domain of X	Good	Bad	Total	Dist. of Good	Dist. of Bad	Category IV	WoE
X_1	523	414	937	0.3381	0.2576	0.0165	−0.2337
X_2	753	805	1558	0.4867	0.5204	0.0022	0.0667
X_3	271	328	599	0.1752	0.2120	0.0070	0.1909
Total	3,094	1,547	1,547	1.0000	1.0000	**0.0257**	

To do a fair evaluation, we trained the same algorithm, with the different subsets of features selected using the different feature selection techniques. As a classification algorithm, we used logistic regression. This algorithm has proven to provide accurate predictions with this type of data in the past [10]. For this experimentation, we trained the logistic regression with penalty set to 'L1', the parameter C equal to 1.0, a $SAGA$ solver, for *multi_class* we used *'one-vs-rest'* which refers to a binary target as in our case. Finally, we used the same random state for the training phase.

To maintain consistency we selected the ten most relevant features selected by each method. We selected only ten in order to compare with the features that had a strong or medium relationship with the target based on IV (IV $>$ 0.1) which are also ten.

For metrics of evaluation, we used the Area under Receiver Operating Characteristic (AUROC) and the Kolmogorov-Smirnov (K-S). These metrics are commonly used to evaluate the performance of a scoring algorithm as seen in [4,18], among others. We used 10-fold cross-validation to test the performance of the algorithms using a given dataset. This validation technique helps us to maximize the use of the data as it uses each case both for training and validating.

5 Analysis of Results

As explained in the previous section, the results will be compared with the results obtained by using the IV and the rules proposed by [15].

In Table 3, we present the results for the logistic regression using the different subsets of features generated using the different feature selection techniques.

Table 3. Table of the performance of the different feature selection methods.

Selection method	Kolmogorov-Smirnov	AUROC
Feature importance	0.2500	0.6540
Extra-Trees	0.3103	**0.7016**
RFE	0.2802	0.6707
L1	**0.3491**	0.6975
Information value	0.2004	0.6327

The bottom line of the table shows the results of using the rules presented on [15] and is the basis for comparison. We can observe that all the remaining methods achieved better performance both in K-S and AUROC when compared to the IV. The ETFS presented the best performance on the AUROC with 0.7016, which represents an increase of 20.28% to the previous result obtained with IV. However, when comparing the K-S, the L1 outperformed the other methods by achieving a K-S of 0.3491, a 74.20% increase relative to the IV performance. This means that the scoring model built with the features selected by L1 is able to differentiate better the distributions of the classes of the target. Therefore, the percentage of correct predictions within the good and bad class is more similar. In our case, the primary goal was to reduce miss-classification of bad cases which in this context is to classify a customer with bad credit as a customer with good credit, therefore granting a loan. We would prefer to have a model with better prediction of the bad customers even if this means to lose some prediction power in the good customers since the cost of default overweight the cost of opportunity. This being said, the model built with the features selected by L1 provided the best accuracy for both the good and the bad class. However, notice that the performance of these models is around 70% overall accuracy. Therefore, these models are only useful to display the difference between the selection methods. In order to build more robust models, we should consider a higher number of features since we capped the number of features to the first ten selected by each method. As for the features selected, 25 unique features were selected by the 5 different feature selection methods. The only feature to be selected as relevant by all the methods was *"number_of_apps_installed"*. This feature refers to the number of different apps installed on the mobile phone of the applicant. Only 11 features were selected by at least 2 feature selection methods and 14 features selected by only 1 method. The top five relevant features selected by ETFS matches the top five selected by L1. This explains why the results of these two methods are similar in terms of AUROC and K-S.

6 Conclusions

As we have shown in the previous section, we can build more robust scoring models by transforming the original dataset with the WoE and then applying a feature selection method. Not only the models generated with the assembled method are more accurate but also the distance between the classes to predict is higher as indicated by the results above. However, some of the feature selection methods have different stopping criteria for selecting the number of features. For this study, we used the ten with a higher score on the respective feature selection method.

For future work, we will focus on using different stopping criteria for the different methods. We believe this can be a determining factor of the performance scoring algorithm as we might be losing relevant features in some methods due to the limitation of using the top ten features. Furthermore, we will study the optimal stopping criteria for the different methods. As some of them do not have a clear threshold to cut as the IV. We will also explore how the different methods for binning the numerical features with the WoE transformation can affect the features predictive power.

Acknowledgment. This article is a result of the project Risk Assessment for Microfinance, supported by Norte Portugal Regional Operational Programme (NORTE 2020), under the PORTUGAL 2020 Partnership Agreement, through the European Regional Development Fund (ERDF)

References

1. Barnes, P.: E-commerce in emerging markets: the biggest growth opportunity (2016). http://marketrealist.com/2016/12/emerging-markets-better-demographics/. Accessed 20 Nov 2019
2. Biçer, I., Sevis, D., Bilgic, T.: Bayesian credit scoring model with integration of expert knowledge and customer data. In: International Conference 24th Mini EURO Conference "Continuous Optimization and Information-Based Technologies in the Financial Sector" (MEC EurOPT 2010), pp. 324–329. Vilnius Gediminas Technical University Publishing House "Technika" (2010)
3. Bjorkegren, D., Grissen, D.: Behavior revealed in mobile phone usage predicts loan repayment (2015)
4. Blanco, A., Pino-Mejías, R., Lara, J., Rayo, S.: Credit scoring models for the microfinance industry using neural networks: evidence from Peru. Expert Syst. Appl. **40**(1), 356–364 (2013)
5. De Cnudde, S., et al.: Who cares about your Facebook friends? Credit scoring for microfinance. Technical report (2015)
6. Fifer Mandell, A., Strawther, M., Zhu, J.: Inventure: building credit scoring tools for the base of the pyramid (2015)
7. Kursa, M.B., Rudnicki, W.R., et al.: Feature selection with the Boruta package. J. Stat. Softw. **36**(11), 1–13 (2010)
8. Madhavi, A.V.: Improving the credit scoring model of microfinance institutions by support vector machine (2014)

9. Poushter, J.: Financial inclusion: helping countries meet the needs of the under-banked and under-servedk (2016). http://www.pewglobal.org/2016/02/22/smartphone-ownership-and-internet-usage-continues-to-climb-in-emerging-economies/. Accessed 20 Nov 2019
10. Ruiz, S., Gomes, P., Rodrigues, L., Gama, J.: Credit scoring for microfinance using behavioral data in emerging markets. Intell. Data Anal. **23**(6), 1355–1378 (2019)
11. Pedro, J.S., Proserpio, D., Oliver, N.: MobiScore: towards universal credit scoring from mobile phone data. In: Ricci, F., Bontcheva, K., Conlan, O., Lawless, S. (eds.) UMAP 2015. LNCS, vol. 9146, pp. 195–207. Springer, Cham (2015). https://doi.org/10.1007/978-3-319-20267-9_16
12. Schreiner, M.: A scoring model of the risk of costly arrears for loans from affiliates of women's world banking in Colombia. Women's World Banking (2000)
13. Schreiner, M.: The risk of exit for borrowers from a microlender in Bolivia. Center for Social Development, Washington University in St. Louis, gwbweb. wustl. edu/users/schreiner (1999)
14. Schreiner, M., et al.: A scoring model of the risk of costly arrears at a microfinance lender in Bolivia. Center for Social Development, Washington University in St. Louis, gwbweb. wustl. edu/users/schreiner (1999)
15. Siddiqi, N.: Credit Risk Scorecards. Devoloping and Implementing Intelligent Credit Scoring. Wiley, New Jersey (2006)
16. Sousa, M.R., Gama, J., Brandão, E.: A new dynamic modeling framework for credit risk assessment. Expert Syst. Appl. **45**, 341–351 (2016)
17. Tibshirani, R.: Regression shrinkage and selection via the Lasso. J. Royal Stat. Soc. Ser. B (Methodological) **58**(1), 267–288 (1996)
18. Van Gool, J., Verbeke, W., Sercu, P., Baesens, B.: Credit scoring for microfinance: is it worth it? Int. J. Fin. Econ. **17**(2), 103–123 (2012)
19. Zhu, J., Hastie, T.: Classification of gene microarrays by penalized logistic regression. Biostatistics **5**(3), 427–443 (2004)

Learning Surrogates of a Radiative Transfer Model for the Sentinel 5P Satellite

Jure Brence[1,2(✉)], Jovan Tanevski[1,3], Jennifer Adams[4], Edward Malina[5], and Sašo Džeroski[1,2,4]

[1] Department of Knowledge Technologies, Jožef Stefan Institute, Ljubljana, Slovenia
jure.brence@ijs.si
[2] Jozef Stefan International Postgraduate School, Ljubljana, Slovenia
[3] Institute for Computational Biomedicine, Faculty of Medicine, Heidelberg University Hospital and Heidelberg University, Heidelberg, Germany
[4] Φ-lab, ESA/ESRIN, Frascati, Italy
[5] Earth and Mission Science Division, ESA/ESTEC, Noordwijk, The Netherlands

Abstract. Surrogate models approximate the predictions of other models. The motivation for learning surrogate models can come from computational concerns, when the predictions of the original model are computationally expensive to obtain. In contrast, the surrogate models are computationally efficient.

In this paper, we propose a framework for machine learning of surrogate models, which operate on the same input and output spaces as their original models. Instead of learning direct mappings from the input to the output space (and vice versa), we first assess the intrinsic dimensionality of the input and output spaces and reduce it appropriately, by using PCA and autoencoders. Predictive models are learned on the reduced spaces by the use of neural networks and their predictions are mapped to the original spaces.

We apply the framework to learn a surrogate model for a complex radiative transfer model RemoTeC, designed and built at SRON in the Netherlands. The original model predicts shortwave infrared (SWIR) spectra, for a given state vector of atmospheric parameters, representative of any geo-location that the Sentinel 5P satellite may encounter. The results indicate a low dimensionality of both the input and the output space and are accurate in both the forward and reverse direction.

Keywords: Surrogate models · Dimensionality reduction · Radiative transfer · Satellite data · Neural networks · Forward and reverse models

J. Brence and J. Tanevski—These authors contributed equally.

A. Appice et al. (Eds.): DS 2020, LNAI 12323, pp. 217–230, 2020.
https://doi.org/10.1007/978-3-030-61527-7_15

1 Introduction

The TROPOspheric Monitoring Instrument (TROPOMI) onboard the Copernicus Sentinel-5 Precursor (S5P) satellite is an important step forward in Earth Observation. TROPOMI provides global information on air quality and greenhouse gases such as methane, carbon monoxide and water vapour, as well as many others. TROPOMI products, such as gas concentrations, are generated through schemes knows as atmospheric retrievals. The retrieval schemes of some of these products, for example methane, typically rely on "optimal estimation methods", which although well proven, require large processing resources due to the running of forward models.

In physical sciences and engineering, theories are often tested by comparing measured or experimental data with the results of computational methods. These methods are expensive to compute, especially when the parameter space is large or an iterative approach is used. In such cases, the simulation or a part of it, can be replaced by a surrogate model - a computationally efficient approximation of a computationally complex function. Machine learning models are typically computationally expensive during the training process, but cheap when used for prediction. The use of a surrogate model speeds up the simulation, which allows researchers to spend their computational time on exploring larger parts of the parameter space, or improving the accuracy of their methods through more iterations [14].

In the case of TROPOMI trace gas retrievals, surrogate models that emulate the full forward physical model are gaining interest recently, since running the complex forward models is a bottleneck in the provision of operational and/or near-real time products. Being able to reduce computational and timing costs provides an opportunity for using more advanced physics for trace gas retrievals, without the overhead penalty on operational retrievals.

In this study we present our work on implementing a surrogate model that would emulate the simulations of the RemoTeC algorithm (currently the main source of TROPOMI trace gas concentrations) accurately and in a computationally efficient way. In addition to forward models learning to predict the output of the simulation, we also train reverse models to predict simulation parameters, based on the outputs. Reverse models can be considered as surrogates for the task of estimating the parameters of a complex system, given real measurements or simulation.

Emulating the RemoTeC algorithm presents unique challenges, because the algorithm features a large number of dimensions in both input and output spaces, compounded by the requirement of developing both forward and backward models. Many machine learning algorithms have trouble with predictions in settings like this due to the curse of dimensionality. Dimensionality reduction is a family of unsupervised learning methods that can be helpful in approaching such problems. These methods try to find an embedding - a projection of the data to a space of fewer dimensions that preserves as much information as possible [4]. Some well known algorithms include principal component analysis (PCA) [13], kernel PCA [4], t-SNE [11], UMAP [12] and autoencoders [3]. We investigate different methods of dimensionality reduction for both input and output spaces and evaluate their effect on the performance of the predictive models as surrogates for the task of simulation and parameter estimation.

2 RemoTec Radiative Transfer Model for Sentinel 5P TROPOMI Simulations

2.1 Copernicus Sentinel-5 Precursor

The S5P satellite was launched in October 2017 with the aim to provide global information on air quality and greenhouse gases. S5P is a joint venture between the European Space Agency (ESA) and the Netherlands, and is the first of several planned missions for air quality monitoring in the ESA/European Commission Copernicus program. Onboard S5P is the TROPOMI instrument which is an imaging spectrometer, providing data in a number of wavebands, of which the SWIR band (2300–2380 nm) is the subject of this work. The SWIR band (typically known as SWIR3) is focused on providing data on atmospheric concentrations of methane, carbon monoxide and water vapour, all of which are important in the context of a changing global climate [9].

2.2 RemoTec Code and LINTRAN RTM

A "retrieval algorithm" is used to convert the spectra captured by TROPOMI (known as Level 1 data) into trace gas concentrations (known as Level 2 data), for example the RemoTeC algorithm, the details of which are covered in [2,7]. RemoTec simulates a realistic approximation of the instrument response of S5P SWIR band, given a wide range of atmospheric parameters, including scattering by aerosols and variations in surface albedo and solar zenith angle, and is one of the current key methods for methane retrievals from S5P/TROPOMI. The core of the RemoTeC algorithm is the LINTRAN radiative transfer model [6], which calculates synthetic spectra based on a set of input atmospheric, spectroscopic, surface and instrument properties/assumptions. The whole retrieval process is computationally intensive, and the speed up of these algorithms is the subject of much work in the atmospheric communities.

The RemoTeC algorithm is split into two components, operational and synthetic, where the operational aspect deals with the active retrievals of methane from TROPOMI [8] and is not the subject of this work. The synthetic component was designed to test the RemoTeC algorithm prior to the launch of S5P, and is based on the retrievals of synthetically generated atmospheric scenarios, where the synthetic scenarios are used to generate synthetic spectra using LINTRAN. These scenarios form the basis of the training data set used in this study, and are described in more detail below.

Table 1. Input atmospheric parameters for the RemoTec model, value distribution and source of information.

Parameter	Variation/distribution	Source
SZA	0–70 deg	
Albedo	0.01, 0.1, 0.3, 0.5, 0.8	ADAM database
CH4 profile	Arctic, mid-latitude and tropical conditions	TM5 model
CO profile	Arctic, mid-latitude and tropical conditions	TM5 model
H20 profile	Arctic, mid-latitude and tropical conditions	ECMWF
Aerosols	Five different aerosol types	ECHAM-HAM
Temperature	Arctic, mid-latitude and tropical conditions	ECMWF
Pressure	Arctic, mid-latitude and tropical conditions	ECMWF

2.3 Inputs: Atmospheric Parameters

The training dataset is generated using the RemoTeC tool, provided by Dr. J. Landgraf at the Dutch Space Research Organisation (SRON). The atmospheric parameters input into RemoTec are designed to cover the range of atmospheric conditions that S5P/TROPOMI is expected to encounter, in order to develop a surrogate model capable of approximating both realistic atmospheric conditions and S5P/TROPOMI measurements. The input state vectors are generated from a combination of chemistry transport models. Table 1 outlines each of the atmospheric parameter inputs, their possible associated values and the source of information (either from chemistry transport models, meta-data or explicitly defined). A combination of values, one for each parameter shown in Table 1, is fed into RemoTeC/LINTRAN as a state vector.

In total, a dataset of 50,000 input state vectors is generated. This dataset comprises 10,000 individual measurement points, split into a global dataset of 3×3 degree bins, representing one day from each season, i.e. January, April, July and October. The dataset is generated by varying the surface albedo conditions between 0.01 and 0.8 (as shown in Table 1) for each of 10k simulated measurement points. Land conditions are considered only, and no sea environments are included in the dataset.

2.4 Outputs: S5P TROPOMI L1 Synthetic Spectra

Given each set of state vectors, synthetic Level 1 radiance (defined as the radiant flux emitted, reflected, transmitted or received by a given surface, per unit angle per unit projected area) spectra are simulated using the RemoTec RTM in the S5P/TROPOMI Shortwave InfraRed (SWIR3) band. An example of the SWIR-3 band spectra produced by RemoTec is given in Fig. 1. In total, 50,000 synthetic spectra were generated under a given state vector of atmospheric parameters, outlined in the Section above.

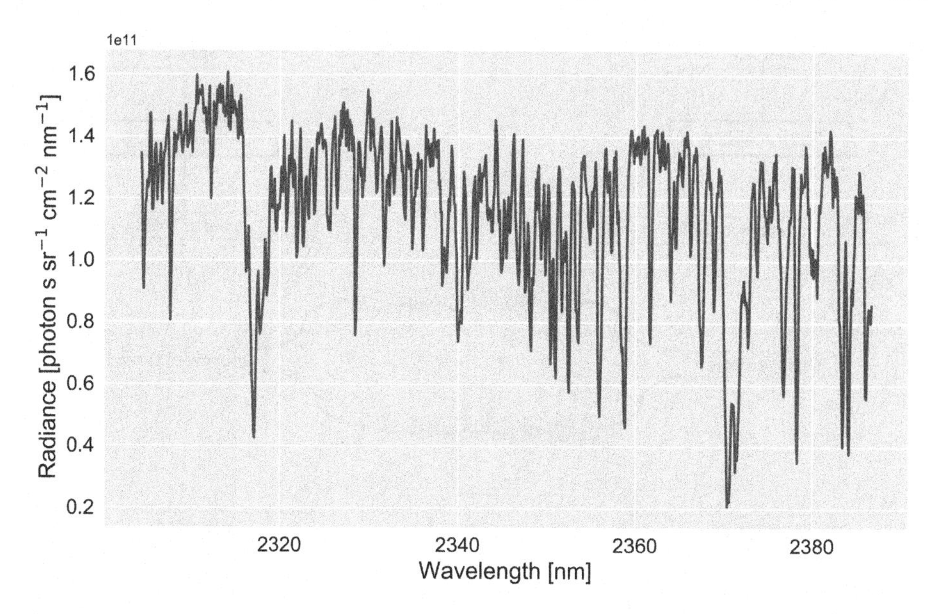

Fig. 1. An example SWIR3 spectrum from RemoTeC, assuming a realistic atmospheric profile and TROPOMI instrument characteristics.

3 Surrogate Model Learning

Surrogate models are commonly used to replace computationally expensive simulations of complex models. However, constructing surrogate models and making predictions can still be computationally complex. This is the case when the dimensionality of the input or the output space is large, increasing the complexity of model construction, as well as the computation time required to make predictions. One way to address this issue is through the use of methods for dimensionality reduction.

In our framework, depicted in Fig. 2, preprocessing and dimensionality reduction is performed on the input and the output spaces, i.e. on atmospheric parameters and spectra, separately. The forward model takes an embedded representation of the parameters as input and predicts the representation of spectra as output. The predictions in the embedding of the output space are then inversely transformed to obtain predictions in the original space of spectra. For the backward model, the roles of input and output are reversed, with the representation of spectra acting as input and the representation of atmospheric parameters acting as output to be predicted.

3.1 Data Exploration and Preprocessing

Initial exploration of the dataset revealed three features with constant values among the atmospheric parameters, which we removed from the data. The remaining 122 features have magnitudes on different scales. The distributions

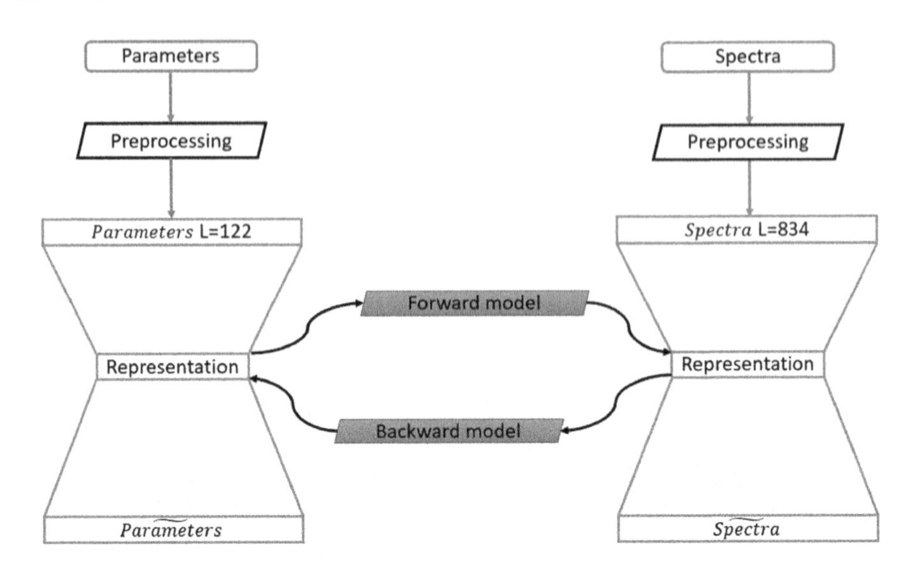

Fig. 2. The architecture of our framework for learning surrogate models.

of the variables are either bimodal or heavy-tailed unimodal, with high density around the modes. We normalized the atmospheric parameter data to a standard distribution by applying the transformation $\frac{x-\mu}{\sigma}$.

The spectral space contains 834 targets, all on a similar scale. Their distributions are predominantly exponential. We divided all values by 10^{11} and applied a logarithmic transformation, followed by normalization to a standard distribution.

3.2 Dimensionality Reduction

We have studied and compared two methods of dimensionality reduction within our framework: principal component analysis (PCA) and autoencoders. PCA is a popular method that is easy to implement and cheap to compute. Autoencoders feature a number of hyperparameters and are computationally more expensive, but have the potential to find better embeddings than PCA, due to the nonlinear transformations they make.

Principal Component Analysis. is a linear method of dimensionality reduction that finds projections into lower-dimensional subspaces, so that variance in the data is maximized. We can gain some insight about the intrinsic dimensionality of our data by looking at how the cumulative relative variance depends on the number of principal components, as depicted in Fig. 3. For the 122-dimensional atmospheric parameter space, we need:

- 23 dimensions to explain 95% of the variance,
- 45 dimensions to explain 99% of the variance, and
- 73 dimensions to explain 99.9% of the variance.

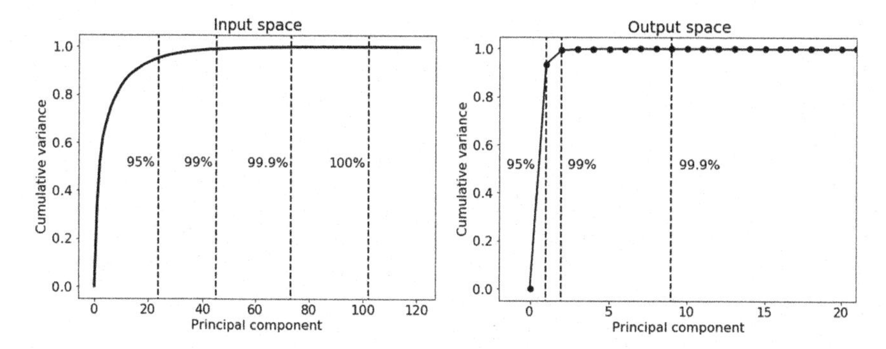

Fig. 3. Dependence of the cumulative relative variance on the number of principal components for both the input and the output space.

In the 834-dimensional spectral space,

- 1 dimension is enough for 95% of the variance,
- 2 dimensions are enough for 99% of the variance, and
- 9 dimensions are enough for 99.9% of the variance.

The spectral space has peculiar properties and an extremely low intrinsic dimensionality. In particular, the first principle component is very high, which might indicate an artifact in the simulations or in the sampling of the dataset. The atmospheric parameter space appears to be overdetermined as well, allowing a reduction from 122 dimensions to 45, while keeping 99% of the variance. The observed properties of the spectral space are worrying and warrant further investigation by domain experts. Nevertheless, the dataset presents an interesting opportunity for the study of the impact of dimensionality reduction on the prediction task at hand.

Autoencoders. are a type of neural network that is used to learn low dimensional embeddings. The network is trained to reproduce the input data on the output layer, with the defining characteristic that the network architecture features a bottleneck - the embedding layer. A concern when designing autoencoders is that a network with sufficient capacity would memorize the entire dataset and simply learn the identity function to satisfy the reproduction loss function. To combat this issue and improve the ability of an autoencoder to capture important information and learn richer representations, different methods of regularization are employed. Some common methods include batch normalization, adding a sparsity term to the loss function and adding noise to the input.

We performed dimensionality reduction using a denoising autoencoder [5]. Adding some amount of Gaussian noise to the input data forces the autoencoder to learn meaningful features. We treated the variance of the noise as a dimensionality reduction hyperparameter and tested several options. Note that the noise is only added during training.

We experimented with a number of autoencoder architectures and settled on a model with a total of 7 layers, including input and output. The structure of the autoencoder for the parameter space can be summarized as follows:

1. input layer of size $N_0 = 122$ + Gaussian noise layer,
2. fully connected layer of size $N_1 < N_0$ and activation ReLu,
3. fully connected layer of size $N_2 = \frac{1}{2}N_1$ and activation ReLu,
4. fully connected (embedding) layer of size $N_3 < N_2$ and linear activation,
5. fully connected layer of size N_2 and activation ReLu,
6. fully connected layer of size N_1 and activation ReLu,
7. output layer of size $N_0 = 122$ and linear activation.

For the spectral space, the architecture is the same, with $N_0 = 834$. The models were trained using the Adam optimizer [10] and a MSE loss function. The models were implemented with Tensorflow 2.0 [1].

3.3 Learning Forward/Backward Models on Reduced Spaces

Our method of choice for learning the predictive models were feedforward neural networks. In our experiments, we used a network with two fully connected hidden layers and ReLu activation functions on every layer, except the output layer, which used a linear activation function. Since we investigated the effect of different levels of dimensionality reduction on both the input and the output space, the number of neurons in each layer varied and was treated as a hyperparameter to be optimized.

The models were trained on the preprocessed training set with reduced dimensions. The hyperparameters defining the network architecture were optimized by using a validation set. The test set was used to compute the prediction errors for evaluating the model and for comparing different combinations of dimensionality reduction and prediction models.

4 Experiments and Results

We tested and evaluated two methods for dimensionality reduction and prediction to identify the best choice for our particular problem. We compared the properties of different methods of dimensionality reduction and how well they are able to reconstruct the original data. We then implemented predictive models that map between the reduced input and output spaces and compared their predictive performance. The workflow can be summarized as follows:

1. Randomly split the data into a training set (80%), validation set (10%) and test set (10%).
2. Preprocess the features and targets.
3. Compute a dimensionality reduction projection on the feature space of the training set, then transform all three sets using the learned projection. Evaluate by using the error of reconstruction on validation set as a measure of quality. Repeat for the target space.

4. Train the neural network to predict from the reduced feature space to the reduced target space. Optimize the hyperparameters of the network by using the validation set. Vice-versa for the backward model.
5. Evaluate the forward and the backward model on the test set.

Note that step 1 - the splitting of the dataset - was done only once, with all subsequent experiments working on the same split.

4.1 Dimensionality Reduction

We compared the different methods for dimensionality reduction on both the atmospheric parameter and the spectral space separately. We reduced the dimensionality of the parameter space from 123 down to 45 dimensions and the dimensionality of the spectral space from 834 to 2. The lowest numbers of dimensions, 45 and 2, correspond to the numbers of principle components in the parameter and the spectral space, respectively, that are needed to explain 99% of the variance.

The reconstruction errors, i.e., the root mean squared errors (RMSE), of the models constructed using reduced dimensions obtained with autoencoders and PCA are shown in Table 2. When training the denoising autoencoder, we added Gaussian noise with mean $\mu = 0$ and different values for the standard deviation σ. The models were trained on the train set and evaluated on the validation set. Unsupervised learning normally does not require a train-test split. Here, we train the autoencoders on the train set only, since the dimensionality reduction models are used to inverse transform the predictions from the embedded space back into the original space. The error of this reconstruction is more objectively estimated by evaluating it using unseen data from the validation set.

Table 2. The reconstruction error (RMSE) of an autoencoder with three different input noise levels (σ) compared to PCA. For the atmospheric parameter space (X), we reduce the dimensions from 123 to 45. For the spectral space (Y), the dimensions are reduced from 834 to 2.

σ	Autoencoder			PCA
	0	0.1	0.5	
X	0.58	0.71	1.51	0.62
Y	1.51	1.52	1.57	1.67

We observed that adding noise resulted in worse reconstruction error on the test set for high noise values. The reconstruction error of the autoencoder on the parameter space is lower than the error of PCA for no added noise, and higher for both nonzero levels of noise. In the spectral space, the autoencoder reconstruction error is lower than the error of PCA for all levels of noise.

4.2 Predictive Models

We compared the prediction error of the models, mapping between the embedding spaces, created by different methods of dimensionality reduction, using the root mean squared error of the predictions. The error was computed in the original space, after inversely transforming the predictions. To provide a reference for the quality of our models, we compared the prediction errors to the error of a conservative baseline model that always predicts the average value of the target from the train set.

Table 3. Comparison of the prediction error of the predictive models with different reduced input and output spaces. The reduced dimensions are 45 for the atmospheric parameter space and 2 for the spectral space, in all four cases.

σ	Autoencoder			PCA	Baseline
	0	0.1	0.5		
Forward	3.8	2.2	2.7	2.4	33.5
Backward	11.4	11.4	11.8	11.4	16.5

Table 3 shows the performance of the predictive models using different dimensionality reduction methods. For the forward model, the prediction using autoencoder representations works best with a noise level of $\sigma = 0.1$. Regularization by adding noise improved the performance of prediction. PCA reduction outperforms reduction with autoencoders without noise, as well as high noise. For the backward model, the performances are similar.

We further compared the effect of input and output space dimensionality on the predictive performance of the learned models. Considering the results of the previous step, we chose $\sigma = 0.1$ for the noise level. We compared the prediction error for different combinations of embedding dimensionality as $dim(X) - dim(Y)$:

- 45-2, which explains 99% of variance in PCA,
- 73-9, which explains 99.9% of variance in PCA, and
- 102-50, which explains nearly 100% of variance in PCA.

We chose 50 as the largest output dimension, because it resulted in best performance and at the same time represents a large reduction in dimensionality. In Table 4, we show the results and compare the performance of the predictive models using different levels of dimensionality reduction, as well as to the performance of the predictive model without any dimensionality reduction.

Models using PCA for dimensionality reduction outperform models using autoencoders for all pairs of reduced dimensions, except for 45-2. For both the forward and the backward model, all combinations of dimensionality reduction outperform the baseline predictor. After dimensionality reduction, the predictive model can have an error, higher than the error of a model constructed on data

Table 4. Comparison of the prediction error on the test set for different dimensions of the atmospheric parameter and spectral space representations. The noise level of the autoencoder was set to $\sigma = 0.1$.

Forward model				
Dimensions (X-Y)	Autoencoder	multicolumn1l—PCA	No DR	Baseline
45-2	2.2	2.4		
73-9	1.9	1.1		
102-50	0.95	**0.85**		
All			0.98	33.5
Backward model				
Dimensions (X-Y)	Autoencoder	PCA	No DR	Baseline
45-2	11.4	11.4		
73-9	7.0	5.0		
102-50	5.9	**4.2**		
All			4.7	16.5

with no dimensionality reduction. Interestingly, the predictive performance for the forward model is improved by using an embedding of 102 dimensions in the parameter space and 50 dimensions in the spectral space, as compared to using the full spaces.

The results for the backward models are similar, with one important difference: the performance of the backward models is generally lower than the performance of the forward models. While still better than the baseline predictor, the difference is not large. The best performing combination is the same as in the forward model: PCA with 102 dimensions in the parameter space and 50 dimensions in the spectral space. Using autoencoders results in lower performance than using PCA. Models using dimensionality reduction outperform models working in the original space.

To confirm the effect of dimensionality reduction, we took the best performing models using dimensionality reduction, as well as models working in the original space, and re-evaluated them by 10-fold cross validation. We compare the coefficient of determination

$$R^2 = 1 - \frac{\text{MSE(model)}}{\text{MSE(baseline)}}$$

in Table 5.

After re-evaluation, the forward models in the original space achieve the same R^2 as forward models in reduced space, up to the fourth decimal. However, the backward models experience a great improvement in performance by reducing the dimensionality, bringing the coefficient of determination from 0.906 to 0.973.

Table 5. Coefficient of determination for the best performing models using dimensionality reduction (PCA 102-50) and the models with no dimensionality reduction, estimated by 10-fold cross validation.

	DR	No DR
Forward	0.9999	0.9999
Backward	0.9733	0.9060

5 Discussion

The results of our study are very promising. The RemoTeC simulation can be emulated well by predictive models based on neural networks in both the forward and backward directions. The use of dimensionality reduction on the parameter and spectral spaces considerably improves the performance of the backward model and does not degrade the performance of the forward model. Furthermore, the predictive model can be simpler when mapping between spaces with fewer dimensions, leading to improved computational efficiency. In our experiments, making predictions using the neural network was approximately 33% faster when mapping between the reduced spaces than when working in the original space. Both options need only a few tenths of a second to make predictions for 5000 examples, while the original simulation requires hours or days of computation. The main motivation for developing surrogate models is to improve computational efficiency and the developed models proved to be very successful at that task.

Our analysis of the dataset revealed an extremely high first principal component in the spectral space and low intrinsic dimensionality of the spectral, as well as the parameter space. This property is likely linked to limitations in the set up of the training dataset and associated sampling of certain parameters. It is unclear how the presence of this artifact affects the performance of the presented methodology.

A large part of this study was devoted to investigating how techniques for dimensionality reduction can be used to improve the performance and computational efficiency of surrogate models. However, the choice of algorithm for prediction was given relatively little attention. Further work will likely include a comparison of different models, with a focus on ensembles of regression trees - the random forest family of methods.

The eventual aim of this study is to input the surrogate model developed here into retrieval algorithms for various trace gases, that currently rely on the full forward physical models. This would allow us to firstly determine the uncertainty of the surrogate model compared to the full physical forward model at the trace gas level with respect to uncertainty requirements developed for the mission. Secondly, given this uncertainty information with respect to requirements, we can assess the trade-off between reduction in computational and timing costs of the surrogate model with potential losses in accuracy, compared with the original physical model.

One of the key advantages of using surrogate over physical models, is the ability to retrieve trace gases using more accurate but computationally expensive models that make use of more complex physics. In operational or real-time use, employing more complex physics (for example retrievals of individual aerosol types, or employing line-by-line spectroscopic parameter calculation) is often not possible. However, surrogate models that learn from more accurate models but perform computations in a fraction of the time (given a possible loss in accuracy) can be extremely useful. Accordingly, further work intends to explore the learning of surrogate models from either more complex physical models or models that include more advanced physics. With large numbers of Earth Observation missions planned in the coming decades, huge volumes of additional data will need to be analysed. By using the methods highlighted in this work, it may be possible to speed up this analysis, while retaining high levels of accuracy without making the assumptions and sacrifices made by current retrieval algorithms in order to remain quick.

References

1. Abadi, M., et al.: TensorFlow: a system for large-scale machine learning. In: 12th USENIX Symposium on Operating Systems Design and Implementation (OSDI 2016), pp. 265–283 (2016)
2. Butz, A., Galli, A., Hasekamp, O., Landgraf, J., Tol, P., Aben, I.: TROPOMI aboard Sentinel-5 Precursor: prospective performance of CH4 retrievals for aerosol and cirrus loaded atmospheres. Remote Sens. Environ. **120**, 267–276 (2012). https://doi.org/10.1016/j.rse.2011.05.030
3. Charte, D., Charte, F., García, S., del Jesus, M.J., Herrera, F.: A practical tutorial on autoencoders for nonlinear feature fusion: taxonomy, models, software and guidelines. Inf. Fusion **44**, 78–96 (2018)
4. Friedman, J., Hastie, T., Tibshirani, R.: The Elements of Statistical Learning. Springer Series in Statistics, vol. 1. Springer, New York (2001). https://doi.org/10.1007/978-0-387-84858-7
5. Goodfellow, I., Bengio, Y., Courville, A.: Deep Learning. MIT Press (2016). http://www.deeplearningbook.org
6. Hasekamp, O.P., Landgraf, J.: A linearized vector radiative transfer model for atmospheric trace gas retrieval. J. Quant. Spectrosc. Radiat. Transf. **75**(2), 221–238 (2002). https://doi.org/10.1016/S0022-4073(01)00247-3
7. Hu, H., et al.: The operational methane retrieval algorithm for TROPOMI. Atmos. Meas. Tech. **9**, 5423–5440 (2016). https://doi.org/10.5194/amt-9-5423-2016. www.atmos-meas-tech.net/9/5423/2016/
8. Hu, H., et al.: Toward Global Mapping of Methane With TROPOMI: First Results and Intersatellite Comparison to GOSAT, April 2018. https://doi.org/10.1002/2018GL077259. http://doi.wiley.com/10.1002/2018GL077259
9. IPCC: Fifth Assessment Report - Impacts, Adaptation and Vulnerability (2014). http://www.ipcc.ch/report/ar5/wg2/
10. Kingma, D.P., Ba, J.: Adam: A method for stochastic optimization. arXiv preprint arXiv:1412.6980 (2014)
11. van der Maaten, L., Hinton, G.: Visualizing data using t-SNE. J. Mach. Learn. Res. **9**, 2579–2605 (2008)

12. McInnes, L., Healy, J., Melville, J.: UMAP: Uniform manifold approximation and projection for dimension reduction. arXiv preprint arXiv:1802.03426 (2018)
13. Pearson, K.: LIII. on lines and planes of closest fit to systems of points in space. London Edinburgh Dublin Philos. Mag. J. Sci. **2**(11), 559–572 (1901)
14. Tanevski, J., Džeroski, S., Todorovski, T.: Meta-model framework for surrogate-based parameter estimation in dynamical systems. IEEE Access **99** (2019)

Nets Versus Trees for Feature Ranking and Gene Network Inference

Nicolas Vecoven(✉), Jean-Michel Begon, Antonio Sutera, Pierre Geurts, and Vân Anh Huynh-Thu

Department of Electrical Engineering and Computer Science, University of Liège, Liège, Belgium
{nvecoven,jm.begon,a.sutera,p.geurts,vahuynh}@uliege.be

Abstract. We investigate several global variable importance measures derived from artificial neural networks (ANN) to address the challenging problem of feature ranking in high-dimensional unstructured problems. While several ANN (local) importance measures have been validated in the context of computer vision or natural language processing tasks, it is not clear how these methods perform on unstructured problems where many variables are expected to be irrelevant. We empirically compare these ANN measures with one standard and state-of-the-art Random forests (RF) importance measure on several artificial and real datasets. These experiments show that ANN measures can achieve performance similar to the RF measure, sometimes outperforming it. On some problems however, the feature rankings returned by ANN are not as good as the ones returned by RF, despite significantly better predictive performance. Importantly, reaching the best performance with the ANN-based methods often comes at the cost of introducing a so-called selection layer at the beginning of the network. Using this specific neural architecture has proven to be critical both in terms of feature ranking and predictive performance on datasets with many irrelevant variables. Finally, we evaluate these methods on the problem of gene network inference, where they yield decent performance, without however outperforming RF.

Keywords: Feature ranking · Deep learning · Neural networks · Gene regulatory network inference · Random forests

1 Introduction

In many supervised learning applications, one is more interested by the interpretability of the trained model than by its actual predictive performance. One way to gain such interpretability is through the application of feature selection (or ranking) techniques, which aim at identifying the most relevant input features for predicting a given output. Recently, motivated by the advent of deep learning, there has been a resurgence of interest towards (old and new) techniques to derive variable importance scores from artificial neural networks (ANN) [1,7,16]. However, these methods have been mostly evaluated in the context of computer vision or text mining applications, where input relevance is typically computed

A. Appice et al. (Eds.): DS 2020, LNAI 12323, pp. 231–245, 2020.
https://doi.org/10.1007/978-3-030-61527-7_16

on a per sample basis (usually referred to as local importance measure) and can be assessed easily by human inspection. It is hence not clear how these ANN-based methods perform on unstructured, high-dimensional datasets where many variables are expected to be irrelevant. Furthermore, only very few works have focused on deriving feature importances in a global way, i.e., at the dataset level.

Feature selection techniques are traditionally divided into three families: filter, wrapper and embedded approaches [5]. Filters perform feature selection by looking at the intrinsic properties of the data. They are simple and fast, but they however work independently of the training of the predictive model. On the other hand, wrapper and embedded methods depend on the trained model. Wrappers use the model as a black box and exploit the predictions returned by the black box for deriving feature importance scores, e.g., by measuring how the output changes when perturbing the input or by training different models on different subsets of features. Embedded approaches directly incorporate feature selection in the training of the model and compute variable importance scores using formulae based on the internal parameters of the trained model. Like wrappers, embedded methods have the advantage to interact with the learning algorithm, while being typically much less computationally intensive.

In this paper, we focus our analysis on embedded approaches for ANN. We extend several existing approaches to derive global variable importance scores from ANN and compare them to a popular embedded approach, which derives variable importance scores from Random forests (RF) [12]. We carry out experiments on benchmark datasets, which show that ANN measures can achieve performance similar to the RF measure, sometimes outperforming it. On some problems however, the feature rankings returned by ANN are not as good as the ones returned by RF, despite significantly better predictive performance. Our experiments also show that the introduction of a selection layer at the beginning of the neural network allows to strongly improve the performance, both in terms of prediction and feature ranking. Finally, we compare ANN and RF on a challenging task in computational biology, namely the inference of gene regulatory networks, where RF are currently amongst the state-of-the-art approaches. While the ANN-based approaches do not outperform RF, they nevertheless return very promising results.

In the following, Sect. 2 formalizes the feature ranking problem, Sect. 3 introduces three ANN-based global variable importance measures, and Sects. 4 and 5 show our results on benchmark problems and on the gene network inference task respectively.

2 Problem Definition

In this paper, we tackle the problem of identifying *relevant* variables in a high-dimensional dataset comprising many *irrelevant* variables. We consider that a variable is relevant if it conveys information about the output, either in isolation or in conjunction with other relevant variables. More formally, let us consider a set $\mathbf{X} = \{X_1, \ldots, X_p\} \in \mathbb{R}^p$ of p input random variables and an output Y

that can be either continuous (regression problem) or categorical (classification problem). Let $\mathbf{X}^{-i}$ be the set $\mathbf{X}\backslash X_i$. Following the standard definition in [9], a variable X_i is said to be relevant if there exists a subset $\mathbf{B} \subseteq \mathbf{X}^{-i}$ (possibly empty) such that X_i brings additional information about Y conditionally to $\mathbf{B}$, i.e. if there exist some values x_i and $\mathbf{b}$ with probability $P(X_i = x_i, \mathbf{B} = \mathbf{b}) > 0$ and a value y such that[1]:

$$P(Y = y|X_i = x_i, \mathbf{B} = \mathbf{b}) \neq P(Y = y|\mathbf{B} = \mathbf{b}). \tag{1}$$

A variable is thus irrelevant if it is independent of the output conditionally to any subset of the other variables.

We assume that we have at our disposal a training set of N instances of input-output pairs, drawn from the unknown probability distribution $P(\mathbf{X}, Y)$:

$$S = \{(\mathbf{x}^k, y^k)\}_{k=1}^N. \tag{2}$$

A feature ranking method is then defined here as a procedure that exploits S to associate an importance score $Imp(X_i)$ to each input variable $X_i (i \in \{1, \ldots, p\})$, with the aim of assigning the highest scores to the relevant variables. Note that while our goal is to rank all the relevant variables above the irrelevant ones, we do not evaluate the ranking of the relevant variables among them.

3 Variable Importances from Neural Networks

In the past few years, numerous approaches have been developed for explaining predictions returned by ANN, among which are embedded methods that compute importance measures for input features. These approaches usually provide local importance scores, measuring the relevance of each input feature for a given individual prediction. They can be broadly divided into two families: gradient-based methods (e.g., [3,22]), which compute the gradient of the output with respect to the input, and decomposition-based methods (e.g., [2,21,24]), which decompose the output prediction (or the difference with respect to a baseline) into a sum of contributions from the different input features. Both gradient-based and decomposition-based methods are backpropagation approaches that propagate the importance signal from an output neuron to the input neurons through each layer of the network. Note that there exists a third category of embedded approaches, which identify the input pattern that activates the neurons in the different ANN layers (e.g., [8,23,26]). These approaches were however specifically developed for visualizing the predictions of a convolutional neural network in the context of image classification, and are therefore out of the scope of our analysis.

Our goal in this paper is to study how ANN-based importance measures perform with respect to standard measures derived from Random forests, rather

[1] Note that this definition applies to discrete variables, but can be easily extended to continuous variables by changing probabilities $P(X_i = x_i)$ to $P(X_i \leq x_i)$.

than comparing the existing ANN-based measures among them. We thus analyze in this paper only one representative method from each family. For a detailed discussion and comparison of the different existing approaches for ANN, the reader can refer to [1,7,16]. As representatives of the gradient-based and decomposition-based approaches, we choose, respectively, a standard approach that computes the absolute value of the derivative of the output with respect to each input feature (called GRAD in the following) [10,22] and the layer-wise relevance propagation (LRP) technique [2]. Both GRAD and LRP can be used with any pre-trained network with an arbitrary feed-forward structure. These methods (and some of their variants) have been previously discussed in the context of image classification, where they were shown to be able to identify the pixels that are useful for classifying a given image [2,22]. As mentioned above, GRAD and LRP provide local importance measures. Although local interpretation can provide fine-grained information about individual predictions, our goal here is different as we would like to derive a global measure of importance over all the instances, in order to get insights into the learned input-output relationships at the dataset level. To compute such global importance, we simply sum (or average) the local measures over all the training samples. In addition to GRAD and LRP, we also study a third approach based on the introduction of a so-called selection layer (SL), which thus requires to train a specific network structure.

Note that since we are targeting high-dimensional unstructured datasets, we focus on fully-connected feed-forward neural networks, although every method investigated in this work can be extended to more complex architectures in a straightforward way. We also assume here for simplification that all the hidden neurons use ReLU activations [17] and we expose each method in a (multi-output) regression setting. In the case of classification, a softmax layer is added and variable importances are derived by considering the inputs of the softmax layer—the logits—as multiple regression outputs. Python source codes of the different ANN-based methods presented below are available at https://github.com/nvecoven/ann_fsl.

3.1 Gradient (GRAD)

Let $f(\mathbf{x})$ be the ANN output for a given sample $\mathbf{x}$. A standard variable importance measure for the input x_i is given by the absolute (or squared) value of the derivative $\frac{\partial f(\mathbf{x})}{\partial x_i}$ [10,22]. This importance score thus measures how much the network output for the sample $\mathbf{x}$ changes regarding an infinitesimal change in x_i, and can be efficiently computed using back-propagation. To obtain a global importance score, we extend this approach by simply taking the sum of the derivatives over all the instances of the training set:

$$Imp(X_i) = \sum_{k=1}^{N} \left| \frac{\partial f(\mathbf{x}^k)}{\partial x_i} \right|, \tag{3}$$

When there are multiple outputs, the importance score can be summed over the different outputs. Note that when using ReLU activation functions, one can show that:

$$\frac{\partial f(\mathbf{x})}{\partial x_i} = \sum_{\forall P \in \Gamma_i(\mathbf{x})} \prod_{w \in \mathbf{w}_P} w, \tag{4}$$

where $\Gamma_i(\mathbf{x})$ is the set of all non-blocked paths from input x_i to the ANN output for the example $\mathbf{x}$ and $\mathbf{w}_P$ is the set of all the weights along the path P. A path is said to be blocked if it passes through an inactive neuron (i.e. a ReLU neuron that has a negative input).

3.2 Layer-Wise Relevance Propagation (LRP)

Although commonly used, the gradient method has the drawback that it does not explain the output of the network but rather how the output varies when the input is changed [16]. Clearly, an input could be relevant even if (3) is zero at a given point. Several alternative importance measures have been proposed to circumvent these limitations. As a representative of these methods, we use below a particular instance of the generic LRP method proposed in [2].

Let us consider a rectified fully-connected feedforward network and let $a_i^{(j)}$ be the activation (i.e. output) of the ith neuron in the jth layer ($0 \leq j \leq L$, $1 \leq i \leq p_j$). In such network, $a_i^{(0)} = x_i$ is the ith input (with $p_0 = p$). For $j \geq 1$, $a_i^{(j)} = ReLU\left(\sum_{k=1}^{p_{j-1}} a_k^{(j-1)} w_{ik}^{(j)} + b_i^{(j)}\right)$, where weights $w_{ik}^{(j)}$, corresponding to connections from the kth neuron of the $j-1$th layer to the ith neuron of the jth layer, and biases $b_i^{(j)}$ are learnable parameters. In classification, the outputs of the Lth layer are called the logits, which can be turned into probabilities by running them through a softmax function. The relevance of the ith neuron of the jth layer can be computed with

$$Imp_i^{(j)}(\mathbf{x}) = \begin{cases} |a_i^{(L)}| & \text{if } j = L \\ \sum_{k=1}^{p_{j+1}} \frac{a_i^{(j)} ReLU\left(w_{ki}^{(j+1)}\right)}{\sum_{l=1}^{p_j} a_l^{(j)} ReLU\left(w_{kl}^{(j+1)}\right)} Imp_k^{(j+1)} & \text{otherwise} \end{cases} \tag{5}$$

This propagation rule corresponds to the LRP rule with $\alpha = 1$ and $\beta = 0$ [2]. In the case of ReLU activations, it can be shown that applying this rule at a given layer can be viewed as a Taylor decomposition of the importance at that layer onto the lower layer [16]. The importance of an input X_i is eventually computed as the sum of $Imp_i^{(0)}(\mathbf{x})$ over all the training examples $\mathbf{x}$. In essence, while GRAD uses the samples to identify the (in)active paths, LRP looks at the activations.

3.3 Selection Layer (SL)

This method is inspired by sparse linear regression [25]. As illustrated in Fig. 1, a one-to-one connected layer with linear activations and no bias, called here *selection layer*, is introduced between the inputs and the first hidden layer of the network [11]. As all the other weights of the network, the weights of SL are

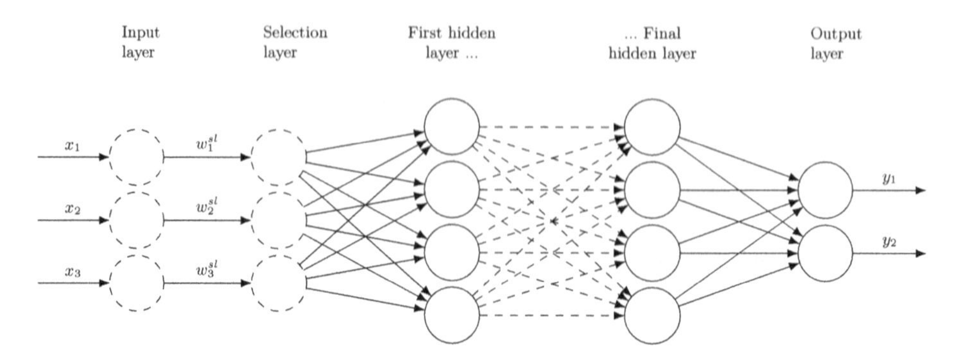

Fig. 1. Example of selection layer architecture. The selection layer consists of a one-to-one connected layer between the input variables x_i and the first hidden layer of the network. Dashed (resp. plain) circles represent neurons with linear (resp. non-linear) activations.

initialized with random values drawn from a truncated normal distribution with 0 mean and 0.1 standard deviation, and the network is trained while penalizing them to ensure that only useful information goes through the network. The importance $Imp(X_i)$ is then simply set to the weight $\left|w_i^{sl}\right|$ of X_i in the selection layer. Penalization can be achieved through elastic net [27], where the overall loss function is of the form:

$$\mathcal{L}_{\theta,sl}(B) = \frac{1}{|B|} \sum_{(\mathbf{x},y)\in B} \mathcal{C}_\theta(\mathbf{x}, y) + \frac{\alpha_1}{p} \sum_{i=1}^{p} |w_i^{sl}| + \frac{\alpha_2}{p} \sum_{i=1}^{p} \left(w_i^{sl}\right)^2 \tag{6}$$

where $\mathcal{L}_{\theta,sl}$ is the loss of the network indexed by parameters θ, $\mathcal{C}_\theta(\mathbf{x}, y)$ is the regular cost function (cross-entropy for classification problems or least-square error for regression problems) over batch $B \subseteq S$, $|B|$ is the batch size, and $\alpha_1 \geq 0$ and $\alpha_2 \geq 0$ are hyper-parameters balancing the penalty terms. Unless otherwise mentioned, in our experiments we focus on a L1 penalty only, i.e. we set $\alpha_2 = 0$.

3.4 Hybrid Methods

Below, we also experiment with mixed strategies, called SL+GRAD and SL+LRP, that train the network using the selection layer but compute the variable importances using the GRAD and LRP techniques respectively. We will show in Sect. 4 that these hybrid methods allow to strongly increase the performance on benchmark datasets.

4 Experiments on Benchmark Problems

We use datasets with a known ground-truth (i.e. known relevant features) in order to evaluate the five ANN-based approaches (GRAD, LRP, SL, SL+GRAD,

SL+LRP) introduced in the previous section. We compare them to standard RF mean decrease impurity (MDI) score, using as impurity measure the Gini index in classification and the variance in regression [12]. Since the relevant features are known, the variable rankings are assessed using the area under the precision-recall curve (AUPR). The AUPR will be equal to 1 if the ranking is perfect, i.e., if all the relevant variables receive a higher importance than the irrelevant ones, while the AUPR will be close to the proportion of relevant variables for a random ranking. In all the experiments, the inputs are centered and re-scaled according to their standard deviation prior to training. RF models are composed of 1000 unpruned trees. The parameter α_1 of SL (and α_2 when a L2 penalty is used), as well as the main parameter of RF (i.e., the number K of randomly chosen variables at each tree node) are tuned to minimize the generalization error (misclassification rate for classification, mean squared error for regression), estimated either using a validation set or by cross-validation (details are given in each respective section).[2]

4.1 Simulated Problems

We consider four different simulated problems:

- LR A linear regression problem generated using the *make_regression* function in scikit-learn [18]. Output Y is computed as $\sum_{i=1}^{25} w_i X_i$, where weights w_i are randomly and uniformly selected in $[0, 100]$, and inputs X_i are $\mathcal{N}(0,1)$ distributed.
- LC A linear, binary classification problem generated by thresholding the LR problem output so that the two classes are perfectly balanced.
- NLR A non-linear regression problem generated using the *make_friedman1* function in scikit-learn, which generates the following problem: $Y = 10 * sin(\pi * X_1 * X_2) + 20 * (X_3 - 0.5)^2 + 10 * X_4 + 5 * X_5 + 0.1 * \epsilon$, where ϵ is a $\mathcal{N}(0,1)$ noise and the inputs X_i are uniformly distributed in $[0, 1]$.
- NLC A non-linear, binary classification problem generated using the *make_classification* function of scikit-learn with 25 relevant features. Briefly, one of the two classes is associated randomly to each vertex of a hypercube of dimension 25 and training examples of the corresponding class are generated in the neighbourhood of each vertex by using a normal distribution centered on the vertex (with $\Sigma = I$).

For each problem, we generate 10 datasets with 2000 training samples, 1000 validation samples and 8000 test samples, and we add in each dataset a varying number of irrelevant features. These irrelevant features are generated using the same type of distribution as for the relevant features (i.e. $\mathcal{N}(0,1)$ for LR, LC and NLC and $\mathcal{U}(0,1)$ for NLR). In all the experiments, unless otherwise stated, each ANN is composed of 3 hidden layers of respectively 300, 150 and 75 ReLU

[2] Unless otherwise stated, values of the parameter K of RF are optimized in $\{\sqrt{p}, \log(p), p/3, p/2, p\}$, where p is the number of inputs.

Table 1. AUPR and ER/MSE for the four simulated problems, with 5000 variables in total in each problem. Values indicate means and standard deviations computed over 10 datasets. The best results are indicated in bold.

		SL+GRAD	SL+LRP	SL	GRAD	LRP	RF
LC	ER	**0.057 ± 0.011**			0.364 ± 0.006		0.239 ± 0.014
	AUPR	**0.902 ± 0.044**	0.881 ± 0.040	0.855 ± 0.048	0.730 ± 0.037	0.729 ± 0.038	0.724 ± 0.038
NLC	ER	**0.049 ± 0.007**			0.390 ± 0.025		0.186 ± 0.021
	AUPR	0.945 ± 0.026	0.941 ± 0.025	0.896 ± 0.039	0.599 ± 0.074	0.604 ± 0.074	**0.996 ± 0.008**
LR	MSE	**0.007 ± 0.003**			0.740 ± 0.018		0.618 ± 0.018
	AUPR	**0.976 ± 0.026**	0.969 ± 0.028	0.967 ± 0.029	0.864 ± 0.064	0.860 ± 0.075	0.815 ± 0.079
NLR	MSE	**0.152 ± 0.044**			0.862 ± 0.010		0.237 ± 0.008
	AUPR	0.860 ± 0.091	0.860 ± 0.091	0.860 ± 0.091	0.800 ± 0.000	0.800 ± 0.000	**1.000 ± 0.000**

Table 2. Results on the NLC problem with an increasing number of irrelevant features (from 25 to 9975 irrelevant features, in addition to the 25 relevant ones).

# feat.		SL+GRAD	SL+LRP	SL	GRAD	LRP	RF
50	ER	**0.039 ± 0.011**			0.040 ± 0.005		0.094 ± 0.014
	AUPR	0.998 ± 0.003	0.995 ± 0.005	0.986 ± 0.011	0.999 ± 0.001	**1.000 ± 0.000**	**1.000 ± 0.000**
2500	ER	**0.051 ± 0.010**			0.352 ± 0.025		0.171 ± 0.017
	AUPR	0.960 ± 0.025	0.955 ± 0.043	0.895 ± 0.000	0.595 ± 0.051	0.606 ± 0.052	**0.997 ± 0.005**
5000	ER	**0.049 ± 0.007**			0.390 ± 0.025		0.186 ± 0.021
	AUPR	0.945 ± 0.026	0.941 ± 0.025	0.896 ± 0.039	0.599 ± 0.074	0.604 ± 0.074	**0.996 ± 0.008**
10000	ER	**0.065 ± 0.016**			0.418 ± 0.027		0.193 ± 0.037
	AUPR	0.905 ± 0.052	0.916 ± 0.052	0.897 ± 0.064	0.603 ± 0.086	0.607 ± 0.085	**0.988 ± 0.015**

neurons, and is trained for 30000 steps on batches of size 50 using dropout and AdamOptimiser with a learning rate of 10^{-3}. Values of the SL parameter α_1 are optimized in $\{10, 100, 1000\}$ on the validation set.

Table 1 reports the AUPR for all the methods on the four benchmark datasets with 4975 irrelevant variables for LC, LR, and NLC and 4995 for NLR (for a total of 5000 variables in each dataset), and Table 2 shows the impact of the number of irrelevant variables on the NLC problem. We also report the predictive performance of each model, i.e., the error/ misclassification rate (ER) in classification and the mean squared error (MSE) in regression, computed on the independent test set. The results clearly show the lack of robustness of standard neural networks (i.e., without any selection layer) in the presence of a large number of irrelevant features. Without SL, ANN are usually worse than RF along both ER/MSE and AUPR, while adding SL allows to strongly increase the performance along both criteria in high-dimensional datasets. Compared to RF, ANN with SL yield higher performance in terms of ER/MSE on all the problems, as well as higher performance in terms of AUPR on the linear problems (LC and LR). RF are better at highlighting the relevant variables on the non-linear problems, despite worse predictive performance. Among the three SL methods, SL+GRAD and SL+LRP yield equivalent AUPR while SL returns inferior results, showing that the weights of SL are not enough to measure feature importances (see also Fig. 2).

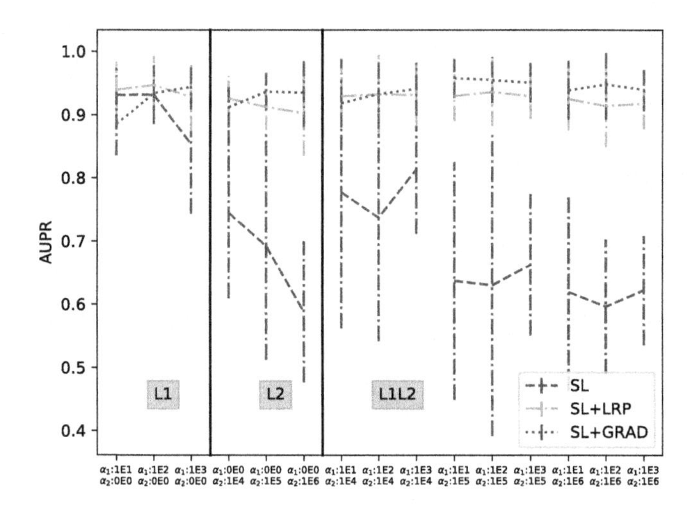

Fig. 2. Impact of α_1 and α_2 for NLC with 4975 irrelevant variables. The figure plots the means and standard deviations of the AUPR over the ten different NLC datasets.

The results in Tables 1 and 2 were obtained using a L1 regularization (i.e., with α_2 in Eq. (6) set to 0) on the weights of SL. Other regularization schemes could be used instead, such as a L2 regularization (with a corresponding regularization coefficient $\alpha_2 > 0$) or a combination of both. However, as shown in Fig. 2, these other regularization schemes do not yield better results than L1 for SL+GRAD and SL+LRP and return lower AUPR for SL.

The network architecture has also a great impact on the AUPR and ER/MSE. For example, we observe in Fig. 3 that networks with three or four hidden layers tend to yield the best results on the NLC problem. The figure also shows that although the ER/MSE and the AUPR are not perfectly correlated, for a single dataset a lower ER/MSE generally corresponds to a higher AUPR. An unbiased estimate of ER/MSE using a validation set (or by cross-validation) thus seems to be a good indicator for automatically selecting the hyper-parameter values.

4.2 HIV Drug Resistance

We also apply the ANN and RF approaches for identifying mutations in the Human Immunodeficiency Virus Type 1 (HIV-1) that are associated with drug resistance [20]. Datasets are available for respectively 16 drugs from 3 different classes (called PI, NRTI and NNRTI, respectively).[3] In each dataset, the output is the drug resistance level and each input feature indicates the presence or absence of a particular mutation at a specific genotype position. Following the same protocol as in [4,13], for each dataset we remove the samples with missing drug resistance information, we keep only the mutations appearing more than

[3] The HIV-1 datasets are available at: https://hivdb.stanford.edu/pages/published_analysis/genophenoPNAS2006/.

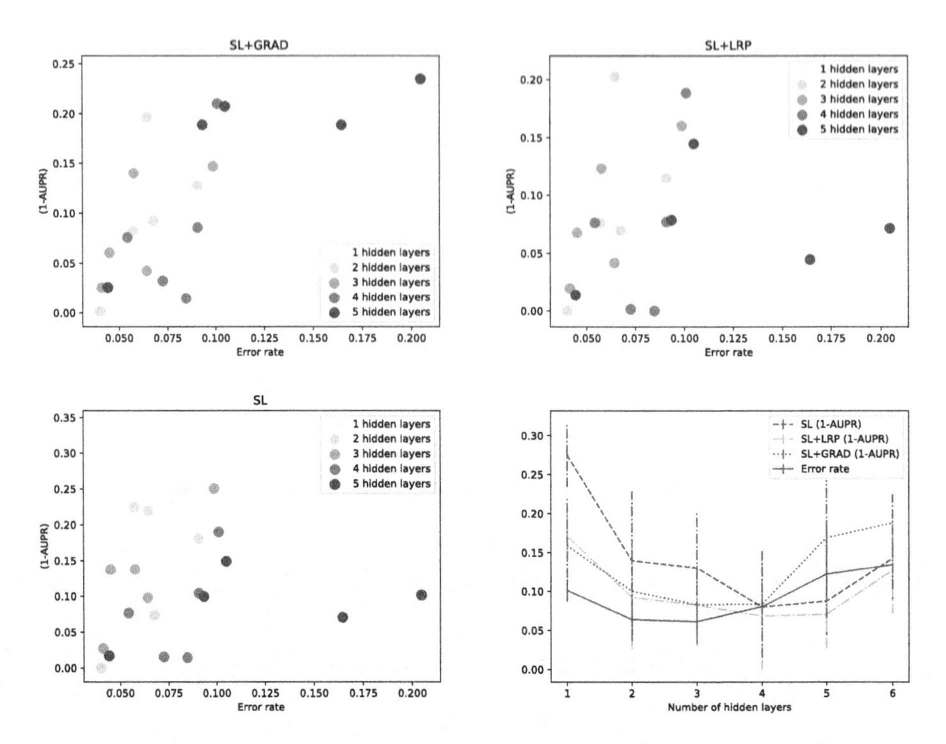

Fig. 3. Impact of the number of hidden layers (150 neurons each) for the NLC dataset with 4975 irrelevant variables. Each scatter plot shows the performance metrics of one feature ranking method for five datasets. The bottom right figure shows the means and standard deviations over the five datasets.

three times in the dataset and we remove duplicated input columns to allow for identifiability. Resulting datasets have thus different sizes, with ∼800 samples and ∼200 mutations for the datasets of the PI class, ∼600 samples and ∼300 mutations for the datasets of the NRTI class, and ∼700 samples and ∼300 mutations for the datasets of the NNRTI class. To evaluate the mutation rankings returned by ANN and RF, we use the genotype positions found, in a separate study, to be associated with a treatment by each class of drug [19]. Still following the protocol used in [4,13], for each drug we consider that a mutation is relevant if it is located at a position found to be associated with the treatment by the corresponding drug class. This results in ∼120, ∼70 and ∼50 relevant mutations for the PI, NRTI and NNRTI classes respectively.

Table 3 shows the results obtained with the different methods. Again, we see here that the addition of the selection layer allows to improve the predictive performance, yielding lower MSE than RF. The selection layer also allows to improve the AUPR: SL+LRP outperforms LRP on all the datasets, while SL+GRAD outperforms GRAD on 11 datasets (out of 16). These improvements are however less impressive than the ones observed on the simulated problems (Sect. 4.1). This can probably be explained by the fact that the proportion of relevant variables is much higher in the HIV-1 datasets compared to the simulated problems. Among the three

Table 3. AUPR and MSE obtained on the HIV-1 datasets. Best results are shown in bold. For each dataset, we use an ANN composed of 3 hidden layers of respectively 100, 50 and 25 ReLU neurons, trained for 10000 steps on batches of size 50 using dropout and AdamOptimiser with a learning rate of 10^{-3}. Values of the hyper-parameter α_1 of SL are tuned in $\{2, 5, 10, 20, 50, 100\}$ by cross-validation, and variable importance scores are computed from the ANN trained on the whole dataset using the optimized α_1 value. Likewise, the hyper-parameter K of RF is tuned in $\{\sqrt{p}, \log(p), p/3, p/2, p\}$, where p is the number of inputs. The generalization error (MSE) of ANN without SL is estimated by cross-validation while the MSE of ANN with SL and of RF are estimated by a double cross-validation loop (using the inner loop for hyper-parameter tuning and the outer loop for error estimation).

PI							
Dataset		GRAD	LRP	SL+GRAD	SL+LRP	SL	RF
APV	AUPR	0.678	0.598	0.687	0.617	0.686	**0.759**
	MSE	0.401 ± 0.469		**0.341 ± 0.448**			0.608 ± 0.793
ATV	AUPR	0.686	0.597	**0.750**	0.696	0.703	0.739
	MSE	0.410 ± 0.478		**0.358 ± 0.403**			0.788 ± 0.876
IDV	AUPR	0.741	0.605	**0.745**	0.690	0.724	0.686
	MSE	0.420 ± 0.563		**0.364 ± 0.388**			0.651 ± 0.646
LPV	AUPR	**0.772**	0.640	0.728	0.666	0.701	0.726
	MSE	0.367 ± 0.400		**0.186 ± 0.108**			0.448 ± 0.502
NFV	AUPR	0.739	0.618	0.765	0.719	**0.771**	0.748
	MSE	0.464 ± 0.648		**0.423 ± 0.627**			0.712 ± 1.112
RTV	AUPR	0.721	0.609	0.757	0.645	0.743	**0.771**
	MSE	0.347 ± 0.266		**0.259 ± 0.194**			0.500 ± 0.347
SQV	AUPR	0.683	0.628	0.720	0.709	0.735	**0.761**
	MSE	**0.409 ± 0.353**		0.481 ± 0.382			0.603 ± 0.460
NRTI							
Dataset		GRAD	LRP	SL+GRAD	SL+LRP	SL	RF
3TC	AUPR	0.331	0.349	0.337	0.357	0.336	**0.376**
	MSE	0.036 ± 0.039		**0.032 ± 0.038**			0.036 ± 0.028
ABC	AUPR	**0.482**	0.254	0.415	0.356	0.374	0.423
	MSE	0.409 ± 0.860		**0.353 ± 0.806**			0.740 ± 1.769
AZT	AUPR	0.381	0.298	0.341	0.346	0.418	**0.519**
	MSE	0.439 ± 0.284		**0.325 ± 0.267**			0.526 ± 0.313
D4T	AUPR	0.480	0.292	0.470	0.383	0.432	**0.495**
	MSE	0.273 ± 0.289		**0.261 ± 0.251**			0.403 ± 0.359
DDI	AUPR	**0.476**	0.294	0.416	0.359	0.400	0.459
	MSE	0.240 ± 0.237		**0.095 ± 0.131**			0.196 ± 0.219
TDF	AUPR	0.289	0.308	0.473	0.389	0.446	**0.529**
	MSE	**0.490 ± 1.450**		0.726 ± 1.541			1.028 ± 2.863
NNRTI							
Dataset		GRAD	LRP	SL+GRAD	SL+LRP	SL	RF
DLV	AUPR	0.463	0.196	0.473	0.301	**0.474**	0.399
	MSE	0.365 ± 0.182		**0.204 ± 0.135**			0.324 ± 0.182
EFV	AUPR	0.513	0.193	**0.546**	0.326	0.534	0.540
	MSE	0.515 ± 0.211		**0.287 ± 0.184**			0.447 ± 0.203
NVP	AUPR	0.543	0.228	**0.586**	0.385	0.580	0.570
	MSE	0.323 ± 0.125		**0.149 ± 0.082**			0.284 ± 0.143
No. wins	AUPR	3	0	4	0	2	7
	MSE	2		14			0

SL-based methods, SL+GRAD tends to be the best performer, having the highest AUPR on 11 datasets. Compared to RF, the performance of SL+GRAD is however rather disappointing, with a higher AUPR than RF on 7 datasets only, despite a better predictive performance.

5 Application to Gene Network Inference

An open problem in computational biology is the reconstruction of gene regulatory networks (GRNs) from gene expression data. A GRN aims at explaining the joint variability in the expression levels of a group of genes through a directed graph, where an edge e_{ij} going from gene g_i to gene g_j indicates that g_i regulates the expression of g_j. Often, the aim is to reconstruct a *weighted* network, where each putative edge is associated with a confidence weight. One approach to the reconstruction of weighted GRNs consists in solving one regression problem for each gene g_j in turn, with the expression of g_j as output variable and the expressions of the other genes as input variables. The variable importance score of gene g_i in the model predicting the expression of g_j is then used as weight for the edge e_{ij}. Using this framework, the RF are currently one of the state-of-the-art approaches for GRN inference [6].

We use the ANN-based variable importance scores to reconstruct the five artificial networks of the DREAM4 multifactorial challenge and the real *Escherichia coli* network [15] used in the DREAM5 challenge [14]. Each DREAM4 network is composed of 100 genes, for which the simulated expressions in 100 samples are available. The *E. coli* dataset contains the expression levels of 4511 genes in 805 experimental conditions. In this dataset, 334 genes are known to be transcription factors, and thus we use only those genes as input variables. A gold standard network is available for each dataset, allowing the evaluation of a predicted ranking of edges in the form of an AUPR. Note that while the DREAM4 gold standard networks are the true (artificial) networks, the *E. coli* gold standard was built from experimentally confirmed interactions and is thus not perfect.

Table 4 shows the AUPR for the five DREAM4 networks. Results for ANN are shown for a fixed architecture and parameter α_1 (columns 1–3 of Table 4) and when both of them are tuned by five-fold cross validation (columns 4–6). To save some computing time, we did not use a standard cross-validation scheme, but rather adopted the following strategy: the samples were divided in five folds $S_1, \dots, S_5$ and the genes were also divided in five folds $G_1, \dots, G_5$. For each gene in subset G_i, the ANN was trained on the learning set composed of the four subsets $S_{j \neq i}$ and its MSE was evaluated using subset S_i. The mean MSE across all genes/ANN was then used to select the ANN architecture and α_1 value. As shown in Table 4, the cross-validation procedure allows to improve the AUPR, but however does not allow the ANN to outperform, and even be competitive with, the state-of-the-art RF. When the number of genes is high, the cross-validation becomes too computationally expensive, and hence could not be applied for the inference of the *E. coli* network. Results with a fixed architecture are shown in Fig. 4, where we see that, for this specific network, the ANN-based approaches are competitive with RF.

Table 4. AUPR for the five DREAM4 networks. The first three columns indicate the results obtained when using a fixed ANN architecture with three layers of 75, 50 and 25 neurons respectively and setting the regularization parameter $\alpha_1 = 10$. Columns 4–6 indicate the results when we select the ANN architecture and the parameter α_1 that optimize the prediction error (MSE) computed using cross-validation. The number of hidden layers was optimized in {2,3}, the number of neurons per layer was optimized in {50,150} and the value of α_1 was optimized in {0,5,60,300,800,1500}. Each network was trained for 10000 steps on batches of size 35, with a learning rate of 10^{-4}. The parameter K of RF was optimized in $\{\sqrt{p}, \log(p), p/3, p/2\}$, where p is the number of inputs (99 in this case). The highest AUPR are indicated in bold.

	Fixed architecture and α_1			Tuned architecture and α_1			RF	Random
	SL+GRAD	SL+LRP	SL	SL+GRAD	SL+LRP	SL		
Net 1	0.127	0.148	0.137	0.148	0.143	0.126	**0.171**	0.018
Net 2	0.079	0.095	0.087	0.109	0.101	0.121	**0.156**	0.025
Net 3	0.147	0.169	0.145	0.178	0.193	0.191	**0.262**	0.020
Net 4	0.141	0.155	0.134	0.184	0.172	0.192	**0.240**	0.021
Net 5	0.116	0.142	0.114	0.187	0.180	0.166	**0.231**	0.020

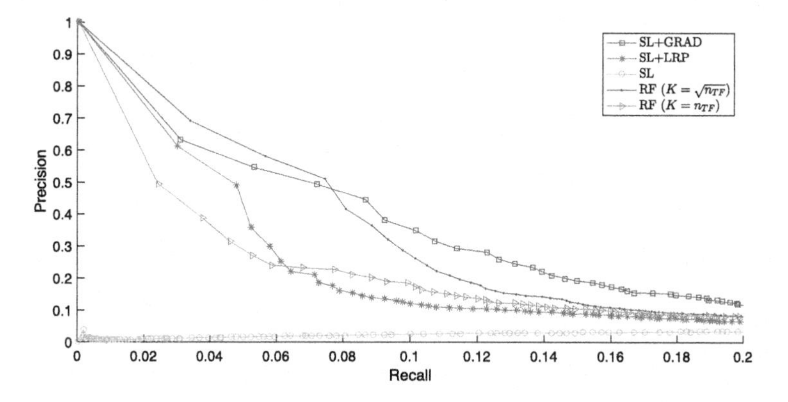

Fig. 4. Precision-recall curves obtained for the *E. coli* network, when using an ANN architecture with 4 layers of 100 neurons each and $\alpha_1 = 10$. The RF performance is shown for two values of the parameter K ($K = \sqrt{n_{TF}}$ and $K = n_{TF}$, where $n_{TF} = 334$ is the number of transcription factors).

6 Conclusion

We evaluated several feature ranking techniques based on ANN and compared them on several problems with RF, chosen as a state-of-the-art reference. While the ANN importance measures can yield performance similar to the RF measure, they remain outperformed by RF on most problems we studied, despite having significantly better predictive performance. Importantly, for datasets with a large number of irrelevant features, reaching good performance, both in terms of feature ranking (AUPR) and generalization error, comes at the cost of introducing a selection layer within the neural network architecture, with a corresponding tuning of the regularization hyper-parameter. Regarding the problem of gene network inference, ANN are competitive with RF on the real *E. coli* network, but are inferior on the artificial DREAM4 networks, even after an extensive tuning of the ANN architecture and regularization parameter.

The fact that the ANN approaches yield better predictive models than RF but not as good feature rankings suggest that the studied ANN-based importance measures could potentially be improved. Future works will include the analysis of other ANN-based importance scores, possibly with different ANN architectures, as well as a better characterization of these different scores.

References

1. Ancona, M., Ceolini, E., Öztireli, C., Gross, M.: Towards better understanding of gradient-based attribution methods for deep neural networks. In: Proceedings of ICLR 2018 (2018)
2. Bach, S., Binder, A., Montavon, G., Klauschen, F., Müller, K.R., Samek, W.: On pixel-wise explanations for non-linear classifier decisions by layer-wise relevance propagation. PLoS ONE **10**(7), e0130140 (2015)
3. Baehrens, D., Schroeter, T., Harmeling, S., Kawanabe, M., Hansen, K., Müller, K.R.: How to explain individual classification decisions. J. Mach. Learn. Res. **11**, 1803–1831 (2010)
4. Barber, R.F., Candès, E.J.: Controlling the false discovery rate via knockoffs. Ann. Stat. **43**(5), 2055–2085 (2015)
5. Guyon, I., Elisseeff, A.: An introduction to variable and feature selection. J. Mach. Learn. Res. **3**, 1157–1182 (2003)
6. Huynh-Thu, V.A., Irrthum, A., Wehenkel, L., Geurts, P.: Inferring regulatory networks from expression data using tree-based methods. PLoS ONE **5**(9), e12776 (2010)
7. Kindermans, P.-J., et al.: The (un)reliability of saliency methods. In: Samek, W., Montavon, G., Vedaldi, A., Hansen, L.K., Müller, K.-R. (eds.) Explainable AI: Interpreting, Explaining and Visualizing Deep Learning. LNCS (LNAI), vol. 11700, pp. 267–280. Springer, Cham (2019). https://doi.org/10.1007/978-3-030-28954-6_14
8. Kindermans, P.J., et al.: Learning how to explain neural networks: pattern net and pattern attribution. ArXiv e-prints (2017)
9. Kohavi, R., John, G.H., et al.: Wrappers for feature subset selection. Artif. Intell. **97**(1–2), 273–324 (1997)
10. Leray, P., Gallinari, P.: Feature selection with neural networks. Behav. Metrika **26**(1), 145–166 (1999)
11. Li, Y., Chen, C.Y., Wasserman, W.W.: Deep feature selection: theory and application to identify enhancers and promoters. In: Proceedings of RECOMB 2015, pp. 205–217 (2015)
12. Louppe, G., Wehenkel, L., Sutera, A., Geurts, P.: Understanding variable importances in forests of randomized trees. In: Advances in Neural Information Processing Systems 26, pp. 431–439 (2013)

13. Lu, Y., Fan, Y., Lv, J., Stafford Noble, W.: DeepPINK: reproducible feature selection in deep neural networks. In: Bengio, S., Wallach, H., Larochelle, H., Grauman, K., Cesa-Bianchi, N., Garnett, R. (eds.) Advances in Neural Information Processing Systems 31, pp. 8676–8686. Curran Associates, Inc. (2018)
14. Marbach, D., et al.: Wisdom of crowds for robust gene network inference. Nat. Methods **9**(8), 796–804 (2012)
15. Marbach, D., Schaffter, T., Mattiussi, C., Floreano, D.: Generating realistic in silico gene networks for performance assessment of reverse engineering methods. J. Comput. Biol. **16**(2), 229–239 (2009)
16. Montavon, G., Samek, W., Müller, K.R.: Methods for interpreting and understanding deep neural networks. Digit. Signal Process. **73**(Suppl. C), 1–15 (2018)
17. Nair, V., Hinton, G.: Rectified linear units improve restricted Boltzmann machines. In: Proceedings of the 27th International Conference on Machine Learning (ICML 2010), pp. 807–814 (2010)
18. Pedregosa, F., et al.: Scikit-learn: Machine learning in Python. J. Mach. Learn. Res. **12**, 2825–2830 (2011)
19. Rhee, S.Y., et al.: HIV-1 protease and reverse-transcriptase mutations: correlations with antiretroviral therapy in subtype B isolates and implications for drug-resistance surveillance. J. Infect. Dis. **192**(3), 456–465 (2005)
20. Rhee, S.Y., Taylor, J., Wadhera, G., Ben-Hur, A., Brutlag, D.L., Shafer, R.W.: Genotypic predictors of human immunodeficiency virus type 1 drug resistance. Proc. Natl. Acad. Sci. **103**(46), 17355–17360 (2006)
21. Shrikumar, A., Greenside, P., Kundaje, A.: Learning important features through propagating activation differences. In: Proceedings of the 34th International Conference on Machine Learning, pp. 3145–3153 (2017)
22. Simonyan, K., Vedaldi, A., Zisserman, A.: Deep inside convolutional networks: visualising image classification models and saliency maps. ArXiv e-prints (2014)
23. Springenberg, J.T., Dosovitskiy, A., Brox, T., Riedmiller, M.: Striving for simplicity: the all convolutional net. ArXiv e-prints (2014)
24. Sundararajan, M., Taly, A., Yan, Q.: Axiomatic attribution for deep networks. In: Proceedings of the 34th International Conference on Machine Learning, pp. 3319–3328 (2017)
25. Tibshirani, R.: Regression shrinkage and selection via the lasso. J. Roy. Stat. Soc.: Ser. B (Methodol.) **58**(1), 267–288 (1996)
26. Zeiler, M.D., Fergus, R.: Visualizing and understanding convolutional networks. In: Fleet, D., Pajdla, T., Schiele, B., Tuytelaars, T. (eds.) ECCV 2014. LNCS, vol. 8689, pp. 818–833. Springer, Cham (2014). https://doi.org/10.1007/978-3-319-10590-1_53
27. Zou, H., Hastie, T.: Regularization and variable selection via the elastic net. J. Roy. Stat. Soc. Ser. B (Stat. Methodol.) **67**(2), 301–320 (2005)

Pathway Activity Score Learning for Dimensionality Reduction of Gene Expression Data

Ioulia Karagiannaki[1(✉)], Yannis Pantazis[2], Ekaterini Chatzaki[3,4], and Ioannis Tsamardinos[1,2,5(✉)]

[1] Department of Computer Science, University of Crete, Heraklion, Greece
jkarayan@csd.uoc.gr, tsamard.it@gmail.com
[2] Institute of Applied and Computational Mathematics, Foundation for Research and Technology - Hellas, Heraklion, Greece
pantazis@iacm.forth.gr
[3] Laboratory of Pharmacology, Medical School, Democritus University of Thrace, Alexandroupoli, Greece
achatzak@med.duth.gr
[4] Institute of Agri-food and Life Sciences, University Research Center, Hellenic Mediterranean University, 71410 Heraklion, Crete, Greece
[5] Gnosis Data Analysis PC, Heraklion, Crete, Greece

Abstract. Molecular gene-expression datasets consist of samples with tens of thousands of measured quantities (e.g., high dimensional data). However, there exist lower-dimensional representations that retain the useful information. We present a novel algorithm for such dimensionality reduction called Pathway Activity Score Learning (PASL). The major novelty of PASL is that the constructed features directly correspond to known molecular pathways and can be interpreted as pathway activity scores. Hence, unlike PCA and similar methods, PASL's latent space has a relatively straight-forward biological interpretation. As a use-case, PASL is applied on two collections of breast cancer and leukemia gene expression datasets. We show that PASL does retain the predictive information for disease classification on new, unseen datasets, as well as outperforming PLIER, a recently proposed competitive method. We also show that differential activation pathway analysis provides complementary information to standard gene set enrichment analysis. The code is available at https://github.com/mensxmachina/PASL.

Keywords: Pathway activity · Dimensionality reduction · Disease classification · Differential activation analysis

1 Introduction

Molecular data, such as gene expressions, are often very high dimensional, measuring tens of thousands molecular quantities. For example, the Affymetrix micro-array platform GPL570 for humans measures the expressions of 54675

A. Appice et al. (Eds.): DS 2020, LNAI 12323, pp. 246–261, 2020.
https://doi.org/10.1007/978-3-030-61527-7_17

probe-sets, corresponding to all known human genes. As such, visually inspecting the data, understanding the multivariate gene correlations, and biologically interpreting the measurements is challenging. To address this problem, several methods have appeared that reduce the dimensionality of the data. Dimensionality reduction (a.k.a. latent representation learning) constructs new dimensions (features, quantities, variables). The purpose is to reduce the number of features making them amenable to inspection while maintaining all "useful" information. For example, consider the representation of music. The raw data (original measured quantities) correspond to the sound spectrum which is visually incomprehensible to humans. However, music at each time-point can be represented as a sum of prototypical states (notes) and musical scores, which are much more intuitive. Similarly, we can ask the questions: Are there prototypical cell states whose sum can represent any cell state (e.g., gene expression profile)? What are the "notes" of biology? How can we learn such representations automatically?

Numerous dimensionality reduction techniques have been proposed. Some of the most prevalent ones are arguably the PCA, Kernel PCA [15], t-SNE [11], and Neural Network autoencoders. All of these methods learn a lower dimensional space (latent space) of newly constructed features and represent the data as a linear combination of those. The projection aims to retain the data variance and exhibit a low data reconstruction error. However, the data representation in the new feature space is biologically unintepretable. To improve interpretability other methods introduce sparsity to the latent space in the sense that new features are constructed as linear combinations of only a few of the original molecular quantities. Such methods are the Sparse PCA [20] and sparse variants of Non-negative Matrix Factorization [10] for molecular data [4,6]). The new constructed features are sometimes called *meta-genes* [3]. Any clustering method could also be defined as creating meta-genes and new features. However, *the meta-genes are still hard to interpret biologically as they do not directly correspond to the known biological pathways or other known gene sets.*

In this work, we develop a novel method for unsupervised feature construction and dimensionality reduction based on the availability of prior knowledge, called Pathway Activity Score Learning or **PASL**. PASL aims at a trade-off between biological interpretability, and computational performance. PASL accepts as input a collection of predefined sets of genes, hereafter called **genesets**, such as molecular pathways or gene ontology groups. It has two phases, the *inference phase* and the *discovery phase.* During the inference phase, **PASL constructs new features that are constrained to directly correspond to the available genesets.** The new features could be thought as **activity scores** of the corresponding genesets. The inference phase ends when it has captured as much information as possible (maximum explained variance) given only the provided genesets. However, a large percentage of the measured quantities is not mapped to any known genesets. In the discovery phase, PASL constructs features that are not constrained to correspond to the given genesets trying to capture the remaining information (variance) in the data.

We evaluate PASL in two sets of computational experiments. (a) We use two collections of real micro-array gene expression datasets, one for Breast Cancer and one for Leukemia. *It is shown that PASL learns latent representations that allow it to perform predictive modeling based on the novel features. The computational experiments are performed on test datasets never seen by PASL during feature construction.* Predictive modeling uses an AutoML platform for molecular data called Just Add Data Bio or **JADBIO** [17] that searches thousands of machine learning pipelines to identify the optimally predictive one and estimates the out-of-sample predictive performance of the final model in a conservative fashion. *Analysis in the new feature space is orders of magnitude faster than the one performed using the original feature space.* In addition, the resulting predictive models are on par and often outperform the ones constructed using the original molecular quantities. PASL is compared against PLIER [12], arguably the algorithm closer in spirit to PASL. PASL outperforms PLIER in terms of predictive performance.

In the second set of computational experiments, (b) we show that PASL's constructed features can complement standard gene set enrichment analysis (**GSEA**). Specifically, the geneset activity scores output by PASL can be employed to perform differential activation analysis (**DAA**) and identify the genesets that behave differently between two different classes (e.g., cases vs controls, or treatment vs controls). Conceptually, this is equivalent to gene differential expression analysis that identifies genes whose expression behaves differently in two classes. Our experiments indicate that DAA complements GSEA: it can identify genesets that are not identified by GSEA as statistically significant. Moreover, DAA has larger statistical power than GSEA and, in general, it identifies the affected genesets with lower p-values than GSEA.

2 Pathway Activity Score Learning Algorithm

2.1 Preliminaries

The PASL algorithm accepts as input two 2D matrices X and G. Matrix $X \in \mathbb{R}^{n \times p}$ contains the molecular measurements, where n is the number of samples and p the number of features. Typically $n \ll p$. For micro-array gene expression data, the rows of X correspond to molecular profiles while the columns to the gene expressions of the probe-sets. Hereafter, *we will refer to probe-sets as genes for simplicity, unless otherwise noted; however, the reader is warned that there is not a one-to-one correspondence between probe-sets and genes.* PASL also accepts a gene membership matrix $G \in \{0,1\}^{g \times p}$ with g being the number of predefined groups of genes. Each row of G, denoted by $\mathbf{g}_i$ for the i-th row, corresponds to a molecular pathway, gene ontology set, or any other predefined gene collection of interest called **geneset** hereafter. We set $G_{ij} = 1$ if gene j belongs to the i-th geneset, and 0 otherwise.

PASL assumes the data X can be decomposed as: $\boxed{X = L \cdot D + \sigma I}$, where $D \in \mathbb{R}^{a \times p}$ is a sparse matrix. In other words, each molecular profile at row j of X is a linear combination of rows of D with coefficients in the jth row of L with an isotropic noise added to it. D is called the **dictionary** and its rows the dictionary **atoms**, denoted with $\mathbf{d}_i$. Given training data X, PASL outputs the two matrices D and L. D is the concatenation of two sub-dictionaries D_1 and D_2 ($D = [D_1; D_2]$) with dimensions $a_1 \times p$ and $a_2 \times p$, respectively (hence, $a = a_1 + a_2$). *D_1 is a dictionary where each atom $\mathbf{d}_i$ is constrained to correspond to only one geneset of the matrix G*, in the sense that the non-zero elements of $\mathbf{d}_i$ correspond to the genes in the particular geneset. Thus, D_1 is the part of the dictionary that is biological interpretable. D_2 is just a sparse dictionary meant to explain the remaining variance of the data and suggest the existence of yet-to-be-discovered genesets. D_1 is the outcome of the first phase of PASL, called **inference phase**, while D_2 is the outcome of the second phase, called the **discovery phase**. $L \in \mathbb{R}^{n \times a}$ is the representation of the data in the latent feature space (*PASL scores*). It provides the optimal projection of X on the row space of D and it is computed by minimizing the Frobenius norm between X and $L \cdot D$.

2.2 Inference Phase

One approach to extract the genesets with the highest variance in the dataset is to restrict the data matrix to the features that correspond to a pathway, estimate the first principal component, repeat the same for all pathways and then keep the principal component with the highest variance *(dynamic approach)*. We mathematically formulate this problem as

$$i^* = \arg\max_{i=1,\dots,g} \max_{\mathbf{d} \in \mathbb{R}^{||\mathbf{g}_i||_0}} ||X(:, \mathbf{g}_i)\mathbf{d}||_2^2 \tag{1}$$

where $X(:, \mathbf{g}_i)$ denotes the data matrix restricted by the i-th geneset. Then, we add the i^* principal component to the dictionary, remove its contribution from the dataset and repeat the same procedure until a pre-specified criterion is met. The described algorithm is guaranteed to return an ordered dictionary whose atoms have the highest variance. Nevertheless, it can be prohibitively expensive in terms of computational cost since at each iteration it computes thousands of principal components that are discarded. In order to remedy the computational burden, one solution could be to pre-compute the principal components for all restricted-to-the-pathways data matrices, then, order them and keep the principal components with the highest variance (*static approach*). Despite being relatively computationally efficient, this approach does not necessarily lead to an optimal solution. Specifically, the ordering of the genesets is fixed, but at each iteration the data matrix changes because the contribution of each new atom is removed from it. This might affect the actual ordering of the variance, hence the optimality of the solution.

Algorithm 1 Pathway Activity Score Learning

Input: Data $X_{n \times p}$, Geneset Matrix $G_{g \times p}$

Output: Dictionary $D_{a \times p}$, Representation of data in D: $L_{n \times a}$

1: //Inference Phase
2: $X_z \leftarrow zscore(X)$
3: $X \leftarrow X_z$
4: $i \leftarrow 1$, $i' \leftarrow 1$ //i: running geneset index, i': atom counter
5: $[i_{\bar{G}}, v_{\bar{G}}] \leftarrow$ OrderOfGenesets(X, G) //$v_{\bar{G}}$: pre-computed variance
6: $\bar{G} \leftarrow G(i_{\bar{G}}, :)$ //$\bar{G}$: ordered geneset matrix
7: **while** $i' \leq a_1$ **do**
8: $X_r \leftarrow X(:, \bar{\mathbf{g}}_i)$
9: $[\mathbf{d}_r, v_r] \leftarrow pca(X_r,\ \#pc = 1)$ //v_r : current variance
10: **if** $\frac{v_r}{v_{\bar{G}}(i)} \leq t$ **then** //how close is v_r to $v_{\bar{G}}(i)$
11: $[i_{\bar{G}}, v_{\bar{G}}] \leftarrow$ OrderOfGenesets(X, G)
12: $\bar{G} \leftarrow G(i_{\bar{G}}, :)$
13: $i \leftarrow 1$ //Reset counter
14: $X_r \leftarrow X(:, \bar{\mathbf{g}}_i)$
15: $[\mathbf{d}_r, v_r] \leftarrow pca(X_r, \#pc = 1)$
16: **end if**
17: $D_1 \leftarrow [D_1; expand(\mathbf{d}_r; \mathbf{g}_i)]$ //Insert the new atom in D_1
18: $X \leftarrow X\big(I - D_1(i, \mathbf{g}_i)^T D_1(i, \mathbf{g}_i)\big)$ //Remove the contribution
19: $v_z \leftarrow ||X_z(I - D_1^+ D_1)||_F^2 \ / \ ||X_z||_F^2$
20: **if** $|v_z - v_{z-1}| < tol$ **then** break **end if**
21: $i \leftarrow i + 1$, $i' \leftarrow i' + 1$
22: **end while**
23: //Discovery Phase
24: $X_z \leftarrow zscore(X)$
25: $D_2 \leftarrow spca(X_z,\ \#pc = a_2,\ \#nz = m)$ //$a_2 = a - i'$
26: $D \leftarrow [D_1; D_2]$
27: $L \leftarrow X_z D^+$
28: **return** D, L

29: **function** OrderOfGenesets(X, G)
30: $v_G \leftarrow \emptyset$, $i_G \leftarrow \emptyset$
31: **for** $i \leftarrow 1$ **to** g **do**
32: $X_r \leftarrow X(:, \mathbf{g}_i)$
33: $[\sim, v_r] \leftarrow pca\big(X_r,\ \#pc = \min(n, ||\mathbf{g}_i||_0)\big)$
34: $v_G \leftarrow \left[v_G; \frac{\lambda \cdot v_r}{(||\mathbf{g}_i||_0^{\lambda} - 1)}\right]$ //Box-Cox normalization
35: $i_G \leftarrow [i_G|i|...|i]$ //Insert $\min(n, ||\mathbf{g}_i||_0)$ elements
36: **end for**
37: $[v_{\bar{G}}, j] \leftarrow \mathrm{sort}(v_G)$
38: $i_{\bar{G}} \leftarrow i_G(j)$
39: **return** $i_{\bar{G}}, v_{\bar{G}}$ //ordered genesets ids and their corresponding variance
40: **end function**

The inference phase of PASL shown in Algorithm 1 (lines 1–22 and 29–40) balances between the dynamic and the static approach. As in the static approach, it computes the ordering of the principal components' variance (lines 5 and 29–40) and iteratively select the atoms based on this ordering (while loop;

lines 7–22). The difference between PASL and the static approach is that PASL checks how close is the current variance from the expected pre-computed variance (line 10). If the relative change is below a threshold then PASL recomputes the ordering of the principal components' variance (lines 11–15). The hyper-parameter t, which takes values between $[0, 1]$, controls how often the variance reordering is performed henceforth the proximity to optimality. The higher the value of t the more often the evaluation of the ordering is happening thus more accurate the dictionary in terms of explained variance is on the cost of being computationally more expensive.The stopping criterion asserts that the inference phase of PASL stops when there is no further decrease in the relative reconstruction error (i.e., the variance of the normalized residual error) (line 20). Finally, we remark that the variance values are normalized before they are ordered (line 34). This is absolutely necessary due to the wide variation of the number of genes in each geneset which varies from few dozens to few thousands of genes. We choose as normalization method the Box-Cox transformation on the number of genes and optimize over its hyper-parameter λ.

2.3 Discovery Phase

After the inference phase where we extracted as much as possible variance from prior knowledge, we will distill the remaining variance of the data without restrictions on the location of the non-zero elements of the dictionary atoms using a sparse –hence, interpretable– dimensionality reduction technique aiming to reveal new potential pathways which were previously unknown. Based on its generality, efficiency and speed, we employ in our experiments Sparse Principal Component Analysis (SPCA) [20] (line 25 in Algorithm 1). We note though that any sparse dimensionality reduction technique can be utilized. SPCA applies both l_1 and l_2 penalties in order to regularize and enforce sparsity.

However, we do not tune the respective hyper-parameters, instead, we require the SPCA algorithm to return a fixed number of non-zero elements per atom. We denote this number with m and we set it to 2000 in our experiments.

2.4 Selection of the Hyper-Parameters' Value

Effect of t on the explained variance and the execution time. The most time-consuming part of PASL is the execution of the function OrderOfGenesets in Algorithm 1 due to the large number of PCA calculations (one for each geneset). Hyper-parameter t controls how often the function OrderOfGenesets will be called. When $t = 1$ then it is called at every iteration while it is called once at the beginning and never again when $t = 0$. In order to determine the optimal value for t, we perform an experiment with a merged collection of microarray datasets where the total number of samples is $n = 4235$, the number of genes $p = 54675$ and a fixed number of atoms $a_1 = 200$. Figure 1(a) demonstrates the explained variance as a function of the execution time for different values of t. Based on this plot, we set t to be equal to 0.9 (cyan star symbol in Fig. 1(a)).

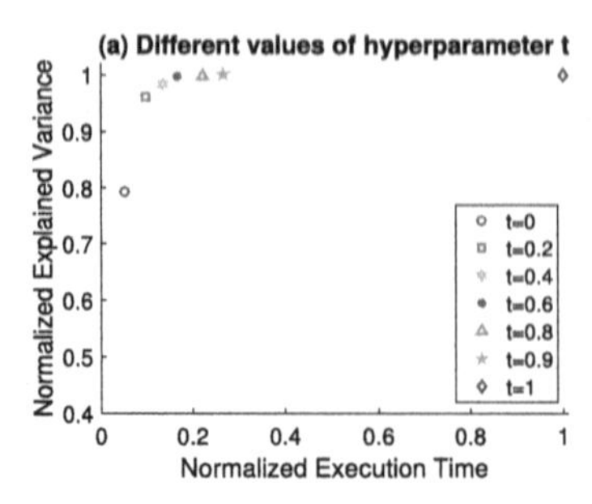

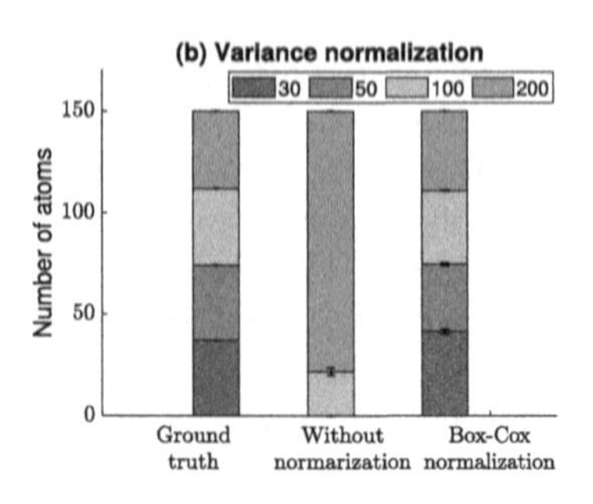

Fig. 1. **(a)** The explained variance (y-axis) as a function of the execution time (x-axis) is shown for different values of t. For $0.4 \leq t \leq 0.9$, the execution time is reduced by a percentage between 65% and 85% with minimal impact on the explained variance. **(b)** The simulated dictionary (ground truth; left bar) consists of equally distributed pathways with 30, 50, 100, 200 genes. The middle bar shows the distribution of selected pathways when PASL is applied without normalization while the right bar shows the selected pathways when Box-Cox normalization is applied with $\lambda = 1/3$. Apparently, the normalization of the variance is necessary for PASL in order to avoid being biased towards selecting genesets with a larger number of genes.

Box-Cox Normalization of the Variance. The number of genes, i.e., the number of non-zero elements in each row of the geneset matrix G, varies from few dozens to several thousands making the geneset ordering based on variance susceptible to such variations. Indeed, we experimentally observe that genesets with more genes tend to be selected frequently while genesets with a low number of genes were rarely selected (see also the middle plot of Fig. 1(b)). Therefore, it is essential to normalize the variance of each geneset relative to the number of genes it contains. We propose to normalize the variance using the Box-Cox transformation [2] on the number of genes (i.e., on $\|\mathbf{g}_i\|_0$) which is given by

$$y' = \begin{cases} (y^\lambda - 1)/\lambda & \text{if } \lambda \neq 0 \\ \log(y) & \text{if } \lambda = 0 \end{cases} \tag{2}$$

where λ is a tunable hyper-parameter which controls the power scaling on y.

The value of λ is determined by a targeted experiment using simulated data which are generated using genesets with both small and large numbers of genes. Simulated data are generated by first creating the prior information matrix G consisting of equally distributed genesets with specific number of genes. Then, we construct a dictionary using randomly selected genesets which are also equally distributed. Specifically, we create $n = 400$ samples with $p = 500$ features while the numbers of genes per geneset take the values 30, 50, 100, 200.

After extensive tests with different values of Box-Cox transformation hyper-parameter, we set $\lambda = 1/3$. The geneset selection results obtained with PASL are presented in Fig. 1(b). Evidently, the use of Box-Cox transformation with $\lambda = 1/3$ (right bar) produced results similar to the ground truth (left bar) while PASL without normalization failed to correctly infer the true dictionary (middle bar).

3 PASL Evaluation on Real Gene Expression Data

Dataset Collections. For our experiments we downloaded microarray datasets available in the Biodataome database [9]. Specifically, we downloaded all the available Breast cancer and Leukemia datasets as of May 2020 measured with the Affymetrix Human Genome U133 Plus 2.0 - GPL570 platform, each having at least 20 samples. The datasets form the *Breast Cancer collection* and *Leukemia collection*. For each collection we select 80% of the datasets to pool together and use them as training data. PASL and PLIER dimensionality reduction algorithms are applied on this training set to learn a dictionary matrix D of atoms (Fig. 2(a)). The remaining 20% of the available datasets are employed as test dataset and are *not seen by neither PASL or PLIER during training*. The selection of datasets used for the train or the test set is random, with the restriction that test datasets have to be accompanied by a discrete outcome (phenotype) for each sample, e.g., disease or mutation status or multiple phenotypes related to the diseases (e.g. rapid/slow early responder). The outcome is either binary or multiclass. The training set for the Breast cancer and the Leukemia collection contains 4200 and 5600 unique gene-expression profiles respectively.

Provided Genesets. In all experiments with real data, the gene membership matrix G includes 1974 pathways found in KEGG [7], Reactome [5] and Biocarta [14] which were downloaded from Molecular Signatures Database (MSigDB) of the Broad Institute [16].

Constructing a Latent Feature Space with PASL and PLIER. Applied to a training dataset X_{train}, PASL learns a transformation to a new feature space given data X_{train} and a geneset matrix G. Subsequently, PASL learns a dictionary D and scores L_{train} such that $X'_{train} \approx L_{train} \cdot D$. Each atom (row) in D corresponds to only one geneset in G or a newly discovered geneset (Fig. 2(a)). To apply the transformation to new test data X_{test} one projects them to the row space of D by computing $L_{test} = X_{test} \cdot D^{+}$ (Fig. 2(b)). An important detail is that both train and test data are first standardized using the means and standard deviations of the training data; thus, the transformation does not require to estimate any quantity from the test data. This is important to avoid information when evaluating predictive performance on the transformed data.

We comparatively evaluate PASL against a recently introduced algorithm called PLIER [12]. Like PASL, PLIER learns a latent feature space that corresponds to known genesets. PLIER also accepts as input data X and a geneset matrix G. Similarly to PASL, it returns the scores L and the dictionary D, such that $X \approx L \cdot D$. PLIER accepts several hyper-parameters. The maxpath hyper-parameter indicates how many genesets an atom of D is supposed to correspond to. We set maxpath = 1 requesting that each atom in D corresponds to one and only geneset, so that the output is comparable to PASL. Unfortunately, *PLIER treats maxpath as indicative; atoms in D may correspond to the union of several genesets, even when maxpath = 1*. In that sense, the atoms in D are not as easy to interpret as the ones returned by PASL. PLIER also ignores genesets with fewer features than minGenes. We set minGenes = 1 so that no

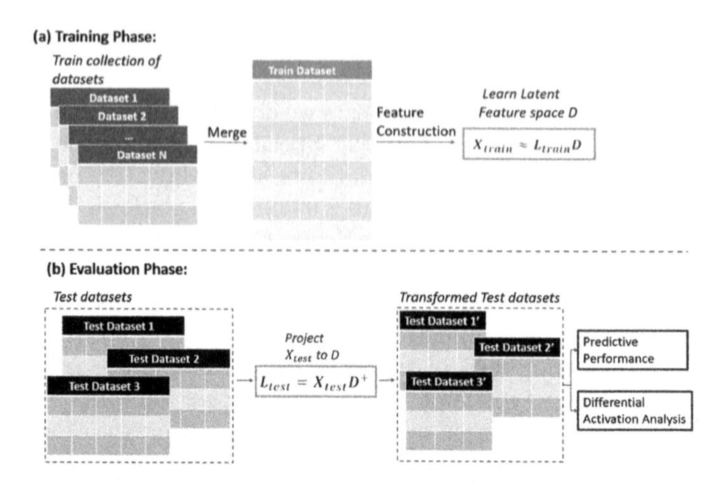

Fig. 2. Experimental Setup. For the construction of the latent feature space, the methods are trained on a collection of gene expression datasets. The evaluation is performed on new unseen test datasets, where the recostruction ability, predictive performance and the significance of the pathways of the latent feature space are examined.

genesets are ignored. Finally, we note that in PLIER the scores L are computed as $X \cdot D^T \cdot (DD^T + \lambda_2 I)^{-1}$, where λ_2 is a parameter learned by the algorithm.

The atoms of PLIER are not as sparse as the ones output by PASL. For example, for the Breast Cancer collection analysis, the mean number of non-zero coefficients in each atom of PLIER is 25833 (almost half of the original feature size), while for PASL it is 1329. For the same number of atoms, PLIER uses more degrees of freedom (non-zero coefficients) to find a suitable transformation to a latent space. For a fair comparison in the subsequent experiments, we impose the restriction that the learned dictionaries D_{PLIER} and D_{PASL} have approximately the same number of non-zero elements. To this end, we first run PLIER allowing it to construct a large number of atoms and estimate the number of atoms a required to reach approximately the same number of non-zeros as PASL. Then, we re-run PLIER constrained to produce only a atoms. Specifically, when PASL is restricted to 500 atoms, its dictionary contains 664695 and 700020 non-zeros for the Breast Cancer and the Leukemia collections, respectively. PLIER is limited to 29 and 30 atoms instead, producing dictionaries with 699976 and 782114 non-zeros, respectively.

3.1 Predictive Performance in Latent Feature Space

This set of experiments examines the following research question: *does the transformation to the latent feature space capture all important information*, defined as the information required to classify to typical outcomes (phenotypes) such as the disease state. To this end, we employ predictive modeling on the **test datasets** and estimate the predictive performance of the best identified model. Each test

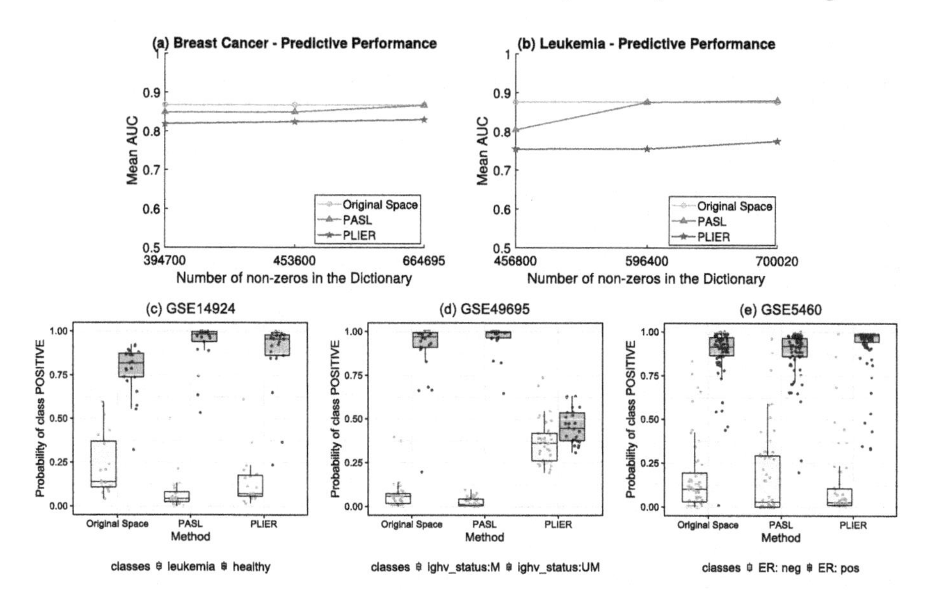

Fig. 3. **(a)**, **(b)** Mean AUC of Breast Cancer and Leukemia test datasets **Lower row:** Out-of-sample probability of selected results of **(c)** The best visualization for PASL vs Original, **(d)** The best visualization for PASL vs PLIER (The outcome stands for the mutation status of immunoglobulin heavy chain (IGHV) gene) and **(e)** The best visualization for PLIER vs PASL.

dataset's outcome leads to binary or multiclass classification tasks. For the classification, we employ an automated machine learning architecture (AutoML), called **JADBIO** (Just Add Data Bio, www.jadbio.com), version 1.1.21. JADBIO has been developed specifically for small-sample, high-dimensional data, such as multi-omics data. The use of JADBIO is meant to ensure that (a) out-of-sample AUC estimates are accurate, and (b) performance does not depend on a single classifier tried with just the default hyper-parameters. Instead, for classification, JADBIO uses the SES feature selection algorithm [8], combined with ridge logistic regression, decision trees, random forests, and SVMs for modelling. It automatically tunes the hyper-parameters of the algorithms, trying thousands of combinations of algorithms and hyper-parameters. It estimates the performance of the final winning model produced by the best configuration (pipeline of algorithms and hyper-parameter values) using the BBC-CV protocol [19]. The latter is a version of cross-validation that adjusts the estimate of performance of the winning configuration for multiple tries to provide conservative AUC estimates. A detailed description of the platform along with a massive evaluation on hundreds of omics datasets is included in [17]. JADBIO has produced novel scientific results in nanomaterial prediction [18], suicide prediction [1] and others.

We performed classification analysis using JADBIO on 13 and 15 test datasets for Breast Cancer and Leukemia, respectively. The analysis uses the original feature space, as well as the PLIER and PASL feature spaces, for different

Table 1. AUC of the test datasets for PASL, PLIER and Original space (initial test datasets). PASL and PLIER are tested for approximately equal number of non-zero entries in the dictionary matrix. For Breast cancer data PASL's latent space consists of 500 dimensions-664695 non-zeros. PLIER's latent space consists of 29 dimensions of 699976 non-zeros. For Leukemia, PASL's latent space consists of 500 dimensions of 700020 non-zeros. PLIER's latent space consists of 30 dimensions of 782114 non-zeros.

Breast Cancer				Leukemia			
Data ID	PASL	PLIER	Original	Data ID	PASL	PLIER	Original
54002	0.999	1	0.995	15434	0.985	0.747	0.987
5460	0.952	0.958	0.96	14924	0.996	0.987	0.91
36771	0.935	0.933	0.963	23025	0.762	0.766	0.741
66161	0.664	0.486	0.579	21029	0.95	0.694	0.966
76124	0.976	0.98	0.97	28654	0.767	0.616	0.762
66159	0.759	0.506	0.776	14671	0.59	0.674	0.625
66305	0.513	0.569	0.535	7440	0.73	0.52	0.736
10780	0.976	0.995	0.962	66006	0.926	0.792	0.952
27562	0.835	0.776	0.914	28460	0.719	0.542	0.697
27830	0.725	0.671	0.759	26713	0.998	0.997	0.952
36769	0.953	0.963	0.96	31048	0.984	0.981	0.99
29431	0.997	0.982	0.991	39411	0.997	0.956	0.985
42568	0.991	0.975	0.927	49695	1	0.612	0.998
				50006	0.979	0.994	0.983
				61804	0.823	0.744	0.869
Mean	**0.8673**	**0.830**	**0.868**	**Mean**	**0.8804**	**0.7748**	**0.876**
Median	**0.952**	**0.958**	**0.96**	**Median**	**0.95**	**0.747**	**0.952**

dimensionalities. For PASL, the number of atoms to learn take the values 250, 400, and 500. The number of atoms with approximately the same number of non-zeros in the dictionary of PLIER is 20, 25, and 30. Thus, there are 7 analyses for each dataset, and $91 + 105$ analyses in total.*For the Breast Cancer (Leukemia) datasets 860002 (983425) classification models were trained in total by JADBIO with different combinations of algorithms and hyper-parameter values on different subsets of the input data (cross-validation).*

Regarding the execution time, the analysis in the space of PASL or PLIER takes about **1 order of magnitude less time** than in the original space. The exact execution time in JADBIO depends on several factors, such as the load of the Amazon servers on which the platform runs, and thus exact timing results are meaningless. Indicatively, we mention a typical case: the analysis of GSE61804 for the original space took 1.15 h, 9 min and 5 min for PASL and PLIER respectively. Figure 3(a),(b) shows the average AUC over all test datasets for each disease for increasing number of non-zeros. **PASL outperforms PLIER and**

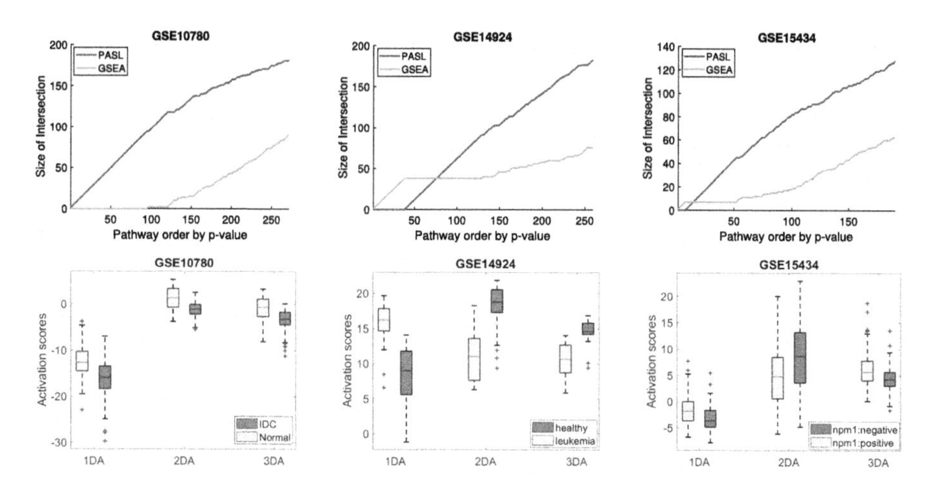

Fig. 4. Upper row: Interaction plots of DAA and GSEA. The x-axis represents the total number of significant genesets. The y-axis represents the number of significant genesets that come from DAA and GSEA. **Lower row:** Box-plots of the activation scores that correspond to the first, second, third differentially activated PASL feature/pathway. It is verified that the differentially activated pathways behave differently between the phenotypes. The outcome of GSE10780 stands for Invasive Ductal Carcinoma/Unremarkable breast ducts, and the outcome of GSE15434 stands for the mutation status of Nucleophosmin 1 (NPM1).

it is on par with analyses on the original space. Thus, the learned dictionary by PASL generalizes to new test data and captures the important information to perform classification with various disease-related outcomes. At the same time, *PASL achieves 2-orders of magnitude dimensionality reduction by a sparse matrix whose atoms directly correspond to known genesets (pathways).*

We now focus on the experiments for the largest dimension of PASL and PLIER. The number of atoms in PASL is set to 500 (664695 non-zeros for Breast Cancer, 700020 non-zeros for Leukemia). PLIER's latent space consists of 29 (699976 non zeros) and 30 (782114 non-zeros) atoms for Breast Cancer and Leukemia respectively. Table 1 contains the detailed results for each dataset and method. The worst case (best case) for PASL is dataset with ID 27562 (14924) where it achieves 8 AUC points (8 AUC points) lower (higher) performance vs no dimensionality reduction. In contrast, there are several datasets (IDs 66161, 66159, 27562, 15434, 21029, 7440, 66006, 28460, 28460, 49695, 61804) where PLIER's performance is lower than 10 or more AUC points.

In the lower row of Fig. 3 we visually demonstrate the ability of PASL to lead to highly predictive models. Each panel corresponds to a different test dataset. Specifically, we chose to present the visualizations from datasets that lead to the "best" visual differences for PASL vs the original space, PASL vs PLIER, and PLIER vs PASL, in Fig. 3(c)–(e), respectively. Each panel shows the box-plots of the *out-of-sample probability* of each molecular profile to belong to the positive class for the models produced in the original, PASL, and PLIER feature space. The out-of-sample predictions are calculated by JADBIO during the cross-validation of the winning model and thus, they do not correspond to the fitting of the samples used for training. The larger the separation of the distribution of the predicted probabilities, the larger the AUC.

3.2 From Gene Set Enrichment Analysis to Differential Activation Analysis

The biological interpretability of PASL's feature space is demonstrated in the following experiments. Since the constructed features correspond to the genesets (atoms of D), we can use their values (stored in the columns of L) to find which genesets behave differently under two conditions, e.g., disease vs. healthy or treatment vs. control. In other words, we can perform **Differential Activation Analysis (DAA)** in a similar fashion that differential expression analysis identifies the genes that behave differently. A current standard alternative method that provides insight into the underlying biology is to use Gene Set Enrichment Analysis (**GSEA**). GSEA first summarizes the probesets that correspond to the same gene e.g. by taking the minimum, maximum or average expression value. Inherently, GSEA loses information by applying this summarization and by not taking into account the covariances of the gene expressions. Subsequently, the null hypothesis is that the p-values of the genes in a pathway have the same distribution as the p-values of the genes that do not belong to the pathway.

We next examine the ability of PASL to identify genesets (pathways) that behave differently between two classes and compare it against GSEA. We employ the GSEA v4.0.3 tool from https://www.gsea-msigdb.org/gsea/index.jsp [13, 16]. We run GSEA on the test datasets in the original feature space using 10000 phenotype permutations for the permutation-based statistical test employed in the package. The input genesets are the same as the ones provided to PASL in the geneset matrix G. We also perform DAA on the test datasets projected to the latent space of PASL (activity scores) using the Matlab's t-test function *mattest* with 10000 permutations. The list of p-values from DAA and GSEA can then be used to identify the affected pathways.

Figure 4 (upper row) shows the number of pathways identified by each method (y-axis) in the top k (lowest p-value) pathways, for each k (x-axis). Each panel corresponds to a different test dataset. We observe that the pathways identified by PASL have lower p-values and are encountered first on the list; PASL has higher statistical power in identifying some genesets that behave differently. PASL's features correspond to pathways. The statistically significant ones are referred as *differentially activated.* Figure 4 (bottom row) visualizes why the PASL features are identified as *differentially activated.* Each panel shows the box-plots for the activation scores corresponding to the first, second, and third most statistically significant PASL feature/pathway (denoted with names 1DA, 2DA, and 3DA, respectively).

Specifically, the top 3 differentially activated pathways of GSE10780 are the "Reactome signaling by GPCR", "Reactome Fructoce Catabolism" and "Reactome Hemostasis". The top 3 differentially activated pathways of GSE14924 is the "Reactome metabolism of Lipids", "Reactome Chromatin Organization" and "Reactome Gene Expression Transcription". The top 3 differentially activated pathways of GSE15434 are the "Reactome Transport of Small Molecules", "Reactome Developmental Biology", "Reactome Post Translational Protein Modification". *It is visually verified that the scores are different between the phenotypes in an easy to understand and intuitive plot.*

While DAA using PASL seems to offer several advantages (lower p-values, intuitive visualization), it also has a major limitation. PASL requires a training set that is related to the application (test) set. It learns atoms that only pertain to capturing information regarding the train data. For example, DAA using PASL cannot be applied to a schizophrenia dataset, before we construct a sufficiently large training dataset for the disease. As such, we consider DAA and GSEA complementary and synergistic.

4 Conclusions

Molecular omics and multi-omics data are notoriously high-dimensional. Statistical or machine learning analysis of such data could hit computational obstacles due to the high dimensionality; results may be hard to interpret (e.g. interpreting thousands of differentially expressed genes or pair-wise correlations and covariances). As a result, several dimensionality reduction methods for such data have been proposed, but usually end up with an unintepretable new feature space. To the extent of our knowledge, PASL is the first technique where the new features directly correspond to prior knowledge about genesets. PASL is relatively computationally efficient by relying on a greedy, yet effective heuristic to construct the next atom. PASL projects the data to a new feature space that maintains the predictive information for a wide range of outcomes, e.g., disease or mutation status, dietary restrictions and others. The classification models created on this space outperform the ones created on the PLIER space and are on par with the ones using the original features. Classification analysis is one order of magnitude faster in PASL space than in the original space. PASL's learned features can be used for Differential Activation Analysis identifying the pathways that behave differently between the phenotypes. This analysis is synergistic to gene set enrichment analysis, it is intuitively visualized, and often produces smaller p-values. Based on these promising results, in a future work PASL will be applied on a much larger corpus of gene expression data, spanning a wide plethora of diseases and conditions.

Acknowledgements. This research has been co-financed by the European Regional Development Fund of the European Union and Greek national funds through the Operational Program Competitiveness, Entrepreneurship and Innovation, under the call RESEARCH–CREATE–INNOVATE (project code:T1EDK-00905) and the

European Research Council under the European Union's Seventh Framework Programme (FP/2007-2013) / ERC Grant Agreement n. 617393.

References

1. Adamou, M., et al.: Toward automatic risk assessment to support suicide prevention. Crisis J. Crisis Interv. Suicide Prevent. (2018)
2. Box, G.E., Cox, D.R.: An analysis of transformations. J. Roy. Stat. Soc. Ser. B (Methodol.) **26**(2), 211–243 (1964)
3. Brunet, J.P., Tamayo, P., Golub, T.R., Mesirov, J.P.: Metagenes and molecular pattern discovery using matrix factorization. Proc. Natl. Acad. Sci. **101**(12), 4164–4169 (2004)
4. Carmona-Saez, P., Pascual-Marqui, R.D., Tirado, F., Carazo, J.M., Pascual-Montano, A.: Biclustering of gene expression data by non-smooth non-negative matrix factorization. BMC Bioinform. **7**(1), 78 (2006)
5. Croft, D., et al.: The reactome pathway knowledgebase. Nucleic Acids Res. **42**(D1), D472–D477 (2014)
6. Fertig, E.J., Ding, J., Favorov, A.V., Parmigiani, G., Ochs, M.F.: CoGAPS: an r/c++ package to identify patterns and biological process activity in transcriptomic data. Bioinformatics **26**(21), 2792–2793 (2010)
7. Kanehisa, M., Goto, S.: KEGG: Kyoto encyclopedia of genes and genomes. Nucleic Acids Res. **28**(1), 27–30 (2000)
8. Lagani, V., Athineou, G., Farcomeni, A., Tsagris, M., Tsamardinos, I.: Feature selection with the r package MXM: discovering statistically-equivalent feature subsets. arXiv preprint arXiv:1611.03227 (2016)
9. Lakiotaki, K., Vorniotakis, N., Tsagris, M., Georgakopoulos, G., Tsamardinos, I.: Biodataome: a collection of uniformly preprocessed and automatically annotated datasets for data-driven biology. Database **2018** (2018)
10. Lee, D.D., Seung, H.S.: Learning the parts of objects by non-negative matrix factorization. Nature **401**(6755), 788–791 (1999)
11. van der Maaten, L., Hinton, G.: Visualizing data using T-SNE. J. Mach. Learn. Res. **9**, 2579–2605 (2008)
12. Mao, W., Zaslavsky, E., Hartmann, B.M., Sealfon, S.C., Chikina, M.: Pathway-level information extractor (plier) for gene expression data. Nat. Methods **16**(7), 607–610 (2019)
13. Mootha, V.K., et al.: PGC-1α-responsive genes involved in oxidative phosphorylation are coordinately down regulated in human diabetes. Nat. Genet. **34**(3), 267–273 (2003)
14. Nishimura, D.: Biocarta. Biotech Softw. Internet Rep. Comput. Softw. J. Sci. **2**(3), 117–120 (2001)
15. Schölkopf, B., Smola, A., Müller, K.R.: Nonlinear component analysis as a kernel eigenvalue problem. Neural Comput. **10**(5), 1299–1319 (1998)
16. Subramanian, A., et al.: Gene set enrichment analysis: a knowledge-based approach for interpreting genome-wide expression profiles. Proc. Natl. Acad. Sci. **102**(43), 15545–15550 (2005)
17. Tsamardinos, I., et al.: Just add data: automated predictive modeling and biosignature discovery. bioRxiv (2020)

18. Tsamardinos, I., Fanourgakis, G.S., Greasidou, E., Klontzas, E., Gkagkas, K., Froudakis, G.E.: An automated machine learning architecture for the accelerated prediction of metal-organic frameworks performance in energy and environmental applications. Microporous Mesoporous Mater., 110160 (2020)
19. Tsamardinos, I., Greasidou, E., Borboudakis, G.: Bootstrapping the out-of-sample predictions for efficient and accurate cross-validation. Mach. Learn. **107**(12), 1895–1922 (2018). https://doi.org/10.1007/s10994-018-5714-4
20. Zou, H., Hastie, T., Tibshirani, R.: Sparse principal component analysis. J. Comput. Graph. Stat. **15**(2), 265–286 (2006)

Distributed Processing

Balancing Between Scalability and Accuracy in Time-Series Classification for Stream and Batch Settings

Apostolos Glenis(✉) and George A. Vouros

Department of Digital Systems, University of Piraeus, Piraeus, Greece
apostglen46@gmail.com, georgev@unipi.gr

Abstract. As big data sources providing time series increase, and data is provided in increased velocity and volume, we need to efficiently recognize data provided, classifying it according to their type, origin etc. This is a first important step in doing analytics on data provided from disparate data sources, such as archival sources, multiple sensors, or social media feeds. Time series classification is the task labeling time series using a set of predefined labels.

In this paper we present the K-BOSS-VS algorithm for time series classification. The proposed algorithm is based on state-of-the-art symbolic time series classification algorithms, and aims to achieve high accuracy, balancing with computational efficiency. K-BOSS-VS exploits K representatives of each time series class to classify new series. This provides opportunities for representing intra-class differences, thus increasing the classification accuracy, while incurring a small performance overhead compared to methods using one class representative. Additionally, K-BOSS-VS offers a solution for classifying time-series in batch and streaming settings, due to the opportunities for increasing computational efficiency and the low memory requirements.

Keywords: Time series classification · Distributed processing · Streaming data

1 Introduction

As more and more devices are getting smarter, and sensors become ubiquitous, in conjunction to the increase of media channels and their users, time series data, i.e. data that is tagged with time-stamps, become bigger and bigger. This necessitates to recognize time series flowing into a system from disparate data sources efficiently, classifying data according to their type, origin, quality, trustworthiness etc. with high accuracy. This is a first important step in doing analytics on data provided from disparate data sources, such as archival data sources, multiple sensors, or social media feeds. The task, named time series classification

A. Appice et al. (Eds.): DS 2020, LNAI 12323, pp. 265–279, 2020.
https://doi.org/10.1007/978-3-030-61527-7_18

has become an important task in data analytics pipelines, classifying the type of time series flowing into a system using a set of predefined labels. Time series classification has gained popularity in a variety of fields such as signal processing, environmental sciences, health care, and chemistry.

In this paper we address the problem of time series classification both in batch and streaming settings, with the objective to balance between computational efficiency and accuracy. Efficiency is important in dealing with big data sources (either due to volume or velocity), while accuracy is needed for automating the task: Addressing this trade-off is the objective of recent works in time-series classification, e.g. in [3,20,23].

In this paper we present the K-BOSS-VS algorithm for time series classification, which is based on state-of-the-art symbolic time series classification algorithms, such as BOSS-VS [20]. In contrast to these, K-BOSS-VS exploits K representatives of each class to classify new series. This provides opportunities for increasing the degrees of parallelism used - although it incurs a computational overhead compared to methods that use a single representative, while increasing the classification accuracy, due to addressing intra-class time series modalities, in contrast to methods using a single representative per class (e.g. class centroid).

More specifically, our proposed algorithm achieves high accuracy, which is comparable, if not higher, to state of the art algorithms. Additionally, it offers a solution for classifying time-series in batch and streaming settings, due to the opportunities for distributing the task in multiple workers and due to the low memory requirements compared to methods comparing a new series to each class member. However, as already pointed out, it incurs a small performance overhead compared to methods using one class representative, which is due to the need to compare with K representatives per class.

The contributions made in this work are as follows:

- We introduce a new method that scales for big data sources, without sacrificing accuracy on time series classification.
- We evaluate and compare the proposed method against state of the art algorithms, showing its ability to achieve highly accurate results, with a small performance overhead that can be absorbed in distributed settings.
- We show the efficacy of the proposed method in batch and streaming settings.

The structure of the paper is as follows: Sect. 2 describes the time series classification problem, while Sect. 3 describes related work and explains the contributions made in this paper. Section 4 gives a high-level description of the proposed algorithm and Sect. 5 describes the actual implementation on top of Apache Spark. Section 6 provides experimental results of the proposed algorithm, compared against state of the art algorithms. Finally, Sect. 7 concludes the paper.

2 Problem Formulation

In our problem definition a time-series T is defined as a series of values ordered by their timestamp, i.e. $T = t_0, t_1 ..., t_{m-1} ...$.

Given a set of n labels $L = \{l_0, l_1 ..., l_{n-1}\}$ the goal is to train a classifier $f : \mathbf{T} \rightarrow L$ to label time-series in $\mathbf{T}$ (target time series) using a label in L.

In this paper we treat all data sources as sources providing batches of time series data: This means that given a time horizon H, we consider all values that are provided within that time horizon. Thus, the time series provided within H time instants from a specific source is of specific length. Of course, this length may vary between data sources, depending on data source velocity and frequency of sampling. In batch settings we fetch series of values within time windows of duration H, while in streaming settings we get values from a starting time point t_0, until $t_0 + H$. The time horizon may be tuned in different cases, depending on the data sources used, and according to domain-specific requirements concerning the classification speed. We do not deal with this problem in this paper.

Having fixed the time horizon H, the classification task follows the λ architecture [17] paradigm, where, while sources might very well be streams, a preprocessing step converts part of the stream (i.e. the values provided within a specified time horizon) to a batch dataframe.

3 Related Work

As we are focusing on symbolic representations of time series, below we provide state of the art methods in this line of research: Starting from SAX and SAX-VSM, SFA, BOSS, BOSS-VS and WEASEL.

In [23] the authors present the SAX-VSM classification algorithm that uses SAX representation and tf-idf weighting. SAX transforms a series of values into a word. The range of values is first divided into segments usually following the Gaussian distribution and then each segment is mapped to a letter from a given alphabet. A tf-idf vector is created for each class of the training, after the training set has been transformed into SAX words. Then the set of target time series is transformed into SAX words and the Term Frequency (TF) vector is created for each time-series in the target set. Finally, to classify a time series, the cosine similarity between the TF vectors of that series and of classes' centers is computed.

In [19] the authors introduce the Bag of SFA Symbols Ensemble classifier (BOSS) that uses Symbolic Fourier Approximation (SFA) [21] to classify a time series. To compute SFA words, a number of Fourier coefficients are computed, which are grouped based on common prefixes, building histograms per group, discretized, and mapped to an alphabet. Thus, the SFA approximation, and thus BOSS, uses a symbolic representation based on the frequency domain, providing information about the whole series. Properties of this representation lead to significant lower training times compared to using the SAX representation. The BOSS ensemble classifier is based on 1-NN classification using multiple BOSS models at different time series substructural sizes. However, BOSS requires the entire training set to be available while classifying target time series. Because the training set is large, the memory requirements, in addition to the computational complexity incurred, prevent BOSS from being a candidate for big data time series classification, although it is very accurate.

In [20] the authors present the BOSS-VS classification algorithm where each data point in the training set is transformed into SFA words. Then a centroid is created for each class and the cosine similarity is computed between centroids and target series, as in [23]. This significantly reduces computational complexity and the memory footprint of the classification algorithm, since now each target time-series is only compared to each class centroid. This makes BOSS-VS suitable for big data and streaming data sources, but it is less accurate than BOSS. This trade-off between computational complexity/scalability and accuracy of classification motivates our work, addressing issues concerning bid data, streaming and batch data sources.

In [22] the authors propose WEASEL as a middle ground between BOSS-VS [20] and BOSS [19] for time series classification, balancing between accuracy and scalability. It uses SFA, but it does a few novel things: First, WEASEL considers differences between classes during feature discretisation, second it uses windows of variable lengths, also considering the order of windows, and finally it uses statistical feature selection, leading to significantly reduced runtime.

There is not much available work on distributed time series classification. A work that is close to our aims is [3]; where the authors present a distributed algorithm that uses shapelets and a random forest classifier. Their algorithm scales well compared to the centralized version and achieves an average accuracy of 82% for one of the data sets and 99% for the other.

Concerning classification of time series in streaming settings, in [15] the authors present a method for classifying time series data using Time Series Bitmaps (TSBs) based on SAX, which are shown to be maintained in constant time. Given that TSBs are very close to a normalized Term Frequency vector, this work is considered to be using a compact signature of the training set's time series to classify a streaming time series. While TSBs are robust to concept drift and spotting new behaviour, authors in [15] use a single centroid per class, resulting to accuracy that is not as high as that achieved by state of the art methods mentioned above.

Finally, in [16] the authors propose a method based on Piecewise Linear Approximation (PLA). Each streaming time series is transformed to a vector by means of a PLA technique. The PLA vector is a sequence of symbols denoting the trend of the series (either UP or DOWN), and it is constructed incrementally. The author proposes efficient in-memory methods in order to a) determine the class of each streaming time series, and b) determine the streaming time series that comprise a specific trend class. In contrast to that approach we do not explicitly model trends, we use SFA for the symbolic representation of time series, and we do not compute things incrementally. Modelling trends and incremental computations are useful features that we shall consider in future work.

In addition to the above approaches, there are recent proposals for time series classification using deep learning methods:

While deep learning approaches, such as [7,10,13,14,24,25], report results that are better than baseline approaches and close to the state of the art time series classification methods, their computational complexity and sample-efficiency while training, together with their memory requirements and accuracy scores, impose specific limitations.

Other methods, such as the Leveraging Bagging method in [5], combining bagging with randomization to the input and output of the classifiers using the ADWIN [4] change detector, the Adaptive Random Forests (ARF) in [12], including a theoretically sound resampling method and adaptive operators that can cope with different types of concept drift, Kappa Updated Ensemble (KUE) [6] using an ensemble classification algorithm for drifting data streams, address explicitly the concept drift problem that we do not address in this work.

4 Algorithm Description

As described in Sect. 3, BOSS is highly accurate but requires that the entire training set is compared to each target time-series using an 1-NN classifier. This, implying that the entire training set is available during classification, incurs specific scalability limitations, which are crucial for big data, batch and streaming settings. On the other hand, BOSS-VS and SAX-VSM use a single centroid for each class label. This makes them more suitable for big data and streaming settings since the computational complexity in terms of comparisons needed, as well as their memory footprint, is significantly reduced compared to BOSS. The problem with using a single centroid to represent a class label is the reduction on the accuracy achieved, given that a single centroid may not be representative of all time series patterns in a class (this is apparent when comparing the accuracy of BOSS-VS and SAX-VSM against BOSS).

The main intuition behind the method proposed here is that instead of having a single centroid to represent each class label, we can have K representatives per class. Our approach requires that K is a bound constant so that the memory and computation time remains close to BOSS-VS and SAX-VSM, while at the same time the use of K representatives preserves, or even increases, the classification accuracy of BOSS-VS. We choose SFA as the symbolic representation for our algorithm, similarly to BOSS, given its representation flexibility and superiority compared to SAX. In addition, BOSS-VS has been proved superior to SAX-VSM in terms of accuracy, something which is also shown in our experiments.

We apply K-means to each class label of the training set to obtain K representatives per class. After we have obtained the representatives for each class, to classify a target time series we compute the cosine similarity between the normalized term frequency vector of that series and the normalized term frequency vector of each of these representatives. The target time series is assigned the label of the closest class representative, using an 1-NN classifier.

5 Implementation

We implemented the proposed method in Apache Spark [29] using Spark's MLLIB [18] for computing the tf-idf vector of each time series, as well as for determining the K representatives per class using K-means. To implement the

1-NN classification step we use the Dataframe API. For the streaming use case we used Spark Streaming, more specifically the Spark Structured Streaming [1] API. The Structured Streaming API makes Dataframe operations of Apache Spark available in a streaming environment.

Before delving into the implementation, we provide very succinctly some preliminaries on Apache Spark.

Apache Spark [29] is a unified engine for distributed data processing. Spark uses the MapReduce [8] programming model, but extends it with an abstraction called Resilient Distributed data sets (RDDs) [27]. RDDs provide a distributed memory abstraction that allows in-memory computations on large clusters in a fault-tolerant manner. Using RDDs, Spark can express a vast array of workloads that needed separate processing engines. Example workloads include SQL [2,9], streaming workloads [28], machine learning [18] and graph processing [26]. Apart from RDDs another Spark abstraction that is relevant to our case is Spark DataFrames which are RDDs of records with a known schema. Spark DataFrames get most of their functionality from the DataFrame abstraction for tabular data in R and Python, and essentially model a database table. Dataframes provide methods for filtering data, computing new columns, and aggregating data. In Spark Dataframes, operations map down to operations of the Spark SQL engine and as such, they use all available optimizations.

Coming to the implementation of K-BOSS-VS, Algorithm 1 shows the auxilliary function for data pre-processing. Line 1 first groups the data from each data source to H-time instances' chunks to create the time series. Then it sorts the target column (the column containing the measurements) by the column containing the time stamp, to account for data points that are out of (temporal) order in the data set. This is important, because real-world time series might contain out-of-order values. It must be noted that we have set an horizon of 24 h, that fits better in the data sets used in our experiments. Different time horizons could be applied for other data sets.

Lines 2 to 8 transform time series to SFA words and then create windows of the SFA representation (SFA windows: In our experiments the SFA window length is 4 h). Finally, the algorithm returns a dataframe with the transformed column together with its label.

Algorithm 2 shows the algorithm for computing the representatives per class. In line 2 the algorithm pre-processes the data set using the auxiliary functions provided above. Then in line 3 it creates the hashing term frequency for each document corresponding to the SFA windows of a single time-series of time horizon H. Then in line 4 each term frequency vector is normalized using L2-normalization. The next step of the algorithm is to compute the representatives for each class label (column) of the data set (line 5) using MLLIB's K-means. After that, each column (class) of the data set has K class representatives.

Algorithm 1: processColumnTrain Algorithm

```
Input  : timestampColumn: The name of the column containing the timestamp
Input  : inputDF: The input Dataframe to be processed
Input  : field_groupedBy: The field to group by the measurement, usually the
         data source ID
Input  : outputFunction: The function that transforms the time-series into the
         symbolic representation
Input  : outputColumnName: The column of the input Dataframe containing
         the measurements
Output: someDF: the Dataframe with the transformed column and associated
         label
1 grouped2 ← inputDF. groupBy (inputDF.col (field_groupedBy), window
  (inputDF.col (timestampColumn),"1 day")) .agg ( collect_list
  (struct(inputDF.col (timestampColumn), inputDF.col (outputColumnName)))
  .as ("columnSorted")) .withColumn
  (outputColumnName+"SortedByTimestamp", getValueFromTuple(sort_array
  ("columnSorted", asc = false)))
2 withAppendedColumnsRdd ← grouped2.rdd.map (row => {
3 myArray ← row.get(outputColumnName+"SortedByTimestamp")
4 newColumn2 ← outputFunction(myArray)
5 getColumns ← List(row.get(field_groupedBy))
6 Row.fromSeq(getColumns :+newColumn2) } )
7 someDF ← withAppendedColumnsRdd.toDF
8 return someDF.withColumn ("variable_name", lit(outputColumnName))
```

To classify a set of target time series, we use the following process:

1. The data set is grouped into intervals of length H and then converted into SFA windows similarly to the training set.
2. Each group of SFA windows is converted to a term frequency vector and normalized as in the training set.
3. The similarity between the normalized term frequency vector of the target time series and the class representatives is computed.

The proposed K-BOSS-VS approach borrows from BOSS-VS and SAX-VSM the low memory footprint, as it uses only a subset of training examples as class representatives. Indeed, it uses a number of representatives per class to balance between scalability (of BOSS-VS) and accuracy (of BOSS): The number of class representatives is small compared to all the series available in the training set, but can be large enough to represent different intra-class modalities and provide accuracy improvement compared to BOSS-VS and SAX-VSM. Furthermore, the additional number of class representatives must be small enough that it can be stored in memory.

It must be noted that both, BOSS-VS and the K-BOSS-VS implementations are inspired by the Lambda Architecture [17], where models are trained in batch setting and tested on an setting that can be either batch or streaming.

Algorithm 2: ComputeRepresentatives Algorithm

```
Input  : training: The Dataframe containg the training set
Input  : columns: List of column names
Input  : timestampColumn: The name of the column containing the timestamp
Input  : fieldGroupBy: The field to group by the measurement, usually the
         data source ID
Input  : windowsColumnName: The label of the column to contain the result
         of the output function
Input  : outputFunction: The function that transforms the time-series into the
         symbolic representation
Output: retVector: The class representatives together with their label
1 for column ← columns do
2 |   currDataframe ← processColumnTrain(timestampColumn, training,
  |   fieldGroupBy, outputFunction, windowsColumnName, column)
3 |   rescaledData2 ← hashingTF.transform (currDataframe)
4 |   l2NormTrain ← normalizer.transform (rescaledData2)
5 |   model ← kmeans.fit(l2NormTrain)
6 |   retVector += (column -> model.classRepresentatives)
7 end for
8 return retVector
```

6 Evaluation

In our evaluation we examine SAX-VS, BOSS, BOSS-VS, K-BOSS-VS, in terms or their accuracy, execution time and scalability. We implemented parallel versions of SAX-VSM, BOSS and BOSS-VS in Apache Spark [29] using Spark's MLLIB [18] for tf-idf and term frequency weighting, so as to have a fair comparison with the parallel version of the proposed K-BOSS-VS method.

For the batch setting we compared our algorithm against BOSS (since it provides state-of-the-art classification accuracy), SAX-VSM and BOSS-VS (since they are more efficient and scalable than BOSS). For the streaming case we compared K-BOSS-VS with BOSS-VS (since, it outperforms SAX-VSM in all the cases in the batch setting, and it is more suitable for streaming settings), and we evaluated the execution time and scalability of the algorithms for both the training phase and the testing phase. As part of the evaluation we measured both the total execution time and speedup, while for the streaming case we also provide results concerning the testing time. For the K-BOSS-VS method the scalability and execution time measurements both in the batch and in the streaming use-case are performed with K equal to 16. This is further justified below.

6.1 Experimental Setup and Data Sets

We tested all algorithms in a cluster with 10 computing nodes in total, where one node of the cluster is reserved as the driver-node in Spark, and the remaining 9 nodes are used as workers. The hardware specifications for each computing node is as follows: 2*XEON e5-2603v4 6-core 1.7 GHz with 15 MB cache, with 128 GB of RAM and 256 GB SSD.

Details of the configurations of parallelism are depicted in Table 1.

Table 1. Configurations of parallelism

Number of executors	Number of cores per executor	Parallelism
9	4	36
4	4	16
1	4	4
1	1	1

To evaluate the algorithms we have used the following data sets:

1. The *power measurement* data set[1] used in [11]. This data set contains two classes, one for watts and another for temperature.
2. The Intel Data Lab Sensor (*lab data*) data set[2]. This data set contains 2.3 Million readings from sensors with classes temperature, humidity, light and voltage.

The time horizon in our experiments for both data sets is equal to 24 in order to create time-series per day. This created time series of different length for each data data set: In Intel Lab data set we have about 58415 data points per day for a total of 38 days (0.67 datapoints per second). In power measurements data we have a data point every 10 s which means we have 8640 points per day.

In our tests 80% of each data set is randomly assigned to the training set, and the remaining 20% to the test set.

In the streaming case we used the training set to obtain the representative vectors per class, and then we used the test set to simulate a streaming setting: For this purpose the test set in JSON was fed into the Apache Spark structured streaming readStream function.

6.2 Evaluation Results

As we can see from Fig. 1a the K-BOSS-VS method is a middle ground between BOSS and BOSS-VS in terms of accuracy, while it outperforms SAX-VSM. This

[1] https://github.com/UniSurreyIoT/KAT/raw/master/logic/data.csv.
[2] http://db.csail.mit.edu/labdata/labdata.html.

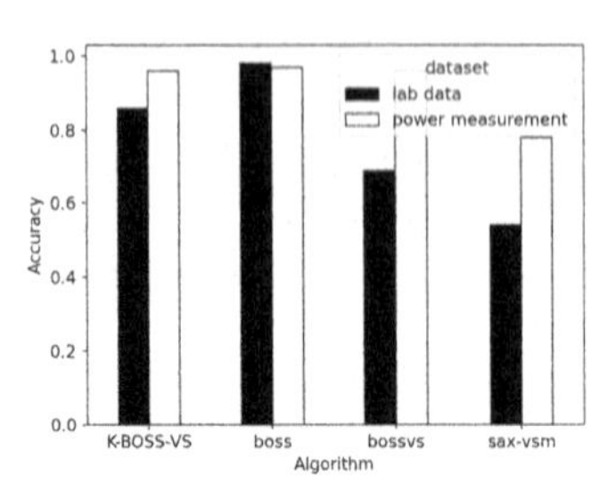

(a) Accuracy of the methods

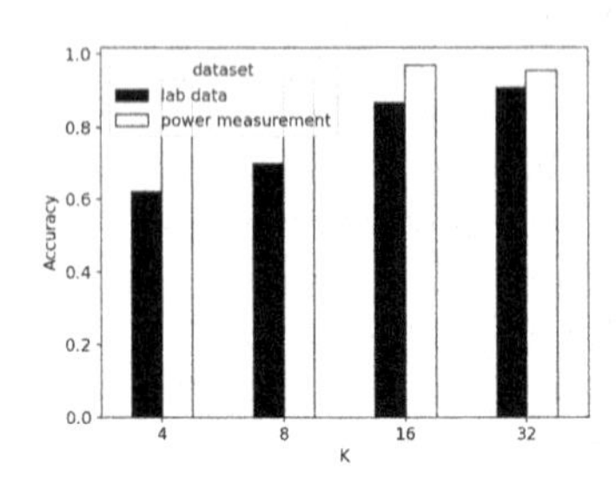

(b) Accuracy of K-BOSS-VS varying K

Fig. 1. Accuracy results

proves that having multiple class representatives per class (in contrast to BOSS-VS and SAX-VSM) provides increased accuracy especially on classification tasks with multiple labels, as in the Intel Data Lab Sensor data set. All the algorithms provide high accuracy in the power measurement data set, but the performance of the algorithms differs in the lab data case. The K-BOSS-VS algorithm provides an accuracy of 96% for the power measurement data set and 86% for the lab data data set for K = 16 compared to 96% and 98% accuracy scores, respectively, from BOSS.

Figure 1b shows the accuracy of K-BOSS-VS for varying values of parameter K. This algorithm reaches high accuracy for the power measurement data set even for low values of K. For the lab data data set the increase of K proves beneficial. For K = 16 K-BOSS-VS uses 41% of the time series per class for the lab data and 48% of the time series per class for the power measurements data set. While increasing K would result in accuracy similar to BOSS, the comparisons needed per target time series would also be similar to BOSS, as the percentage of representatives per class would increase. These results supports our intuition that a sufficient number of class representatives is beneficial to

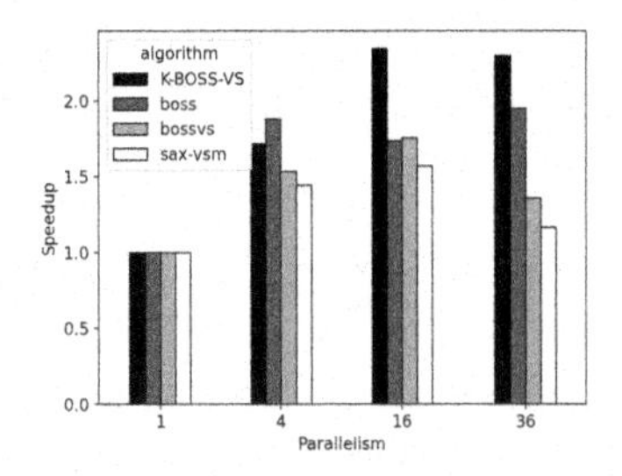

(a) power measurements data set

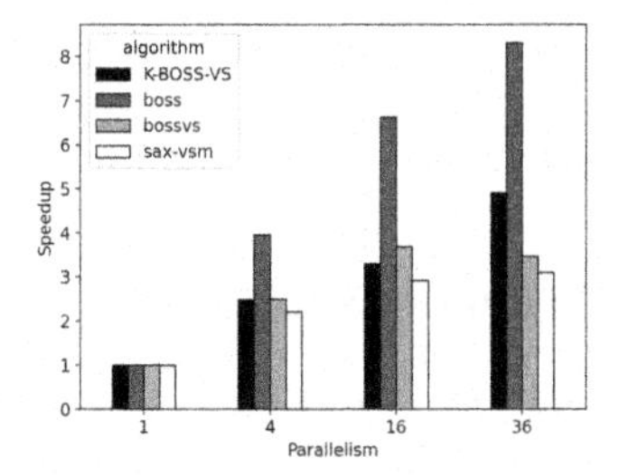

(b) lab data set

Fig. 2. Speedup of methods in the batch setting, per data set.

the accuracy of the algorithm, while we can effectively reduce the comparisons needed per target case.

Figures 2a and 2b show the speedup for the batch settings for all four algorithms on the two data sets. As we can see, based on our implementations, BOSS scales better than SAX-VSM and BOSS-VS as parallelism increases, given that it can distribute comparisons of test cases with each of the class members effectively. For the power measurement data set the K-BOSS-VS method provides the best speedup, while is has the second best speedup after BOSS for the lab data set.

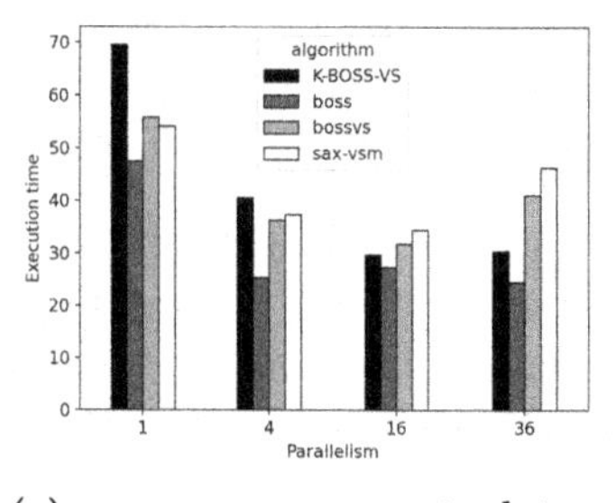

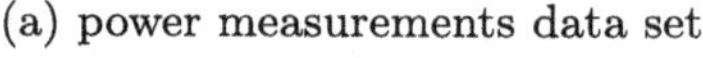
(a) power measurements data set

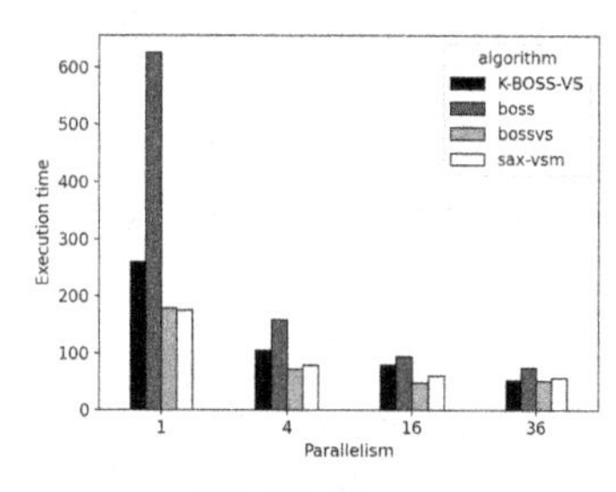

(b) lab data set

Fig. 3. Total methods' execution time in the batch setting, per data set.

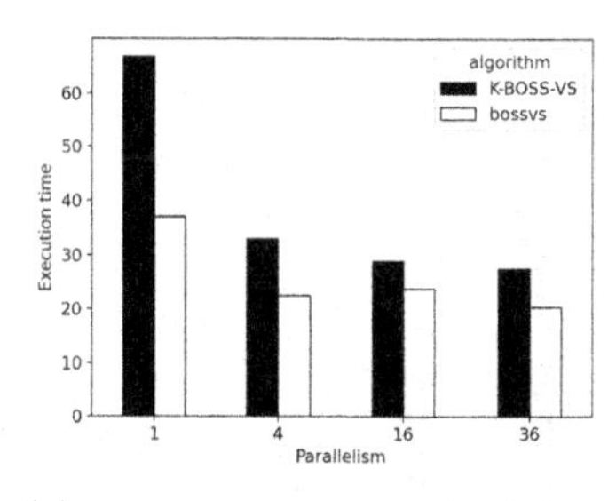

(a) power measurements data set

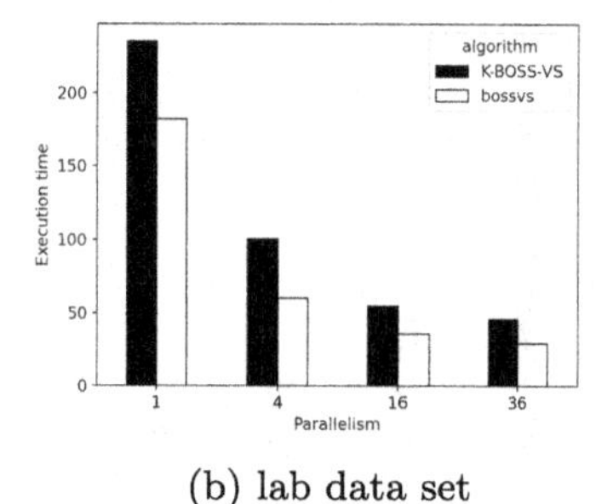

(b) lab data set

Fig. 4. Training time in the streaming setting for K-BOSS-VS and BOSS-VS, per data set.

Figures 3a and 3b show the total execution time (i.e., the time for the training and test-phase of the algorithms) in the batch setting. On the power measurements data set the results are not what we would expect, especially when comparing BOSS with BOSS-VS and SAX-VSM (and with K-BOSS-VS): This may be due to the fact that the data set is small and auxiliary operations add non-trivial execution time. On the larger and more complex data set the picture is different. BOSS, as expected, has the largest execution time in the low parallelism settings, but as parallelism increases it becomes competitive to the other algorithms. The K-BOSS-VS algorithm reports high execution time in low parallelism settings, compared to BOSS-VS and SAX-VSM, although it is much faster than BOSS. In high parallelism setting K-BOSS-VS matches the performance of BOSS-VS and SAX-VSM. It must be noted that for parallelism equal to 36 the training time for K-BOSS-VS is 47 s and the test time is 6 s. More results regarding the test time are provided in the streaming settings, where these are more relevant.

Figures 4a and 4b show the execution time for training on the two data sets for BOSS-VS and the K-BOSS-VS method for the streaming case. As we can see the K-BOSS-VS algorithm takes longer time to train, as expected, due to the cost incurred on computing the K representatives per class. However, it becomes more competitive with increased parallelism.

Figures 5a and 5b show the execution time for the test part of the algorithms in streaming settings. The time needed for online classification is important - especially in the streaming case, since, the training phase of the streaming algorithm can be performed offline in the fashion of the Lambda Architecture [17]. As we can see, although the K-BOSS-VS algorithm is competitive to BOSS-VS, it is still slightly slower in low parallelism settings, as it incurs additional cost for testing with multiple representatives per class. The performance benefits of BOSS-VS are less apparent with increased parallelism and the two algorithms perform identically for parallelism equal to 36 (given that K = 16), although K-BOSS-VS achieves similar (in the power measurements data set) or better (in the lab data set) accuracy scores, as shown above.

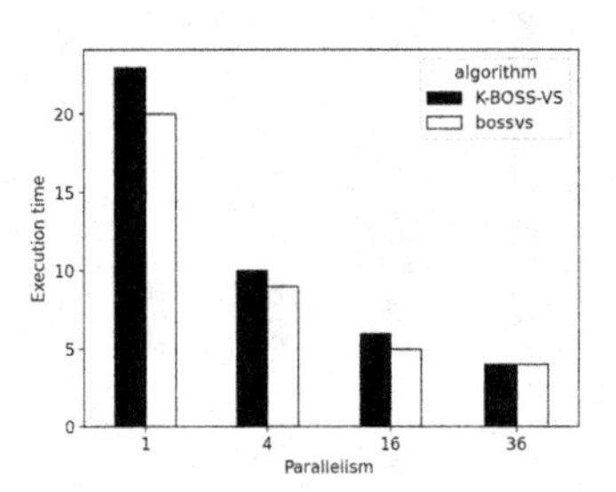

(a) power measurements data set

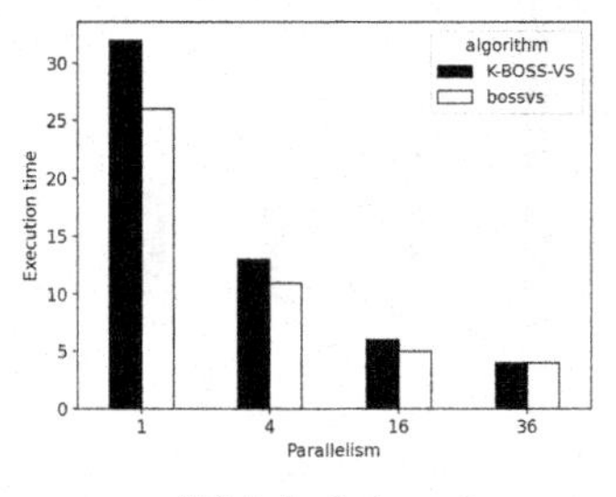

(b) lab data set

Fig. 5. Testing time in the streaming setting, for K-BOSS-VS and BOSS-VS, per data set.

Figures 6a and 6b show the scalability of the test phase in the streaming setting. As we can see the K-BOSS-VS method scales better than BOSS-VS due to higher opportunities for parallelism, given that it can compare each target time series to K representatives for each of the classes, in parallel.

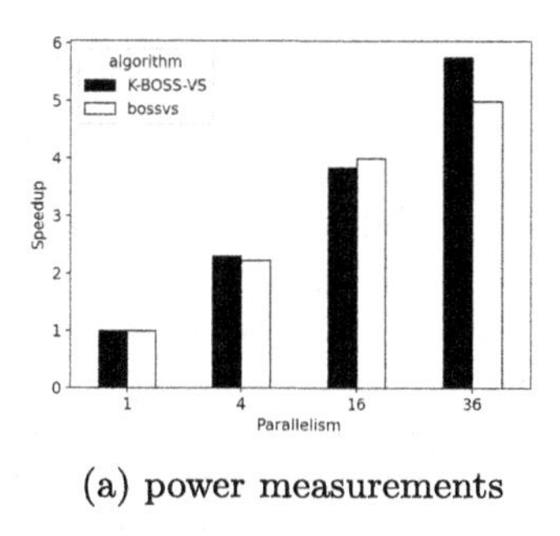

(a) power measurements

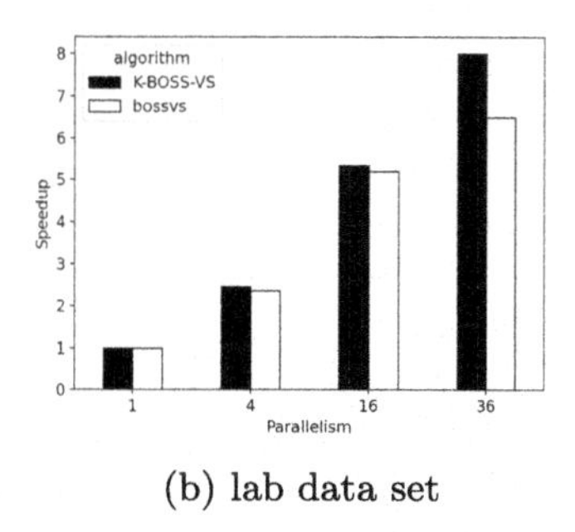

(b) lab data set

Fig. 6. Speedup for testing in the streaming setting, for K-BOSS-VS and BOSS-VS, per data set.

Figures 7a and 7b compare the scalability of BOSS-VS with the scalability of the K-BOSS-VS algorithm in the training phase. As we can see, in the smaller data set K-BOSS-VS scales better, whereas in the larger data set BOSS-VS scales better, given the cost incurred by K-BOSS-VS in comparing each target series with K representatives per class. However, despite the fact that the algorithms are close in terms of scalability, K-BOSS-VS is significantly more accurate than BOSS-VS in the later case.

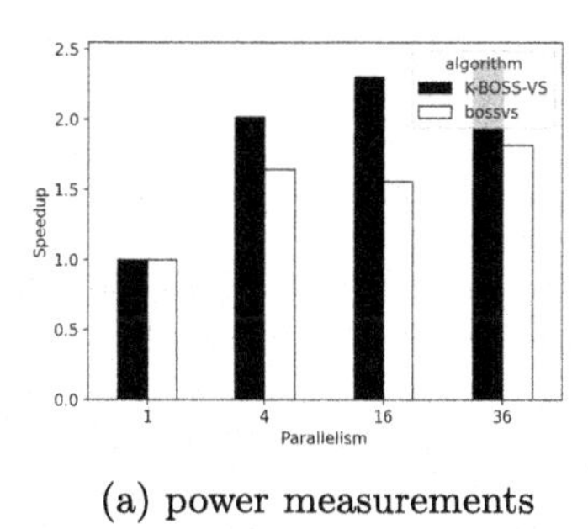

(a) power measurements

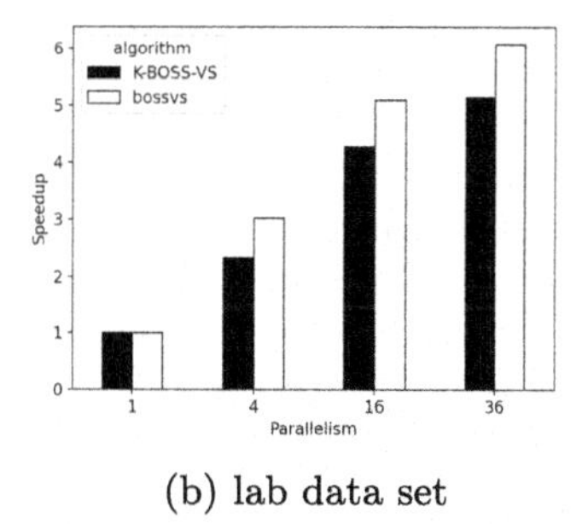

(b) lab data set

Fig. 7. Speedup for training in the streaming setting, for K-BOSS-VS and BOSS-VS, per data set.

7 Conclusions

In this paper we have introduced the K-BOSS-VS algorithm for time-series classification in batch and streaming settings. Our proposed algorithm uses K representatives per class to compare each test case using an 1-NN classifier, providing accuracy similar or better than state of the art classifiers that compare each test case with every member of each class. On the contrary, although the algorithm incurs a performance overhead compared to algorithms using a single representative (centroid) per class, maintaining K representatives per class, achieves higher accuracy. However, on high parallelism settings it matches the performance of algorithms that use a single centroid due to better scalability. The small memory footprint of the algorithm, the high accuracy, and the efficiency of the classification despite the performance overhead that it incurs - mainly in the training phase, make it suitable for streaming applications.

As future work we plan to investigate trends' representations by means of class representatives, as well as incremental computations, aiming to further increase the computational efficiency of the method while increasing further its accuracy. We also plan to investigate time series classification methods following paradigms that are different to the methods used here, such as decision trees, also addressing the concept drift problem. Finally, we plan to investigate time series classification purely on a stream setting without having to train on batch.

Acknowledgement. This work is partially supported by the University of Piraeus Research Center.

References

1. Armbrust, M., et al.: Structured streaming: a declarative API for real-time applications in apache spark. In: Proceedings of the 2018 International Conference on Management of Data, pp. 601–613 (2018)
2. Armbrust, M., et al.: Spark SQL: relational data processing in spark. In: Proceedings of the 2015 ACM SIGMOD International Conference on Management of Data, pp. 1383–1394. ACM (2015)
3. Baldán, F.J., Benítez, J.M.: Distributed fastshapelet transform: a big data time series classification algorithm. Inf. Sci. **496**, 451–463 (2019)
4. Bifet, A., Gavalda, R.: Learning from time-changing data with adaptive windowing. In: Proceedings of the 2007 SIAM International Conference on Data Mining, pp. 443–448. SIAM (2007)
5. Bifet, A., Holmes, G., Pfahringer, B.: Leveraging bagging for evolving data streams. In: Balcázar, J.L., Bonchi, F., Gionis, A., Sebag, M. (eds.) ECML PKDD 2010. LNCS (LNAI), vol. 6321, pp. 135–150. Springer, Heidelberg (2010). https://doi.org/10.1007/978-3-642-15880-3_15
6. Cano, A., Krawczyk, B.: Kappa updated ensemble for drifting data stream mining. Mach. Learn. **109**(1), 175–218 (2020)
7. Cui, Z., Chen, W., Chen, Y.: Multi-scale convolutional neural networks for time series classification. arXiv preprint arXiv:1603.06995 (2016)
8. Dean, J., Ghemawat, S.: Mapreduce: simplified data processing on large clusters. Commun. ACM **51**(1), 107–113 (2008)
9. Engle, C., et al.: Shark: fast data analysis using coarse-grained distributed memory. In: Proceedings of the 2012 ACM SIGMOD International Conference on Management of Data, pp. 689–692. ACM (2012)
10. Fawaz, H.I., Forestier, G., Weber, J., Idoumghar, L., Muller, P.A.: Transfer learning for time series classification. In: 2018 IEEE International Conference on Big Data (Big Data), pp. 1367–1376. IEEE (2018)

11. Ganz, F., Barnaghi, P., Carrez, F.: Automated semantic knowledge acquisition from sensor data. IEEE Syst. J. **10**(3), 1214–1225 (2014)
12. Gomes, H.M., et al.: Adaptive random forests for evolving data stream classification. Mach. Learn., 1469–1495 (2017). https://doi.org/10.1007/s10994-017-5642-8
13. Hüsken, M., Stagge, P.: Recurrent neural networks for time series classification. Neurocomputing **50**, 223–235 (2003)
14. Karim, F., Majumdar, S., Darabi, H., Chen, S.: LSTM fully convolutional networks for time series classification. IEEE Access **6**, 1662–1669 (2018)
15. Kasetty, S., Stafford, C., Walker, G.P., Wang, X., Keogh, E.: Real-time classification of streaming sensor data. In: 2008 20th IEEE International Conference on Tools with Artificial Intelligence, vol. 1, pp. 149–156. IEEE (2008)
16. Kontaki, M., Papadopoulos, A.N., Manolopoulos, Y.: Continuous trend-based classification of streaming time series. In: Eder, J., Haav, H.-M., Kalja, A., Penjam, J. (eds.) ADBIS 2005. LNCS, vol. 3631, pp. 294–308. Springer, Heidelberg (2005). https://doi.org/10.1007/11547686_22
17. Marz, N., Warren, J.: Big Data: Principles and Best Practices of Scalable Realtime Data Systems. Manning Publications Co., New York (2015)
18. Meng, X., et al.: MLlib: machine learning in apache spark. J. Mach. Learn. Res. **17**(1), 1235–1241 (2016)
19. Schäfer, P.: The boss is concerned with time series classification in the presence of noise. Data Min. Knowl. Disc. **29**(6), 1505–1530 (2015)
20. Schäfer, P.: Scalable time series classification. Data Min. Knowl. Disc. **30**(5), 1273–1298 (2015). https://doi.org/10.1007/s10618-015-0441-y
21. Schäfer, P., Högqvist, M.: SFA: a symbolic Fourier approximation and index for similarity search in high dimensional datasets. In: Proceedings of the 15th International Conference on Extending Database Technology, pp. 516–527. ACM (2012)
22. Schäfer, P., Leser, U.: Fast and accurate time series classification with weasel. In: Proceedings of the 2017 ACM on Conference on Information and Knowledge Management, pp. 637–646. ACM (2017)
23. Senin, P., Malinchik, S.: SAX-VSM: interpretable time series classification using SAX and vector space model. In: 2013 IEEE 13th International Conference on Data Mining, pp. 1175–1180. IEEE (2013)
24. Smirnov, D., Nguifo, E.M.: Time series classification with recurrent neural networks
25. Wang, Z., Yan, W., Oates, T.: Time series classification from scratch with deep neural networks: a strong baseline. In: 2017 International Joint Conference on Neural Networks (IJCNN), pp. 1578–1585. IEEE (2017)
26. Xin, R.S., Gonzalez, J.E., Franklin, M.J., Stoica, I.: GraphX: a resilient distributed graph system on spark. In: First International Workshop on Graph Data Management Experiences and Systems, p. 2. ACM (2013)
27. Zaharia, M., et al.: Resilient distributed datasets: a fault-tolerant abstraction for in-memory cluster computing. In: Proceedings of the 9th USENIX Conference on Networked Systems Design and Implementation, p. 2. USENIX Association (2012)
28. Zaharia, M., Das, T., Li, H., Hunter, T., Shenker, S., Stoica, I.: Discretized streams: fault-tolerant streaming computation at scale. In: Proceedings of the Twenty-Fourth ACM Symposium on Operating Systems Principles, pp. 423–438. ACM (2013)
29. Zaharia, M., et al.: Apache spark: a unified engine for big data processing. Commun. ACM **59**(11), 56–65 (2016)

DeCStor: A Framework for Privately and Securely Sharing Files Using a Public Blockchain

Maria Siopi, George Vlahavas(✉), Kostas Karasavvas, and Athena Vakali

Aristotle University of Thessaloniki, Thessaloniki, Greece
gvlahavas@csd.auth.gr

Abstract. Cloud services have become increasingly popular during the past few years. Through these services, users can store their data remotely and access them any time and from anywhere. These services are offered by centralized systems where an organization or company usually offers their resources to users. The centralized nature of these systems causes several problems; a single point of failure exists, security issues might provide unwarranted access to intruders and there are privacy issues to consider as well. A solution to these problems is the decentralization of the system. A core technology that can help in this respect is the blockchain. It does not require any centralized control and its security model is based on the nodes of the blockchain network to share and verify transactions. This work aims to develop a secure decentralized cloud service, which does not expose the users' personal data. To this effect, a framework that implements a cloud service using the Ethereum blockchain ecosystem and the Swarm decentralized storage platform was developed. In this, file access is provided through user-specific decryption keys. By developing a decentralized cloud, using a secure encryption model for the data, a service which is more secure, and where the users have full control over their data is possible.

Keywords: Blockchain · Cloud storage · Decentralization · Security · Privacy

1 Introduction

Advances in networking technologies in recent years, along with an ever increasing need for computing resources has prompted individuals and organizations around the world to outsource their storage needs to cloud storage providers. These providers, offer on-demand network access to a shared pool of configurable computing resources that can be rapidly provisioned and released with minimal management effort or service provider interaction [22] and provide benefits like flexibility, convenience, reliability, access to resources from anywhere and at any time etc. In most cases, in addition to providing an online storage space, they also provide the means for file sharing and collaboration. Their business model

A. Appice et al. (Eds.): DS 2020, LNAI 12323, pp. 280–293, 2020.
https://doi.org/10.1007/978-3-030-61527-7_19

usually dictates a free service tier, which most individuals find adequate for their needs, while anyone with increased needs can purchase additional resources at a cost effective level. These fall into the category of public cloud service providers. While generally convenient and cost effective, they introduce significant security and privacy risks. A potential breach in the security of the cloud storage provider may allow the attacker access to private data. This issue is important for individuals, organizations and enterprises as well. Cloud service providers usually implement strong security policies, and the potential security risks may be negligible, but they certainly exist. Privacy may be an even more important concern though. It is widely known that most cloud service providers access user files, stored on their premises, with the intent to harvest information about the users. They can then use that information to direct the user to other paid services, or simply target them with more effective advertisements of third party products and services. While this may be an acceptable trade-off for a large number of individuals, it is certainly not acceptable by everyone. And especially for enterprises and government organizations, this access to potentially mission-critical data may not be acceptable at all.

To counter these problems, private cloud solutions have been developed. In these, infrastructure is managed by the user, organization or enterprise. While these may address most of the privacy related issues with public cloud services, the cost of supporting their infrastructure falls on the end user. What's more important, they are also more susceptible to security issues. This is due to the fact that a small organization or an individual may lack the resources for proper maintenance of the infrastructure and related software, leaving known security issues unpatched for large periods of time.

In order to address these concerns, we argue for designing a framework which allows the users to privately and securely share files using public blockchain technologies. Blockchain technologies have become increasingly popular since the introduction of Bitcoin [23]. The blockchain is an immutable ledger of interlinked blocks. In each block, transactions between its users are recorded. Apart from their primary use as a value store and currency, they can also be used for recording arbitrary data. In later blockchain technologies, such as Ethereum [11], it is also possible to execute computer code in a decentralized manner. However, it is not cost-effective to store large amounts of data on the blockchain itself. For that reason, it is most often the case that only metadata are stored. Recently, several decentralized storage solutions have emerged, mostly based on blockchain technologies, such as IPFS [1], Sia [3], Storj [4] and Swarm [5]. Our proposed framework is based on Ethereum and Swarm technologies. Ethereum has been chosen as it is the most mature public smart contract platform at this point and Swarm is a technology that is closely linked to Ethereum. However, the same methodology that is used with our proposed framework could easily be used with other technologies. Key characteristics of our proposal, mostly inherited by the underlying technologies, include:

- reliability: all data is redundantly stored in multiple locations

- confidentiality: nobody other than the user the files belong to may learn any information about their data
- integrity: no unauthorized modification of data may take place
- availability: data should be available from anywhere with an Internet access, at all times
- cost effectiveness: access to the provided services should not incur prohibitive costs
- data sharing: users may share their data with other users that they choose to, and those users only

The added value provided by our proposal, includes the combination of these underlying technologies in a novel way, coupled with an encryption scheme that allows for private and data sharing that is isolated for individual users.

The structure of this paper is as follows: in the next section, related work is outlined. In Sect. 3, the DeCStor framework is presented in detail. In Sect. 4, a cost analysis with respect to our proposal's use is performed. Finally, conclusions are drawn in the last section, potential limitations of our work and possible mitigations actions are also listed and ways of possibly extending this work are indicated.

2 Related Work

There has been extensive work pertaining to the security and privacy of centralized cloud solutions during the past years. In [17], several architectures that combine recent and non-standard cryptographic primitives in order to provide security on public cloud infrastructure, are considered. A review of cryptographic techniques that are used on existing cloud storage, their adoption and role are analyzed in [28]. Moreover, the security issues that are the major concerns in a cloud environment, along with issues related to data location, storage, availability and integrity are noted in [18]. Security in public cloud storage solutions is also discussed in [16], followed by possible mitigation techniques.

Users' privacy concerns with respect to cloud storage solutions are recorded in [15], indicating that their privacy requirements are different than those of companies. Risks, solutions, and open problems related to ensuring privacy of users accessing services or resources in the cloud, sensitive information stored at external parties, and access to such information is discussed in [12].

A comprehensive review of existing privacy and security issues in cloud computing, along with the relationships between them, potential vulnerabilities that may be exploited by attackers, threat models as well as existing defense strategies appear in [33]. The unique issues of cloud computing that exacerbate security and privacy challenges in clouds along with various approaches to address these challenges are illustrated in [31].

The ways that security, trust and privacy issues occur in the context of cloud computing and how these may be addressed are presented in [27]. It is noted that in order to provide and support trustworthy and innovative cloud computing services that are useful for a range of different situations, problems with privacy and security should be addressed.

Unsolicited government access to end user data in cloud storage solutions is discussed in depth in [30], together with potential mitigations that include data encryption and strict adherence to no-logging policies.

Several cloud computing system providers are investigated with respect to their concerns on security and privacy issues in [34]. The researchers find that these impose strong barriers for users' adoption of cloud systems and cloud services and conclude that more security strategies should be deployed in the cloud environment and that privacy acts should be altered to that end.

The privacy challenges that software engineers face when targeting the cloud as their production environment are outlined in [26].

With respect to blockchain based solutions, a modified version of the Inter-Planetary File System (IPFS) that leverages Ethereum smart contracts to provide access controlled file sharing is described in [10].

A blockchain based data provenance architecture to provide assurance of data operations in a cloud storage application, while enhancing privacy and availability is presented in [21]. A decentralized document version control system that uses the Ethereum blockchain [11] together with IPFS [1] is described in [25], with the authors claiming that their proposed solution is free of commonly known security vulnerabilities and attacks.

A blockchain-based security architecture for distributed cloud storage, where users can divide their own files into encrypted data chunks, and upload those data chunks randomly into the P2P network nodes is introduced in [20]. The authors employ a genetic algorithm to solve the file block replica placement problem and find that the proposed architecture outperforms the traditional cloud storage architectures in terms of file security.

In [19], a data-sharing mechanism that enables users to share their encrypted data under a blockchain-based decentralized storage architecture is presented. Encrypted data are stored in dedicated storage nodes and a proxy re-encryption mechanism is used to ensure secure data-sharing in this untrusted environment.

The present work aims at improving the security and privacy implications of using public decentralized storage, through the use of strong encryption ciphers and user-specific decryption keys. Our solution involves the Ethereum blockchain and the Swarm decentralized storage platform with a smart contract controlling the upload and sharing of files that are stored on Swarm.

3 The DeCStor Framework

In this section, the DeCStor framework is outlined. The most important technologies, that DeCStor builds upon, are presented, accompanied by a brief description of how they are used to overcome different problems. The encryption model that is used by DeCStor is analyzed in depth and the way account management functions is presented. The role of smart contracts is detailed and the different components of a proof-of-concept implementation are described.

The most important aims of the DeCStor framework design are complete decentralization and data protection. Technological solutions that have been adopted and design details have been carefully chosen with those in mind.

3.1 Building Blocks

The following technologies are key parts of the DeCStor framework, as it was designed and implemented. A brief introduction to these technologies follows:

Ethereum provides a public blockchain platform, accessible to anyone, in which smart contracts can be executed. It was first released in 2013, with the goals of improving the scripting and other restrictions of bitcoin and other cryptocurrencies. Smart contracts are usually written using Ethereum's dedicated programming language, Solidity. These are executed in the Ethereum Virtual Machine (EVM), which is a decentralized Turing complete virtual machine. Using the EVM, smart contracts in Ethereum are possible to carry arbitrary state and perform any arbitrary computation. The code that is included with a smart contract is executed on all participating nodes of the Ethereum network, as part of the block creation process [11].

Swarm is a "distributed storage platform and content distribution service, a native base layer service of the ethereum web3 stack that aims to provide a decentralized and redundant store for Distributed Application (DApp) code, user data, blockchain and state data" [5]. Swarm is a service available to the Ethereum blockchain. It essentially allows users to store and distribute DApp code and data utilizing its peer-to-peer data sharing network. In this, files are addressed by the hash value of their content. Instead of storing files in individual servers, these are hosted in this peer-to-peer data storage network. Implementing cross-node replication and erasure coding ensures data availability. These traits makes Swarm fault-tolerant and censorship resistant. Its decentralized nature also means that it is resistant to DDoS (distributed denial of service) types of attacks, with effectively zero-downtime. It is also designed to use an incentive system, so that the network's viability is ensured [13].

Public Key Infrastructure is a system that enables users of an insecure public network to securely and privately exchange data through the use of private and public cryptographic key pairs. These key pairs are commonly obtained and shared through a trusted authority, although self-creation is also quite common. PKI assumes the use of public key cryptography, a cryptographic technique that reliably verifies the identity of an entity via digital signatures. The main features that PKI offers are non-repudiation, privacy, integrity, accountability and trust [14,32].

The Advanced Encryption Standard (AES) cipher is an iterative cipher that implements a symmetric encryption algorithm. It is widely adopted for the secure and private communication of data and is supported in both software and hardware. Until now, there exist no practical attacks against AES. It can use keys of variable length, which allows a degree of "future-proofing" against progress in the ability to perform exhaustive key searches.

3.2 Encryption Model

The most important feature of the DeCStor framework is data security. User uploaded data should remain private and accessible only to specific users that the owner provides access to. Therefore, the selection and design of a suitable encryption model, that implements strong cryptographic principles, is of paramount importance.

The encryption model that DeCStor uses, combines two cryptographic primitives that are commonly used and are currently impossible to break, in practical terms. The first one is the Advanced Encryption Standard (AES) cipher [7], which is a symmetric-key algorithm. That means that the same key is used for both the encryption and the decryption of data. AES has a block size of 128 bits, and the configuration that is used by DeCStor uses a key length of 256 bits, commonly called the AES-256 algorithm. The second is Public Key Infrastructure (PKI) [9]. In public-key cryptography, a pair of keys is used. This key pair consists of a "private key" and a "public key". The private key is known only to the owner of the key pair, while the public key, as the name suggests, may be disseminated publicly. The public key is actually derived from the private key by applying certain cryptographic functions. This is performed in a manner, so that someone can encrypt a message with a public key provided to them, with only the owner of the corresponding private key being able to decrypt the message [24]. The most typically used algorithm used in PKI is RSA [29], but other algorithms can be used as well.

Before any file operations are performed, a private/public key pair is issued for each user. When a user uploads a new file, a "File-key" that corresponds to that specific file is generated. This File-key is simply a 32-byte randomly generated number that is then encoded as a base-64 ASCII string. The file is then encrypted with the AES-256 cipher, using the File-key as the encryption key. The encrypted file is then uploaded to Swarm and a hash that corresponds to the file's location in Swarm is returned to the file owner.

In order to provide access to the encrypted file to another user, the owner of the file needs to obtain the potential recipient's public key. The owner then encrypts the File-key with that public key, thus creating a "Share-key". The Share-key is then sent to the recipient, along with the Swarm hash that indicates the location of the file.

When the recipient wants to download the encrypted file, they use their own private key to decrypt the Share-key, recreating the File-key in the process. They can then download the AES-256 encrypted file using its hash from Swarm and decrypt it using the File-key they obtained.

3.3 Account Management

A user account management system typically identifies users in a centralized cloud based storage platform. Since the adoption of such an account management system would add a centralized component to DeCStor, the adoption of

Ethereum addresses as the user accounts was preferred. This way, user management is decoupled from the DeCStor framework and a completely decentralized solution can be implemented. An additional benefit is that a cryptographically secure solution that is already used widely is adopted.

An Ethereum address is derived from a user's private key. Specifically, for a given private key, an Ethereum address (a 160-bit number) to which it corresponds is defined as the right most 160-bits of the Keccak hash of the corresponding ECDSA public key [11]. The same private key that is used to derive the Ethereum address, is also used for PKI operations in the DeCStor framework.

In order to use Ethereum addresses for user identification, an Ethereum "wallet" application is required. In general, and despite the name, a wallet application for any cryptocurrency does not hold any amount of currencies. It just stores and manages user accounts in the form of private/public key pairs. For this purpose, a web-based browser plugin wallet application, such as Metamask [2] may be chosen. This has the additional benefit that a web application can interface with the plugin by using a library such as web3.js [6].

3.4 Smart Contract

A smart contract, deployed on the Ethereum blockchain is used to manage file access, including file access request actions and file sharing actions.

When a file is uploaded, through the smart contract, to Swarm, the smart contract stores the file's Swarm hash, along with the owner's address to the Ethereum blockchain (Fig. 1).

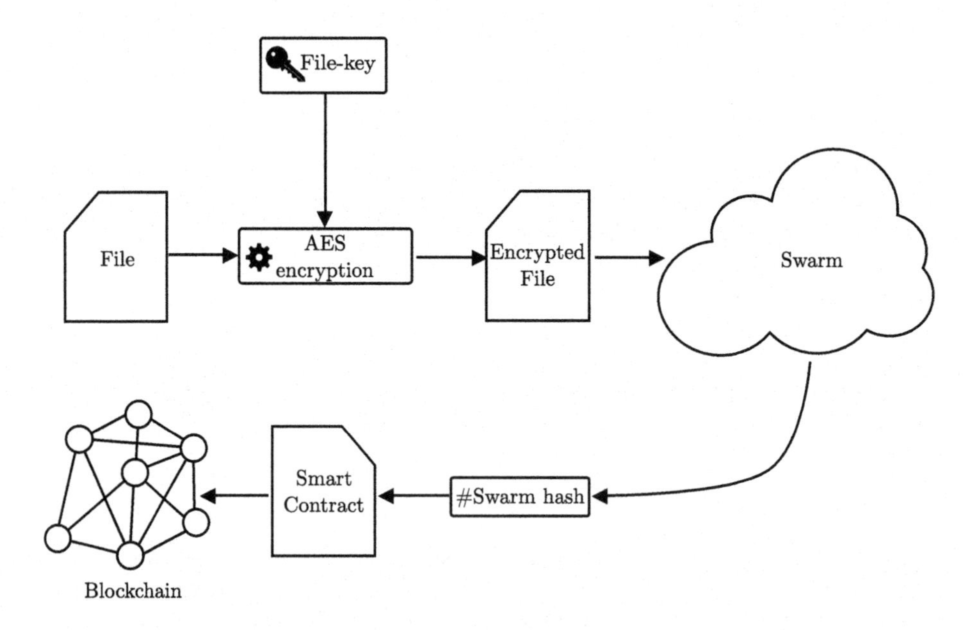

Fig. 1. File upload process

During a file access request by a user, an appropriate method in the smart contract is initiated (Fig. 2). The information that is passed to this method includes the recipient's address in the Ethereum network, as well as the public key that is attached to that address. These will be used by the file owner to encrypt the File-key before providing access to the file itself. The smart contract also notifies the owner for the file access request by means of emitting an appropriate message in the Ethereum event logging system. This can be typically read by a monitoring application so that the owner is aware of the access request.

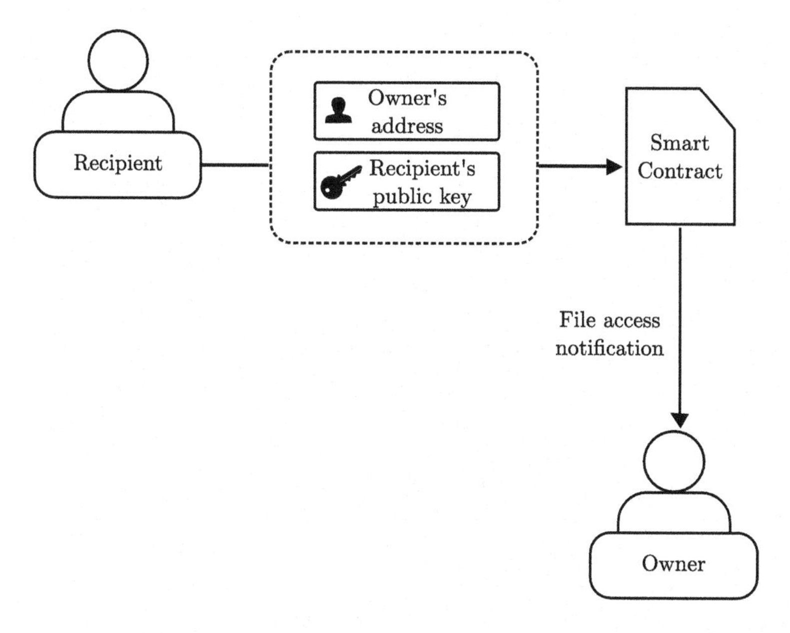

Fig. 2. File access request process

When a file owner responds to a file access request, they use the recipient's public key to encrypt the File-key to a Share-key that is specific to the recipient (Fig. 3). That way, only the recipient, using their own private key, is able to later decrypt the Share-key. Having the Share-key, they can then get back the File-key, and in turn decrypt the AES-encrypted file to its unencrypted form. The smart contract is used to store the file's Swarm hash, that is the address of the file in Swarm, the recipient's address as well as the corresponding public key. In that way, the smart contract acts as a logging mechanism of all people that have been provided access to the file.

During a file download (Fig. 4), no data is stored by the smart contract. Only data that has been previously shared between the file's owner and recipient is used to download the file from Swarm and decrypt it using the File-key.

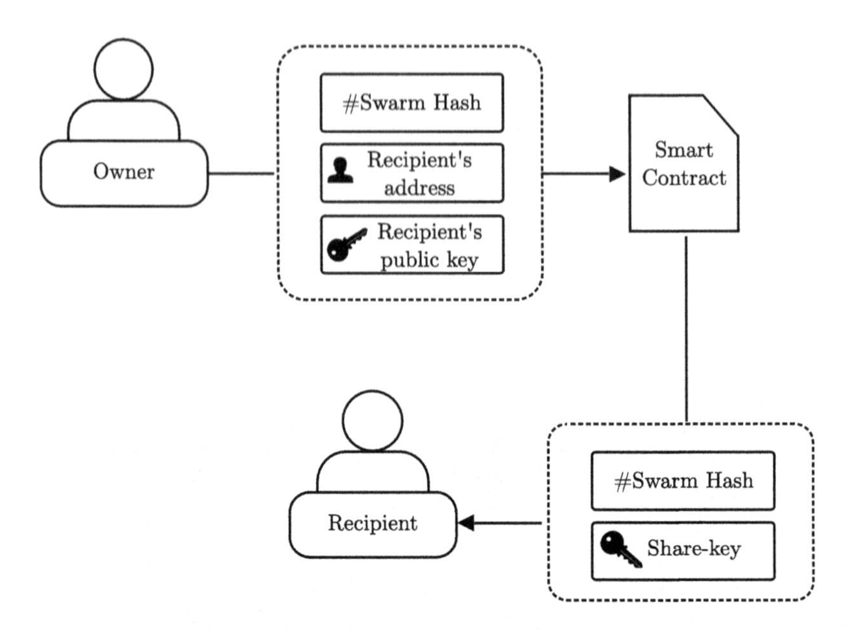

Fig. 3. File sharing process

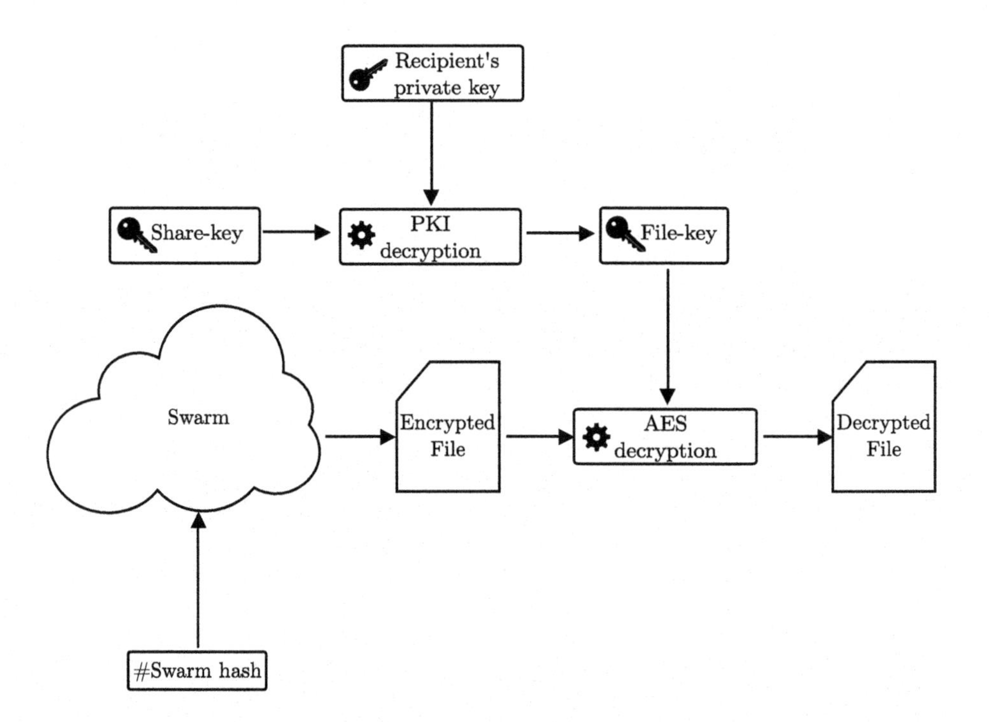

Fig. 4. File download process

3.5 Implementation

A proof-of-concept implementation has been developed, in order to showcase that the DeCStor framework can be practically used. The implementation consists of different components:

- a smart contract,
- a web3 javascript library and
- an HTML5 front-end

All of these components are available, in source code form, under a free software license[1].

The smart contract is written in the Solidity programming language [8], which is the most common programming language for developing smart contracts for the Ethereum blockchain. This essentially functions as the "back-end" part of the proof-of-concept application. Access to the methods that the smart contract includes, is provided by an Application Binary Interface (ABI). Using this ABI, users are allowed to interact with the Ethereum blockchain network in the manner specified by the smart contract. The smart contract is responsible for storing and retrieving data to and from the Ethereum blockchain. It stores data with respect to the identities of users (in the form of Ethereum addresses) that upload files, request access to files and ultimately share files with other users. This information is stored using so called "mappings", which are in essence key-value data stores. In any case, the key is always an Ethereum address and the value is a struct. A struct is a composite data type, similar to that found in other common programming languages, such as C. A struct defines a grouped list of related variables under one name. In this case, structs contain the owner's and recipient's Ethereum addresses as well as the Swarm hash of the uploaded files and access information according to the action that is being requested by the user (Figs. 1-3). In addition, it is designed to emit the respective events for any changes that might occur in the data that the smart contract manages.

The web3 [6] javascript library interfaces with the web3.js libraries, that use a generic JSON-RPC interface to connect with one or more active nodes in the Ethereum network. Users may select to connect to a full node, locally or remotely, or they may use a lightweight client, such as the Metamask web browser plugin to establish a connection with the Ethereum network. The DeCStor library includes functionality that facilitates the following actions:

- Data upload. The function encrypts the contents of a user provided file and uploads the encrypted file to Swarm. It is also responsible for uploading Share-keys when the file's owner elects to do so.
- Data re-encryption. The function receives the encrypted data from Swarm, decrypts them and in turn encrypts them again using new encryption keys. All relevant information is stored back on the blockchain.
- Request for access permission. This is a small function that stores the request on the blockchain.

[1] https://gitlab.com/datalab-auth/blockchain/decstor.

- Data download. This function facilitates the data download from Swarm. The downloaded data are stored as a file locally in the user's device, which is decrypted using the respective Share-key. In contrast to all previous functions, it does not interact with the blockchain in any way.

The front-end web application is a simple implementation, written using HTML5/Javascript technologies. It includes forms for file upload, file access request, file sharing and file download providing simple and user-friendly interfaces to all functionality.

All of these components are actually parts of a Decentralized Application (DApp). With DApps, in general, the "back-end" part of the application runs in a completely decentralized manner, on blockchain networks, such as the Ethereum network. Therefore, DApps inherit all characteristics of blockchain networks, such as immutability, increased security, fault tolerance and zero downtime. The "front-end" part of a DApp can be a Web UI, built with common web technologies, as in this case. However, there is no need to host this Web UI in a centralized manner. In fact, there can be multiple instances of the front-end code running simultaneously on different web servers. Front-end code can also be ran as mobile phone or desktop computer applications, with no restriction to the number of running instances. It is also possible to have the front-end code stored in a decentralized manner on decentralized storage, such as Swarm or IPFS, that way making the entire DApp completely decentralized.

4 Cost Analysis

In this section, costs associated with using the DeCStor framework are going to be calculated and compared to respective costs of traditional cloud providers.

Cost analysis for the DeCStor implementation was held in May 2020. At that time, the recommended gas price was approximately 15 gwei (1 gwei = 10^9 ETH), while 1 ETH cost approximately $180. Due to the volatility of cryptocurrency market prices, calculations will probably differ for different timeframes.

Deployment of the smart contract that has been developed for DeCStor, to the Ethereum blockchain, incurs a cost of approximately 30000 gas, which equals approximately 0.00045 ETH. That translates to $0.081, an amount that can be considered trivial for most. This is also a one-time cost and should not concern users that employ the smart contract to upload and share files.

When users issue a transaction to the smart contract, either for uploading a file, requesting access to a file, or providing access to it, the cost is not proportional to the file's size. This is due to the fact that all the significant calculations that pertain to the file's contents occur in the front-end part of the DApp. The data that is relayed to the smart contract, for each one of the respective functions is always constant. The cost of executing any of the smart contracts methods is between 15–20 gas, which translates to about 0.00000023-0.00000030 ETH, or roughly, $0.00005. It has to be noted, that even this cost relates to file upload, file access request and file sharing actions. The actual file download does not call the smart contract and has no cost whatsoever.

With such small operational costs, the most important costs are likely to be Swarm storage costs. However, Swarm is still in early access stage and it has not been deployed to the main network yet. There is no way to guess how much it will ultimately cost to store data on Swarm, but judging from similar implementations, such as IPFS, that is not expected to be significant. For these reasons, storage costs are not included in our calculations.

5 Conclusions and Future Work

In this section, the general outcome of this work is briefly outlined. Possible limitations are also discussed and ideas for future work are presented.

Our goal was to design a framework for secure and private file access and sharing using public blockchain technologies. Our proposed architecture involves the encryption of data with strong encryption ciphers and decryption keys that are specific to each user that requests access to the data. A smart contract is used for requesting and providing access to data uploaded to the Swarm decentralized network storage platform. A proof-of-concept implementation has also been created in order to prove the practical application of our design. Finally, costs associated with its use have been calculated.

A potential issue with our architecture, is that once a Share-key has been decrypted to the File-key that is used to encrypt and decrypt the data, the File-key, or indeed the entire decrypted file may be shared between third parties, without the original owner being aware of it. While a solution to this issue could involve the use of some access control technologies, such as Digital Rights Management (DRM) in order to restrict access, it is out of scope with respect to this work. Such technologies could be used on top of our proposed framework, without significant, or in fact any, alterations to it.

Access revocation is however something to consider. If a user that has received access to a file is now considered untrustworthy, or the File-key has been otherwise compromised, access to the original file should be revoked. To that effect, our proof-of-concept implementation includes functionality to re-encrypt and re-upload data to Swarm. During this action, the originally uploaded file gets downloaded, a new File-key is generated, the file is encrypted with this new File-key and it is uploaded as a new file to Swarm. If the original file's owner stops providing incentive to other Swarm nodes, in order for them to keep hosting the contents of the original file, this will in time be garbage collected and entirely removed from Swarm. Nevertheless, a window of opportunity for potentially malicious users that want to freely share the file with other parties still exists. Additionally, if the malicious user has already downloaded the full data from Swarm, there is no way to stop them from sharing them using any other means.

As already noted, at the time of writing, the Swarm decentralized storage platform that is used by DeCStor, has yet to be deployed to the main network and is to be considered experimental. Once it is finally released, exact storage costs will be possible to calculate, in proportion to the uploaded files' size. Until that time, it is not possible to know what these costs might be, however these are expected to be similar or lower than offerings by centralized cloud storage providers.

Future improvements and changes may include the potential deployment on different decentralized network storage platforms, such as IPFS, since our solution could ultimately be used with almost any underlying storage platform. Additionally, the development of a more polished front-end application, together with mobile device applications that may facilitate the use of our solution is one of our future goals.

Acknowledgement. This research has been co-financed by the European Union H2020 Research and Innovation Programme under Grant Agreements No. 826404 and No 871403.

References

1. IPFS: a peer-to-peer hypermedia protocol designed to make the web faster, safer, and more open. https://ipfs.io/. Accessed 29 May 2020
2. Metamask: a crypto wallet & gateway to blockchain apps. https://metamask.io. Accessed 21 May 2020
3. Sia: decentralized storage for the post-cloud world. https://sia.tech/. Accessed 28 May 2020
4. Storj decentralized cloud storage. https://storj.io. Accessed 28 May 2020
5. Swarm: storage and communication for a sovereign digital society. https://swarm.ethereum.org/. Accessed 22 May 2020
6. web3js: Ethereum javascript api. https://web3js.readthedocs.io. Accessed 25 May 2020
7. Announcing the advanced encryption standard (aes). Federal Information Processing Standards Publication 197. United States National Institute of Standards and Technology (NIST), November 2001. https://nvlpubs.nist.gov/nistpubs/FIPS/NIST.FIPS.197.pdf
8. Solidity. https://solidity.readthedocs.io (2020). Accessed 08 May 2020
9. Adams, C., Lloyd, S.: Understanding PKI: concepts, standards, and deployment considerations. Addison-Wesley Professional (2003)
10. Benet, J.: IPFS - Content Addressed, Versioned, P2P File System. arXiv e-prints arXiv:1407.3561, July 2014
11. Buterin, V.: A next-generation smart contract and decentralized application platform-ethereum whitepaper, 2014 (2014). https://www.weusecoins.com/assets/pdf/library/Ethereum_white_paper-a_next_generation_smart_contract_and_decentralized_application_platform-vitalik-buterin.pdf. Accessed 24 May 2020
12. De Capitani di Vimercati, S., Foresti, S., Samarati, P.: Managing and accessing data in the cloud: Privacy risks and approaches. In: 2012 7th International Conference on Risks and Security of Internet and Systems (CRiSIS), pp. 1–9 (2012)
13. Dhillon, V., Metcalf, D., Hooper, M.: Unpacking Ethereum, pp. 25–45. Apress, Berkeley, CA (2017). https://doi.org/10.1007/978-1-4842-3081-7_4
14. Faraj Al-Janabi, S.T., Abd-alrazzaq, H.K.: Combining mediated and identity-based cryptography for securing E-mail. In: Ariwa, E., El-Qawasmeh, E. (eds.) DEIS 2011. CCIS, vol. 194, pp. 1–15. Springer, Heidelberg (2011). https://doi.org/10.1007/978-3-642-22603-8_1
15. Ion, I., Sachdeva, N., Kumaraguru, P., Čapkun, S.: Home is safer than the cloud! privacy concerns for consumer cloud storage. In: Proceedings of the Seventh Symposium on Usable Privacy and Security. SOUPS 2011, Association for Computing Machinery, New York (2011). https://doi.org/10.1145/2078827.2078845

16. Ka, S., Jayanthi, S.: A review on cloud data security and its mitigation techniques. Procedia Comput. Sci. **48**, 347–352 (2015)
17. Kamara, S., Lauter, K.: Cryptographic cloud storage. In: Sion, R., Curtmola, R., Dietrich, S., Kiayias, A., Miret, J.M., Sako, K., Sebé, F. (eds.) FC 2010. LNCS, vol. 6054, pp. 136–149. Springer, Heidelberg (2010). https://doi.org/10.1007/978-3-642-14992-4_13
18. Kaur, M., Singh, H.: A review of cloud computing security issues. Int. J. Educ. Manag. Eng. **5**, 32–41 (2015)
19. Li, D., Du, R., Fu, Y., Au, M.H.: Meta-key: a secure data-sharing protocol under blockchain-based decentralized storage architecture. IEEE Netw. Lett. **1**(1), 30–33 (2019)
20. Li, J., Wu, J., Chen, L.: Block-secure: blockchain based scheme for secure p2p cloud storage. Inf. Sci. **465**, 219–231 (2018). https://doi.org/10.1016/j.ins.2018.06.071
21. Liang, X., Shetty, S., Tosh, D., Kamhoua, C., Kwiat, K., Njilla, L.: Provchain: a blockchain-based data provenance architecture in cloud environment with enhanced privacy and availability. In: 2017 17th IEEE/ACM International Symposium on Cluster, Cloud and Grid Computing (CCGRID), pp. 468–477 (2017)
22. Mell, P., Grance, T.: The NIST definition of cloud computing. NIST Special Publication, pp. 800–145 (2011)
23. Nakamoto, S.: Bitcoin: a peer-to-peer electronic cash system (2008)
24. Nechvatal, J.: Public-key cryptography. Technical Report NIST Special Publication 800–2, National Institute of Standards & Technology, Research Information Center, Gaithersburg, MD 20899 (1991)
25. Nizamuddin, N., Salah, K., Azad, M.A., Arshad, J., Rehman, M.: Decentralized document version control using ethereum blockchain and IPFS. Comput. Electr. Eng. **76**, 183–197 (2019). https://doi.org/10.1016/j.compeleceng.2019.03.014
26. Pearson, S.: Taking account of privacy when designing cloud computing services. In: 2009 ICSE Workshop on Software Engineering Challenges of Cloud Computing, pp. 44–52 (2009)
27. Pearson, S.: Privacy, security and trust in cloud computing. In: Pearson, S., Yee, G. (eds.) Privacy and Security for Cloud Computing. CCN, pp. 3–42. Springer, London (2013). https://doi.org/10.1007/978-1-4471-4189-1_1
28. Peng, Y., Zhao, W., Xie, F., Dai, Z.H., Gao, Y., Chen, D.Q.: Secure cloud storage based on cryptographic techniques. J. China Univ. Posts Telecommun. **19**, 182–189 (2012). https://doi.org/10.1016/S1005-8885(11)60424-X
29. Rivest, R.L., Shamir, A., Adleman, L.: A method for obtaining digital signatures and public-key cryptosystems. Commun. ACM **21**(2), 120–126 (1978). https://doi.org/10.1145/359340.359342
30. Soghoian, C.: Caught in the cloud: privacy, encryption, and government back doors in the web 2.0 era. J. Telecommun. High Technol. Law **8**, 359 (2010)
31. Takabi, H., Joshi, J.B.D., Ahn, G.: Security and privacy challenges in cloud computing environments. IEEE Secur. Privacy **8**(6), 24–31 (2010)
32. Vacca, J.R.: Public Key Infrastructure: Building Trusted Applications and Web Services. CRC Press, Boca Raton (2004)
33. Xiao, Z., Xiao, Y.: Security and privacy in cloud computing. IEEE Commun. Surv. Tutorials **15**(2), 843–859 (2013)
34. Zhou, M., Zhang, R., Xie, W., Qian, W., Zhou, A.: Security and privacy in cloud computing: a survey. In: 2010 Sixth International Conference on Semantics, Knowledge and Grids, pp. 105–112 (2010)

Investigating Parallelization of MAML

Jan Bollenbacher[1,2(✉)], Florian Soulier[2], Beate Rhein[2], and Laurenz Wiskott[1]

[1] Institut für Neuroinformatik, Ruhr-University Bochum, 44801 Bochum, Germany
laurenz.wiskott@ini.rub.de

[2] TH Köln (University of Applied Sciences), 50678 Köln, Germany
{jan.bollenbacher,beate.rhein}@th-koeln.de

Abstract. We propose a meta-learning framework to distribute Model-Agnostic Meta-Learning (DMAML), a widely used meta-learning algorithm, over multiple workers running in parallel. DMAML enables us to use multiple servers for learning and might be crucial if we want to tackle more challenging problems that often require more CPU time for simulation. In this work, we apply distributed MAML on supervised regression and image recognition tasks, which are quasi benchmark tasks in the field of meta-learning. We show the impact of parallelization w.r.t. wall clock time. Therefore, we compare distributing MAML over multiple workers and merging the model parameters after parallel learning with parallelizing MAML itself. We also investigate the impact of the hyperparameters on learning and point out further potential improvements.

Keywords: Meta-learning · Distributed learning · Few-shot learning · Parallel computing · MAML

1 Introduction

In machine-learning, models are traditionally learned by applying a vast dataset on a specific task and train a model from scratch. Artificial learners perform poorly when only a small amount of data is available, or the task is changing, and they need to adapt.

In contrast, humans can abstract problems and effectively utilize prior knowledge to learn new skills quickly and can learn with just a few samples. Meta-learning is an emerging trend of research and tackles the problem of learning to learn. Meta-learning can be seen as a generalization of using experience and knowledge acquired earlier and is related to techniques for fine-tuning a model on a task or hyperparameter optimization.

Model-Agnostic Meta-Learning (MAML) is a popular and influential meta-learning algorithm. On a broad set of tasks, the MAML algorithm estimates a set of parameters, which is used as a starting point for fast adaptation so that new tasks can be learned quickly. MAML also has the benefit of being model-agnostic in a way, that it can be applied to many different instances of structurally similar tasks, solvable using gradient descent.

A. Appice et al. (Eds.): DS 2020, LNAI 12323, pp. 294–306, 2020.
https://doi.org/10.1007/978-3-030-61527-7_20

In this paper, we apply MAML in a distributed manner using multiple workers and compare it to a parallelization of MAML itself. We provide an empirical hyperparameter study and show the impact on the learning outcome. We can show that the benefit of parallelization highly depends on how we can sample the tasks for learning, and we recommend how to train MAML accordingly.

2 Related Work

MAML is a highly popular meta-learning algorithm for few-shot learning problems and achieves competitive performance on common benchmark few-shot learning problems [10,14,16,18–20]. It is an optimization-based meta-learning algorithm learning the parameters of a task-specific classifier. Other approaches in the same family of meta-learning are [3,7,11,12,21]. Where, for example, in [21], a concept generator is learned parallel with a meta-learner, the meta-learner learns in a high-level concept space instead of the original representation.

Of all the algorithms in the family of optimization-based meta-learning algorithms, MAML is especially influential and inspired many direct extensions in literature recently [1,2,6,8,13,17,22]. These extensions critically rely on the core structure of the MAML algorithm, using an outer-learning loop (for meta-training) and an inner-learning loop (for task-specific adaptation). In [2,13] an approach for implementing regimes to learn the inner learning rate α is proposed and in [15] in the inner-learning loop only the last layer of the model is adapted leading to less parameter updates. [22] divides the model parameters which are learned into *context* and *shared parameters* aiming to make MAML easier to parallelize and more interpretable.

In this work, we also aim to parallelize MAML by distributing learning over multiple workers. This approach is similar to [4], where the learning is performed by individual workers in a federated learning setting. Federated learning aims at training on heterogeneous datasets, whereas distributed learning aims on parallelizing computing power. In this work, we are performing an empirical investigation of MAML in the distributed learning setting.

3 Distributed Model-Agnostic Meta-Learning

3.1 Model-Agnostic Meta-Learning

The main goal of MAML introduced by [5] is to find model parameters f_θ which serve as a good starting point for training the model on new instances of tasks by applying just a few gradient steps

We sample training tasks $\mathcal{T}_i$ from a distribution of tasks $p(\mathcal{T})$. A task $\mathcal{T}_i$ consists of a support and query set $\mathcal{T}_i = (\mathcal{D}_i^s, \mathcal{D}_i^q)$:

1. The support set is used to fit the model parameters for solving the task by using a high inner-loop learning rate α and
2. the query set is used for slow training the meta-model parameters using a lower meta-learning rate β.

In MAML, we apply learning by two nested loops:

Inner-learning loop: Using the support set $\mathcal{D}_i^s$ of a task $\mathcal{T}_i$, we calculate the gradient of the loss function $\mathcal{L}$ w.r.t. the parameters of f_θ and store these in γ_i (Eq. 1).

$$\gamma_i \leftarrow \alpha \nabla_\theta \mathcal{L}_{\mathcal{D}_i^s}(f_\theta) \tag{1}$$

Outer-learning loop: Using the Adam optimizer, we update the meta-model parameters θ in a way that the error loss using the query set $\mathcal{D}_i^q$ is minimized for all task specific values γ_i (Eq. 2).

$$\theta \leftarrow \theta - \beta \sum_{\mathcal{T}_i} \nabla_\theta \mathcal{L}_{\mathcal{D}_i^q}(f_{\theta-\gamma_i}) \tag{2}$$

The quantity of data points and the structure of the sets are highly dependent on the task domain. We evaluate the domains supervised regression and image recognition; the tasks are described in detail in Sect. 4.

3.2 Distributed Learning

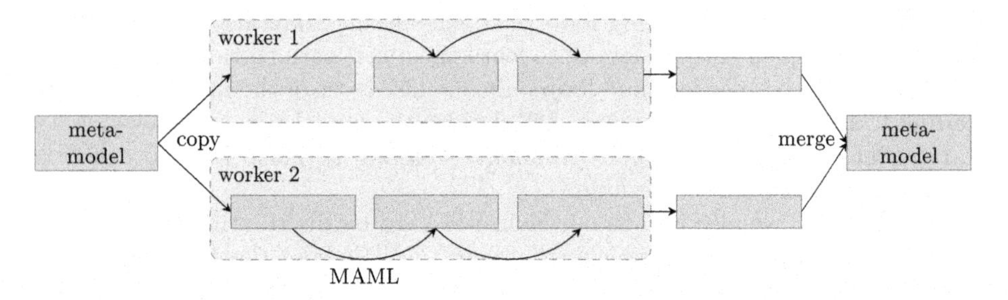

Fig. 1. One complete DMAML run: The central learner sends a copy of the meta-model to both workers. The workers perform 2 MAML steps and send the model parameters back to the central learner where they are merged.

We parallelize learning by distributing the learning over several workers. The central learner holds the meta-model parameters and sends a copy to the workers. Each worker is applying MAML steps to its model parameters in parallel. One MAML step is one update of the model parameters, as described in the previous Sect. 3.1. After the workers performed their MAML steps, the model parameters are sent back to the central learner. The central learner averages all parameters and starts over again. We call this procedure from copying the model parameters to the workers until averaging the adapted parameters one run. One Distributed Model-Agnostic Meta-Learning (DMAML) run is shown on Fig. 1.

The central learner receives the updated model parameters of each worker along with the following metrics: loss and accuracy. Therefore, strategies can be implemented on how to merge the model parameters. It is possible to drop model parameters of underperforming workers or give parameters of a well-performing worker a greater weight.

Workers can be run in parallel and are not increasing wall clock time used, while the consecutive MAML steps are executed sequentially and increase wall clock time linearly.

4 Experiments

The goal of our experimental evaluation is to answer the following questions:

1. Which impact does parallelization have on learning?
2. Which impact do the number of workers and MAML steps have, and is it beneficial to use more tasks during one MAML step?
3. Which parameters are best if tasks are either perfectly parallelized or serialized?

We compare our model to plain MAML, which is equivalent to DMAML using only one worker. In all of our experiments, we use TensorFlow 2 (without GPU support). All parameters used for training are listed in the Appendix B.

4.1 Regression Task

The task domain of tasks T_i is to perform a regression to data points produced by a sine function $y(x) = A \cdot sin(\omega x - b)$ with constant frequency $\omega = 1$ but random amplitude $A \in [0.1, 5.0]$ and phase $b \in [0, \pi]$.

The regressor $f_\theta(x)$ is a simple neural network with parameters θ consisting of one input neuron, 2 hidden layers of size 40, and ReLU nonlinearities followed by an output layer with size one following [5]. The loss is the mean-squared error between $f_\theta(x)$ and the true value $y(x)$.

For each experiment we apply 10,000 consecutive MAML steps on the model parameters θ and use $n \in \{1, 4, 16, 64\}$ workers. Validation is performed by optimizing the model parameter θ for a given validation task T_i and applying 5 gradient steps. As stated in [5], we use $K = 10$ points for training, randomly chosen from $x \in [-5.0, 5.0]$ and $K \in \{5, 10, 20\}$ for validation.

Results show that all models trained with more than one worker perform better (Fig. 2). Table 1 shows that we achieve similar results for more than 4 workers and comparable results with even more data points for validation. We compare the mean loss and standard deviation after 600 tasks and with a confidence interval of 95%. We assume that the increase in performance is a result of the quantity of tasks the meta-learner observes.

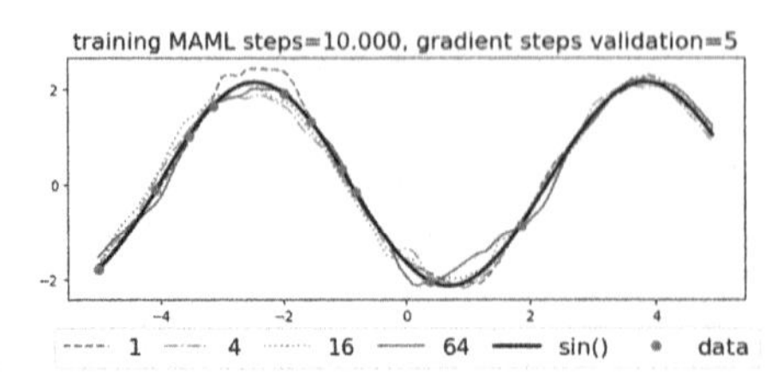

Fig. 2. The data points used for the gradient steps for validation are all sampled on the left side, so the model needs to extrapolate the values on the right side.

Table 1. Validation loss for n workers after training a meta-learner for 10,000 MAML steps and one gradient step w.r.t. the validation task

n	K		
	5	10	20
1	1.24 ± 0.1	0.78 ±0.05	0.52 ±0.04
4	1.02 ±0.08	0.63 ±0.05	0.43 ±0.04
16	**0.83 ±0.07**	**0.53 ±0.05**	**0.32 ±0.03**
64	0.9 ±0.08	**0.54 ±0.05**	**0.33 ±0.03**

4.2 Image Recognition Task

We evaluate DMAML on Omniglot, and miniImageNet image recognition tasks, which are the most common recently used few-shot learning benchmarks [16,18,20].

We split the 50 available Omniglot alphabets into 30 alphabets for training and 20 for testing, as proposed by [18]. The dataset is augmented by rotating each instance by 90°, and every rotated class is treated as a new class. The training dataset consists of 3.760 classes, the validation dataset of 2.732 classes. Every class consists of 20 images, which are resized to 28×28 pixels and are treated as grayscale images.

The miniImageNet dataset proposed by [16] consists of 64 training, 12 validation, and 24 test classes. As we are just doing validation, we aggregate the validation and test classes.

As proposed by [20], we apply the N-way, K-shot task as follows:

1. We apply fast learning using a support-set consisting of N (5 or 20) different classes and K (1 or 5) instances per class and
2. evaluate the models' ability to classify new instances of the N classes using a query-set consisting of K (Omniglot $K = 1$, miniImagenet $K = 15$) unseen instances of the N classes.

The models used for training have a simple architecture following [20]:

For Omniglot image recognition task we use a model with an input shape of (28, 28, 1), four modules with a 3×3 convolution with 64 filters, and a stride of 2. Each module followed by batch normalization [9] and a ReLU nonlinearity.

The miniImageNet model follows a similar architecture: The model has an input shape of (84, 84, 3) and consists of four modules with a 3×3 convolution with 32 filters, followed by a batch normalization, ReLU nonlinearity, and 2×2 max-pooling.

For all models, the output layer has a size of N and a softmax nonlinearity. The loss function is the cross-entropy error between the predicted and true class.

The following experiments were performed on the Omniglot image recognition task as a 5-way, 1-shot problem. In our experiments, we focus on the first 1,000 runs, which show a comparable trend in learning without training the algorithm exceedingly long. All experiments run with 3 different seeds, recognizing the need for more runs for a confidential result. The validation accuracy is the average over 10 different validation tasks, taken after fitting the model parameters to each task for 3 (Omniglot) and 10 (miniImageNet) gradient steps. We show the standard deviation of all 3 models and the output of a polynomial model fitted to the experiment data. The detailed results for both data sets are shown in the Appendix A.

4.3 Parallel Workers

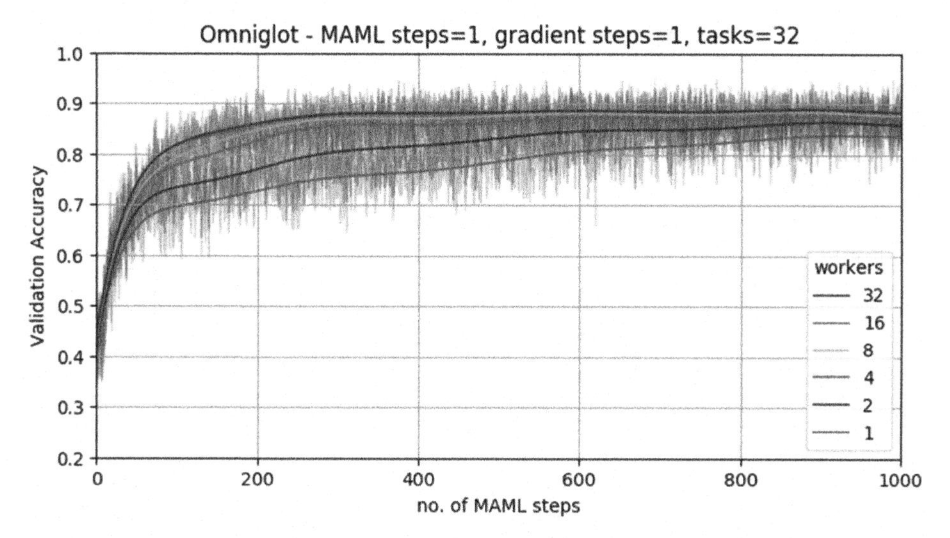

Fig. 3. Validation accuracy using n workers for learning. Learning is accelerated, when more workers are used. There also seems to be a breakeven point where more workers do not result in faster learning.

We run DMAML using $n \in \{1, 2, 3, 8, 16, 32\}$ parallel workers for learning. On the image recognition task, a small increase of workers leads to a steeper slope and faster convergence on a higher accuracy level (Fig. 3). The difference in accuracy between 16 and 64 workers is much smaller than the difference between 1 and 4 workers. This small difference leads us to the assumption that there must be a breakeven point, where increasing the parallel workers n does not increase the outcome significantly. The maximum tradeoff between CPU-time and performance in this example is 2 or 4 workers. We run our experiments also for 10,000 MAML steps and can observe an acceleration of learning for the first 1,000 MAML steps. If n is higher, we can observe a divergence of the validation accuracy after 1,500 MAML steps (Appendix A). This divergence reinforces our assumption that 2 or 4 workers might be optimal for this setting. Also, a form of schedule to decrease n over time or a strategy to drop the parameters of underperforming workers might be helpful.

4.4 MAML Steps per Run

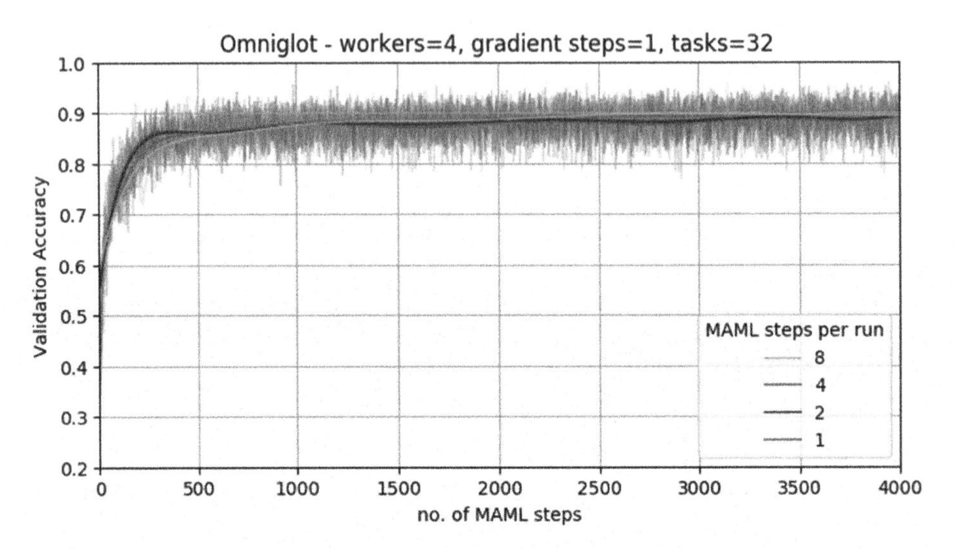

Fig. 4. The performance is shown depending on MAML steps. There seems to be no advantage in increasing the consecutive MAML steps in one run.

In this experiment, we increase the number of consecutive MAML steps per run. One worker performs multiple MAML steps before the model parameters are merged in the central learner. The resulting graphs for 1, 2, 4 and 8 consecutive MAML steps are shown on Fig. 4. Less consecutive MAML steps increase the acceleration of learning at first but converge to the same level of accuracy. This convergence leads to the conclusion that there is no advantage in increasing the number of MAML steps in one run.

4.5 Tasks per MAML Step

In the meta-learning setting, tasks are costly, as most of the time, time-consuming simulations need to be run. Therefore, we try to minimize overall tasks needed for training. In this experiment, we also want to determine how we can obtain the greatest output out of one single task.

On Fig. 5 (left), we assume a perfect level of parallelization of tasks and show that the accuracy rises with an increasing number of tasks observed in one MAML step. If we assume complete serialization of tasks, it is favorable to use

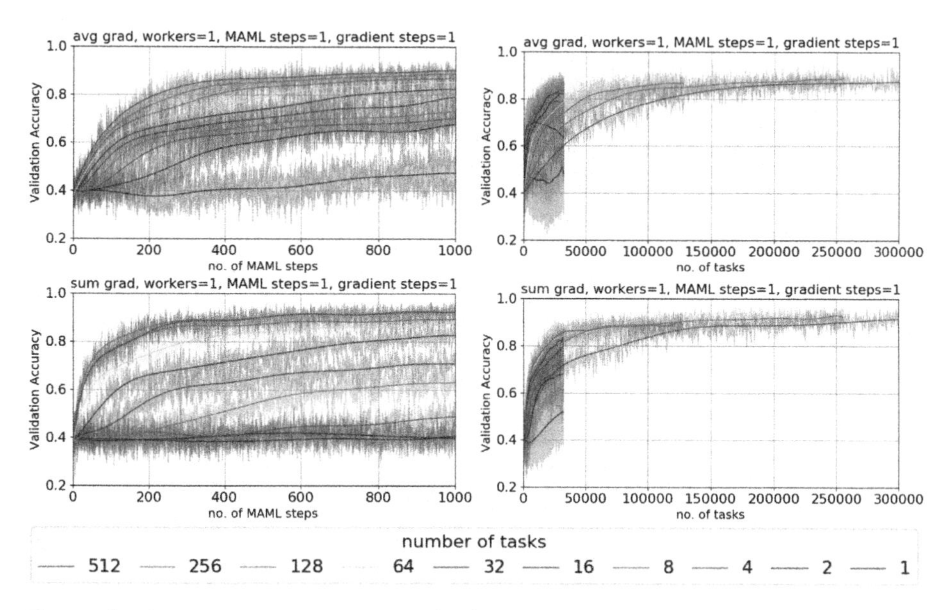

Fig. 5. Applying on Omniglot dataset (left) perfect parallelization: an increase of tasks perceived in one MAML step increases performance. (right) perfect serialization: an increase of tasks perceived in one MAML step increases wall clock time linearly. If we use fewer tasks per MAML step, we achieve a higher performance in a shorter amount of time. (top) we average the gradient over all tasks and apply the same learning rate β. (bottom) we sum over the gradients and apply a learning rate scheme.

less tasks per MAML step (Fig. 5 (right)). In the top section, we show the results when averaging the gradients in the outer-learning loop over all tasks and apply the same learning rate β, resulting in a similar step size for each MAML step.

In contrast, we sum the gradients of all tasks in the outer-learning loop and apply a learning rate scheme (Fig. 5, bottom). We set the learning rate for 32 tasks $\beta = 0.001$ as proposed by [5]. If we double the tasks used in one MAML step, we can be more certain that the gradient is more accurate, so we also double the learning rate and vice versa. Due to this learning rate scheme, learning is stabilized using less than 32 tasks, and learning is accelerated when using more tasks. As more tasks are used to determine the gradient, we can be more confident of the gradient's accuracy and perform a more significant update on the model parameters.

In one MAML step, the meta-model parameters are adapted to perform better for all observed tasks. With the same total number of tasks across all workers, the meta-learner achieves a higher accuracy if a single worker observes more tasks in one MAML step instead of distributing them to more workers (Fig. 6). This observation leads to the conclusion that we should aim at assigning as many tasks per MAML step as possible.

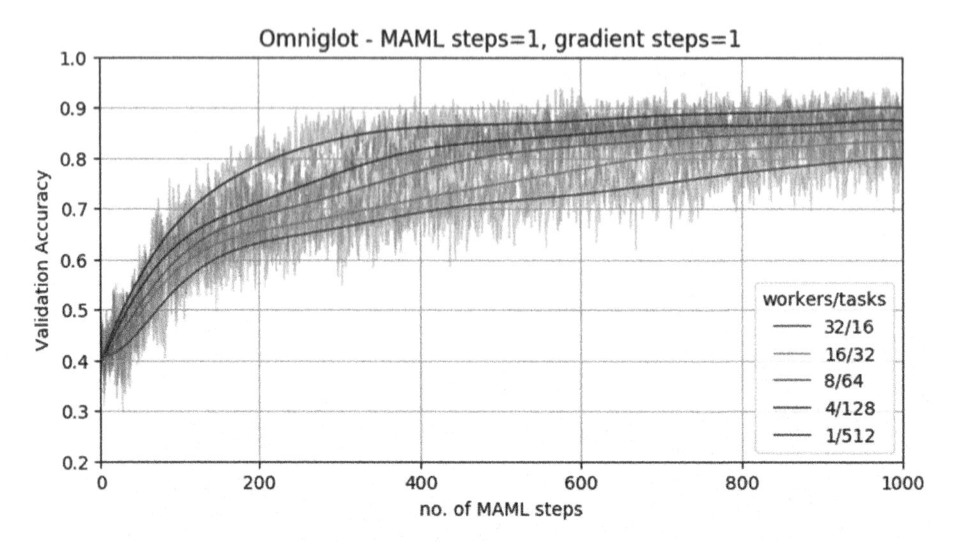

Fig. 6. The same amount of tasks distributed over multiple workers does not result in the same performance as using just one worker.

5 Conclusion

We found two ways to parallelize MAML: 1. Parallelize the computation of the tasks or 2. distribute learning over multiple workers, DMAML.

In DMAML, every worker receives a copy of the model parameters and updates the parameters by applying MAML. We have investigated the impact of DMAML and the influence of hyperparameters on learning (consecutive MAML steps, number of workers and tasks).

In this work, we have shown:

1. If perfect parallelization of tasks is possible, meaning every MAML step takes the same amount of wall clock time, we should aim at using as many tasks as possible in one MAML step to increase performance.
2. If we need to serialize tasks, meaning the sum of computing all training tasks results in the same amount of wall clock time, we should aim for many MAML steps with fewer tasks per step. This results in an acceleration of learning.
3. Distributing learning over multiple workers does not improve learning compared to parallelizing the inner-learning loop of MAML.

Further research can investigate if other strategies of merging the model parameters in the central learner improves learning and suppresses the divergence (e.g., dropping the parameters of a worker with a high loss).

As mentioned, using more tasks in one MAML step is beneficial and, due to the averaging, can be expected to lead to a more accurate gradient. Therefore, it might be possible to increase the learning rate. We also worked with a constant learning rate β, however, if we optimize the learning rate individually, the optimal solution might be shifted. This requires further investigation.

A Detailed Experiment Results

The validation accuracy after training DMAML using n workers for a number of MAML steps is calculated using the model with the highest validation accuracy after 10.000 steps out of three trained models, each using a different seed. We use a confidence interval of 95% and use 600 tasks for validation.

The result for Omniglot few-shot tasks are shown in Tables 2, 3 and 4, and the results for miniImageNet few-shot tasks are shown in Table 5.

Table 2. Comparison of n workers training for a number of MAML steps on the Omniglot image recognition task (5-way, 1-shot)

n	500	1000	2000	4000	8000	10000
1	0.81 ± 0.006	0.82 ± 0.007	0.87 ± 0.005	0.88 ± 0.006	0.85 ± 0.007	0.86 ± 0.007
2	0.84 ± 0.006	0.86 ± 0.006	0.87 ± 0.006	0.89 ± 0.005	0.89 ± 0.005	0.9 ± 0.005
4	0.87 ± 0.005	0.88 ± 0.005	0.88 ± 0.005	0.89 ± 0.006	0.91 ± 0.005	0.93 ± 0.004
8	0.88 ± 0.005	0.88 ± 0.005	0.86 ± 0.006	0.83 ± 0.007	0.88 ± 0.006	0.91 ± 0.005
16	0.88 ± 0.005	0.89 ± 0.005	0.84 ± 0.006	0.77 ± 0.01	0.86 ± 0.006	0.87 ± 0.007
32	0.89 ± 0.005	0.89 ± 0.005	0.85 ± 0.006	0.81 ± 0.009	0.75 ± 0.016	0.9 ± 0.005

Table 3. Comparison of n workers training for a number of MAML steps on the Omniglot image recognition task (5-way, 5-shot).

n	500	1000	2000	4000	8000	10000
1	0.96 ± 0.002	0.96 ± 0.002	0.97 ± 0.002	0.97 ± 0.002	0.97 ± 0.002	0.97 ± 0.003
2	0.95 ± 0.003	0.95 ± 0.003	0.96 ± 0.003	0.97 ± 0.002	0.97 ± 0.002	0.97 ± 0.002
4	0.96 ± 0.002	0.97 ± 0.002	0.97 ± 0.002	0.94 ± 0.005	0.97 ± 0.002	0.97 ± 0.004
8	0.96 ± 0.002	0.97 ± 0.002	0.97 ± 0.003	0.92 ± 0.009	0.85 ± 0.016	0.72 ± 0.021
16	0.96 ± 0.002	0.97 ± 0.002	0.97 ± 0.002	0.92 ± 0.008	0.84 ± 0.015	0.95 ± 0.007
32	0.97 ± 0.002	0.97 ± 0.002	0.97 ± 0.002	0.94 ± 0.005	0.67 ± 0.020	0.71 ± 0.019

Table 4. Comparison of n workers training for a number of MAML steps on the Omniglot image recognition task (20-way, 1-shot).

n	500	1000	2000	4000	8000	10000
1	0.59 ± 0.004	0.68 ± 0.004	0.73 ± 0.004	0.75 ± 0.004	0.76 ± 0.004	0.76 ± 0.004
2	0.66 ± 0.004	0.73 ± 0.004	0.77 ± 0.004	0.78 ± 0.004	0.78 ± 0.004	0.79 ± 0.004
4	0.71 ± 0.004	0.79 ± 0.004	0.81 ± 0.004	0.81 ± 0.004	0.81 ± 0.003	0.78 ± 0.004
8	0.74 ± 0.003	0.82 ± 0.003	0.84 ± 0.003	0.85 ± 0.002	0.75 ± 0.004	0.73 ± 0.003
16	0.77 ± 0.003	0.83 ± 0.003	0.83 ± 0.003	0.8 ± 0.004	0.77 ± 0.004	0.77 ± 0.003
32	0.76 ± 0.003	0.82 ± 0.003	0.81 ± 0.003	0.71 ± 0.004	0.76 ± 0.004	0.77 ± 0.003

Table 5. Comparison of n workers training for a number of MAML steps on the miniImageNet image recognition task (5-way, 1-shot and 5-way, 5-shot).

n	5-way, 1-shot			5-way, 5-shot		
	500	1000	2000	500	1000	2000
1	0.21 ± 0.003	0.21 ± 0.003	0.23 ± 0.004	0.3 ± 0.004	0.35 ± 0.005	0.39 ± 0.006
2	0.27 ± 0.005	0.27 ± 0.005	0.29 ± 0.005	0.38 ± 0.005	0.43 ± 0.006	0.46 ± 0.006
4	0.31 ± 0.005	0.32 ± 0.005	0.34 ± 0.006	0.44 ± 0.006	0.46 ± 0.006	0.5 ± 0.006
8	0.33 ± 0.005	0.36 ± 0.006	0.37 ± 0.006	0.46 ± 0.006	0.49 ± 0.006	0.52 ± 0.006
16	0.34 ± 0.006	0.37 ± 0.006	0.39 ± 0.006	0.47 ± 0.006	0.5 ± 0.006	0.53 ± 0.006
32	0.35 ± 0.006	0.36 ± 0.006	0.4 ± 0.006	0.48 ± 0.006	0.51 ± 0.006	0.53 ± 0.006

B Training Parameters

The parameters used for training DMAML are shown in Table 6.

Table 6. Experiment parameters

Experiment parameter	Omniglot	MiniImageNet	Regression
Support-set samples (k)	1	1	10
Query-set samples (k')	1	15	10
Inner-loop learning rate (α)	0.4	0.4	0.01
Meta-learning rate (β)	0.01	0.01	0.01
Tasks (i)	32	4	10
MAML steps	1	1	1
Gradient steps	1	5	1
Validation steps	3	10	1
Validation tasks	10	10	10

References

1. Antoniou, A., Storkey, A., Edwards, H.: How to train your MAML. In: 7th International Conference on Learning Representations, ICLR 2019. International Conference on Learning Representations, ICLR (2019). http://arxiv.org/abs/1810.09502
2. Behl, H.S., Baydin, A.G., Torr, P.H.S.: Alpha MAML: adaptive model-agnostic meta-learning, May 2019. http://arxiv.org/abs/1905.07435
3. Bertinetto, L., Torr, P.H., Henriques, J., Vedaldi, A.: Meta-learning with differentiable closed-form solvers. In: 7th International Conference on Learning Representations, ICLR 2019. International Conference on Learning Representations, ICLR, May 2019. http://arxiv.org/abs/1805.08136

4. Chen, F., Luo, M., Dong, Z., Li, Z., He, X.: Federated Meta-Learning with Fast Convergence and Efficient Communication, February 2018. http://arxiv.org/abs/1802.07876
5. Finn, C., Abbeel, P., Levine, S.: Model-agnostic meta-learning for fast adaptation of deep networks. In: 34th International Conference on Machine Learning, ICML 2017, vol. 3, pp. 1856–1868 (2017)
6. Finn, C., Xu, K., Levine, S.: Probabilistic model-agnostic meta-learning. In: Advances in Neural Information Processing Systems. vol. 2018-Decem, pp. 9516–9527. Neural Information Processing Systems Foundation, June 2018. http://arxiv.org/abs/1806.02817
7. Gordon, J., Bronskill, J., Nowozin, S., Bauer, M., Turner, R.E.: Meta-learning probabilistic inference for prediction. In: 7th International Conference on Learning Representations, ICLR 2019. International Conference on Learning Representations, ICLR, May 2019. http://arxiv.org/abs/1805.09921
8. Grant, E., Finn, C., Levine, S., Darrell, T., Griffiths, T.: Recasting gradient-based meta-learning as hierarchical bayes. In: 6th International Conference on Learning Representations, ICLR 2018 - Conference Track Proceedings, January 2018. http://arxiv.org/abs/1801.08930
9. Ioffe, S., Szegedy, C.: Batch normalization: Accelerating deep network training by reducing internal covariate shift. In: 32nd International Conference on Machine Learning, ICML 2015, vol. 1, pp. 448–456. International Machine Learning Society (IMLS), February 2015
10. Koch, G.: Siamese neural networks for one-shot image recognition. In: ICML Deep Learning Workshop (2015). http://www.cs.toronto.edu/~gkoch/files/msc-thesis.pdf
11. Lee, K., Maji, S., Ravichandran, A., Soatto, S.: Meta-learning with differentiable convex optimization. In: Proceedings of the IEEE Computer Society Conference on Computer Vision and Pattern Recognition, pp. 10649–10657, April 2019. https://doi.org/10.1109/CVPR.2019.01091, http://arxiv.org/abs/1904.03758
12. Lee, Y., Choi, S.: Gradient-based meta-learning with learned layerwise metric and subspace. In: 35th International Conference on Machine Learning, ICML 2018, pp. 4574–4586, January 2018. http://arxiv.org/abs/1801.05558
13. Li, Z., Zhou, F., Chen, F., Li, H.: Meta-SGD: learning to learn quickly for few-shot learning, July 2017. http://arxiv.org/abs/1707.09835
14. Nichol, A., Achiam, J., Schulman, J.: On First-Order Meta-Learning Algorithms, March 2018. http://arxiv.org/abs/1803.02999
15. Raghu, A., Raghu, M., Bengio, S., Vinyals, O.: Rapid Learning or Feature Reuse? Towards Understanding the Effectiveness of MAML, September 2019. http://arxiv.org/abs/1909.09157
16. Ravi, S., Larochelle, H.: Optimization as a model for few-shot learning. In: 5th International Conference on Learning Representations, ICLR 2017 - Conference Track Proceedings, November 2019
17. Rusu, A.A., et al.: Meta-learning with latent embedding optimization. In: 7th International Conference on Learning Representations, ICLR 2019, July 2019. http://arxiv.org/abs/1807.05960
18. Santoro, A., Bartunov, S., Botvinick, M., Wierstra, D., Lillicrap, T.: Meta-learning with memory-augmented neural networks. In: 33rd International Conference on Machine Learning, ICML 2016, pp. 2740–2751 (2016). http://proceedings.mlr.press/v48/santoro16.pdf

19. Snell, J., Swersky, K., Zemel, R.: Prototypical networks for few-shot learning. In: Advances in Neural Information Processing Systems, pp. 4078–4088 (2017). http://arxiv.org/abs/1703.05175
20. Vinyals, O., Blundell, C., Lillicrap, T., Kavukcuoglu, K., Wierstra, D.: Matching networks for one shot learning. In: Advances in Neural Information Processing Systems, pp. 3637–3645 (2016). http://arxiv.org/abs/1606.04080
21. Zhou, F., Wu, B., Li, Z.: Deep meta-learning: learning to learn in the concept space, February 2018. http://arxiv.org/abs/1802.03596
22. Zintgraf, L., Shiarlis, K., Kurin, V., Hofmann, K., Whiteson, S.: Fast context adaptation via meta-learning. In: 36th International Conference on Machine Learning, ICML 2019, pp. 13262–13276, October 2019. http://arxiv.org/abs/1810.03642

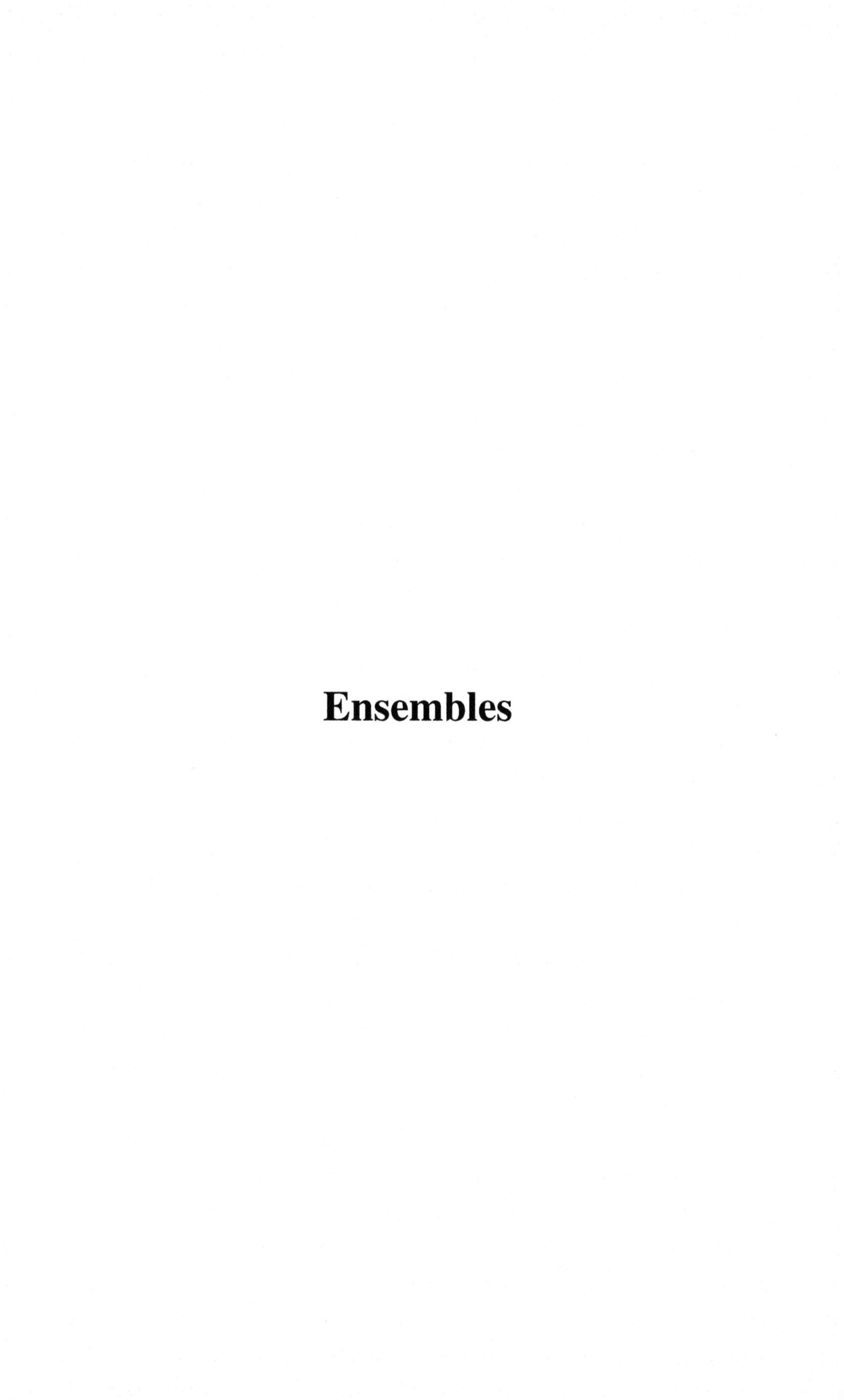

Ensembles

Extreme Algorithm Selection with Dyadic Feature Representation

Alexander Tornede(✉), Marcel Wever, and Eyke Hüllermeier

Heinz Nixdorf Institute and Department of Computer Science, Paderborn University, Warburger Str. 100, 33100 Paderborn, Germany
{alexander.tornede,marcel.wever,eyke}@uni-paderborn.de

Abstract. Algorithm selection (AS) deals with the automatic selection of an algorithm from a fixed set of candidate algorithms most suitable for a specific instance of an algorithmic problem class, e.g., choosing solvers for SAT problems. Benchmark suites for AS usually comprise candidate sets consisting of at most tens of algorithms, whereas in algorithm configuration (AC) and combined algorithm selection and hyperparameter optimization (CASH) problems the number of candidates becomes intractable, impeding to learn effective meta-models and thus requiring costly online performance evaluations. In this paper, we propose the setting of extreme algorithm selection (XAS), which, despite assuming limited time resources and hence excluding online evaluations at prediction time, allows for considering thousands of candidate algorithms and thereby facilitates meta learning. We assess the applicability of state-of-the-art AS techniques to the XAS setting and propose approaches leveraging a dyadic representation, in which both problem instances and algorithms are described in terms of feature vectors. We find this approach to significantly improve over the current state of the art in various metrics.

Keywords: Extreme algorithm selection · Dyadic ranking · Surrogate model

1 Introduction

Algorithm selection (AS) refers to a specific recommendation task, in which the choice alternatives are algorithms: Given a set of candidate algorithms to choose from, and a specific instance of a problem class, such as SAT or integer optimization, the task is to select or recommend an algorithm that appears to be most suitable for that instance, in the sense of performing best in terms of criteria such as runtime, solution quality, etc. Hitherto practical applications of AS, as selecting a SAT solver for a logical formula, typically comprise candidate sets consisting of at most tens of algorithms, and this is also the order of magnitude that is found in standard AS benchmark suites such as ASlib [2].

This is in contrast with the problem of combined algorithm selection and hyperparameter optimization (CASH) [24] as considered in automated machine

A. Appice et al. (Eds.): DS 2020, LNAI 12323, pp. 309–324, 2020.
https://doi.org/10.1007/978-3-030-61527-7_21

learning (AutoML), where the number of potential candidates is very large and potentially infinite [6,16,24]. Corresponding methods heavily rely on computationally extensive search procedures combined with costly online evaluations of the performance measure to optimize for, since learning effective meta models for an instantaneous recommendation becomes infeasible.

In this paper, we propose *extreme algorithm selection* (XAS) as a novel setting in-between traditional AS and AC/CASH, which is motivated by application scenarios characterized by

- the demand for prompt recommendations in quasi real time,
- an extremely large (though still finite) set of candidate algorithms.

An example is the scenario of "On-the-fly computing" [10], including "On-the-fly machine learning" [17] as one of its instantiations, where users can request online (machine learning) software services customized towards their needs. Here, users are unwilling to wait for several hours until their service is ready, but rather claim a result quickly. Hence, for providing a first version of an appropriate service, costly search and online evaluations are not affordable. As will be seen, XAS offers a good compromise solution: Although it allows for the consideration of extremely many candidate solutions, and even offers the ability to recommend configurations that have never been encountered so far, it is still amenable to AS techniques and avoids costly online evaluations.

In a sense, XAS relates to standard AS as the emerging topic of extreme classification (XC) [1] relates to standard multi-class classification. Similar to XC, the problem of learning from sparse data is a major challenge for XAS: For a single algorithm, there are typically only observations for a few instances. In this paper, we propose a benchmark dataset for XAS and investigate the ability of state-of-the-art AS approaches to deal with this sparsity and to scale with the size of candidate sets. Furthermore, to support more effective learning from sparse data, we propose methods based on "dyadic" feature representations, in which both problem instances and algorithms are represented in terms of feature vectors. In an extensive experimental study, we find these methods to yield significant improvements.

2 From Standard to Extreme Algorithm Selection

In the standard (per-instance) algorithm selection setting, first introduced in [20], we are interested in finding a mapping $s : \mathcal{I} \longrightarrow \mathcal{A}$, called algorithm selector. Given an instance i from the instance space $\mathcal{I}$, the latter selects the algorithm a^* from a set of candidate algorithms $\mathcal{A}$, optimizing a performance measure $m : \mathcal{I} \times \mathcal{A} \longrightarrow \mathbb{R}$. Furthermore, m is usually costly to evaluate. The optimal selector is called *oracle* and is defined as

$$s^*(i) := \arg\max_{a \in \mathcal{A}} \mathbb{E}\big[\, m(i,a) \,\big] \tag{1}$$

for all $i \in \mathcal{I}$. The expectation operator $\mathbb{E}$ accounts for any randomness in the application of the algorithm—in the non-deterministic case, the result of applying a to i, and hence the values of the performance measure, are random variables.

Most AS approaches leverage machine learning techniques, in one way or another learning a surrogate (regression) model $\widehat{m} : \mathcal{I} \times \mathcal{A} \longrightarrow \mathbb{R}$, which is fast to evaluate and thus allows one to compute a selector $\widehat{s} : \mathcal{I} \longrightarrow \mathcal{A}$ by

$$\widehat{s}(i) := \arg\max_{a \in \mathcal{A}} \widehat{m}(i, a) . \tag{2}$$

In order to infer such a model, we usually assume the existence of a set of training instances $\mathcal{I}_D \subset \mathcal{I}$ for which we have instantaneous access to the associated performances of some or often all algorithms in $\mathcal{A}$ according to m.

The XAS setting distinguishes itself from the standard AS setting by two important properties. Firstly, we assume that the set of candidate algorithms $\mathcal{A}$ is *extremely* large. Thus, approaches need to be able to scale well with the size of $\mathcal{A}$. Secondly, due to the size of $\mathcal{A}$, we can no longer reasonably assume to have evaluations for each algorithm on each training instance. Instead, we assume that the training matrix spanned by the training instances and algorithms is only sparsely filled. In fact, we might even have algorithms without any evaluations at all. Hence, suitable approaches need to be able to learn from very few data and to tackle the problem of "zero-shot learning" [29].

Similarly, the XAS setting differs from the AC and CASH settings in two main points. Firstly, dealing with real-valued hyperparameters, the set of (configured) algorithms $\mathcal{A}$ is generally assumed to be *infinite* in both AC and CASH, whereas $\mathcal{A}$ is still finite (even if extremely large) in XAS. More importantly, in both AC and CASH, one usually assumes having time to perform online evaluations of solution candidates at recommendation time. However, as previously mentioned, this is not the case in XAS, where instantaneous recommendations are required. Hence, the XAS setting significantly differs from the AS, AC, and CASH settings. A summary of the main characteristics of these settings is provided in Table 1.

Table 1. Overview of the characteristics of the problem settings we distinguish.

Characteristics/Setting	AS	XAS	AC	CASH
Size of $\mathcal{A}$	at most tens	extremely many	potentially infinite	potentially infinite
Training data	complete	sparse	mostly not present	mostly not present
Online evaluations	no	no	yes	yes

3 Exploiting Instance Features

Instance-specific AS is based on the assumption that instances can be represented in terms of feature information. For this purpose, $f_I : \mathcal{I} \longrightarrow \mathbb{R}^k$ denotes a function representing instances as k-dimensional, real-valued feature vectors, which can be used to learn a surrogate model (2). This can be done based on different types of data and using different loss functions.

3.1 Regression

The most common approach is to tackle AS as a regression problem, i.e., to construct a regression dataset for each algorithm, where entries consist of an instance representation and the associated performance of the algorithm at question. Accordingly, the dataset associated with algorithm $a \in \mathcal{A}$ consists of tuples of the form $\big(f_I(i), m(i,a)\big)$, created for those instances $i \in \mathcal{I}_D$ to which a has been applied, so that a performance evaluation $m(i,a) \in \mathbb{R}$ is available. Using this dataset, a standard regression model $\widehat{m}_a$ can be learned per algorithm a, and then used as a surrogate. The model can be realized as a neural network or a random forest, and trained on loss functions such as root mean squared or absolute error. For an overview of methods of this kind, we refer to Sect. 6.

This approach has two main disadvantages. Firstly, it is not well suited for the XAS setting, as it requires learning a huge number of surrogate models, one per algorithm. Although these models can usually be trained very quickly, the assumption of sparse training data in the XAS setting requires them to be learned from only a handful of training examples—it is not even uncommon to have algorithms without any performance value at all. Accordingly, the sparser the data, the more drastically this approach drops in performance, as will be seen in the evaluation in Sect. 5. Secondly, it requires precise real-valued evaluations of the measure m as training information, which might be costly to obtain. In this regard, one may also wonder, whether regression is not solving an unnecessarily difficult problem: Eventually, AS is only interested in finding the best algorithm for a given problem instance, or, more generally, in ranking the candidate algorithms in decreasing order of their expected performance. An accurate prediction of absolute performances is a *sufficient* but not a *necessary* condition for doing so.

3.2 Ranking

As an alternative to regression, one may therefore think of tackling AS as a *ranking* problem. More specifically, the counterpart of the regression approach outlined above is called *label ranking* (LR) in the literature [28]. Label ranking deals with learning to rank choice alternatives (referred to as "labels") based on given contexts represented by feature information. In the setting of AS, contexts and labels correspond to instances and algorithms, respectively. The type of training data assumed in LR consists of rankings π_i associated with training instances $i \in \mathcal{I}_D$, that is, order relations of the form $(f_I(i), a_{i,1}) \succ \ldots \succ (f_I(i), a_{i,l_i})$, in which $\succ$ denotes an underlying preference relation; thus, $(f_I(i), a) \succ (f_I(i), a')$ means that, for instance i represented by features $f_I(i)$, algorithm a is preferred to (better than) algorithm a'. If i is clear from the context, we also represent the ranking by $a_1 \succ \ldots \succ a_{l_i}$. Compared to the case of regression, a ranking dataset of this form can be constructed more easily, as it only requires qualitative comparisons between algorithms instead of real-valued performance estimates.

A common approach to label ranking is based on the so-called Plackett-Luce (PL) model [4], which specifies a parameterized probability distribution on rankings over labels (i.e., algorithms in our case). The underlying idea is to associate each algorithm a with a latent utility function $\widehat{m}_a : \mathcal{I} \longrightarrow \mathbb{R}_+$ of a context (i.e., an instance), which estimates how well an algorithm is suited for a given instance. The functions $\widehat{m}_a$ are usually modeled as log-linear functions

$$\widehat{m}_a(i) = \exp\left(\boldsymbol{\theta}_a^\top f_I(i)\right), \tag{3}$$

where $\boldsymbol{\theta}_a \in \mathbb{R}^k$ is a real-valued, k-dimensional vector, which has to be fit for each algorithm a. The PL model specifies a probability distribution on rankings: given an instance $i \in \mathcal{I}$, the probability of a ranking $a_1 \succ \ldots \succ a_z$ over any subset $\{a_1, \ldots, a_z\} \subseteq \mathcal{A}$ is

$$\mathbb{P}(a_1 \succ \ldots \succ a_z \mid \boldsymbol{\Theta}) = \prod_{n=1}^{z} \frac{\widehat{m}_{a_n}(i)}{\widehat{m}_{a_n}(i) + \ldots + \widehat{m}_{a_z}(i)} . \tag{4}$$

A probabilistic model of that kind suggests learning the parameter matrix $\boldsymbol{\Theta} = \{\boldsymbol{\theta}_a \mid a \in \mathcal{A}\}$ via maximum likelihood estimation, i.e., by maximizing

$$L(\boldsymbol{\Theta}) = \prod_{i \in \mathcal{I}_D} \mathbb{P}(\pi_i \mid \boldsymbol{\Theta})$$

associated with (4); this approach is explained in detail in [4]. Hence, the associated loss function under which we learn is now of a probabilistic nature (the logarithm of the PL-probability). It no longer focuses on the difference between the approximated performance $\widehat{m}_a(i)$ and the true performance $m(i, a)$, but on the ranking of the algorithms with respect to m—putting it in the jargon of preference learning, the former is a "pointwise" while the latter is a "listwise" method for learning to rank [3].

This approach potentially overcomes the second problem explained for the case of regression, but not the first one: It still fits a single model per algorithm a (the parameter vector $\boldsymbol{\theta}_a$), which essentially disqualifies it for the XAS setting.

3.3 Collaborative Filtering

This may suggest yet another approach, namely the use of collaborative filtering (CF) [8], in the setting of AS originally proposed by [23]. In CF for AS, we assume a (usually sparse) performance matrix $R^{|\mathcal{I}_D| \times |\mathcal{A}|}$, where an entry $R_{i,a} = m(i, a)$ corresponds to the performance of algorithm a on instance i according to m if known, and $R_{i,a} = ?$ otherwise. CF methods were originally designed for large-scale settings, where products (e.g. movies) are recommended to users, and data to learn from is sparse. Hence, they appear to fit well for our XAS setting.

Similar to regression and ranking, model-based CF methods also learn a latent utility function. They do so by applying matrix factorization techniques to the performance matrix R, trying to decompose it into matrices $U \in \mathbb{R}^{|\mathcal{I}_D| \times t}$ and $V \in \mathbb{R}^{t \times |\mathcal{A}|}$ w.r.t. some loss function $L(R, U, V)$, such that

$$R \approx \widehat{R} = UV^\top , \tag{5}$$

where U (V) can be interpreted as latent features of the instances (algorithms), and t is the number of latent features. Accordingly, the latent utility of a known algorithm a for a known instance i can be computed as

$$\widehat{m}_a(i) = U_{i,\bullet} V_{\bullet,a}^\top \,, \tag{6}$$

even if the associated value $R_{i,a}$ is unknown in the performance matrix used for training. The loss function $L(R, U, V)$ depends on the exact approach used—examples include the root mean squared error and the absolute error restricted by some regularization term to avoid overfitting. In [15], the authors suggest a CF approach called Alors, which we will use in our experiments later on. It can deal with unknown instances by learning a feature map from the original instance to the latent instance feature space. Alors leverages the CF approach CoFi^{RANK} [31] using the normalized discounted cumulative gain (NDCG) [30] as loss function $L(R, U, V)$. Since the NDCG is a ranking loss, it focuses on decomposing the matrix R so as to produce an accurate ranking of the algorithms. More precisely, it uses an exponentially decaying weight function for ranks, such that more emphasis is put on the top and less on the bottom ranks. Hence, it seems particularly well suited for our use case.

4 Dyadic Feature Representation

As discussed earlier, by leveraging instance features, or learning such a representation as in the case of Alors, the approaches presented in the previous section can generalize over instances. Yet, none of them scales well to the XAS setting, as they do not generalize over algorithms; instead, the models are algorithm-specific and trained independently of each other. For the approaches presented earlier (except for Alors), this does not only result in a large number of models but also requires these models to be trained on very few data. Furthermore, it is not uncommon to have algorithms without any observation. A natural idea, therefore, is to leverage feature information on algorithms as well.

More specifically, we use a feature function $f_A : \mathcal{A} \longrightarrow \mathbb{R}^d$ representing algorithms as d-dimensional, real-valued feature vectors. Then, instead of learning one latent utility model per algorithm, the joint feature representation of a "dyad" consisting of an instance and an algorithm, allows us to learn a single joint model

$$\widehat{m} : f_I(\mathcal{I}) \times f_A(\mathcal{A}) \longrightarrow \mathbb{R} \,, \tag{7}$$

and hence to estimate the performance of a given algorithm a on a given instance i in terms of $\widehat{m}(f_I(i), f_A(a))$.

4.1 Regression

With the additional feature information at hand, instead of constructing one dataset per algorithm, we resolve to a single joint dataset comprised of examples

$\Big(\psi(f_I(i), f_A(a)), m(i,a)\Big)$ with dyadic feature information for all instances $i \in \mathcal{I}_D$ and algorithms $a \in \mathcal{A}$ for which a performance value $m(i,a)$ is known. Here,

$$\psi : \mathbb{R}^k \times \mathbb{R}^d \longrightarrow \mathbb{R}^q \tag{8}$$

is a joint feature map that defines how the instance and algorithm features are combined into a single feature representation of a dyad. What is sought, then, is a (parametrized) latent utility function $\widehat{m}_{\boldsymbol{\theta}} : \mathbb{R}^q \longrightarrow \mathbb{R}$, such that

$$\widehat{m}_{\boldsymbol{\theta}}\Big(\psi(f_I(i), f_A(a))\Big) \tag{9}$$

is an estimation of the performance of algorithm a on instance i. Obviously, the choice of ψ will have an important influence on the difficulty of the regression problem and the quality of the model (9) induced from the data $\mathcal{D}^{REG}$. The regression task itself comes down to learning the parameter vector $\boldsymbol{\theta}$. In principle, this can be done exactly as in Sect. 3.1, also using the same loss function. Note that this is a generalization of the approach used by SMAC [11] for predicting performances across instances in algorithm configuration. We allow for a generic joint feature map ψ and an arbitrary model for $\widehat{m}_{\boldsymbol{\theta}}$, whereas SMAC limits itself to a concatenation of features and trains a random forest for modeling $\widehat{m}_{\boldsymbol{\theta}}$. Once again, it is noteworthy that SMAC by itself is not applicable in the XAS setting, as it relies on costly online evaluations.

4.2 Ranking

A similar adaptation can be made for the (label) ranking approach presented in Sect. 3.2 [25]. Formally, this corresponds to a transition from the setting of label ranking to the setting of *dyad ranking* (DR) as recently proposed in [21]. The first major change in comparison to the setting of label ranking concerns the training data, where the rankings π_i over subsets of algorithms $\{a_{i,1}, \dots, a_{i,l_i}\} \subseteq \mathcal{A}$ for instance i are now of the form

$$\psi\big(f_I(i), f_A(a_{i,1})\big) \succ \dots \succ \psi\big(f_I(i), f_A(a_{i,l_i})\big)\,. \tag{10}$$

Thus, we no longer represent an algorithm a simply by its label (a) but by features $f_A(a)$. Furthermore, like in the case of regression, we no longer learn one latent utility function per algorithm, but a single model of the form (9) based on a dyadic feature representation. In particular, we model $\widehat{m}_{\boldsymbol{\theta}}$ as a feed-forward neural network, where $\boldsymbol{\theta}$ represents its weights, which, as shown in [21], can be learned via maximum likelihood estimation on the likelihood function implied by the underlying PL model. Note that the use of a neural network is of particular interest here, since it allows one to learn the underlying joint feature map ϕ implicitly. Although both instance and algorithm features are simply fed as a concatenated vector into the network, it can recombine these features due to its structure and thus implicitly learn such a joint feature representation.

In contrast to the methods presented in the previous section, the methods based on dyadic feature information are capable of assigning a utility to unknown algorithms. Thus, they are well suited for the XAS setting and in principle even applicable when $\mathcal{A}$ is infinite, as long as a suitable feature representation f_A is available. Furthermore, as demonstrated empirically in Sect. 5, the dyadic feature approaches are very well suited for dealing with sparse performance matrices that are typical of the XAS setting.

5 Experimental Evaluation

In our experiments, we evaluate well established state-of-the-art approaches to algorithm selection as well as the proposed dyadic approaches in the XAS setting. More specifically, we consider the problem of selecting a machine learning classifier (algorithm) for a new classification dataset (instance) as a case study related to the "on-the-fly machine learning" scenario [17]. Please note that this is just one amongst many conceivable instantiations of the XAS setting, which is supposed to demonstrate the performance of the presented methods. To this end, we first generate a benchmark and then use this benchmark for comparison. The generated benchmark dataset as well as the implementation of the approaches including detailed documentation is provided on GitHub[1].

5.1 Benchmark Dataset

In order to benchmark the generalization performance of the approaches presented above in the XAS setting, we consider the domain of machine learning. More precisely, the task is to select a classification algorithm for an (unseen) dataset. Therefore, a finite set of algorithms $\mathcal{A}$ for classification and a set of instances $\mathcal{I}$ corresponding to classification datasets need to be specified. Furthermore, a performance measure is needed to score the algorithms' performance.

The set of candidate algorithms $\mathcal{A}$ is defined by sampling up to 100 different parameterizations of 18 classification algorithms from the machine learning library WEKA [7], ensuring these parameterizations not being too similar. An overview of the algorithms, their parameters and the number of instantiations contained in $\mathcal{A}$ is given in Table 2. This yields $|\mathcal{A}| = 1,270$ algorithms in total. The last row of the table sums up the items of the respective column, providing insights into the dimensionality of the space of potential candidate algorithms.

The set of instances $\mathcal{I}$ is taken from the OpenML CC-18 benchmarking suite[2] [27], which is a curated collection of various classification datasets that are considered interesting from a model selection resp. hyperparameter optimization point of view. This property makes the datasets particularly appealing for the XAS benchmark dataset, as it ensures more diversity across the algorithms.

Accordingly, the total performance matrix spanned by the algorithms and classification datasets in principle features $1,270 \cdot 71 = 88,900$ entries for which the benchmark contains 55,919 actual values and the rest are unknown.

[1] https://github.com/alexandertornede/extreme_algorithm_selection.

[2] https://docs.openml.org/benchmark/#openml-cc18 (Excluding datasets 554, 40923, 40927, 40996 due to technical issues.).

In the domain of machine learning, one is usually more interested in the generalization performance of an algorithm than in the runtime. Therefore, m is chosen to assess the solution performance of an algorithm. To this end, we carry out a 5-fold cross validation and measure the mean accuracy across the folds[3]. As the measure of interest, accuracy is a reasonable though to some extent arbitrary choice. Note that in principle any other measure could have been used for generating the benchmark as well.

Table 2. The table shows the types of classifiers used to derive the set $\mathcal{A}$. Additionally, the number of numerical parameters (#num.P), categorical parameters (#cat.P), and instantiations (n) is shown.

Learner	0R	1R	BN	DS	DT	IBk	J48	JR	KS	L	LMT	MP	NB	PART	REPT	RF	RT	SMO
#num.P	0	1	0	0	1	1	2	2	1	1	2	2	0	2	3	3	4	1
#cat.P	0	0	2	0	3	3	6	2	2	0	5	6	2	2	2	2	4	2
n	1	30	12	1	45	89	100	100	99	100	100	100	3	91	100	99	100	100

Training data for CF and regression-based approaches can then be obtained by using the performance values as labels. In contrast, for training ranking approaches, the data is labeled with rankings derived by ordering the algorithms in a descending order w.r.t. their performance values. Note that information about the exact performance value itself is lost in ranking approaches.

We would like to note that the problem underlying this benchmark dataset could of course be cast as an AC or CASH problem. However, here we make the assumption that there is no time for costly online evaluations due to the on-the-fly setting and hence standard AC and CASH methods are not applicable.

Instance Features. For the setting of machine learning, the instances are classification datasets and associated feature representations are called meta-features [18]. To derive a feature description of the datasets, we make use of a specific subclass of meta-features called *landmarkers*, which are performance scores of cheap-to-validate algorithms on the respective dataset. More specifically, we use all 45 landmarkers as provided by OpenML [27], for which different configurations of the following learning algorithms are evaluated based on the error rate, area under the (ROC) curve, and Kappa coefficient: Naive Bayes, One-Nearest Neighbour, Decision Stump, Random Tree, REPTree and J48. Hence, in total we use 45 features to represent a classification dataset.

Algorithm Features. The presumably most straight-forward way of representing an algorithm in terms of a feature vector is to use the values of its hyperparameters. Thus, we can describe each individual algorithm by a vector of their hyperparameter-values. Based on this, the general feature description

[3] The standard deviation of the performance values per dataset is on average 0.101, minimum 0.0064 and maximum 0.33.

is obtained by concatenation of the vectors. As already mentioned, the neural network-based dyad ranking approach implicitly learns a more sophisticated joint feature map. Due to the way in which we generated the set of candidate algorithms $\mathcal{A}$, we can compress the vector sharing features for algorithms of the same type. Additionally, we augment the vector by a single categorical feature denoting the type of algorithm. Given any candidate algorithm, its feature representation is obtained by setting the type of algorithm indicator feature to its type, each element of the vector corresponding to one of its hyperparameters to the specific value, and other entries to 0. Categorical parameters, i.e. features, are one-hot encoded yielding a total of 153 features to represent an algorithm.

5.2 Baselines

To better relate the performance of the different approaches to each other and to the problem itself, we employ various baselines. While `RandomRank` assigns ranks to algorithms simply at random, `AvgPerformance` first averages the observed performance values for each candidate algorithm and predicts the ranking according to these average performances. k-NN LR retrieves the k nearest neighbors from the training data, averages the performances and predicts the ranking which is induced by the average performances. Since `AvgRank` is commonly used as another baseline in the standard AS setting, we note that we omit this baseline on purpose. This is because meaningful average ranks of algorithms are difficult to compute in the XAS setting, where the number of algorithms evaluated, and hence the length of the rankings of algorithms, vary from dataset to dataset.

5.3 Experimental Setup

In the following experiments, we investigate the performance of the different approaches and baselines in the setting of XAS for the example of the proposed benchmark dataset as described in Sect. 5.1.

We conduct a 10-fold cross validation to divide the dataset into 9 folds of known and 1 fold of unknown instances. From the resulting set of known performance values, we then draw a sample of 25, 50, or 125 pairs of algorithms for every instance under the constraint that the performances of the two algorithms is not identical. Thus, a maximum fill degree of 4%, 8% respectively 20% of the performance matrix is used for training, as algorithms may occur more than once in the sampled pairs. The sparse number of training examples is motivated by the large number of algorithms in the XAS setting. The assumption that performance values are only available for a small subset of the algorithms is clearly plausible here. Throughout the experiments, we ensure that all approaches are provided the same instances for training and testing, and that the label information is at least based on the same performance values.

In the experiments, we compare various models with each other. This includes two versions of `Alors`, namely `Alors (REGR)` and `Alors (NDCG)` optimizing for a regression respectively ranking loss. Furthermore, we consider a state-of-the-art regression approach learning a RandomForest regression model per algorithm (PAReg). Note that for those algorithms with no training data at all, we make PAReg predict a performance of 0, as recommending such an algorithm does not seem reasonable. Lastly, we consider two approaches leveraging a dyadic feature representation, internally fitting either a RandomForest for regression (DFReg) or a feed-forward neural network for ranking (DR). For both dyadic approaches, the simple concatenation of instance and algorithm features is used as a feature map. In contrast to the other methods, the ranking model is only provided the information which algorithm of a sampled pair performs better, as opposed to the exact performance value that is given to other methods. A summary of the type of features and label information used by the different approaches/baselines is given on the left side of Table 3.

Table 3. Overview of the data provided to the approaches and their applicability to the considered scenarios.

	Approach	f_I	f_A	Label		Approach	f_I	f_A	Label
approaches	Alors (REGR)	✓	✗	m	baselines	RandomRank	✗	✗	
	Alors (NDCG)	✓	✗	m		AvgPrfm	✗	✗	m
	PAReg	✓	✗	m		AvgRank	✗	✗	π
	DFReg	✓	✓	m		k-NN LR	✓	✗	m
	DR	✓	✓	π					

The test performance of the approaches is evaluated by sampling 10 algorithms for every (unknown) instance to test for. The comparison is done with respect to different metrics detailed further below, and the outlined sampling evaluation routine is repeated 100 times.

Statistical significance w.r.t performance differences between the best method and any other method is determined by a Wilcoxon rank sum test with a threshold of 0.05 for the p-value. Significant improvements of the best method over another one is indicated by •.

Experiments were run on nodes with two Intel Xeon Gold "Skylake" 6148 with 20 cores each and 192 GB RAM.

5.4 Performance Metrics

On the test data, we compute the following performance metrics measuring desirable properties of XAS approaches.

regret@k is the difference between the performance value of the best algorithm within the predicted top-k of algorithms and the actual best algorithm. The domain of regret@k is $[0, 1]$, where 0 is the optimum meaning no regret.

NDCG@k is a position-dependent ranking measure (*n*ormalized *d*iscounted *c*umulative *g*ain) to measure how well the ranking of the top-k algorithms can be predicted. It is defined as

$$\text{NDCG@}k(\pi, \pi^*) = \frac{\text{DCG@}k(\pi)}{\text{DCG@}k(\pi^*)} = \frac{\left(\sum_{n=1}^{k} \frac{2^{m(i,\pi_n)-1}}{\log(n+2)}\right)}{\left(\sum_{n=1}^{k} \frac{2^{m(i,\pi_n^*)-1}}{\log(n+2)}\right)},$$

where i is a (fixed) instance, π is a ranking and π^* the optimal ranking, and π_n gives the algorithm on rank n in ranking π. The NDCG emphasizes correctly assigned ranks at higher positions with an exponentially decaying importance. NDCG ranges in $[0, 1]$, where 1 is the optimal value.

Kendall's τ is a rank correlation measure. Given two rankings (over the same set of elements) π and π', it is defined as

$$\tau(\pi, \pi') = \frac{C - D}{\sqrt{(C + D + T_\pi) \cdot (C + D + T_{\pi'})}} \tag{11}$$

where C/D is the number of so-called *concordant*/*discordant* pairs in the two rankings, and $T_\pi/T_{\pi'}$ is the number of ties in π/π'. Two elements are called a concordant/discordant pair if their order within the two rankings is identical/different, and tied if they are on the same rank. Intuitively, this measure determines on how many pairs the two rankings coincide. It takes values in $[-1, 1]$, where 0 means uncorrelated, -1 inversely, and 1 perfectly correlated.

Table 4. Results for the performance metrics Kendall' tau (τ), NDCG@k (N@3, N@5), and regret@k (R@1, R@3) for varying number of performance value pairs used for training. The best performing approach is highlighted in bold, the second best is underlined, and significant improvements of the best approach over others is denoted by •.

Approach	4% fill rate / 25 performance value pairs					8% fill rate / 50 performance value pairs				
	τ	N@3	N@5	R@1	R@3	τ	N@3	N@5	R@1	R@3
PAReg	0.1712 •	0.9352 •	0.9433 •	0.0601 •	0.0185 •	0.2537 •	0.9453	0.9594	0.0493	0.0136
Alors (NDCG)	0.0504 •	0.9205 •	0.9223 •	0.0686 •	0.0225	0.0472 •	0.9155 •	0.9164 •	0.0614 •	0.0208
Alors (REGR)	0.0303 •	0.9117 •	0.9191 •	0.0794 •	0.0190 •	0.0807 •	0.9172 •	0.9304 •	0.0754 •	0.0285 •
DR	0.3445	0.9523	0.9604	0.0381	0.0089	**0.3950**	**0.9584**	**0.9685**	0.0322	**0.0087**
DFReg	**0.3819**	**0.9564**	**0.9652**	**0.0302**	**0.0079**	0.3692	0.9573	0.9661	**0.0300**	0.0123
RandomRank	-0.0038 •	0.8933 •	0.9105 •	0.0878 •	0.0272 •	-0.0038 •	0.8933 •	0.9105 •	0.0878 •	0.0272 •
AvgPerformance	0.1384 •	0.9388 •	0.9433 •	0.0337	0.0090	0.2083 •	0.9355 •	0.9508 •	0.0493 •	0.0199 •
1-NN LR	0.1227 •	0.9290 •	0.9310 •	0.0733 •	0.0230 •	0.1059 •	0.9246 •	0.9296 •	0.0564 •	0.0209
2-NN LR	0.1303 •	0.9278 •	0.9310 •	0.0642 •	0.0193 •	0.0874 •	0.9269 •	0.9343 •	0.0541 •	0.0206

Approach	20% fill rate / 125 performance value pairs				
	τ	N@3	N@5	R@1	R@3
PAReg	0.3003 •	0.9525	0.9632	0.0395	0.0107
Alors (NDCG)	0.0540 •	0.9220 •	0.9242 •	0.0542 •	0.0228 •
Alors (REGR)	0.1039 •	0.9160 •	0.9329 •	0.0604 •	0.0222 •
DR	**0.4507**	**0.9696**	0.9715	**0.0241**	**0.0055**
DFReg	0.4264	0.9629	**0.9720**	0.0292	0.0071
RandomRank	-0.0038 •	0.8933 •	0.9105 •	0.0878 •	0.0272 •
AvgPerformance	0.2541 •	0.9437 •	0.9536 •	0.0523 •	0.0084
1-NN LR	0.1152 •	0.9245 •	0.9318 •	0.0594 •	0.0249 •
2-NN LR	0.1142 •	0.9292 •	0.9350 •	0.0412	0.0176 •

5.5 Results

The results of the experiments are shown in Table 4. It is clear from the table that the methods for standard algorithm selection tend to fail especially in the scenarios with only few algorithm performance values per instance. This includes the approach of building a distinct regression model for each algorithm (PAReg) as well as for the collaborative filtering approach Alors, independently of the loss optimized for, even though the NDCG variant has a slight edge over the regression one. Moreover, Alors even fails to improve over simple baselines, such as AvgPerformance and k-NN LR. With an increasing number of training examples, PAReg improves over the baselines and also performs better than Alors, but never yields the best performance for any of the considered settings or metrics.

In contrast to this, the proposed dyadic feature approaches clearly improve over both the methods for the standard AS setting and the considered baselines for all the metrics. Interestingly, DFReg performs best for the setting with only 25 performance value pairs, while DR has an edge over DFReg for the other two settings. Still, the differences between the dyadic feature approaches are never significant, whereas significant improvements can be achieved in comparison to the baselines and the other AS approaches.

Moreover, our study demonstrates the heterogeneity of the benchmark dataset. As described in [22], a relevant measure for heterogeneity is the per-instance potential for improvement over a solution that is static across instances, i.e., what is often called the single best algorithm or solver (SBS). In this case study, the SBS is represented by the AvgPerformance baseline, which is always worse than the oracle with respect to all measures and in particular the regret@k measures. Hence, as the superior performances of our approach compared to the AvgPerformance demonstrate, the benchmark dataset offers a potential for per-instance algorithm selection.

The results of our study show that models with strong generalization performance can be obtained despite the small number of training examples. Moreover, the results suggest that there is a need for the development of specific methods addressing the characteristics of the XAS setting. This concerns the large number of different candidate algorithms as well as the sparsity of the training data.

6 Related Work

As most closely related work, we subsequently highlight several AS approaches to learning latent utility functions. For an up-to-date survey, we refer to [12].

A prominent example of a method learning a regression-based latent utility function is [32], which features an empirical hardness model per algorithm for estimating the runtime of an algorithm, i.e., its performance, for a given instance based on a ridge regression approach in the setting of SAT solver selection. Similarly, [13] learn per-algorithm hardness models using statistical (non-)linear regression models for algorithms solving the winner determination problem. Depending on whether a given SAT instance is presumably satisfiable or not, conditional runtime prediction models are learned in [9] using ridge linear regression.

In [5], a label-ranking-based AS approach for selecting collaborative filtering algorithms in the context of recommender systems is presented leveraging nearest neighbor and random forest label rankers.

Similar to our work, [19] leverages algorithm features in the form of a binary vector indicating which algorithm is considered to learn a probabilistic ranking model considering up to tens of algorithms. AS was first modeled as a CF problem in [23], using a probabilistic matrix factorization technique to select algorithms for the constraint solving problem. Assuming a complete performance matrix for training, low-rank latent factors are learned in [14] using singular value decomposition to obtain a selector en par with the oracle. Lastly, in [26] a decision-theoretic approach is proposed leveraging survival analysis to explicitly acknowledge timeouts of algorithms in the learning process.

7 Conclusion

In this paper, we introduced the extreme algorithm selection (XAS) setting and investigated the scalability of various algorithm selection approaches in this setting. To this end, we defined a benchmark based on the OpenML CC-18 benchmark suite for classification and a set of more than 1,200 candidate algorithms. Furthermore, we proposed the use of dyadic approaches, specifically dyad ranking, taking into account feature representations of both problem instances (datasets) and algorithms, which allows them to work on very few training data. In an extensive evaluation, we found that the approaches exploiting dyadic feature representations perform particularly well according to various metrics on the proposed benchmark and outperform other state-of-the-art AS approaches developed for the standard AS setting.

The currently employed algorithm features allow for solving the cold start problem only to a limited extent, i.e., only algorithms featuring known hyperparameters can be considered as new candidate algorithms. Investigating features to describe completely new algorithms is a key requirement for the approaches considered in this paper, and therefore an important direction for future work.

Acknowledgements. This work was supported by the German Research Foundation (DFG) within the Collaborative Research Center “On-The-Fly Computing” (SFB 901/3 project no. 160364472) and by the Paderborn Center for Parallel Computing (PC^2) through computing time.

References

1. Bengio, S., Dembczynski, K., Joachims, T., Kloft, M., Varma, M.: Extreme classification (dagstuhl seminar 18291). Dagstuhl Reports **8**(7), 62–80 (2018)
2. Bischl, B., et al.: Aslib: a benchmark library for algorithm selection. Artif. Intell. **237**, 41–58 (2016)
3. Cao, Z., Qin, T., Liu, T., Tsai, M., Li, H.: Learning to rank: from pairwise approach to listwise approach. In: ICML (2007)

4. Cheng, W., Hüllermeier, E., Dembczynski, K.J.: Label ranking methods based on the plackett-luce model. In: ICML (2010)
5. Cunha, T., Soares, C., de Carvalho, A.C.: A label ranking approach for selecting rankings of collaborative filtering algorithms. In: SAC (2018)
6. Feurer, M., Klein, A., Eggensperger, K., Springenberg, J., Blum, M., Hutter, F.: Efficient and robust automated machine learning. In: NIPS (2015)
7. Frank, E., Hall, M., Witten, I.: The WEKA Workbench. Data Mining (2016)
8. Goldberg, D., Nichols, D., Oki, B.M., Terry, D.: Using collaborative filtering to weave an information tapestry. Commun. ACM **35**(12), 61–70 (1992)
9. Haim, S., Walsh, T.: Restart strategy selection using machine learning techniques. In: SAT (2009)
10. Happe, M., Meyer auf der Heide, F., Kling, P., Platzner, M., Plessl, C.: On-the-fly computing: a novel paradigm for individualized it services. In: ISORC (2013)
11. Hutter, F., Hoos, H.H., Leyton-Brown, K.: Sequential model-based optimization for general algorithm configuration. In: LION (2011)
12. Kerschke, P., Hoos, H.H., Neumann, F., Trautmann, H.: Automated algorithmselection: Survey and perspectives. ECJ **27**(1), 3–45 (2019)
13. Leyton-Brown, K., Nudelman, E., Shoham, Y.: Learning the empirical hardness of optimization problems: the case of combinatorial auctions. In: CP (2002)
14. Malitsky, Y., O'Sullivan, B.: Latent features for algorithm selection. In: SOCS (2014)
15. Mısır, M., Sebag, M.: Alors: an algorithm recommender system. Artif. Intell. **244**, 219–344 (2017)
16. Mohr, F., Wever, M., Hüllermeier, E.: ML-Plan: automated machine learning via hierarchical planning. Mach. Learn. **107**(8), 1495–1515 (2018). https://doi.org/10.1007/s10994-018-5735-z
17. Mohr, F., Wever, M., Tornede, A., Hüllermeier, E.: From automated to on-the-fly machine learning. In: INFORMATIK (2019)
18. Nguyen, P., Hilario, M., Kalousis, A.: Using meta-mining to support data mining workflow planning and optimization. JAIR **51**, 605–644 (2014)
19. Oentaryo, R.J., Handoko, S.D., Lau, H.C.: Algorithm selection via ranking. In: AAAI (2015)
20. Rice, J.R.: The algorithm selection problem. In: Advances in computers, vol. 15. Elsevier (1976)
21. Schäfer, D., Hüllermeier, E.: Dyad ranking using plackett-lucemodels based on joint feature representations. Mach. Learn. **107**(5), 903–941 (2018)
22. Schneider, M., Hoos, H.H.: Quantifying homogeneity of instance sets for algorithm configuration. In: LION (2012)
23. Stern, D.H., Samulowitz, H., Herbrich, R., Graepel, T., Pulina, L., Tacchella, A.: Collaborative expert portfolio management. In: AAAI (2010)
24. Thornton, C., Hutter, F., Hoos, H.H., Leyton-Brown, K.: Auto-weka: combined selection and hyperparameter optimization of classification algorithms. In: SIGKDD (2013)
25. Tornede, A., Wever, M., Hüllermeier, E.: Algorithm selection as recommendation: from collaborative filtering to dyad ranking. In: CI Workshop, Dortmund (2019)
26. Tornede, A., Wever, M., Werner, S., Mohr, F., Hüllermeier, E.: Run2survive: a decision-theoretic approach to algorithm selection based on survival analysis. CoRR abs/2007.02816 (2020)
27. Vanschoren, J., van Rijn, J.N., Bischl, B., Torgo, L.: Openml: networked science in machine learning. SIGKDD Explorations **15**(2), 49–60 (2013)

28. Vembu, S., Gärtner, T.: Label ranking algorithms: a survey. In: Preference Learning. Springer, Heidelberg (2010). https://doi.org/10.1007/978-3-642-14125-6_3
29. Wang, W., Zheng, V.W., Yu, H., Miao, C.: A survey of zero-shot learning: settings, methods, and applications. ACM TIST (2019)
30. Wang, Y., Wang, L., Li, Y., He, D., Chen, W., Liu, T.: A theoretical analysis of NDCG ranking measures. In: COLT (2013)
31. Weimer, M., Karatzoglou, A., Le, Q.V., Smola, A.J.: Cofi rank-maximum margin matrix factorization for collaborative ranking. In: NIPS (2008)
32. Xu, L., Hutter, F., Hoos, H.H., Leyton-Brown, K.: Satzilla: portfolio-based algorithm selection for sat. JAIR **32**, 565–606 (2008)

Federated Ensemble Regression Using Classification

Oghenejokpeme I. Orhobor[1(✉)], Larisa N. Soldatova[2], and Ross D. King[1,3,4]

[1] Department of Chemical Engineering and Biotechnology, University of Cambridge, Cambridge CB3 0AS, UK
oo288@cam.ac.uk
[2] Department of Computing, Goldsmiths, University of London, London SE14 6AD, UK
[3] The Alan Turing Institute, London NW1 2DB, UK
[4] Department of Biology and Biological Engineering, Chalmers University of Technology, 412 96 Gothenburg, Sweden

Abstract. Ensemble learning has been shown to significantly improve predictive accuracy in a variety of machine learning problems. For a given predictive task, the goal of ensemble learning is to improve predictive accuracy by combining the predictive power of multiple models. In this paper, we present an ensemble learning algorithm for regression problems which leverages the distribution of the samples in a learning set to achieve improved performance. We apply the proposed algorithm to a problem in precision medicine where the goal is to predict drug perturbation effects on genes in cancer cell lines. The proposed approach significantly outperforms the base case.

Keywords: Ensemble learning · Machine learning · Regression · Bioinformatics · Gene expression

1 Introduction

In a standard regression setting, one builds a model on pre-existing learning data with the goal of making predictions on future unseen samples. In this case, a single model is built using a preferred learning algorithm. However, it has been demonstrated that one can improve predictive accuracy even further by aggregating the predictive power of multiple models built using the same learning data [14]. These models can be built in a variety of ways, from varying the attributes used in building the models to using multiple learning algorithms. This is done to ensure heterogeneity in the models, such that given a set of new samples, they are all wrong in different ways and their aggregation leads to improved predictions [5]. The typical approach in the use of a single model or an ensemble is that the distribution of the continuous response one is interested in predicting is often not given much thought. For example, if one imagines that the response for the

A. Appice et al. (Eds.): DS 2020, LNAI 12323, pp. 325–339, 2020.
https://doi.org/10.1007/978-3-030-61527-7_22

samples in a dataset follows a normal distribution. Then it also follows that any model that is naively built using this data is going to be very good at predicting samples that a near the centre of the distribution, but not those at the tails. A close analogy to this phenomenon is the class imbalance problem in a classification setting. Where given a dataset in which one class is over-represented, models built on this dataset using machine learning algorithms typically perform poorly when presented with a sample from the under-represented class [2,15]. Therefore, we hypothesised that predictive performance in a regression setting can be improved by accounting for the distribution of the response.

We take an ensemble learning approach to solving this problem. First, we split the learning data into a pre-specified number of bins using a known discretization technique [9]. We then build a regressor for each bin using only the samples that belong to that bin, each of which generalises on only a restricted portion of the distribution. We then build a classifier for each bin, treating the samples which belong to said bin as the positive samples, and the samples in the other bins as the negative samples. Therefore, there is a classifier-regressor pair for each bin. Given an unseen sample, real-valued predictions are made using the regressor for each bin. The corresponding classifier for each regressor is then used to predict the probability that the unseen sample is similar to the samples used in building the regressor. The predictions are then aggregated by weighting the probabilities and applying them to the predictions. This process is described diagrammatically in Fig. 1.

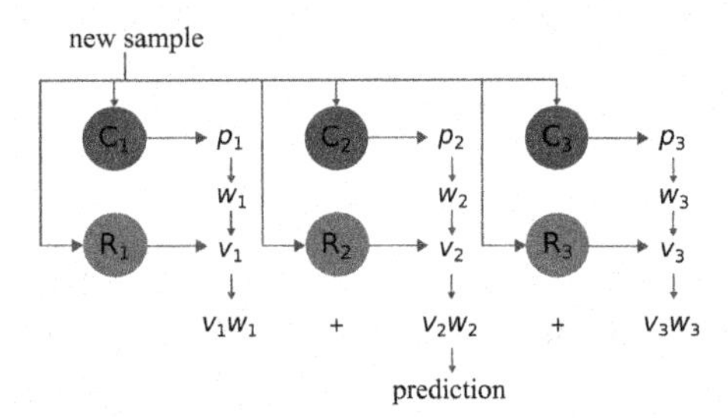

Fig. 1. Representation of the proposed approach when bin size is 3.

This approach is valuable to problems in precision medicine, where tail case prediction is of vital importance. An example of such a problem is the prediction of drug perturbation effects on genes in cancer cell lines, which, with improved predictive accuracy, has the potential to dramatically improve the rate at which new cancer drugs are developed. In our evaluation, we used data from the library of integrated network-based cellular signatures (LINCS) [16], which curates the drug perturbation effects on human genes. Our evaluation shows a significant improvement in performance over the base case. Our contributions are as follows:

1. An ensemble learning approach which considers response distribution for regression problems.
2. An application to a real-world dataset in precision medicine.

2 Related Work

Ensemble learning takes a variety of forms, from bootstrap aggregating (bagging) which is central to popular and robust learning algorithms like random forests [4], to methods like stacking [3]. The proposed approach shares some similarities with both of these methods. Stacking is most commonly used when one intends on aggregating the predictions made by multiple learning algorithms, or if a single learning algorithm is used, multiple models are built using subsets of the feature space [18]. There are three main processes in a stacking procedure: meta-feature generation, pruning, and aggregation [17]. Assume one has a learning and a test set. In the meta-feature generation phase, meta-features are generated for both the learning and test sets, and total to the number of models whose predictive power one wants to aggregate. Pruning is then used to optimise for the best meta-features. Finally, aggregation is done by learning weights using the learning set meta-features and then applying these weights to the test meta-features to form the final prediction.

In contrast to a typical stacking approach which we have described, we do not generate meta-features in our approach. The utility of the meta-features is that they provide a mechanism through which aggregating weights can be learned using a meta-level learning algorithm. Instead, we opt for a scheme where given a new sample, individual classifiers predict how much we can trust the predictions of their corresponding regressor as described in the introduction, which is more closely aligned with the concept of local classifiers in the hierarchical classification literature [20]. This implies that we also do not perform a pruning step. It is worth noting that while aggregation in stacking can be performed using weights learned with a meta-learner, it is also possible to simply average the predictions, we explore this in our evaluation. Other similarities exist. For example, one can argue that our weighting and aggregating procedure is a form of dynamic weighting, where new samples are weighted based on their similarities to samples used in building a model [19]. However, rather than being a separate step, dynamic weighting is implicit in the proposed learning procedure.

Central to the proposed method is the discretization of the continuous response one is interested in learning how to predict. Several methods to perform this task have been proposed, and they have been classed into supervised and unsupervised methods [7,9]. We considered only unsupervised methods in our evaluation. However, the use of supervised methods will be explored in future work. Methods which use classification as a means to perform regression in an ensemble setting have also been proposed. Ahmad et al. proposed the use of extreme randomized discretization to perform regression via classification [12]. In contrast to what we propose, the authors do not use a classifier-regressor pair to estimate the prediction for a new sample. Rather they do this using the minimum or maximum of the training data points and the bin boundary [1,12]. Also closely related to what we propose is work by Gonzalez et al. for problems that involve multi-variate spatio-temporal data [11]. The main differences in our approaches is two-fold. Firstly, they are interested in classifying bands of attributes before performing regression. Secondly, aggregation is done by first selecting the best models using leave-one-out cross-validation and the median predicted values by these models is treated as the final prediction for a new sample.

3 Methodology

3.1 Algorithm

The proposed approach can be split into a training and a prediction phase. An informal description follows, however, a more formal representation is given in Algorithm 1. In the training phase, given a training set with input vectors, a response, and a pre-specified number of bins c:

1. Discretize the response into c bins, forming c datasets.
2. For each c bin, build a regressor R_c and a classifier C_c. The regressor is built using the training samples for the particular bin. Whereas, the classifier is built by treating the samples in the current bin as the positive class and all other samples in the training set as the negative class.

In the prediction phase, given a new sample:

1. With all the R_c regressors, predict values for the new sample.
2. With all the C_c classifiers, predict the probability that the sample belongs in that c bin.
3. Generate weights using the c probabilities such that they sum to 1. This is done by summing the c probabilities and then dividing each c probability by this sum.
4. Get the final prediction by summing the values generated by applying each corresponding c weight to the prediction made by its c regressor in step (1).

3.2 Considerations

When tackling a machine learning problem, the choice of learning algorithm is vital as it plays a crucial role in predictive performance. However, it is clear from the description of the proposed approach outlined above that it is learner agnostic. That is, one can choose to build the classifiers and regressors using their preferred algorithm of choice. This property is particularly useful as one can choose to optimise for different properties using approaches from multiple kernel learning [22] or even stack multiple learning algorithms if they so choose. The choice of discretization technique is also open-ended, where one can choose to use known supervised or unsupervised discretization techniques, or a custom technique tailored to a particular problem.

When the number of bins is greater two, it will generally be the case that there will be some form of class imbalance. This may be in favour of the positive or negative class, and can be quite severe, depending on the distribution of the response variable under consideration, choice of discretization technique, and the number of specified bins. Therefore, it is important that this be taken into consideration, as it is known that class imbalance can have significant effects on predictive accuracy [2]. To combat this, methods which balance an imbalanced dataset such as oversampling methods like the synthetic minority oversampling technique (SMOTE) [6] should be considered. We explore the effects of discretization technique and class imbalance in our evaluation.

Algorithm 1. Federated Ensemble Learning using Classification

Input: Training set matrix $\mathbf{L} \in \mathbb{R}^{m \times b}$, response vector y, c bins, and test set matrix $\mathbf{T} \in \mathbb{R}^{n \times b}$

Output: Test set predictions

Training:

1: Split y into c bins using a discretization technique of choice, producing $\mathcal{L} = (\mathbf{L}_1, \ldots, \mathbf{L}_c)$ and $\mathcal{Y} = (y_1 \ldots y_c)$

2: **for** each c bin in $\mathcal{L}$ and $\mathcal{Y}$ **do**

3: Build a regressor R_c using $\mathbf{L}_c$ and y_c

4: Build a classifier C_c using $\mathbf{L}_c$ as the positive samples and $\mathcal{L} - \mathbf{L}_c$ as negative samples. *Note: class balancing may be required*

5: **end for**

Prediction:

6: **for** each c regressor-classifier pair R_c and C_c **do**

7: Predict the response for $\mathbf{T}$ using R_c

8: Predict the probability that the samples in $\mathbf{T}$ belong in c using C_c

9: **end for**

The process above generates predicted response and probability matrices $\mathbf{R}, \mathbf{P} \in \mathbb{R}^{n \times c}$

10: $v_n = \sum_{j=1}^{c} p_{n,j}$

11: Create weight matrix $\mathbf{W} \in \mathbb{R}^{n \times c}$ by dividing all elements in each row in $\mathbf{P}$ by the value in the corresponding row index in v_n

12: Create weighted response matrix $\mathbf{R}^w \in \mathbb{R}^{n \times c}$ by performing the element-wise multiplication of $\mathbf{R}$ and $\mathbf{W}$

13: The final prediction $T = \sum_{j=1}^{c} r_{n,j}^{w}$

14: **return** T

4 Evaluation Setup

We used data from the general LINCS Phase II dataset with accession code GSE70138. We had 7000 training samples and 3000 test samples. The predictive task is the expression levels of 20 cancer-related genes [8,10] using perturbation conditions as input. We evaluated four bin sizes: 2,3,4, and 5. We also considered four discretization methods. The first involves randomly assigning samples to bins, the second involves splitting samples evenly into bins after sorting, the third and fourth are equal frequency interval and k-means clustering. It worth noting that even splitting and frequency interval are the same in that they discretize a vector of continuous variables evenly given a specified size. However, they differ in that equal frequency does not achieve perfect equally sized groups if there are duplicates, naive even splitting does. For aggregation methods, we considered simple averaging, a case in which no classifiers are used in aggregation. For the cases in which classifiers are involved, we considered one in which class imbalance is ignored, we refer to this simply as imbalanced for the rest of the manuscript. The other classifier approaches used are one in which undersampling is used to reduce the number of samples in one class when it outweighs those in the other, and oversampling, which is the reverse. Undersampling was performed by randomly selecting samples from the over-represented class equal to that of the under-represented class. Oversampling was performed using SMOTE with the `smotefamily` package [21], where $k = 5$. We used random forests as our learning algorithm. All models were built using 1000 trees and default settings with the `ranger` [24] library in `R` [13]. The reported performance metric for regression is the coefficient of determination (R^2), as we are interested in the amount of the observed variance explained by the ensemble. We also report the performance of the classifying aggregators, for these we report accuracy, precision, recall and the F1 score. The dataset used in our experiment is available here http://dx.doi.org/10.17632/8mgyb6dyxv.2, and it is named *base_fp*, and the code is available here https://www.github.com/oghenejokpeme/FERUC.

5 Results

5.1 Overall Performance

We observed that on average multiple combinations of the considered discretizer-aggregator pairs generally outperformed the base case (see Table 1). Certain discretizer-aggregator pairs tended to consistently perform well or poorly. Even split and frequency interval combined with oversampling outperformed all other combination pairs, whereas k-means combined with averaging or undersampling generally underperformed when compared to the others (see Table 2). When paired with oversampling, the even split and frequency interval discretizers both achieved an average percentage performance increase of approximately 100% over the base case. Combined, both of these methods performed best when the number of bins is set to 5 (Table 3). With the assumption that there is no difference in performance between these two combinations and the base case, paired t-tests suggest that the null hypothesis can be rejected with a significance level of 0.01, with p-values of 2.6×10^{-8} and 9.7×10^{-9} respectively. Note that the average percentage performance difference between two competing approaches is calculated by estimating the percentage difference in performance for each gene pair, and then finding the mean.

5.2 Discretizer Effects

Discretization is the first step in the proposed learning algorithm, and the method by which we stratify the distribution of the response one might be interested in predicting into narrow-bins (Algorithm 1). It is clear from Fig. 2 that

Table 1. Mean predictive performance (R^2) of the 20 considered cancer genes for bin sizes 2, 3, 4, and 5. Discretization methods: random, even split, frequency interval, and k-means. Aggregation methods: simple averaging (AVG), class imbalance is ignored (RG), undersampling (US), and oversampling (OS). The best performing method for each bin size is underlined.

Bins	*Base*	Random				Even Split				Frequency Interval				K-means			
		AVG	RG	US	OS	AVG	RG	US	OS	AVG	RG	US	OS	AVG	RG	US	OS
2	0.075	0.079	0.079	0.079	0.079	0.057	0.081	0.081	0.081	0.057	0.082	0.082	0.082	−0.496	0.080	−0.451	0.054
3	0.075	0.081	0.083	0.083	0.085	0.040	0.085	0.088	0.100	0.040	0.085	0.086	0.094	−0.678	0.082	−0.500	0.050
4	0.075	0.082	0.084	0.084	0.088	0.031	0.086	0.084	0.103	0.031	0.086	0.084	0.102	−0.567	0.081	−0.432	0.036
5	0.075	0.082	0.084	0.082	0.091	0.024	0.087	0.078	0.103	0.024	0.087	0.078	0.104	−0.433	0.082	−0.302	0.031

Table 2. Average percentage performance difference of the two best and worst performing discretizer-aggregator combinations compared to the base case for each bin. The percentage increase or decrease is given, followed by the number of genes for which a discretizer-aggregator pair outperforms the base case. The best and worst performers are in boldface.

Bin	Best performers	Worst performers
2	Frequency interval – oversampling 26.5%(17)	k-means – undersampling −1324.2%(3)
	Frequency interval – imbalanced 26.4%(17)	k-means – averaging −1289.0%(1)
3	Even split – oversampling 105.5%(20)	**k-means – averaging** **−1732.5%(0)**
	Frequency interval – oversampling 75.3%(20)	k-means – undersampling −1460.3%(0)
4	**Even split – oversampling** **118.3%(20)**	k-means – averaging −1643.3%(1)
	Frequency interval – oversampling 113.4%(20)	k-means – undersampling −1520.4%(0)
5	Even split – oversampling 117.5%(20)	k-means – averaging −856.5%(1)
	Frequency interval – oversampling 116.7%(20)	k-means – undersampling −864.2%(0)

Table 3. Predictive performance (R^2) of the considered genes when bin size is 5. Shown are the results for the base case, the even split – oversampling pair (ES–OS), and frequency interval – oversampling pair (FRQ–OS). The percentage increase over the base case is also given.

Genes	*Base*	ES–OS	FRQ–OS
AKT1	0.197	0.207 (5.1)	0.208 (5.6)
APOE	0.071	0.088 (23.9)	0.090 (26.8)
BRCA1	0.111	0.152 (36.9)	0.152 (36.9)
CDH3	0.032	0.056 (75.0)	0.058 (81.2)
CDK4	0.291	0.294 (1.0)	0.296 (1.7)
CFLAR	0.017	0.049 (188.2)	0.049 (188.2)
EGF	0.055	0.076 (38.2)	0.076 (38.2)
EGFR	0.051	0.090 (76.5)	0.089 (74.5)
FGFR2	0.059	0.086 (45.8)	0.085 (44.1)
IGF1R	−0.007	0.042 (700.0)	0.041 (685.7)
KIT	0.069	0.096 (39.1)	0.098 (42.0)
LYN	0.083	0.109 (31.3)	0.112 (34.9)
PAX8	0.011	0.041 (272.7)	0.040 (263.6)
PTK2	0.069	0.082 (18.8)	0.082 (18.8)
RAD51C	0.057	0.087 (52.6)	0.087 (52.6)
STK10	0.044	0.066 (50.0)	0.067 (52.3)
TERT	0.073	0.107 (46.6)	0.108 (47.9)
TGFBR2	0.010	0.045 (350.0)	0.044 (340.0)
TNFRSF21	−0.022	0.042 (290.9)	0.042 (290.9)
TP53	0.228	0.245 (7.5)	0.247 (8.3)

the choice of discretizer plays a crucial role in predictive performance. When averaging is used as the aggregator, random sampling outperforms all other discretizers, with even split and frequency interval performing equally well. This is interesting as it shows that without the aggregating classifiers, the regressors built using methods like frequency interval perform worse than those built using random sampling. The reason is because when the response is put into bins using random sampling, the values in each of these bins will generally follow the same distribution as the overall response. Therefore, aggregating the predictions made by regressors built using these bins by averaging will generally yield good results. This is in contrast to when methods like even split or frequency interval are used, as each bin comprises of narrow generally non-intersecting bands of the overall distribution. It is worth noting here though that k-means performs remarkably poorly, producing negative R^2 values, suggesting that it fits worse than the horizontal line. One might be quick to note that this is one of the disadvantages of using R^2 as a performance metric in a regression problem when there is the potential for non-linearity. However, we would argue that for this particular application, it is vital that we have a clear representation of how much of the observed variance is explained by the proposed ensemble.

When class imbalance is ignored as is the case in the imbalanced aggregators, we observed that even split and frequency interval have near identical performance, with k-means and random sampling coming third and fourth depending on bin size. When undersampling is used to balance the dataset before building the classifying aggregators, we observed that as the number of bins increases, random sampling tended to outperform the even split and frequency interval discretizers. This is because as bin size increases, the number of samples in each bin decreases, and by undersampling, the classifying aggregators are built using fewer and fewer samples, making them less powerful. The performance of the random sampling discretizer does not suffer as much from this because its regressors are built using bins which generally represent the overall distribution of the response. We discuss this further when we discuss aggregator effects in the next section.

The performance of the discretizers when oversampling is used to handle class imbalance supports and contrasts with their performance when undersampling is used. We observed that the even split and frequency interval discretizers generally perform vastly better than how they do when undersampling is used to deal with class imbalance. In contrast to undersampling, the classifying aggregators are built using datasets in which the positive class has been oversampled, improving the models which classify new samples into bins. Given that the overall distribution of a response is represented in each bin when random sampling is used, building accurate bin delineating classifiers becomes more difficult as the samples in the positive and negative classes are very much alike. However, the expectation is that these classifiers will essentially predict that a new sample belongs in its bin, and produce a probability based on how closely related it is to the positive samples used in their construction. Therefore, for the random sampling discretizer, one would expect better performance when undersampling is used, which is what we observed (see Fig. 2).

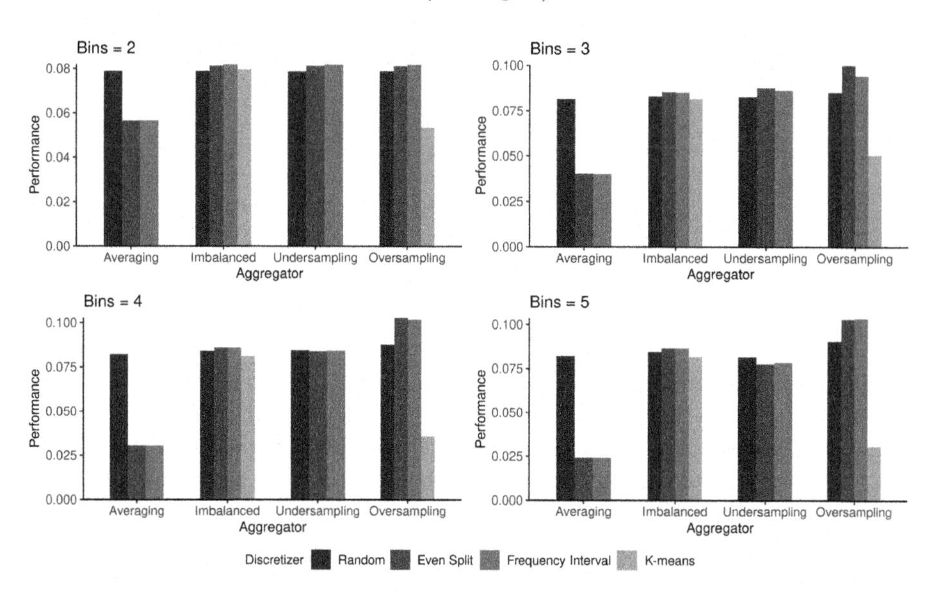

Fig. 2. Average discretizer performance (R^2) for the considered bin sizes across the considered aggregation approaches. Frequency interval is excluded from averaging and undersampling aggregation results because it consistently produced negative R^2 values.

5.3 Aggregator Effects

In the previous section, we discussed the effects the choice of discretizer can have on predictive performance. Although the discretizers were our main focus, it is clear that there is a synergistic effect between the choice of discretizer and aggregator. Figure 2 also shows that the choice of aggregator has a clear effect on predictive performance, with averaging performing worse overall, oversampling outperforming all the others, and undersampling generally performing worse than imbalanced. Here, our primary focus is to discuss why this is the case, especially as it has to do with the classifying aggregators. Table 4 shows the average predictive performance (accuracy, precision, recall, and F1 score) for all discretizer-aggregator pairs, and for all bin sizes we considered. These results explain the observed predictive performance discussed in the previous two sections. Although the accuracy of the classifying models are also reported, our discussion will be mostly centered around the precision and recall metrics, given that we are dealing with input datasets which may be class imbalanced.

Table 4. Average predictive performance of the aggregating classifiers built using datasets whose class representations are imbalanced (RG), undersampled (US), and oversampled (OS) for the considered discretizers. The reported performance metrics are accuracy (Acc), precision (Prec), recall (Rec), and F1 score.

Discretizer	*Aggregators*	Bins															
		2				3				4				5			
		Acc	Prec	Rec	F1	Acc	Prec	Rec	F1	Acc	Prec	Rec	F1	Acc	Prec	Rec	F1
Random	RG	0.50	0.50	1.00	0.67	0.13	0.13	1.00	0.23	0.05	0.05	1.00	0.09	0.02	0.02	1.00	0.03
	US	0.50	0.50	1.00	0.66	0.50	0.50	1.00	0.67	0.50	0.50	1.00	0.67	0.50	0.50	1.00	0.67
	OS	0.50	0.50	1.00	0.67	0.17	0.17	1.00	0.29	0.10	0.10	1.00	0.18	0.06	0.06	1.00	0.11
Even split	RG	0.58	0.58	0.58	0.58	0.65	0.23	0.45	0.30	0.73	0.10	0.37	0.16	0.79	0.06	0.32	0.10
	US	0.58	0.58	0.58	0.58	0.56	0.56	0.39	0.46	0.55	0.55	0.29	0.38	0.54	0.54	0.23	0.32
	OS	0.58	0.57	0.58	0.58	0.65	0.26	0.44	0.33	0.72	0.17	0.35	0.22	0.77	0.12	0.30	0.17
Frequency interval	RG	0.58	0.58	0.58	0.58	0.65	0.23	0.45	0.30	0.73	0.10	0.37	0.16	0.79	0.06	0.32	0.10
	US	0.58	0.58	0.58	0.58	0.56	0.56	0.39	0.46	0.55	0.55	0.29	0.38	0.54	0.54	0.23	0.32
	OS	0.58	0.58	0.58	0.58	0.65	0.25	0.44	0.32	0.72	0.15	0.36	0.21	0.77	0.12	0.30	0.16
K-means	RG	0.80	0.55	0.67	0.55	0.73	0.36	0.52	0.34	0.77	0.24	0.37	0.15	0.80	0.15	0.15	0.06
	US	0.60	0.60	0.55	0.52	0.60	0.60	0.39	0.42	0.60	0.60	0.30	0.34	0.60	0.60	0.24	0.29
	OS	0.79	0.58	0.62	0.58	0.73	0.42	0.47	0.43	0.76	0.32	0.39	0.34	0.79	0.25	0.33	0.27

When random sampling the discretizer, we observed that across all bin sizes, the recall of all the classifying aggregators is exactly 1. This is consistent with previously discussed results. It shows that the classifiers are classifying all the test samples as being similar to those used in their building. This is unsurprising since the samples used in building each bin's classifier follows the same distribution as the original response vector. The precision of the aggregating methods is more nuanced. In the case in which class imbalance is ignored, although the recall maintains its value of 1 as bin size increases, the precision steadily decreases. This makes sense, as the expectation is that the models will consistently become worse at identifying false positives. The results for undersampling and oversampling are contrasting. While recall is also consistently 1 as bin size increases, the precision for undersampling stays at approximately 50%, while like the class imbalance case, the precision for oversampling steadily declines. This is also consistent with expectation. In the case of undersampling, we are building binary bin classifiers using a perfect 50−50 split in class representation, but with fewer samples as bin size increases. It is no surprise that accuracy is also approximately 50%. For oversampling, accuracy and precision both hold 50% when bin size is 2, but steadily declines as it increases. Here, we argue that oversampling the samples in the imbalanced class, which is usually the positive class, makes the classifiers even worse at predicting false positives. This is to be expected, due to the properties of the random sampling discretizer.

For the even split and equal frequency discretizers, all three aggregators have an average value of 58% for accuracy, precision, recall, and F1 when the bin size is 2. This suggests that the models are capable of classifying positive and negative samples equally well. However, this changes as bin size increases. When the input dataset is imbalanced, we observed that both precision and recall steadily decreases, with precision getting remarkably worse-off than recall. The explanation for this is that the class imbalance is exacerbated by the increasing bin size with fewer samples in each bin, making it harder for the models to identify false positives. When undersampling is used, precision generally remains the same as bin size increases but recall decreases. This shows that while the classifiers' false positive prediction rate does not get significantly worse, its number of false negative predictions increases. This phenomenon can be easily explained by the fact that as bin size increases, fewer samples in general are used in building the classifiers. For oversampling, what we observe for recall and precision are in contrast to those of undersampling. Though they both decrease as bin size increases, recall is better than precision. When compared to the imbalanced case, although the recall values are similar, the precision in the oversampling case is generally better, especially as bin size increases. This explains why oversampling outperforms the imbalanced and undersampling cases. The difference in performance between even split and frequency interval as seen in Table 2 can be explained by a slight increase in precision and a slight decrease in recall for even split compared to frequency interval (see Table 4). Therefore, it is worth noting that for the proposed approach, seemingly duplicate values should not be excluded during discretization.

For k-means, the precision and recall values are generally similar to those of even split and frequency interval for the considered bin sizes with the exception of when bin size is 2. However, even with this similarity, we still observed that the undersampling aggregator performed remarkably poorly when paired with k-means (see Tables 1 and 2). Our analysis of the results showed that this is because of a known limitation of k-means discretization, which is that it is very sensitive to outliers [7], which we expect to mainly be at the tails of the response distribution. Individual investigation of classifier performance for each gene showed that this occurs because the models built using samples at the tails have either very low precision and very high recall, or the opposite.

This is in contrast to other discretizers, for which the precision-recall ratio is better balanced. This is evident from the difference in F1 scores between the different discretizers across the considered bin sizes (see Table 4).

5.4 Bin Size Effects

Figure 3 shows how aggregator predictive performance changes as bin size increases for the considered discretizers. For the random discretizer, most aggregators tend to steadily improve as bin size increases. The exception to this is undersampling, which peaks at a bin size of 4. For even split and frequency interval, the four aggregators behave similarly as expected. The imbalanced and oversampling aggregators get better as bin sizes increases, with the imbalanced aggregator doing so at a slower rate. Averaging gets worse as bin size increases as discussed in previous sections. Lastly, the undersampled aggregators reach peak performance at a bin size of 3 and begin to decline. When the k-means discretizer is paired with averaging and undersampling, we see a performance decrease from bin size 2 to 3, then steady increase from 3 to 5. However, as noted in the previous two sections, the performance is still remarkably poor. For oversampling, predictive performance sees a slow decline as bin size increases. Whereas the imbalanced aggregator tends to hold its performance. From these results, it is clear that bin size also plays a crucial role in the performance of the proposed ensemble regression approach. However, to what extent this is the case is beyond the scope of this work and will be the subject of future work.

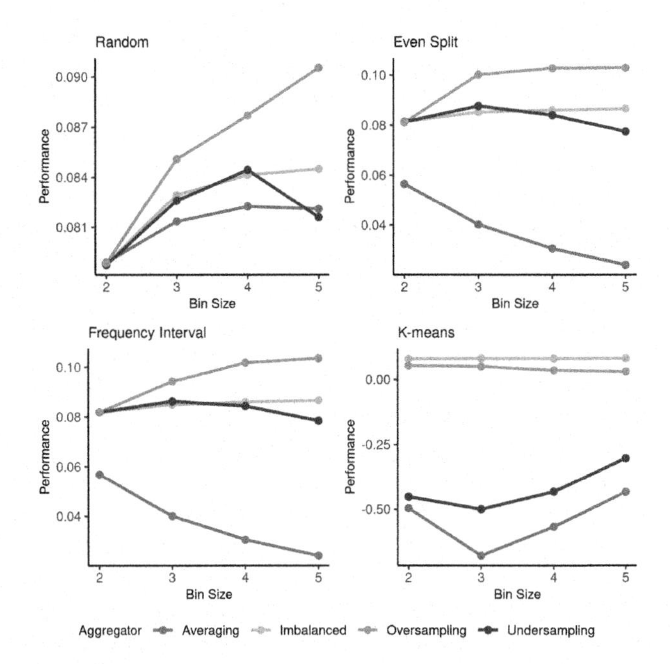

Fig. 3. Average aggregator performance (R^2) for the considered discretizers as bin size increases.

6 Discussion

An important task in the machine learning model building process is the selection of the right parameters. Our results show that the choice of bin size, discretizer, and aggregator all play an important role in predictive performance. Although we do not directly evaluate it here, we argue that these parameters can be easily optimised using the standard model selection approach with cross-validation. Assuming a near optimal bin size has been selected, the proposed ensemble learning algorithm is limited by the fact that it can only do as well the classifier-regressor pairs. Although we used only random forests in our evaluation, which is capable of building both classifiers and regressors, one can choose to use one learning algorithm for the classifiers and another for the regressors. In fact, it is possible to extend what we have proposed using traditional stacking, where multiple learning algorithms are used as classifiers and regressors. Of course this will come with increased cost in the form of computational time complexity. Another obvious extension is in multi-target regression problems. For example, one can imagine using this as the core predictor in an ensemble of regressor chains [23]. All of this, along with evaluations on other datasets will be the subject of future work.

7 Conclusion

We have presented an ensemble learning algorithm for regression using classification which leverages the underlying distribution of the response one is interested in predicting. We evaluated this approach on an important problem in precision medicine, which is the *in silico* estimation of drug perturbation effects on genes in cancer cell lines. We found that this approach significantly outperforms the base case, with several directions for extension which we conjecture will further improve its predictive capabilities.

Acknowledgments. This work was supported by the Engineering and Physical Sciences Research Council (EPSRC) UK through the ACTION on cancer grant (EP/R022925/1, EP/R022941/1). Prof. King acknowledges the support of the Knut and Alice Wallenberg Foundation Wallenberg Autonomous Systems and Software Program (WASP).

References

1. Ahmad, A., Khan, S.S., Kumar, A.: Learning regression problems by using classifiers. J. Intell. Fuzzy Syst. **35**(1), 945–955 (2018)
2. Ali, A., Shamsuddin, S.M., Ralescu, A.L., et al.: Classification with class imbalance problem: a review. Int. J. Adv. Soft Comput. Appl. **7**(3), 176–204 (2015)
3. Breiman, L.: Stacked regressions. Mach. Learn. **24**(1), 49–64 (1996)
4. Breiman, L.: Random forests. Mach. Learn. **45**(1), 5–32 (2001)
5. Caruana, R., Niculescu-Mizil, A., Crew, G., Ksikes, A.: Ensemble selection from libraries of models. In: Proceedings of the Twenty-first International Conference on Machine Learning, p. 18 (2004)

6. Chawla, N.V., Bowyer, K.W., Hall, L.O., Kegelmeyer, W.P.: SMOTE: synthetic minority over-sampling technique. J. Artif. Intell. Res. **16**, 321–357 (2002)
7. Dash, R., Paramguru, R.L., Dash, R.: Comparative analysis of supervised and unsupervised discretization techniques. Int. J. Adv. Sci. Technol. **2**(3), 29–37 (2011)
8. Dolgin, E.: The most popular genes in the human genome. Nature **551**(7681), 427 (2017)
9. Dougherty, J., Kohavi, R., Sahami, M.: Supervised and unsupervised discretization of continuous features. In: Machine Learning Proceedings 1995, pp. 194–202. Elsevier (1995)
10. Futreal, P.A., et al.: A census of human cancer genes. Nat. Rev. Cancer **4**(3), 177–183 (2004)
11. Gonzalez, D.L., et al.: Hierarchical classifier-regression ensemble for multi-phase non-linear dynamic system response prediction: application to climate analysis. In: 2012 IEEE 12th International Conference on Data Mining Workshops, pp. 781–788. IEEE (2012)
12. Halawani, S.M., Albidewi, I.A., Ahmad, A.: A novel ensemble method for regression via classification problems (2011)
13. Ihaka, R., Gentleman, R.: R: a language for data analysis and graphics. J. Comput. Graph. Statist. **5**(3), 299–314 (1996)
14. Jahrer, M., Töscher, A., Legenstein, R.: Combining predictions for accurate recommender systems. In: Proceedings of the 16th ACM SIGKDD International Conference on Knowledge Discovery and Data Mining, pp. 693–702 (2010)
15. Japkowicz, N., Stephen, S.: The class imbalance problem: a systematic study. Intell. Data Anal. **6**(5), 429–449 (2002)
16. Koleti, A., Terryn, R., et al.: Data portal for the library of integrated network-based cellular signatures (LINCS) program: integrated access to diverse large-scale cellular perturbation response data. Nucleic Acids Res. **46**(D1), D558–D566 (2017)
17. Mendes-Moreira, J., Soares, C., Jorge, A.M., Sousa, J.F.D.: Ensemble approaches for regression: a survey. ACM Comput. Surv. **45**(1), 1–40 (2012)
18. Orhobor, O.I., Alexandrov, N.N., King, R.D.: Predicting rice phenotypes with meta-learning. In: Soldatova, L., Vanschoren, J., Papadopoulos, G., Ceci, M. (eds.) DS 2018. LNCS (LNAI), vol. 11198, pp. 144–158. Springer, Cham (2018). https://doi.org/10.1007/978-3-030-01771-2_10
19. Rooney, N., Patterson, D., Anand, S., Tsymbal, A.: Dynamic integration of regression models. In: Roli, F., Kittler, J., Windeatt, T. (eds.) MCS 2004. LNCS, vol. 3077, pp. 164–173. Springer, Heidelberg (2004). https://doi.org/10.1007/978-3-540-25966-4_16
20. Silla, C.N., Freitas, A.A.: A survey of hierarchical classification across different application domains. Data Min. Knowl. Disc. **22**(1–2), 31–72 (2011)
21. Siriseriwan, W.: Smotefamily: a collection of oversampling techniques for class imbalance problem based on smote (2018). http://cran.r-project.org/package=smotefamily
22. Sonnenburg, S., Rätsch, G., Schäfer, C., Schölkopf, B.: Large scale multiple kernel learning. J. Mach. Learn. Res. **7**(Jul), 1531–1565 (2006)
23. Spyromitros-Xioufis, E., Tsoumakas, G., Groves, W., Vlahavas, I.: Multi-label classification methods for multi-target regression. arXiv preprint arXiv:1211.6581, pp. 1159–1168 (2012)
24. Wright, M.N., Ziegler, A.: ranger: a fast implementation of random forests for high dimensional data in C++ and R. arXiv preprint arXiv:1508.04409 (2015)

One-Class Ensembles for Rare Genomic Sequences Identification

Jonathan Kaufmann, Kathryn Asalone, Roberto Corizzo(✉), Colin Saldanha, John Bracht, and Nathalie Japkowicz

American University, Washington 20016, DC, USA
jkaufmann@google.com, ka5144a@student.american.edu,
{rcorizzo,saldanha,jbracht,japkowic}@american.edu

Abstract. The next-generation sequencing revolution has impacted biological research by allowing the collection and analysis of very large datasets. However, despite the large availability of data, current computational methods used by biologists present some limitations in challenging domains, such as extremely imbalanced datasets characterized by almost only negative examples. In this paper, we address the problem of identifying sequences from the zebra finch (songbird) germline-restricted chromosome (GRC), which is present only in reproductive tissues and missing from all other cells. Since the germline contains the GRC in addition to other chromosomes, sequencing germline DNA must be followed by separation into GRC or non-GRC sequences. The complexity of this task depends on the limited availability of known GRC sequences. In this paper, we propose a one-class ensemble learning method to solve this problem, and we compare its performance with state-of-the-art methods for one-class classification. Our results show that the proposed method is able to identify positive sequences with high accuracy, having been trained only with negative sequences, and tuned with a limited number of positive sequences. Moreover, a biological analysis revealed that positive sequences from a verified GRC gene were ranked in the top third of all the sequences, showing that our method is successful in demarcating GRC from non-GRC sequences. Our method thus represents a valuable tool for biologists, since model predictions can allow them to focus their limited resources towards the experimental validation of a subset of higher confidence sequences.

Keywords: Machine learning · One-class learning · Anomaly detection · Genomics · Biology

1 Introduction

The next-generation sequencing revolution has fundamentally impacted biological research, with researchers sequencing and assembling very large datasets. Although biologists have relied on computational methods to handle this massive amount of data for several decades, they face many problems that cannot be tackled by the currently available tools.

A. Appice et al. (Eds.): DS 2020, LNAI 12323, pp. 340–354, 2020.
https://doi.org/10.1007/978-3-030-61527-7_23

In this paper, we focus on the identification of sequences from the zebra finch (songbird) germline-restricted chromosome (GRC). Differently than most organisms, in which every cell contains the same genome [14], the Zebra Finch germline presents an extra "germline-restricted chromosome" (GRC) in the ovaries and the testes [34], that is missing in all its other tissues composed by somatic cells.

For this reason, sequencing germline DNA must be followed by separation between GRC (positive class) sequences and non-GRC (negative class) sequences. The GRC has been found in Bengalese Finch [37] and 14 other examined song bird species, suggesting a wide or even ubiquitous distribution in songbirds [45]. However, only two GRC genes have been identified and sequenced to date. The only known genetic sequences from the GRC are a non-coding repeat [18] and a single coding transcript [9].

The limited number of GRC genes identified so far depends also on the fact that the isolation of GRC sequences from total cell sequences is a challenging scientific problem. Unfortunately, the GRC cannot be physically isolated by conventional methods due to its size [17]. Moreover, the germline genome consists primarily of chromosomes also present in somatic cells, with approximately 10% of sequence complexity deriving from GRC and thus unique to germline cells. It is thought that the GRC is derived by ancient duplication of specific genomic elements, which subsequently diverged along a distinctive evolutionary trajectory, because only mutations obtained in females are inherited [34,35]. For this reason, GRC elements identified to date have been divergent copies of sequences on other chromosomes [9,18] making computational subtraction [5] alone ineffective in distinguishing GRC from non-GRC sequences with high confidence.

Since biological experimental validation through Polymerase Chain Reaction (PCR) can only be performed for a small number of sequences, a machine learning method could provide a valuable computational strategy to rapidly identify bona fide rare GRC sequences on a large scale, and allow researchers to only test a limited number of sequences that the model classifies as GRC with high confidence.

However, the classification task in this domain poses two main challenges: *i)* the large number of data sequences, and *ii)* the rarity of positively labeled data sequences. While the former calls for efficient algorithms that are capable to process large-scale data, the latter requires machine learning approaches that can handle extreme class imbalance.

In this paper, we propose a one-class learning method for sequence classification. Our method trains a model on somatic (non-GRC) negative sequences represented as merged pairs of raw reads, and predicts the class of unseen sequences (GRC or non-GRC) in the germline DNA, consisting of assembled contiguous sequences (contigs). Our method can be applied to any dataset where sequences experience a distinctive mutational pattern relative to other genomic elements, such as the exposure to the germline milieu of only a single sex, as in the case addressed in our study.

The paper is structured as follows. In Sect. 2, we provide an overview of related works in the biological and machine learning fields. In Sect. 3, we present our proposed method. In Sect. 4 we describe the datasets and the experimental results obtained in our study. Finally, Sect. 5 concludes the paper.

2 Background

2.1 Biological Background

Early in embryonic development, the GRC is eliminated from every cell except for the germline, and it is expelled during spermatogenesis in adult males so it is passed down through the maternal lineage only [34,35]. Similar to the mode of inheritance of the mitochondrial DNA in humans, the GRC has a uniparental inheritance from the mother, because any mutations that occur in the males are not passed on to the next generation. However, the GRC is not found in an organelle, making it analogous to the human Y chromosome in terms of its nuclear environment, and suggesting it may display a mutational signature different from other chromosomes [21,39]. The association of the GRC with female biology makes it a particularly intriguing genomic element and an important counterpoint to Y-chromosome analysis. Of particular interest are any genes it may harbor.

Unlike the Y chromosome, where many genes involved in the development and function of the testis are known [26] the function of any potential genes on the GRC are unknown. Given that the GRC is the longest chromosome in zebra finch, encoding an estimated 10% of the total sequence complexity [34] it may encode hundreds of genes.

2.2 Machine Learning Background

The one-class classification concept, also called outlier detection [38] and novelty detection [10], was first proposed by [32] and subsequently analyzed in depth by [19] and [44]. Unlike traditional binary classification approaches which learn to *discriminate* between positive and negative examples, one-class classification methods train on negative (or positive) instances only and learn to *recognize* instance from the class they were trained on. By default, instances not recognized by the method are labeled as negative (positive). While this approach is generally not as effective as binary classification [8], it is sometimes the only approach available given a particular problem. General surveys of one-class classification methods can be found in [22,23], and [36], while surveys of anomaly or outlier detection methods that includes one-class classification can be found in [1,13].

Many one-class classification methods have been proposed including One-Class SVM [40], Autoencoder-based classification [15,16,20], One-Class Local Outlier Factor (LOF) [11], and most recently, Isolation Forests [30]. Other methods are based on clustering [7] and ensembles of Support Vector Machines [47] [48]. In order to improve one-class classification results, many different combinations have been proposed in [2].

3 Method

Given the large availability of true-negative data and the rarity of true-positive data, the problem addressed in this study provides a canonical case for one-class learning. In order to analyze large datasets, it is important to leverage machine learning methods that exhibit low time and space efficiency. Isolation Forest and LOF represent two efficient methods for one class learning, and were shown to provide high accuracy performance in different domains [41,43].

LOF works by comparing the local density of a point to that of its neighbors. If a point has a much lower density than its neighbor, it is considered an outlier. Isolation Forest works quite differently. It consists of building trees in which the data is cut to isolate data points. The fewer cuts it takes to isolate a data point, the higher on the tree the point appears and the more likely it is to be an outlier.

In our method, we adopt Isolation Forest and LOF as base learning methods, and we subsequently combine their outputs in an ensemble. The rationale for our mixed ensemble is that combining the outcome of different learners has the potential to reduce prediction errors, and reduce the bias deriving from single-model specifications.

Each method outputs anomaly scores, where a lower score represents a stronger anomaly. However, Isolation Forest produces anomaly scores between -0.5 and 0.5, whereas LOF produces both positive and negative scores in an unbounded range. For this reason, we adopt a normalization approach. More specifically, we find the value such that 95% of all scores below 0 are greater than that value, and found the multiplier such that the value multiplied by the multiplier is -0.5. Symmetrically, we find the value such that 95% of all scores above 0 are greater than that value, and identify the multiplier such that the value multiplied by the multiplier is 0.5. We perform the same process for both Isolation Forest and LOF models. The multipliers allow us to calculate new normalized transformation scores for all models. We also devise a combination strategy based on model averaging (formally described in Algorithm 1)

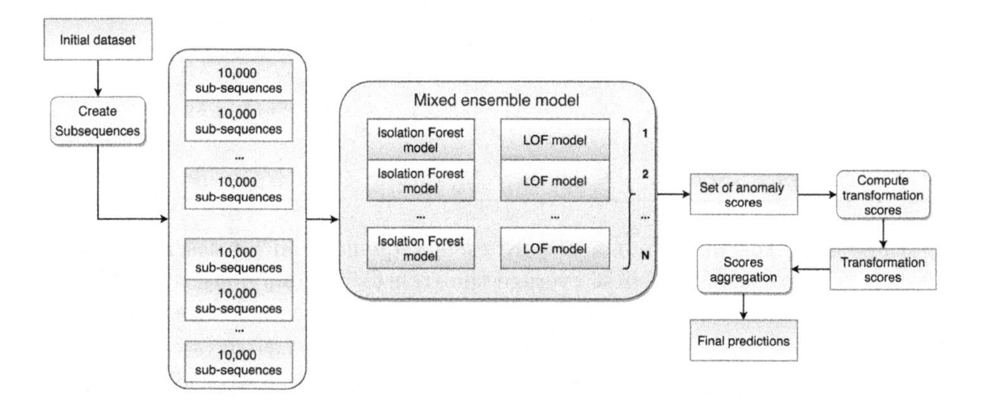

Fig. 1. A graphical representation of our proposed mixed ensemble method.

in order to extract a final anomaly score from the mixed ensemble. A graphical representation of the method is shown in Fig. 1.

Algorithm 1: Combined prediction strategy with mixed ensemble

Input: Sequence S,
Ensemble of Isolation Forest models I,
Isolation Forest transformation scores I_α, I_β
Ensemble of LOF models L,
LOF transformation scores L_α, L_β

Result: Final anomaly score S_{score}

1 $P = \{\}$
2 **for** $s \in S$ **do**
3 $I_S = predict(I, s)$
4 **for** $s \in I_S$ **do**
5 **if** *if* $s < 0$ **then**
6 $s = s \times I_\beta$
7 **else**
8 $s = s \times I_\alpha$
9 **end**
10 **end**
// Average score for Isolation Forest sub-ensemble
11 $\overline{I_S} = \frac{\sum_{s \in I_S}}{|I_S|}$
12
13 $L_S = predict(L, s)$
14 **for** $s \in L_S$ **do**
15 **if** *if* $s < 0$ **then**
16 $s = s \times L_\beta$
17 **else**
18 $s = s \times L_\alpha$
19 **end**
20 **end**
// Average score for LOF sub-ensemble
21 $\overline{L_S} = \frac{\sum_{s \in L_S}}{|L_S|}$
22
23 $\overline{s} = \frac{\overline{I_S} + \overline{L_S}}{2}$
24 $P = P \cup \overline{s}$
25 **end**
26 $S_{score} = \frac{\sum_{\overline{s} \in P}}{|P|}$

Effectively, the final anomaly score for a sequence can be described as an average of averages, as we first average the transformed anomaly scores generated by each model on a particular subsequence, and then average that score across all subsequences. In accordance with the behavior of the base learners, our method returns low scores for high-confidence anomalous sequences, that are consequently labeled as positive.

3.1 Biological Data Preparation

A first step of merging and quality filtering of sequences was performed using the software PEAR v0.9.11 [49] specifying a minimum length of 50 *bp* and a phredscore of 30. This process led to a resulting dataset of 26,198,608 liver and 69,911,879 testis sequences. Subsequently, high-confidence testis merged reads mapping to the alpha-SNAP were identified with BWA v0.7.12 [27]. These high-confidence reads were considered the "true-positive" data for our validation and testing sets, while the liver data was considered "true-negative".

Quality filtered DNA sequences from male testis tissue were corrected for read errors using Karect Master v2 [3]. Corrected data was assembled using SPAdes v3.12.0 [6] using the options `-s` for unpaired reads, `--sc` for single-celled data, and `--only-assembler`.

The error corrected reads were mapped onto two published genome assemblies [25,46] and the SPAdes assembly of testis DNA. BBMAP [12] was used with options: `minid = 0.98` for the minimum alignment identity, `idfilter = 0.98` sets minimum identity separate of minid, and the rest of the options in default. We used 98% identity to allow alleles to map but not paralogs. SNPs were called using SAMtools v1.9 [28] and BCFtools v1.9 [29].

Scaffolds that did not contain any SNPs were isolated and subtractive blast methods were used to remove any remaining somatic sequences that match known somatic references [5]. To ensure that only matches with a greater than 98% identity are included, blast v2.9.0 [4] was run using the `-perc_identity=98`. The two published assemblies [25,46] were masked using RepeatMasker v4.0.5 [31] with the custom repeat library from zebra finch using RepeatModeler v1.0.8 [42]. The masked assemblies were used as references, or databases for the blast search, for the somatic genome. The output after two rounds of subtraction were the hypothetical GRC sequences, a set of 13,818 sequences.

Qualitative polymerase chain reaction (qPCR) was used to validate a contig predicted to be a part of the GRC. This process is a way to validate that the sequences identified by the machine learning approach are correct from a biological viewpoint. The contig was identified by machine learning and through blast search for contigs with the known non-coding repetitive element on the GRC. DNA from brain, heart, and testis using primers specific for a 400 *bp* region of the contig was used that codes for Splicing Factor 38 A. qPCR was run as a two-stage cycle with 95 °C for 10 min for the initial melt, followed by 40 cycles of 95 °C for 30 s, 55 °C for 15 s. Signals from qPCR were measured relative to β-actin using ΔCt. The averages and standard deviations of $2^{-(gene\ Ct\ -\ \beta-actin\Delta Ct)}$ were calculated and statistical significance was measured using an ANOVA test.

Concerning data availability, two of the assemblies used in our dataset are publicly available [24][1] [45][2]. The in-house SPAdes assembly was obtained from raw reads from our laboratory, which are publicly available[3].

[1] https://www.ncbi.nlm.nih.gov/nuccore/MUGN00000000.1/.
[2] https://www.ncbi.nlm.nih.gov/bioproject/?term=ABQF00000000.
[3] https://www.ncbi.nlm.nih.gov/sra/SRR6896648.

3.2 Data Pre-processing

GRC Dataset. Somatic (liver) DNA represent our training data (non-GRC, negative class). For all our analyses we select 10,000 random sequences for each model in the ensemble, and we split the data sequences into subsequences of size 60 to provide a constant size and avoid bias due to variation in sequence length between datasets. For the positive class, we derive 183 high-confidence GRC subsequences from the 109 high-confidence testis DNA reads that map to the single known GRC transcript. We use half of the 183 true-positive GRC subsequences and 10,000 true-negative liver subsequences as our validation dataset. We select the other half of the 183 true-positive GRC subsequences and 10,000 true-negative liver subsequences as our testing data.

Hypothetical GRC Dataset. Based on the methods described in the biological data preparation section above, we construct a second test dataset made of 13,818 sequences, derived by assembling and processing testis DNA, and hypothesize them to be part of the GRC (positive class). For each dataset, we counted the instances of overlapping 4-grams[4] in each sequence as features, providing 256 possible features given that the DNA alphabet contains four characters.

4 Experiments

We present our results in the following four subsections. In the first subsection, we compare the classification performance of our method and competitor methods on the GRC dataset (validation data). For this purpose, we tuned the models and attempted alternative configurations on this dataset. For all methods, we use the Python implementations available in the SciKit-Learn library [33].

Therefore, we construct five candidate model specifications: a single Isolation Forest, a single LOF, an ensemble of 10 Isolation Forest model[5], an ensemble of ten LOFs, and a mixed ensemble of 10 Isolation Forests and 10 LOFs[6].

[4] We also evaluated alternative constructions with 5-grams, which exhibited similar performance but substantial increases in computation time, as the feature space size is 4^n. Larger complexity n-grams were deemed infeasible for this application, due to this exponential feature space growth.

[5] Experimentation yielded that performance was asymptotic to the number of models. 8 models was approximately optimal, but 10 was selected to be conservative and so that we could experiment with dropping out the maximum and minimum score, leaving 8 scores.

[6] Experimentally, we observed that an homogeneous ensemble of size 10 IF or LOF models exhibits roughly equivalent performance to a size 20, within a delta of 0.0001. Therefore, comparing a mixed-model ensemble with 10 IF and 10 LOF models to size 10 IF or LOF homogenous ensemble is equivalent to comparing to a size 20 IF or LOF homogenous ensemble.

Each model was trained on separate, unique sets of 10,000 subsequences with length 60 from the training dataset[7].

In the second subsection, we discuss the results obtained on the GRC dataset (test data). In the third section, we discuss the results we obtained by applying our selected model to the hypothetical GRC dataset. Given the large size of the dataset and the scarcity of positive examples, we used a train-validation-test rather than a cross-validation evaluation scheme. While this scheme may be less reliable than cross-validation, it appears more feasible for this study from a computational cost perspective. A blind test allowed us to further validate the significance of the results obtained from a biological point of view. Following the analysis of the two datasets, the fourth subsection contains the results of our blind test. Finally, the fifth subsection contains a biological analysis of the results obtained.

4.1 Evaluation on the GRC Dataset (Validation Data)

We first evaluated the mixed LOF/Isolation Forest ensemble against homogeneous ensembles of Isolation Forest and LOF, as well as single models of each type, using the validation dataset. After tuning, parameters for Isolation Forest were $n\_estimators = 100$ and $max\_samples = 256$. Parameters for LOF were $n\_neighbors = 25, leaf\_size = 30, p = 2$, and $novelty = True$. Results in Table 1 show that the mixed model exhibits a higher AUC than the other candidate specifications. Alternative configurations, using more data per model (25,000 and 50,000) did not affect significantly the performance of the model, but did substantially increase the computational cost. Similarly, dropping out the highest and lowest scores from the ensembles before averaging did not affect AUC. Lastly, we tested whether summed anomaly scores instead of averaging them was better, but observed no effect on performance in terms of accuracy.

4.2 Evaluation on the GRC Dataset (Test Data)

This subsection presents results of all methods on test data. From the results shown in Table 1 we can observe that most methods present a slightly higher AUC score than the validation data results. However, in accordance with results obtained on the validation data, the proposed method based on mixed ensemble outperforms other methods.

[7] We experimented with larger and smaller training data sizes per model. Smaller size training sets exhibited AUC performance losses, while larger sizes demonstrated no significant performance gains, paired with substantial additional computational expense, particularly for LOF.

Table 1. Classification accuracy results in terms of Area Under The Curve (AUC) obtained by all methods with the GRC dataset on validation and test data. The best result is marked in bold.

Method	AUC (Validation data)	AUC (Test data)
Single Isolation Forest	0.5807	0.6062
Single LOF	0.7441	0.7845
Isolation Forest Ensemble	0.6918	0.6759
LOF Ensemble	0.7655	0.8101
Mixed Ensemble	**0.8107**	**0.8416**

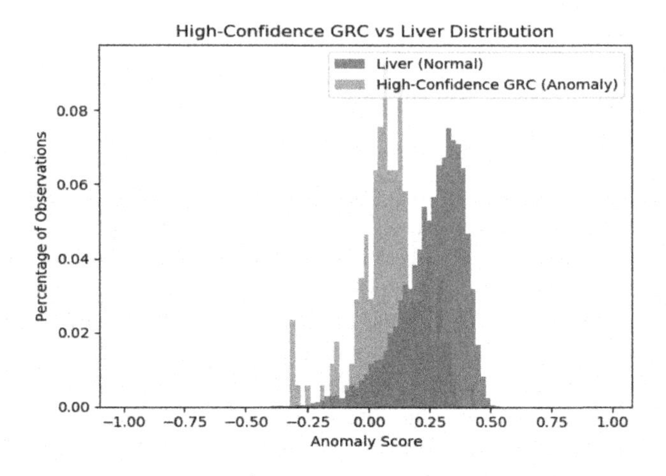

Fig. 2. High-confidence GRC vs Liver Anomaly Scores

In order to visually inspect the results, we plot the anomaly scores of the 183 known-GRC subsequences against the 20,000 somatic subsequences in our testing and validation datasets (Fig. 2). A lower score indicates a more anomalous sequence. In this graph, we merge the scores obtained on both validation and test sets. From the graph it is possible to observe a consistent difference in distribution between the known-GRC and somatic sequences. The somatic distribution appears to be skew-normal, while the GRC appears to be more normal. However, we note additional possible peaks at 0.25 as well as below −0.1 in the known-GRC. This raises the possibility of the true GRC distribution being multi-modal, but it is possible that the limited size of the known-GRC dataset puts too much emphasis on each observation. The clear differences in distribution between known-GRC and liver subsequences provide a benchmark against which we can compare the anomaly score distribution of hypothetical GRC sequences.

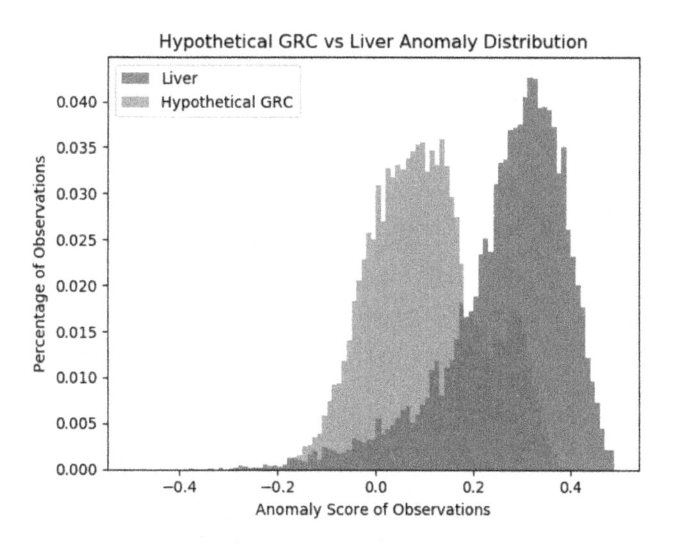

Fig. 3. Hypothetical GRC vs Liver Anomaly Scores

4.3 Evaluation on the Hypothetical GRC Dataset

Next, we evaluate the 13,818 hypothetical GRC sequences, adopting the averaging approach described in Algorithm 1 to generate final anomaly scores. We compare the hypothetical GRC anomaly score to 15,000 randomly selected liver sequences in Fig. 2. We apply the same average of averages prediction approach to generate scores for the liver sequences, by averaging the anomaly scores of their component subsequences. The liver distribution precisely mimics that of the test and validation data liver set. The hypothetical GRC sequences, similar to the known-GRC subsequences, have a much lower average anomaly score. We note that there appears to be a second peak in the hypothetical-GRC distribution around 0.27, close to the liver mean of 0.25. Given that the combination of genomic methods used to isolate the hypothetical GRC are imperfect, the machine learning approach appears to be both confirming the overall distinctness of these data from liver and the presence of some contaminating somatic data remaining to be separated out (Fig. 3).

Finally, we compare the hypothetical GRC to the known-GRC in Fig. 4. Though the known-GRC graph is unfortunately sparse and noisy, it is clear the hypothetical GRC closely follows this distribution. It is unclear if there is support for the secondary hypothetical GRC peak in the known-GRC distribution, as there is possibly a secondary peak in the known-GRC distribution around the same point. However, we cannot rule out this is not merely noise from our limited known-GRC data size.

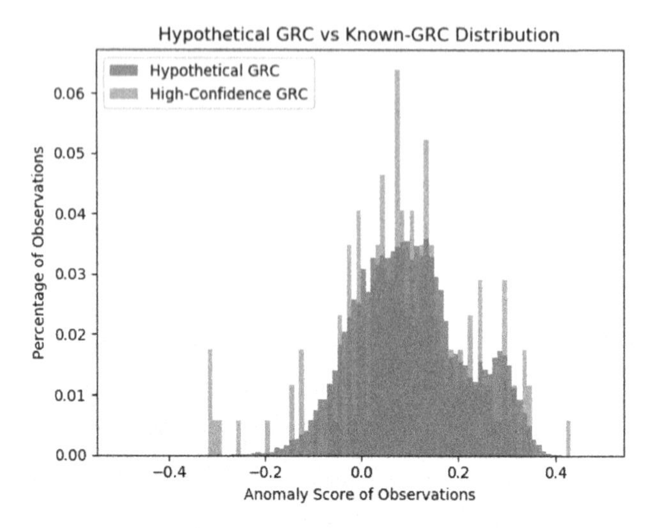

Fig. 4. Known-GRC vs High-Confidence GRC

4.4 Blind Test Results

It is important to independently verify that our classifier correctly identifies GRC sequences. For this reason, we searched for additional validated GRC genes. One of the sequences from the hypothetical GRC set, encoding a splicing factor 38 A gene, was physically linked with a known GRC sequence [18] in one of our assemblies (not shown here), giving us reason to pursue it further. Using qPCR we verified experimentally that this sequence is encoded on the GRC and is present in multiple copies, based on elevated copynumber only observed in testis DNA (Fig. 5).

The splicing factor 38A-containing contig has an anomaly score of 0.027, making it number 3, 753 of 13, 818 total sequences in the hypothetical GRC list (ranked in descending order of anomaly score). This puts it in the top 27% of all hypothetical GRC sequences, suggesting that anomaly scores of this value or lower are likely to be real GRC genes, and that thousands of hypothetical GRC sequences identified in our experiments are likely to be correct. This result, along with the results found in the previous sections, suggests that the machine learning approach proposed in this study is, indeed, successful at predicting GRC sequences and can accelerate the pace of genomic sequencing.

4.5 Biological Discussion

The inheritance of the Y-chromosome in a single sex imparts a distinctive mutational signature relative to the other chromosomes which spend half their time in each sex [21,39]. Mutational rates in males are higher because sperm undergo many more rounds of cell division than eggs, so chromosomes which are inherited only through males average a higher mutational rate than sequences inherited

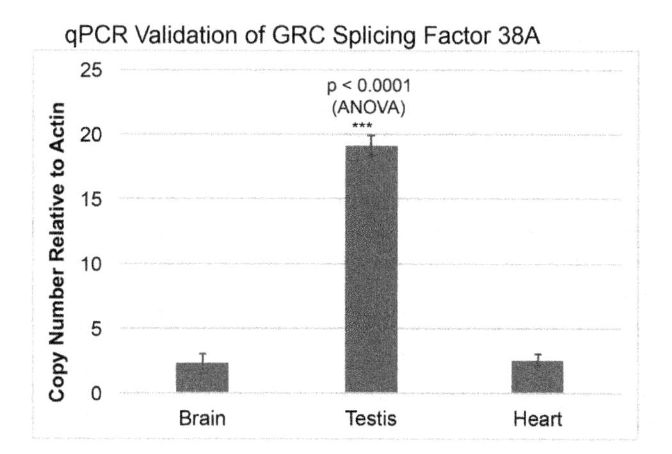

Fig. 5. Confirmation that GRC Splicing Factor 38 A is GRC-encoded by qPCR of the gene in brain, testis or heart relative to β-actin single-copy control.

both through males and females; consequently, sequences inherited through only female lineages should experience the fewest mutations [21,39]. The bird sex-determination system has ZZ males and ZW females, thus a female-only W chromosome, which has a very low mutation rate [24]. We hypothesized that the GRC should similarly reflect a protection from mutations accumulated in males, and this signature should be detectable relative to other chromosomes. By training a one-class anomaly detector we are the first to identify this signature and use it to identify many more high-confidence GRC sequences by anomaly score. Because our second validated GRC gene, a splicing factor 38A gene was number 3,753 on the list by anomaly score, we suggest that there are likely thousands of true-positive GRC sequences in our hypothetical dataset, significantly expanding known GRC sequences. In doing so, we demonstrate that one-class machine learning methods applied to genomic data can assist subject matter experts in identifying anomalous sequences. Specifically, we permit biologists with hypothetical anomalous data to rank-order their sequences, allowing them to be more confident about the anomalous nature of some sequences versus others. This significantly narrows the scope of which sequences must be experimentally verified through PCR, a process which is extremely time intensive. As researchers are likely only able to experimentally PCR test a few dozen sequences, our model permits the biologist to focus their limited resources on the hypothetical sequences that are most in doubt—or on finding the anomaly score threshold that approximately demarcates the line between true-positives and true-negatives.

5 Conclusion

In this study, we proposed an ensemble-based machine learning method to identify rare genomic sequences in imbalanced datasets. In more detail, our study considers the problem of identifying sequences from a specific chromosome present in certain types of songbirds (GRC) from datasets containing mixtures of this chromosome and common chromosomes. The results showed highly accurate classification results, also when compared to state-of-the-art one class machine learning methods.

We found that the anomaly scores our method obtained on the hypothetical GRC data were very close to those obtained on known GRC data. Furthermore, their distribution differed from that of the non-GRC sequences in a way similar to the way in which the known GRC differed from the non-GRC sequences. Our method assigned a high anomaly score to the gene presented as a blind test, thus ranking it as GRC with high confidence. The results extracted can help biologists prioritize thousands of candidate sequences and facilitate discoveries that would, otherwise, be highly laborious or impossible to do experimentally due to financial or temporal constraints.

In the future, we plan to experimentally verify more sequences that we ranked with high confidence to test whether, indeed, our tool is indicative of GRC sequences. We hypothesize that this tool may be useful in examining other anomalous sequences such as the finch W chromosome, another female-specific element, which should reflect altered mutation rates like the GRC. In addition to continuing to verify our computational findings in biology laboratories, we also intend to optimize our approach further, by exploring more complex combination methods. Finally, we will try to quantify the differential in speed of GRC identification with machine learning tools and, on the other hand, proceeding in a traditional manner.

References

1. Aggarwal, C.C.: Outlier analysis. Data Mining, pp. 237–263. Springer, Cham (2015). https://doi.org/10.1007/978-3-319-14142-8_8
2. Aggarwal, C.C.: Outlier ensembles: position paper. ACM SIGKDD Explor. Newslett. **14**(2), 49–58 (2013)
3. Allam, A., Kalnis, P., Solovyev, V.: Karect: accurate correction of substitution, insertion and deletion errors for next-generation sequencing data. Bioinformatics **31**(21), 3421–3428 (2015)
4. Altschul, S.F., Gish, W., Miller, W., Myers, E.W., Lipman, D.J.: Basic local alignment search tool. J. Mol. Biol. **215**(3), 403–410 (1990)
5. Asalone, K.C., Nelson, M.M., Bracht, J.R.: Novel sequence discovery by subtractive genomics. JoVE (J. Vis. Exp.) (143), e58877 (2019)
6. Bankevich, A., et al.: Spades: a new genome assembly algorithm and its applications to single-cell sequencing. J. Comput. Biol. **19**(5), 455–477 (2012)
7. Barracchia, E.P., Pio, G., D'Elia, D., Ceci, M.: Prediction of new associations between ncrnas and diseases exploiting multi-type hierarchical clustering. BMC Bioinform. **21**(1), 1–24 (2020)
8. Bellinger, C., Sharma, S., Japkowicz, N.: One-class versus binary classification: which and when? In: 2012 11th International Conference on Machine Learning and Applications, vol. 2, pp. 102–106 (2012)

9. Biederman, M.K., Nelson, M.M., Asalone, K.C., Pedersen, A.L., Saldanha, C.J., Bracht, J.R.: Discovery of the first germline-restricted gene by subtractive transcriptomic analysis in the zebra finch taeniopygia guttata. Curr. Biol. **28**(10), 1620–1627 (2018)
10. Bishop, C.M.: Novelty detection and neural network validation. IEE Proc. Vis. Image Sign. Process. **141**(4), 217–222 (1994)
11. Breunig, M.M., Kriegel, H.P., Ng, R.T., Sander, J.: LOF: identifying density-based local outliers. In: Proceedings of the 2000 ACM SIGMOD International Conference on Management of Data, pp. 93–104 (2000)
12. Bushnell, B.: BBMap short read aligner (2016)
13. Chandola, V., Banerjee, A., Kumar, V.: Anomaly detection: a survey. ACM Comput. Surv. (CSUR) **41**(3), 15 (2009)
14. Consortium, I.H.G.S.: Initial sequencing and analysis of the human genome. Nature **409**(6822), 860 (2001)
15. Corizzo, R., Ceci, M., Japkowicz, N.: Anomaly detection and repair for accurate predictions in geo-distributed big data. Big Data Res. **16**, 18–35 (2019)
16. Corizzo, R., Ceci, M., Zdravevski, E., Japkowicz, N.: Scalable auto-encoders for gravitational waves detection from time series data. Expert Syst. Appl. **151**, 113378 (2020)
17. Herschleb, J., Ananiev, G., Schwartz, D.C.: Pulsed-field GEL electrophoresis. Nat. Protoc. **2**(3), 677 (2007)
18. Itoh, Y., Kampf, K., Pigozzi, M.I., Arnold, A.P.: Molecular cloning and characterization of the germline-restricted chromosome sequence in the zebra finch. Chromosoma **118**(4), 527–536 (2009)
19. Japkowicz, N.: Concept-learning in the absence of counter-examples: an autoassociation-based approach to classification (1999)
20. Japkowicz, N., Myers, C., Gluck, M.: A novelty detection approach to classification. IJCAI **1**, pp. 518–523 (1995)
21. Johnson, N.A., Lachance, J.: The genetics of sex chromosomes: evolution and implications for hybrid incompatibility. Ann. N. Y. Acad. Sci. **1256**(1), E1–E22 (2012)
22. Khan, S.S., Madden, M.G.: A survey of recent trends in one class classification. In: Irish Conference on Artificial Intelligence and Cognitive Science, pp. 188–197 (2009)
23. Khan, S.S., Madden, M.G.: One-class classification: taxonomy of study and review of techniques. Knowl. Eng. Rev. **29**(3), 345–374 (2014)
24. Kirkpatrick, M., Hall, D.W.: Male-biased mutation, sex linkage, and the rate of adaptive evolution. Evolution **58**(2), 437–440 (2004)
25. Korlach, J., et al.: De novo PacBio long-read and phased avian genome assemblies correct and add to reference genes generated with intermediate and short reads. GigaScience **6**(10), gix085 (2017)
26. Krausz, C., Casamonti, E.: Spermatogenic failure and the Y chromosome. Hum. Genet. **136**(5), 637–655 (2017). https://doi.org/10.1007/s00439-017-1793-8
27. Li, H., Durbin, R.: Fast and accurate short read alignment with Burrows-wheeler transform. Bioinformatics **25**(14), 1754–1760 (2009)
28. Li, H., et al.: The sequence alignment/map format and SAMtools. Bioinformatics **25**(16), 2078–2079 (2009)
29. Li, H.: A statistical framework for SNP calling, mutation discovery, association mapping and population genetical parameter estimation from sequencing data. Bioinformatics **27**(21), 2987–2993 (2011)
30. Liu, F.T., Ting, K.M., Zhou, Z.H.: Isolation forest. In: 2008 Eighth IEEE International Conference on Data Mining, pp. 413–422. IEEE (2008)

31. Tarailo-graovac, M., Chen, N.: Using repeatmasker to identify repetitive elements in genomic sequences. Curr. Protoc. Bioinform. **25**(1), 4 (2009)
32. Moya, M.M., Koch, M.W., Hostetler, L.D.: One-class classifier networks for target recognition applications. NASA STI/Recon Technical Report **93** (1993)
33. Pedregosa, F., et al.: Scikit-learn: Machine learning in Python. J. Mach. Learn. Res. **12**, 2825–2830 (2011)
34. Pigozzi, M., Solari, A.: Germ cell restriction and regular transmission of an accessory chromosome that mimics a sex body in the zebra finch Taeniopygia guttata. Chromosome Res. **6**(2), 105–113 (1998)
35. Pigozzi, M., Solari, A.: The germ-line-restricted chromosome in the zebra finch: recombination in females and elimination in males. Chromosoma **114**(6), 403–409 (2005)
36. Pimentel, M.A., Clifton, D.A., Clifton, L., Tarassenko, L.: A review of novelty detection. Sign. Process. **99**, 215–249 (2014)
37. del Priore, L., Pigozzi, M.I.: Histone modifications related to chromosome silencing and elimination during male meiosis in Bengalese finch. Chromosoma **123**(3), 293–302 (2014). https://doi.org/10.1007/s00412-014-0451-3
38. Ritter, G., Gallegos, M.T.: Outliers in statistical pattern recognition and an application to automatic chromosome classification. Pattern Recogn. Lett. **18**(6), 525–539 (1997)
39. Sayres, M.A.W.: Genetic diversity on the sex chromosomes. Genome Biol. Evol. **10**(4), 1064–1078 (2018)
40. Schölkopf, B., Williamson, R.C., Smola, A.J., Shawe-Taylor, J., Platt, J.C.: Support vector method for novelty detection. In: Advances in Neural Information Processing Systems, pp. 582–588 (2000)
41. Shriram, S., Sivasankar, E.: Anomaly detection on shuttle data using unsupervised learning techniques. In: 2019 International Conference on Computational Intelligence and Knowledge Economy (ICCIKE), pp. 221–225 (2019)
42. Smit, A.F., Hubley, R.: RepeatModeler Open-1 (2008). http://www.repeatmasker.org
43. Tan, Y., Tian, H., Jiang, R., Lin, Y., Zhang, J.: A comparative investigation of data-driven approaches based on one-class classifiers for condition monitoring of marine machinery system. Ocean Eng. **201**, 107174 (2020)
44. Tax, D.: One-class classification; Concept-learning in the absence of counterexamples. Ph. D thesis. Delft University of Technology, ASCI Dissertation Series, 2001 146 (2001)
45. Torgasheva, A.A., et al.: Germline-restricted chromosome (GRC) is widespread among songbirds. Proc. Natl. Acad. Sci. **116**(24), 11845–11850 (2019)
46. Warren, W.C., et al.: The genome of a songbird. Nature **464**, 757 (2010)
47. Wu, R.S., Chung, W.H.: Ensemble one-class support vector machines for content-based image retrieval. Expert Syst. Appl. **36**(3), 4451–4459 (2009)
48. Xing, H.J., Liu, W.T.: Robust adaboost based ensemble of one-class support vector machines. Inf. Fusion **55**, 45–58 (2020)
49. Zhang, J., Kobert, K., Flouri, T., Stamatakis, A.: Pear: a fast and accurate Illumina Paired-End reAd mergeR. Bioinformatics **30**(5), 614–620 (2013)

Explainable and Interpretable Machine Learning

Explaining Sentiment Classification with Synthetic Exemplars and Counter-Exemplars

Orestis Lampridis[1], Riccardo Guidotti[2,3](✉), and Salvatore Ruggieri[2]

[1] Aristotle University of Thessaloniki, Thessaloniki, Greece
lorestis@csd.auth.gr
[2] University of Pisa, Pisa, Italy
{riccardo.guidotti,salvatore.ruggieri}@unipi.it
[3] ISTI-CNR, Pisa, Italy
riccardo.guidotti@isti.cnr.it

Abstract. We present XSPELLS, a model-agnostic local approach for explaining the decisions of a black box model for sentiment classification of short texts. The explanations provided consist of a set of exemplar sentences and a set of counter-exemplar sentences. The former are examples classified by the black box with the same label as the text to explain. The latter are examples classified with a different label (a form of counter-factuals). Both are close in meaning to the text to explain, and both are meaningful sentences – albeit they are synthetically generated. XSPELLS generates neighbors of the text to explain in a latent space using Variational Autoencoders for encoding text and decoding latent instances. A decision tree is learned from randomly generated neighbors, and used to drive the selection of the exemplars and counter-exemplars. We report experiments on two datasets showing that XSPELLS outperforms the well-known LIME method in terms of quality of explanations, fidelity, and usefulness, and that is comparable to it in terms of stability.

Keywords: Explainable sentiment classification · Synthetic exemplars

1 Introduction

Opinions expressed by people in social media are increasingly being collected for several purposes [24]. People look at others' opinions on a product before buying it, on a restaurant or hotel before making a reservation. Managers take decisions supported by consumers' opinions on company brand, products, and services. Public decision makers care for what the citizens in their community want.

The massive amount of online texts (posts, tweets, reviews, etc.) makes it necessary to automate the analyses of such data. *Sentiment classification* is the task of learning a model that is able to predict the sentiment of a given text from labeled examples [31]. These machine learning models are exploited in various applications, e.g., personalization of advertisements, peer suggestion in social

A. Appice et al. (Eds.): DS 2020, LNAI 12323, pp. 357–373, 2020.
https://doi.org/10.1007/978-3-030-61527-7_24

networks, recommendations of news, movies, etc. The analysis of short texts, which abound in micro-blogging sites such as Twitter and in online reviews, is especially challenging, due to their sparsity, non-uniformity, and noisiness. Deep Neural Networks (DNNs) [23,40] and Random Forests (RFs) [6,39], have been shown to be effective in terms of predictive accuracy and robustness to noise. However, the logic learned by a DNN or by a RF to classify a given text remains obscure to human inspection. These inscrutable "black box" models may hide biases learned from data, such as prejudice [2] or spurious correlations [33]. Consequently, they may reproduce and amplify such biases in their predictions [10].

Explainability of black box decisions is nowadays a mandatory requirement [9,11]. Developers need to understand model's decisions for debugging purposes. People subject to black box decisions may inquire to be provided with "meaningful information of the logic involved" (*right to explanation* [26] in the European Union GDPR). For example, if a comment in a social network has been removed because it has been classified as *hate speech*, the author has the right to know *why* the machine learning system has assigned such a label to her comment.

In this paper, we investigate the problem of explaining the decisions of a black box for sentiment classification on a given input (short) text. We design and experiment with a model-agnostic local approach named XSPELLS (eXplaining Sentiment Prediction generating ExempLars in the Latent Space). XSPELLS's explanations for the sentiment $y = b(x)$ assigned by a black box b to a text x consists of set of *exemplar* texts E, a set of *counter-exemplar* texts C, and the most frequent words in each of those sets $W = W_E \cup W_C$. Exemplars are sentences classified by the black box as x and close in meaning to x. They are intended to provide the user with hints about the kind of texts in the neighborhood of x that the black box classifies in the same way as x. Counter-exemplars are sentences that the black box classifies differently from y, but like exemplars, are also close in meaning to x. They are intended to provide the user with hints about the kind of texts in the neighborhood of x that the black box classifies differently from x. The usefulness of *counter-factual reasoning* has been widely recognized in the literature on explainable machine learning [4], particularly as a tool for causal understanding of the behavior of the black box. By contrasting exemplars and counter-exemplars, the user can gain an understanding of the factors affecting the classification of x. To help such an understanding, XSPELLS provides also the most frequent words appearing in E and C.

The main novelty of our approach lies in the fact that the exemplars and counter-exemplars produced by XSPELLS are *meaningful* texts, albeit synthetically generated. We map the input text x from a high-dimensional vector space into a low-dimensional latent space vector z by means of Variational Autoencoders [22], which couple encoding and decoding of texts. Then we study the behavior of the black box b in the neighborhood of z, or, more precisely, the behavior of b on texts decoded back from the latent space. Finally, we exploit a decision tree built from latent space neighborhood instances to drive the selection of exemplars and counter-exemplars. Experiments on two standard datasets and two black box classifiers show that XSPELLS overtakes the baseline method LIME [33] by providing understandable, faithful, useful, and stable explanations.

This paper is organized as follows. Section 2 discusses related work. Section 3 formalizes the problem and recalls key notions for the proposed method, which is described in Sect. 4. Section 5 presents an experimental validation. Finally, Sect. 6 summarizes our contribution, its limitations, and future work.

2 Related Work

Research on interpretability and explainability in machine learning has bloomed over the last few years [17,28]. Explanation methods can be categorized as: *(i) model-specific* or *model-agnostic*, depending on whether or not the approach requires access to the internals of the model; *(ii) local* or *global*, depending on whether the approach explains the prediction for a specific instance or the overall logic of the machine learning model.

XSPELLS, falls into the category of *local*, *model-agnostic* methods which originated with [33] and extended along diverse directions by [12] and by [14,16]. Well known model-agnostic local explanation methods able to also work on textual data include LIME, ANCHOR and SHAP. LIME [33] randomly generates synthetic instances in the neighborhood of the instance to explain. An interpretable linear model is trained from such instances. Feature weights of the linear model are used for explaining the feature importance over the instance to explain. In the case of texts, a feature is associated to each word in a vocabulary. LIME has two main weaknesses. First, the number of top features/words to be considered is assumed to be provided in input by the user. Second, the neighborhood texts are generated by randomly removing words, possibly generating meaningless texts [15]. ANCHOR [34] is developed following principles similar to LIME but it returns decision rules (called anchors) as explanations. It adopts a bandit algorithm that randomly constructs anchors with predefined minimum precision. Its weaknesses include the discretization of continuous features, the need for user-defined precision threshold parameters, and, as for LIME, the usage of meaningless synthetic instances. SHAP [25] relates game theory with local explanations and overcomes some of the limitations of LIME and ANCHOR. Also SHAP audits the black box with possibly meaningless synthetic sentences. The method XSPELLS proposed in this paper recovers from this drawback by generating the sentences for the neighborhood in a latent space by resorting to Variational Autoencoders.

LIONETS, DEEPLIFT and NEUROX are model-specific local explanation methods designed to explain deep neural networks able to work also on textual data. DEEPLIFT [36] decomposes the prediction of neural networks on a specific input by back-propagating the contributions of all neurons in the network to the input features. Then it compares the activation of each neuron to its "reference activation" and it assigns contribution scores according to the difference. NEUROX [7] facilitates the analysis of individual neurons in DNNs. In particular, it identifies specific dimensions in the vector representations learned by a neural network model that are responsible for specific properties. Afterwards, it allows the ranking of neurons and dimensions based on their overall saliency. Finally, LIONETS [29] looks at the penultimate layer of a DNN, which models texts in an alternative representation, randomly permutes the weights of nodes in that layer to generate new vectors, classifies them, observes the classification outcome and returns the explanation using a linear regressor like LIME. Differently from these model-specific methods, XSPELLS is not tied to a specific architecture and it can be used to explain any black box sentiment classifier.

3 Setting the Stage

We address the *black box outcome explanation problem* [17] in the domain of sentiment classification, where machine learning classifiers are trained to predict the class value (sentiment) of a natural language text (simply, a text). We will mainly consider short texts such as posts on social networks, brief reviews, or single sentences, as these are typically the subject of sentiment classification. In this context, a black box model is a non-interpretable or inaccessible sentiment classifier b which assigns a sentiment label y to a given text x, i.e., $b(x) = y$. Example of black box models include Random Forests (RF) and Deep Neural Networks (DNN). We assume that the black box b can be queried at will. We use the notation $b(X)$ as a shorthand for $\{b(x) \mid x \in X\}$. Formally, we have:

Definition 1. *Let b be a black box sentiment classifier, and x a text for which the decision $y = b(x)$ has to be explained. The* black box outcome explanation problem for sentiment classification *consists of providing an explanation $\xi \in \Xi$ belonging to a human-interpretable domain Ξ.*

We introduce next the key tools that will be used in our approach.

3.1 Factual and Counter-Factuals

A widely adopted human-interpretable domain Ξ consists of *if-then* rules. They provide conditions (in the if-part) met by the instance x to be explained, that determined the answer of the black box (then-part). Rules can also be used to provide *counter-factuals*, namely alternative conditions, not met by x, that would determine a different answer by the black box [4]. In our approach, we will build on LORE [14], a local explainer for *tabular data* that learns a decision tree from a given neighborhood Z of the instance to explain. Such a tree is a *surrogate* model of the black box, i.e., it is trained to reproduce the decisions of the black box. LORE provides in output: *(i)* a *factual* rule r, corresponding to the path in the surrogate tree that explains why an instance x has been labeled as y by the black box b; and *(ii)* a set of *counter-factual* rules Φ, explaining minimal changes in the features of x that would change the class y assigned by b. In LORE, the neighborhood Z is synthetically generated using a genetic algorithm that balances the number of instances similar to x and with its same label y, and the number of instances similar to x but with a different label $y' \neq y$ assigned by b.

Algorithm 1: XSPELLS(x, b, ζ, η)

Input : x - text to explain, b - black box, ζ - encoder, η - decoder
Output: ξ - explanation

```
1 z ← ζ(x);                       // encode text into the latent space
2 Z ← neighgen(z, b, ζ, η);       // generate latent neighborhood
3 Z̃ ← η(Z);                       // decode neighborhood
4 Y ← b(Z̃);                       // classify neighborhood
5 ldt ← learnTree(Z, Y);          // learn latent surrogate decision tree
6 r ← rule(z, ldt);               // extract factual latent rule
7 E, C ← explCexpl(r, Z, Z̃, Y);   // select exemplars and counter-exemplars
8 W ← mostCommon(E, C);           // extract most common words
9 return ξ = ⟨E, C, W⟩;           // return explanation
```

3.2 Variational Autoencoder

Local explanation methods audit the behavior of a black box in the neighborhood of the instance to explain. A non-trivial issue with textual data is how to generate *meaningful* synthetic sentences in the neighborhood (w.r.t. semantic similarity) of the instance. We tackle this problem by adopting Variational Autoencoders (VAEs) [22]. A VAE is trained with the aim of learning a representation that reduces the dimensionality from the large m-dimensional space of words to a small k-dimensional space of numbers (*latent space*), also capturing non-linear relationships. An *encoder* ζ, and a decoder *decoder* η are simultaneously learned with the objective of minimizing the *reconstruction loss.* Starting from the reduced encoding $z = \zeta(x)$, the VAE reconstructs a representation as close as possible to its original input $\tilde{x} = \eta(z) \simeq x$. After training, the *decoder* can be used with generative purposes to reconstruct instances never observed by generating vectors in the latent space of dimensionality k. The difference with standard autoencoders [19] is that VAEs are trained by considering an additional limitation on the loss function such that the latent space is scattered and does not contain "dead zones". Indeed, the name *variational* comes from the fact that VAEs work by approaching the posterior distribution with a variational distribution. The encoder ζ emits the parameters for this variational distribution, in terms of a multi-factorial Gaussian distribution, and the latent representation is taken by sampling this distribution. The decoder η takes as input the latent representation and focuses on reconstructing the original input from it. The avoidance of dead zones ensures that the instances reconstructed from vectors in the latent space, e.g., posts or tweets, are semantically meaningful [3].

4 Explaining Sentiment Classifiers

We propose a local model agnostic explainer for sentiment classification of short texts, called XSPELLS (*eXplaining Sentiment Prediction generating ExempLars in the Latent Space*). Given a black box b, a short text x, e.g., a post on a

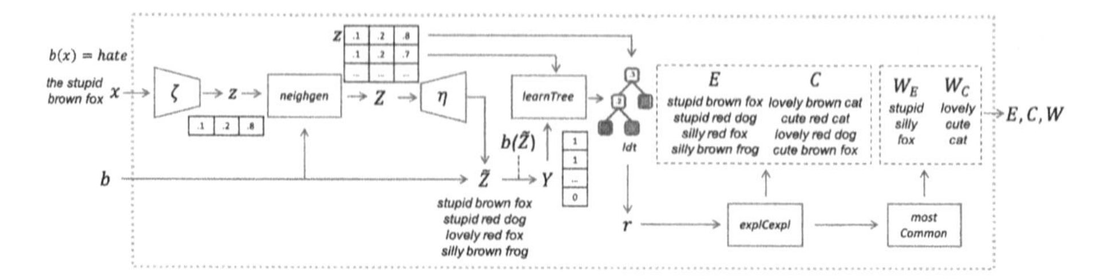

Fig. 1. XSPELLS process on a sample input. XSPELLS takes as input the short text x and the sentiment assigned $b(x)$. The output is a set of exemplars and counter-exemplars, and the most common discriminative words.

social network, and the sentiment label $y = b(x)$ assigned by the black box, e.g., *hate* or *neutral*, the explanation provided by XSPELLS is composed of: *(i)* a set of *exemplar* texts; *(ii)* a set of *counter-exemplar* texts; and, *(iii)* the set of *most common words* in exemplars and counter-exemplars. Exemplar and counter-exemplar texts respectively illustrate instances classified with the same and with a different label than x. Such texts are close in meaning to x, and they offer an understanding of what makes the black box determine the sentiment of texts in the neighborhood of x. Exemplars help in understanding reasons for the sentiment assigned to x. Counter-exemplars help in understanding reasons that would reverse the sentiment assigned. The most common words in the exemplars and counter-exemplars may allow for highlighting terms (not necessarily appearing in x) that discriminate between the assigned sentiment and a different sentiment. These components form the human-interpretable explanation $\xi \in \Xi$ for the classification $y = b(x)$ returned by XSPELLS, whose aim is to satisfy the requirements of counter-factuability, usability, and meaningfulness [4,28,32].

Besides the black box b and the text x to explain, XSPELLS is parametric in: an encoder ζ and a decoder η for representing texts in a compact way in the latent space. Algorithm 1 details XSPELLS, and Fig. 1 shows the steps of the explanation process on a sample input. First, x is transformed into a low-dimensionality vector $z = \zeta(x)$ in the latent space. XSPELLS then generates a neighborhood Z of z, which is decoded back to a set of texts $\tilde{Z}$. The dataset Z and the decisions of the black box on the decoded text $Y = b(\tilde{Z})$ are used to train a surrogate decision tree (in the latent space).

Then, the *explCexpl*() module selects exemplars E and counter-exemplars C from Z by exploiting the knowledge extracted (i.e., the decision tree branches), and decodes them into texts. Finally, the most common words $W = W_E \cup W_C$ are extracted from E and C and the overall explanation ξ is returned. Details of each step are presented in the rest of this section.

4.1 Latent Encoding and Neighborhood Generation

The input text x is first passed to a trained VAE ζ (line 1 of Algorithm 1), thus obtaining the latent space representation $z = \zeta(x)$. The number of latent dimensions k is kept low to avoid dimensionality problems. We capture the sequential information in texts by adopting VAEs based on long short-term memory layers (LSTM) [20] for both the encoder ζ and decoder η (lines 1 and 3). In particular, the decoder η is trained to predict the next characters of the text, given the previous characters of the text. In more detail, it is trained to convert a given text into the same text, but being offset by a time-step in the future.

XSPELLS generates a set Z of n instances in the latent feature space for a given z. The neighborhood generation function *neighgen* (line 2) can be implemented by adopting several different strategies, ranging from a purely random approach like in LIME [33], to using a given distribution and a genetic algorithm maximizing a fitness function like in LORE [14]. XSPELLS adopts a random generation of latent synthetic instances by relying on the fact that the encoder maps uniformly the data distribution over the latent space. XSPELLS guarantees a minimum number n of distinct instances by removing duplicates. Next, XSPELLS uses the synthetically generated instances $\tilde{Z}$ for querying the black box b (line 4). This is made possible by turning back the latent representation to text through the decoder η [3] (line 3). We tackle the requirement of generating *local* instances by randomly generating $N \gg n$ latent instances, and then retaining in Z only the n closest instances to z, i.e., $|Z| = n$. The distance used in the latent space is the Euclidean distance. The neighborhood generation *neighgen* actually returns a set $Z = Z_= \cup Z_{\neq}$ with $z' \in Z_=$ such that $b(\eta(z')) = b(\eta(z))$, and instances $z' \in Z_{\neq}$ such that $b(\eta(z')) \neq b(\eta(z))$. We further consider the problem of imbalanced distributions in Z, which may lead to weak decision trees. Class balancing between the two partitions is achieved by adopting the SMOTE [5] procedure if the proportion of the minority class is less than a predefined threshold τ.

4.2 Local Latent Rules and Explanation Extraction

Given Z and $Y = b(\tilde{Z})$, XSPELLS builds a latent decision tree *ldt* (line 5) acting as a local surrogate of the black box, i.e., being able to locally mime the behavior of b. XSPELLS adopts decision tree because decision rules can be derived from a root-to-leaf path [14]. Indeed, the premise p of the rule $r = p \rightarrow y$ is the conjunction of the split conditions from the root to the leaf of the tree that is followed by features in z. This approach is a variant of LORE (see Sect. 3.1) but in a latent feature space. The consequence y of the rule is the class assigned at that leaf[1].

Given a text x, the explanations returned by XSPELLS are of the form $\xi = \langle E, C, W \rangle$, where: $E = \{e_1^x, \ldots, e_u^x\}$ is the set of *exemplars* ($b(e_i^x) = b(x)\ \forall i \in [1, u]$); $C = \{c_1^x, \ldots, c_v^x\}$ is the set of *counter-exemplars* ($b(c_i^x) \neq b(x)\ \forall i \in [1, v]$); and $W = W_E \cup W_C$ is the set of the h most frequent words in exemplars E and of the h most frequent words in counter-exemplars C. Here, u, v, and h are

[1] In theory, it might happen that $y \neq b(x)$, namely the path followed by z predicts a sentiment different from $b(x)$. In our experiments, this never occurred. In such cases, XSPELLS restarts by generating a new neighborhood and then a new decision tree.

parameters that can be set in XSPELLS. Exemplars are chosen starting from the latent instances in Z which satisfy both the premise p and the consequence y of the rule $r = p \rightarrow y$ above, namely the instances $z' \in Z$ that follow the same path as z in the decision tree, and such that the $b(\eta(z')) = y$. The u instances z' closest to z are selected, using Euclidean distance. They are decoded back to the text space $\eta(z')$ and included in E. Counter-exemplars are chosen starting from the latent instances $z' \in Z$ which do not satisfy the premise p and such that $b(\eta(z')) \neq b(x)$. The v instances closest to z are chosen. They are decoded back to the text space $\eta(z')$ and included in C.

5 Experiments

In this section, we illustrate qualitative/quantitative experimental analyses of faithfulness, usefulness, and stability properties of XSPELLS explanations[2]. The XSPELLS system has been developed in Python, and it relies on the CART decision tree algorithm as implemented by the `scikit-learn` library, and on VAEe implemented with the `keras` library[3].

5.1 Experimental Settings

We experimented with the proposed approach on two datasets of tweets. The *hate speech dataset* (`hate`) [8] contains tweets labeled as hate, offensive or neutral. Here, we focus on the 1,430 tweets that belong to the *hate* class, and on the 4,163 tweets of the *neutral* class. The *polarity dataset* (`polarity`) [30] contains tweets about movie reviews. Half of these tweets are classified as *positive* reviews, and the other half as *negative* ones. These two datasets are remarkable examples where a black box approach is likely to be used to remove posts or to ban users, possibly in automated way. Such extreme actions risk to hurt the free speech rights of people. Explanations of the black box decision are then of primary relevance both to account for the action and to test/debug the black box.

For both datasets, we use 75% of the available data for training a black box machine learning classifier. The remaining 25% of data is used for testing the black box decisions. More specifically, 75% of that testing data is used for training the autoencoder, and 25% for explaining black box decisions (*explanation set*). Datasets details are reported in Table 1 (left).

We trained and explained the following black box classifiers: Random Forest [38] (RF) as implemented by the `scikit-learn` library, and Deep Neural Networks (DNN) implemented with the `keras` library. For the RF, we transformed texts into their TF-IDF weight vectors [38], after removing stop-words, including Twitter stop-words such as "rt", hashtags, URLs and usernames. A randomized cross-validation search was then performed for parameter tuning. Parameters for RF models were set as follows: 100 decision trees, *Gini* split criterion, $\sqrt{m}$ random features where m is the total number of features; no limit on

[2] The source code is available at: *https://github.com/orestislampridis/X-SPELLS*.
[3] https://scikit-learn.org/stable/modules/tree.html, https://keras.io.

Table 1. Datasets description, black box models accuracy, and VAE RMSE.

Dataset	No. tweets	Avg. no words	No. classes	Bb train size	VAE train size	Expl. size	Accuracy		VAE
							RF	DNN	MRE
`hate`	5,593	20.82	2	4,195	1,048	350	.9257	.8485	0.26
`polarity`	10,660	24.87	2	7,995	1,998	666	.6702	.6302	0.59

tree depth. The DNNs adopted have the following architecture. The first layer is a dense embedding layer. It takes as input a sparse vector representation of each text (subject to same pre-processing steps as for the RF, without the TF-IDF representation) obtained by using a Keras tokenizer[4] to turn the text into an array of integers and a padder so that each vector has the same length. This way, we allow the network to learn its own dense embeddings of size 64. The first embedding layer is followed by a dropout layer at 0.25. Afterwards, the DNN is composed by three dense layers with sizes 64, 512 and 128. The central layer is an LSTM [20] that captures the sequential nature of texts and has size 100. After that, there are three dense layers with sizes 512, 64 and 32. The dense layers adopt the *ReLu* activation function. Finally, the *sigmoid* activation function is used for the final classification. We adopted *binary cross-entropy* as loss function and the *Adam* optimizer. We trained the DNN for 100 epochs. Classification performances are reported in Table 1 (center-right).

We designed the VAEs used in experiments with both the encoder ζ and the decoder η consisting of a single LSTM layer. We fed the text into the VAE using a one-hot vectorization that takes an input tensors with dimensions $33 \cdot 5368 = 177,144$ for the `hate` dataset, and $48 \cdot 5308 = 254,784$ for the `polarity` dataset, after stop-words removal. The numbers above represent the maximum text length and the number of distinct words considered. In order to provide to the VAE knowledge also about unseen words with respect to those in its training set, we extended the vocabulary with the 1000 most common English words[5] We considered $k = 500$ latent features for both datasets[6]. Table 1 (right) reports the *Mean Reconstruction Error* (MRE) calculated as the average cosine similarity distance between the original and reconstructed texts when converted to TF-IDF vectors. We set the following XSPELLS hyper-parameters. The neighborhood generation *neighgen* is run with $N = 600$, $n = 200$, $\tau = 40\%$. For the latent decision tree we used the default parameter of the CART implementation. Finally, with regards to the explanation hyper-parameters, we set $u = v = 5$ (counter-)exemplars, and $h = 5$ most frequent words for exemplars and for counter-exemplars.

[4] https://keras.io/preprocessing/text.

[5] https://1000mostcommonwords.com.

[6] Experiments (not reported due to lack of space) show that $k = 500$ is a good compromise between MRE and the reduced dimensionality of the latent space when varying $k \in \{100, 250, 500, 1000, 2500\}$.

In the experiments we compare XSPELLS against LIME [33]. We cannot compare against SHAP [25] and ANCHOR [34] because it is not immediate how to practically employ them to explain sentiment classifiers. Other approaches such as IntGrad [37] or LRP [1] could theoretically be used to explain sentiment classifiers. However, first, they are not agnostic but tied to DNNs, and second, they are typically used for explaining image classifiers.

5.2 Qualitative Evaluation

In this section, we qualitatively compare XSPELLS explanations with those returned by LIME. Tables 2 and 3 show sample explanations for both experimental datasets, and considering the RF black box sentiment classifier.

The first and second tweet in Table 2 belong to the `hate` dataset and are classified as *hate*. Looking at the exemplars returned by XSPELLS, the *hate* sentiment emerges from the presence of the word "hate", from sexually degrading references, and from derogatory adjectives. On the other hand counter-exemplars refer to women and to work with a positive perspective. The second tweet for the `hate` dataset follows a similar pattern. The focus this time is on the word "retard", used here with negative connotations. Differently from XSPELLS, the explanations returned by LIME in Table 3 for the same tweets show that the `hate` sentiment is mainly due to the words "faggot" and "retards" but there are not any further details, hence providing to the user a limited understanding.

The usefulness of the exemplars and counter-exemplars of XSPELLS are even more clear for the `polarity` dataset, where the RF correctly assigns the sentiment *negative* to the sample tweets in Table 2. For the first tweet, XSPELLS recognizes the negative sentiment captured by the RF and provide exemplars containing negative words such as "trash", "imperfect", and "extremely unfunny" as negative synonyms of "eccentric", "forgettable", and "doldrums". The counter-exemplars show the positive connotation and context that words must have to turn the sentiment into *positive*. On the contrary, LIME (Table 3) is not able to capture such complex words and it focuses on terms like "off", "debut", or "enough". For the second tweet, XSPELLS is able to generates exemplar similar in meaning to the tweet investigated: the tweet starts positive (or appear positive), but reveals/hides a negative sentiment in the end. In this case the most frequent words alone are not very useful. Indeed, (the surrogate linear classifier of) LIME mis-classifies the second tweet as positive giving importance to the word "work" that, however, is not the focus of the negative sentiment.

Overall, since LIME extracts words from the text under analysis, it can only provide explanations using such words. On the contrary, the (counter-)exemplars of XSPELLS consist of texts which are close in meaning, but including different wordings that help the user better grasp the reasons behind black box decision.

5.3 Fidelity Evaluation

We evaluate the *faithfulness* [11,17] of the surrogate latent decision tree adopted by XSPELLS by measuring how well it reproduces the behavior of the black box b

Table 2. Explanations returned by XSPELLS for texts classified as *hate* in the `hate` dataset, and as *negative* in the `polarity` dataset. Three exemplars (E) and two counter-exemplars (C) for each tweet. Relative word frequencies in parenthesis.

	Tweet	(Counter-)exemplars	E/C	$W_{=}$	$W_{\neq}$
`Hate`	I dont have any problems with zak, but you seem like a faggot	I hate dumb bitches	E	Hate (.22)	Work (.06)
		I hate fat bitches wear show	E	Bitches (.17)	Love (.06)
		I hate fat bitches	E	Fat (.11)	Wearing (.06)
		This is why i work	C	Dumb (.06)	Fuzzy (.06)
		I really want a girl	C	Wear (.06)	Blankets (.06)
`hate`	California's biggest retards. Don't forget about HOLY who just released an amazing EP	This girl is retarded	E	Retarded (.08)	Im (.14)
		The fucking royals bitch work	E	Hated (.08)	Love (.07)
		Im such a retard sometimes	E	Bitch (.08)	Birds (.07)
		This is why i love birds	C	Fucking (.08)	Brownies (.07)
		Wait did take my brownies	C	Retard (.08)	Sorry (.07)
`polarity`	Eccentric enough to stave off doldrums, caruso's self-conscious debut is also eminently forgettable	It has ever under trash without to a familiar	E	Trash (.05)	Fun (.10)
		This extremely unfunny movie in at 80 min	E	Imperfect (.05)	Remarkable (.07)
		This movie makes for one thing imperfect	E	Unfunny (.05)	Appears (.07)
		A story of musical and character and love	C	Without (.05)	Want (.04)
		It is a movie fun for fans who cant stop	C	Ever (.05)	Love (.04)
`polarity`	While some of the camera work is interesting, the film's mid-to-low budget is betrayed by the surprisingly shoddy makeup work	In the end i kept this one at two stars	E	Bad (.07)	New (.12)
		Odd poetic road movie spiked by jolts of pop	E	Attempt (.07)	Really (.06)
		In attempt to the bad sense with this summer	E	End (.07)	Safe (.03)
		Does what a fine documentary does best	C	Sense (.04)	Fine (.03)
		A film that plays things so nice n safe	C	Odd (.04)	Safe (.03)

Table 3. Explanations returned by LIME for tweets classified as *hate* in the `hate` dataset, and as *negative* in the `polarity` dataset. LIME word importance in parenthesis.

	Tweet	Top features		Tweet	Top features
`hate`	I dont have any problems with zak, but you seem like a faggot	Faggot (−0.62) You (−0.03) Like (0.01) Any (−0.01) Problems (0.01)	`polarity`	Eccentric enough to stave off doldrums, caruso's self-conscious debut is also eminently forgettable	off (−0.30) Debut (0.03) Enough (0.03) Also (0.03) Self (−0.01)
`hate`	California's biggest retards. Don't forget about HOLY who just released an amazing EP	Retards (−0.24) Dont (−0.03) California (−0.01) Who (−0.01) Holy (0.01)	`polarity`	While some of the camera work is interesting, the film's mid-to-low budget is betrayed by the surprisingly shoddy makeup work	Work (0.11) While (0.04) low (−0.04) Some (−0.04) Interesting (−0.03)

Table 4. Mean and standard deviation of fidelity. The higher the better.

	RF		DNN	
	LIME	XSPELLS	LIME	XSPELLS
`hate`	0.62 ± 0.30	0.98 ± 0.01	0.92 ± 0.15	0.98 ± 0.01
`polarity`	0.89 ± 0.14	0.98 ± 0.01	0.91 ± 0.20	0.97 ± 0.01

in the neighborhood of the text x to explain – a metric known as *fidelity*. Let Z be the neighborhood of x in the latent space generated at line 2 of Algorithm 1 and *ldt* be the surrogate decision tree computed at line 5. The fidelity metric is $|\{y \in Z \mid ldt(y) = b(\eta(y))\}|/|Z|$, namely the accuracy of *ldt* assuming as ground truth the black box. The fidelity values over all instances in the explanation set are aggregated by taking their average and standard deviation.

We compare XSPELLS against LIME, which adopts as surrogate model a linear regression over the feature space of words and generates the neighborhood using a purely random strategy. Table 4 reports the average fidelity and its standard deviation. On the `hate` dataset, XSPELLS reaches almost perfect fidelity for both black boxes. LIME performances are markedly lower for the RF black box. On the `polarity` dataset, the difference is less marked, but still in favor of XSPELLS. A Welch's t-test shows that the difference of fidelity between XSPELLS and LIME is statistically significant (p-value < 0.01) in all cases from Table 4.

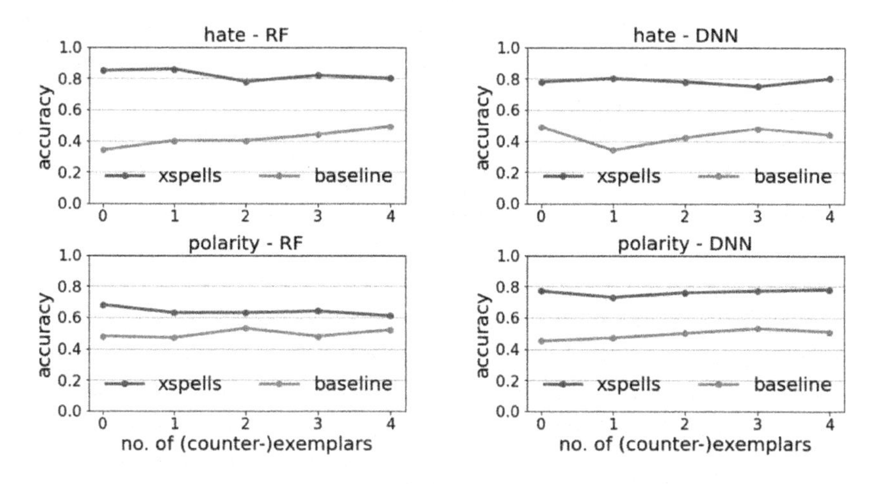

Fig. 2. Usefuless as 1-NN accuracy varying the number of (counter-)exemplars.

5.4 Usefulness Evaluation

How can we evaluate the usefulness of XSPELLS explanations? The gold standard would require to run lab experiments involving human evaluators. Inspired by [21], we provide here an indirect evaluation by means of a k-Nearest Neighbor (k-NN) classifier [38]. For a text x in the explanation set, first we randomly select n exemplars and n counter-exemplars from the output of XSPELLS. Then, a 1-NN classifier[7] is trained over such (counter-)exemplars. Finally, we test 1-NN over the text x and compare the prediction of 1-NN with the sentiment $b(x)$ predicted by the black box. In other words, the 1-NN approximates a human in assessing the (counter-)exemplars usefulness. The accuracy computed over all x's in the explanation set is a proxy measure of how good/useful are (counter-)exemplars at delimiting the decision boundary of the black box. We compare such an approach with a *baseline* (or null) model consisting of a 1-NN trained on n texts per sentiment, selected randomly from the training set and not including x.

The accuracy of the two approaches are reported in Fig. 2 by varying the number n of exemplars and counter-exemplars. XSPELLS neatly overcomes the *baseline*. The difference is particularly marked for when n is small. Even though the difference tend to decrease for large n's, large-sized explanations are less useful in practice due to cognitive limitations of human evaluators. Moreover, XSPELLS performances are quite stable w.r.t. n, i.e., even one or two exemplars and counter-exemplars are sufficient to let the 1-NN classifier distinguish the sentiment assigned to x in an accurate way.

[7] Distance function adopted: cosine distance between the TF-IDF representations.

Table 5. Mean and stdev of the coherence index C_x. The closer to 1 the better.

	RF		DNN	
	LIME	XSPELLS	LIME	XSPELLS
hate	1.10 ± 0.17	1.05 ± 0.25	1.06 ± 0.08	1.12 ± 0.39
polarity	1.05 ± 0.15	1.15 ± 0.20	1.13 ± 0.18	1.09 ± 0.14

5.5 Stability Evaluation

Stability of explanations is a key requirement, which heavily impacts users' trust on explainability methods [35]. Several metrics of stability can be devised [18,27]. A possible choice is to use sensitivity analysis with regard to how much an explanation varies on the basis of the randomness in the explanation process. Local methods relying on random generation of neighborhoods are particularly sensitive to this problem. In addition, our method suffers of the variability introduced by the encoding-decoding of texts in the latent space. Therefore, we measure here stability as a relative notion, that we call *coherence*. For a given text x in the explanation set, we consider its closest text x^c and its k-th closest text x^f, again in the explanation set. A form of Lipschitz condition [27] would require that the distance between the explanations $e(x)$ and $e(x^f)$, normalized by the distance between x and x^f, should not be much different than the distance between the explanations $e(x)$ and $e(x^c)$, again normalized by the distance between x and x^c. Stated in words, normalized distances between explanations should be as similar as possible. Formally, we introduce the following *coherence index*:

$$C_x = \frac{dist_e(e(x^f), e(x))/dist(x^f, x)}{dist_e(e(x^c), e(x))/dist(x^c, x)}$$

where we adopt as distance function $dist$ the cosine distance between the TF-IDF representation of the texts, and as distance function $dist_e$ the Jaccard distance between the 10 most frequent words in each explanation (namely, the W set). In experiments, we set x^f to be the k = 10-closest text w.r.t. x. For comparison, the coherence index is computed also for LIME, with Jaccard similarity calculated between the sets of 10 words (a.k.a. features) that LIME deems more relevant.

Table 5 reports the average coherence over the explanation set. XSPELLS and LIME have comparable levels of coherence, and an even number of cases where one overcomes the other. A Welch's t-test shows that the difference of the coherence indexes between XSPELLS and LIME is statistically significant (p-value < 0.01) in only one case, namely for the polarity dataset and RF black box model.

6 Conclusion

We have presented XSPELLS, a local model-agnostic explanation approach for black box sentiment classifiers. The key feature of XSPELLS is the adoption of variational autoencoders for generating meaningful synthetic texts from a latent space. Such a space reveals essential also for inducing a decision tree which helps in characterizing exemplar and counter-factual exemplar texts. The approach advances over baseline explainers, such as LIME, which only highlight the contribution of words already in the text to explain. Experiments showed that XSPELLS also exhibits better fidelity and usefulness, and comparable stability.

The proposed approach has some clear limitations. *First*, performance is strictly dependent on the VAE adopted: a better autoencoder would lead to more realistic exemplars and counter-exemplars. The structure of the autoencoder needs then to be further explored and evaluated beyond the specific one adopted in this paper. This may also require trading-off quality with computational costs, which may slow down the response time of XSPELLS. *Second*, we will consider extending the explanations returned by XSPELLS with logic rules, which convey information at a more abstract level than exemplars. Such rules can be extracted from the decision tree on the latent space, but have to decoded back to rules on texts – a challenging task. *Third*, XSPELLS could be extended to account for long texts, e.g., by adopting word2vec embeddings [13] for modeling the input/output of the VAE. *Fourth*, we could rely on linguistic resources, such a thesaurus or domain ontologies, to empower both synthetic text generation and to enrich the expressiveness of the (counter-)exemplars. *Fifth*, a human evaluation of XSPELLS would be definitively required, e.g., through crowdsourcing experiments.

Acknowledgements. This work is partially supported by the European Community H2020 programme under the funding schemes: G.A. 871042 *SoBigData++*, (sobigdata.eu), G.A. 952026 *Humane AI-Net*, (humane-ai.eu), G.A. 825619 *AI4EU*, (ai4eu.eu), and G.A. 860630 *NoBIAS* (nobias.eu). The first author would like to thank Ioannis Mollas and Grigorios Tsoumakas for their support.

References

1. Bach, S., et al.: On pixel-wise explanations for non-linear classifier decisions by layer-wise relevance propagation. PloS One **10**(7), e0130140 (2015)
2. Bolukbasi, T., et al. Man is to computer programmer as woman is to homemaker? Debiasing word embeddings. In: NIPS, pp. 4349–4357 (2016)
3. Bowman, S., et al.: Generating sentences from a continuous space. 1511.06349 (2015)
4. Byrne, R.M.: Counterfactuals in explainable artificial intelligence (XAI): evidence from human reasoning. In: IJCAI, pp. 6276–6282 (2019)
5. Chawla, N., et al.: SMOTE: synthetic minority over-sampling technique. JAIR **16**, 321–357 (2002)
6. Da Silva, N.F., Hruschka, E.R., Hruschka Jr., E.R.: Tweet sentiment analysis with classifier ensembles. Decis. Support Syst. **66**, 170–179 (2014)
7. Dalvi, F., et al.: What is one grain of sand in the desert? Analyzing individual neurons in deep NLP models. In: AAAI, vol. 33, pp. 6309–6317 (2019)
8. Davidson, T., et al.: Automated hate speech detection and the problem of offensive language. CoRR, abs/1703.04009 (2017)
9. Doshi-Velez, F., Kim, B.; Towards a rigorous science of interpretable machine learning. arXiv:1702.08608 (2017)

10. Ntoutsi, E., et al.: Bias in data-driven AI systems - an introd. survey. DMKD (2020)
11. Freitas, A.A.: Comprehensible classification models: a position paper. ACM SIGKDD Explor. **15**(1), 1–10 (2014)
12. Frosst, N., et al.: Distilling a neural network. 1711.09784 (2017)
13. Goldberg, Y., Levy, O.: word2vec explained: deriving Mikolov et al'.s negative-sampling word-embedding method. arXiv:1402.3722 (2014)
14. Guidotti, R., et al.: Factual and counterfactual explanations for black box decision making. IEEE Intell. Syst. **34**, 14–23 (2019)
15. Guidotti, R., Monreale, A., Cariaggi, L.: Investigating neighborhood generation methods for explanations of obscure image classifiers. In: Yang, Q., Zhou, Z.-H., Gong, Z., Zhang, M.-L., Huang, S.-J. (eds.) PAKDD 2019. LNCS (LNAI), vol. 11439, pp. 55–68. Springer, Cham (2019). https://doi.org/10.1007/978-3-030-16148-4_5
16. Guidotti, R., Monreale, A., Matwin, S., Pedreschi, D.: Black box explanation by learning image exemplars in the latent feature space. In: Brefeld, U., Fromont, E., Hotho, A., Knobbe, A., Maathuis, M., Robardet, C. (eds.) ECML PKDD 2019. LNCS (LNAI), vol. 11906, pp. 189–205. Springer, Cham (2020). https://doi.org/10.1007/978-3-030-46150-8_12
17. Guidotti, R., Monreale, A., Ruggieri, S., Turini, F., et al.: A survey of methods for explaining black box models. CSUR **51**(5), 93 (2019)
18. Guidotti, R., Ruggieri, S.: On the stability of interpretable models. In: IJCNN, pp. 1–8. IEEE (2019)
19. Hinton, G.E., Salakhutdinov, R.R.: Reducing the dimensionality of data with neural networks. Science **313**(5786), 504–507 (2006)
20. Hochreiter, S., et al.: Long short-term memory. NC **9**(8), 1735–1780 (1997)
21. Kim, B., et al.: Examples are not enough, learn to criticize! In: NIPS (2016)
22. Kingma, D.P., et al.: Auto-encoding variational bayes. arXiv:1312.6114 (2013)
23. Korde, V., Mahender, C.N.: Text classification and classifiers: a survey. Int. J. Artif. Intell. Appl. **3**(2), 85 (2012)
24. Liu, B., Zhang, L.: A survey of opinion mining and sentiment analysis. In: Aggarwal, C., Zhai, C. (eds.) Mining text data, pp. 415–463. Springer, Boston (2012). https://doi.org/10.1007/978-1-4614-3223-4_13
25. Lundberg, S.M., Lee, S.-I.: A unified approach to interpreting model predictions. In: Advances in Neural Information Processing Systems, pp. 4765–4774 (2017)
26. Malgieri, G., Comandé, G.: Why a right to legibility of automated decision-making exists in the GDPR. Int. Data Privacy Law **7**(4), 243–265 (2017)
27. Melis, D.A., Jaakkola, T.: Towards robust interpretability with self-explaining neural networks. In: NIPS (2018)
28. Miller, T.: Explanation in artificial intelligence: insights from the social sciences. Artif. Intell. **267**, 1–38 (2019)
29. Mollas, I., Bassiliades, N., Tsoumakas, G.: LioNets: local interpretation of neural networks through penultimate layer decoding. In: Cellier, P., Driessens, K. (eds.) ECML PKDD 2019. CCIS, vol. 1167, pp. 265–276. Springer, Cham (2020). https://doi.org/10.1007/978-3-030-43823-4_23
30. Pang, B., et al.: Seeing stars: exploiting class relationships for sentiment categorization with respect to rating scales. In: ACL, USA, pp. 115–124 (2005)
31. Pang, B., et al.: Opinion mining and sentiment analysis. FTIR **2**(1–2), 1–135 (2008)
32. Pedreschi, D., et al.: Meaningful explanations of black box AI decision systems. In: AAAI, vol. 33, pp. 9780–9784 (2019)

33. Ribeiro, M.T., Singh, S., Guestrin, C.: Why should I trust you?: explaining the predictions of any classifier. In: KDD, pp. 1135–1144. ACM (2016)
34. Ribeiro, M.T., Singh, S., Guestrin, C.: Anchors: high-precision model-agnostic explanations. In AAAI (2018)
35. Rudin, C.: Stop explaining black box machine learning models for high stakes decisions and use interpretable models instead. NMI **1**(5), 206–215 (2019)
36. Shrikumar, A., et al.: Learning important features through propagating activation differences. In: ICML, pp. 3145–3153. JMLR. org (2017)
37. Sundararajan, M., et al.: Axiomatic attribution for deep networks. In: ICML, pp. 3319–3328. JMLR. org (2017)
38. Tan, P.-N., et al.: Introduction to Data Mining. Pearson Education (2016)
39. Xu, B., Guo, X., Ye, Y., Cheng, J.: An improved random forest classifier for text categorization. JCP **7**(12), 2913–2920 (2012)
40. Zhang, X., Zhao, J., LeCun, Y.: Character-level convolutional networks for text classification. In: NIPS, pp. 649–657 (2015)

Generating Explainable and Effective Data Descriptors Using Relational Learning: Application to Cancer Biology

Oghenejokpeme I. Orhobor[1](✉), Joseph French[5], Larisa N. Soldatova[2], and Ross D. King[1,3,4]

[1] Department of Chemical Engineering and Biotechnology, University of Cambridge, Cambridge CB3 0AS, UK
oo288@cam.ac.uk
[2] Department of Computing, Goldsmiths, University of London, London SE14 6AD, UK
[3] The Alan Turing Institute, London NW1 2DB, UK
[4] Department of Biology and Biological Engineering, Chalmers University of Technology, 412 96 Gothenburg, Sweden
[5] Manchester, UK

Abstract. The key to success in machine learning is the use of effective data representations. The success of deep neural networks (DNNs) is based on their ability to utilize multiple neural network layers, and big data, to learn how to convert simple input representations into richer internal representations that are effective for learning. However, these internal representations are sub-symbolic and difficult to explain. In many scientific problems explainable models are required, and the input data is semantically complex and unsuitable for DNNs. This is true in the fundamental problem of understanding the mechanism of cancer drugs, which requires complex background knowledge about the functions of genes/proteins, their cells, and the molecular structure of the drugs. This background knowledge cannot be compactly expressed propositionally, and requires at least the expressive power of Datalog. Here we demonstrate the use of relational learning to generate new data descriptors in such semantically complex background knowledge. These new descriptors are effective: adding them to standard propositional learning methods significantly improves prediction accuracy. They are also explainable, and add to our understanding of cancer. Our approach can readily be expanded to include other complex forms of background knowledge, and combines the generality of relational learning with the efficiency of standard propositional learning.

Keywords: Relational learning · Inductive logic programming · Gene expression

1 Introduction

Effective data representations are the key to success in machine learning (ML) [28]. Most ML is based on data representations that use tuples of

J. French — Independent researcher

A. Appice et al. (Eds.): DS 2020, LNAI 12323, pp. 374–385, 2020.
https://doi.org/10.1007/978-3-030-61527-7_25

descriptors, i.e. the data can be put into a single table, where the descriptors (attributes) are the columns, and the examples are rows. Descriptors are properties of the examples that are believed to be important: for example if one wishes to classify pictures of animals then image pixel values are useful descriptors. Such tuple-based representations are essentially based on propositional logic [24]. The effectiveness of the propositional descriptors used for learning can vary greatly, and traditionally, most of the effort in ML went into hand-crafting effective descriptors. This has changed with the success of deep neural-networks (DNNs), which has been based on their capacity to utilize multiple neural network layers, and large amounts of data, to learn how to convert raw propositional descriptors (e.g., image pixel values) into richer internal representations that are effective for learning. Thanks to this ability DNNs have succeeded in domains that had previously proved recalcitrant to ML, such as face recognition and learning to play Go. The archetypal success is face recognition, which was once considered to be intractable, but can now be solved with super-human ability on certain limited problems [20]. *Therefore, a key lesson of the success of DNNs is: use ML to learn better data representations for ML.*

For many problems the standard propositional representation of data is problematic, as such a representation cannot efficiently express all the known relational structure (background knowledge) in the data. In some cases this structure can be encoded for using special purpose methods. For example convolutional neural networks encode relational information about the position of descriptors in the structure of the net. Similarly, recurrent neural networks encode information about temporal structure in the net, graph neural networks encode graphical information, etc. In many cases such special purpose methods can work very well. However, these methods must be redesigned for each new type of problem, and the structure encoded in the learning process is not explicit. It would be more beneficial (and elegant) if the learning biases in DNNs were explicit, and not inherent in the structure of the network. A more general approach to encoding known structure in data is to use logic programs [21] to represent the data – relational learning (RL) [24]. Such programs can express spatial, temporal, graphical structure, etc. using a single formalism, and, crucially, this structure is explicit instead of being implicit (e.g. in the connection of neurons). Logic programs provide a unified way of representing the relations between objects. They also promote explainable ML, as it is usually straightforward to translate logic programs into a series of easily understandable sentences that can be interpreted by domain experts. More formally, logic programs are a subset of 1st-order predicate logic, and therefore more general that propositional representations.

The main disadvantages of using a relational representation compared to a standard propositional one are that RL is more computationally expensive and difficult, as the search space of possible models is much larger, and that RL technology is much less developed. This suggests a hybrid strategy where RL is used to learn effective descriptors, and then standard ML is used to learn the final model [6]. This hybrid approach is particularly suited to problems where the data is semantically complicated, and where symbolic explainable models are required. In such problems RL has the potential to effectively learn new descriptors that are understandable to domain experts. Many biomedical ML problems are potentially suitable for a hybrid RL approach, such as understanding the mechanism of cancer drugs. In this problem one needs to encode background knowledge (problem structure) about gene/protein function, associated pathways, known drug targets, cancer cell type, the molecular structure of drugs, etc. We took data on this problem from the Library of Integrated Network-based Cellular Signatures [18] (LINCS). Specifically, we used the Phase II data, which consists of gene expression levels for 978 landmark human genes under perturbation conditions, making this a regression problem. The perturbation conditions consist of a cancer drug added to a cancer cell line, and it is worth noting here that only the response gene expression values are provided, and one would need to independently construct the input variables from the provided metadata.

We hypothesized that we could improve both ML model explainability, and predictive accuracy, by including additional background knowledge in the learning process using a hybrid RL approach. A key source of this background knowledge was the Stanford Biomedical Network Dataset Collection [22] (SBND). Using RL we mined frequent patterns about each drug found in relation to additional background knowledge. These patterns are expressed in Datalog [8] and are explainable to domain experts. They can also be used as binary descriptors in standard ML methods. It is worth noting that ML model explainability heavily depends on the learning algorithm, as some learning algorithms are more interpretable than others. However, we argue that the descriptors generated using the hybrid RL approach will generally be more interpretable than their propositional counterparts.

We evaluated the predictive performance of the newly learnt RL descriptors versus the standard descriptors, both when used by standard ML in isolation, and in combination. We compared two approaches to combining sets of descriptors: one in which the features from both representations are concatenated to form a single dataset, and another where predictions are stacked [2]. We found that the standard descriptors generally outperform the RL descriptors when used in isolation. However, the RL descriptors significantly improve predictive performance when used in combination. Moreover, these new effective RL descriptors are understandable by domain experts. The main contributions of the paper are as follows:

1. Demonstration of the effectiveness of hybrid RL learning on an important real-world problem.
2. Learnt explainable patterns underlying common cancer drugs.
3. A fully integrated biomedical knowledge base in Datalog.

2 Related Work

The problem of building models to predict drug effects has been widely studied, from potential adverse effects [33] and drug-drug interactions [32] to cancer cell sensitivity [23]. One such task is the learning of quantitative structure activity relationships (QSARs), where one is interested in predicting the effect of a drug or chemical compound from its molecular structure [25]. Molecular structure is usually represented using molecular fingerprints, which are tuples of Boolean descriptors [5]. However, several other approaches also exist. For example, some authors have used the 3-dimensional structure of chemicals [35], while others have extracted molecular vector embeddings using graph neural networks [15]. In our evaluation, we used the most widely adopted molecular fingerprint representation as the propositional approach. The LINCS data has been used in several studies, e.g. for the task of predicting gene expression levels using perturbation conditions [4]. In contrast to our evaluation, the authors do not utilise background knowledge in the learning process.

Several techniques have been applied to interconnected knowledge bases for various problems in biology [1,37]. RL in particular has been used in problems such as predicting gene function [16], gene regulation [9] and QSAR-related problems [31]. RL algorithms such as WARMR, which we use in our evaluation, have been shown to be successful in identifying relationships in linked data [17]. However, there are other algorithms like AMIE [10] which have also been shown to perform remarkably well. Furthermore, there exist several other approaches for learning representations from graph or inherently relational data [3,13] with varying levels of predictive performance and interpretability. We argue that our decision to use WARMR in our evaluation offers a good foundation from which all of these other methods can be explored in tackling the stated problem as part of future work.

One can think of the Boolean molecular fingerprint and RL representations of the drugs in the stated problem as views in multi-view learning, as both of these representations offer different perspectives in what constitutes the known properties of a drug. In a standard multi-view learning problem the views are typically distinct, meaning that special consideration is made as to the learning algorithm used in building a model for a particular view. Multiple kernel learning [29], which is essentially a form of stacking, has been proposed for such a scenario, where a kernel that is best suited for a particular view is used and the predictions from all views are then combined to form the final prediction. This is in contrast to how we perform our evaluation, because though we considered multiple learning algorithms, a specific learning algorithm is not used for a particular representation.

3 Methodology

The LINCS Phase II dataset with accession code GSE70138 provides the expression levels for 978 landmark human genes for 118,050 perturbation conditions. In the metadata, the perturbation conditions are described by their cell line, cell site, drug dosage, drug timepoint, and the applied drug. The Broad Institute identifiers along with their canonical smiles are also provided for the drugs. We were able to map 1,089 of the applied drugs to their DrugBank [36] and ChEMBL

[11] identifiers. This is relevant because the SBND knowledge graph uses DrugBank identifiers. The drugs we could map across these databases were applied to only 57,749 of the 118,050 perturbation conditions. For all the aforementioned perturbation condition properties but the drugs, we engineered features for the perturbation conditions using one-hot encoding, and treat them as *base* features. For the propositional representation of the drugs, we converted them into molecular fingerprints using RDKit [19], with 1024 bits, a radius of 4, and `useFeatures` set to `True`.

For the RL representation of the drugs, we formalised the following relations from the SBND: drug-drug (ChCh-Miner), drug-gene (ChG-Miner), gene-function (GF-Miner), disease-drug (DCh-Miner), and disease-function (DF-Miner). Additionally, we included relationships between functions from Gene Ontology [12], such as `is_a` and `part_of`. Furthermore, we included the chemical properties of each drug, such as the presence of rings. In total, this Datalog knowledge base contains 11,175 drugs, 6,869 genes, 45,089 functions and 5,941 diseases. It is available for download at https://github.com/oghenejokpeme/RLCBkb. The hypothesis language we used is given in Fig. 1.

```
drug_drug(+drug, -drug)
drug_disease(+drug, -disease)
disease_function(+disease, -function)
drug_gene(+drug, -gene)
gene_function(+gene, -function)
has_functional_group(+drug, #group)
has_ring(+drug, #ringtype)
has_group_count(+drug, #group, #count)
has_ring_count(+drug, #ringtype, #count)
has_group_ring_attachment(+drug, #group, #ringtype)
is_a(+function, #function)
part_of(+function, #function)
has_part(+function, #function)
regulates(+function, #function)
neg_regulates(+function, #function)
pos_regulates(+function, #function)
```

Fig. 1. Permitted relations in the body of valid clauses.

Using this knowledge base, we learned 1,024 frequent patterns using the WARMR algorithm in the Aleph inductive logic programming engine [30] for each drug, which we then use as binary features. WARMR is a levelwise RL algorithm based on a breath-first search of the input knowledge base, as relations are structured as a lattice. It allows for the learning of frequent patterns present in a knowledge base within pre-specified constraints, such as the proportion of the sample space a learned pattern must cover [17]. In Aleph, we used a minimum cover of 5%, a maximum clause length of 20, and 5000 nodes. We should note that there is evidence that suggests that beyond a certain point, increasing the number of learned features leads to no performance gains [26]. With the base features, we then created three data descriptors for the perturbation conditions. One with the base and RL features, another with the base and fingerprint features, and finally, one with the base, RL and fingerprint features. We refer to these datasets as RL, FP, and RL+FP for the remainder of this paper, all of which at this point, have 57,749 samples.

4 Evaluation Setup

We used a train-test split in our evaluation and selected only a subset of the samples due to computational complexity. The selection procedure entailed an initial random split of the 57,749 samples into training and testing buckets, 70%-30%. We then randomly selected 7,000 samples from the training bucket for the training set and 3,000 samples from the testing bucket for the test set. This was performed exactly once, and the dataset is available here: http://dx.doi.org/10.17632/8mgyb6dyxv.2. One might argue that this is a paired-input problem, as we are predicting gene expression on pairs of drug perturbation conditions and cancer cell lines. Therefore, we should extend our evaluation to take this into account. We expect that the naive train-test split evaluation approach we have taken will produce more optimistic results than an evaluation procedure for which each entity in a pair present in the test set is also not present in the training set [27]. However, we argue that for the purposes of evaluating standard propositional and RL data descriptors, such an evaluation setting will suffice.

As we mentioned previously, the LINCS dataset contains the gene expression levels of 978 genes. We selected genes that are dissimilar from one another by associated function using Gene Ontology associations at a tree depth of 1, where 0 are the base nodes. This process selected 46 genes, which we used in our experiments. We did this to reduce computational complexity and to select genes that are uncorrelated on a functional level in order to get performance estimates that are generally representative of the complete set of genes. For learning algorithms, we used the least absolute shrinkage and selection operator (LASSO) [34], ridge regression (RR) [14], and random forests (RF) in our evaluation. For LASSO and RR the regularization parameter was chosen using internal 10-fold cross-validation, and the RF models were built with 1000 trees and default settings. The performance metric reported is the coefficient of determination (R^2), as we are most interested in the amount of variance explained by the built models. Apart from the standard regression experiments using all three datasets, we also evaluated integrating the predictions made by the RL and FP representations using simple averaging, which is a form of stacking [2]. In this case, we averaged the predictions made using the RL and FP representations. We refer to these results as AVG in the discussion of the evaluation results. All code used for this experiment is available at https://github.com/oghenejokpeme/RLCBexp.

5 Results

5.1 Predictive Performance

We observed that on average the RL representation consistently performs worse than all the other representations (see Table 1). For the approaches which combine the RL and FP representations, we found that RL+FP consistently outperforms RL and does not strictly outperform FP on any of the learners. However, RL+FP and FP perform equally well on LASSO. Like RL+FP, AVG also consistently outperforms RL, but is outperformed by FP on both LASSO and RR, but not on RF. These results suggests that the effect the RL representations have when used to augment FP representations depends on two things; the choice of learning algorithm and how the representations are combined. From the mean performance results in Table 1, one might argue that overall, the performance of the representations is generally low. While this is true, we would argue that this is to be expected, as we are attempting to recreate laboratory conditions *in silico*, and predict the expression of 46 genes which often vary in concert and not in isolation of each other. Furthermore, it is worth pointing out that the representations perform reasonably well on some genes, with a maximum R^2 of 0.366 when the RL and FP predictions for RF are averaged (Table 1).

Table 1. The predictive performance (R^2) of the engineered datasets (RL, FP and RL+FP) and the aggregation by mean of the predictions made by RL and FP (AVG) on the learning algorithms. We show the mean with the minimum and maximum performance for the 46 considered genes. The best performing descriptor for each learner is in boldface.

Learner	R^2	RL	FP	RL+FP	AVG
LASSO	Mean	0.028	**0.086**	**0.086**	0.081
	min – max	−0.001–0.106	**0.004–0.319**	**0.002–0.320**	0.013–0.271
RR	Mean	0.030	0.090	**0.089**	0.084
	min – max	0.002–0.105	0.009–0.330	**0.008–0.330**	0.013–0.289
RF	Mean	0.068	0.094	0.089	**0.114**
	min – max	−0.012–0.238	−0.047–0.364	−0.050–0.360	**−0.006 – 0.366**

Given the difference in performance between the different methods, we tested for statistical significance using sign tests and paired t-tests. For LASSO, Table 2 shows that RL+FP underperforms when compared to FP, with a ratio of 18–28 of the 46 considered genes and a 0.12% average performance decrease from FP to RL+FP. However, this difference in performance is not statistically significant for both sign test and paired t-test. When AVG is compared to FP, we found that FP performed better on more genes, with a ratio of 17–29. However, we found that when compared to FP, AVG achieves a 9.6% average percentage performance increase over FP, with statistical significance according to the paired t-tests. Further investigation showed that although FP outperformed AVG on

Table 2. Performance (R^2) comparisons between the different datasets for the learning algorithms we considered. The comparisons are structured as approach A/B. For each compared pair, the number of genes for which one strictly outperforms the other is given. Additionally, an asterisk ($*$) and a dagger ($\dagger$) are used to indicate a statistically significant difference in performance with a significance level of 0.05 for a sign test and a paired t-test respectively. Lastly, the average percentage performance increase or decrease when approach A is compared to B is given. It is worth noting that this average percentage performance is calculated by taking the mean percentage difference in performance of genes between A and B, and not simply the percentage difference in mean performance given in Table 1.

Comparison	LASSO	RR	RF
FP/RL	$43/3^{*\dagger}$ (386.5%)	$46/0^{*\dagger}$ (310.9%)	$33/13^{*\dagger}$ (63.7%)
RL+FP/RL	$43/3^{*\dagger}$ (380.9%)	$44/2^{*\dagger}$ (301.5%)	$30/16^{\dagger}$ (26.3%)
RL+FP/FP	18/28 (−0.12%)	$13/33^{*\dagger}$ (−1.4%)	$0/46^{*\dagger}$ (−23.4%)
AVG/RL	$46/0^{*\dagger}$ (355.3%)	$46/0^{*\dagger}$ (276.6%)	$46/0^{*\dagger}$ (191.5%)
AVG/FP	$17/29^{\dagger}$ (9.6%)	$14/32^{*\dagger}$ (−0.07%)	$43/3^{*\dagger}$ (178.7%)
AVG/RL+FP	$18/28^{\dagger}$ (19.7%)	$14/32^{*\dagger}$ (2.0%)	$43/3^{*\dagger}$ (117.6%)

more genes, AVG tended to do a lot better than FP on the genes it outperformed FP on, explaining the percentage performance increase. For RR, both RL+FP and AVG both see a statistically significant decrease in average percentage performance when compared to FP. For RF, RL+FP significantly underperforms when compared to FP, but the reverse is true for AVG, with an average percentage performance increase of 178.7%. These results show that for two of the learners we considered, the RL representations can significantly improve predictive performance when used to augment the traditional RL representations. Having established that how the RL and FP representations are combined plays a crucial role in predictive performance, we conjecture that techniques from the multiple kernel learning literature might further improve predictive performance.

5.2 Explainability

RL enables the introduction of additional background knowledge to the model building process, and it can improve both understandability and predictive performance. In our experiments we were interested in learning frequent patterns present in the knowledge base for the considered drugs. We learned such rules as:

```
frequent_pattern(B):-
  has_functional_group(B,oxide),
  has_ring_count(B,benzene_ring,2).
```

and

```
frequent_pattern(B):-
  drug_gene(B,C),
  gene_function(C,D),
  pos_regulates(D,'GO:0008152').
```

The former rule can be interpreted as *a frequent pattern is for a drug to have an oxide group and two benzene rings.* Note that the standard fingerprint representation of molecules cannot express the simple concept of a molecule having two benzene rings unless a special descriptor 'two benzene rings' is pre-generated. Nor can it express the concept of a drug having an oxide group and two benzene rings unless it is pre-generated. To pre-generate all possible descriptors would produce an exponential number of descriptors.

The latter rule can be interpreted as *a frequent pattern is for a drug to target a gene that positively regulates a metabolic process.* Note that is a second-order pattern, the drug targets a gene that in turn regulates metabolism. Most drugs inhibit their targets, and in this pattern the overall result is likely to be decrease in a metabolic process, which is generally desirable in cancer therapy. These examples show that rules are easily understandable by a human reader. One can conjecture that if feature selection is performed when such rules are used as features in a predictive problem, the why of the observed variance in the target could be explained easier. However, it is beyond the scope of this work.

6 Discussion

The great success of DNNs is based on their ability *to learn* how to transform a simple input data representation into an effective internal representation. The limitations of the DNN approach are that it requires a large amount of data, the internal representation is obscure, and there is not a general way to encode known problem structure and background knowledge. In many biomedical problems, such as understanding the effect of anti–cancer drugs, it is required to encode a large amount of background knowledge. In this paper we have shown that a hybrid RL approach can *learn* new descriptors that are effective and explainable. The limitations of the hybrid RL approach are that it is a two stage approach rather than end–to–end learning (it is computationally efficient to learn frequent patterns, but they are not necessarily effective), and that the learning model is not differentiable, which makes it more difficult to find model improvements. The main criticism of RL in the past was that it was too inefficient to be applied. However, now, given the vast resources used to train DNNs this no longer applies. It is therefore interesting to consider whether there is a more general way of learning how to improve data representations that combines the advantages of DNNs and the hybrid RL approach, as is the case with deep relational machines [7]. Furthermore, it is worth noting that though the RL representations might be explainable, the interpretability of the models built using them will vary based on the learning algorithm. For example, one might conceivably inspect the important variables of a random forest model, but will find this far more challenging in a deep attention neural network.

7 Conclusion

In this paper we report the use of RL representations to enhance the predictive accuracy of traditional propositional data representations for a relevant problem in cancer biology. Apart from improved predictive accuracy, we also learnt explainable patterns underlying common anti–cancer drugs, and built a fully integrated biomedical knowledge base in Datalog which is now publicly available. We intend to investigate other forms of RL as part of future work.

Acknowledgments. This work was supported by the Engineering and Physical Sciences Research Council (EPSRC) UK through the ACTION on cancer grant (EP/R022925/1, EP/R022941/1). Prof. King acknowledges the support of the Knut and Alice Wallenberg Foundation Wallenberg Autonomous Systems and Software Program (WASP).

References

1. Barracchia, E.P., Pio, G., D'Elia, D., Ceci, M.: Prediction of new associations between ncRNAs and diseases exploiting multi-type hierarchical clustering. BMC Bioinf. **21**(1), 1–24 (2020)
2. Breiman, L.: Stacked regressions. Mach. Learn. **24**(1), 49–64 (1996)
3. Ceci, M., Appice, A.: Spatial associative classification: propositional vs structural approach. J. Intell. Inf. Syst. **27**(3), 191–213 (2006)
4. Chen, Y., Li, Y., Narayan, R., et al.: Gene expression inference with deep learning. Bioinformatics **32**(12), 1832–1839 (2016)
5. Cherkasov, A., Muratov, E.N., Fourches, D., et al.: QSAR modeling: where have you been? Where are you going to? J. Med. Chem. **57**(12), 4977–5010 (2014)
6. Clare, A., King, R.D.: Machine learning of functional class from phenotype data. Bioinformatics **18**(1), 160–166 (2002)
7. Dash, T., Srinivasan, A., Vig, L., Orhobor, O.I., King, R.D.: Large-scale assessment of deep relational machines. In: Riguzzi, F., Bellodi, E., Zese, R. (eds.) ILP 2018. LNCS (LNAI), vol. 11105, pp. 22–37. Springer, Cham (2018). https://doi.org/10.1007/978-3-319-99960-9_2
8. Dehaspe, L., Toivonen, H.: Discovery of frequent datalog patterns. Data Min. Knowl. Disc. **3**(1), 7–36 (1999)
9. Fröhler, S., Kramer, S.: Inductive logic programming for gene regulation prediction. Mach. Learn. **70**(2–3), 225–240 (2008)
10. Galárraga, L.A., Teflioudi, C., Hose, K., Suchanek, F.: AMIE: association rule mining under incomplete evidence in ontological knowledge bases. In: Proceedings of the 22nd international conference on World Wide Web, pp. 413–422 (2013)
11. Gaulton, A., Bellis, L.J., Bento, A.P., et al.: ChEMBL: a large-scale bioactivity database for drug discovery. Nucleic Acids Res. **40**(D1), D1100–D1107 (2011)
12. Gene Ontology Consortium: The Gene Ontology (GO) database and informatics resource. Nucleic Acids Res. **32**, D258–D261 (2004)

13. Hamilton, W.L., Ying, R., Leskovec, J.: Representation learning on graphs: methods and applications. arXiv preprint arXiv:1709.05584 (2017)
14. Hoerl, A.E., Kennard, R.W.: Ridge regression: biased estimation for nonorthogonal problems. Technometrics **12**(1), 55–67 (1970)
15. Jeon, W., Kim, D.: FP2VEC: a new molecular featurizer for learning molecular properties. Bioinformatics **35**(23), 4979–4985 (2019)
16. King, R.D.: Applying inductive logic programming to predicting gene function. AI Mag. **25**(1), 57–57 (2004)
17. King, R.D., Srinivasan, A., Dehaspe, L.: Warmr: a data mining tool for chemical data. J. Comput. Aided Mol. Des. **15**(2), 173–181 (2001)
18. Koleti, A., Terryn, R., et al.: Data portal for the library of integrated network-based cellular signatures (LINCS) program: integrated access to diverse large-scale cellular perturbation response data. Nucleic Acids Res. **46**(D1), D558–D566 (2017)
19. Landrum, G.: RDKit: open-source cheminformatics (2006)
20. LeCun, Y., Bengio, Y., Hinton, G.: Deep learning. Nature **521**(7553), 436 (2015)
21. Lloyd, J.W.: Foundations of Logic Programming. Springer, Heidelberg (2012)
22. Marinka Zitnik, Rok Sosič, S.M., Leskovec, J.: BioSNAP datasets: stanford biomedical network dataset collection, August 2018. http://snap.stanford.edu/biodata
23. Menden, M.P., et al.: Machine learning prediction of cancer cell sensitivity to drugs based on genomic and chemical properties. PLoS One **8**(4), e61318 (2013)
24. Muggleton, S., et al.: ILP turns 20. Mach. Learn. **86**(1), 3–23 (2011). https://doi.org/10.1007/s10994-011-5259-2
25. Olier, I., et al.: Meta-QSAR: a large-scale application of meta-learning to drug design and discovery. Mach. Learn. **107**(1), 285–311 (2017). https://doi.org/10.1007/s10994-017-5685-x
26. Orhobor, O.I.: A general framework for building accurate and understandable genomic models: a study in rice (Oryza sativa). Ph.D. thesis, The University of Manchester (United Kingdom) (2019)
27. Park, Y., Marcotte, E.M.: Flaws in evaluation schemes for pair-input computational predictions. Nat. Methods **9**(12), 1134 (2012)
28. Russell, S.J., Norvig, P.: Artificial Intelligence: A Modern Approach. Pearson, London (2016)
29. Sonnenburg, S., Rätsch, G., Schäfer, C., Schölkopf, B.: Large scale multiple kernel learning. J. Mach. Learn. Res. **7**(Jul), 1531–1565 (2006)
30. Srinivasan, A.: The aleph manual (2001)
31. Srinivasan, A., Page, D., Camacho, R., King, R.: Quantitative pharmacophore models with inductive logic programming. Mach. Learn. **64**(1–3), 65 (2006)
32. Takeda, T., Hao, M., Cheng, T., et al.: Predicting drug-drug interactions through drug structural similarities and interaction networks incorporating pharmacokinetics and pharmacodynamics knowledge. J. Cheminform. **9**(1), 16 (2017)
33. Tatonetti, N.P., Patrick, P.Y., Daneshjou, R., et al.: Data-driven prediction of drug effects and interactions. Sci. Transl. Med. **4**(125), 25ra31–125ra31 (2012)
34. Tibshirani, R.: Regression shrinkage and selection via the lasso. J. Roy. Stat. Soc.: Ser. B (Methodol.) **58**(1), 267–288 (1996)
35. Verma, J., Khedkar, V.M., Coutinho, E.C.: 3D-QSAR in drug design-a review. Curr. Top. Med. Chem. **10**(1), 95–115 (2010)
36. Wishart, D.S., Knox, C., Guo, A.C., et al.: DrugBank: a comprehensive resource for in silico drug discovery and exploration. Nucleic Acids Res. **34**, D668–D672 (2006)
37. Zitnik, M., Agrawal, M., Leskovec, J.: Modeling polypharmacy side effects with graph convolutional networks. Bioinformatics **34**(13), i457–i466 (2018)

Interpretable Machine Learning with Bitonic Generalized Additive Models and Automatic Feature Construction

Noëlie Cherrier[1,2(✉)], Michael Mayo[3], Jean-Philippe Poli[1], Maxime Defurne[2], and Franck Sabatié[2]

[1] CEA, LIST, 91191 Gif-sur-Yvette Cedex, France
jean-Philippe.poli@cea.fr
[2] Irfu, CEA, Université Paris-Saclay, 91191 Gif-sur-Yvette Cedex, France
{noelie.cherrier,maxime.defurne,franck.sabatie}@cea.fr
[3] Department of Computer Science, University of Waikato, Hamilton, New Zealand
michael.mayo@waikato.ac.nz

Abstract. In many machine learning applications, interpretable models are necessary for the sake of trust or for further understanding the patterns in the data. In particular, scientists often want models that elucidate knowledge and therefore may lead to new discoveries. Currently, Generalized Additive Models (GAM) are gaining interest in other application domains because of their ability to fit the data well while at the same time being intelligible. Moreover, prior domain-specific knowledge is often valuable to guide the learning. In this work, extensions and generalizations of GAM are proposed to incorporate prior knowledge during the learning phase. Specifically, the fitting method for GAM is modified so that it can fit the data with bitonic functions. In physics for instance, the most discriminative variables often present specific distributions with respect to the target variable, especially peaking (i.e. bitonic) distributions. An algorithm is also described to build automatically bitonic high-level features to be used in the GAM terms. Experiments on three physics datasets are used to validate these ideas in conjunction with physics scientists.

Keywords: Bitonicity · Generalized additive models · Experimental physics

1 Introduction

A common obstacle in machine learning (ML) comes from a tradeoff between performance and interpretability. Several application domains require interpretability to be able to use ML models, for instance in healthcare [1]. In scientific domains in particular, interpretable models are needed for validation on real data and hopefully for knowledge discovery. Arrieta et al. [2] notably provide a definition for interpretability as "the ability to explain or to provide the meaning in understandable terms to a human". This work focuses on producing models interpretable by an expert.

A. Appice et al. (Eds.): DS 2020, LNAI 12323, pp. 386–402, 2020.
https://doi.org/10.1007/978-3-030-61527-7_26

Generalized Additive Models [3] (GAM) are often considered as both intelligible and well performing models [4]. The predicted variable in a GAM is a sum of smooth functions of the input variables. The final model is interpreted by observing the inferred smoothed functions of each input variable independently. However, GAM must meet a few requirements to remain interpretable, as they may still be overly complex [2]. According to Arrieta et al. the variables and smooth functions of the GAM must be constrained "within human capabilities for understanding". To fulfill this need, prior knowledge should be incorporated about the problem.

For illustrative purposes, this work focuses on high-energy physics (HEP) problems. Generally, a preprocessing step of feature engineering is performed manually based on domain expertise. In HEP, quantities related to energy, mass or momentum balances which are dependent on the process of interest are derived from base variables. Although understandable and analyzable by construction, nothing guarantees that these quantities are optimized for the analysis of the process of interest. The field of feature construction (FC) aims at automating the feature engineering step. In this way, interpretable FC is performed in this work to determine the discriminative variables of interest to be used in GAM.

In addition, prior knowledge on the expected distributions of the features can be integrated during the inference of GAM terms. A monotonicity assumption is often made in the literature [5–7]: one or several input variables are assumed to be monotonic with respect to the target variable. The ML model is then constrained to respect this assumption such that the predicted value of the model should be monotonic with respect to the input variable(s). However, we observe that in the HEP field in particular, the most frequently used high-level variables often present a local extremum (see Fig. 1 for instance). Other applications can also benefit from bitonicity constraints, such as dose–response analysis [8]. Bitonicity is introduced in this work as an extension to monotonicity and enforced in GAM terms in the context of HEP applications.

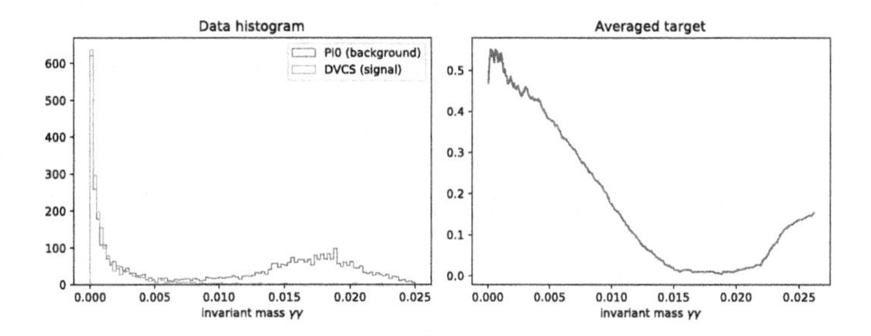

Fig. 1. Invariant mass $\gamma\gamma$, a high-level variable often used to recognize π^0 production events. On the left, the unnormalized distributions of the two classes with respect to the invariant mass $\gamma\gamma$. On the right, the averaged target (corresponding to the ratio between numbers of signal and background instances per bin).

The contributions of this work can be summarized as follows: firstly, a definition of bitonicity and an algorithm to verify the bitonicity of a distribution is presented (Sect. 2); secondly, a method to constrain GAM terms to be bitonic is described in Sect. 3, including a bitonic FC algorithm. Several experiments are performed in Sect. 4 to validate the method. The overall interpretability of the generated models is discussed in Sect. 5.

2 Definition and Verification of Bitonicity

A function f of one real variable is commonly said to be monotonic if and only if $f(y) \geq f(x)$ (resp. $f(y) \leq f(x)$ for the decreasing case) for any (x, y) in $\mathbb{R}$ such that $y \geq x$. Canini et al. [5] define a multi-variable function to be monotonic with respect to feature d if and only if $f(y) \geq f(x)$ (resp. $f(y) \leq f(x)$) for any two feature vectors x, y in $\mathbb{R}^D$ such that $y[d] \geq x[d]$ and $y[m] = x[m]$ for $m \neq d$ with $m, d \in \{1, 2, ..., D\}$.

Definition 1. *f is positively (resp. negatively) bitonic w.r.t. feature d if and only if for each set of values $[x_m]_{m \neq d}$ (setting all values of input X except feature d), it exists at most one x_d^* in the domain of feature d such that these two conditions are satisfied:*

- *$f(X) \geq f(X')$ (resp. $f(X) \leq f(X')$) with $X = (x_1, ..., x_d, ..., x_D)$ and $X' = (x_1, ..., x'_d, ..., x_D)$ for each x_d and x'_d such that $x_d \leq x'_d < x_d^*$,*
- *$f(X) \leq f(X')$ (resp. $f(X) \geq f(X')$) with $X = (x_1, ..., x_d, ..., x_D)$ and $X' = (x_1, ..., x'_d, ..., x_D)$ for each x_d and x'_d such that $x_d^* < x_d \leq x'_d$.*

In contrast to the usual definition of bitonicity in the context of bitonic sorters, the circular shifts are here not taken into account. In addition, this definition includes fully monotonic functions, e.g.. if the value of x_d^* is beyond the range of feature d. Quasi-convex and quasi-concave functions are also bitonic functions (the reciprocal is false for $D > 1$). *Unimodality* is a similar term mostly used for distributions. Figure 2 displays two examples of univariate bitonic functions, one example of a bivariate bitonic function and one bivariate non bitonic function.

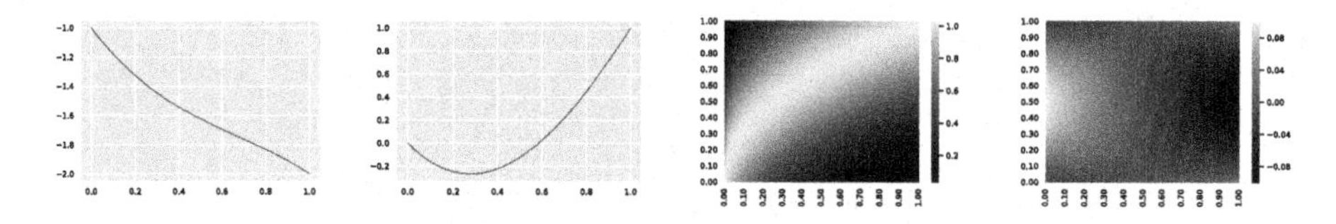

Fig. 2. First two plots: two bitonic univariate functions. Third plot: bivariate bitonic function (i.e. bitonic w.r.t its two variables). Fourth plot: non bitonic bivariate function, since the y variable is increasing then decreasing for low x and the opposite for high x.

The bitonicity of a function is often hard to prove with an analytical method, except for simple functions [9]. To numerically quantify the non-bitonicity degree of a function, it is sampled into an ordered vector (sequence noted s), then compared to a close bitonic sequence. Monotonicity can be assessed by comparing the sequence to its cumulative maximum (for a non-decreasing sequence) or to its cumulative minimum (non-increasing). The ith-component of the cumulative maximum is the maximum value taken by the sequence between components 0 and i. A bitonic sequence is thus equal to its cumulative maximum in its increasing part and to its cumulative minimum in its decreasing part. Other techniques could be considered such as unimodal regression [10], but the previously described method using cumulatives is preferred because it is simpler to implement and since the objective is to rank the features with respect to each other, not to get the closest bitonic approximation. To summarize, the procedure is as follows:

1. Find the point in the sequence leading to the best bitonic approximation using cumulative minimum and maximum on each side of the point;
2. Compute the integral of the absolute difference between the sequence and its bitonic approximation and normalize it by the length and amplitude of the sequence.

The bitonicity penalty of a sequence varies consequently from 0 to 0.5, the worst case being a sequence of alternating zeros and ones. In practice, the maximum is not reached in the experiments since a prior smoothing of the sequences is performed. The bitonicity of a data feature is checked by looking at the variation of the target values along this feature. However, applying directly the procedure detailed above to evaluate bitonicity can be troublesome: first, data are often noisy; second, the target values take a finite number of values in a classification task. To ensure robustness, the data feature is preprocessed as follows (numbers have been determined empirically):

1. Take the target vector $r = (r_1, ..., r_n)$ sorted along the evaluated feature $f = (f_1, ..., f_n)$. Average the values of r where f takes the same values.
2. A moving average box of size $\frac{n}{10}$ is propagated through the r vector.
3. If $n > 1000$, a median filter of size $\frac{n}{100}$ is applied to r.
4. Then if $n > 10000$, a median filter of size $\frac{n}{1000}$ is applied to r.

Finally, the bitonicity is evaluated on the smoothed r sequence. Figure 6 in Subsect. 3.5 illustrates examples of this procedure.

3 Bitonic Functions and Features for GAM

This section recalls the background on GAM and details the method to enforce bitonicity before presenting a summary of the overall approach and describing bitonic FC.

3.1 Enforcing Bitonicity of Shape Functions in GAM

Generalized Additive Models (GAM) [11] are of the form: $g(\hat{y}) = c_0 + \sum_{i \in \mathcal{S}} f_i(x_i)$, where g is the link function and $\mathcal{S}$ the set of features. The shape functions f can be modeled in different ways, for instance using splines [12] fitted with backfitting [3] or penalized iteratively reweighted least squares [13]. Other approaches fit tree ensembles with gradient boosting [4,14].

A method is proposed hereafter to enforce bitonicity for two types of shape functions without loss of generality: splines and functions learnt by a neural network. The idea in both cases is based on the exploitation of the regularization parameter.

Splines as commonly used in GAM are written $f(x) = \sum_k \beta_k b_k(x)$ with b_k basis functions (B-splines) and β_k the parameters to fit. The fitting of f is done by minimizing the penalized sum of squares: $\min_\beta \left\{ \left\| y - B^T \beta \right\|^2 + \lambda \beta^T P \beta \right\}$ with B the vector of $b_k(x)$ and $\lambda \beta^T P \beta$ a penalty term. P can for instance penalize the differences between adjacent β_k, or the second derivative of the shape function. The larger the λ parameter, the smoother the final function. This smoothing parameter is usually optimized (with respect to a performance metric) by generalized cross-validation (GCV) or restricted maximum likelihood (REML).

The following procedure permits to obtain a bitonic shape function:

1. Fit the shape function f with GCV or REML and retrieve smoothing parameter λ_0.
2. Check the bitonicity of f by applying it to a regular test sequence s spanning the range of x and assess bitonicity of $f(s)$ using the procedure detailed in Sect. 2.
3. If $f(s)$ is bitonic, then accept function f as bitonic. Else set $\lambda = \lambda\mu$ with $\mu > 1$, refit f with imposed λ and go back to step 2.

As long as the test sequence s is large enough to account for the complexity of f, the proposed procedure permits to obtain a bitonic function at the cost of multiple refits. Moreover, this procedure will always converge since an infinite λ leads to a linear function. The choice of μ balances between speed and performance: the performance tends to decrease as λ increases and moves off the optimum found by GCV or REML.

The procedure is similar for shape functions learnt by a neural network. The weights of the network must minimize a cost function plus a regularization term expressed as $\lambda \mathcal{R}(W)$, with $\mathcal{R}$ for instance a $\mathcal{L}^1$ or $\mathcal{L}^2$ regularization of the weights. The intuition behind λ is the same as for spline fitting: the larger the λ, the smoother the resulting function. To enforce bitonicity of the learnt function, the same procedure than for splines is applied but setting the first parameter λ_0 as a hyperparameter (small enough).

An experimental validation of this approach is conducted in the next subsection.

Related Work on Unimodal Regression with Splines. Some previous works constrain the shape of splines by adding linear constraints to the optimization problem, producing functions that are increasing, decreasing, convex, concave among others [15]. In [8], data are fitted by unimodal B-splines, namely a function with a single local maximum. However, it requires knowing beforehand the location of the maximum to formulate the linear constraint. One must either try all possible locations of the maximum (because of possible noise, the global maximum may not be the proper one for unimodal regression), or perform a more computationally intensive Bayesian approach. Moreover, the authors do not consider other forms of bitonicity including monotonic functions and decreasing-increasing functions. In contrast, our approach does not require prior knowledge on the type of bitonicity nor on the optimum location. Taking their approach is more time-consuming because of multiple REML computations: two for each monotonicity type, and twice the number of knots for the two other bitonicity types (the knots correspond to the possible locations of the optimum). Our approach computes REML once and then only solves consecutive penalized least squares problems by increasing λ while the function is not bitonic. Moreover, it can be used with any shape function that supports penalization.

3.2 Validation of the Principle to Enforce Bitonicity of Shape Functions

The previously presented method is tested on a toy dataset of 1000 instances displayed on Fig. 3 (data generated with the `make_classification` function of scikit-learn [16]). Experiments are conducted by fitting a bivariate GAM term, either with splines or with a multilayer perceptron (MLP), while trying different values for the regularization parameter λ. The evolution of the F1 score and the bitonicity penalty obtained by the fitted function is plotted on Fig. 4. Figure 5 depicts the fitted terms at small, large, and optimal λ (for which the bitonicity penalty is 0).

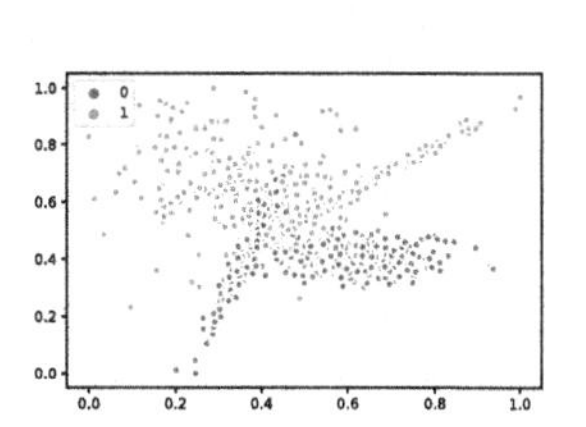

Fig. 3. Toy dataset

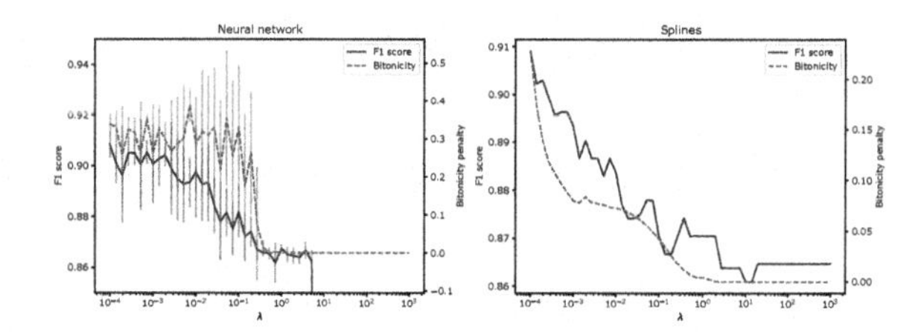

Fig. 4. Evolution of F1 score and function bitonicity against regularization parameter λ.

These plots show an experimental validation of the proposed method, as the bitonicity decreases until reaching 0 definitively as λ increases. Moreover, these plots give an interesting comparison between spline and neural network fitting

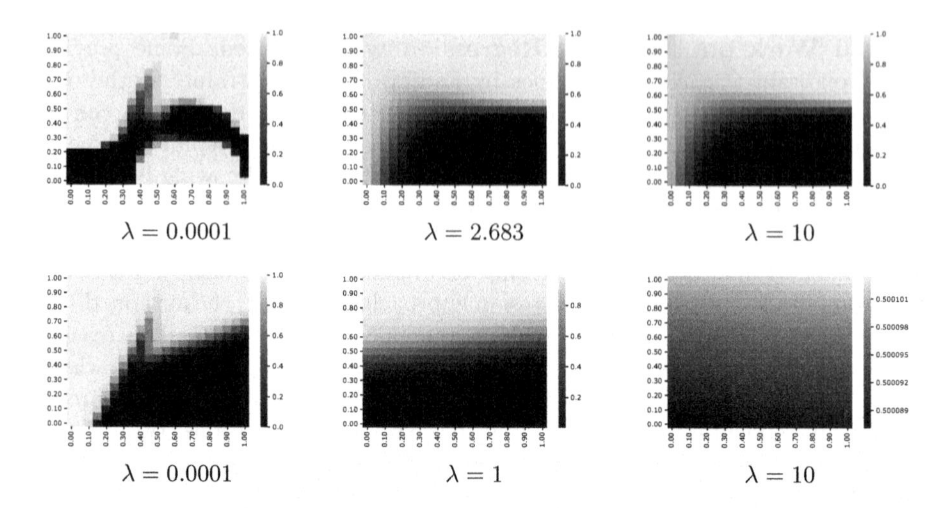

Fig. 5. Top: spline models fitted on toy dataset. $\lambda = 2.683$ corresponds to the minimal λ value for which the bitonicity equals 0. Bottom: MLP models fitted on toy dataset. $\lambda = 1$ corresponds to the minimal λ value for which the bitonicity equals 0.

methods: while the spline terms never obtain a score below 0.8, the score of the neural network fitted terms finally drops to 0 (outside the scope of the plot) shortly after reaching bitonicity.

3.3 Building a Complete Model with Gradient Boosting

The following algorithm builds a complete GAM with bitonic shape functions. For a binary classification task, a list of inputs X and a target vector y are considered. A standard GAM can be modeled by a sigmoid of the sum $F(x)$ of the GAM terms. In the proposed method, the GAM is built iteratively, building a feature at each step. A first prediction model predicts p_0 for each x in X, p_0 being the proportion of the majority class. The objective is to minimize the cross-entropy between the target y and the prediction $p(x)$, noted p:

$$\mathcal{L}(F(x)) = -\left(y\log(p) + (1-y)\log(1-p)\right), \quad p(x) = \frac{1}{1+e^{-F(x)}}. \tag{1}$$

A gradient boosting method is applied to iteratively improve this convex loss function. At the n-th step, a term $h_n(x)$ is added to the current GAM: it is fitted to the current objective while enforcing its bitonicity using the set $\{(x_i, r_{i,n})\}$.

$$r_{i,n} = -\frac{\partial \mathcal{L}_i(F_{n-1}(x))}{\partial F_{n-1}(x)} = y_i - p_{i,n-1}, \quad F_n(x) = F_{n-1}(x) + \alpha_n h_n(x). \tag{2}$$

A gradient descent with its own learning rate β determines the learning rate α_n to minimize the averaged cross-entropy. α_n is directly linked to the importance of the new term in the global result.

Algorithm 1: FCGAM algorithm

Input: *data* used to build the features, of size (m, d), y target vector of length m
n the number of GAM terms to learn

Initialization:

$p \leftarrow p_0$ proportion of the majority class in y, $F \leftarrow \log\left(\frac{p}{1-p}\right)$, $r \leftarrow y - p$

for $i \leftarrow 0$ **to** n **do** `// one iteration builds one term of the GAM`

(1) Build one single feature z using FC algorithm with r as the target for the fitness function. Bitonicity may or may not be enforced at this step.

(2) Train a single GAM term on the built feature z with target r. Bitonicity may or may not be enforced at this step.

$h \leftarrow predict(z)$, $g \leftarrow +\infty$, $\alpha \leftarrow$ random(0,100)

while $|g| > \epsilon$ **do**

$\widetilde{F} \leftarrow F + \alpha h$, $\widetilde{p} \leftarrow \frac{1}{1+e^{-\widetilde{F}}}$, $g \leftarrow \frac{1}{D}\sum_k h(\widetilde{p} - y)$, $\alpha \leftarrow \alpha - \beta g$

$F \leftarrow F + \alpha h$, $p \leftarrow \frac{1}{1+e^{-F}}$, $r \leftarrow y - p$

The remaining concern is to carefully select the involved variable x in each term h_n. In HEP, relevant variables are often high-level combinations of some base variables, as stated in the introduction. At each step of the boosting algorithm, one feature is built adapted to the ongoing regression problem with the FC algorithm presented in next subsection. The overall process is summarized in Algorithm 1.

3.4 Automatic Construction of Bitonic Features

This subsection proposes an adapted bitonic feature construction (BFC) algorithm. The literature in automatic FC is very abundant and a survey can be found in [17] or [18]. A constrained genetic programming algorithm is proposed in [19] to build interpretable features for HEP applications. One of the contributions is to use a grammar to enforce the respect of physical units during the construction of a new feature.

The constrained FC method of [19] is reused to build features for GAM terms. At a given step n, a single GAM term that fits the target $r_n = y - p_{n-1}$ is added to the model. The fitness function for FC is adapted to better suit the tackled problem of this work. A shallow decision tree with maximum four leafs is trained on the set $\{(z_i, r_{i,n})\}$, z_i being the candidate feature. Thus, the decision tree can only perform cuts on the candidate feature and observe its discriminating power. The fitness of the candidate feature is minus the RMS (Root Mean Squared) error between the prediction of the shallow tree and the target r_n. To enforce bitonicity, a bitonicity penalty term is added to the fitness of z_i. This penalty term b is the result of applying the procedure of Sect. 2 on the sequence of residuals r_n, sorted along z_i. In the end, the fitness of the candidate feature z_i is $f = -(\text{RMS} + b)$, to be maximized during the evolution process.

Table 1. Four variants of Algorithm 1. FCGAM_b_{max} makes the shape functions bitonic if and only if the feature is itself bitonic, i.e. if and only if the bitonicity b of the feature is below b_{max}.

Name	Bitonicity enforced in FC	Bitonicity enforced in shape functions
FCGAM_∅	No	No
FCGAM_b_{max}	No	Yes if $b \leq b_{max}$
FCGAM_∞	No	Yes
BFCGAM	Yes	Yes

3.5 Algorithm Variants and Bitonicity Threshold

We consider four variants of Algorithm 1 regarding the bitonicity constraints (summarized in Table 1). Bitonicity can be enforced or not during FC (label (1) in Algorithm 1) or for shape functions (label (2)).

To activate the bitonicity constraint for shape functions, a parameter b_{max} is set: if the bitonicity of a built feature is below b_{max}, the associated shape function will be forced to be bitonic. Looking at various features from the three datasets used in the experiments, a near-optimal bitonicity threshold b_{max} can be set at 0.04 (see Fig. 6).

4 Experiments

4.1 Experimental Setup

Datasets. Three HEP binary classification problems are considered in this study, with very different experimental setups and studied processes. However the objective is the same: to isolate signal instances from one or several background sources. Our focus on HEP problems with information about the input variables complicates experiments on a larger number of datasets. The DVCS dataset comprises 30 raw features for 14730 instances, the Higgs dataset [20] 17 raw features for 100000 used instances, the MAGIC dataset [21] 10 raw features for 19020 instances. Detailed descriptions of the DVCS and Higgs datasets are provided in Sect. 5 along with an interpretability discussion.

Shape Functions Parameters. The GAM version with neural network as shape function uses a MLP regressor with two hidden layers of size 100 each, Adam optimizer with a constant learning rate of 0.001 and rectified linear unit as activation function. The shape (100,100) of the network is not fine-tuned since the objective is to have sufficient degrees of freedom to handle the dimensionality of the problem, while letting the optimization of the regularization parameter (through bitonicity requirement) compensate the potential overfitting. These parameters were found by quick testing on the datasets.

As for the terms modeled with splines, penalized B-splines are used with 14 knots spaced along the quantiles of the feature.

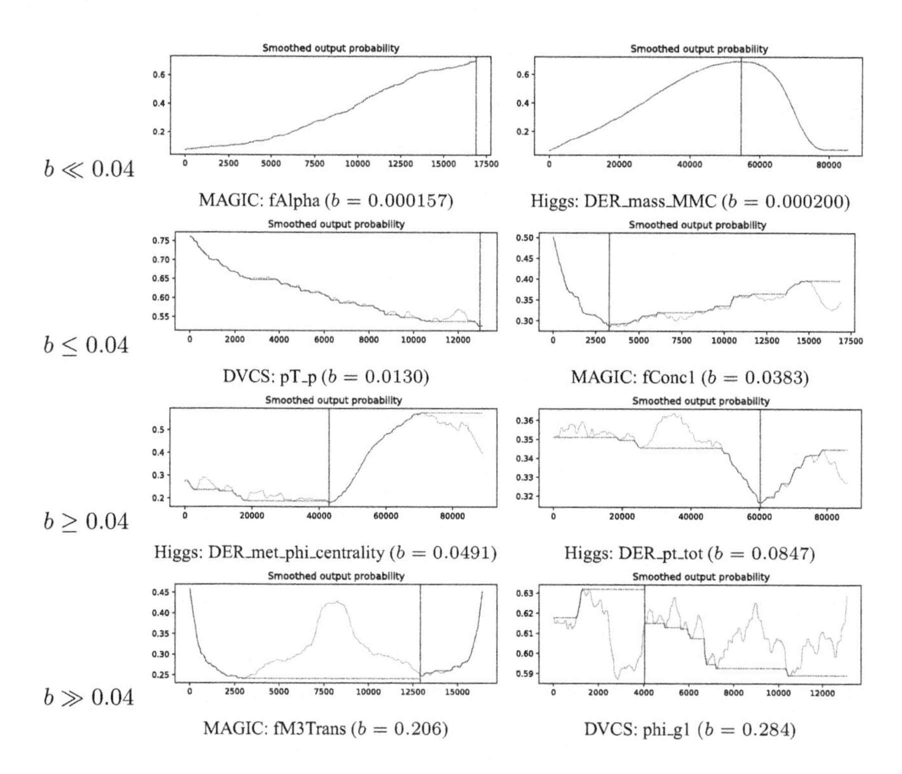

Fig. 6. Feature bitonicity examples on all three datasets, from smaller to higher bitonicity penalties. On the graph is plotted the output probability after smoothing (i.e. the vector used for the computation of the bitonicity penalty). In dotted black is the cumulative minimum/maximum that is the reference to compute the difference and get the bitonicity penalty corresponding to the area between the dotted line and the orange feature data. The red vertical line marks the hypothesis of the algorithm for the extremum.

Other Hyperparameters. In the FC algorithm, we set the population size to 500 and the number of generations to 70. The multiplying factor μ to increase the regularization λ if the resulting shape function is not bitonic is arbitrarily set to $\mu = \sqrt{10}$. The β parameter for the gradient descent of the learning rates α_n is set to 30 and the demanded maximum ϵ for the gradient is set to 10^{-5}. These parameters have been experimentally proven to lead to convergence in almost all cases for the experimental datasets. In the case an α_n rate was not found, the current iteration n of the FCGAM algorithm is dropped and recomputed (a new FC and shape function fit are done). 10 independent runs for each configuration are performed because of the stochastic nature of genetic programming in FC. The mean and standard deviation of the F1 score are presented for each dataset and algorithm configuration, averaged over the 10 runs and doing 5-fold cross-validation (50 runs in total for each numerical result).

Table 2. F1 score (mean and standard deviation over 10 runs for each fold in a 5-fold cross-validation) of the proposed methods on three datasets, compared with a few baselines. The best score for each dataset is in bold font, while the best variant for splines and for MLP is underlined.

	DVCS	Higgs	MAGIC
Baselines			
(100,100) MLP	0.751 ± 0.007	**0.673 ± 0.022**	0.795 ± 0.012
XGBoost	0.772 ± 0.005	0.587 ± 0.002	0.797 ± 0.009
GAM with boosted DT	0.748 ± 0.004	0.539 ± 0.002	0.781 ± 0.008
FCGAM with splines			
FCGAM_∅	0.792 ± 0.012	<u>0.626 ± 0.015</u>	<u>0.806 ± 0.011</u>
FCGAM_b_{max}	<u>**0.793 ± 0.006**</u>	0.625 ± 0.013	<u>0.806 ± 0.011</u>
FCGAM_∞	0.788 ± 0.007	0.609 ± 0.025	0.804 ± 0.012
BFCGAM	0.789 ± 0.007	0.545 ± 0.031	0.792 ± 0.012
FCGAM with (100,100) MLP			
FCGAM_∅	<u>0.792 ± 0.009</u>	0.627 ± 0.013	<u>**0.808 ± 0.012**</u>
FCGAM_b_{max}	<u>0.792 ± 0.008</u>	<u>0.628 ± 0.013</u>	0.807 ± 0.011
FCGAM_∞	<u>0.792 ± 0.006</u>	0.625 ± 0.013	0.807 ± 0.012
BFCGAM	0.788 ± 0.005	0.529 ± 0.057	0.785 ± 0.012

4.2 Performance Comparison

Table 2 presents the results obtained while applying the four variants detailed in Subsect. 3.5 on the three datasets using either splines or MLP as GAM shape functions, along with three baselines. A first observation is that the FCGAM algorithm gives better results than the baselines on DVCS and MAGIC datasets. One could have expected the opposite, since more complex and less interpretable algorithms such as a neural network or XGBoost are supposed to perform better on complex problems, which is the case for the Higgs dataset.

Apart from the baselines comparison, the fourth version (BFCGAM) always gives worse scores than the no bitonicity FCGAM_∅ version, in a significant manner for the Higgs dataset in particular. In all cases, letting the FC free and enforcing bitonicity only on relevant shape functions (i.e. those for which the feature is actually bitonic) in the FCGAM_b_{max} variant improves the score at the level of the FCGAM_∅ version. Even if not significantly, the FCGAM_b_{max} version with $b_{max} = 0.004$ threshold sometimes gets better results than the FCGAM_∅. One conclusion for this is that forcing bitonicity may be good for interpretability (this will be discussed in Sect. 5), but can be too restrictive: some really discriminative features are not bitonic and are indispensable to get a good score. Next subsection discusses the bitonicity potential of the datasets and why some datasets behave well under bitonicity constraint while others do not.

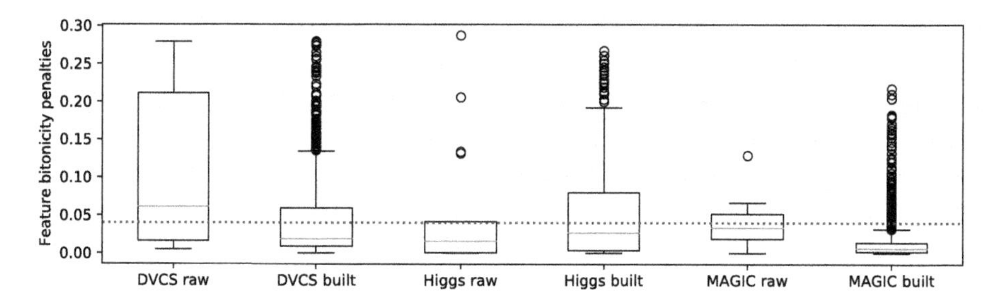

Fig. 7. Boxplots of features bitonicities. The box indicates the first and third quantiles with an orange line at the median, while the whiskers extend to the farthest data point within 1.5 IQR (interquartile range) after the box. The horizontal dotted red line represents the bitonicity threshold b_{max} set to 0.04 to trigger the bitonicity constraint on the shape function. (Color figure online)

4.3 Bitonicity Potential of the Different Datasets

Enforcing bitonicity on built features or shape functions will only be beneficial if there exist a discriminative set of bitonic features for a given dataset. Figure 7 displays the boxplots of the features bitonicities: those already present in the dataset (raw features) and those which have been found to be discriminative through the FC process (built features) without the bitonicity constraint. Therefore, the built features boxplots represent well the distribution of the most discriminative features for a given dataset. These plots have been made with all the features built in the FCGAM_∅ configuration, so around 1000 built features and around 10 raw features for each dataset.

The bitonicity of the raw features seems not to influence the bitonicity of the built high-level features nor the potential of a dataset to get good results while enforcing bitonicity. Indeed, the Higgs dataset presents mainly bitonic raw features whereas the scores are impaired under bitonicity constraints, when it is the opposite for DVCS.

The large majority of the features built for the MAGIC dataset are bitonic, without the need to add a bitonicity penalty term during the FC process. It is then logical that the scores of the MAGIC dataset are not penalized when adding this constraint which is already satisfied. However, the most discriminative features found for the Higgs dataset (i.e. the built features) are often not bitonic, hence the decrease in score when trying to force the bitonicity. Not all the features built for the DVCS dataset are bitonic, however the bitonicity penalty does not impair the performance on this dataset. Some redundancy may indeed be present between the built features, hence all the required information to perform a proper classification can be contained in the subsample of bitonic features.

5 Discussion on Interpretability

The goal here is to assess the interpretability of the trained models from the point of view of experts in the field. Notably, an explicit textual explanation of the model is not needed. Indeed, expert physicists are able to interpret high-level built features (physical formulas) and shape functions (can be viewed as function curves).

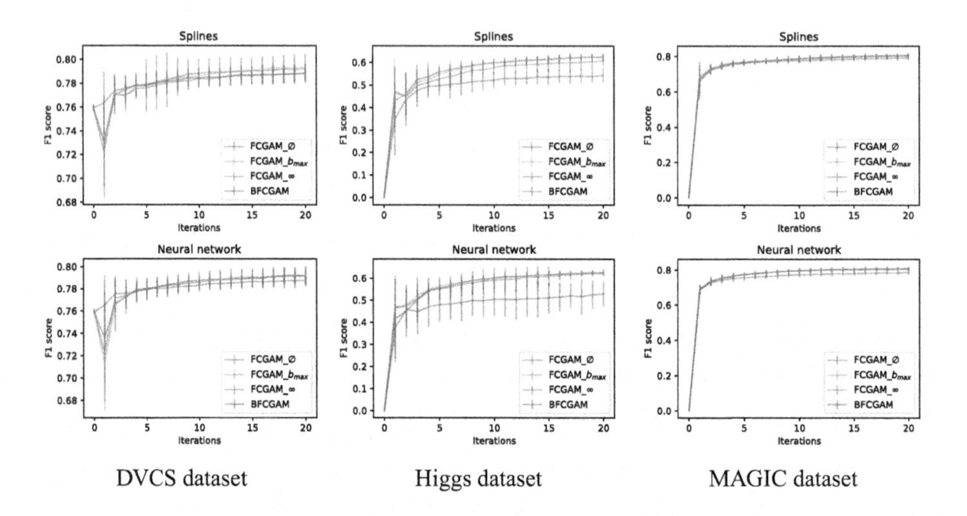

Fig. 8. Convergence curves for all three datasets with splines (on the top) or neural network (on the bottom) as shape function. The evolution of the F1 score is plotted against the number of iterations in the boosting algorithm (Algorithm 1).

The global GAM consists of 20 terms, namely 20 shape functions each associated to a built feature. Figure 8 displays the evolution of the classification score for each dataset against the number of GAM terms. Interpretability decreases with increasing number of GAM terms hence increasing number of features. This is a trade-off that experts must consider depending on their own criteria.

The focus now is on the DVCS and Higgs datasets only since we are not experts in the MAGIC classification problem. An analysis of one feature and the associated fitted shape function for each of the DVCS and Higgs datasets is performed hereafter. Figures 9 and 10 are presented in the same way: the left plot is the target vector binned along the built feature (so the y value on the plot is the averaged target for a bin), the central plot is a GAM term learnt for this feature using splines without bitonicity enforcement, the right plot is a GAM term learnt using splines and with bitonicity enforcement. For the DVCS and Higgs datasets, the physical problem is first explained followed by an interpretation of a frequent GAM term.

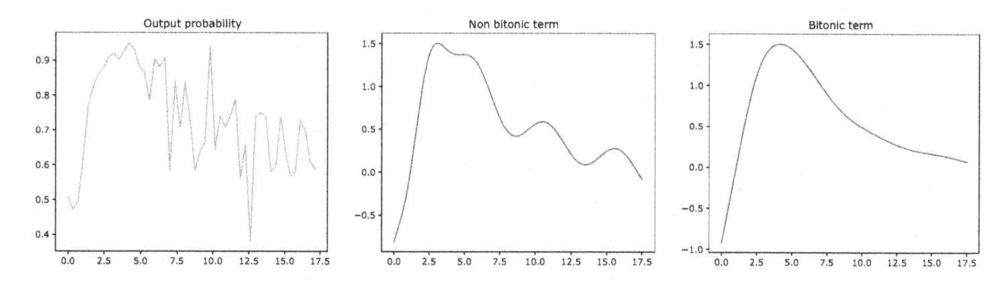

Fig. 9. DVCS: $\angle(p_{\gamma_2}, p_{\gamma_1} + p_{\gamma_2})$. Lower value means higher probability to have a signal event.

DVCS Dataset. At Jefferson Laboratory, an electron beam scatters off protons. The objective is to discriminate between the γ^{DVCS}-events whose final state is composed of an electron, a proton and a photon noted γ, and the $\gamma\gamma^{\pi^0}$-events which have a similar final state, except that two correlated γ photons are produced. The three-dimensional momentum (i.e. mass times speed) and angles are available for each identified particle.

Figure 9 illustrates a DVCS built feature: the angle between γ_2 the lowest energetic photon and the sum of two detected photons $\gamma_1 + \gamma_2$. A signal γ^{DVCS}-event involves a single γ photon. But an uncorrelated photon from background may be simultaneously detected. It then resembles the major background being $\gamma\gamma^{\pi^0}$-events. The two γ photons of a $\gamma\gamma^{\pi^0}$-event are correlated since produced by the decay of a same particle. Therefore, the distribution of this angle is not random and presents a peak around 5°. However, the oscillations in the non bitonic term are probably learnt from the noise present in the data. The bitonic term permits to solve this irregularity: experts can visually tell that it generalizes better and is more consistent with their expectations.

Higgs Dataset [20]**.** At CERN, Higgs particles are notably produced out of the collisions of two protons. The objective of the dataset is to detect Higgs bosons decaying into two τ-particles. Geometrical features are available for each detected particle.

The angle between the lepton and a hypothetical missing particle illustrated by Fig. 10 is one of the most common feature built by a FCGAM for the Higgs dataset. Indeed this missing momentum actually relates to undetectable particles called neutrinos. In signal events, the neutrinos are in majority emitted in the same direction than the lepton. However, in several background processes, only one neutrino is emitted in the opposite direction of the lepton (see Fig. 11). Therefore, the probability to have a signal event is higher at 0° and at its lowest at ±180°. This feature is highly discriminative but not bitonic. Enforcing bitonicity on this feature is counterproductive.

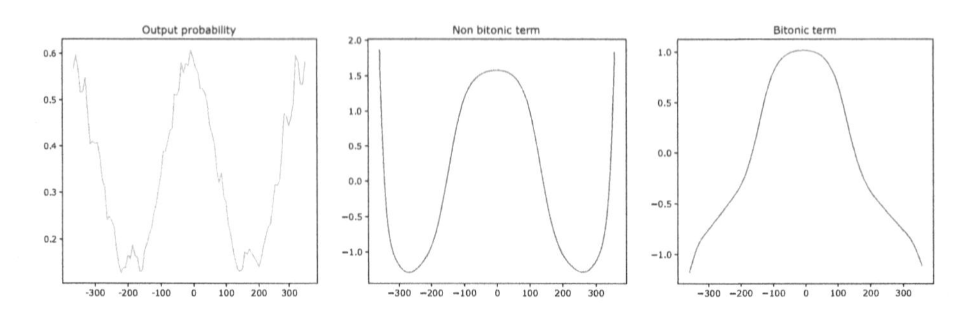

Fig. 10. Higgs dataset: $\phi_{lep} - \phi_{missingtE}$. Higher value means higher probability to have a signal event.

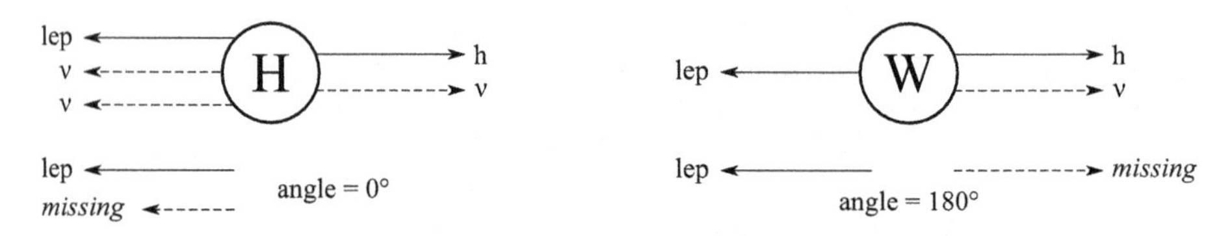

Fig. 11. Illustration of signal events on the left and one type of background events on the right.

6 Conclusion

GAM are widely considered as intelligible models, suitable for applications where transparency and expert interpretation is needed. In this work, bitonicity is introduced to take into account prior knowledge about HEP applications. A method is proposed to test the bitonicity of a feature and to enforce it when fitting shape functions. Feature construction is also incorporated in the process since raw variables in HEP are often not the most relevant for classification purposes and since interpretable models often lack of sufficiently complex internal representation of data.

Experiments on three HEP datasets show that enforcing bitonicity on terms associated with bitonic features increases the interpretability potential and generalization power of the global model, with a performance classification score comparable to the score obtained without constraint, if not greater. However, some datasets have shown to be more adapted to the bitonicity approach, depending on the bitonicity degree of the most discriminative features for the classification problem.

In future work, we plan to deepen our studies on the reasons why bitonicity is working better on some datasets than on others. In addition, we started conducting experiments including 2D terms that involve pairwise interactions between features to complete our existing studies on univariate GAM terms.

Acknowledgments. We would like to thank the CLAS12 collaboration for the simulation software.

References

1. Gilpin, L.H., Bau, D., Yuan, B.Z., Bajwa, A., Specter, M., Kagal, L.: Explaining explanations: an overview of interpretability of machine learning. In: 2018 IEEE 5th International Conference on Data Science and Advanced Analytics (DSAA), pp. 80–89. IEEE (2018)
2. Arrieta, A.B., et al.: Explainable artificial intelligence (XAI): concepts, taxonomies, opportunities and challenges toward responsible AI. Inf. Fusion **58**, 82–115 (2020)
3. Hastie, T., Tibshirani, R.: Generalized additive models. Stat. Sci. **1**(3), 297–310 (1986)
4. Lou, Y., Caruana, R., Gehrke, J.: Intelligible models for classification and regression. In: Proceedings of the 18th ACM SIGKDD International Conference on Knowledge Discovery and Data Mining, pp. 150–158 (2012)
5. Fard, M.M., Canini, K., Cotter, A., Pfeifer, J., Gupta, M.: Fast and flexible monotonic functions with ensembles of lattices. In: Advances in Neural Information Processing Systems, pp. 2919–2927 (2016)
6. Gupta, M., et al.: Monotonic calibrated interpolated look-up tables. J. Mach. Learn. Res. **17**(1), 3790–3836 (2016)
7. Nguyen, A.P., Martínez, M.R.: Mononet: towards interpretable models by learning monotonic features. arXiv preprint arXiv:1909.13611 (2019)
8. Köllmann, C., Bornkamp, B., Ickstadt, K.: Unimodal regression using bernstein-schoenberg splines and penalties. Biometrics **70**(4), 783–793 (2014)
9. Barthelemy, T.: On the unimodality of METRIC Approximation subject to normally distributed demands (2015)
10. Stout, Q.F.: Unimodal regression via prefix isotonic regression. Comput. Stat. Data Anal. **53**(2), 289–297 (2008)
11. Lou, Y., Caruana, R., Gehrke, J., Hooker, G.: Accurate intelligible models with pairwise interactions. In: Proceedings of the 19th ACM SIGKDD International Conference on Knowledge Discovery and Data Mining, pp. 623–631 (2013)
12. Wood, S.N.: Thin-plate regression splines. J. Roy. Stat. Soc. (B) **65**(1), 95–114 (2003)
13. Wood, S.: Generalized Additive Models: An Introduction with R, 2nd edn. Chapman and Hall/CRC, Boca Raton (2017)
14. Friedman, J.H.: Greedy function approximation: a gradient boosting machine. Ann. Stat. **29**(5), 1189–1232 (2001)
15. Pya, N., Wood, S.N.: Shape constrained additive models. Stat. Comput. **25**(3), 543–559 (2014). https://doi.org/10.1007/s11222-013-9448-7
16. Pedregosa, F., et al.: Scikit-learn: machine learning in Python. J. Mach. Learn. Res. **12**, 2825–2830 (2011)
17. Sondhi, P.: Feature construction methods: a survey. sifaka. cs. uiuc. edu **69**, 70–71 (2009)

18. Swesi, I.M.A.O., Bakar, A.A.: Recent developments on evolutionary computation techniques to feature construction. In: Huk, M., Maleszka, M., Szczerbicki, E. (eds.) ACIIDS 2019. SCI, vol. 830, pp. 109–122. Springer, Cham (2020). https://doi.org/10.1007/978-3-030-14132-5_9
19. Cherrier, N., Poli, J.P., Defurne, M., Sabatié, F.: Consistent feature construction with constrained genetic programming for experimental physics. In: 2019 IEEE Congress on Evolutionary Computation (CEC), pp. 1650–1658. IEEE (2019)
20. Adam-Bourdarios, C., Cowan, G., Germain, C., Guyon, I., Kegl, B., Rousseau, D.: Learning to discover: the Higgs boson machine learning challenge - Documentation (2014)
21. Dua, D., Graff, C.: UCI machine learning repository (2017). http://archive.ics.uci.edu/ml

Predicting and Explaining Privacy Risk Exposure in Mobility Data

Francesca Naretto[1], Roberto Pellungrini[2](✉), Anna Monreale[2], Franco Maria Nardini[3], and Mirco Musolesi[4,5]

[1] Scuola Normale Superiore, Pisa, Italy
francesca.naretto@sns.it
[2] University of Pisa, Pisa, Italy
{roberto.pellungrini,anna.monreale}@di.unipi.it
[3] ISTI CNR, Pisa, Italy
francomaria.nardini@isti.cnr.it
[4] University College London, London, UK
m.musolesi@ucl.ac.uk
[5] University of Bologna, Bologna, Italy

Abstract. Mobility data is a proxy of different social dynamics and its analysis enables a wide range of user services. Unfortunately, mobility data are very sensitive because the sharing of people's whereabouts may arise serious privacy concerns. Existing frameworks for privacy risk assessment provide tools to identify and measure privacy risks, but they often (i) have high computational complexity; and (ii) are not able to provide users with a justification of the reported risks. In this paper, we propose EXPERT, a new framework for the prediction and explanation of privacy risk on mobility data. We empirically evaluate privacy risk on real data, simulating a privacy attack with a state-of-the-art privacy risk assessment framework. We then extract individual mobility profiles from the data for predicting their risk. We compare the performance of several machine learning algorithms in order to identify the best approach for our task. Finally, we show how it is possible to explain privacy risk prediction on real data, using two algorithms: SHAP, a feature importance-based method and LORE, a rule-based method. Overall, EXPERT is able to provide a user with the privacy risk and an explanation of the risk itself. The experiments show excellent performance for the prediction task.

Keywords: Privacy risk assessment · Privacy risk prediction · Explainability

1 Introduction

There is a growing research interest in mobility data analysis, since it is a key enabler of a new wave of knowledge-based services and applications. However, the use of human mobility data raises concerns associated to the potential leakage of personal sensitive information as mobility data analysis might reveal details

A. Appice et al. (Eds.): DS 2020, LNAI 12323, pp. 403–418, 2020.
https://doi.org/10.1007/978-3-030-61527-7_27

of people's private life. For example, de Montjoye *et al.* [17] showed that four spatio-temporal points can be enough to uniquely identify 95% of the individuals in a mobility dataset. The existence of these privacy issues has led researchers to develop techniques to mitigate the privacy risks while preserving mobility data [4,15,26]. For enabling a practical application of these techniques, Pratesi *et al.* proposed PRUDEnce [21], a framework for a systematic assessment of individual privacy risk in a mobility dataset. PRUDEnce helps data controllers being compliant with the new EU General Data Protection Regulation (GDPR)[1]. However, PRUDEnce is characterized by a high computational complexity, because it requires the computation of the maximum risk of re-identification (or privacy risk) given an external knowledge that a malicious adversary might use for an attack [20]. The high computational complexity becomes a non-negligible practical limitation in some online user-centric applications where it is useful to have a continuously up-to-date indicator of privacy exposure. In user-centric applications, providing users with an explanation of the reasons of the identified privacy risk might contribute to raise their self-awareness.

In this paper, to overcome the computational complexity drawback and to increase users' awareness, we propose EXPERT, an EXplainable Privacy ExposuRe predicTion framework that exploits *(i)* machine learning (ML) models for predicting a user's individual privacy risk and *(ii) local* explainers for producing explanations of the predicted risk. First, EXPERT extracts from human mobility data an individual mobility profile describing the mobility behavior of any user. Second, for each user it exploits PRUDEnce to compute the associated privacy risk. Third, it uses the mobility profiles of the users with their associated privacy risks to train a ML model. For the prediction task, EXPERT exploits tree-based ensemble models to effectively handle the class-imbalance problem, i.e., a high number of risky users vs a low number of non-risky ones, that is typical of the data in this context. The aim is to have a predictor that preserves the privacy of risky users while providing the freedom of using data-driven services to users with low privacy risk. For a new user, along with the prediction of risk, EXPERT also provides an explanation of the predicted risk. EXPERT exploits two state-of-the-art explanation techniques, i.e., SHAP [13] and LORE [11]. The two methods produce explanations based on feature importance and logic rules, respectively. The goal of explanations is to provide users with insights on which mobility behavior contributes to their privacy risk. We evaluate EXPERT on real-world mobility data showing the effectiveness of the framework. Results show that the proposed framework is able to classify the privacy risk level of unseen users in the urban areas. Moreover, we observe a high recall on the high-risk users, meaning that the probability of misclassifying a high-risk user as low-risk is negligible, while achieving good performance in classifying low-risk users.

The paper is organized as follows. Sect. 2 discusses related work. In Sect. 3, we briefly discuss PRUDEnce, the framework we used for the privacy risk assessment. Section 4 introduces our novel EXPERT framework. In Sect. 5, we report

[1] EU GDPR can be found at the following link: http://bit.ly/1TlgbjI.

the results of a comprehensive experimental evaluation of EXPERT on mobility data. Finally, Sect. 6 concludes the work and discusses future work.

2 Related Work

Our framework leverages the privacy risk assessment framework PRUDEnce [21], which allows for the systematic calculation of the empirical privacy risk. Another risk management framework is LINDDUN [8] useful for modeling privacy threats in software-based systems, but lacks a quantitative evaluation of privacy risk. Some works [23,25] propose to evaluate the privacy risk by a unicity measure computed as the number of records uniquely identified. Armando *et al.* [2] proposed a risk-aware framework for information disclosure supporting runtime risk assessment where access-control decisions are based on the disclosure-risk associated with a data access request and adaptive anonymization is used as a risk-mitigation method.

In the context of mobility analysis, an overview on problems, techniques and methodologies can be found in [28]. Human mobility analysis can reveal personal sensitive information and habits leading to possible privacy violation. Thus, many techniques for privacy-preserving analysis have shown that we can design data-driven mobility services where the quality of results coexists with the privacy protection. Some works, e.g., [4,16], are based on the differential privacy model [9] while others, e.g., [15,26], are based on the k-anonymity model [24].

Our work can be seen as an extension of the prediction methodology proposed by Pellungrini *et al.* [20], showing how it is possible to predict privacy risk in mobility data with a feature based approach. We extend it by providing a unified framework that provides both prediction and explanation about the individual privacy risk. Moreover, our proposal is based on a prediction module that is able to handle the high class imbalance of the data typical of this domain [29].

The importance of interpretability in machine learning has led to an increasing research work in this field. An overview of explainable machine learning models can be found in [12]. This survey identifies two main families of approaches: *local* and *global* explainers. The first category aims at explaining the reason for a specific instance classification [11,13,22], while the goal of the second one is to explain the logic of the "machine learning black-box" as a whole [5–7].

3 Background

Human mobility data contain information about the movement of individuals during a given period of observation. They are typically collected by electronic devices, such as mobile phones and GPS devices installed in vehicles [28]. All the movements of a user in the period of observation are described using a sequence of spatio-temporal data points, i.e., a trajectory. In other words, each sequence item is a pair composed of a geographic location, often expressed in coordinates (generally latitude and longitude), and a timestamp indicating when the user stopped in or went through that location.

Definition 1 (Trajectory). *A human mobility trajectory is a temporally ordered sequence of pairs,* $T_u = (l_1, t_1), (l_2, t_2), \ldots, (l_m, t_m)$*, where* $l_i = \langle x_i, y_i \rangle$ *is the location identified by the latitude* x_i *and longitude* y_i*, while* t_i $(i = 1, \ldots, m)$ *denotes the corresponding timestamp such that* $\forall 1 \leq i \leq m \ t_i < t_{i+1}$*.*

We denote by $\mathcal{D} = T_1, \ldots, T_n$ the *mobility dataset* that describes the complete history of movements of n individuals, in a specific period of observation.

3.1 Privacy Risk Assessment Framework

In this paper, we consider the framework PRUDEnce [21], which allows for a systematic assessment of the privacy risk inherent to human mobility data. It considers a scenario where a Service Developer (SD) requests data from a Data Provider (DP) to develop services or perform an analysis. In order to guarantee the right to privacy of individuals, the DP has to assess their privacy risk before the data sharing. Once assessed the privacy risk, the DP can choose how to protect the data before sharing them, selecting the most appropriate privacy-preserving technology. Taking into account the data requirements of the SD, the DP aggregates, selects, and filters the dataset $\mathcal{D}$ to meet its requirements and on top of it performs a privacy risk assessment. This operation requires the definition of a set of possible attacks that an adversary might conduct on the data, and their simulation. The user's privacy risk is related to her probability of re-identification in a dataset with respect to a set of attacks. An attack assumes that an adversary gets access to a dataset, then, using some previously obtained background knowledge, i.e., the knowledge of a portion of an individual's mobility data, the adversary tries to re-identify all the records in the dataset regarding that individual. An attack is defined by a matching function, which represents the process with which an adversary exploits the background knowledge to find the corresponding individual in the data. As far as the attack definition is concerned, PRUDEnce is based on the notions of background knowledge category, configuration and instance. The first one denotes the type of information known by the adversary about a specific set of dimensions of an individual's mobility data: e.g.., a subset of the locations visited by a user (spatial dimension) or the specific times a user visited those locations (spatial and temporal dimensions). The number of the elements known by the adversary is called background knowledge configuration. An example is the adversary knowledge of $h = 2$ locations visited by an individual. Finally, an instance of background knowledge is defined as the specific information known by the adversary, such as a visit in a specific location. Consider a trajectory from D: $T_u = \langle (l_1, t_1), (l_2, t_2), (l_3, t_3), (l_4, t_4) \rangle$ of an individual u. Based on T_u the DP can generate all the possible instances of a background knowledge configuration that an adversary might use to re-identify the whole T_u. If the adversary knows the ordered subsequences of locations and $h = 2$, we obtain the background knowledge configuration: $B_2 = \{((l_1, t_1), (l_2, t_2)),\ ((l_1, t_1), (l_3, t_3)),\ ((l_1, t_1), (l_4, t_4)),\ ((l_2, t_2), (l_3, t_3)),\ ((l_2, t_2), (l_4, t_4)),\ ((l_3, t_3), (l_4, t_4))\}$. The adversary might know instance $b = ((l_1, t_1), (l_4, t_4)) \in B_{h=2}$ and aims at detecting all the records in D regarding u, in order to reconstruct the whole trajectory T_u.

The definition of privacy risk is based on these notions and on the following definition of probability of re-identification.

Definition 2. *Given an attack and its function* $matching(T, b)$ *indicating if a record* $T \in \mathcal{D}$ *matches the instance of background knowledge configuration* $b \in B_h$*, and a function* $M(\mathcal{D}, b) = \{T \in \mathcal{D} | matching(T, b) = True\}$*, we define the probability of re-identification of an individual* u *in dataset* $\mathcal{D}$ *as:* $PR_{\mathcal{D}}(T = u|b) = \frac{1}{|M(\mathcal{D},b)|}$ *that is the probability to associate a record* $T \in \mathcal{D}$ *to an individual* u*, given instance* $b \in B_h$*.*

Since each instance $b \in B_h$ has its own probability of re-identification, the risk of re-identification of an individual is defined as the maximum probability of re-identification over the set of instances of a background knowledge configuration:

Definition 3. *The risk of re-identification (or privacy risk) of an individual* u *given a background knowledge configuration* B_h *is her maximum probability of re-identification* $Risk(u, \mathcal{D}) = \max PR_{\mathcal{D}}(T = u|b)$ *for each* $b \in B_h$*.*

4 Explainable Privacy Risk Prediction Framework

PRUDEnce [21] assumes a worst case scenario approach for the privacy risk computation and therefore, it evaluates all the possible background knowledge configurations for a potential adversary generating them with a combinatorial approach directly from the data of a user. While the framework provides a comprehensive methodology for worst-case privacy risk assessment, its computational complexity is high. Moreover, PRUDEnce is designed for supporting data providers (companies) in identifying portions of data with high privacy risk by simulations of the attacks. The computation requires the availability of the entire dataset, like that stored in the servers of the companies. In other words, PRUDEnce is not suited for providing personalized recommendations in terms of risks associated to sharing personal trajectories. Indeed, for any new user requiring risk evaluation, the system should re-compute the privacy risk against the whole dataset. Moreover, it does not provide any explanation of the privacy risk derived by the system. In this paper we present an explainable framework for the *individual* prediction of a user's privacy risk, in order to increase privacy risk awareness, by also providing an explanation of the derivation of the risk associated to sharing sensitive location information. The idea is inspired by the explainable privacy-preserving system theorized in [3]. To this end, we propose EXPERT which, given a user's trajectory, predicts the privacy risk associated with it. The explanation provided to the users is based on their trajectory given in input. Figure 1 depicts the architecture of EXPERT which is composed of two main modules: the *privacy risk prediction* module which takes as input the user's trajectory and, exploiting a trained ML model, predicts the privacy risk level of that user, and the *explanation* module which produces the explanation of the predicted risk. The ML model is the result of several steps: *(i)* the empirical computation of the individual privacy risk, *(ii)* the extraction of individual mobility profiles from human mobility data, summarizing users' mobility behavior, and *(iii)* the training of a ML model.

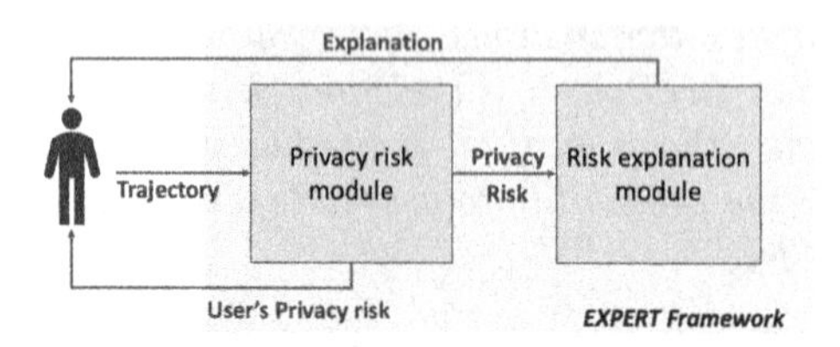

Fig. 1. The general structure of the proposed framework EXPERT.

4.1 Learning a Prediction Model for Individual Privacy Risk

The basic idea is to train a ML model to predict the privacy risk level of users based solely on their individual mobility profile. Thus, given a human mobility dataset of n user trajectories, we propose to derive the training dataset $\langle M, \Gamma \rangle$, where M is a set of n individual mobility profiles, and Γ is the vector of their associated privacy risk levels. Since, the privacy risk is related to a specific attack (see Sect. 3.1), the procedure for *building a training dataset* depends on the adversary attack modelling. As a consequence, given a specific attack, characterized by a background knowledge configuration B_h, the procedure performs the following two steps:

- *Mobility Profile Extraction*: Given a mobility dataset $\mathcal{D}$, for every user trajectory T_u we propose to extract a mobility profile in order to characterize her mobility behavior. To this end, we propose to derive a set of well-known mobility features (presented in the next section). We denote by $M_u \in M$ the mobility feature vector of a user.
- *Privacy Risk Computation*: For each user u a privacy risk value is computed by simulating an attack with background knowledge configuration B_h on the mobility dataset $\mathcal{D}$. Since the goal is to predict the privacy risk level, the privacy risk vector is discretized to get a set of risk classes[2], and the vector of n user's privacy risk levels $\mathit{\Gamma}$.

After the execution of the above two steps, we get a training set $\langle M, \mathit{\Gamma} \rangle$. The derived training dataset $\langle M, \Gamma \rangle$ is used to train a predictive model which will be used within EXPERT to immediately estimate the privacy risk level of previously unseen users, whose data were not used in the learning process. Clearly, in prediction time, in order to predict the privacy risk of a new trajectory instance the process requires, first the computation of the mobility profile for that user and

[2] In our experiments we discretize the risk in two main classes: low risk (privacy risk ≤ 0.5) and high risk (privacy risk > 0.5).

then, the application of the predictive model. Among the different ML methods, we propose to employ models able to handle classification tasks with imbalanced data. Indeed, as we show in our experiments, one of the characteristics of our training data is that most of the users have high privacy risk. Our goal is to get a predictor able to guarantee the privacy protection of risky users while providing the freedom of using data-driven services to users with low privacy risk. Thus, the optimal predictor should be characterized by a low probability of misclassifying a high risk user as a low risk one, while maintaining also good performance with respect to the classification of low risk users. In this paper, we propose to apply the GCFOREST model [29], a decision tree ensemble approach with performance highly competitive to deep neural networks in a broad range of tasks. It is especially suitable to handle highly extra-imbalanced data [27]. GCFOREST relies on multiple layers of parallel forests of trees whose output is then concatenated to re-represent data to subsequent layers. In our experiments we compare GCFOREST against models such as decision tree, logistic regression, and random forest.

Mobility Profile Extraction. The goal of this step is to construct the matrix M representing the set of individual mobility profiles, expressed by a set of mobility features that describe and summarize the mobility behavior of an individual. In our setting, we employ measures widely used in the literature [18,20]. Some of them describe only the mobility behaviour of an individual, while others describe an individual mobility behaviour in relation to collective mobility characteristics. Table 1 reports all the mobility measures used in the study. First of all, we define V as the number of visits of a user, it corresponds to the total number of locations in the user's trajectory. To quantify the erratic behaviour of a user during the day we compute the average number of daily visits $\overline{V}$, dividing V by the total number of days in the period of observation. *Locs*, instead, is the number of distinct locations visited by a user during the period of observation, while $Locs_{ratio}$ represents the fraction of locations covered by a user. We compute it by dividing *Locs* by the total number of locations available in the territory. We also evaluated some measures about the distances travelled by the users. We define D_{max} as the maximum distance travelled by each user, i.e. the longest trip for each user. This measure is then employed for the computation of D_{max}^{trip}: it is the ratio between the maximum distance travelled D_{max} and the maximum distance that is possible to travel in the area of observation. We also consider D_{sum}, i.e., the sum of all the distances travelled by a user. This value is then used in the definition of $\overline{D_{sum}}$, which is the average of D_{sum} over the period of observation (expressed in days). We also consider the *radius of gyration* [19] representing the characteristic distance travelled by a user during the period of observation and is defined as $r_g = \sqrt{\frac{1}{N}\sum_{i \in L} w_i(r_i - r_{cm})^2}$, in which $i \in L$ is the visited location by a user, w_i represents a user's frequency of visits at a location i, r_i denotes the geographical description of the location i and it is a bi-dimensional vector, while r_{cm} is the center of mass of the user under consideration. Mathematically, the latter is defined as $r_{cm} = \frac{1}{V}\sum_{1 \in L} r_i$. We also measure the *mobility entropy* E as the predictability of a user's trajectory. We employ the Shannon entropy measure [10]: $E = -\sum_{i \in L} p_i \log_2 p_i$, in which p_i is the probability of the location i for the user under analysis. For each user, we also consider three locations that characterize a user's mobility: the most visited location, the second most visited location and the least visited location. Typically, the most visited location corresponds to user's home, while the second most visited location is users' work place. For each one of these locations, we evaluate the frequency of visits during the period of observation w_i, where i represents the specific location under analysis. We also define $\overline{w_i}$ as the daily average of the frequency of visits at the location i for the user under analysis. Then, we denote by w_i^{pop} the frequency of visits divided by the popularity of the location, i.e. the total frequency of the location in the dataset. In this way, we normalize the frequency of the user for a particular location considering the behaviour of all the users in the dataset. For these three locations, we also consider U_i, i.e., the number of distinct users that visited the location i in the period of observation. Out of U_i, we also compute U_i^{ratio}, in which the number of distinct users that visited the location i is divided by the total number of users in the dataset. The last measure we consider for each of the three locations is the entropy. In this case, we compute a *location entropy* E_i, that represents the predictability of a visit at the location i defined as: $E = -\sum_{u \in U_i} p_u \log_2 p_u$, where U_i is the set of users that visited the location i and p_u is the probability that a user u visited the location i. When working with trajectories, we have also a temporal information: each trajectory is composed by $\langle l_i, t_i \rangle$, in which t_i is the timestamp corresponding to time of arrival of a user at a location l_i. We exploit this information to compute the *path time* [18], i.e., the time occurring between the first and last visit of a trajectory.

Table 1. Mobility features of the individual mobility profile.

Notation	Description	Notation	Description
V	visits	$\overline{V}$	daily visits
D_{max}	max distance	D_{sum}	sum distances
D_{max}^{tot}	max distance over total max distance for a user	$\overline{D}_{sum}$	D_{sum} per day
D_{max}^{trip}	D_{max} over area	$Locs$	distinct locations
$Locs_{ratio}$	$Locs$ over area	R_g	radius of gyration
E	mobility entropy	E_i	location entropy
U_i	individuals per location	U_i^{ratio}	U_i over individuals
w_i	location frequency	w_i^{pop}	w_i over the total frequency of location i
$\overline{w}_i$	daily location frequency	PT_j	Path time per user

Privacy risk computation. The goal of this module is to compute for each user trajectory in $\mathcal{D}$ a privacy risk value by using a re-identification algorithm. We propose to apply the PRUDEnce framework (Sect. 3.1) that enables the definition and simulation of any desired privacy attacks over the entire dataset. Several attacks might be defined on the basis of the type of background knowledge possessed by an adversary [20,21]. In this paper we instantiate our risk computation using the location sequence attack, introduced in [14,15], where the adversary knows a subset of the locations visited by the individual and the temporal ordering of the visits. Given an individual u, we denote by $L(T_u)$ the sequence of locations $l_i \in T_u$ visited by u. The background knowledge category of a location sequence attack is defined as follows:

Definition 4. *Let h be the number of locations l_i of an individual u known by the adversary. The Location Sequence background knowledge is a set of configurations based on h locations, defined as $B_h = L(T_u)^{[h]}$, where $L(T_u)^{[h]}$ denotes the set of all the possible h-subsequences of the elements in the set $L(T_u)$.*

We indicate with $a \preceq b$ that a is a subsequence of b. Each instance $b \in B_h$ is a location subsequence $X_u \preceq L(T_u)$ of length h. Given a record $T \in \mathcal{D}$ we define the matching function as: $matching(T, b) = true$ if $b \preceq L(T)$, *false* otherwise. PRUDEnce uses this function to compute the probability of re-identification for any instance of background knowledge (Definition 2) enabling the privacy risk computation for each trajectory (Definition 3).

4.2 Risk Explanation Module

The last module of EXPERT is the *explainer* aiming at providing the end-user with an explanation for the predicted risk label. The objective is to increase users' awareness about the privacy risks. EXPERT is modular with respect to the explainer allowing the use of any explanation method suitable to tabular data. Since the goal is to explain a specific decision, *local* methods [11,13,22] are more suitable for this task. The main difference between them is the type of explanation returned. LIME [22] and SHAP [13] are mainly based on the notion of feature importance and LORE [11] instead provides a logical rule-based explanation for the prediction. In our experiments we considered LORE and SHAP as explainers. Given our ML model and an individual trajectory belonging to a user u, transformed into the mobility profile M_u and labeled with a specific privacy risk level r_u by our model, LORE (LOcal Rule-based Explanation) builds a simple, interpretable predictor by first generating a balanced set of neighbor instances of the given M_u through an ad-hoc genetic algorithm, and then extracting from such a set a decision tree classifier. A *local explanation* is then extracted from the obtained decision tree. The local explanation is a pair composed by *(i)* a *logic rule*, corresponding to the path in the tree that explains why M_u has been labeled as r_u by the predictor, and *(ii)* a set of *counterfactual rules*, explaining which changes in M_u would invert the risk class assigned. SHAP (SHapley Additive exPlanations) is a local approach for interpreting model predictions that assigns to each feature an importance value for a particular prediction. Moreover, for each model's prediction SHAP defines an *explanation* model. The main idea is that the explanation model is an interpretable approximation of the original model and works with simplified input data. SHAP exploits the collaborative game theory to determine the importance value of a feature for the instance prediction.

5 Experiments

We experimentally validate the different components of our framework by analyzing the performance of: *i)* the prediction module implemented with different machine learning models by varying their complexity; and *ii)* the explanation module by comparing two state-of-the-art approaches.

Data. We use data containing GPS tracks of private vehicles in Tuscany (Italy) provided by Octo Telematics. We selected trajectories from an area comprising two major urban centers, Prato and Pistoia, considering the period from 1st May to 31st May 2011, for a total of 8651 distinct vehicles. We performed two different transformations of the original data in order to obtain two different datasets. In the first dataset, called `istat`, trajectory points are generalized according to the geographical tessellation provided by the Italian National Statistics Bureau (ISTAT): each point is substituted with the centroid of the geographical cell to which it belongs. We then remove redundant points, i.e., points mapped to the same cell at the same time, obtaining 2274 different locations with an average length of 31.9 points per trajectory. With respect to the second dataset, called `voronoi`, we first apply a data-driven Voronoi tessellation of the territory [1], taking into consideration the traffic density of an area, and then we used the cells of this tessellation to increase the granularity of the original trajectories. The algorithm also performs interpolation between non adjacent points[3]. We obtained 1473 different locations with an average length of 240.2 points per trajectory. For both datasets we computed the mobility features M for extracting the users' mobility profiles and the privacy risk according to the simulation of the location sequence attack (Sect. 4.1) with four background knowledge configurations B_h using $h = 2, 3, 4, 5$, getting four different risk datasets, $\Gamma_{h=2,3,4,5}$. We discretized the risk values in intervals: $[0, 0.5]$ and $(0.5, 1]$ named *low* and *high* risk class, respectively. Then, we built our classification datasets merging each risk dataset with the feature-based mobility profiles: $\langle M, \Gamma_h \rangle$, as explained in Sect. 4.1. To better handle the imbalance in the data, we learned our predictive models using stratified sampling, undersampling and 5-fold cross-validation. Tables 3 and 2 report the class balance after under-sampling the majority class. We also performed hyper-parameter tuning by grid search in the parameter space[4].

Predicting Risk. We validate the effectiveness of the prediction module of EXPERT by comparing four different ML models: Decision Tree (DT), Logistic Regression (LR), Random Forest (RF)[5], and GCFOREST (GC)[6]. Decision Tree and Logistic Regression are two well-known, white-box models. Random Forest and GCFOREST [29] are ensemble models proven to be effective when dealing with imbalanced data. This task is characterized by strong imbalance of the two risk classes, therefore being a challenging machine learning problem, where

[3] Voronoi tessellation obtained by using: http://geoanalytics.net/V-Analytics/.
[4] Hyper-parameter settings: https://github.com/francescanaretto/prp.
[5] https://scikit-learn.org/stable/.
[6] https://github.com/kingfengji/gcForest.

Table 2. Predictive models evaluation on mobility profiles derived from `istat`.

B_h	Class Balance	Under-sampling	Metric	DT	LR	RF	GC
h=2	High=77 Low=23	High=40 Low=60	$F1_{high}$	0.92 (0.00)	0.92 (0.00)	**0.94** (0.00)	**0.94** (0.02)
			P_{high}	0.90 (0.01)	0.91 (0.01)	0.91 (0.00)	**0.92** (0.01)
			R_{high}	0.93 (0.01)	**0.96** (0.00)	0.95 (0.00)	**0.96** (0.00)
			$F1_{low}$	0.69 (0.02)	0.71 (0.01)	**0.75** (0.01)	**0.75** (0.01)
			P_{low}	0.73 (0.02)	0.77 (0.01)	0.81 (0.01)	**0.82** (0.01)
			R_{low}	0.66 (0.02)	0.42 (0.03)	**0.70** (0.09)	**0.70** (0.02)
h=3	High=93 Low=7	No under-sampling	$F1_{high}$	0.96 (0.00)	0.92 (0.00)	**0.97** (0.00)	**0.97** (0.03)
			P_{high}	0.95 (0.01)	0.94 (0.01)	**0.96** (0.00)	**0.96** (0.00)
			R_{high}	0.96 (0.00)	**0.98** (0.00)	**0.98** (0.00)	**0.98** (0.00)
			$F1_{low}$	0.70 (0.02)	0.71 (0.01)	0.75 (0.01)	**0.79** (0.03)
			P_{low}	0.72 (0.02)	0.77 (0.03)	0.83 (0.03)	**0.84** (0.03)
			R_{low}	0.70 (0.06)	0.41 (0.03)	0.70 (0.04)	**0.74** (0.05)
h=4	High=95 Low=5	No under-sampling	$F1_{high}$	0.96 (0.00)	0.96 (0.00)	**0.97** (0.00)	**0.97** (0.00)
			P_{high}	0.96 (0.05)	0.95 (0.00)	0.96 (0.00)	**0.97** (0.00)
			R_{high}	0.97 (0.00)	**0.98** (0.00)	**0.98** (0.00)	**0.98** (0.00)
			$F1_{low}$	0.73 (0.02)	0.70 (0.02)	0.77 (0.02)	**0.80** (0.02)
			P_{low}	0.75 (0.02)	0.80 (0.01)	**0.85** (0.02)	**0.85** (0.09)
			R_{low}	0.70 (0.01)	0.45 (0.03)	0.74 (0.05)	**0.76** (0.03)
h=5	High=96 Low=4	No under-sampling	$F1_{high}$	0.96 (0.04)	0.96 (0.00)	**0.97** (0.00)	**0.97** (0.00)
			P_{high}	0.96 (0.04)	0.95 (0.00)	**0.97** (0.00)	**0.97** (0.00)
			R_{high}	0.96 (0.01)	**0.98** (0.00)	**0.98** (0.00)	**0.98** (0.00)
			$F1_{low}$	0.73 (0.03)	0.70 (0.03)	0.78 (0.02)	**0.80** (0.02)
			P_{low}	0.72 (0.03)	0.80 (0.05)	0.83 (0.02)	**0.85** (0.02)
			R_{low}	0.70 (0.03)	0.46 (0.03)	0.75 (0.04)	**0.76** (0.03)

the classifier performance in terms of accuracy is less significant due to the dominance of the majority class on the metric.

Indeed, as discussed in Sect. 4.1, our desiderata is a classifier with a conservative approach with respect to high risk users, to avoid their misclassification as low risk users. On the other hand, we aim at achieving high precision and recall for both high and low risk users. As a consequence, for the performance evaluation of the machine learning models, we select the following indicators: *i)* precision (P_{high}) and recall (R_{high}) on high risk; *ii)* precision (P_{low}) and recall (R_{low}) on low risk; and *iii)* the two corresponding *F1-Score* for low ($F1_{low}$) and high ($F1_{high}$) risk. In a setting where the size of high risk class is larger than that of the low risk one, achieving good performance for the low risk users is difficult. The results for the two datasets are shown in Tables 2 and 3. We note that `istat` represents a typical situation in the privacy context, where a high number of risky users exists. We also built `voronoi` to present a balanced situation and to verify how our models behave in such a case. In general, we found that the ensemble methods have good performance in terms of both *F1-Score* on high risk and *F1-Score* on low risk. This means that these models are suitable for our target. More precisely, we observe that, although GC and RF have comparable performance, for `istat`, that is extra imbalanced, GC performs slightly better than RF on the low risk class. Moreover, ensemble methods also outperform the white-box classifiers and again, their advantage is more evident in `istat`; especially, they considerably improve the classification scores for the more difficult category of low-risk users. Indeed, we found that GC increases of 0.04–0.06 (0.09–0.13) points the R_{low} (P_{low}) of DT and of 0.28–0.33 (0.05–0.07) points the R_{low} (P_{low}) of LR. Clearly, these results contribute to have GC with the best $F1_{low}$ for every value of h, while still maintaining a conservative behaviour highlighted by the high values of recall on high risk class (R_{high}). Regarding `voronoi`, we further notice that, although the data are more balanced, the ensemble methods always maintain the conservative approach for high risk users (high R_{high}) while improving the overall classification for low risk users ($F1_{low}$). Overall, these results suggest that GC is the most suitable option for our specific predictive task with RF as a close second one.

Table 3. Predictive models evaluation on mobility profiles derived from `voronoi`.

B_h	Class Balance	Under-sampling	Metric	DT	LR	RF	GC
h=2	High=28 Low=72	High=30 Low=70	$F1_{high}$	0.71 (0.02)	0.65 (0.07)	0.75 (0.02)	**0.80** (0.01)
			P_{high}	0.73 (0.01)	0.73 (0.02)	0.78 (0.01)	**0.79** (0.01)
			R_{high}	0.74 (0.04)	0.77 (0.03)	0.72 (0.02)	**0.80** (0.03)
			$F1_{low}$	0.87 (0.00)	0.86 (0.01)	**0.89** (0.01)	**0.89** (0.00)
			P_{low}	0.70 (0.01)	0.89 (0.01)	0.87 (0.01)	**0.90** (0.02)
			R_{low}	0.85 (0.01)	0.82 (0.02)	**0.91** (0.01)	0.86 (0.01)
h=3	High=55 Low=45	No under-sampling	$F1_{high}$	0.88 (0.01)	0.88 (0.01)	**0.92** (0.01)	**0.92** (0.01)
			P_{high}	0.89 (0.01)	0.88 (0.01)	**0.91** (0.00)	**0.91** (0.00)
			R_{high}	0.86 (0.02)	0.89 (0.03)	**0.92** (0.01)	**0.92** (0.01)
			$F1_{low}$	0.84 (0.02)	0.82 (0.01)	**0.87** (0.01)	**0.87** (0.01)
			P_{low}	0.80 (0.02)	0.83 (0.03)	**0.88** (0.09)	**0.88** (0.01)
			R_{low}	**0.89** (0.02)	0.81 (0.02)	0.87 (0.01)	0.86 (0.01)
h=4	High=57 Low=43	High=40 Low=60	$F1_{high}$	0.91 (0.00)	0.90 (0.00)	**0.93** (0.00)	**0.93** (0.00)
			P_{high}	0.91 (0.01)	0.88 (0.00)	0.92 (0.00)	**0.94** (0.01)
			R_{high}	0.91 (0.02)	**0.92** (0.01)	**0.92** (0.01)	0.91 (0.01)
			$F1_{low}$	0.84 (0.01)	0.80 (0.01)	**0.87** (0.01)	**0.87** (0.01)
			P_{low}	0.84 (0.03)	0.84 (0.01)	**0.85** (0.01)	**0.85** (0.01)
			R_{low}	0.84 (0.02)	0.77 (0.03)	**0.88** (0.01)	**0.88** (0.02)
h=5	High=62 Low=38	High=50 Low=50	$F1_{high}$	0.93 (0.01)	0.93 (0.01)	**0.94** (0.00)	**0.94** (0.01)
			P_{high}	0.92 (0.03)	0.90 (0.01)	0.94 (0.01)	**0.95** (0.02)
			R_{high}	0.93 (0.02)	0.93 (0.02)	**0.94** (0.01)	**0.94** (0.01)
			$F1_{low}$	0.83 (0.01)	0.80 (0.03)	**0.86** (0.01)	**0.86** (0.02)
			P_{low}	0.83 (0.03)	0.83 (0.03)	**0.86** (0.03)	**0.86** (0.02)
			R_{low}	0.84 (0.03)	0.84 (0.03)	**0.87** (0.02)	0.86 (0.03)

Explaining Risk. Regarding the explanation task in our experiments, we employed LORE [11] and SHAP [13]. We followed the experimental methodology proposed in [11]: we selected the best models from the k-fold validation presented in Sect. 5 and its associated train and test datasets. In particular, we used a RF and a GC model for $h = 2$ on the `istat` dataset. For SHAP we trained the *Kernel Explainer* on the training dataset. For LORE, we chose a genetic generation of the neighborhood and the Euclidean distance as distance among the neighbors. We performed a comparative analysis to evaluate the compactness and comprehensibility of returned explanations. To this end, we considered the diversity of the explanation structure provided by the two methods: LORE outputs rules with premises of variable lengths, while SHAP, outputs the importance of each feature in the data. Thus, we considered two different settings: i) *no-zero features*, where in the SHAP result we only keep features with importance values different from zero; and, ii) *top-k features*, that tries to automatically identify the k features with highest importance values. The value k depends on the record explanation under analysis. To detect the best k for each explanation, we used an elbow-like approach which, given the SHAP result, first sorts in descending order the importance values and then, calculates the segment s bounded by the biggest and the smallest importance values. At this point, it selects the importance value m with the maximum distance from the segment s. Thus, only features with importance values greater than or equal to m are kept. For analyzing the compactness of the explanations we considered their average lengths: LORE explanations have an average length of 2.9 ± 1.3 (RF) and 3.8 ± 1.4 (GC), against the average lenghts of paths of the decision tree of 7.8 ± 1.5. SHAP explanations have an average length of 17.1 ± 3.1(RF) and 16.2 ± 3.2 (GC) for the *no-zero features* setting, which decrease to 9.8 ± 6.3 (RF) and 8.3 ± 7.1 (GC) for the *top-k features* setting. Hence, LORE provides more compact explanations with respect to the paths of the decision tree and the SHAP importance values. We also compare the two explanation types in terms of semantic coherence. To this end, we propose to use the *Jaccard similarity* to highlight the degree of common features used for the explanations and *coherence* measure aiming at capturing the percentage of features used in LORE explanations which are important also in SHAP explanations. The *Jaccard similarity* measure, is defined as $\frac{1}{n}\sum_{i=1}^{n}\frac{F_i^{lore} \cap F_i^{shap}}{F_i^{lore} \cup F_i^{shap}}$ while the *coherence* is defined as $\frac{1}{n}\sum_{i=1}^{n}\frac{F_i^{lore} \cap F_i^{shap}}{|F_i^{lore}|}$. Here, F_i refers to the set of features included in the explanation for the record i.

Table 4 reports the results of the coherence analysis. Regarding the *no-zero features* setting, we found out that the Jaccard similarity is close to zero, highlighting that the intersection of the two feature sets is quite small compared to their union. Concerning the coherence, a value equal to 1 means that all the features of LORE are also in SHAP explanations. Results highlight that SHAP explanations contain the majority of the features used by LORE. In the *top-k features* setting, we observe a general decrease in the values of both measures. This means that the majority of the features that LORE uses in its rules are actually among the least important features of SHAP. Thus, when considering only the *top-k* features the discrepancy between SHAP important values and LORE increases. Our analysis highlights that the two methods consider different important features for providing explanations. LORE explanations tend to be more compact and easy to understand due to the logic structure of the rules. SHAP outputs a visualization and a large amount of information, which might potentially be difficult for a user to navigate. Indeed, a large number of the

Table 4. SHAP vs LORE in the **istat** dataset with $h = 2$.

Setting		Jaccard	Coherence
Top-k Features	RF	0.133 ± 0.063	0.472 ± 0.381
	GC	0.096 ± 0.101	0.393 ± 0.038
No-zero Features	RF	0.133 ± 0.063	0.816 ± 0.250
	GC	0.165 ± 0.072	0.767 ± 0.232

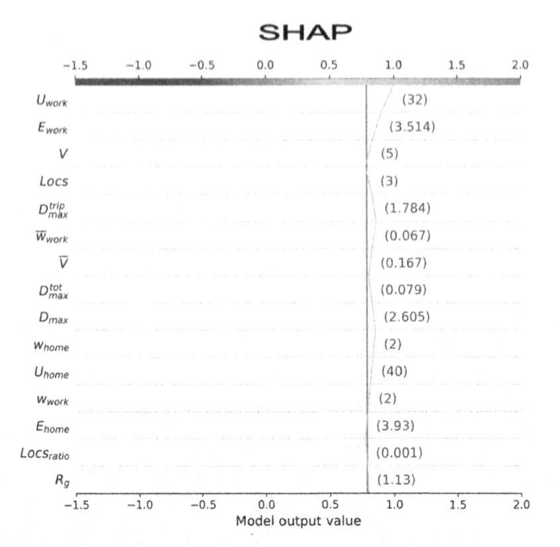

LORE $\overline{w}^{pop}_{home} \leq 0.36, U_{home} \leq 1722, E \leq 1.09, \overline{w}_{work} \leq 0.82 \implies HighRisk$

LORE $\overline{w}^{pop}_{home} \leq 0.36, U_{home} \leq 1722, E \leq 1.09, \overline{w}_{work} \leq 0.82 \implies HighRisk$

Fig. 2. SHAP vs LORE: Table 4 quantifies the similarity between the two explanations. SHAP visualization (right) and the LORE rule (left) represent the explanations for a specific record classified as *high risk* by GCFOREST.

values of the importance features are close to zero. Moreover, given a feature used in an explanation, LORE provides a richer information that could help in understanding more about certain mobility habits that contribute to a specific risk value. For example, let us analyze Fig. 2, where we provide SHAP (right) and LORE (left) explanations for a high risky user according to GCFOREST. With SHAP a user can only understand which feature (with its specific value indicated between parentheses) is important or not for classification, while the LORE rule provides a user with a more detailed motivation, which includes the set of conditions on features that a user satisfies. For example, for the LORE explanation a user can understand that their risk depends on the fact that she travelled more than 0.09 km (D_{max}), their home location is visited by less than 1772 distinct users, and their work location is not enough popular in the data. This reasoning is not supported by the SHAP result. After the local explanation evaluation, we also performed a comparative analysis of global feature importance among all the ML models (Table 5). An interesting result is that the number of locations (*Locs*) is the most important feature for LR, DT and GC, while for RF it is in the second position. Moreover, LR is the only one which considers the entropy of locations (home and work) as important features.

Table 5. Global top-5 most important features of machine learning models.

DT	LR	RF	GC
$Locs$ (0.45)	$Locs$ (0.35)	D_{sum} (0.15)	$Locs$ (0.07)
D_{max} (0.10)	E_{home} (0.14)	$Locs$ (0.13)	U_{work} (0.04)
U_{work} (0.06)	E_{work} (0.12)	$Locs_{ratio}$ (0.08)	$Locs_{ratio}$ (0.03)
$\overline{D}_{sum}$ (0.06)	W_{work} (0.10)	$\overline{D}_{sum}$ (0.07)	U_{home} (0.03)
U_{home} (0.06)	$\overline{D}_{sum}$ (0.08)	U_{work} (0.07)	D^{trip}_{max} (0.02)

6 Conclusions

We have presented EXPERT, a framework for predicting and explaining users' privacy risk associated to the analysis of mobility data. EXPERT exploits ML techniques that are suitable to handle extra-imbalanced data and local explainers to provide users with meaningful explanations about the predicted privacy risk. The empirical evaluation of EXPERT using real-world data demonstrate its effectiveness in predicting privacy risk and in increasing users' self-awareness in relation to potentially risky mobility behavior. The main limitation of the framework is that it requires domain expertise for extracting users' profiles for the prediction. Our future research agenda includes the substantiation of the prediction module by a ML model that does not require the extraction of mobility features. This work could also be extended to generic sequential data.

Acknowledgments. This work has been funded by the European projects SoBigData-PlusPlus (Grant Id 871042), XAI (Grant Id 834756) and HumanE-AI-Net (Grant Id 952026).

References

1. Andrienko, N.V., Andrienko, G.L.: Spatial generalization and aggregation of massive movement data. IEEE Trans. Vis. Comput. Graph. **17**(2), 205–219 (2011)
2. Armando, A., et al.: Risk-based privacy-aware information disclosure. Int. J. Secur. Softw. Eng. **6**(2), 70–89 (2015)
3. Baron, B., Musolesi, M.: Interpretable machine learning for privacy-preserving pervasive systems. IEEE Pervasive Comput. **19**(1), 73–82 (2020)
4. Cormode, G., Procopiuc, C.M., Srivastava, D., Tran, T.T.L.: Differentially private summaries for sparse data. In: ICDT 2012, pp. 299–311 (2012)
5. Craven, M., Shavlik, J.W.: Extracting tree-structured representations of trained networks. In: NIPS, pp. 24–30 (1996)
6. Craven, M.W., Shavlik, J.W.: Using sampling and queries to extract rules from trained neural networks. In: JMLR, pp. 37–45. Elsevier (1994)
7. Deng, H.: Interpreting tree ensembles with intrees. Int. J. Data Sci. Anal. **7**(4), 277–287 (2019). https://doi.org/10.1007/s41060-018-0144-8
8. Deng, M., et al.: A privacy threat analysis framework: supporting the elicitation and fulfillment of privacy requirements. Requir. Eng. **16**(1), 3–32 (2011). https://doi.org/10.1007/s00766-010-0115-7

9. Dwork, C., McSherry, F., Nissim, K., Smith, A.: Calibrating noise to sensitivity in private data analysis. In: Halevi, S., Rabin, T. (eds.) TCC 2006. LNCS, vol. 3876, pp. 265–284. Springer, Heidelberg (2006). https://doi.org/10.1007/11681878_14
10. Eagle, N., Pentland, A.S.: Eigenbehaviors: identifying structure in routine. Behav. Ecol. Sociobiol. **63**, 1057–1066 (2009). https://doi.org/10.1007/s00265-009-0739-0
11. Guidotti, R., et al.: Factual and counterfactual explanations for black box decision making. IEEE Intell. Syst. **34**(6), 14–23 (2019)
12. Guidotti, R., et al.: A survey of methods for explaining black box models. ACM Comput. Surv. **51**, 1–42 (2019)
13. Lundberg, S.M., Lee, S.I.: A unified approach to interpreting model predictions. In: NIPS, pp. 4765–4774 (2017)
14. Mohammed, N., et al.: Walking in the crowd: anonymizing trajectory data for pattern analysis. In: CIKM, pp. 1441–1444. ACM (2009)
15. Monreale, A., et al.: Movement data anonymity through generalization. TDP **3**(2), 91–121 (2010)
16. Monreale, A., et al.: Privacy-preserving distributed movement data aggregation. In: Vandenbroucke, D., Bucher, B., Crompvoets, J. (eds.) Geographic Information Science at the Heart of Europe. Lecture Notes in Geoinformation and Cartography. Springer, Cham (2013). https://doi.org/10.1007/978-3-319-00615-4_13
17. de Montjoye, Y.A., et al.: Unique in the crowd: the privacy bounds of human mobility. Sci. Rep. **3**, 1376 (2013)
18. Muntean, C.I., et al.: On learning prediction models for tourists paths. ACM Trans. Intell. Syst. Technol. **7**(1), 8:1–8:34 (2015)
19. Pappalardo, L., et al.: Returners and explorers dichotomy in human mobility. Nat. Commun. **6**, 1–8 (2015)
20. Pellungrini, R., et al.: A data mining approach to assess privacy risk in human mobility data. ACM TIST **9**(3), 31:1–31:27 (2018)
21. Pratesi, F., et al.: Prudence: a system for assessing privacy risk vs utility in data sharing ecosystems. Trans. Data Priv. **11**(2), 139–167 (2018)
22. Ribeiro, M.T., et al.: "Why should I trust you?": explaining the predictions of any classifier. In: ACM SIGKDD, pp. 1135–1144 (2016)
23. Rossi, L., Musolesi, M.: It's the way you check-in: identifying users in location-based social networks. In: COSN, pp. 215–226. ACM (2014)
24. Samarati, P., Sweeney, L.: Generalizing data to provide anonymity when disclosing information (abstract). In: PODS, p. 188. ACM (1998)
25. Song, Y., et al.: Not so unique in the crowd: a simple and effective algorithm for anonymizing location data. In: International Workshop on Privacy-Preserving IR: When Information Retrieval Meets Privacy and Security, pp. 19–24 (2014)
26. Terrovitis, M., Mamoulis, N.: Privacy preservation in the publication of trajectories. In: MDM, pp. 65–72 (2008)
27. Zhang, Y.L., et al.: Distributed deep forest and its application to automatic detection of cash-out fraud. ACM Trans. Intell. Syst. Technol. **10**(5), 1–9 (2019)
28. Zheng, Y.: Trajectory data mining: an overview. ACM TIST **6**(3), 29:1–29:41 (2015)
29. Zhou, Z.H., Feng, J.: Deep forest: towards an alternative to deep neural networks. In: IJCAI, pp. 3553–3559 (2017)

Graph and Network Mining

Maximizing Network Coverage Under the Presence of Time Constraint by Injecting Most Effective k-Links

Kouzou Ohara[1](✉), Takayasu Fushimi[2], Kazumi Saito[3,4], Masahiro Kimura[5], and Hiroshi Motoda[6]

[1] College of Science and Engineering, Aoyama Gakuin University, Sagamihara, Japan
ohara@it.aoyama.ac.jp
[2] School of Computer Science, Tokyo University of Technology, Hachioji, Japan
takayasu.fushimi@gmail.com
[3] Faculty of Science, Kanagawa University, Hiratsuka, Japan
k-saito@kanagawa-u.ac.jp
[4] Center for Advanced Intelligence Project, RIKEN, Tokyo, Japan
kazumi.saito@riken.jp
[5] Faculty of Advanced Science and Technology, Ryukoku University, Otsu, Japan
kimura@rins.ryukoku.ac.jp
[6] Institute of Scientific and Industrial Research, Osaka University, Suita, Japan
motoda@ar.sanken.osaka-u.ac.jp

Abstract. We focus on a class of link injection problem of spatial network, *i.e.*, finding best places to construct k new roads that save as many people as possible in a time-critical emergency situation. We quantify the network performance by node coverage under the presence of time constraint and propose an efficient algorithm that maximizes the marginal gain by use of lazy evaluation making the best of time constraint. We apply our algorithm to three problem scenarios (disaster evacuation, ambulance call, fire engine dispatch) using real-world road network and geographical information of actual facilities and demonstrate that 1) use of lazy evaluation can achieve nearly two orders of magnitude reduction of computation time compared with the straightforward approach and 2) the location of new roads is intuitively explainable and reasonable.

1 Introduction

There is pressing need for understanding the structure and functions of large complex networks in many different fields of science such as sociology, biology, physics and computer science [20] as we challenge to optimize complex systems be they physical or cyber. One common approach to analyze large complex networks is investigating their characteristics through a measure called centrality. Various kinds of centralities are proposed and used according to our objectives. For example, if our goal is to know the topological characteristics of a network, degree, closeness, and betweenness centralities [14] are the candidates. If it is to know the importance of nodes that constitute a network, HITS [6] and PageRank [4] centralities are the candidates. Influence degree

A. Appice et al. (Eds.): DS 2020, LNAI 12323, pp. 421–436, 2020.
https://doi.org/10.1007/978-3-030-61527-7_28

centrality [17] is yet another one to measure the importance of nodes. Traditionally centrality measures have been applied to individual nodes, whereas a notion of group centralities have been proposed [12] to cope with situations such as information diffusion from multiple source nodes over a social network and emergency escape to evacuation facilities over a spatial network. It is natural to consider a group of nodes instead of individual nodes in such situations.

We address the problem of injecting k new links in a spatial network under a time critical situation setting and propose a new group centrality called time-bounded group link-injection centrality that can be maximized to solve this problem. This is similar to the problem of identifying critical links where the focus is to identify links that are already there in a network [1,16,29,30,32]. A critical link in a network is such a link that exerts a substantial effect on the network performance when the link fails to function properly. Thus, given a network structure and a performance measure, critical links can be found by solving the corresponding optimization problem of finding a link such that its blocking maximally degrades the network performance. Saito et al. [31] adopted as the performance measure the marginal loss of group closeness centrality when they are blocked under the constraint of time-bound. Our problem, on the other hand, is link injection where the task is to find a new link (or more generally k new links) that, if added, maximally increases the network performance. So far, link injection problem has been also explored from various angles [10,22,25,27,28]. Among them, closest to our work is Ohara et al. [22]. They consider the marginal gain of group closeness centrality with actual link distance under the constraint of the length of links to be added.

Our problem here is different from [22,31] in that we impose the time bound constraint in link injection problem. The network performance is measured by the weighted sum of nodes that are reachable to the target nodes (group of nodes) within the given time-bound, *i.e.*, we are maximizing the network coverage. As mentioned earlier we focus on time critical situations. We assume that each node in the network represents the population around it, the traveling time (or the real distance) is assigned to each link, and there exists a set of target nodes. One such example of time critical situations is disaster evacuation we consider in this paper, in which case, the target nodes $\mathcal{U}$ consists of the evacuation facilities, and people living in the neighborhood of a node v evacuate to the facility $u \in \mathcal{U}$ nearest to v. Most importantly there is a time limit set to each case. Then, the measure to be maximized, that is, the marginal gain of the weighted sum of nodes corresponds to the number of people who become able to escape to one of the facilities within the maximum permissible time when a link is injected. This measure is referred to as time-bounded group link-injection centrality, a new centrality proposed in this work. We note that the gain is attained only when a node that is not reachable to $u \in \mathcal{U}$ due to the time constraint before link injection becomes reachable after the injection. This motivated us to devise a very efficient algorithm that prunes unnecessary search by estimating the upper limits of reachable nodes from every single node first and use them later (lazy evaluation). We apply our algorithm to three emergency problem scenarios (disaster evacuation, ambulance call, and fire engine dispatch) in Tokyo area using the real road network and the geographical information of actual facilities and demonstrate that 1) use of lazy evaluation can achieve nearly two orders of magnitude reduction of computation time compared with the straightforward approach and 2) the location of new roads is intuitively explainable and reasonable.

The paper is organized as follows. After briefly describing the related work in Sect. 2, we mathematically formulate our link injection problem in Sect. 3. We present the proposed method in Sect. 4, and report the experimental results in Sect. 5. We give our conclusion in Sect. 6.

2 Related Work

The problem we pose in this paper is related to defining a new centrality measure for complex networks. Many centrality measures have been introduced to quantify the importance of each node or link in a network [3]. As the size of a network is increasing, it gets more difficult to compute centrality measures that have to be derived using global network structure (e.g., closeness centrality). Thus, several approximation approaches have been proposed to compute such centrality measures for a huge network [2,8,9]. Unlike those studies, we focus on exactly computing our new centrality measure. Time-bound constraint and an efficient search algorithm have made this possible. Thus, there is no need to evaluate the approximation error.

In this paper, we especially give a new centrality measure defined for links. Grady et al. [15] presented the notion of link salience as a link centrality measure to find a network's skeleton by improving link betweenness centrality, and Fang et al. [13] examined link capacity allocation methods under limited investment costs to prevent cascading failures for power transmission networks. Moreover, a problem of finding critical links in a network was addressed in various scenarios [1,33], and several detection algorithms have been proposed to solve the problem [16,29,30,32]. In particular, Saito et al. [31] defined the concept of time-bounded criticalness centrality as a link centrality measure for spatial networks, and presented an effective method of identifying the critical links. Unlike those critical link detection problems, we tackle a link injection problem in this paper. We also note that although our new centrality measure can be regarded as a kind of vitality index [18] in centrality indices of complex networks, it is completely different from the original one. In fact, our centrality definition requires not only network topology but also the notions of traveling time and time bound constraint.

Link injection problem, that is, a problem of finding appropriate k new links to inject into a network for improving network performance has also been tackled from various angles. Aiming to minimize the average shortest path distance over all pairs of nodes in a network, a problem of discovering optimal k new shortcut links was investigated [25, 27]. Crescenzi et al. [10] and Parotsidis et al. [28] dealt with the problem of selecting k new links to be added into a network for maximizing the closeness centrality of a specific node. A problem of promoting information diffusion in a social network by creating new links was explored [7,34]. Unlike our problem, these studies neither were intended to a spatial network with geodesic distance between nodes nor have taken group centrality into account. Ohara et al. [22] examined the problem of creating k new links satisfying a length constraint in order to maximize the gain of the group closeness centrality in a spatial network with actual link distance. As described in Sect. 1, our problem is mainly different from their problem in terms of the objective function to evaluate network performance. In addition, they do not consider time constraint. To the best of our knowledge, our work is the first to address the problem of creating new links for maximizing the time-bounded group link-injection centrality in a spatial network.

Our problem is also related to analyzing and characterizing the structure and functions of large spatial networks such as urban streets. Conventional centrality measures were often utilized [11,26] to investigate the structural properties of road networks. Montis et al. [19], Opsahl et al. [24] and Ohara et al. [21] extended use of the conventional centrality measures and explored weighted spatial networks based on road usage frequency and real distance for urban streets. From functional points of view, traffic usage patterns in urban streets were widely examined [5,35]. Also, using a transportation simulation software, the vulnerability of a road network was investigated [23]. Unlike these existing investigations, we focus on realistic emergency situations, and aim at improving the structure and functions of spatial road networks by newly introducing the concept of time-bounded group link-injection centrality.

3 Problem Formulation

Let $G = (\mathcal{V}, \mathcal{E})$ be a given simple connected, undirected (or bidirectional) network without self-loops, where $\mathcal{V} = \{v, w, x, \cdots\}$ and $\mathcal{E} = \{e, \cdots\}$ are sets of nodes and undirected links, respectively. We also express each link e as a pair of nodes, *i.e.*, $e = (v, w)$. For each link $e = (v, w) \in \mathcal{E}$, we assign its traveling time $t(v, w; G)$ between these nodes. For each pair of nodes that does not have the direct connection, *i.e.*, $(v, x) \notin \mathcal{E}$, we define its traveling time $t(v, x; G)$ by the minimum traveling time over all possible paths between them.

In our problem setting, we assume a fixed group of nodes $\mathcal{U} \subset \mathcal{V}$ such as evacuation facilities, emergency hospitals or fire stations on a road network. Here, for each node $v \in \mathcal{V}$, we can compute the minimum traveling time $f(v; \mathcal{U}, G)$ between v and some node $w \in \mathcal{U}$ as follows:

$$f(v; \mathcal{U}, G) = \min_{w \in \mathcal{U}} t(v, w; G). \tag{1}$$

For each node $v \in \mathcal{V}$, we assume that v has some weight denoted by $\rho(v)$, which is intended to represent the number of residences or houses around node v in a road network. Then, the following group closeness centrality measure for minimizing the average arrival time has been proposed and studied by Ohara et al. [22]

$$\tilde{g}(\mathcal{U}; G) = \sum_{v \in \mathcal{V}} f(v; \mathcal{U}, G)\rho(v). \tag{2}$$

and they investigated the problem of injecting k new links to minimize $\tilde{g}(\mathcal{U}; G)$.

Now, in case of a disaster such as tsunami right after a large-scale earthquake, people living in the flooded area must evacuate to some facility before the time of tsunami arrival. Let τ be such a maximum permissible time, and then we propose the following cover-based group closeness centrality measure with time-bound τ for maximizing the weighted sum of nodes whose traveling times from/to some facility are less than a maximum permissible time.

$$g(\mathcal{U}; G, \tau) = \sum_{\{v \in \mathcal{V} \mid f(v; \mathcal{U}, G) \leq \tau\}} \rho(v). \tag{3}$$

For a given time-bound τ, we compute the ratios of uncovered nodes and weight as $1 - |\{v \in \mathcal{V} \mid f(v; \mathcal{U}, G) \leq \tau\}|/|\mathcal{V}|$ and $1 - g(\mathcal{U}; G, \tau)/\sum_{v \in \mathcal{V}} \rho(v)$, respectively.

For each graph G, let $G(\{e\})$ be a graph constructed by injecting a link $e = (v, w)$, *i.e.*, $G(\{e\}) = (\mathcal{V}, \mathcal{E} \cup \{e\})$. Here we assume that the traveling time for a new link $t(v, w)$ can be computed from their locations. Then, we can define the following marginal gain of injecting a link $e \in \mathcal{H}$ over G, where $\mathcal{H} = \{e = (v, w) \in (\mathcal{V} \times \mathcal{V}) \setminus \mathcal{E} \mid v \neq w\}$.

$$h(e; \mathcal{U}, G, \tau) = g(\mathcal{U}; G(\{e\}), \tau) - g(\mathcal{U}; G, \tau). \tag{4}$$

Note that the value $h(e; \mathcal{U}, G, \tau)$ can be interpreted as the weighted sum of nodes, *i.e.*, people, who become able to move to one of these evacuation facilities within the maximum permissible time when a link e is injected in case of evacuation scenario. Similar interpretation is possible for other scenarios. Hereafter, the measure defined in Eq. (4) is referred to as time-bounded link-injection centrality.

Let $G(\mathcal{R}) = (\mathcal{V}, \mathcal{E} \cup \mathcal{R})$ be the network constructed by newly injecting a set of links $\mathcal{R} \subset \mathcal{H}$ to the original network G. The number of links to be injected is set to k, *i.e.*, $|\mathcal{R}| = k$. Then, we can define the following marginal gain based on our time-bounded link-injection centrality $h(e; \mathcal{U}, G, \tau)$ as our objective function to be maximized with respect to $\mathcal{R}$:

$$h(\mathcal{R}; \mathcal{U}, G, \tau) = g(\mathcal{U}; G(\mathcal{R}), \tau) - g(\mathcal{U}; G, \tau). \tag{5}$$

Then, based on Eq. (5), our time-bounded k-links injection problem is formulated as the problem of finding the following optimal set of new links denoted by $\mathcal{R}^*$:

$$\mathcal{R}^* = \underset{\mathcal{R} \subset \mathcal{H},\ |\mathcal{R}| = k}{\operatorname{argmax}}\ h(\mathcal{R}; \mathcal{U}, G, \tau). \tag{6}$$

Here note that the time constraint effectively imposes the constraint on the length of the roads to be injected which was considered in [22]. Hereafter, the measure defined in Eq. (5) is referred to as time-bounded group link-injection centrality.

4 Proposed Method

For a given spatial network $G = (\mathcal{V}, \mathcal{E})$ with a weight $\rho(v)$ and location for each node $v \in \mathcal{V}$ and a traveling time $t(v, y; G)$ for each link $(v, y) \in \mathcal{E}$, together with a fixed group of nodes $\mathcal{U} \subset \mathcal{V}$ and a maximum permissible time τ, we describe our proposed algorithm for efficiently computing the link $\hat{e} \in \mathcal{H}$ which maximizes the time-bounded link-injection centrality value $h(e; \tau, \mathcal{U}, G)$ defined in Eq. (4). Then, we can easily obtain a reasonably good solution $\mathcal{R}$ for our time-bounded k-links injection problem by the following greedy algorithm.

G1: Initialize $\mathcal{R} \leftarrow \emptyset$ and $j \leftarrow 0$.
G2: Compute $\hat{e} = \operatorname{argmax}_{e \in \mathcal{H} \setminus \mathcal{R}} h(e; \mathcal{U}, G(\mathcal{R}), \tau)$, and set $\mathcal{R} \leftarrow \mathcal{R} \cup \{e\}$ and $j \leftarrow j + 1$.
G3: Go to G2 if $j < k$; otherwise output the set of k-links $\mathcal{R}$ and then terminate.

Evidently, for this basic algorithm, we can improve the quality of solutions by introducing some techniques such as local improvement.

In order to compute the best link $\hat{e}$ which maximizes the centrality value $h(e; \mathcal{U}, G, \tau)$, among all links $e \in \mathcal{H}$, we consider injecting a link denoted by $e = (u(y), y)$. Here, as studied by Ohara et al. [22], by selecting the parent node of y as

$$u(y) = \underset{v \in \mathcal{V}}{\operatorname{argmin}}\{f(v; \mathcal{U}, G) + t(v, y)\}, \tag{7}$$

we can maximize the gain with respect to the descendant nodes of y obtained by the best-first search described below, in terms of the maximum reduction of the minimum traveling times after injecting the link $e = (u(y), y)$. Note that we obtain a positive gain $\rho(z)$ for some node z in case that $f(z; \mathcal{U}, G) > \tau$ and $f(z; \mathcal{U}, G(\{(u(y), y)\})) < \tau$. Hereafter, we denote the centrality value with respect to each node $y \in \mathcal{V} \setminus \mathcal{U}$ as

$$\psi(y) = h((u(y), y); \mathcal{U}, G(\{(u(y), y)\}), \tau), \tag{8}$$

In this paper, based on the best-first search strategy, we propose our algorithm consisting of the following three steps:

S1: Compute the minimum traveling time $f(v; \mathcal{U}, G)$ for every node $v \in \mathcal{V}$;
S2: Compute the upper bound value $\phi(y)$ of $\psi(y)$ for every node $y \in \mathcal{V} \setminus \mathcal{U}$;
S3: Compute the node $\hat{y}$ which maximizes the centrality value $\psi(y)$ based on a lazy evaluation technique, and output the best link as $\hat{e} = (u(\hat{y}), \hat{y})$.

Below we describe the details of these three steps.

For a subset of nodes $\mathcal{W} \subset \mathcal{V}$, let $\mathcal{I}(\mathcal{W})$ be a set of incident links from $\mathcal{W}$, *i.e.*, $\mathcal{I}(\mathcal{W}) = \{(w, x) \in \mathcal{E} \mid w \in \mathcal{W}, x \notin \mathcal{W}\}$. Then, as the first step S1, we compute the minimum traveling time $f(v; \mathcal{U}, G)$ for every node $v \in \mathcal{V}$ as follows.

S1-1: Initialize $\mathcal{W} \leftarrow \mathcal{U}$ and $f(w; \mathcal{U}, G) \leftarrow 0$ for each $w \in \mathcal{U}$.
S1-2: Select the best-first link $(\hat{w}, \hat{x}) \leftarrow \operatorname{argmin}_{(w,x) \in \mathcal{I}(\mathcal{W})}\{f(w; \mathcal{U}, G) + t(w, x; G)\}$, and set $\mu \leftarrow f(\hat{w}; \mathcal{U}, G) + t(\hat{w}, \hat{x}; G)$.
S1-3: Set $\mathcal{W} \leftarrow \mathcal{W} \cup \{\hat{x}\}$ and $f(\hat{x}; \mathcal{U}, G) \leftarrow \mu$.
S1-4: Go to S1-2 if $\mathcal{W} \neq \mathcal{V}$; otherwise output $f(v; \mathcal{U}, G)$ for every node $v \in \mathcal{V}$ and then terminate.

Here note that after obtaining $f(v; \mathcal{U}, G)$ for every $v \in \mathcal{V}$, we can easily compute our cover-based group closeness centrality measure with time-bound τ defined in Eq. (3).

Let $\mathcal{Z} = \{z \in \mathcal{V} \mid f(z; \mathcal{U}, G) > \tau \wedge \rho(z) > 0\}$ be the set of uncovered nodes with positive weight. In order to obtain a positive gain by injecting a link $e = (u(y), y)$, the traveling time from y to $z \in \mathcal{Z}$ must be smaller than τ. Therefore, as the second step S2, we consider computing the upper bound value $\phi(y)$ of $\psi(y)$ for every node $y \in \mathcal{V} \setminus \mathcal{U}$ as follows.

S2-0: Initialize $\phi(y) \leftarrow 0$ for every $y \in \mathcal{V} \setminus \mathcal{U}$ and after iterating the following steps for every $z \in \mathcal{Z}$, output $\phi(y)$ for $y \in \mathcal{V} \setminus \mathcal{U}$.
S2-1: Initialize $\mathcal{W} \leftarrow \{z\}$, $f(z; \{z\}, G) \leftarrow 0$, and $\phi(z) \leftarrow \phi(z) + \rho(z)$.
S2-2: Select the best-first link $(\hat{w}, \hat{x}) \leftarrow \operatorname{argmin}_{(w,x) \in \mathcal{I}(\mathcal{W})}\{f(w; \{z\}, G) + t(w, x; G)\}$, and set $\mu \leftarrow f(\hat{w}; \{z\}, G) + t(\hat{w}, \hat{x}; G)$.
S2-3: Terminate z's iteration if $\mu > \tau$,

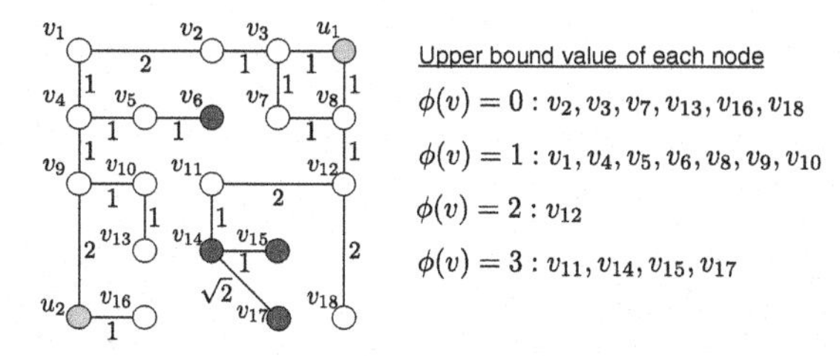

Fig. 1. An example of spatial network having 4 uncovered nodes (red nodes) in the case of $\tau = 4$ and upper bound values of each node.

S2-4: Set $\mathcal{W} \leftarrow \mathcal{W} \cup \{\hat{x}\}$, $f(\hat{x}; \{z\}, G) \leftarrow \mu$, $\phi(\hat{x}) \leftarrow \phi(\hat{x}) + \rho(z)$ and go to S2-2.

Here note that in case of general and practical problem settings, we can assume $|\mathcal{Z}| \ll |\mathcal{V}|$ and generally expect that this step works with small amount of computation.

Let $\mathcal{Y} = \{y_1, \cdots, y_M\}$ be a set of indexed nodes arranged according to their upper bound values, *i.e.*, $\phi(y_i) \geq \phi(y_{i+1})$ for $1 \leq i < M$, where $M = |\{y_i\}|$ for $\phi(y_i) > 0$. Then, based on the idea of lazy evaluation, during our iteration from $i = 1$ to M, it is guaranteed that at the i-th step, the current best node $\hat{y}$ is the optimal one when satisfying $\psi(\hat{y}) > \phi(y_i)$. Here, for an injection candidate link $e \in \mathcal{H}$, let $\mathcal{J}(\mathcal{W}; e)$ be a set of incident links from $\mathcal{W}$ whose adjacent nodes have improved traveling times, *i.e.*, $\mathcal{J}(\mathcal{W}; e) = \{(w, x) \in \mathcal{I}(\mathcal{W}) \mid f(x; \mathcal{U}, G(\{e\})) < f(x; \mathcal{U}, G)\}$. Then, as the third step S3, we compute the node $\hat{y}$ which maximizes the centrality value $\psi(y)$ as follows.

S3-0: Initialize $\psi(\hat{y}) \leftarrow 0$, iterate the following steps from $i \leftarrow 1$ to M while $\psi(\hat{y}) < \phi(y_i)$, and then output $\hat{e} = (u(\hat{y}), \hat{y})$.
S3-1: Initialize $\mathcal{W} \leftarrow \{y_i\}$, $e = (u(y_i), y_i)$, $f(y_i; \mathcal{U}, G(\{e\})) \leftarrow f(u(y_i); \mathcal{U}, G) + t(u(y_i), y_i)$, and $\psi(y_i) \leftarrow \rho(y_i)$ if $f(y_i; \mathcal{U}, G) > \tau$ and $f(y_i; \mathcal{U}, G(\{e\})) \leq \tau$; otherwise $\psi(y_i) \leftarrow 0$.
S3-2: Terminate y_i's iteration if $\mathcal{J}(\mathcal{W}; e) = \emptyset$; otherwise select the best-first link $(\hat{w}, \hat{x}) \leftarrow \text{argmin}_{(w,x)\in\mathcal{J}(\mathcal{W};e)}\{f(w; \mathcal{U}, G(\{e\})) + t(w, x; G)\}$, and set $\mu \leftarrow f(\hat{w}; \mathcal{U}, G(\{e\})) + t(\hat{w}, \hat{x}; G)$.
S3-3: Terminate y_i's iteration if $\mu > \tau$,
S3-4: Set $\mathcal{W} \leftarrow \mathcal{W} \cup \{\hat{x}\}$, $f(\hat{x}; \mathcal{U}, G(\{e\})) \leftarrow \mu$, $\psi(y_i) \leftarrow \psi(y_i) + \rho(\hat{x})$ if $f(\hat{x}; \mathcal{U}, G) > \tau$, $\hat{y} \leftarrow y_i$ if $\psi(y_i) > \psi(\hat{y})$, and go to S3-2.

For example, suppose the spatial network shown in Fig. 1 is given, in which the number assigned to each link is the traveling time between its two terminal nodes and $\mathcal{U} = \{u_1, u_2\}$. We assume $\rho(v) = 1$ for every node v for simplicity. In this case, we can find nodes v_6, v_{14}, v_{15}, and v_{17} are not reachable from any of u_1 and u_2 when $\tau = 4$. For every node other than u_1 and u_2, according to S2-1 to S2-4, we can obtain the upper bound values shown in Fig. 1. Based on these values, we first apply steps S3-1 to S3-4 to v_{11} and obtain $\psi(v_{11}) = 1$ by injecting a new link (u_1, v_{11}) whose traveling time is $\sqrt{8} \approx 2.83$. As $\psi(v_{11}) < \phi(v_{14})$ holds, we apply these steps to v_{14}, which results in $\psi(v_{14}) = 3$ due to a new link (u_2, v_{14}) with traveling time of $\sqrt{5} \approx 2.24$. Since $\psi(v_{14}) < \phi(v_{15})$ does not hold, we do not have to consider the remaining nodes any more, and can prune further steps for them. Note that node v_6 will be covered if $k > 1$.

As the novel characteristics, our proposed algorithm employs the lazy evaluation technique based on the upper bounds of actual centrality values. In contrast, by straightforwardly adopting the method proposed by Ohara et al. [22], we can obtain a conventional method without lazy evaluation as follows.

C1: Compute the minimum traveling time $f(v; \mathcal{U}, G)$ for every node $v \in \mathcal{V}$ by performing the steps from S1-1 to S1-4.
C2: Compute the node $\hat{y}$ which maximizes the centrality value $\psi(y)$ by performing the steps from S3-1 to S3-4 for every $y \in \mathcal{Y}$, and output the best link as $\hat{e} = (u(\hat{y}), \hat{y})$.

Evidently, the efficiency of our proposed algorithm in comparison to the above conventional method is affected by several factors including the maximum permissible time τ, the number of nodes in $\mathcal{U}$ and so on. Thus, we evaluate the performance of our proposed algorithm in our computational experiments.

5 Experiments

Using real data of road network $G = (\mathcal{V}, \mathcal{E})$ and facilities $\mathcal{U}$, we evaluated the effectiveness of the proposed algorithm.

5.1 Experimental Settings

In our experiments, we used the actual road network of Tokyo in Japan as $G = (\mathcal{V}, \mathcal{E})$, and considered three different realistic scenarios, *i.e.*, disaster evacuation, ambulance call, and fire engine dispatch, as studied by Saito et al. [31]. Below we briefly describe the experimental settings for the sake of readers' convenience. The spatial road network of Tokyo was extracted from the Open Street Map data[1], *i.e.*, the spatial network was constructed by regarding the ends, intersections, and curve-fitting-points as nodes and the streets between them as links. The resulting network consists of $|\mathcal{V}| = 6,571,077$ nodes and $|\mathcal{E}| = 7,312,007$ links. As for the facilities $\mathcal{U}$ of each scenario, geographical information about evacuation facilities, emergency hospitals and fire stations was gathered from the site of National Land Information Division of Ministry of Land, Infrastructure, Transport and Tourism (MLIT) of Japan[2], respectively, and each of these facilities was mapped to the nearest node in the spatial network. The numbers of evacuation facilities, emergency hospitals, and fire engines are 3,919, 55, and 318, respectively. Further, it is assumed that a person moves at 1 m per second (3.6 km/h) on foot in the case of disaster evacuation, and that both an ambulance and a fire engine move at 10 m per second (36 km/h). Each traveling time $t(v, w; G)$ was computed by dividing the distance $dist(v, w)$ by the velocity corresponding to each scenario for each link $(v, w) \in \mathcal{E}$. Here it should be emphasized that we can arbitrary change $t(v, w; G)$ according to the other conditions such as road width. Weight $\rho(v)$ setting of each node was determined by using the 2015 census population aggregation data [3]. More specifically, for a given population number n of each small region containing a subset of nodes $\mathcal{X} \subset \mathcal{V}$, the average number $\rho(v) = n/|\mathcal{X}|$ was assigned to each node $v \in \mathcal{X}$ as its weight.

[1] https://openstreetmap.jp/.
[2] http://nlftp.mlit.go.jp/ksj-e/index.html.
[3] https://www.e-stat.go.jp/gis/.

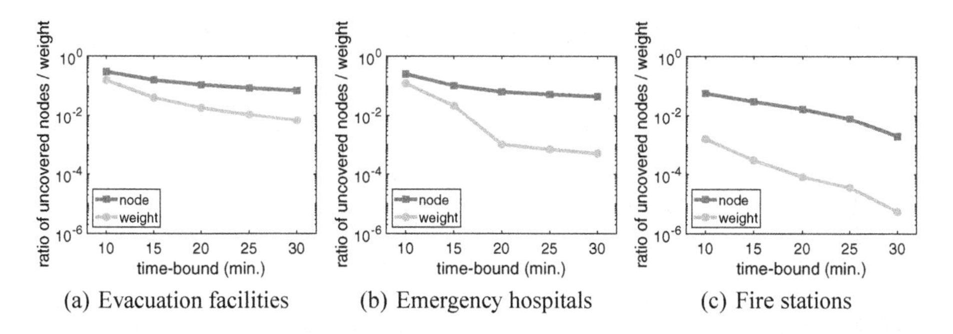

(a) Evacuation facilities (b) Emergency hospitals (c) Fire stations

Fig. 2. The ratios of uncovered nodes and weight as a function of maximum permissible time.

As basic characteristics of our three scenarios, we investigated the ratios of uncovered nodes and weight defined in Sect. 3, *i.e.*, $1 - |\{v \in \mathcal{V} \mid f(v; \mathcal{U}, G) \leq \tau\}|/|\mathcal{V}|$ and $1 - g(\mathcal{U}; G, \tau)/\sum_{v \in \mathcal{V}} \rho(v)$. Figure 2 shows the experimental results, where Figs. 2(a), 2(b), and 2(c) are cases using evacuation facilities, emergency hospitals, and fire stations as $\mathcal{U}$, respectively. In this investigation, we varied the maximum permissible time τ from 10 (min.) to 30 (min.) by 5 (min.). From these results, we can observe that both the ratios of uncovered nodes and weight monotonically decrease as the maximum permissible time τ becomes longer. For all the cases, we can see that the ratios of uncovered weight are substantially smaller than those of uncovered nodes although the degree of difference depends on the problems and settings of the maximum permissible times. The ratios of uncovered nodes and weight for $\tau = 10$ are more than 0.1 in Fig. 2(a). This is because we assumed that a person moves at 1 m per second (60 m per minute) on foot in the scenario of disaster evacuation, and thus, it is difficult for some people to reach the nearest evacuation facility within 10 min. This difficulty is alleviated by setting τ to a larger value. However, even in case of $\tau = 30$, we still have room to improve the ratio of uncovered weight by injecting effective links. Similar tendency can be observed in the cases of ambulance call and fire engine dispatch in Figs. 2(b) and 2(c), where we assumed that both an ambulance and a fire engine move at 10 m per second (600 m per minute). But, we note that the ratios of uncovered weight for these scenarios when $\tau = 10$ are quite different, more than 0.1 for ambulance call as shown in Fig. 2(b) and about 0.001 for fire engine dispatch as shown in Fig. 2(c), which comes from the difference in the number of facilities, *i.e.*, 55 for emergency hospitals and 318 for fire stations. Thus, the situation becomes more critical for the time limitation of 10 min in case of ambulance call. Further notable difference we see is that the ratios of uncovered weight for fire engine dispatch are much smaller than those for the other scenarios for larger values of the maximum permissible time τ. We will show this reason in our later experiment.

5.2 Experimental Results

First, we evaluated the computational efficiency of our proposed algorithm with the lazy evaluation technique, in comparison to the conventional method constructed by straightforwardly adopting the method proposed by Ohara et al. [22], which is described in the

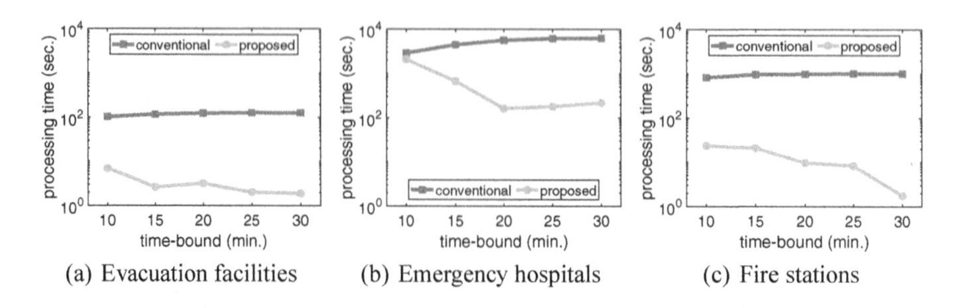

(a) Evacuation facilities (b) Emergency hospitals (c) Fire stations

Fig. 3. Evaluation of computational efficiency as a function of maximum permissible time.

previous section. For this purpose, we focused on the processing time to compute the optimal injection link $\hat{e} \in \mathcal{V} \times \mathcal{V} \setminus \mathcal{E}$ according to the marginal gain $h(e; \mathcal{U}, G, \tau)$ defined in Eq. (4). Recall that $\hat{e}$ thus chosen is the best link in terms of time-bounded link-injection centrality. The programs to compute the best link $\hat{e}$ were implemented in C and run on a computer with a single thread (Xeon X5690 3.47GHz CPUs) with a 192GB main memory capacity. Figure 3 shows the resultant processing time (seconds), where Figs. 3(a), 3(b), and 3(c) are cases using evacuation facilities, emergency hospitals, and fire stations as $\mathcal{U}$, respectively. In this experiment, we also varied the maximum permissible time τ from 10 (min.) to 30 (min.) by 5 (min.). From these results, we can see that for most of our experimental settings, the proposed method works about 2 orders of magnitude faster than the conventional method. Namely, these results support that our proposed lazy evaluation technique works practically effective.

More specifically, we can see that the processing time of the conventional method for each scenario slightly increases as the maximum permissible time τ gets longer. It should be noted that the computational complexity of performing the steps from S3-1 to S3-4 can be roughly approximated by $O(|\tilde{\mathcal{W}}(\tau)|)$, where $|\tilde{\mathcal{W}}(\tau)|$ stands for the size of the final node set $\mathcal{W}$ when this iteration terminates. Thus, the above observation can be naturally explained by the fact that the size $|\tilde{\mathcal{W}}(\tau)|$ slightly increases as the maximum permissible time τ gets longer. Moreover, noting that $|\tilde{\mathcal{W}}(\tau)|$ is roughly regarded to be the size of the descendant nodes of best-first search from some starting node, the size $|\tilde{\mathcal{W}}(\tau)|$ becomes smaller as the number of facilities $|\mathcal{U}|$ gets larger. Actually, from Fig. 3, we can see that relatively larger computation times are required for emergency hospitals, but relatively smaller ones for evacuation facilities, where recall that the numbers of evacuation facilities, emergency hospitals, and fire engines are $3,919$, 55, and 318, respectively. In contrast, we can see that the processing time of the proposed method for each scenario substantially decreases as the maximum permissible time τ gets longer. It should be noted that we can naturally suppose that these processing times decrease as the size of uncovered nodes $|\mathcal{Z}|$ decreases. In fact, we can find that the uncovered ratios shown in Fig. 2 somewhat correlate to the processing times of the proposed method shown in Fig. 3.

Second, we investigated how much the network coverage improves by varying the maximum permissible time τ and the number k of links to be injected. Figure 4 shows the experimental results, in which the marginal gain $h(\mathcal{R}; \mathcal{U}, G, \tau)$ defined in Eq. (5)

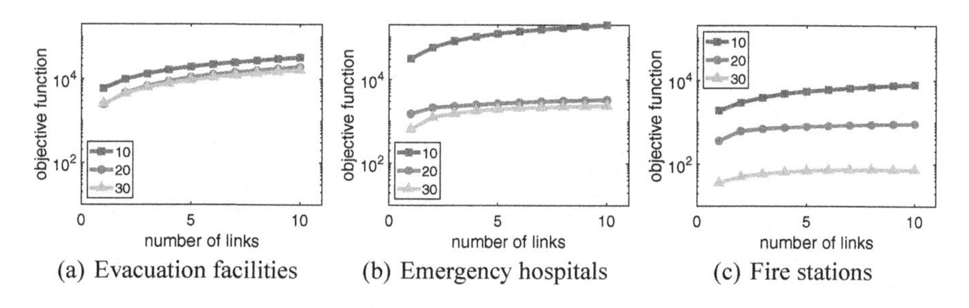

(a) Evacuation facilities (b) Emergency hospitals (c) Fire stations

Fig. 4. Evaluation of time-bounded group link-injection centrality as a function of the number of links to be injected.

with respect to the number k of links to be injected by varying from $k = 1$ to 10 are depicted. Here, we plotted the case that the maximum permissible time τ is set to $\tau = 10$, $\tau = 20$ or $\tau = 30$ in each figure. Recall that this gain is also referred to as time-bounded group link-injection centrality. From these results, we can see that the centrality values for each scenario steadily increase as the number k of injected links gets larger. On the other hand, the difference in the value of the objective function between the different values of maximum permissible times τ depends on the scenario, *i.e.*, quite small for all the settings in the case of disaster evacuation, substantially large only between $\tau = 10$ and $\tau = 20$ in the case of ambulance call, and substantially large for all the settings in the case of fire engine dispatch. Here we should note that these tendencies basically coincide with those for the improvement ratios of uncovered weight shown in Fig. 2, which is explained later in Table 1.

Third, we investigated how differently the injected links in each scenario are distributed. To this end, we marked them on an actual map of Tokyo as shown in Fig. 5, namely, we plotted each node $u \in \mathcal{U}$ corresponding to the location of each facility as a star shaped marker, and then depicted each node $y \in \mathcal{V}$ obtained as $e = (u(y), y) \in \mathcal{R}$ as a red position marker. We show our results for each scenario of disaster evacuation, ambulance call, and fire engine dispatch with the maximum permissible time set to $\tau = 10$ in Figs. 5(a), 5(c) and 5(e), and those with the maximum permissible time set to $\tau = 30$ in Figs. 5(b), 5(d), and 5(f), respectively, where the number of links to be injected was set to $k = 10$. From these figures, we can see a tendency that the injected link positions move from the city area (east) to the mountain region (west), especially for scenarios of ambulance call and fire engine dispatch. Here note that the distance corresponding to the maximum permissible time $\tau = 30$ is 1.8 km on foot in the scenario of disaster evacuation, and 18 km by vehicle in the scenario of ambulance call or fire engine dispatch, where the east-west and north-south distances of Tokyo are roughly 100km and 25km, respectively. Thus, from Fig. 5(b), we can see that the injected link positions are located not so far from those of evacuation facilities even in case of $\tau = 30$. In contrast, from Figs. 5(d) and 5(f), we can see that most of the injected link positions are located substantially far from those of evacuation facilities in case of $\tau = 30$. Notably, they are located in a quite far-west mountain region in the case of fire engine dispatch because there exist only a few fire stations in the mountain region, as shown in Figs. 5(f). This fact can explain that both the ratio of uncovered weight and the network coverage are

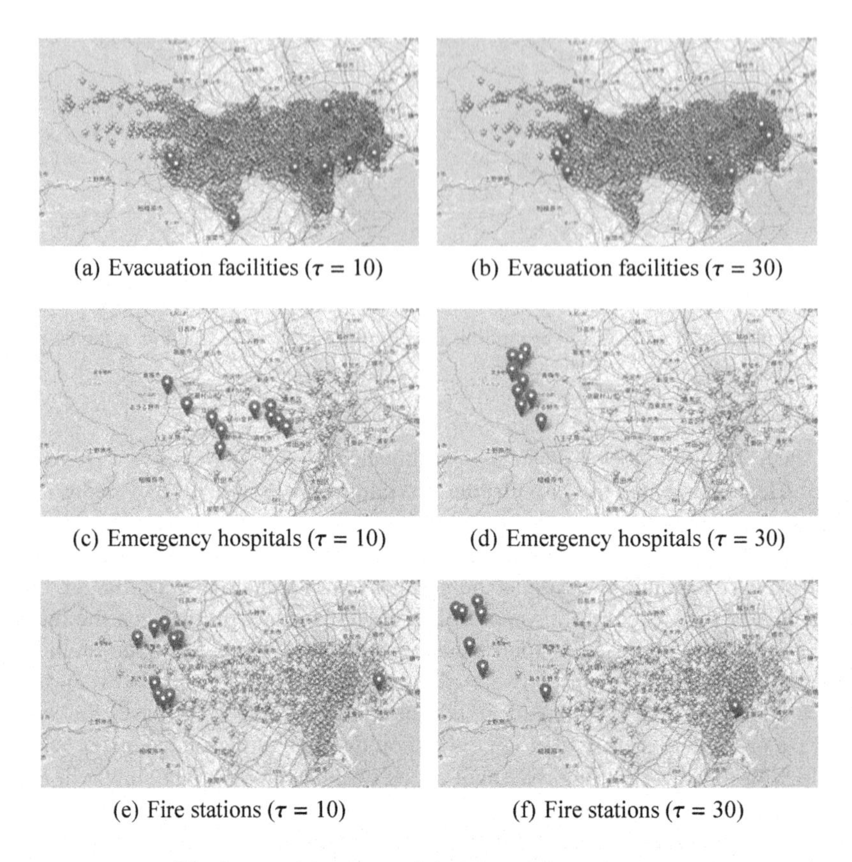

(a) Evacuation facilities (τ = 10) (b) Evacuation facilities (τ = 30)

(c) Emergency hospitals (τ = 10) (d) Emergency hospitals (τ = 30)

(e) Fire stations (τ = 10) (f) Fire stations (τ = 30)

Fig. 5. Actual locations of the injected links (k = 10).

substantially improved only for the case of fire engine dispatch, as shown in Figs. 2(c) and 4(c).

Finally, we further discuss our findings reported separately above. We showed that the time-bounded group link-injection centrality is a new centrality that can effectively be applied to three different problem scenarios: disaster evacuation, ambulance call, fire engine dispatch. The improvement in the coverage differs from problem to problem as shown in Table 1, which summarizes the ratio of uncovered weight (uncovered), the ratio of covered weight by link injection (covered+), and the improved ratio defined as covered+/uncovered (improved) in the cases of $\tau = 10, 20, 30$ and $k = 1, 2$. To get a rough idea we focus on the case of $\tau = 30$ and $k = 10$. The population of Tokyo is about $13,500,000$. Under these settings the number of people uncovered is initially $91,000$, $6,900$ and 74 for disaster evacuation, ambulance call and fire engine dispatch, respectively. The corresponding number of people newly covered by link injection (road constructions) is $16,400$, $2,400$ and 74, respectively. These amount to the improvement of 18%, 35% and 100%, respectively. It is understandable that the amount of improvement is smallest for disaster evacuation because there are already many evacuation facilities and the distance people can walk is limited. On the other hand,

Table 1. The ratio of uncovered weight (uncovered), the ratio of covered weight by link injection (covered+), and the improved ratio defined as covered+/uncovered (improved) in the case of varying $\tau = 10, 20, 30$ for each scenario in the cases of $k = 1$ and $k = 10$. (%)

	τ	$k = 1$			$k = 10$		
		Uncovered	Covered+	Improved	Uncovered	Covered+	Improved
Evacuation facilities	10	15.5939	0.0444	0.2850	15.5939	0.2414	1.5481
	20	1.8222	0.0183	1.0063	1.8222	0.1435	7.8762
	30	0.6767	0.0191	2.8228	0.6767	0.1213	17.9219
Emergency hospitals	10	12.0646	0.2322	1.9246	12.0646	1.4717	12.1982
	20	0.1019	0.0114	11.1957	0.1019	0.0248	24.2979
	30	0.0508	0.0048	9.4518	0.0508	0.0179	35.1508
Fire stations	10	0.1585	0.0142	8.9468	0.1585	0.0588	37.1164
	20	0.0082	0.0027	32.3497	0.0082	0.0068	83.2203
	30	0.0005	0.0003	48.5574	0.0005	0.0005	100.0000

the amount of improvement is largest for fire engine dispatch and 100% is covered. This is because the number of fire stations in the mountain region is small and link injection in these region is very effective. The improvement is in between for hospital call. This is attributed to the fact that the number of emergency hospitals is far smaller than the number of fire stations, especially in the mountain region. The drop of uncovered weight ratio in Fig. 2(b) from $\tau = 15$ to $\tau = 20$ is also caused by the shortage of the emergency hospitals. There are areas of dense population that are not reachable from the nearest hospitals within $\tau = 15$ but reachable within $\tau = 20$. Overall, we confirm that the proposed method can inject links in appropriate locations in reasonable time if the target facilities are appropriately located. We note that the population density ρ is sensitive in the results. The current method uses average density for local regions and its estimate needs improvement for large local regions that have to be represented by small number of nodes.

6 Conclusion

In this paper, we addressed the problem of injecting k new links in a spatial network so that they can improve the network performance measured by the weighted sum of nodes reachable to one of target nodes within the given time-bound. To solve this problem, we proposed a new group centrality called time-bounded group link-injection centrality, which is defined as the marginal gain of the network performance obtained by injecting k new links, and we devised an efficient algorithm to solve the optimization problem that can maximize the proposed group centrality, *i.e.*, can maximize the node coverage. The proposed algorithm fully utilizes the time-bound to prune the search space consisting of potential links based on the lazy evaluation technique. Through the empirical evaluation using a real road network and three kinds of facility locations each corresponding to actual time-critical problem scenarios, that is, disaster evacuation, ambulance call, and fire engine dispatch, we demonstrated that the proposed algorithm can achieve nearly two orders of magnitude reduction of computation time compared with the straightforward algorithm and the resulting location of new links (roads) is intuitively explainable and reasonable. The amount of improvement in the network performance, that is, the coverage of nodes by injecting k new links depends on the problem in hand. To further improve it and solve actual time-critical problems such as disaster evacuation by optimally allocating urban resources, it would be valuable to solve not only link injection, but also critical node identification in a cooperative way. This is one of our immediate future work.

Acknowledgments. This material is based upon work supported by JSPS Grant-in-Aid for Scientific Research (C) (No. 20K11940).

References

1. Akram, V.K., Dagdeviren, O.: Breadth-first search-based single-phase algorithms for bridge detection in wireless sensor networks. Sensors **13**(7), 8786–8813 (2013)
2. Boldi, P., Vigna, S.: In-core computation of geometric centralities with hyperball: a hunderd billion nodes and beyond. In: Proceedings of the 2013 IEEE 13th International Conference on Data Mining Workshops (ICDMW 2013), pp. 621–628 (2013)
3. Brandes, U., Erlebach, T. (eds.): Network Analysis. LNCS, vol. 3418. Springer, Heidelberg (2005). https://doi.org/10.1007/b106453
4. Brin, S., Page, L.: The anatomy of a large-scale hypertextual web search engine. Comput. Netw. ISDN Syst. **30**, 107–117 (1998)
5. Burckhart, K., Martin, O.J.: An interpretation of the recent evolution of the city of Barcelona through the traffic maps. J. Geogr. Inf. Syst. **4**(4), 298–311 (2012)
6. Chakrabarti, S., et al.: Mining the web's link structure. IEEE Comput. **32**, 60–67 (1999)
7. Chaoji, V., Ranu, S., Rastogi, R., Bhatt, R.: Recommendations to boost content spread in social networks. In: Proceedings of the 21th International Conference on World Wide Web (WWW 2012), pp. 529–538 (2012)
8. Chierichetti, F., Epasto, A., Kumar, R., Lattanzi, S., Mirrokni, V.: Efficient algorithms for public-private social networks. In: Proceedings of the 21st ACM SIGKDD International Conference on Knowledge Discovery and Data Mining (KDD 2015), pp. 139–148 (2015)
9. Cohen, E.: All-distances sketches, revisited: hip estimators for massive graphs analysis. In: Proceedings of the 33rd ACM SIGMOD-SIGACT-SIGART Symposium on Principles of Database Systems, pp. 88–99 (2014)
10. Crescenzi, P., D'angelo, G., Severini, L., Velaj, Y.: Greedily improving our own closeness centrality in a network. ACM Trans. Knowl. Discov. Data **11**(1), 1–32 (2016)
11. Crucitti, P., Latora, V., Porta, S.: Centrality measures in spatial networks of urban streets. Phys. Rev. E **73**(3), 036125 (2006)
12. Everett, M.G., Borgatti, S.P.: Extending centrality. In: Carrington, P.J., Scott, J., Wasserman, S. (eds.) Models and Methods in Social Network Analysis, pp. 57–76. Cambridge University Press, New York (2005)
13. Fang, Y.P., Pedroni, N., Zio, E.: Comparing network-centric and power flow models for the optimal allocation of link capacities in a cascade-resilient power transmission network. IEEE Syst. J. **99**, 1–12 (2014)
14. Freeman, L.: Centrality in social networks: conceptual clarification. Soc. Netw. **1**, 215–239 (1979)
15. Grady, D., Thiemann, C., Brockmann, D.: Robust classification of salient links in complex networks. Nature Commun. **3**(864), 1–10 (2012)
16. Kimura, M., Saito, K., Motoda, H.: Blocking links to minimize contamination spread in a social network. ACM Trans. Knowl. Discov. Data **3**, 9:1–9:23 (2009)

17. Kimura, M., Saito, K., Ohara, K., Motoda, H.: Speeding-up node influence computation for huge social networks. Int. J. Data Sci. Anal. **1**(1), 3–16 (2015). https://doi.org/10.1007/s41060-015-0001-y
18. Koschützki, D., et al.: Centrality indices. In: Brandes, U., Erlebach, T. (eds.) Network Analysis. LNCS, vol. 3418, pp. 16–61. Springer, Heidelberg (2005). https://doi.org/10.1007/978-3-540-31955-9_3
19. Montis, D.A., Barthelemy, M., Chessa, A., Vespignani, A.: The structure of interurban traffic: a weighted network analysis. Environ. Plan. **34**(5), 905–924 (2007)
20. Newman, M.: The structure and function of complex networks. SIAM Rev. **45**, 167–256 (2003)
21. Ohara, K., Saito, K., Kimura, M., Motoda, H.: Accelerating computation of distance based centrality measures for spatial networks. In: Calders, T., Ceci, M., Malerba, D. (eds.) DS 2016. LNCS (LNAI), vol. 9956, pp. 376–391. Springer, Cham (2016). https://doi.org/10.1007/978-3-319-46307-0_24
22. Ohara, K., Saito, K., Kimura, M., Motoda, H.: Maximizing network performance based on group centrality by creating most effective *k*-links. In: Proceedings of the 4th IEEE Data Science and Advanced Analytics (DSAA2017), pp. 561–570. IEEE Explore (2017). https://doi.org/10.1109/DSAA.2017.44
23. Oliveira, E.L., Portugal, L.S., Junior, W.P.: Determining critical links in a road network: vulnerability and congestion indicators. Proc. Soc. Behav. Sci. **162**, 158–167 (2014)
24. Opsahl, T., Agneessens, F., Skvoretz, J.: Node centrality in weighted networks: generalizing degree and shortest paths. Soc. Netw. **32**(3), 245–251 (2010)
25. Papagelis, M.: Refining social graph connectivity via shortcut edge addition. ACM Trans. Knowl. Discov. Data **10**(2), 1–35 (2015)
26. Park, K., Yilmaz, A.: A social network analysis approach to analyze road networks. In: Proceedings of the ASPRS Annual Conference 2010 (2010)
27. Parotsidis, N., Pitoura, E., Tsaparas, P.: Selecting shortcuts for a smaller world. In: Proceedings of the 2015 SIAM International Conference on Data Mining (SDM 2015), pp. 28–36 (2015)
28. Parotsidis, N., Pitoura, E., Tsaparas, P.: Centrality-aware link recommendations. In: Proceedings of of the Ninth ACM International Conference on Web Search and Data Mining (WSDM 2016), pp. 503–512 (2016)
29. Saito, K., Ohara, K., Kimura, M., Motoda, H.: Accurate and efficient detection of critical links in network to minimize information loss. J. Intell. Inf. Syst. **51**(2), 235–255 (2018). https://doi.org/10.1007/s10844-018-0523-6
30. Saito, K., Ohara, K., Kimura, M., Motoda, H.: Efficient detection of critical links to maintain performance of network with uncertain connectivity. In: Geng, X., Kang, B.-H. (eds.) PRICAI 2018. LNCS (LNAI), vol. 11012, pp. 282–295. Springer, Cham (2018). https://doi.org/10.1007/978-3-319-97304-3_22
31. Saito, K., Ohara, K., Kimura, M., Motoda, H.: Efficient identification of critical links based on reachability under the presence of time constraint. In: Nayak, A.C., Sharma, A. (eds.) PRICAI 2019. LNCS (LNAI), vol. 11671, pp. 404–418. Springer, Cham (2019). https://doi.org/10.1007/978-3-030-29911-8_31
32. Shen, Y., Nguyen, N.P., Xuan, Y., Thai, M.T.: On the discovery of critical links and nodes for assessing network vulnerability. IEEE/ACM Trans. Netw. **21**(3), 963–973 (2013)
33. Stojmenovic, I., Simplot-Ryl, D., Nayak, A., Velaj, Y.: Toward scalable cutvertex and link detection with applications in wireless ad hoc networks. IEEE Netw. **25**(1), 44–48 (2011)

34. Tong, H., Prakash, B.A., Eliassi-Rad, T., Faloutsos, M., Faloutsos, C.: Gelling, and melting, large graphs by edge manipulation. In: Proceedings of the 21st ACM international conference on Information and knowledge management (CIKM 2012), pp. 245–254 (2012)
35. Wang, P., Hunter, T., Bayen, A.M., Schechtner, K., Gonzalez, M.C.: Understanding road usage patterns in urban areas. Sci. Rep. **2**, 1001:1–1001:6 (2012)

On the Utilization of Structural and Textual Information of a Scientific Knowledge Graph to Discover Future Research Collaborations: A Link Prediction Perspective

Nikolaos Giarelis, Nikos Kanakaris, and Nikos Karacapilidis(✉)

Industrial Management and Information Systems Lab, MEAD, University of Patras, 26504 Rio Patras, Greece

giarelis@ceid.upatras.gr, nkanakaris@upnet.gr, karacap@upatras.gr

Abstract. We consider the discovery of future research collaborations as a link prediction problem applied on scientific knowledge graphs. Our approach integrates into a single knowledge graph both structured and unstructured textual data through a novel representation of multiple scientific documents. The Neo4j graph database is used for the representation of the proposed scientific knowledge graph. For the implementation of our approach, we use the Python programming language and the scikit-learn ML library. We benchmark our approach against classical link prediction algorithms using accuracy, recall, and precision as our performance metrics. Our initial experimentations demonstrate a significant improvement of the accuracy of the future collaboration prediction task. The experimentations reported in this paper use the new COVID-19 Open Research Dataset.

Keywords: Link prediction · Research knowledge graphs · Natural language processing · Document representation · Future research collaborations

1 Introduction

In recent years, we have witnessed an increase in the adoption of graph-based approaches for predicting future research collaborations (Nathani et al. 2019; Vahdati et al. 2018). In these approaches, a collaboration between two researchers is generally denoted by a scientific article written by them (Ponomariov and Boardman 2016). Graph-based approaches (particularly those concerning knowledge graphs) build on concepts and methods from graph theory (e.g. node centrality, link prediction and node similarity measures) to discover hidden knowledge from the structural characteristics of the corresponding research graph (Wang et al. 2017). However, despite their broad adoption, existing graph-based approaches aiming to discover future research collaborations utilize only the structural characteristics of a research graph (Veira et al. 2019). In cases where unstructured textual data is available (e.g. graph nodes that correspond to scientific articles), existing approaches are incapable of simultaneously exploiting both the structural and the textual information of the graph.

A. Appice et al. (Eds.): DS 2020, LNAI 12323, pp. 437–450, 2020.
https://doi.org/10.1007/978-3-030-61527-7_29

To remedy the above weakness, this paper proposes the construction and utilization of a scientific knowledge graph where structured and unstructured data co-exist (e.g. document, author and word nodes). Building on our previous work, we represent the documents of a scientific graph as a *graph-of-docs* (Giarelis et al. 2020a; Giarelis et al. 2020b). This enables us to exploit both the structural and textual characteristics of a research graph towards building a novel link prediction algorithm for discovering future collaborations. The proposed approach uses the Neo4j graph database (https://neo4j.com) for the representation of the knowledge graph. For the implementation of our experiments, we use the Python programming language and the scikit-learn ML library (https://scikit-learn.org).

To evaluate the outcome of this paper, we benchmark our approach against different combinations of link prediction measures, which utilize only the structural information of a research graph. Our performance metrics include the accuracy, the precision, and the recall for each of the Machine Learning (ML) models considered. For our experiments, we use the COVID-19 Open Research Dataset (CORD-19). To examine whether our approach is affected by the size of the dataset (e.g. overfits or underfits), we extract and consider nine different well-balanced datasets. The experimental results show a significant improvement of the accuracy of the link prediction problem.

The remainder of the paper is organized as follows. Section 2 introduces background concepts and related work. Our approach is thoroughly presented in Sect. 3. Section 4 reports on the experiments carried out to evaluate the proposed approach. Limitations of our approach, future work directions and concluding remarks are outlined in Sect. 5.

2 Background Issues

For the discovery of future research collaborations, the proposed approach exploits a set of natural language processing (NLP), graph-based text representation, graph theory and knowledge graph techniques.

2.1 Graph Measures and Indices

Diverse graph measures and indices to capture knowledge related to the structural characteristics of a graph have been proposed in the literature (Vathy-Fogarassy and Abonyi 2013). Below, we mention a small subset of them, which is used in our approach.

The *Common Neighbors* measure, denoted by $CN(a, b)$, calculates the number of nodes that are common neighbors for a pair of nodes a and b (Li et al. 2018). It is defined as:

$$CN(a, b) = |\Gamma(a) \cap \Gamma(b)| \tag{1}$$

where $\Gamma(x)$ denotes the set of neighbors of a node x.

The *Total Neighbors* measure, denoted by $TN(a, b)$, takes into consideration all neighbors of a pair of nodes a and b (and not only the common ones as is the case in the previous measure). It is defined as:

$$TN(a, b) = |\Gamma(a) \cup \Gamma(b)| \tag{2}$$

The *Preferential Attachment* measure, denoted by *PA(a, b)*, calculates the product of the in-degree values of a pair of nodes a and b (Albert and Barabási 2001). This measure assumes that two highly connected nodes are far more likely to be connected in the future, in contrast to two loosely connected ones. This measure is defined as:

$$PA(a,\ b)\ =\ |\Gamma(a)| * |\Gamma(b)| \tag{3}$$

The *Adamic Adar* measure, denoted by *AA(a, b)*, calculates the sum of the inverse logarithm of the degree of the set of neighbors shared by a pair of nodes a and b (Adamic and Adar 2003). This measure assumes that nodes of a low degree are more likely to be influential in the future. It is defined as:

$$AA(a, b) = \sum_{c \epsilon \Gamma(a) \cap \Gamma(b)} \left(\frac{1}{log|c|} \right) \tag{4}$$

Finally, the *Jaccard Coefficient* index, denoted by *J(a, b)*, resembles the CN measure mentioned above; however, it differs slightly in that, for a pair of nodes a and b, it considers the amount of the intersection of their neighbor nodes over the union of them (Jaccard 1901). It is defined as:

$$J(a, b) = \frac{|\Gamma(a) \cap \Gamma(b)|}{|\Gamma(a) \cup \Gamma(b)|} \tag{5}$$

2.2 Graph-Based Text Representations

The *graph-of-words* textual representation (Rousseau et al. 2013) represents each document of a corpus as a single graph. In particular, each graph node corresponds to a unique word of a document and each edge denotes the co-occurrence between two words within a sliding window of text. Rousseau et al. (2015) suggest that a window size of four seems to be the most appropriate value, in that it does not sacrifice either the performance or the accuracy of the ML models. Compared to the *bag-of-words* representation, it enables a more sophisticated feature engineering process due to the fact that it takes into consideration the co-occurrence between the terms. In any case, the limitations of the graph-of-words text representation are that: (i) it is unable to assess the importance of a word for a whole set of documents; (ii) it does not allow for representing multiple documents in a single graph, and (iii) it is not easily expandable to support more complicated data architectures.

To remedy the shortcomings of the graph-of-words representation, Giarelis et al. (2020b) have proposed the *graph-of-docs* representation, which depicts and elaborates multiple textual documents as a single graph. This last representation: (i) enables the investigation of the importance of a term into a whole corpus of documents, and (ii) it allows multiple node types to co-exist in the same graph, thus being easily expandable and adaptable to more complex data. In this paper, we utilize the graph-of-docs model to represent the textual data of a knowledge graph.

2.3 Related Work

As far as the discovery of future research collaborations using link prediction techniques is concerned, works closest to ours are those of Liben-Nowell and Kleinberg (2007), Sun et al. (2011), Guns and Rousseau (2014), Huang et al. (2008), and Yu et al. (2014). Specifically, Liben-Nowell and Kleinberg (2007) rely only on network topology aspects of a co-authors network, and the proximity of a pair of nodes to calculate the probability of future research collaborations between them. Sun et al. (2011) propose the use of structural properties to predict future research collaborations in heterogeneous bibliographic networks, where multiple types of nodes (e.g. venues, topics, papers, authors) and edges (e.g. publish, mention, write, cite, contain) co-exist. They exploit the relationships between the papers to improve the accuracy of their link prediction algorithm.

Guns and Rousseau (2014) recommend potential research collaborations using link prediction techniques and a random forest classifier. For each pair of nodes of a co-authorship network, they calculate a variety of topology-based measures such as Adamic Adar and Common Neighbors, and they combine them with location-based characteristics related to the authors. Hence, they propose future collaborations based on the location of the authors and their position on the co-authorship network. Huang et al. (2008) construct a co-authorship network for the Computer Science field that represents research collaborations from 1980 to 2005. They rely on classical statistical techniques and graph theory algorithms to describe the properties of the constructed co-authorship network. The dataset used contains 451,305 papers from 283,174 authors.

Yu et al. (2014) utilize link prediction algorithms to discover future research collaborations in medical co-authorship networks. For a given author, they attempt to identify potential collaborators that complement her as far as her skillset is concerned. They calculate common topological and structural measures for each pair of author nodes, including Adamic Adar, Common Neighbors and Preferential Attachment. ML models are used for the identification of possible future collaborations.

For a broader link prediction perspective, we refer to (Fire et al. 2011), (Julian and Lu 2016) and (Panagopoulos et al. 2017); these works describe approaches concerning the task of predicting possible relationship types between nodes (e.g. friendships in social networks).

3 Our Approach

Our approach first constructs a scientific knowledge graph that contains both structured and unstructured textual data. The integration of the unstructured textual data into the knowledge graph is accomplished through a graph-based text representation, namely *graph-of-docs* (see Sect. 2.2). Then, it employs graph measures and graph similarity techniques to extract features associated to both structural and textual information concerning the entities of a knowledge graph. Finally, it utilizes the produced features to build an ML model, which discovers future research collaborations by mapping the whole problem to a link prediction task. A detailed description of the abovementioned steps appears in (Giarelis et al. 2020a; Giarelis et al. 2020b).

3.1 The Scientific Knowledge Graph

Our knowledge graph allows diverse types of entities and relationships to co-exist in a the same graph data schema, including entity nodes with types such as 'Paper', 'Author', 'Laboratory', 'Location', 'Institution' and 'Word', and relationship edges with types such as 'is_similar', 'cites', 'writes', 'includes', 'connects', 'co_authors' and 'affiliates_with' (see Fig. 1).

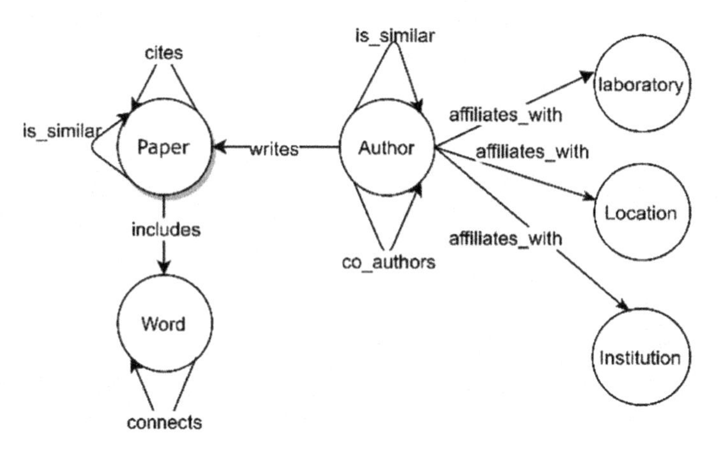

Fig. 1. The data schema of the scientific knowledge graph.

A 'Paper' entity represents a scientific paper or document. An 'Author' entity represents an author of a scientific paper or document. The 'Laboratory' entity represents the laboratory of an author. The 'Location' entity represents the location of a laboratory. The 'Institution' entity represents the institution of an author. Each 'Word' entity corresponds to a unique word of a scientific paper or document.

An 'includes' relationship connects a 'Word' with a 'Paper' entity. It marks the presence of a specific word to a certain paper. A 'connects' relationship is only applicable between two 'Word' entities and denotes their co-occurrence within a predefined sliding window of text. The subgraph constructed by the 'Word' and 'Paper' entities, as well as the 'includes', 'connects' and 'is_similar' relationships, corresponds to the graph-of-docs representation of the textual data of the available papers (see Fig. 2).

An 'is_similar' relationship links either a pair of 'Paper' or 'Author' nodes. In the former case, it denotes the graph similarity of the graph-of-docs representation of each paper. In the latter, it denotes the graph similarity between the graph-of-docs representations associated to the two authors. The subgraph that consists of the 'Author' entities and the 'is_similar' relationships corresponds to the authors similarity subgraph.

A 'cites' relationship links two 'Paper' nodes. A 'writes' relationship links an 'Author' with a 'Paper' entity. An 'affiliates_with' relationship

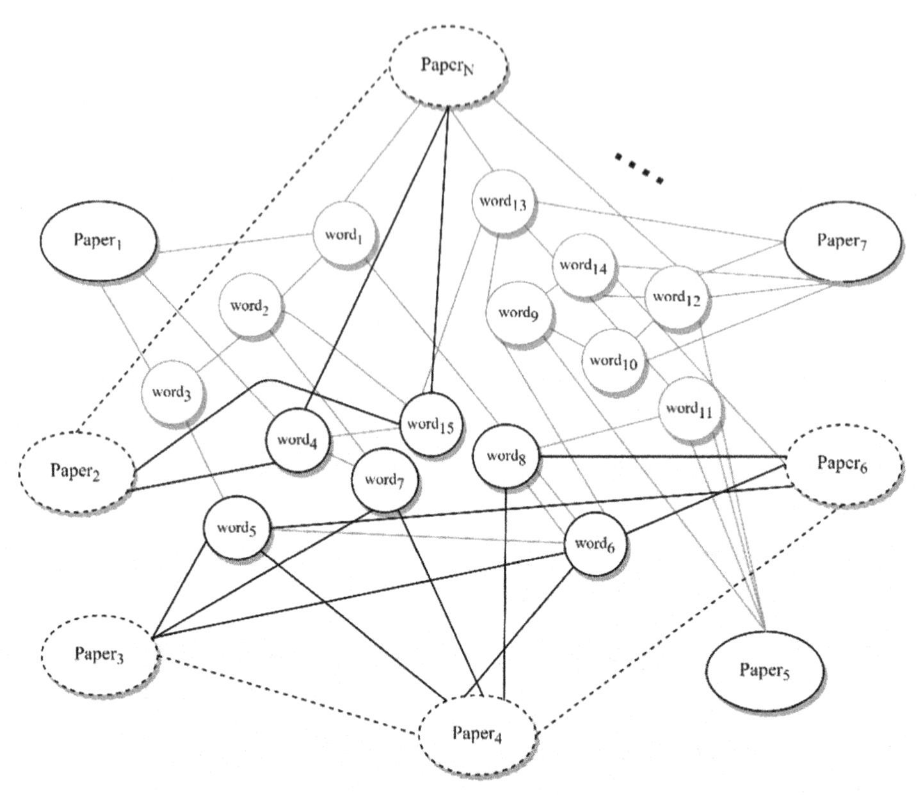

Fig. 2. Representing textual data of papers using the graph-of-docs model (relationships between papers are denoted with dotted lines). The graph-of-docs representation is associated to the `'Paper'` and `'Word'` entities, and the `'includes'`, `'connects'` and `'is_similar'` relationships of the scientific knowledge graph.

connects an `'Author'` entity with a `'Laboratory'`, `'Location'` or `'Institution'` entity. A `'co_authors'` relationship denotes a research collaboration between the connected `'Author'` entities. The subgraph constructed of the available `'Author'` entities and the `'co_authors'` relationships corresponds to the co-authors' subgraph.

The produced knowledge graph enables the utilization of well-studied graph algorithms, which in turn assists in gaining insights about various tasks, such as finding experts nearby based on the `'Location'` entities, recommending similar research work, and discovering future research collaborations; this paper focuses on the last of these tasks.

3.2 Discovery of Future Research Collaborations Using a Link Prediction Approach

For the discovery of future research collaborations, we employ various link prediction and ML techniques. Particularly, we reduce the problem of predicting future research collaborations to the common binary classification problem. By using a binary classifier, we are able to predict the presence or the absence of a `'co_authors'` relationship between two `'Author'` entities, and thus build a link prediction algorithm for the discovery of future research collaborations. Available binary classifiers include logistic regression, k-nearest neighbors, linear support vector machines, decision tree, and neural networks (Aggarwal 2018).

4 Experiments

For the implementation and evaluation of our approach, we used the Python programming language and the scikit-learn ML library (https://scikit-learn.org). The Neo4j graph database (https://neo4j.com) has been utilized for the representation of the graph-of-docs and the corresponding knowledge graph. The full code, datasets, and evaluation results of our experiments are freely available at https://github.com/NC0DER/CORD19_GraphOfDocs.

4.1 CORD-19

The COVID-19 Open Research Dataset (CORD-19) (Wang et al. 2020) contains information about 63,000 research articles, related to COVID-19, SARS-CoV-2 and other similar coronaviruses. It is freely distributed from the Allen Institute for AI and Semantic Scholar (https://www.semanticscholar.org/cord19). The articles in CORD-19 have been collected from popular scientific repositories and publishing houses, including Elsevier, bioRxiv, medRxiv, World Health Organization (WHO) and PubMed Central (PMC). Each scientific article in CORD-19 has a list of specific attributes, namely 'citations', 'publish time', 'title', 'abstract' and 'authors', while the majority of the articles (51,000) also includes a 'full text' attribute. Undoubtfully, the CORD-19 dataset is a valuable source of knowledge as far as the COVID-19-related research is concerned; however, the fact that the majority of the data included is unstructured text renders a set of limitations in its processing. As advocated in the literature, the exploitation of a graph-based text representation in combination with a knowledge graph seems to be a promising step towards structuring this data (Veira et al. 2019; Wang et al. 2017; Wang et al. 2016). For the construction of our scientific knowledge graph, we utilize the 'abstract', 'authors' and 'publish time' attributes of each scientific article. We do not exploit the 'full text' attribute due to hardware limitations; however, we assume that the abstract of a paper consists a representative piece of its full text.

4.2 Experimental Setup

Selection of measures and metrics. To construct the authors similarity subgraph and to populate the edges of the `'Author'.'is_similar'` type, we use the Jaccard similarity index, since it deals only with the percentage of common set of words versus all words, ignoring their document frequency.

Construction of datasets for the link prediction problem. To test whether our approach performs well and does not overfit, regardless of the sample size of the dataset, we extract nine different datasets from the original one, corresponding to different volumes of papers (ranging from 1,536 to 63,023). For the creation of a sample creation, we utilize (i) the authors similarity subgraph, and (ii) the co-authors subgraph (i.e. the subgraph generated from the `'co_authors'` edges; it is noted that edges also store the year of the first collaboration between authors, as a property). The features of a sample encapsulate either structural or textual characteristics of the whole knowledge graph (e.g. the similarity between the papers of two authors). Furthermore, each sample describes the relationship between two `'Author'` nodes of the knowledge graph.

The features of a sample are analytically described in Table 1. Each of the nine datasets consists of a different number of randomly chosen samples. All datasets are balanced, in that the number of positive and negative samples are equal (see Table 2). To examine whether the features taken into account each time affect the efficiency of the ML models, we execute a set of experiments with different combinations of selected features (see Table 3). Finally, it is noted that the samples for the training subset are selected from an earlier instance in time of the co-authors subgraph, which is created from `'co_authors'` edges first appeared within or before the year of 2013; respectively, the samples of the testing subset include `'co_authors'` edges created after 2013. This separation in time ensures that we avoid any data leakage between the training and testing subsets (Liben-Nowell and Kleinberg 2007).

Table 1. A detailed explanation of the features of a sample. Each feature is associated to either a structural or a textual relationship between two given `'Author'` nodes.

Feature	Description	Type
`adamic_adar`	The sum of the inverse logarithm of the degree of the set of common neighbor `'Author'` nodes shared by a pair of nodes	Structural
`common_neighbors`	The number of neighbor `'Author'` nodes that are common for a pair of `'Author'` nodes	Structural
`preferential_attachment`	The product of the in-degree values of a pair of `'Author'` nodes	Structural
`total_neighbors`	The total number of neighbor `'Author'` nodes of a pair of `'Author'` nodes	Structural
`similarity`	The textual similarity of the graph-of-docs graphs of two `'Author'` nodes. The Jaccard index is used to calculate the similarity	Textual
`label`	The existence or absence of a `'co_authors'` edge between two `'Author'` nodes. A positive label (1) denotes the existence, whereas the absence is denoted by a negative label (0)	Class

Table 2. Number of samples (|samples|), number of positive (|positive|) and negative (|negative|) samples of the training and testing subsets of each dataset. A positive sample denotes the existence of a 'co_authors' edge between two 'Author' nodes, while a negative sample denotes the absence of such an edge.

	Training subset			Testing subset		
	|samples|	|positive|	|negative|	|samples|	|positive|	|negative|
Dataset 1	668	334	334	840	420	420
Dataset 2	858	429	429	1566	783	783
Dataset 3	1726	863	863	2636	1318	1318
Dataset 4	3346	1673	1673	7798	3899	3899
Dataset 5	5042	2521	2521	12976	6488	6488
Dataset 6	5296	2648	2648	16276	8138	8138
Dataset 7	6210	3105	3105	25900	12950	12950
Dataset 8	8578	4289	4289	34586	17293	17293
Dataset 9	13034	6517	6517	49236	24618	24618

Table 3. Combinations of features aiming to test how different set of features affect the performance of an ML model.

Combination name	Features included
structural characteristics (STRS)	adamic_adar, common_neighbors, preferential_attachment, total_neighbors
structural and textual characteristics (ALL)	adamic_adar, common_neighbors, preferential_attachment, total_neighbors, similarity
adamic adar and authors similarity (AA-SIM)	adamic_adar, similarity
adamic adar (AA)	adamic_adar

4.3 Evaluation

To evaluate the effectiveness of our approach, we assess how the performance of various binary classifiers is affected by the 'similarity' feature. The list of the binary classifiers considered in this paper includes: logistic regression (LR), k-nearest neighbors (50NN), linear support vector machines (LSVM), decision tree (DT) and neural networks (NN). An extensive list of experiments using various classifiers along with different hyperparameter configurations can be found on the GitHub repository of this paper (https://github.com/NC0DER/CORD19_GraphOfDocs). Our performance metrics include the *accuracy*, *precision* and *recall* of the binary classifiers. The Friedman

Table 4. Mean (AVG), minimum (MIN), maximum (MAX) and standard deviation (SD) of accuracy, precision and recall metrics per text classifier for each combination of selected features on the nine different datasets. Bold font indicates the best method for each ML model as far as the mean and the standard deviation value of each individual metric are concerned.

		Accuracy				Precision				Recall			
Method		AVG	MIN	MAX	SD	AVG	MIN	MAX	SD	AVG	MIN	MAX	SD
LR	STRS	0.955	0.938	0.965	0.007	0.955	0.907	0.981	0.022	0.955	0.933	0.976	0.013
	ALL	0.959	0.942	0.968	0.007	0.961	0.909	0.984	0.023	0.957	0.940	0.982	0.013
	AA-SIM	**0.972**	0.964	0.981	0.005	**0.977**	0.962	0.991	0.011	0.969	0.955	0.981	0.007
	AA	0.971	0.964	0.978	0.005	0.968	0.945	0.985	0.014	**0.974**	0.964	0.986	0.008
50NN	STRS	0.956	0.925	0.975	0.014	0.928	0.872	0.965	0.025	0.988	0.982	0.997	0.005
	ALL	0.959	0.925	0.976	0.015	0.933	0.872	0.966	0.026	0.989	0.982	0.996	0.004
	AA-SIM	**0.967**	0.948	0.979	0.010	**0.944**	0.908	0.969	0.018	**0.993**	0.990	0.996	0.002
	AA	0.957	0.941	0.969	0.010	0.927	0.895	0.947	0.019	0.992	0.987	0.999	0.004
LSVM	STRS	0.959	0.931	0.971	0.012	0.941	0.884	0.972	0.026	0.979	0.969	0.993	0.008
	ALL	0.963	0.936	0.976	0.012	0.946	0.888	0.975	0.026	0.983	0.973	0.996	0.007
	AA-SIM	**0.973**	0.957	0.981	0.009	**0.957**	0.926	0.977	0.017	**0.989**	0.985	0.994	0.003
	AA	0.968	0.953	0.980	0.009	0.952	0.920	0.975	0.018	0.987	0.981	0.994	0.004

(continued)

Table 4. (*continued*)

Method		Accuracy				Precision				Recall			
		AVG	MIN	MAX	SD	AVG	MIN	MAX	SD	AVG	MIN	MAX	SD
DT	STRS	0.922	0.826	0.979	0.057	0.878	0.742	0.967	0.084	0.994	0.989	1	0.004
	ALL	**0.933**	0.837	0.980	0.046	**0.891**	0.755	0.969	0.070	0.994	0.991	0.999	0.003
	AA-SIM	0.931	0.836	0.972	0.045	0.887	0.754	0.955	0.068	**0.995**	0.991	0.999	0.002
	AA	0.879	0.660	0.955	0.094	0.825	0.595	0.922	0.108	0.994	0.989	1	0.004
NN	STRS	0.928	0.801	0.982	0.061	0.888	0.715	0.975	0.087	0.993	0.988	0.999	0.003
	ALL	0.938	0.807	0.979	0.054	0.902	0.721	0.971	0.078	0.993	0.988	0.999	0.003
	AA-SIM	**0.965**	0.943	0.979	0.012	**0.941**	0.898	0.968	0.022	**0.994**	0.992	0.999	0.002
	AA	0.956	0.899	0.976	0.024	0.928	0.834	0.968	0.041	0.991	0.985	0.999	0.004

test and the post-hoc test of Nemenyi (alpha value 0.05) are also used to calculate the significant importance between the evaluated approaches.

The obtained results indicate that the inclusion of the 'similarity' feature (i) increases the average accuracy, precision and recall scores, and (ii) decreases the standard deviation of the aforementioned scores (Table 4). The decrement of the standard deviation in the accuracy score indicates that our approach is reliable regardless of the size of the given dataset. Furthermore, by comparing the average precision score to the average recall score, we conclude that our approach predicts most of the future collaborations correctly. The best average accuracy score is achieved by the LSVM classifier, using the 'adamic_adar' and the 'similarity' features. Hence, the combination of these two features seems to be the most appropriate one. On the contrary, features such as 'common_neighbors', 'preferential_attachment' and 'total_neighbors' add noise to the overall link prediction process.

Our approach differs from existing ones in that it considers both the textual similarity between the abstracts of the papers for each pair of authors and the structural characteristics of the associated 'Author' nodes, aiming to predict a future collaboration between them. The utilization of the textual information in combination with the structural information of a scientific knowledge graph results in better and more reliable ML models, which are less prone to overfitting. Contrary to existing algorithms for the discovery of future research collaborations, our approach exploits structural characteristics and does not ignore the importance of the information related to the unstructured text of papers written by authors. Finally, existing approaches that concentrate only on the exploitation of unstructured textual data rely heavily on NLP techniques and textual representations, which in turn necessitate the generation of sparse feature spaces; hence, in such approaches, the effects of the 'curse-of-dimensionality' phenomenon re-emerge.

5 Conclusions

This paper considers the problem of discovering future research collaborations as a link prediction problem applied on scientific knowledge graphs. The proposed approach integrates into a single knowledge graph both structured and unstructured textual data using the graph-of-docs text representation. For the required experimentations, we generated nine different datasets using the CORD-19 dataset. For evaluation purposes, we benchmarked our approach against several link prediction settings, which use various combinations of a set of available features. The evaluation results demonstrated (i) an improvement of the average accuracy, precision and recall of the future collaborations prediction task, and (ii) a mitigation of the effects of the 'curse-of-dimensionality' phenomenon.

In any case, our approach has a performance issue, since the time required to build the scientific knowledge graph increases exponentially with the number of graph nodes. Aiming to address the above limitation, while also enhancing the performance and advancing the applicability of our approach, our future work directions include: (i) the utilization of in-memory graph databases in combination with Neo4j; (ii) the experimentation with word, node and graph embeddings (Mikolov et al. 2013; Nikolentzos et al. 2017; Hamilton et al. 2017); (iii) the integration of other scientific research graphs such as OpenAIRE (Manghi et al. 2019) and Microsoft Academic Graph (Arnab et al. 2015), and (iv) the integration and meaningful exploitation of our approach into collaborative research environments (Kanterakis et al. 2019).

Acknowledgments. The work presented in this paper is supported by the OpenBio-C project (www.openbio.eu), which is co-financed by the European Union and Greek national funds through the Operational Program Competitiveness, Entrepreneurship and Innovation, under the call RESEARCH – CREATE – INNOVATE (Project id: T1EDK- 05275). The authors would also like to thank Stamatis Karlos for his assistance with the statistical analysis of the data.

References

Adamic, L.A., Adar, E.: Friends and neighbors on the Web. Soc. Networks **25**, 211–230 (2003)

Aggarwal, C.C.: Machine Learning for Text. Springer, Cham (2018). https://doi.org/10.1007/978-3-319-73531-3

Albert, R., Barabási, A.: Statistical mechanics of complex networks. ArXiv, cond-mat/0106096 (2001)

Arnab, S., Zhihong, S., Yang Song, H.M., Darrin Eide, B.H., Kuansan, W.: An overview of microsoft academic service (MAS) and applications. In: Proceedings of the 24th International Conference on World Wide Web (WWW 2015 Companion), pp. 243–246. ACM, New York (2015)

Fire, M., et al.: Link prediction in social networks using computationally efficient topological features. In: 2011 IEEE Third International Conference on Privacy, Security, Risk and Trust and 2011 IEEE Third International Conference on Social Computing, pp. 73–80 (2011)

Giarelis, N., Kanakaris, N., Karacapilidis, N.: An innovative graph-based approach to advance feature selection from multiple textual documents. In: Maglogiannis, I., Iliadis, L., Pimenidis, E. (eds.) AIAI 2020. IAICT, vol. 583, pp. 96–106. Springer, Cham (2020a). https://doi.org/10.1007/978-3-030-49161-1_9

Giarelis, N., Kanakaris, N., Karacapilidis, N.: On a novel representation of multiple textual documents in a single graph. In: Czarnowski, I., Howlett, Robert J., Jain, Lakhmi C. (eds.) IDT 2020. SIST, vol. 193, pp. 105–115. Springer, Singapore (2020b). https://doi.org/10.1007/978-981-15-5925-9_9

Guns, R., Rousseau, R.: Recommending research collaborations using link prediction and random forest classifiers. Scientometrics **101**(2), 1461–1473 (2014). https://doi.org/10.1007/s11192-013-1228-9

Hamilton, W., Ying, Z., Leskovec, J.: Inductive representation learning on large graphs. In Advances in neural information processing systems, pp. 1024–1034 (2017)

Huang, J., Zhuang, Z., Li, J., and Giles, C. L.: Collaboration over time: characterizing and modeling network evolution. In: Proceedings of the 2008 international conference on web search and data mining, pp. 107–116 (2008)

Jaccard, P.: Étude comparative de la distribution florale dans une portion des Alpes et des Jura. Bull Soc Vandoise Sci Nat **37**, 547–579 (1901)

Julian, K., Lu, W.: Application of machine learning to link prediction (2016)

Kanterakis, A., et al.: Towards reproducible bioinformatics: the OpenBio-C scientific workflow environment. In: Proceedings of the 19th IEEE International Conference on Bioinformatics and Bioengineering (BIBE), Athens, Greece, pp. 221–226 (2019)

Li, S., Huang, J., Zhang, Z., Liu, J., Huang, T., Chen, H.: Similarity-based future common neighbors model for link prediction in complex networks. Sci. Rep. **8**, 1–11 (2018)

Liben-Nowell, D., Kleinberg, J.M.: The link-prediction problem for social networks. J. Am. Soc. Inform. Sci. Technol. **58**, 1019–1031 (2007)

Manghi, P., et al.: OpenAIRE Research Graph Dump (Version 1.0.0-beta) [Data set]. Zenodo. (2019). http://doi.org/10.5281/zenodo.3516918

Mikolov, T., Sutskever, I., Chen, K., Corrado, G.S., Dean, J.: Distributed representations of words and phrases and their compositionality. In Advances in neural information processing systems (NeurIPS), pp. 3111–3119 (2013)

Nathani, D., Chauhan, J., Sharma, C., Kaul, M.: Learning attention-based embeddings for relation prediction in knowledge graphs. In: Proceedings of the 57th Annual Meeting of the Association for Computational Linguistics (ACL), pp. 4710–4723 (2019)

Nikolentzos, G., Meladianos, P., Vazirgiannis, M.: Matching node embeddings for graph similarity. In: Thirty-First AAAI Conference on Artificial Intelligence (2017)

Panagopoulos, G., Tsatsaronis, G., Varlamis, I.: Detecting rising stars in dynamic collaborative networks. J. Informetrics **11**, 198–222 (2017)

Ponomariov, B., Boardman, C.: What is co-authorship? Scientometrics **109**(3), 1939–1963 (2016). https://doi.org/10.1007/s11192-016-2127-7

Rousseau, F., Kiagias, E., Vazirgiannis, M.: Text categorization as a graph classification problem. In: Proceedings of the 53rd Annual Meeting of the Association for Computational Linguistics and the 7th International Joint Conference on Natural Language Processing, vol. 1, pp. 1702–1712 (2015)

Rousseau, F., Vazirgiannis, M.: Graph-of-word and TW-IDF: new approach to ad hoc IR. In: Proceedings of the 22nd ACM International Conference on Information & Knowledge Management, pp. 59–68, ACM Press (2013)

Sun, Y., Barber, R., Gupta, M., Aggarwal, C.C., Han, J.: Co-author relationship prediction in heterogeneous bibliographic networks. In: 2011 International Conference on Advances in Social Networks Analysis and Mining, pp. 121–128 IEEE (2011)

Vahdati, S., Palma, G., Nath, R.J., Lange, C., Auer, S., Vidal, M.-E.: Unveiling scholarly communities over knowledge graphs. In: Méndez, E., Crestani, F., Ribeiro, C., David, G., Lopes, J.C. (eds.) TPDL 2018. LNCS, vol. 11057, pp. 103–115. Springer, Cham (2018). https://doi.org/10.1007/978-3-030-00066-0_9

Vathy-Fogarassy, Á., Abonyi, J.: Graph-based clustering and data visualization algorithms. Springer, London (2013). https://doi.org/10.1007/978-1-4471-5158-6

Veira, N., Keng, B., Padmanabhan, K., Veneris, A.: Unsupervised embedding enhancements of knowledge graphs using textual associations. In: Proceedings of the 28th International Joint Conference on Artificial Intelligence, pp. 5218–5225. AAAI Press (2019)

Wang, L., et al.: CORD-19: The Covid-19 Open Research Dataset. arXiv preprint arXiv:2004.10706 (2020)

Wang, Q., Mao, Z., Wang, B., Guo, L.: Knowledge graph embedding: a survey of approaches and applications. IEEE Trans. Knowl. Data Eng. **29**(12), 2724–2743 (2017)

Wang, Z., Li, J., Liu, Z., Tang, J.: Text-enhanced representation learning for knowledge graph. In: Proceedings of International Joint Conference on Artificial Intelligent (IJCAI), pp. 4–17 (2016)

Yu, Q., Long, C., Lv, Y., Shao, H., He, P., Duan, Z.: Predicting co-author relationship in medical co-authorship networks. PLoS ONE **9**(7), 101214 (2014)

Simultaneous Process Drift Detection and Characterization with Pattern-Based Change Detectors

Angelo Impedovo(✉), Paolo Mignone, Corrado Loglisci, and Michelangelo Ceci

Department of Computer Science, University of Bari "Aldo Moro", 70125 Bari, Italy
{angelo.impedovo,paolo.mignone,corrado.loglisci,
michelangelo.ceci}@uniba.it

Abstract. Traditional process mining approaches learn process models assuming that processes are in steady-state. This does not comply with the flexibility and adaptation often requested for information systems and business models. In fact, these approaches should discover variations to adapt to new circumstances, which is a peculiarity that conventional change analysis based on time-series, could not provide, because the processes are complex artifacts. This problem can be handled with change-aware structured representations, such as those typically used for network data. In this paper, we propose a novel pattern-based change detection (PBCD) algorithm for discovering and characterizing changes in event logs encoded as dynamic networks. In particular, PBCDs are unsupervised change detection methods, based on observed changes in sets of patterns observed over time, which are able to simultaneously detect and characterize changes in evolving data. Experimental results, on both real and synthetic data, show the usefulness and the increased accuracy with respect to state-of-the-art solutions.

1 Introduction

The aim of the process mining techniques is learning models (for instance, in the form of Petri nets or heuristic maps) from collections of traces recording observed process executions. Thus, the models can be seen as an abstract form of the really-performed processes and can therefore be used for predictive problems, such as the prediction of outcomes and for conformance checking, that is, the adherence of new traces to the models.

A common assumption of many process mining algorithms is the "invariability" of the process model, meaning that the traces are in a steady-state, that is, they should obey the configuration dictated by the models, without any deviation with respect to the reference process. This aspect has been investigated by methods which recognize variations present in the traces and learn process variants [2]. In many information systems this is not sufficient because the traces might present frequent or regularly repeated changes. A change becomes necessary whenever there is a need for people and institutions to adapt their ordinary behavior to changing circumstances and environments. Various examples

A. Appice et al. (Eds.): DS 2020, LNAI 12323, pp. 451–467, 2020.
https://doi.org/10.1007/978-3-030-61527-7_30

can be found both in society and nature. For example, new regulations and laws require citizens and organizations to change their processes. In a dynamic market, flexible organizations should quickly adapt their internal and external operating procedures to natural disasters as well as to the introduction of new laws and regulations. Therefore, the presence of substantial changes could make the process model inconsistent with respect to the (actual) instances. In order to effectively deal with this, we should revise the working hypothesis and consider the processes as non-stationary, allowing for abrupt or gradual changes exhibited over time. Consequently, the process modeling approaches should react to such process *drifts* by quickly detecting and understanding them [5].

Existing methodologies suffer from several drawbacks. In particular, they work on an over-simplified data representation which does not account for the traces as complex artifacts. This leads to considering only one set of numerical features [4] of the executed traces, while neglecting the temporal component associated to the activities and interactions among the activities, actors and resources, which are sources of information able to explain drifts between traces of the same process model. These representational forms often limit the task of drift detection to a mere quantification of the magnitude of the change between different traces, without providing an explanation of the nature of the change. Thus, any attempt to explain or characterize the changes requires the intervention of the human process modeler or reference knowledge to identify the components of a trace which determine the changes [14].

In this work, we simultaneously solve the problems of process drift detection and characterization with Pattern-based Change Detectors (PBCDs hereafter). PBCDs refer to a class of change detection algorithms in which *i)* the change is detected on patterns discovered from the data over time, and *ii)* the patterns responsible for a given change already constitute an off-the-shelf descriptive model of the change. They have been exploited to study changes on dynamic networks [11] thanks to the peculiarities to identify sub-graphs related to the changes, associate changes to variations of the occurrences of the sub-graphs and quantify the magnitude order of the changes with frequency-based quantitative measures. Thus, our intuition is that of encoding process traces (from the event log) into a graph-based representation and detecting process drifts through PBCDs. This perspective offers several advantages: *i)* the use of an established unsupervised approach to simultaneously solve the problems of drift detection and characterization, *ii)* a computational solution able to account for the temporal order of the activities, *iii)* a method able to determine the most promising set of features mirroring the changes and represent them in form of sub-graph patterns, *iv)* the possibility to capture both gradual changes and sudden changes, which, thus, would appear as mild frequency-based variations and strong frequency-based variations respectively.

The manuscript is organized as follows. Firstly, we introduce some related works in process drift detection and motivate the adoption of PBCDs. Then, we discuss some preliminary notions about processes and dynamic networks, so as to explain how event logs can be transformed into dynamic networks.

The adopted PBCD methodology is then discussed by emphasizing how process drifts are detected and characterized. Then experimental results on both synthetic and real-world event logs are illustrated before drawing some conclusions.

2 Related Works

The adoption of PBCDs for detecting and characterizing the process drifts in event logs occurs at the intersection of two research directions: one concerning pattern-based learning of process models and the other concerning process drift detection methods.

A well-known result in process mining is that *frequent sequential patterns* offer an alternative way of representing process models instead of Petri nets, discovered by the traditional α algorithm [1], or heuristic maps learned by the HeuristicMiner algorithm [17]. Specifically, while *sequential patterns* model the contiguous sequence of executed activities, *frequent sequential patterns* are used to discover statistical evident paths of executions in an event log seen as a database of sequence. Hence, sequential pattern mining algorithms can be used to learn process models as done in [7,8]. An aspect worth mentioning is that frequent patterns effectively model stable features of the process over time. Consequently, our claim is to effectively leverage such features when executing PBCDs on event logs. Unfortunately, to the best of our knowledge, no PBCD based on sequential patterns exists.

As for the process drift detection methods in process mining, different methodologies have been proposed, although none of them is pattern-based. The first is proposed in [4] and implemented in ProM[1], in which the change detection approach is able to detect drifts, via statistical significance testing, by considering a set of four numeric global and local features. In this approach, the event log is transformed into a multivariate time-series, and, hence, changes are detected in such an intermediate representation in which the original control-flow perspective is lost. The second method is the ProDrift algorithm defined in [13] and implemented in the Apromore framework[2]. ProDrift also performs a statistical significance test on a run-based encoding of the traces, obtained prior to the detection phase. Both methods adopt the sliding window model. In particular, the statistical significance test is assessed by comparing the populations of two sliding windows, the reference and the detection windows, that slide over the data whenever a new trace is observed. Both the methods are parametric change detection algorithms, working on an intermediate representation of traces and, lastly, they do not characterize the detected changes.

3 Background

Let A be the set of activities, then an event log over A is defined as the time series of n traces $E = \{T_t\}_{t=1}^{n}$. Each trace $T_i = \langle a_1, \ldots, a_k \rangle$ captures the sequence of k activities $a_i \in A$ as executed at the time point t_i in a given process instance.

[1] https://svn.win.tue.nl/trac/prom/browser/Packages/ConceptDrift.
[2] https://apromore.org/platform/tools/.

Since PBCDs leverage differences between patterns exhibited by the data over time for detecting changes, the principal requirement for their use is the existence of a pattern mining methodology that best suits the data representation at hand. In their natural formulation, event logs are not immediately compatible with traditional pattern mining methods. On the other hand, various existing PBCDs are specifically designed for dynamic networks. Therefore, we encode event logs in the form of dynamic networks as an intermediate representation compatible with existing graph-based PBCDs. Let N be the set of nodes, L be the set of edge labels and $I = N \times N \times L$ the alphabet of all the possible labeled edges. A dynamic network is defined as the time series of n graph snapshots $G = \{G_t\}_{t=1}^n$. Each snapshot $G_i \subseteq I$ is a set of edges denoting a directed graph observed in t_i allowing self-loops and multiple edges with different labels.

3.1 From Event Logs to Dynamic Networks

Encoding the event log $E = \{T_t\}_{t=1}^n$ as the dynamic network $G = \{G_t\}_{t=1}^n$ is done by transforming every trace T_i into the associated graph snapshot $G_i = g(T_i)$. The map $g(T_i)$ allows us to consider the dynamic network $G = \{g(T_t)\}_{t=1}^n$ in place of the initial event log. In particular, let $T = \langle a_1, \ldots, a_k \rangle$ be a trace, the graph $G = g(T)$ is built by considering i) the set of edge labels $L = \{a_1, \ldots, a_k\}$, ii) the set of nodes $N = \{0, 1, \ldots, n\}$ and iii) $I = N \times N \times L$. Then, $G = g(T) = \{(i-1, i, a_i) \in I \mid a_i \in T\} \subseteq I$ is a labeled graph in which edge labels denote activities, and nodes denote natural numbers. This graph-based representation of traces keeps the temporal ordering of activities in a trace, as shown in Fig. 1, and this is a necessary condition to preserve the process control-flow perspective in the drift detection activity.

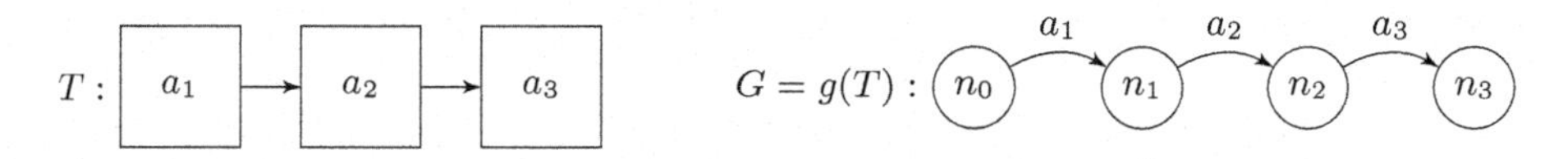

Fig. 1. Example of a trace T made of 3 activities (a_1, a_2 and a_3) represented as the graph snapshot $G = g(T)$. Activity names in T become edge labels in G.

3.2 Frequent and Emerging Subgraph Discovery

The representation of event logs as dynamic networks fits the one adopted in transactional data mining, meaning that it is possible to discover interesting sub-graphs with traditional sub-graph mining algorithms designed for dynamic networks. In the transactional setting, a snapshot $G_{tid} \in G$ is a transaction uniquely identified by *tid*, whose items are labeled edges from I. A sub-graph $S \subseteq I$, with *length* $|S|$, can be seen as a word $S = \langle i_1 \ldots i_n \rangle$ of n lexicographic sorted items, with prefix $P = \langle i_1 \ldots i_{n-1} \rangle$ and suffix i_n.

For this work *frequent connected sub-graphs* (FCSs hereafter) are deemed to be interesting, as they denote stable features that are useful for the drift detection step. FCSs are discovered from graph snapshots belonging to time windows. A window $W = [t_i, t_j]$, with $t_i < t_j$, is the sequence of snapshots $\{G_i, \ldots, G_j\} \subseteq G$. Consequently, the width $|W| = j - i + 1$ is equal to the number of snapshots collected in W. Let S be a sub-graph, then the *tidset* of S in the window W is defined as $tidset(S, W) = \{tid \mid \exists G_{tid} \in W \wedge S \subseteq G_{tid}\}$, while the *support* of S in W is $sup(S, W) = \frac{|tidset(S,W)|}{|W|}$. S is *frequent* in W if $sup(S, W) > minSUP$, where $minSUP \in [0, 1]$. We term F_W the set of all the FCSs in the window W.

Once detected a process drift needs to be characterized. While the FCSs support the drift detection by capturing statistically evident parts of the process, as observed in a time window, they do not characterize drifts. To this end, we deem interesting the *emerging connected sub-graphs* (ESs hereafter), discovered between two time windows by evaluating the growth-rate of sub-graphs. Let S be a sub-graph, W and W' two consecutive time windows, then the *growth-rate* of S between W and W' is $gr(S, W, W') = \frac{max(sup(S,W), sup(S,W'))}{min(sup(S,W), sup(S,W'))}$. S is *emerging* between W and W' if $gr(S, W, W') > minGR$, where $minGR > 1$. We term $es(W, W')$ the set of the ESs between W and W' according to $minGR$.

The ESs are the building blocks of the change characterizations. However, i) the combinatorial explosion of the ESs worsens the readability of characterizations, and ii) ESs singularly add small contributions to the characterizations. To tackle these problems, we only consider the *maximal emerging connected sub-graphs* (MESs hereafter). Let $S \in es(W, W')$, then S is maximal if there is not another sub-graph $Q \in es(W, W')$ such that $S \subset Q$. We term $ms(W, W')$ the set of all the MESs between W and W' according to $minGR$.

3.3 Problem Statement

Let $E = \{T_t\}_{t=1}^{n}$ be an event log, $minSUP \in [0, 1]$ be the minimum support threshold, $minMC \in [0, 1]$ the minimum change threshold, $minGR > 1$ be the minimum growth-rate threshold. Then:

- the *dynamic network* $G = \{g(T_t)\}_{t=1}^{n}$ of E is built as a pre-processing step.
- *pattern-based change detection* finds pairs of windows $W = [t_b, t_e]$ and $W' = [t_{e+1}, t_c]$ from D, where $t_b \leq t_e < t_{e+1} \leq t_c$. Each pair of windows denotes a change which is:
  - quantified by the pattern dissimilarity score $d(F_W, F_{W'}) > minMC$
  - explained by the maximal emerging sub-graphs $ms(F_W, F_{W'})$ discovered according to $minGR$

where F_W ($F_{W'}$) denote the FCSs discovered on W (W') according to $minSUP$.

Process drifts are detected on the dynamic network encoding of the event log. Specifically, a drift is detected every time a relevant difference between the set of FCSs F_W and $F_{W'}$ is measured. Finally, the drift is explained by the MESs, as they describe the (appearing or disappearing) sequences of activities involved in the change of the underlying process model.

3.4 Computational Approach

The afore-mentioned change detection and explanation problem can be solved by various computational solutions. Among them, we mention the class of pattern-based change detection algorithms (PBCD). In general, a PBCD forms a two-step approach in which: i) a pattern mining algorithm extracts the set of patterns observed from the incoming data, and ii) the amount of change is quantified by adopting a dissimilarity measure defined between sets of patterns. More specifically, a PBCD is an iterative algorithm that consumes data coming from a data source, in our case a dynamic network, and produces quantitative measures of changes. For instance, the KARMA algorithm proposed in [11] is a PBCD for detecting and characterizing changes in network data. KARMA is based on the exhaustive mining of FCSs, whose general workflow can be seen in Figure 2. The algorithm iteratively consumes blocks Π of graph snapshots coming from D (Step 2) by using two successive landmark windows W and W' (Step 3). Thus, it mines the complete sets of FCSs, F_W and $F_{W'}$, which are necessary for the detection steps (Steps 4–5). The window grows ($W = W'$) with new graph snapshots, and the associated set of FCSs is kept updated (Steps 8–9) until the change score $d(F_W, F_{W'})$ exceeds β and a change is detected. In that case, the algorithm drops the content of the window by retaining only the last block of transactions ($W = \Pi$, Steps 6–7). Then the analysis restarts.

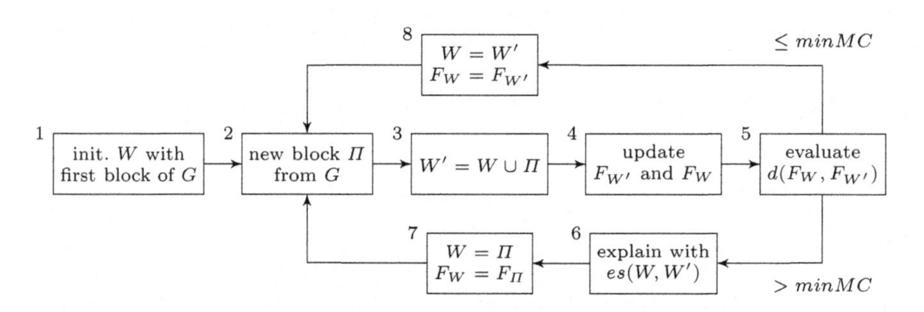

Fig. 2. The KARMA algorithm flowchart

However, KARMA does not naturally fit the given problem statement and is not the optimal solution. Firstly, while KARMA relies on successive landmark windows of increasing size, our problem statement compares two successive non-overlapping windows of different size. Secondly, KARMA discovers FCSs on data represented in a more general representation than the dynamic network encoding of event logs in the form of sequences of graph chains. Consequently, no FCS, which is not a simple chain, would be returned by the mining algorithm: chains are only discovered when mining FCSs in sequences of chains. However, although this solution is always able to discover FCSs that are also chains, it is also inefficient: the mining algorithm would also generate and discard the FCSs which are not chains. Therefore, we restrict the pattern language to *frequent subtrees*

(FSs hereafter), that is, FCSs in which every node is connected to a parent node, except for the root node. To meet these requirements, we adapt the KARMA algorithm to the KARMATree approach depicted in Fig. 3. In this case, an alternative time window model is used to arrange incoming blocks of transactions (Steps 3, 7 and 8). Then sets F_W and $F_{W'}$ of FSs are discovered instead of FCSs (Step 4). Lastly, changes are characterized by discovering the maximal emerging subtrees (Step 6) instead of the emerging ESs by KARMA.

Let G be a dynamic network over $|I| = k$ possible edges, with n snapshots and $m = \frac{n}{|\Pi|}$ blocks of size $|\Pi|$. KARMA requires time proportional to $O(m \cdot |FCSs|)$ in the worst case scenario [11], while KARMATree requires $O(m \cdot |FSs|)$ where $|FSs| << |FCSs| < e^k$, since the number of subtrees is lower than the number of subgraphs in a network. However, KARMATree is an exhaustive PBCD, relying on complete mining of FSs, which could not work well in limited memory scenarios. As a solution, a non-exhaustive variant could be obtained by equipping KARMATree with the heuristic mining approach shown in [10].

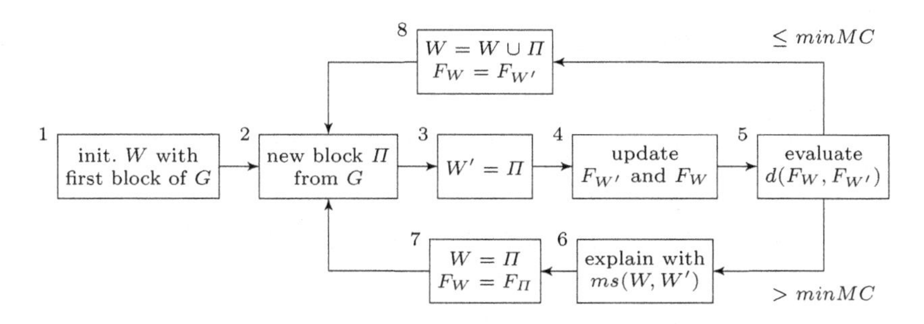

Fig. 3. The KARMATree algorithm flowchart

4 Experiments

The experiments are organized according to different perspectives concerning both synthetic and real-world processes. In particular, we answer the following research questions: **Q1)** Is the proposed PBCD approach *more accurate* than existing process drift detection approaches when detecting changes on synthetic processes? **Q2)** Is the proposed PBCD approach *more efficient* than existing process drift detection approaches when detecting changes on synthetic processes? **Q3)** Do the characterizations describe changes in real-world process?

Assessing the accuracy of the proposed approach compared with competitor methodologies is not an easy task. Although the process evolution is a well-established concept in process mining, to the best of our knowledge, no proper ground truth for process drift detection is known. The main consequence is the difficulty of measuring the accuracy on real-world datasets. Moreover, existing synthetic log generators are not flexible enough. For instance, the one proposed in [6] randomly builds and evolves single process models and simulates their execution to synthesize event logs.

The major limitation is that the simulation is based only on a single process model, and hence none of its evolutions is considered. This means that i) traces in the resulting event log conform to the process model used in the simulation, and ii) consequently no evident change is injected into the resulting log. To overcome this limitation, we extended the process log generator[3] to i) build a chain of n process models where the first is randomly generated and the others are subsequent random evolutions, and ii) generate the complete event log by simulating an equal number of traces for every process model in the chain. We synthesized 10 event logs, each built by considering a chain made of 20 evolving process models. In particular, each model has been used to simulate a block made of 100 traces, for a total number of 2,000 traces per each log, thus ensuring a change between every pair of subsequent blocks.

By so doing, every generated log can be used as a ground-truth about the presence of changes when evaluating the accuracy of the proposed PBCD approach.

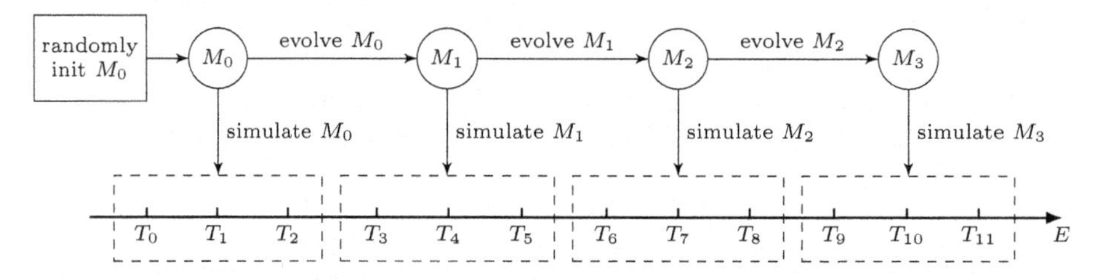

Fig. 4. Synthetic event log E built by simulating the execution of a process model M_0 which evolves 3 times. Each model evolution produces a block of 3 traces.

To answer the afore-mentioned research questions, we first discuss the results of a comparative evaluation between our approach and existing drift detectors for process data. Then a case study is shown to illustrate the usefulness of KARMATree for simultaneously detecting and characterizing changes in real world datasets. In particular we compare our proposed KARMATree PBCD approach with two state-of-the-art process drift detection algorithms, the *ProDrift* [13] available in the Apromore framework and the drift detector by Bose et al. [4] available in the ProM framework, respectively. Both the competitors are parametric change detection algorithms specifically designed for process data. They are built so as to embed the ADWIN [3] algorithm, therefore, they discover changes in event logs by scanning them through adaptively-sized time windows (we term these two algorithms *ProDrift (adwin)* and *Bose et al. (adwin)*). The *ProDrift* algorithm can also be used with fixed-size windows, termed as *ProDrift (fixed)*. Another difference between KARMATree and both ProDrift

[3] https://bitbucket.org/carbonkid/process/.

and Bose et al. is their detection method. In fact, while KARMATree is a non-parametric drift detection approach, relying on pattern-set dissimilarities, both ProDrift and Bose et al. seek changes by performing statistical hypothesis testing, at a given p-value, between the data population in the time windows. Specifically, Bose et al. employ two global features which are defined over an event log, and two local features, which are defined at a trace level by considering a fixed-size window, while ProDrift works only at trace level by considering an event log as a continuous stream of traces and it is designed to adaptively identify the right window size.

4.1 The Most Accurate Process Drift Detection Approach

In this set of experiments we executed KARMATree and the three competitor algorithms on 10 synthetic event logs, generated according to the procedure previously described, and collected their accuracies, false positive rates (FPRs) and false negative rates (FNRs). We fixed the initial size of time windows to 20 in every considered approach, as for KARMATree, we fixed the minimum support threshold to $minSUP = 0.1$ and the minimum growth-rate threshold to $minGR = 1.0$. On the contrary, we tuned the minimum change threshold as $minMC = \{0.5, 0.6, 0.7, 0.8, 0.9\}$. On the other hand, we fixed a critical p-value of 0.95 for both ProDrift and Bose et al.

Table 1. Accuracy of KARMATree against ProDrift (fixed, adwin) and Bose et al. (adwin) when tuning *minMC* on 10 synthetic event logs.

dataset	Accuracy @ *minMC*							
	KARMATree					ProDrift	ProDrift	Bose et al.
	0.5	0.6	0.7	0.8	0.9	(fixed)	(adwin)	(adwin)
synth-log-01	0.989	0.989	0.989	0.959	0.918	1	0.846	0.959
synth-log-02	0.959	0.959	0.948	0.928	0.867	0.989	0.877	0.858
synth-log-03	1	1	1	0.979	0.959	1	0.816	0.909
synth-log-04	1	1	0.989	0.938	0.857	1	0.846	0.898
synth-log-05	0.959	0.969	0.959	0.948	0.918	0.959	0.857	0.929
synth-log-06	0.989	0.979	0.959	0.908	0.867	0.928	0.846	0.898
synth-log-07	0.989	0.979	0.979	0.938	0.908	1	0.816	0.929
synth-log-08	0.969	0.969	0.948	0.928	0.887	1	0.846	0.939
synth-log-09	0.979	0.969	0.928	0.908	0.857	1	0.826	0.959
synth-log-10	0.969	0.969	0.969	0.959	0.938	1	0.826	0.939

As for the accuracy (Table 1), we report that KARMATree always outperforms ProDrift (adwin) for every value of *minMC*. The same is not true for Bose et al. (adwin), which is outperformed by KARMATree when $minMC \leq 0.8$, and outperforms KARMATree when $minMC = 0.9$. On the contrary, ProDrift (fixed) is a top competitor based on time windows of fixed size, differently from KARMATree and every adwin-based competitor adopting time windows of dynamic size. Specifically, since i) ProDrift (fixed) consumes blocks of 20 traces, and ii) the synthetic datasets are generated so to report a change once every 100 traces, the algorithm compares two clearly distinct group of traces once every 5 windows, on which a change is detected. However, knowing in advance the temporal distribution of changes (once every 100 traces) requires prior knowledge on the observed process, which could not always be available. In this perspective, differently from the remaining adwin-based competitors, KARMATree still outperforms ProDrift (fixed) on synth-log-03/05/06 (for $minMC \leq 0.7$), and synth-log-04 (for $minMC \leq 0.6$).

This analysis is confirmed by the false positive rates (FPRs) and false negative rates (FNRs). As expected, ProDrift (fixed) exhibits both FPRs and FNRs approximately equal to 0 on every dataset. Also, ProDrift (adwin) exhibits very low FPRs but moderately high FNRs. On the contrary Bose et al. exhibits the worst FPR on almost every dataset (except for synth-log-09) and remarkable FNRs, which in turn are no worse than ProDrift (adwin). As for KARMATree, the algorithm always outperforms the competitors with respect to their FPRs. As for the FNRs, KARMATree outperforms every competitor for low values of *minMC*. From these results, two tendencies arise: i) both FPRs and FNRs decrease with *minMC*, and ii) the accuracy increases for low values of *minMC* (Table 2).

4.2 The Most Efficient Process Drift Detection Approach

In this set of experiments we compared the running times (seconds) of KARMATree against the ones of the three competitors on 10 synthetic event logs (Table 3). As before, we fixed the initial size of time windows to 20 in every considered approach. As for KARMATree we fixed the minimum support to $minSUP = 0.1$ and the minimum growth-rate to $minGR = 1.0$. On the contrary, we tuned the minimum change threshold as $minMC = \{0.5, 0.6, 0.7, 0.8, 0.9\}$. We fixed a critical p-value of 0.95 for ProDrift and Bose et al. No clear tendency emerges when looking at decreasing values of $minMC$ for KARMATree. This is an expected result, since $minMC$ does not influence the running times, which are strongly determined by the mining step in the PBCD pipeline. However, when comparing KARMATree with respect to both ProDrift fixed and adwin-based,

Table 2. False positive rate (FPR) and False negative rate (FNR) of KARMATree against ProDrift (fixed, adwin) and Bose et al. (adwin) when tuning *minMC* on 10 synthetic event logs.

dataset	False positive rate @ *minMC*							
	KARMATree					ProDrift (fixed)	ProDrift (adwin)	Bose et al. (adwin)
	0.5	0.6	0.7	0.8	0.9			
synth-log-01	0	0	0	0	0	0	0	0
synth-log-02	0	0	0	0	0	0.013	0	0.087
synth-log-03	0	0	0	0	0	0	0.013	0.062
synth-log-04	0	0	0	0	0	0	0	0.087
synth-log-05	0.013	0	0	0	0	0.051	0.013	0.0375
synth-log-06	0	0	0	0	0	0.051	0.063	0.075
synth-log-07	0	0	0	0	0	0	0	0
synth-log-08	0	0	0	0	0	0	0	0.05
synth-log-09	0.013	0	0	0	0	0	0	0
synth-log-10	0	0	0	0	0	0	0	0
dataset	False negative rate @ *minMC*							
	KARMATree					ProDrift (fixed)	ProDrift (adwin)	Bose et al. (adwin)
	0.5	0.6	0.7	0.8	0.9			
synth-log-01	0.053	0.053	0.053	0.211	0.421	0	0.789	0.211
synth-log-02	0.211	0.211	0.263	0.368	0.684	0	0.632	0.368
synth-log-03	0	0	0	0.105	0.211	0	0.895	0.211
synth-log-04	0	0	0.053	0.316	0.737	0	0.789	0.158
synth-log-05	0.158	0.158	0.211	0.263	0.421	0	0.684	0.211
synth-log-06	0.053	0.105	0.211	0.474	0.684	0.158	0.526	0.211
synth-log-07	0.053	0.105	0.105	0.316	0.474	0	0.947	0.368
synth-log-08	0.158	0.158	0.263	0.368	0.579	0	0.789	0.105
synth-log-09	0.053	0.158	0.368	0.474	0.737	0	0.895	0.211
synth-log-10	0.158	0.158	0.158	0.211	0.316	0	0.895	0.316

our approach is more efficient than the two competitors (except for synth-log-03 when $minMC \leq 0.7$). Moreover, KARMATree is more efficient than Bose et al. by at most two orders of magnitude. Therefore, we conclude that KARMATree is able to devise more accurate and more efficient drift detection on almost every considered dataset.

Table 3. Running times (seconds) of KARMATree against ProDrift (fixed, adwin) and Bose et al. (adwin) when tuning *minMC* on 10 synthetic event logs.

dataset	Running times (seconds) @ *minMC*							
	KARMATree					ProDrift	ProDrift	Bose et al.
	0.5	0.6	0.7	0.8	0.9	(Fixed)	(Adwin)	(Adwin)
synth-log-01	0.86	0.874	0.891	1.106	0.856	3.38	3.453	28.284
synth-log-02	1.499	1.625	1.688	1.782	1.58	3.908	3.411	47.411
synth-log-03	6.107	6.155	5.932	4.966	5.735	5.005	5.054	187.65
synth-log-04	3.188	3.257	3.392	3.203	3.705	4.932	4.753	92.523
synth-log-05	1.471	1.452	1.592	1.532	1.437	4.172	3.558	47.311
synth-log-06	0.658	0.628	0.629	0.642	0.606	2.79	2.795	18.789
synth-log-07	3.047	2.764	2.954	3.066	3.065	4.4	4.251	96.541
synth-log-08	2.489	2.496	2.717	2.584	2.562	4.777	4.449	70.165
synth-log-09	3.85	3.504	4.43	4.353	5.105	6.243	5.421	101.7
synth-log-10	1.436	1.375	1.445	1.512	1.367	3.83	3.714	42.003

4.3 Case Study

We illustrate a case study in which KARMATree is used to detect and characterize the changes in a real process. When a change is detected between two windows, the reference window and the target window, it is reasonable to expect some differences between the two associated process models. Intuitively, the change can be characterized by listing the modifications necessary to transform the process model, learned from traces in the reference window, into the one learned from traces in the target window. We recall that KARMATree characterizes changes by listing the maximal emerging subtrees between the reference and the target windows. Two considerations arise: first, it is possible that a subtree which was frequent in the reference window becomes infrequent in the target window, and second, a subtree which was infrequent in the reference window may become frequent in the target window. Since an emerging subtree denotes an appearing (disappearing) sequence of activities, then the associated activities will (will not) be included in the process models. Furthermore, since KARMATree discovers *maximal emerging subtrees*, the change is characterized in terms of the longest sequences of activities which appear or disappear over time.

The real process we consider is the hospital billing process collected in [15]. The event log collects events related to the billing of medical services as provided by a regional hospital. The dataset was collected from the financial modules of the ERP system of the hospital. Specifically, the event log contains 100.000 anonymized traces recorded over a period of three years. Our purpose is to detect and characterize the changes in the hospital billing process. With this objective in mind, we first encoded the dataset as a dynamic network by following the procedure described in Sect. 3. Then we executed the KARMATree algorithm on the resulting dynamic network by fixing the input parameters as follows: the minimum support threshold was fixed at $minSUP = 0.1$, the minimum change threshold at $minMC = 0.5$ and the minimum growth-rate threshold at $minGR = 1.0$. Once it had been executed, the algorithm was able to detect and characterize 95 change points out of 1000 blocks.

We report two changes detected by KARMATree, by focusing on the associated characterizations. In particular, given a change detected between a reference window W and a target window W', we match the MESs against the two heuristic maps discovered by running the HeuristicMiner algorithm (available in ProM [17]) on traces from the two windows, respectively. Thus, we show how MESs, discovered by KARMATree, mirror the differences between the two heuristic maps. We note that heuristic maps highlight the frequently executed parts of the process in black, while the less executed ones appear in gray.

The first change is detected when KARMATree consumes 6000 traces. Specifically, the change score amounts to 54% and is spotted between the reference window containing the first 5900 traces ($W = [1, 5900]$) and the target window containing the remaining 100 ($W' = [5901, 6000]$). The billing process in W is depicted in Fig. 5. First, a new billing is created and the associated diagnosis is set. Then, the fine is created and released. Consequently, the bill is closed only when the fine is released with success. Occasionally, the diagnosis can be changed multiple times before deleting the billing prematurely (due to errors, for example). Errors can also affect the fine, in that case a new one is created by returning to the create fine activity. However, the billing process looks different when observed in W' (Fig. 6). The change diagnosis activity is less frequently executed than in W, and may cause the deletion of the bill. Consequently, a new fine is created right after the billing process starts. When comparing this heuristic map to the previous one, it emerges that the billing process has been shortened by avoiding the change diagnosis activity.

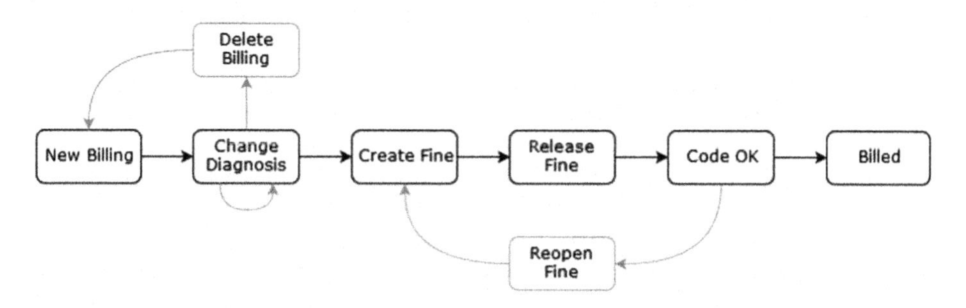

Fig. 5. Heuristic map for the billing process in the *reference window* $W = [1, 5900]$.

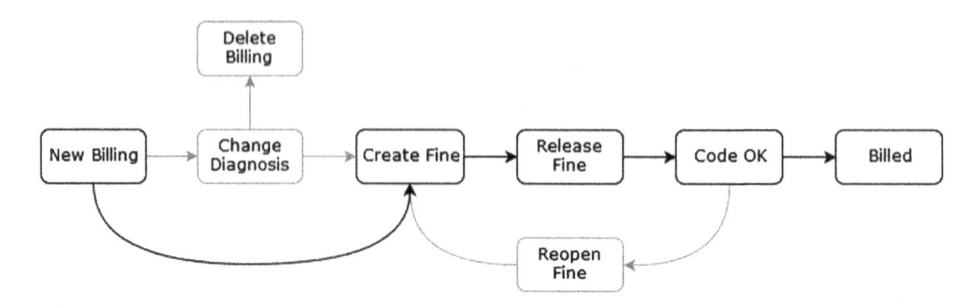

Fig. 6. Heuristic map for the billing process in the *target window* $W' = [5901, 6000]$.

KARMATree explains the change with a single *maximal emerging pattern* S which is frequent in W, with a support of 39%, and infrequent in W' with a support of 7%. Therefore, S emerges with a growth-rate of 558%:

$$S = \{(0, 1, \text{new billing}), (1, 2, \text{change diagnosis}), (2, 3, \text{create fine}), (3, 4, \text{release fine}), (4, 5, \text{code ok}), (5, 6, \text{billed})\}$$

Since S is emerging and infrequent in W', it suggests that the billing process, as performed in W, is not compliant with the process model observed on W'. This is an expected result, since the heuristic map discovered on W' does not depict the billing process as S does. On the contrary, S is compliant with the heuristic map discovered on W. Clearly, every subtree $S' \subset S$ involving the change diagnosis activity also characterizes the change occurring between W and W' (for example, $S' = \{(0, 1, \textit{new billing}), (1, 2, \textit{change diagnosis}), (2, 3, \textit{create fine})\}$). Indeed, the same change could have been represented by various emerging subtrees, each of which adds a small contribution to the remaining ones. The usefulness of discovering *maximal emerging subtrees* is precisely the characterizing of changes in a succinct way, that is, by only considering the longest sequences of activities.

KARMATree detects a second change (54%) immediately after the arrival of 100 new traces. In this case, the reference window is $W = [5901, 6000]$ and is equivalent to the target window of the previous example, while the current target window is $W' = [6001, 6100]$. Consequently, the heuristic map associated to W is depicted in Fig. 6, and the one associated to W' is depicted in Fig. 7. This new heuristic map specifies that the billing process can be alternatively completed by either changing the diagnosis associated to the billing or not. Moreover, the map also states that i) new billings have been immediately deleted after their creation and ii) no fine is reopened. The change is characterized by two maximal emerging subtrees S_1 and S_2 which are infrequent in W (support of 7% and 4%, resp.) and frequent in W' (support of 42% and 12%, resp.).

$$S_1 = \{(0, 1, \text{new billing}), (1, 2, \text{change diagnosis}), (2, 3, \text{create fine}), (3, 4, \text{release fine}), (4, 5, \text{code ok}), (5, 6, \text{billed})\}$$

$$S_2 = \{(0, 1, \text{new}), (1, 2, \text{delete})\}$$

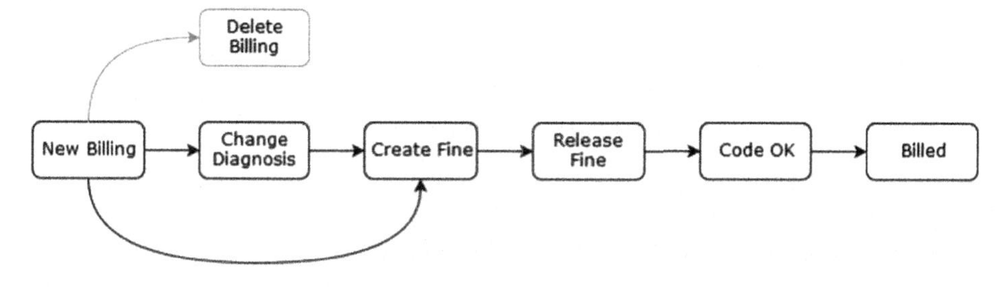

Fig. 7. Heuristic map for the billing process in the *reference window* $W = [6001, 6100]$.

Since both the subtrees are frequent and emerging in W', they denote novel parts of the process, as executed according to the heuristic map on W'. In particular, while S_1 reintroduces the change diagnosis activity in the billing process, S_2 states that new billings are immediately deleted right after their creation. We note that the two subtrees are compliant with the heuristic map learned on W'.

5 Conclusions

We have presented the KARMATree for simultaneously detecting and characterizing process drifts. KARMATree detects changes in an intermediate representation of event logs in the form of dynamic networks. Specifically, changes are i) sought by tracking variations in the frequent subtrees observed over time on non-overlapping time windows and ii) characterized with maximal emerging subtrees. Experiments have shown that KARMATree is more efficient and more accurate than existing state-of-the-art process drift detection algorithms. Furthermore, a case study on real world data has shown that the characterizations provided by KARMATree spot parts of the process involved in a given change. As to future research directions, we plan to i) improve the efficiency through the use of filter-and-refinement techniques, already explored on spatio-temporal data [16], ii) work on the conciseness of the changes through condensed representations of the patterns [9], iii) study the process drift over a longer temporal horizon through evolution chains [12].

Acknowledgments. We acknowledge the support of the MIUR - Ministero dell'Istruzione dell'Universitàe della Ricerca through the project "TALIsMan - Tecnologie di Assistenza personALizzata per il Miglioramento della quAlità della vitA" (Grant ID: ARS01_01116), funding scheme PON RI 2014–2020. We would also like to thank Lynn Rudd for her help in reading the manuscript.

References

1. van der Aalst, W.M.P., Weijters, T., Maruster, L.: Workflow mining: discovering process models from event logs. IEEE Trans. Knowl. Data Eng. **16**(9), 1128–1142 (2004). https://doi.org/10.1109/TKDE.2004.47
2. Assy, N., van Dongen, B.F., van der Aalst, W.M.P.: Discovering hierarchical consolidated models from process families. Adv. Inf. Syst. Eng. - CAiSE **2017**, 314–329 (2017). https://doi.org/10.1007/978-3-319-59536-8_20
3. Bifet, A., Gavaldà, R.: Learning from time-changing data with adaptive windowing. In: Proceedings of the Seventh SIAM International Conference on Data Mining, pp. 443–448 (2007).https://doi.org/10.1137/1.9781611972771.42
4. Bose, R.P.J.C., van der Aalst, W.M.P., Zliobaite, I., Pechenizkiy, M.: Handling concept drift in process mining. In: Advances Information Systems Engineering, pp. 391–405 (2011). https://doi.org/10.1007/978-3-642-21640-4_30
5. Bose, R.P.J.C., van der Aalst, W.M.P., Zliobaite, I., Pechenizkiy, M.: Dealing with concept drifts in process mining. IEEE Trans. Neural Networks Learn. Syst. **25**(1), 154–171 (2014). https://doi.org/10.1109/TNNLS.2013.2278313
6. Burattin, A.: PLG2: multiperspective process randomization with online and offline simulations. BPM Demo Track **2016**, 1–6 (2016)
7. Ceci, M., Lanotte, P.F., Fumarola, F., Cavallo, D.P., Malerba, D.: Completion time and next activity prediction of processes using sequential pattern mining. In: Discovery Science - 17th International Conference, pp. 49–61 (2014). https://doi.org/10.1007/978-3-319-11812-3_5
8. Hassani, M., Siccha, S., Richter, F., Seidl, T.: Efficient process discovery from event streams using sequential pattern mining. In: IEEE Symposium on Computer Intelligence 2015, pp. 1366–1373 (2015). https://doi.org/10.1109/SSCI.2015.195
9. Impedovo, A., Loglisci, C., Ceci, M., Malerba, D.: Condensed representations of changes in dynamic graphs through emerging subgraph mining. Eng. Appl. Artif. Intell. **94** (2020). https://doi.org/10.1016/j.engappai.2020.103830
10. Impedovo, A., Ceci, M., Calders, T.: Efficient and accurate non-exhaustive pattern-based change detection in dynamic networks. In: Discovery Science - 22nd International Conference, DS 2019, Split, Croatia, 28–30 October 2019, Proceedings, pp. 396–411 (2019). https://doi.org/10.1007/978-3-030-33778-0_30
11. Loglisci, C., Ceci, M., Impedovo, A., Malerba, D.: Mining microscopic and macroscopic changes in network data streams. Knowl. Based Syst. **161**, 294–312 (2018)
12. Loglisci, C., Ceci, M., Malerba, D.: Discovering evolution chains in dynamic networks. In: New Frontiers in Mining Complex Patterns - First International Workshop, NFMCP 2012, Held in Conjunction with ECML/PKDD 2012, Bristol, UK, 24 September 2012, Revised Selected Papers, pp. 185–199 (2012). https://doi.org/10.1007/978-3-642-37382-4_13
13. Maaradji, A., Dumas, M., Rosa, M.L., Ostovar, A.: Fast and accurate business process drift detection. In: Business Process Management - 13th International Conference, pp. 406–422 (2015). https://doi.org/10.1007/978-3-319-23063-4_27
14. Maaradji, A., Dumas, M., Rosa, M.L., Ostovar, A.: Detecting sudden and gradual drifts in business processes from execution traces. IEEE Trans. Knowl. Data Eng. **29**(10), 2140–2154 (2017). https://doi.org/10.1109/TKDE.2017.2720601
15. Mannhardt, F., de Leoni, M., Reijers, H.A., van der Aalst, W.M.P.: Data-driven process discovery - revealing conditional infrequent behavior from event logs. In: Advances Information Systems Engineering - 29th International Conference, pp. 545–560 (2017). https://doi.org/10.1007/978-3-319-59536-8_34

16. Vieira, M.R., Bakalov, P., Tsotras, V.J.: On-line discovery of flock patterns in spatio-temporal data. In: 17th ACM International Symposium on Advances in Geographic Information Systems, pp. 286–295 (2009). https://doi.org/10.1145/1653771.1653812
17. Weijters, A.J.M.M., Ribeiro, J.T.S.: Flexible heuristics miner (FHM). In: Proceedings of the IEEE Symposium on Computational Intelligence and Data Mining, CIDM 2011, part of the IEEE Symposium Series on Computational Intelligence 2011, France, pp. 310–317 (2011). https://doi.org/10.1109/CIDM.2011.5949453

Multi-target Models

Extreme Gradient Boosted Multi-label Trees for Dynamic Classifier Chains

Simon Bohlender(✉), Eneldo Loza Mencía, and Moritz Kulessa

Knowledge Engineering Group, Technische Universität Darmstadt, Hochschulstr. 10, 64289 Darmstadt, Germany
simon.bohlender@gmail.com, research@eneldo.net, mkulessa@ke.tu-darmstadt.de

Abstract. Classifier chains is a key technique in multi-label classification, since it allows to consider label dependencies effectively. However, the classifiers are aligned according to a static order of the labels. In the concept of dynamic classifier chains (DCC) the label ordering is chosen for each prediction dynamically depending on the respective instance at hand. We combine this concept with the boosting of extreme gradient boosted trees (XGBoost), an effective and scalable state-of-the-art technique, and incorporate DCC in a fast multi-label extension of XGBoost which we make publicly available. As only positive labels have to be predicted and these are usually only few, the training costs can be further substantially reduced. Moreover, as experiments on eleven datasets show, the length of the chain allows for more control over the usage of previous predictions and hence over the measure one wants to optimize.

Keywords: multi-label classification · classifier chains · gradient boosted trees

1 Introduction

Classical supervised learning tasks deal with the problem to assign a single class label to an instance. Multi-label classification (MLC) is an extension of these problems where each instance can be associated with multiple labels from a given label space [18]. A straight-forward solution, referred to as binary relevance decomposition (BR), learns a separate classification model for each of the target labels. However, it neglects possible interactions between the labels. Classifier chains (CC) similarly learn one model per label, but these take the predictions of the previous models along a predetermined sequence of the labels into account [13]. It was shown formally that CC is able to capture local as well as global dependencies and that these are crucial if the goal is to predict the correct label combinations, rather than each label for itself [2]. However, in practice the success of applying CC highly depends on the order of the labels along the chain. Finding a good sequence is a non-trivial task. First, the number of possible sequences to consider grows exponentially with the number of labels.

A. Appice et al. (Eds.): DS 2020, LNAI 12323, pp. 471–485, 2020.
https://doi.org/10.1007/978-3-030-61527-7_31

Second, even though a sequence might exist which is optimal w.r.t. some global dependencies in the data, local dependencies make it necessary to consider different chains for different instances. For instance, in a driving scene scenario it is arguable easier to detect first a car and then infer its headlights during the day, whereas it is easier to first detect the lights and from that deduce the car during the night. Roughly speaking, each instance has its own sequence of best inferring its true labels.

Dynamic chain approaches address the problem of finding a good sequence for a particular instance. For instance, Kulessa and Loza Mencía [5] propose to build an ensemble of random decision trees (RDT) with special label tests at the inner nodes. The approach predicts at each iteration the label for which the RDT is most certain and re-uses that information in subsequent iterations. Despite the appealing simplicity due to the flexibility of RDT, it comes at the expense of predictive performance since RDT are not trained in order to optimize a particular measure.

The Extreme Gradient Boosted Trees *(XGBoost)* approach [1], instead, is a highly optimized and efficient tree induction method which has been very successful recently in international competitions. Similarly to CC and dynamic chain approaches, XGBoost refines its predictions in subsequent iterations by using boosting. This served as inspiration to the proposed approach *XDCC*,[1] which integrates ***D****ynamic* ***C****lassifier* ***C****hains* into the e**x**treme boosting structure of gradient boosted trees. XDCC's optimization goal in each boosting round is to predict only a single label for which it is the most certain. This label can be different for each training instance and depends on the given data, label dependencies and previous predictions for the instance at hand. The information about the predicted labels is carried over to subsequent rounds.

A key advantage of the proposed approach is the reduced run time in comparison to classifier chains. This is due to the fact that though the total number of labels can be quite high in MLC, the number of actually relevant labels for each instance is relatively low, usually below 10. Hence, only few rounds are potentially enough if only the positive labels are predicted, whereas CC-based approaches have to still make predictions for each of the existing labels.

2 Preliminaries

This section provides a short overview of the notations used. Additionally an insight to XGBoosts basic functionality is given, i.e., to the tree boosting and classification process as well as the way it can deal with multiple targets.

2.1 Multilabel classification

Multilabel classification (MLC) is the task of predicting for a finite set of N unique class labels $\Lambda = \{\lambda_1, \ldots, \lambda_N\}$ whether they are relevant/positive, i.e., $y_j = 1$ if λ_j is relevant, or $y_j = 0$ if λ_j is irrelevant/negative, for a given

[1] Publicly available at https://github.com/keelm/XDCC

instance. The training set consists of training examples $\mathbf{x}_i \in \mathcal{X}$ and associated label sets $\mathbf{y}_i \in \mathcal{Y} = \{0,1\}^N$, $1 \leq i \leq M$, which can be represented as matrices $X = (x_{ik}) \in A^{M \times K}$ and $Y = (y_{ij}) \in \{0,1\}^{M \times N}$, where features x_{ij} can be represented as continuous, categorical or binary values. An MLC classifier $f : \mathcal{X} \rightarrow \mathcal{Y}$ is trained on the training set in order to learn the mapping between input features and output label vector. The prediction of f for a test example $\mathbf{x}$ is a binary vector $\hat{\mathbf{y}} = f(\mathbf{x})$. An extensive overview over MLC is provided by Tsoumakas and Katakis [18].

The simplest solution to MLC is to learn a binary classifier f_j for each of the labels λ_j using the corresponding column in Y as target signal. The approach is referred to as binary relevance decomposition (BR) and disregards dependencies between the labels. For instance, BR might predict contradicting label combinations (for a specific dataset) since the labels are predicted independently from each other.

2.2 (Dynamic) Classifier Chains

The approach of classifier chains [13] overcomes the disadvantages of BR as it neither assumes full label independence nor full dependence. As in BR, a set of N binary classifiers is trained, but in order to being able to consider dependencies, the classifiers are connected in a chain according to the Bayesian chain rule and pass their predictions along a chain. Each classifier then takes the predictions of all previous ones as additional features and builds a new model.

More specifically each f_j is trained on the augmented training data $X \times Y_{\bullet,1} \times \ldots \times Y_{\bullet,j-1}$ to predict the j-th column $Y_{\bullet,j}$ of Y based on previous predictions $\hat{y_1}, \ldots, \hat{y}_{j-1}$ as follows

$$\hat{y}_j = f_j(\mathbf{x}, \hat{y_1}, \ldots, \hat{y}_{j-1}) \tag{1}$$

with $\hat{y}_1 = f_1(\mathbf{x})$ and assuming for convenience an ascending order on the labels.

As further research revealed, CCs are able to capture global dependencies as well as dependencies appearing only locally in the instance space [2]. However, the ability of the CC approach to capture dependencies is determined by the chosen ordering of the labels. A common approach is to set the order of the labels randomly. Early experimental results already revealed that the ordering has an obvious effect on the predictive performance [13,16]. A solution is to use ensembles of CCs with different orderings [6,13], but creating and maintaining an ensemble of CCs is not always feasible [4] and comes with further issues on combining the predictions. An alternative way to handle the label ordering problem is to determine a good chain sequence in advance. For this purpose methods such as genetic algorithms [4], Bayesian networks [17] or double Monte Carlo optimization technique [12] have been used.

Apart from the computational disadvantages of exploring different label sequences, which often leads to just choosing a random ordering in practice, another issue is the underlying assumption that there is one unique, globally optimal ordering which fits equally to all instances. Instead, dynamic approaches choose the label ordering depending on the test instance at hand. For instance, Silva et al. [16] determine the classification order on the fly by looking at the nearest neighbors of the test instance and using the label ordering which works well on the neighbors. However, the approach is computationally expensive since new CC models have to be build during prediction. Nam et al. [10] use recurrent neural networks to predict the positive labels as a sequence. Nam et al. [9] further use reinforcement learning to determine a different, best fitting sequence over the positive labels of each training instance. Predicting only the positive labels has the advantage of considerably lower computational costs during prediction, since the number of relevant labels is usually low in comparison to the total number of available labels. The advantage comes at the expense of ignoring dependencies to negative labels. Predicting the absence of a label is often much easier than finding positive ones and the knowledge about the absence of a label might be very useful to predict a positive label. Despite the induction of the positive labels does not depend on the number of labels, the approaches of Nam et al. [9,10] can still be computationally very demanding due to the complex neural architectures needed, especially regarding the usage of reinforcement learning which actually has again to explore many possible label sequences during training.

Kulessa and Loza Mencía [5] propose to integrate dynamic classifier chains in random decision trees (RDT) [20]. In contrast to the common induction of decision trees or to random forests, RDTs are constructed completely at random without following any predictive quality criterion. These special RDTs place tests on the labels at the inner nodes, which they can turn on and off without altering the original target of the RDT since it is only specified during prediction by the way of combing the statistics in the leaves. Hence, it is possible to simulate any binary base classifier of a CC in any possible chain sequence. In an iterative process, the same RDTs is queried subsequently to determine the next most certain (positive or negative) label. In the respective next iteration the predicted label is added to the input features like in CC and the respective label tests are turned on. The results of their experimental evaluation show that the dynamic classification achieves a major improvement over static label orderings. However, the lack of any optimization may lead to an insuperable gap to state-of-the-art methods. In fact, the results also show that RDT are inherently not suitable for sparse data like text.

2.3 Extreme Gradient Boosted Trees

Extreme Gradient Boosted Trees *(XGBoost)* [1] is a versatile implementation of gradient boosted trees. One of the reasons for its success is the very good scalability due to the specific usage of advanced techniques for dealing with large scale data. XGBoost was originally designed for dealing with regression problems, but different objectives can be defined by correspondingly adapting the objective function and the interpretation of the numeric estimates. Each model consists of a predefined number of decision trees. These trees are built using gradient boosting, i.e., the model is step-wise adding trees which further minimize the training loss. They are constructed recursively, starting at the root node, by adding feature tests on the inner nodes. At each inner node, all possible feature tests are evaluated according to the gain obtained by applying the split on the data. The test candidate returning the highest gain score is then taken and both children are further split up until the maximum depth is reached or the gains stay below a certain threshold. A prediction can be calculated by passing an instance through all trees and summing up their respective leaf scores.

Boosted Optimization. We refer to [1] for a more detailed description of XGBoost. An XGBoost model consists of a sequence of T decision trees $f_1, \ldots, f_T$. Each tree returns a numeric estimate $f_c(\mathbf{x})$ for a given instance $\mathbf{x}$. Predictions are generated by passing an instance through all trees and summing up their leaf scores. The model is trained in an additive manner and each boosting round adds a new tree that improves the model most. For the t-th tree the loss to minimize becomes

$$L^{(t)} = \sum_{i=1}^{M} l\left(y_i,\ (\hat{y}_i^{(t-1)} + f_t(\mathbf{x}_i))\right) + \Omega(f_t), \tag{2}$$

where $\hat{y}_i^{(t-1)} = \sum_{k=1}^{t-1} f_k(\mathbf{x}_i)$ is the prediction of the tree ensemble so far, $l(y, \hat{y})$ is the loss function for each individual prediction and Ω is an additional term to regularize the tree. Combined with a convex differential loss function the objective can be simplified by taking the second-order approximation which gives us the final objective to optimize:

$$obj^{(t)} = \sum_{v=1}^{T} [G_v w_v + \frac{1}{2}(H_v + \epsilon) w_v^2] + \gamma T \quad \text{with} \quad G_v = \sum_{i \in I_v} g_i, H_v = \sum_{i \in I_v} h_i,$$

where I_v is the set of indices for all data points in leaf v, G_v defines the sum of the gradients for all instances I_v in leaf v, H_v is the corresponding sum of the Hessians (cf. also Sect. 3.1), w_v is the vector of leaf scores and ϵ and γ are regularization terms derived from Ω. With the optimal weights w_v^* for leaf v the objective becomes

$$obj^* = -\frac{1}{2} \sum_{v=1}^{T} \frac{G_v^2}{H_v + \epsilon} + \gamma T \qquad \text{with} \qquad w_v^* = -\frac{G_v}{H_v + \epsilon}. \tag{3}$$

These weights finally lead to the gain function used to evaluate different splits. The indices L and R for G and H refer to the proposed right and left child candidates:

$$L_{split} = \frac{1}{2} \left[\frac{G_L^2}{H_L + \epsilon} + \frac{G_R^2}{H_R + \epsilon} - \frac{(G_L + G_R)^2}{H_L + H_R + \epsilon} \right] - \gamma. \tag{4}$$

There are only few special adaptations of the gradient boosting approach to MLC in the literature and they mainly deal with computational costs. Both Si et al. [15] and Zhang and Jung [21] propose to exploit the sparse label structure which they try to transfer to the gradient and Hessian matrix by using $L0$ regularization. These approaches are limited to decomposable evaluation measures, which roughly speaking means that, opposed to the classifier chains approaches, they are tailored towards predicting the labels separately rather than jointly. Moreover, different technical improvements regarding parallelization and approximate split finding are proposed which could also be applied to the proposed technique in the following. Recently, Rapp et al. [11] proposed to use gradient boosting in order to induce classification rules. Instead of predicting the labels in sequence, the rules predict all labels at once, which allows for minimizing also non-decomposable losses. On the other hand, previous predictions can only be exploited indirectly.

3 Learning a Dynamic Chain of Boosted Tree Classifiers

Instead of learning a static CC that predicts labels in a predefined rigid order, we introduce a dynamic classifier chain (DCC) where each chain-classifier predicts only a single label which is not predetermined and can be different for each sample. To prevent learning bad label dependencies, the base-classifiers are built to maximize the probability of only a single label. On the one hand this allows to exploit more complex label dependencies, and on the other side to massively reduce the length of the chain, while still being able to predict all labels. Given a dataset with 100 labels and a cardinality of five, a DCC of length five has the ability to predict all labels, whereas a CC would have to train 100 models. Because of XGBoosts highly optimized boosting-tree architecture, we decided to use it as base-classifiers in our chain. Therefore we have to modify it and make it capable of building multilabel-trees which can deal with an arbitrary number of labels.

3.1 Multi-label XGBoost

Since XGBoost only supports binary classification with its trees in the original implementation, the underlying tree structure had to be adapted in order to support multi-label targets.

The first modification is to calculate leaf weights and gradients over all class labels instead of only a single one. More specifically, $G_{j,v} = \sum_{i \in I_v} g_{j,i}$ and $H_{j,v} = \sum_{i \in I_v} h_{j,i}$ extend to the labels $1 \leq j \leq N$. In consequence, the objective (3) and gain functions (4) have to be adapted to consider gradient and hessian values from all classes. A common approach in multi-variate regression and multi-target classification is to compute the average loss of the model over all targets [19]. Adapted to our XGBoost trees, this corresponds to the sum of $\frac{G_j^2}{H_j + \epsilon}$ over all labels (cf. Table 1). We refer to it as the ***sumGain*** split method. We use cross entropy as our loss, as it has demonstrated to be appropriate practically and also theoretically for binary and especially multi-label classification tasks [2,8]. Hence, the loss is computed as (shown here only for a single label)

$$l_{ce}(y, \hat{y}) = -y \log(\hat{y}) + (1 - y) \log(1 - \hat{y}). \tag{5}$$

In order to get $\hat{y}$ as a probability between zero and one, a sigmoid transformations has to be applied to the summed up raw leaf predictions $\tilde{y} = \sum_{t=1}^{T} f_t(\mathbf{x})$, returned from all boosting trees, where $\hat{y} = \text{sigmoid}(\tilde{y}) = \frac{1}{1+e^{-\tilde{y}}}$. This is also beneficial for calculating g and h, since the gradients of the loss function simply become

$$g = g_{ce} = \nabla_{\hat{y}} l_{ce}(y, \hat{y}) = \hat{y} - y \quad \text{and} \quad h = h_{ce} = \nabla^2_{\hat{y}} l_{ce}(y, \hat{y}) = \hat{y} \cdot (1 - \hat{y}). \quad (6)$$

One might not expect a very different prediction from the combined formulation than from minimizing the loss for each label separately by separate models (as by BR). However, as Waegeman et al. [19] note fitting one model to optimize the average label loss has a regularization effect that stabilizes the predictions, especially for infrequent labels. In addition, only one model has to be inferred in comparison to N, which has a major implication on the computational costs. This is especially an advantage in the case of a large number of labels and our proposed dynamic approach can directly benefit from it.

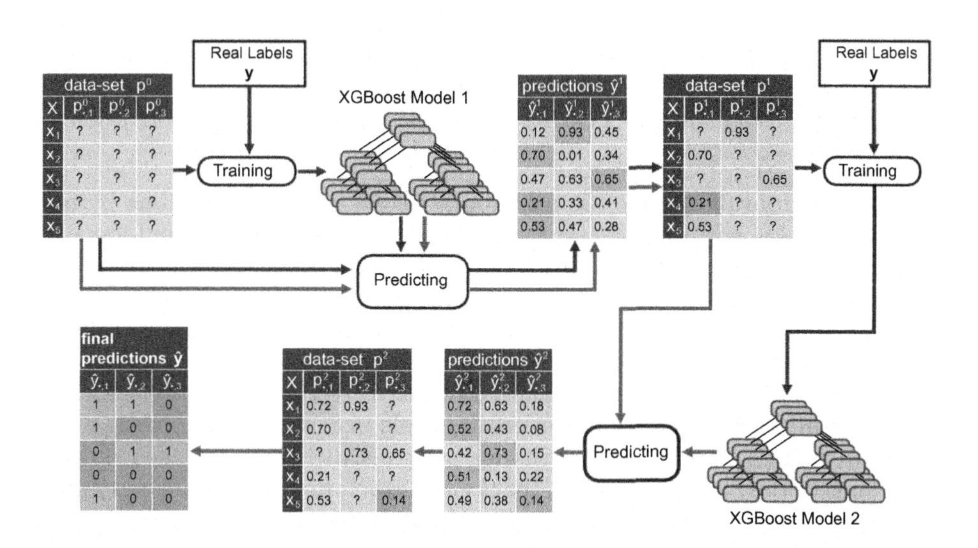

Fig. 1. Dynamic Chain: training pipeline (blue arrows) & prediction pipeline (red arrows).

3.2 Extreme Dynamic Classifier Chains

After introducing the ML-XGBoost models, which can deal with multiple labels, the next step is to modify the tree construction to align it with our goal of predicting a single label per instance.

Table 1 shows the proposed split functions and an example for each one to demonstrate the gain calculations. We assume to have a single instance with four different target labels $\mathbf{y} \in [0, 1]^4$ and their corresponding predictions $\hat{\mathbf{y}}$. g and h are calculated according to Eq. (6) and we get $G = (-0.2, -0.8, 0.9, 0.1)$. We have

Table 1. Proposed split gain calculations with a simplified example calculation for the predicted scores $\hat{\mathbf{y}} = (0.8, 0.2, 0.9, 0.1)$ of the previous trees and given true labels $\mathbf{y} = (1, 1, 0, 0)$. For convenience, we assume $H_j + \epsilon = 1$.

Gain	Formula	Example	Gain	Formula	Ex.
sumGain	$\sum_{j=1}^{N} \left(\frac{G_j^2}{H_j + \epsilon} \right)$	$0.2^2 + 0.8^2 + 0.9^2 + 0.1^2$	*maxGain*	$\max_{1 \leqslant j \leqslant N} \left(\frac{G_j^2}{H_j + \epsilon} \right)$	0.9^2
sumSigned	$\sum_{l=1}^{N} \left(\frac{-G_j}{H_j + \epsilon} \right)$	$0.2 + 0.8 - 0.9 - 0.1$	*maxSigned*	$\max_{1 \leqslant j \leqslant N} \left(\frac{-G_j}{H_j + \epsilon} \right)$	0.8
sumAbsG	$\sum_{l=1}^{N} \left(\left| \frac{-G_j}{H_j + \epsilon} \right| \right)$	$0.2 + 0.8 + 0.9 + 0.1$	*maxAbsG*	$\max_{1 \leqslant j \leqslant N} \left(\left| \frac{-G_j}{H_j + \epsilon} \right| \right)$	0.9

focused on different characteristics for each function.The *max* versions focus on optimizing a tree for predicting only a single label, whereas *sum* functions aim for finding a harmonic split that generates predictions with high probabilities over all labels. The *signed* variants focus on directly optimizing the tree outputs and hence prefer positive labels, while the *gain* splits stay close to XGBoosts original gain calculation and try to optimize positive and negative labels to the same extend. Hereinafter we give a more detailed description and motivation for each gain function:

Maximum default gain over all labels XDCC predicts labels one by one. It hence does not need to find a split which increases the expected loss over all labels (such as *sumGain*), but only one. Hence, ***maxGain*** is tailored to find the label with maximal gain, which corresponds to the label for which the previous trees produced the largest error. In the example in Table 1, this corresponds to λ_3 for which a change of 0.9^2 w.r.t. cross entropy was computed if the prediction is changed to the correct one.

Sum and maximum gradients over all labels In contrast to *maxGain*, ***sumSigned*** aims at good predictions for positive labels only and hence corresponds to the idea of predicting the positive labels first. Positive labels obtain positive scores, whereas negative labels obtain negative scores. The variant ***maxSigned*** chooses the positive label for which the greatest improvement is possible and only goes for the best performing negative label if there are no true positive labels in the instance set. In the example, λ_2 is chosen since the improvement is greater than for λ_1, and definitely greater as for the negative labels.

Sum and maximum absolute gradients over all labels In contrast to *sumGain* and *maxGain*, the measures *sumSigned* and *maxSigned* not only favour positive labels but also take the gradients linearly instead of quadratically into account. This might, for instance, reduce the sensitivity to outliers. Hence, we also include two variants ***sumAbsG*** and ***maxAbsG*** which encourage to predict the labels where the model would improve the most, regardless whether it is positive and negative, but which similarly to *sumSigned* and *maxSigned* use a linear scale on the gradients.

Even though DCC's original design is to predict a single positive label per round, good overall predictions might be required from the beginning for instance in the case of shorter chains. Therefore, we use the split-method as an additional hyper-parameter to choose it individually for different XDCC variants and datasets.

Training Process A schematic view for training the dynamic chain with a length of two is shown in Figure 1 following the blue lines. In a first preprocessing step, the training datasets have to be adapted. For each label λ_j a new *label-feature* p_j^0, initialized as unknown (?), is added to the original features resulting in the augmented space $(\mathbf{x}, \mathbf{p}) \in \mathcal{X} \times [0,1]^N$. While proceeding through the chain, these "?" values are replaced with predicted label probabilities out of $\hat{\mathbf{y}}^r = \text{sigmoid}\left(\sum_{t=1}^{T} f_t^r((\mathbf{x}, \mathbf{p}^{r-1}))\right)$ in round r. As soon as these feature columns begin to be filled with values, following classifiers may detect dependencies and base their predictions on them. Each round r, for $1 \leq r \leq N$, starts with training a new ML-XGBoost model by passing the train set combined with the additional label-features $\mathbf{p}^{r-1}$ and the target label matrix $\mathbf{y}$ to it. Afterwards, the model is used to generate predictions $\hat{\mathbf{y}}^r$ on the same data used to train it, shown in the *predictions* tables. In the last step these predictions are then propagated to the next chain classifier by replacing the corresponding label features with the predicted probabilities $\mathbf{p}^r$. Three different cases can occur during this process:

- At least one label, that was not propagated previously, has a probability ≥ 0.5: The label with the highest probability is propagated.
- All labels, that were not propagated previously, have probabilities < 0.5: The label with the lowest probability is propagated.
- Otherwise, no additional label is propagated.

They can be formalized where $p_{i,j}^r$ denotes the added label feature and $\hat{y}_{i,j}^r$ the corresponding predictions for label λ_j of an instance $\mathbf{x}_i$ in training round r.

$$p_{i,j}^r = \begin{cases} \hat{y}_{i,j}^r & \text{if } p_{i,j}^{r-1} = ? \text{ and } \max_m \hat{y}_{i,m}^r \geq 0.5 \text{ and } \hat{y}_{i,j}^r = \max_m \hat{y}_{i,m}^r \\ \hat{y}_{i,j}^r & \text{if } p_{i,j}^{r-1} = ? \text{ and } \max_m \hat{y}_{i,m}^r < 0.5 \text{ and } \hat{y}_{i,j}^r = \min_m \hat{y}_{i,m}^r \\ p_{i,j}^{r-1} & \text{otherwise} \end{cases} \tag{7}$$

In all cases where labels are propagated, later classifiers are not allowed to change these labels from positive to negative or the other way around, based on the assumption that later classifiers tend to have a higher error rate, since their decisions are based on previous predictions [14].

Prediction Process The prediction process is similar to the training process. Instead of training a model in each step, we reuse the models from the training phase to generate predictions on the test set. After all predictions are propagated, the propagated labels are mapped to label predictions, where probabilites $p_{i,j} <$ 0.5 or equal to ? are interpreted as negative labels and probabilities $p_{i,j} \geq 0.5$ as positive labels. The process is depicted in Figure 1 following the red lines.

3.3 Refinements to the chain

In this section we shortly describe two problems we faced during development of the DCC approach and propose two crucial methods to tackle them.

Separate and Conquer Consecutive models in the chain tend to select the same splits and therefore predict the same labels, especially ones which are easy to learn, i.e. if they clone existing features. We solve this problem by introducing an approach similar to separate-and-conquer from rule learning [3]. The *separating* step turns all gradient and hessian values of previous predicted labels for an instance to zero. Thereby they are no longer considered during split score calculation in the *conquering* step and other splits become more likely since scores for already used splits are lower.

Cumulated Predictions A second observation was that final predictions, after traversing the chain, contain too little positive labels. Analyzing the chain models showed that especially early models predict multiple positive labels, but are only allowed to propagate the one with the highest probability. Therefore we introduce *cumulated predictions* to preserve these otherwise forgotten positive predictions. The idea is to save all predictions of each chain classifier and merge them afterwards with the chain predictions of the unmodified DCC using the following heuristic. The final cumulated prediction $c_{i,j}$ for label λ_j and instance $\mathbf{x}_i$ is computed as

$$c_{i,j} = \begin{cases} p^N_{i,j} & \text{if } p^N_{i,j} \neq ? \\ \max(\hat{y}^1_{i,j}, ..., \hat{y}^N_{i,j}) & \text{otherwise} \end{cases} \tag{8}$$

Table 2. Datasets, # of instances, labels, cardinality, # of distinct label combinations.

Dataset	Instances	Labels	Cardinality	Distinct	Dataset	Instances	Labels	Cardinality	Distinct
EMOTIONS	593	6	1.869	27	GENBASE	662	27	1.252	32
SCENE	2407	6	1.074	15	MEDICAL	978	45	1.245	94
FLAGS	194	7	3.392	54	ENRON	1702	53	3.378	753
YEAST	2417	14	4.237	198	BIBTEX	7395	159	2.402	2856
BIRDS	645	19	1.014	133	CAL500	502	174	26.044	502
TMC2007	28596	22	2.158	1341					

4 Experiments

The experiments were evaluated for the following models, where **BR**, **CC** and **RDT** serve as baselines:

- **BR**: Binary Relevance with default XGBoost models for binary-classification.
- **CC**: Classifier Chains with a random order and default XGBoost as base-models.

- **RDT-DCC**: Dynamic Classifier Chains using Random Decision Trees [5].
- **ML-XGB**: A single multi-label XGBoost model introduced in Section 3.1.
- **XDCC$_{\text{std}}$**: Our Dynamic Classifier Chain with ML-XGB models as base classifiers.
- **XDCC$_{\text{cum}}$**: The cumulated version of DCC introduced in Section 3.2.

The evaluated datasets in Table 2 cover a wide variety of application areas for multi-label classification. All datasets came with predefined train-tests splits which were used for the final evaluation. Parameters were tuned in terms of obtaining best F1 on a randomly chosen 20% of the training set.[2]

4.1 Evaluation Measures

From the large variety of evaluation measures that exist for MLC the most interesting ones for analyzing our proposed methods are *Hamming accuracy* (*HA*) and *subset accuracy* (*SA*). Hamming accuracy denotes the accuracy of predicting individual labels averaged over all labels, whereas subset accuracy measures the ability of a classifiers of predicting exactly the true label combination for an instance. In the case of predicting a large amount of labels, subset accuracy is often of limited use since it often evaluates to zero. Hence, we additionally consider *example-based F1* (*F1*) as measure especially for the parameter tuning. It can be considered as a compromise between *HA* and *SA* and was also used by Nam et al. [9,10] as surrogate loss for *SA* More formally, the comparison between true $\mathbf{y}$ and predicted $\hat{\mathbf{y}}$ for a test instance $\mathbf{x}$ is evaluated to (with $\mathbb{I}$ as indicator function)

$$SA = \mathbb{I}\left[\mathbf{y} = \hat{\mathbf{y}}\right] \quad HA = \frac{1}{N}\sum_{j=1}^{N}\mathbb{I}\left[y_j = \hat{y}_j\right] \quad F1 = \frac{2\sum_{j=1}^{N} y_j\hat{y}_j}{\sum_{j=1}^{N} y_j + \sum_{j=1}^{N}\hat{y}_j}$$

As Dembczyński et al. [2] indicate, *HA* and *SA* are orthogonal to each other. From a probabilistic perspective, to predict the true label combination requires to find the mode of the joint label distribution, whereas it is sufficient to find the modes of the marginal label distributions for *HA*. If there are dependencies between labels, both modes do not have to coincide. In consequence, an approach such as binary relevance is sufficient if one is interested in good *HA* (or there are no dependencies). CC, especially if using the same base learner and configuration as its BR counterpart, cannot be expected to improve over BR regarding *HA*. On the other hand, the reverse behaviour can be expected for *SA*. Hence, the trade-off between both measures and the relation to BR can serve us to assess the ability of considering label dependencies.

[2] The following parameters were tuned by grid-search: number of trees {100, 300, 500}, max tree depth {5, 10, 20, 30, 50}, percentage labels {0.1, 0.2, 0.3} for RDT, max tree depth {5, 10, 20, 50, 100}, number of boosting rounds {10, 20, 50, 100}, learning rate {0.1, 0.2, 0.3} for XDCC, ML-XGB, BR, CC, split methods in Table 1 for XDCC, ML-XGB.

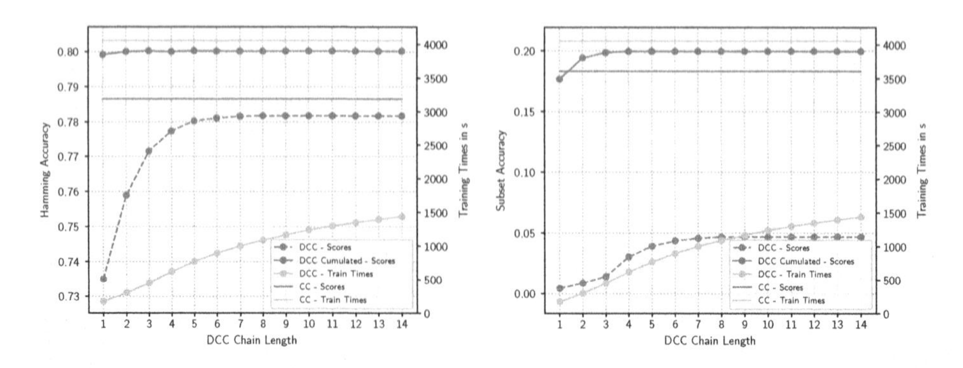

Fig. 2. Comparison with respect to length of the chain on YEAST w.r.t. *HA* and *SA*.

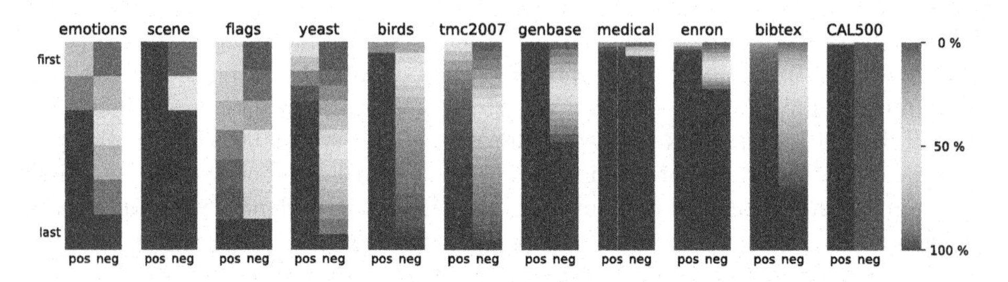

Fig. 3. Heat maps of the development of the predictions of positive and negative labels (left and right side of the bar, respectively) from the first (top row) to last round (bottom row) given as fraction (color level) of the total number of positive and negative predictions on the respective dataset.

4.2 Results

As described in Section 3.2, XDCC can provide a meaningful prediction after each round, which is a major advantage over CC in terms of computational costs. Moreover, by subsequently refining its predictions based on previous predictions, we expected to advance especially in terms of *SA*. Figure 2 shows measures *HA*, *SA* and the time for training for different lengths of the chain on YEAST. CC and XDCC were trained with optimal parameters for CC for a fair comparison of the computational times. Note that length 1 corresponds to ML-XGB when the same parameter were used. The first observation is that, as expected, *HA* and *SA* increase with increasing length for the standard XDCC variant until a little bit further than the average cardinality of 4 of the dataset. If we add the cumulated predictions, the performances converge much faster. Yet, there is a clear improvement visible for *SA*, which indicates that XDCC_{cum} is able to directly benefit from the previous predictions in order to match the correct label combinations. The cumulated predictions are also decisive to surpass CC. Interestingly, the training costs of CC are also never reached although the same XGBoost parameters were used.

Table 3. Predictive performance and training times comparison. Shown are the average ranks over the 10 datasets and the ranks over these in brackets.

Method	*HA*	*SA*	*F1*	Train time
BR	2.20 (1)	3.45 (4)	3.00 (2)	3.00 (3)
CC	3.05 (3)	2.60 (1)	3.30 (3)	3.30 (4)
RDT-DCC	5.10 (6)	4.05 (5)	3.60 (5)	–
ML-XGB	2.90 (2)	3.20 (3)	3.35 (4)	1.10 (1)
$XDCC_{cum}$	3.15 (4)	3.05 (2)	2.45 (1)	2.60 (2)
$XDCC_{std}$	4.60 (5)	4.65 (6)	5.30 (6)	2.60 (2)

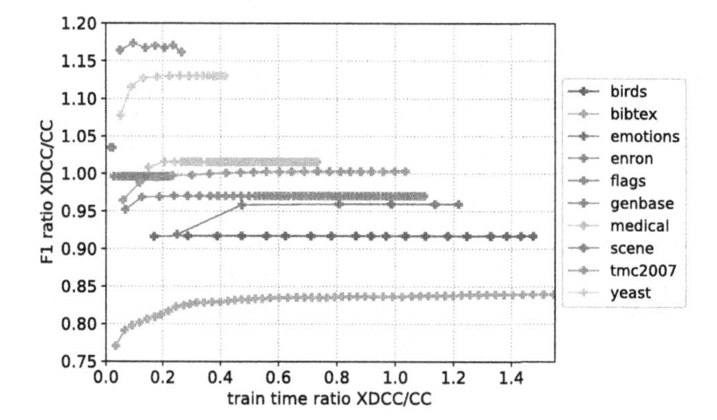

Fig. 4. Train time ratios between $XDCC_{cum}$ and CC in relation to their ratio with respect to F1 for nine datasets. For instance, all points below $x = 1$ and $y = 1$ indicate $XDCC_{cum}$ models which consume less training time but perform worse than CC.

The point where train times of CC are reached by XDCC are further investigated in Figure 4. It shows the ratio of $XDCC_{cum}$ to CC for the different datasets (connected lines) and chain lengths. CAL500 around (0.1,1.4) is not shown for convenience and BIBTEX would continue to (5.3,0.84). The diagram shows that XDCC only takes longer than CC on four datasets and only for the last rounds. For three of these datasets XDCC does not reach CC's *F1*. As already shown in Figure 2, $XDCC_{cum}$ only improves in the first rounds, and sometimes there is even a tendency to decrease. The progress of predicting the labels is also depicted in Figure 3. It visualizes that positive labels are generally predicted in earlier rounds, as expected from the design of the split functions. As shown previously, this behaviour is decisive for the fast convergence and hence the possibility to end the training and prediction processes already in early rounds.

Table 3 also includes a comparison to the RDT-DCC baseline. The first observation is the strong baseline achieved by BR regarding Hamming, as partially expected in Section 4.1. In the same way, CC is best in terms of *SA*. However, ML-XGB performs second regarding *HA* and $XDCC_{cum}$ is second regarding *SA*, which suggests that the proposed approach is able to trade-off between both extremes. This is also confirmed by the best position in terms of *F1*. RDT is the worst performing approach and is even sometimes surpassed by XDCC_{std} which is only included for showing the effect of cumulative predictions.

5 Conclusions

We have proposed in this work XDCC, an adaptation of extreme gradient boosted trees which integrate dynamic chain classifier. XDCC predicts labels along the chain in a dynamic order which adapts to each test instance individually. It was shown that the positive labels are predominantly predicted at the beginning of the process, which allows XDCC to achieve its maximum performance already after a few rounds. This allows XDCC to reduce the length of the chain, which together with the multi-target formulation of XDCC leads to substantial improvements in comparison to binary relevance and classifier chains regarding computational costs, often even if the full chain is processed. The length of the chain also trades-off between the two orthogonal objectives of BR and CC, leading to in average the best results in terms of *F1*.

We will consider in the future to specifically adapt our approach to the setting of large number of labels, e.g. by integrating some of the sparse techniques proposed in [15,21]. Since the number of associated labels per instance is usually not affected by the increasing number of labels, it will be interesting to see how XDCC will behave with respect to computational costs, but also regarding the exploitation of label dependencies since the size of the (dependency) chains should remain of the same size. In order to actually benefit computationally from these short chains, we are planning to include a virtual label which indicates the end of the training and prediction process, similar to the idea of the calibrating label in pairwise learning [7].

References

1. Chen, T., Guestrin, C.: Xgboost: A scalable tree boosting system. In: Proc. of the 22nd SIGKDD Int. Conf. on Knowledge Discovery and Data Mining. pp. 785–794. ACM (2016)
2. Dembczyński, K., Waegeman, W., Cheng, W., Hüllermeier, E.: On label dependence and loss minimization in multi-label classification. Machine Learning **88**(1–2), 5–45 (2012)
3. Fürnkranz, J.: Separate-and-conquer rule learning. Artificial Intelligence Review **13**(1), 3–54 (1999)
4. Goncalves, E.C., Plastino, A., Freitas, A.A.: A Genetic Algorithm for Optimizing the Label Ordering in Multi-label Classifier Chains. In: Proceedings of the IEEE 25th International Conference on Tools with Artificial Intelligence. pp. 469–476 (2013)
5. Kulessa, M., Loza Mencía, E.: Dynamic classifier chain with random decision trees. In: Proceedings of the 21st International Conference of Discovery Science (DS-18) (2018)

6. Li, N., Zhou, Z.: Selective Ensemble of Classifier Chains. In: Multiple Classifier Systems: 11th International Workshop on Multiple Classifier Systems, pp. 146–156 (2013)
7. Loza Mencía, E., Park, S.H., Fürnkranz, J.: Efficient voting prediction for pairwise multilabel classification. Neurocomputing **73**(7–9), 1164–1176 (2010)
8. Nam, J., Kim, J., Loza Mencía, E., Gurevych, I., Fürnkranz, J.: Large-scale multi-label text classification - revisiting neural networks. In: Proceedings of the European Conference on Machine Learning (ECML-PKDD-14). pp. 437–452 (2014)
9. Nam, J., Kim, Y., Loza Mencía, E., Park, S., Sarikaya, R., Fürnkranz, J.: Learning context-dependent label permutations for multi-label classification. In: Proceedings of the 36th International Conference on Machine Learning (ICML-19). pp. 4733–4742 (2019)
10. Nam, J., Loza Mencía, E., Kim, H.J., Fürnkranz, J.: Maximizing subset accuracy with recurrent neural networks in multi-label classification. In: Advances in Neural Information Processing Systems 30 (NIPS-17). pp. 5419–5429 (2017)
11. Rapp, M., Loza Mencía, E., Fürnkranz, J., Nguyen, V.L., Hüllermeier, E.: Learning gradient boosted multi-label classification rules. In: Proceedings of the European Conference of Machine Learning (ECML2020) (2020)
12. Read, J., Martino, L., Luengo, D.: Efficient Monte Carlo methods for multi-dimensional learning with classifier chains. Pattern Recognition **47**(3), 1535–1546 (2014)
13. Read, J., Pfahringer, B., Holmes, G., Frank, E.: Classifier chains for multi-label classification. Machine Learning **85**(3), 333–359 (2011)
14. Senge, R., Del Coz, J.J., Hüllermeier, E.: On the problem of error propagation in classifier chains for multi-label classification. In: Data Analysis, Machine Learning and Knowledge Discovery, pp. 163–170 (2014)
15. Si, S., Zhang, H., Keerthi, S.S., Mahajan, D., Dhillon, I.S., Hsieh, C.J.: Gradient boosted decision trees for high dimensional sparse output. In: Proceedings of the 34th International Conference on Machine Learning. Proceedings of Machine Learning Research, vol. 70, pp. 3182–3190 (06–11 Aug 2017)
16. Silva, P.N.d., Gonçalves, E.C., Plastino, A., Freitas, A.A.: Distinct Chains for Different Instances: An Effective Strategy for Multi-label Classifier Chains, pp. 453–468 (2014)
17. Sucar, L.E., Bielza, C., Morales, E.F., Hernandez-Leal, P., Zaragoza, J.H., Larrañaga, P.: Multi-label Classification with Bayesian Network-based Chain Classifiers. Pattern Recognition Letters **41**, 14–22 (2014)
18. Tsoumakas, G., Katakis, I.: Multi-label classification: An overview. International Journal of Data Warehousing and Mining **3**(3), 1–17 (2007)
19. Waegeman, W., Dembczyński, K., Hüllermeier, E.: Multi-target prediction: a unifying view on problems and methods. Data Mining and Knowledge Discovery **33**(2), 293–324 (2019)
20. Zhang, X., Yuan, Q., Zhao, S., Fan, W., Zheng, W., Wang, Z.: Multi-label Classification without the Multi-label Cost. In: Proceedings of the Society for Industrial and Applied Mathematics International Conference on Data Mining. pp. 778–789 (2010)
21. Zhang, Z., Jung, C.: GBDT-MO: Gradient Boosted Decision Trees for Multiple Outputs. ArXiv preprints arXiv:1909.04373 [cs.CV] (2019)

Hierarchy Decomposition Pipeline: A Toolbox for Comparison of Model Induction Algorithms on Hierarchical Multi-label Classification Problems

Vedrana Vidulin(✉) and Sašo Džeroski

Jožef Stefan Institute, Jamova cesta 39, 1000 Ljubljana, Slovenia
vedrana.vidulin@gmail.com, saso.dzeroski@ijs.si

Abstract. Hierarchical multi-label classification (HMC) is a supervised machine learning task, where each example can be assigned more than one label and the possible labels are organized in a hierarchy. HMC problems emerge in domains like functional genomics, habitat modelling, text and image categorization. They can be addressed with global model induction algorithms, which induce a single model that predicts the complete hierarchy, as well as with local algorithms, which induce multiple models that predict different segments of the hierarchy. However, there is no consensus about which of these approaches perform the best over different domains, especially in the setting of learning ensembles.

We introduce the hierarchy decomposition pipeline, a publicly available toolbox for comparison of model induction algorithms on HMC problems in an ensemble setting. The pipeline includes five algorithms, including the algorithm that predicts the complete hierarchy, and algorithms that perform partial and complete hierarchy decompositions. One of these algorithms is the novel "label specialization" algorithm that constructs a local multi-label classification model for each parent label in a hierarchy that simultaneously predicts the respective children labels.

We apply the pipeline on ten HMC data sets from four domains, which have both tree and directed acyclic graph label hierarchies, and confirm that there is no single best algorithm for all HMC problems. This finding shows that there exists a need for such a pipeline that enables a user to choose the best performing algorithm for his/her HMC data set. Finally, we show that the choice can be narrowed to a specific type of algorithm, based on the characteristics of the label hierarchy and the data set label cardinality.

Keywords: Hierarchical multi-label classification · Hierarchy decomposition · Structured prediction

1 Introduction

Hierarchical multi-label classification (HMC) is a supervised machine learning task, where each example can be assigned more than one label and the possible labels are

A. Appice et al. (Eds.): DS 2020, LNAI 12323, pp. 486–501, 2020.
https://doi.org/10.1007/978-3-030-61527-7_32

organized in a hierarchy [1]. The hierarchy can be in the shape of a tree, where each label has exactly one parent label, or in the form of a directed acyclic graph (DAG), where a label can have multiple parent labels. Label assignments follow the hierarchy constraint: When a label is assigned to an example, all labels on all possible paths from that label to the root of the hierarchy must be assigned too.

Many real life problems are best represented with HMC data sets [2–10]. An example of a HMC problem is gene function prediction, which aims to predict the biological functions of genes. Examples of gene function are the tree shaped hierarchy of FunCat [11] and the more comprehensive DAG shaped hierarchy of the Gene Ontology [12]. The latter is composed of three domains – molecular function, biological process and cellular component – and a single gene can be assigned with multiple functions from each of the three domains [13].

Model induction algorithms for HMC problems can be divided into two groups [14, 15]. Global algorithms induce a single model that predicts complete hierarchy. They can exploit dependencies among labels during a model induction phase to improve model's predictive performance. An example of the global algorithms is Clare and King [16] adaptation of the decision tree algorithm C4.5 [17], which predicts labels on different levels of a hierarchy by assigning a larger cost to misclassification high up in the hierarchy. Another example is the predictive clustering tree (PCT) algorithm, a generalization of the decision tree algorithm that predicts labels from both tree [18–20] and DAG hierarchies [1]. Local algorithms induce multiple models that predict a different part of a hierarchy, and then combine the predictions of those models. Some examples of a local algorithm construct an SVM model for each label and then combine predictions so as to satisfy the hierarchy constraint [21–24].

Levatić et al. [25] compare the predictive performance of four model induction algorithms over HMC problems from different domains. They compare two global and two local algorithms, where one in each group exploits the hierarchical dependencies among labels when constructing model(s) and the other does not. Both global algorithms construct one multi-label model, while both local algorithms construct many single-label classification models. All four approaches construct single PCT models of (random forest and bagging) ensembles of PCTs. When a single PCT models are constructed, the algorithms that exploit hierarchical dependencies outperform those that do not. However, when PCT ensembles are constructed, it is less clear what is the best performing algorithm.

We introduce the hierarchy decomposition pipeline, a publicly available toolbox for comparison of model induction algorithms for HMC problems in the ensemble setting (https://github.com/vedranav/hierarchy-decomposition-pipeline). The pipeline includes five algorithms, beginning with an algorithm that predicts the complete hierarchy in one shot, and following with four algorithms that perform partial and complete hierarchy decompositions. Partial decomposition algorithms construct models that predict the presence of one or more edges of the hierarchy, while complete decomposition algorithms construct model(s) that predict the presence of individual or all hierarchy nodes. We propose a novel partial decomposition algorithm, called the "label specialization", that constructs a multi-label classification model for each parent label in a hierarchy, which predicts the presence of its children labels. The algorithm is an extension of the hierarchical single label classification algorithm [1] that constructs a single-label classification model for each parent-child pair in a hierarchy, where collection of such models for a given parent can be viewed as a binary relevance classifier. Apart from the mentioned algorithms, the pipeline contains tools for performance-based comparison of the algorithms.

We applied the pipeline on ten HMC data sets from four domains. In the text categorization domain, we use the Enron data set that categorizes e-mails from the Enron corporation officials [4] and the Reuters data set that categorizes Reuters stories [5]. In the image categorization domain, we use two data sets from the 2007 CLEF image retrieval campaign that categorize medical X-ray images [7]. In the habitat modelling domain, we use the Danish farms data set that models the habitats of soil microarthropods [6] and the Slovenian rivers data set that models the habitats of aquatic organisms [2]. In the functional genomics domain, we use two data sets intended for predicting biological functions of genes in two model organisms: the plant *Arabidopsis thaliana* and the baker's or brewer's yeast *Saccharomyces cerevisiae* [3]. In addition, we use two data sets intended for predicting functions of genes in thousands of bacterial and archaeal organisms [8, 9]. In the first eight data sets, the labels are interconnected in tree shaped hierarchies, while in the last two the labels form a DAG.

The results of the performance comparison confirm that there is no single best model induction algorithm for all HMC data sets in the ensemble setting. There is no significant difference in the predictive performance of the five algorithms over the ten data sets. This finding shows that there exists a need for the proposed pipeline, which enables a user to find the best performing algorithm for his/her custom HMC data set. Finally, the results show that the search for the best performing algorithm can be narrowed to a specific type of an algorithm based on the characteristics of the hierarchy and the data set cardinality.

The remainder of the paper is organized as follows. In Sect. 2, we describe the hierarchy decomposition pipeline. Section 3 describes the experimental setup, including values of the pipeline's parameters and the HMC data sets. The results of the performance analysis are presented in Sect. 4. We conclude the paper with Sect. 5.

2 Hierarchy Decomposition Pipeline

The hierarchy decomposition pipeline is a toolbox for comparing model induction algorithms for HMC problems in the ensemble setting. The pipeline takes as input a HMC data set specified by a user and applies five model induction algorithms, beginning with the algorithm that induces a model predicting the complete hierarchy and continuing with partial and complete hierarchy decomposition algorithms. The performance-based evaluation tool computes the areas under the average precision-recall curves and performs a statistical test. The components of the pipeline are shown in Fig. 1.

The pipeline receives two **input** files: an HMC data set and a settings file (Fig. 1A). A description of the input file formats is available on the repository.

The **cross-validation module** takes the data set and divides its examples into folds (Fig. 1B). For each fold, it sends a training set to the hierarchy decomposition and the model induction modules, and a test set to the annotation module.

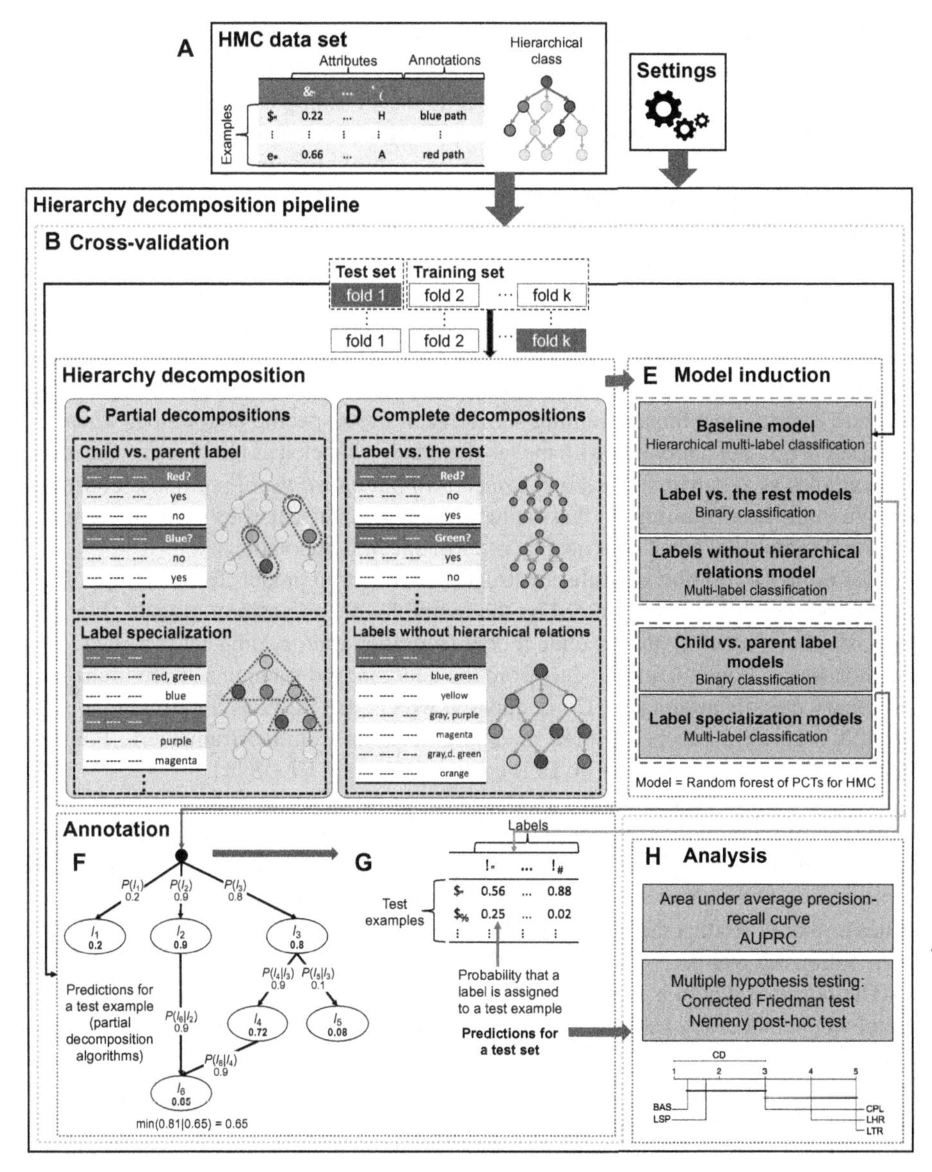

Fig. 1. The hierarchy decomposition pipeline.

The **hierarchy decomposition module** transforms an input training set into multiple training sets by applying two types of decomposition:

Partial decompositions construct multiple training sets representing different edges of a hierarchy (Fig. 1C). The first partial decomposition "*child vs. parent label*" constructs a binary training set for each child-parent label pair in a hierarchy, composed of the training examples originally labeled with the parent label. In a newly created training set, the examples originally labeled with the child label are labeled as positive, while the rest of the examples are labeled as negative. The second partial decomposition, "*label specialization*", constructs a multi-label training set for each parent-children group of labels in a hierarchy, where the training set contains all examples originally labeled with the parent label, now only labeled with the applicable children labels.

Complete decompositions construct one or multiple training sets representing the nodes of the hierarchy, and ignoring the edges (Fig. 1D). The training set(s) contain the same examples as the input training set, but annotated with the labels that belong to the subset of most specific annotations. For example, if an example is originally labeled with two paths "root > 1 > 1.1" and "root > 2 > 2.1 > 2.1.1", the example will be newly labeled with the most specific annotations 1.1 and 2.1.1. Accordingly, the subset of most specific annotations contains the labels that appear at least once as the most specific annotation in the input data set. The first complete decomposition "*label vs. the rest*" constructs a binary training set for each most specific annotation, where the examples originally annotated with the label are newly labeled as positive and the rest of the examples as negative. The second "*labels without hierarchical relations*" constructs a single multi-label training set that captures label cooccurrences by labeling examples with one or multiple labels that qualify as most specific annotations.

The **model induction module** constructs classification models from the input training set and the training sets created by the hierarchy decomposition module (Fig. 1E). The task of constructing the baseline model from the input training set is a HMC task, and the tasks of constructing models from the decomposed training sets are multi-label and binary classification tasks. Consequently, we choose PCTs as a base model, since the PCT algorithm covers all three modeling tasks in a unified framework. For each training set, a random forest of PCTs is constructed using CLUS [1].

Annotation module classifies an input test set by using the models created by the model induction module. It combines the predictions from multiple models and applies the hierarchy constraint (Fig. 1F, G). For each of the five model induction algorithms it outputs a table with predictions, where rows are test examples, columns labels and values probabilities that the labels are assigned to the examples. The tables are obtained in the following manner:

The *baseline* model is the global model that implicitly enforces the hierarchy constraint when annotating test examples. We use the outputted predictions as given by the model.

The *child vs. parent label* model collection is composed of multiple binary classification models, one for each non-root label l_j that outputs a conditional probability $P(l_j|\text{parent}(l_j))$. To make a prediction for a test example e_i and a label l_j, the product rule $P(l_j) = P(l_j|\text{parent}(l_j)) \cdot P(\text{parent}(l_j))$ is applied recursively, beginning with the model where $\text{parent}(l_j)$ is the root of a hierarchy. The procedure is illustrated with an example in Fig. 1F. To compute the probability that l_5 is assigned to e_i, we first use the model for the label l_3 to predict $P(l_3)$, which is 0.8. Then, we use the model for the label l_5 that predicts $P(l_5|l_3)$, which is 0.1. Finally, $P(l_5)$ is computed by applying the product rule $P(l_5|l_3) \cdot P(l_3)$, which is 0.08. By using the product rule we enforce the hierarchy constraint, ensuring that the probabilities decrease with increasing depth within the hierarchy. The presented example shows the case with a single path from a label node to root of a hierarchy. When a class is a DAG, there can be multiple paths from the label node to the root node. In this case, the probability is computed for each path, and the minimal observed probability is considered (P(l_6) in Fig. 1F).

The *label specialization* model collection is composed of multiple multi-label classification models, one for each parent node in the hierarchy, which output conditional probabilities P(l_j|parent(l_j)) for children labels. When the output space is a DAG, a label l_j can have multiple parents and, consequently, multiple multi-label models can output a conditional probability P(l_j|parent(l_j)). In this case, we consider as P(l_j|parent(l_j)) the maximal predicted conditional probability. To predict a label l_j for a test example e_i, we then use the product rule as in the case of "child vs. parent label" model.

The *label vs. the rest* model collection is composed of multiple binary classification models, one for each label l_j that qualifies as most specific annotation. For a test example e_i and a label l_j from the subset of most specific annotations, an l_j specific model outputs the probability P(l_j) that the label is assigned to the example. In the case of labels that do not belong to the subset of most specific annotations, the probability is zero.

The *labels without hierarchical relations* model is a single multi-label classification model that can output probabilities that labels from the subset of most specific annotations are assigned to an example.

The **analysis module** compares the performances of the five model induction algorithms, based on the predictions output by the annotation module. An algorithm's performance is measured as the area under average precision-recall curve (AUPRC) [1]. The statistical significance of AUPRC differences is assessed by using the corrected Friedman test and the Nemenyi post-hoc test [26] (Fig. 1H).

AUPRC is a threshold independent performance measure, where precision and recall points are obtained by changing the value of the threshold t from zero to one with the step of 0.01. For each value of t, precision and recall values are micro-averaged: $\overline{precision_t} = \frac{\sum_{i=1}^{p} TP_i}{\sum_{i=1}^{p} TP_i + \sum_{i=1}^{p} FP_i}$, $\overline{recall_t} = \frac{\sum_{i=1}^{p} TP_i}{\sum_{i=1}^{p} TP_i + \sum_{i=1}^{p} FN_i}$, where p is the number of labels that qualify as most specific annotations, TP are true positives, FP false positives and FN false negatives.

The statistical test is performed on a r by k matrix, where r is the number of model induction algorithms (five in our case), k is the number of cross-validation folds and the values in the matrix are AUPRCs. The corrected Friedman test determines if there is at least one algorithm with significantly different performance. For each fold, the test ranks the algorithms in decreasing order of AUPRC. In case of a tie, an average rank is assigned. Next, the test averages ranks over the k folds and calculates the Friedman statistic Q, distributed according to the χ^2 distribution with $r - 1$ degrees of freedom. The p-value is defined as $P\left(\chi_{r-1}^2 \geq Q\right)$. If, according to the p-value, the difference is significant, the Nemenyi post-hoc test is used for pairwise comparisons among the algorithms. The performance of two algorithms is significantly different if their average ranks differ by more than a critical distance. The critical distance is computed from r, k and a critical value for a given significance level (a Studentized range statistic).

3 Experimental Setup

We applied the hierarchy decomposition pipeline on ten data sets, using a unified experimental setup. The unified setup means that all models are random forests of 500 PCTs.

The size of random subspaces considered at each node is equal to the square root of the number of attributes, and the five model induction algorithms are evaluated by performing 10-fold cross-validation.

The ten data sets used are described below. They represent four domains: text categorization, image categorization, habitat modelling and functional genomics. Eight of them have a tree-shaped label hierarchy and two a DAG shaped label hierarchy. Data set statistics are presented in Table 1.

Table 1. Data set statistics. Columns: n = number of examples; a_d = discrete attributes; a_n = numeric attributes; h_n = hierarchy nodes; h_l = hierarchy leaves; c-h_l = cardinality accounting for labels that are hierarchy leaves (cardinality is an average number of labels per example); p = labels that qualify as most specific annotations; c-p_c = cardinality accounting for most specific annotations available to the complete decomposition algorithms; c-p_h = cardinality accounting for most specific annotations available to the hierarchical algorithms; d = maximal depth of the hierarchy; type = tree or DAG hierarchy.

Data set	n	a_d/a_n	h_n	h_l	c-h_l	p	c-p_c	c-p_h	d	Type
Enron	1,648	1,001/0	56	52	2.85	53	2.87	3.37	3	Tree
Reuters	6,000	0/47,236	100	79	1.19	99	1.46	3.13	4	Tree
ImCLEF07A	11,006	0/80	96	63	1.00	63	1.00	1.00	3	Tree
ImCLEF07D	11,006	0/80	46	26	1.00	26	1.00	1.00	3	Tree
Danish farms	1,893	132/5	70	35	6.27	39	6.74	7.08	3	Tree
Slo. rivers	1,060	0/16	724	492	24.56	637	33.04	50.67	4	Tree
ExprYeast	3,788	4/547	417	161	2.28	194	2.29	4.00	4	Tree
SeqAra	3,718	2/4,448	196	148	0.94	194	1.30	3.32	4	Tree
PP	15,313	2,071/0	1,260	377	0.89	947	2.59	16.67	14	DAG
MPP-I	3,531	0/4,777	826	220	1.32	620	3.30	20.49	13	DAG

Text categorization is a problem of automatic annotation of textual documents with one or several categories. The Enron data set contains bag-of-words descriptions of e-mails from the labeled subset of the Enron corpus [4]. Hierarchically organized categories define genre, emotional tone and topic. Reuters data set contains tf-idf descriptions of stories from the "Topics" category of the Reuters Corpus Volume I (RCV1) [5]. Hierarchically organized categories are topic-based, e.g., economics, industrial or government.

Image categorization annotates images with categories that represent visual concepts the images contain. ImCLEF07A and ImCLEF07D represent medical X-ray images annotated with parts of the human anatomy and orientations of body parts [7]. The images are described with edge histograms indicating a frequency and a directionality of brightness changes in an image.

Habitat modelling studies relationships between environmental variables and the presence of plants and animals in the environment. The Danish farms data set represents habitats of soil microarthropods on Danish experimental and organic farms [6]. The Slovenian rivers data set represents habitats of aquatic organisms in Slovenian rivers [2]. The tree-shaped hierarchies in both data sets represent parts of the taxonomic hierarchy that contain habitat-specific species.

Functional genomics annotates genes with their biological functions. The ExprYeast data set represents *Saccharomyces cerevisiae* (baker's yeast) microarray gene expression levels measured under various experimental conditions, such as heat shock or nitrogen depletion [3]. The SeqAra data set contains attributes derived from amino acid sequences of the *Arabidopsis thaliana* plant genes, such as amino acid ratios, molecular weight and sequence length [3]. The PP data set represents phyletic profiles, i.e., presence and absence patterns of gene families (clusters of genes that share function) in 2,071 bacterial and archaeal genomes [8]. The MPP-I data set represents metagenome phyletic profiles, i.e., relative abundances of gene families in metagenomes obtained from the IMG database [9]. The first two data sets are annotated with functions from the tree shaped hierarchy of FunCat [11] and the last two with functions from the DAG of the Gene Ontology [12].

4 Results

The analysis has three goals. First, to clarify how model induction algorithms for HMC problems can be compared. Second, to examine whether there exists a single best performing algorithm for all ten HMC problems. Third, to investigate whether specific properties of HMC data sets can be related to a type of the best performing algorithm.

4.1 How Model Induction Algorithms for HMC Problems Can Be Compared?

The performance-based comparison of model induction algorithms should be ideally based on annotations available to all of the algorithms. The pipeline contains two types of algorithms. The complete decomposition algorithms use only the most specific annotations, while the hierarchical algorithms use additional annotations obtained by enforcing the hierarchy constraint. To compare the two types of algorithms we aggregate AUPRC over the subset of labels common to both types, that is, over the labels that qualify as most specific annotations. However, this step does not guarantee that the comparison is performed on the common set of annotations.

Distributions of common labels are not necessarily the same in the training sets created by the complete decomposition algorithms and the hierarchical algorithms. This property is best illustrated with an example. Suppose that we have a data set annotated with labels from the hierarchy in Fig. 2A and a derived training set composed of five examples annotated with most specific annotations as presented in Fig. 2B. The training set, which illustrates the annotations available to complete decomposition algorithms, shows that six labels qualify as most specific annotations: "1", "1.1", "1.1.1", "1.1.2", "1.2" and "2.2". When the hierarchical algorithms apply the hierarchy constraint on those six labels, the training set will look like the one in Fig. 2C. The difference in distribution of common labels is considerable: the cardinality doubled.

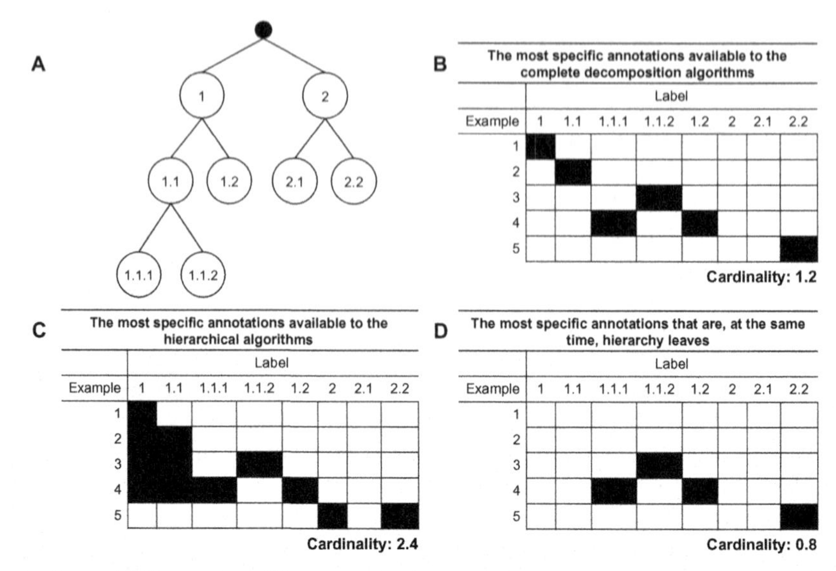

Fig. 2. Example illustrating differences in distribution of labels common to the complete decomposition and hierarchical algorithms.

The difference in distribution of common labels can give an advantage to the hierarchical algorithms due to additional information from the hierarchy. This property affects eight data sets with an exception of the two data sets from image categorization domain (see columns $c\text{-}p_c$ and $c\text{-}p_h$ in Table 1) and tends to have higher impact on the data sets with larger hierarchies.

The presented problem can be addressed by considering only those annotations that are, at the same time, hierarchy leaves. This approach would, however, ignore many annotations (Fig. 2D, see columns $c\text{-}h_l$ and $c\text{-}p_c$ in Table 1). For example, the cardinality of the SeqAra and PP data sets would fall below one, although both data sets have at least one label per example (column $c\text{-}h_l$ in Table 1).

4.2 Is There a Single Best Model Induction Algorithm Across All HMC Data Sets?

To answer this question we: (1) measure AUPRC_m aggregated over the labels that qualify as most specific annotations (Table 2); (2) measure AUPRC_l aggregated over the labels that qualify as most specific annotations and are, at the same time, hierarchy leaves (Table 2); and (3) examine whether the differences in AUPRC_m and AUPRC_l among the five model inductions algorithms are statistically significant at the significance threshold of 0.05.

As a statistical significance test, we use the corrected Friedman test on the matrix where rows are the ten data sets, columns are the five algorithms and values are AUPRCs. We apply the test separately for each type of AUPRC, using the two matrices in Table 2. The p-value for AUPRC_m is 0.029 and for AUPRC_l 0.458. At a significance threshold of 0.05, there are no significant differences in performance considering AUPRC_l.

Table 2. Area under average precision-recall curve aggregated over the labels that qualify as most specific annotations ($AUPRC_m$) and labels that are hierarchy leaves ($AUPRC_l$). AUPRCs of the best performing algorithms are shown in bold. Abbreviations: BAS = baseline, CPL = child vs. parent label, LSP = label specialization, LHR = labels without hierarchical relations, LTR = label vs. the rest, alg. = algorithms, d. = decomposition algorithms.

Data set	$AUPRC_m$					$AUPRC_l$				
	Hierarchical alg.			Complete d.		Hierarchical alg.			Complete d.	
	BAS	CPL	LSP	LHR	LTR	BAS	CPL	LSP	LHR	LTR
Enron	0.646	0.648	**0.657**	0.533	0.532	0.596	**0.601**	0.600	0.594	0.595
Reuters	0.798	**0.816**	0.797	0.446	0.462	0.668	**0.692**	0.661	0.632	0.692
ImCLEF07A	0.886	0.891	0.889	0.888	**0.898**	0.886	0.891	0.889	0.888	**0.898**
ImCLEF07D	0.872	0.871	0.870	0.872	**0.882**	0.872	0.871	0.870	0.872	**0.882**
Danish farms	0.824	0.815	0.825	0.816	**0.827**	0.828	0.819	0.828	**0.830**	0.828
Slo. rivers	**0.658**	0.642	0.657	0.456	0.432	0.504	0.495	**0.510**	0.509	0.486
ExprYeast	0.465	0.449	**0.489**	0.407	0.320	0.372	0.381	**0.415**	0.401	0.347
SeqAra	0.498	**0.524**	0.488	0.238	0.230	0.381	0.395	0.385	0.381	**0.410**
PP	0.341	**0.349**	0.345	0.095	0.097	0.127	0.130	0.129	**0.157**	0.151
MPP-I	0.497	**0.511**	0.507	0.348	0.342	0.301	**0.472**	0.427	0.392	0.398

For $AUPRC_m$, we proceed to the post hoc test (Fig. 3). At the significance level of 0.05, the post hoc test shows that none of the algorithms perform significantly better that the rest. Given the results of the statistical test, we confirm the hypothesis that there is no single best model induction algorithm across all HMC data sets.

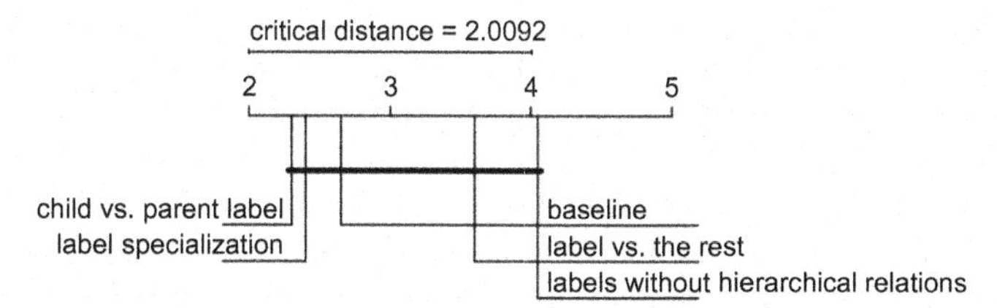

Fig. 3. Average ranks diagram comparing predictive performance, measured as $AUPRC_m$, of the five model induction algorithms over the ten HMC data sets. The numbers on the line represent ranks of the algorithms averaged over the data sets. Better performing algorithms are on the left-hand side. The algorithms with average ranks that differ by less than the critical distance for a p-value of 0.05 are connected with a line.

4.3 Can We Relate Properties of HMC Data Sets to a Type of the Best Performing Model Induction Algorithm?

The best performing algorithm for a data set is the one that receives the highest rank in the statistical test (described in Sect. 2, analysis module), and is significantly better than at least one other algorithm at the significance level of 0.05 (Fig. 4). The second criterion is not satisfied by any of the five algorithms on the PP and MPP-I data sets (Fig. 4I, J). Since we cannot determine the best performing algorithm on the two data sets, they are not going to be used in the analysis. We characterize the best performing algorithm along each of the two dimensions: single- or multi-label classification algorithm, and hierarchical or complete decomposition algorithm. The former dimension indicates whether the model(s) constructed by the algorithm perform(s) single or multi-label classification. The latter indicates whether hierarchical constraint is applied.

Data sets are characterized with two groups of properties, the first describing the hierarchy of labels and the second describing the density of annotations. The hierarchy is described in terms of the number of nodes and leaves, and a branching factor. Annotations are described through cardinality computed both for annotations available to the hierarchical and complete decomposition algorithms. The number of annotations available to the hierarchical, but not to the complete decomposition algorithms is measured as a difference between the two cardinalities. Finally, we measure a share of incomplete annotations in most specific annotations. The most specific annotation is incomplete if it is not a leaf label and can, consequently, be further specialized.

Multi-label classification algorithms perform best on the data sets with large hierarchies: they perform the best on the two data sets with the largest hierarchies, ExprYeast and Slovenian rivers (417 and 724 nodes, Table 3). An average branching factor, which is an indicator of complexity, is, however, not related in the same way. We expected that multi-label classification algorithms would perform best on data sets with high cardinality, but this is not the case. The ExprYeast and Enron data sets have moderate cardinality (from 2.29 to 4), and a multi-label classification algorithm performs best on the former and a single-label classification algorithm performs best on the latter. Similarly, the Danish farms and Slovenian rivers data sets have high cardinality (from 6.74 to 50.67), and a single-label classification algorithm performs best on the former and a multi-label classification algorithm on the latter. An exception are the data sets with low cardinality (less than two) where a single-label classification algorithm always performs best.

Hierarchical algorithms perform better on data sets where they can obtain additional annotations (by applying the hierarchy constraint), as compared to the complete decomposition algorithms. They profit even when only half of an annotation on average is added to examples (Table 4). Interestingly, when the best performing algorithm is a hierarchical algorithm, it performs significantly better than both complete decomposition algorithms (Fig. 4A, B, F–H). Furthermore, the presence of incomplete annotations in a data set is related to the emergence of a hierarchical algorithm as the best performing: When at least 1% of the annotations in a data set are incomplete, a hierarchical algorithm is the best choice. Hierarchical algorithms also perform the best on class hierarchies with more than 100 nodes and an average branching factor higher than three.

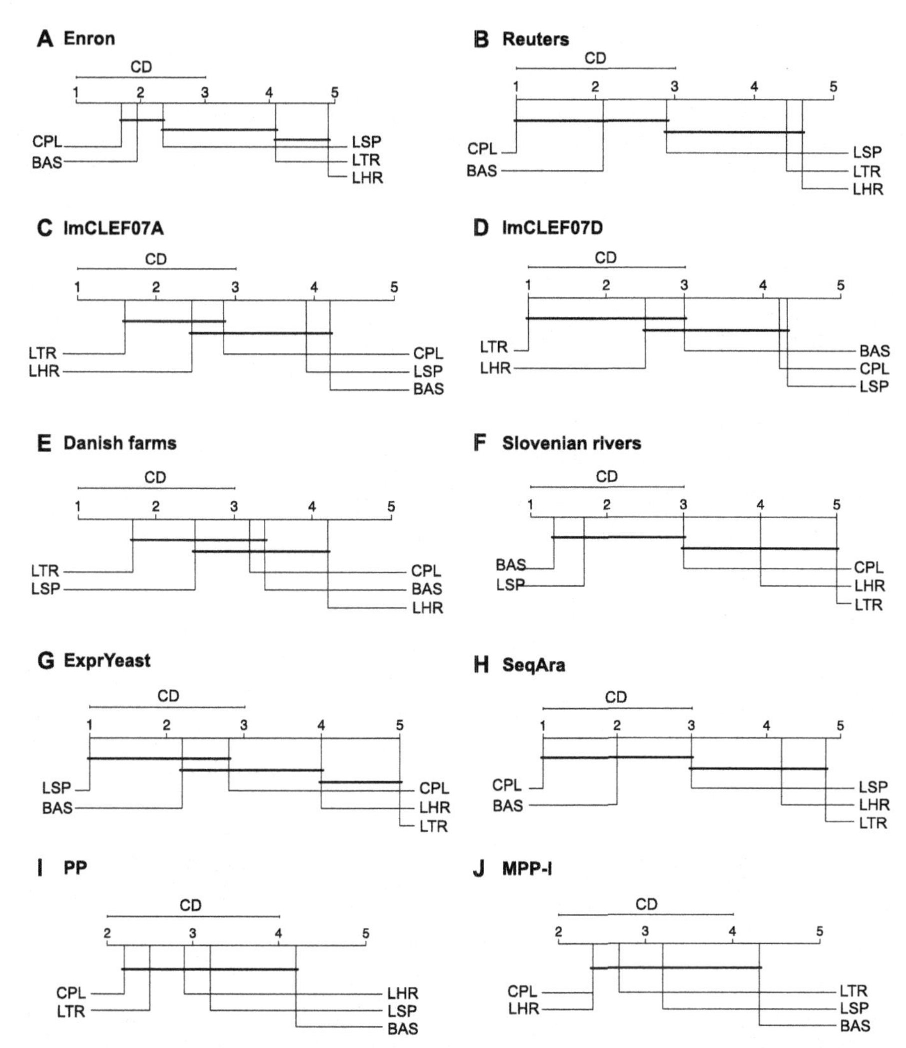

Fig. 4. Average ranks diagrams comparing predictive performance, measured as $AUPRC_m$, of the five model induction algorithms for each of the ten HMC data sets. The numbers on the line represent ranks of the algorithms averaged over the ten cross-validation folds. Better performing algorithms are on the left-hand side. The algorithms with average ranks that differ by less than the critical distance for a p-value of 0.05 are connected with a line. CD = critical distance = 2.0092. Abbreviations: BAS = baseline, CPL = child vs. parent label, LSP = label specialization, LHR = labels without hierarchical relations, LTR = label vs. the rest.

Table 3. The relation between the size of a hierarchy and emergence of a single- or a multi-label classification algorithm as the best performing. Columns: h_n = hierarchy nodes; h_l = hierarchy leaves; $\overline{b}$ = average branching factor; $c\text{-}p_c$ = cardinality accounting for most specific annotations available to the complete decomposition algorithms; $c\text{-}p_h$ = cardinality accounting for most specific annotations available to the hierarchical algorithms; Alg. = algorithm (for the abbreviations of algorithm names, we refer to Fig. 4).

Data set	Hierarchy			Annotations		Best performing algorithm		
	h_n	h_l	$\overline{b}$	$c\text{-}p_c$	$c\text{-}p_h$	Alg.	Multi-label?	Hierarchical?
ImCLEF07D	**46**	26	2.19	1.00	1.00	LTR	**No**	No
Enron	**56**	52	11.20	2.87	3.37	CPL	**No**	Yes
Danish farms	**70**	35	1.94	6.74	7.08	LTR	**No**	No
ImCLEF07A	**96**	63	2.82	1.00	1.00	LTR	**No**	No
Reuters	**100**	79	4.55	1.46	3.13	CPL	**No**	Yes
SeqAra	**196**	148	4.00	1.30	3.32	CPL	**No**	Yes
ExprYeast	**417**	161	1.62	2.29	4.00	LSP	**Yes**	Yes
Slo. rivers	**724**	492	3.11	33.04	50.67	BAS	**Yes**	Yes

Table 4. The relation between the amount of additional annotations available to the hierarchical algorithms and the emergence of a hierarchical or a complete decomposition algorithm as the best performing. Columns: diff = $c\text{-}p_h$ - $c\text{-}p_c$; ia = percentage of incomplete annotations. For a description of the rest of the abbreviations, we refer to Table 3.

Data set	Annotations		Hierarchy			Best performing algorithm		
	diff	ia	h_n	h_l	$\overline{b}$	Alg.	Multi-label?	Hierarchical?
ImCLEF07A	**0**	0%	96	63	2.82	LTR	No	**No**
ImCLEF07D	**0**	0%	46	26	2.19	LTR	No	**No**
Danish farms	**0.34**	0%	70	35	1.94	LTR	No	**No**
Enron	**0.50**	1%	56	52	11.20	CPL	No	**Yes**
Reuters	**1.67**	18%	100	79	4.55	CPL	No	**Yes**
ExprYeast	**1.71**	1%	417	161	1.62	LSP	Yes	**Yes**
SeqAra	**2.02**	27%	196	148	4.00	CPL	No	**Yes**
Slo. rivers	**17.63**	26%	724	492	3.11	BAS	Yes	**Yes**

5 Conclusions and Discussion

We introduced the hierarchy decomposition pipeline, a publicly available software toolbox for comparison of model induction algorithms for hierarchical multi-label classification (HMC) problems in the ensemble setting. The pipeline contains five algorithms: the algorithm that constructs a global model, which predicts the complete hierarchy, two partial decomposition algorithms that construct local models, which predict different edges of a hierarchy, and two complete decomposition algorithms that construct one or multiple models, which predict subset(s) of hierarchy nodes. The pipeline also contains tools for performance-based comparison of the algorithms, which compute the area under the average precision-recall curve and perform a statistical test of the differences in performance.

We applied the pipeline on ten HMC data sets and draw the following conclusions:

First, by comparing the algorithms on a set of common labels, we cannot guarantee that they will be compared on a set of common annotations. The set of labels common to all algorithms is composed of those labels that are assigned to at least one example as the most specific annotation. While the complete decomposition algorithms use only the most specific annotations, the hierarchical algorithms may assign additional annotations for the common labels, simply by applying the hierarchy constraint. This issue can be addressed by comparing the algorithms on a set of common labels that are, at the same time, hierarchy leaves. However, we should have in mind that by making this choice we may omit many annotations. The middle ground is to perform both types of comparisons considering their advantages and disadvantages.

Second, there exists a need for the proposed pipeline, since there is no single best algorithm for all HMC problems.

Third, the properties of a HMC data set can be related to the type of best performing algorithm on that data set. Multi-label classification algorithms perform best on data sets with large hierarchies. Interestingly, high cardinality is not strongly related to the advantage of multi-label classification algorithms. Hierarchical algorithms perform best on data sets from which they can obtain additional annotations compared to the complete decomposition algorithms, simply by applying the hierarchy constraint. They also perform best on data sets with large and complex hierarchies.

The limitation of the analysis that relates the properties of a HMC data set to the type of best performing algorithm is the small number of data sets in the study. The limitation can be addressed by performing a simulation that: (1) generates hundreds of artificial HMC data sets with predefined properties; (2) applies the proposed pipeline on the data sets to determine the best performing algorithm; and (3) uses the collected data for meta learning to produce a classifier relating dataset properties to the type of best performing algorithm. The simulation should consider data sets with both a tree shaped and a DAG shaped label hierarchy. In this study, we had only two data sets with a DAG shaped label hierarchy and none of the algorithms performed significantly better than the rest on those two data sets. They were, consequently, left out of the analysis.

The pipeline can be improved in several directions. While it has been developed in the ensemble setting to maximize predictive performance on complex HMC data sets (e.g., functional genomics data sets), it can be adapted to construct single models. Furthermore, it can be modified to construct ensembles other than random forests, e.g., bagging or boosting. Finally, additional research need to be performed to understand whether it is possible to add stratified cross-validation as an option.

References

1. Vens, C., Struyf, J., Schietgat, L., Džeroski, S., Blockeel, H.: Decision trees for hierarchical multi-label classification. Mach. Learn. **73**, 185–214 (2008)
2. Džeroski, S., Demšar, D., Grbović, J.: Predicting chemical parameters of river water quality from bioindicator data. Appl. Intell. **13**(1), 7–17 (2000)
3. Clare, A.: Machine learning and data mining for yeast functional genomics. Ph.D. thesis, University of Wales Aberystwyth, Aberystwyth, UK (2003)
4. Klimt, B., Yang, Y.: The enron corpus: a new dataset for email classification research. In: Boulicaut, J.-F., Esposito, F., Giannotti, F., Pedreschi, D. (eds.) ECML 2004. LNCS (LNAI), vol. 3201, pp. 217–226. Springer, Heidelberg (2004). https://doi.org/10.1007/978-3-540-30115-8_22
5. Lewis, D.D., Yang, Y., Rose, T.G., Li, F.: RCV1: a new benchmark collection for text categorization research. J. Mach. Learn. Res. **5**, 361–397 (2004)
6. Demšar, D., et al.: Using multi-objective classification to model communities of soil. Ecol. Modell. **191**(1), 131–143 (2006)
7. Dimitrovski, I., Kocev, D., Loskovska, S., Džeroski, S.: Hierchical annotation of medical images. In: Proceedings of the 11th International Multiconference - Information Society, pp. 174–181. JSI, Ljubljana (2008)
8. Vidulin, V., Šmuc, T., Supek, F.: Extensive complementarity between gene function prediction methods. Bioinformatics **32**(23), 3645–3653 (2016)
9. Vidulin, V., Šmuc, T., Džeroski, S., Supek, F.: The evolutionary signal in metagenome phyletic profiles predicts many gene functions. Microbiome **6**(1), 129 (2018)
10. Madjarov, G., Vidulin, V., Dimitrovski, I., Kocev, D.: Web genre classification with methods for structured output prediction. Inf. Sci. **503**, 551–573 (2019)
11. Ruepp, A., et al.: The FunCat, a functional annotation scheme for systematic classification of proteins from whole genomes. Nucleic Acids Res. **32**(18), 5539–5545 (2004)
12. Ashburner, M., et al.: Gene ontology: tool for the unification of biology. Nat. Genet. **25**(1), 25 (2000)
13. Zhou, N., et al.: The CAFA challenge reports improved protein function prediction and new functional annotations for hundreds of genes through experimental screens. Genome Biol. **20**(1), 1–23 (2019)
14. Bakır, G.H., Hofmann, T., Schölkopf, B., Smola, A.J., Taskar, B., Vishwanathan, S.V.N. (eds.): Predicting Structured Data. The MIT Press, Cambridge (2007)
15. Silla, C., Freitas, A.: A survey of hierarchical classification across different application domains. Data Min. Knowl. Disc. **22**(1–2), 31–72 (2011)
16. Clare, A., King, R.D.: Predicting gene function in Saccharomyces cerevisiae. Bioinformatics **19**(S2), ii42–ii49 (2003)
17. Quinlan, J.R.: C4.5: Programs for Machine Learning. Morgan Kaufmann, San Francisco (1993)
18. Blockeel, H.: Top-down induction of first order logical decision trees. Ph.D. thesis, Katholieke Universiteit Leuven, Leuven, Belgium (1998)
19. Blockeel, H., Bruynooghe, M., Džeroski, S., Ramon, J., Struyf, J.: Hierarchical multi-classification. In: Proceedings of the ACM SIGKDD Workshop on Multi-Relational Data Mining, pp. 21–35 (2002)
20. Blockeel, H., Schietgat, L., Struyf, J., Džeroski, S., Clare, A.: Decision trees for hierarchical multilabel classification: a case study in functional genomics. In: Fürnkranz, J., Scheffer, T., Spiliopoulou, M. (eds.) PKDD 2006. LNCS (LNAI), vol. 4213, pp. 18–29. Springer, Heidelberg (2006). https://doi.org/10.1007/11871637_7

21. Obozinski, G., Lanckriet, G., Grant, C., Jordan, M.I., Noble, W.S.: Consistent probabilistic outputs for protein function prediction. Genome Biol. **9**(S1), S6+ (2008)
22. Barutcuoglu, Z., Schapire, R.E., Troyanskaya, O.G.: Hierarchical multi-label prediction of gene function. Bioinformatics **22**(7), 830–836 (2006)
23. Guan, Y., Myers, C.L., Hess, D.C., Barutcuoglu, Z., Caudy, A., Troyanskaya, O.: Predicting gene function in a hierarchical context with an ensemble of classifiers. Genome Biol. **9**(S1), S3+ (2008)
24. Valentini, G.: True path rule hierarchical ensembles for genome-wide gene function prediction. IEEE ACM Trans. Comput. Biol. **8**(3), 832–847 (2011)
25. Levatić, J., Kocev, D., Džeroski, S.: The importance of the label hierarchy in hierarchical multi-label classification. J. Intell. Inf. Syst. **45**(2), 247–271 (2014). https://doi.org/10.1007/s10844-014-0347-y
26. Demšar, J.: Statistical comparisons of classifiers over multiple data sets. J. Mach. Learn. Res. **7**, 1–30 (2006)

Missing Value Imputation with MERCS: A Faster Alternative to MissForest

Elia Van Wolputte[1,2](✉) and Hendrik Blockeel[1,2]

[1] Department of Computer Science, KU Leuven, 3000 Leuven, Belgium
{elia.vanwolputte,hendrik.blockeel}@kuleuven.be
[2] Leuven.AI - KU Leuven Institute for AI, 3000 Leuven, Belgium

Abstract. Fundamentally, many problems in Machine Learning are understood as some form of *function approximation*; given a dataset $\mathcal{D}$, learn a function $f_{\boldsymbol{X} \rightarrow \boldsymbol{Y}}$. However, this overlooks the ubiquitous problem of missing data. E.g., if afterwards an unseen instance has missing input variables, we actually need a function $f_{\boldsymbol{X}' \rightarrow \boldsymbol{Y}}$ with $\boldsymbol{X}' \subset \boldsymbol{X}$ to predict its label. Strategies to deal with missing data come in three kinds: naive, probabilistic and iterative. The naive case replaces missing values with a fixed value (e.g. the mean), then uses $f_{\boldsymbol{X} \rightarrow \boldsymbol{Y}}$ as if nothing was ever missing. The probabilistic case has a generative model $\mathcal{M}$ of $\mathcal{D}$ and uses probabilistic inference to find the most likely value of $\boldsymbol{Y}$, given values for any subset of $\boldsymbol{X}$. The iterative approach consists of a loop: according to some model $\mathcal{M}$, fill in all the missing values based on the given ones, retrain $\mathcal{M}$ on the completed data and redo your predictions, until these converge. `MissForest` is a well-known realization of this idea using Random Forests. In this work, we establish the connection between `MissForest` and `MERCS` (a multi-directional generalization of Random Forests). We go on to show that under certain (realistic) conditions where the retraining step in `MissForest` becomes a bottleneck, `MERCS` (which is trained only once) offers at-par predictive performance at a fraction of the time cost.

Keywords: Missing value imputation · Ensemble methods · Multi-directional models · Decision trees

1 Introduction

Many machine learning methods assume there are no missing values in the data, or missing values are relatively infrequent. Under this assumption, a variety of techniques has been proposed to handle missing data. It is useful to maintain a clear distinction between two cases: *missing values at training time* (relevant during learning) and *missing values at prediction time* (making a prediction, using a given model, for an instance that lacks certain information needed by the model). First, when missing values occur at training time, the learning procedure may deal with them by ignoring all instances with missing values, ignoring

Code available at `github.com/eliavw/missmercs`.

A. Appice et al. (Eds.): DS 2020, LNAI 12323, pp. 502–516, 2020.
https://doi.org/10.1007/978-3-030-61527-7_33

attributes that have missing values, guessing the missing value (imputation) before proceeding with the computations, or using other techniques. The second case, missing values at prediction time, is quite a different problem: a model is given, but the model needs information that is not available. Nevertheless, some techniques for handling missing values during prediction resemble those for the training phase, e.g. imputation can be used, if some model for imputation is available.

In this paper, we focus specifically on missing values at prediction time. There are contexts where missing values at prediction time may be much more frequent, and possibly also more systematic, than typically assumed by many learners. To illustrate, consider two practical examples;

- First, **machine learning in industrial contexts** often depends on sensor data. Consider an AI-system (e.g. a predictive maintenance application) which makes automatic decisions based upon input information coming from sensors. When a single sensor breaks down and no longer provides information, the AI-system needs to carry on and perform as well as possible, although less input information is now available.
- Second, consider a common **spreadsheet**. Suppose a user filling in data in a spreadsheet or a web form: ML methods exist to assist users by predicting information to be inserted in certain cells. Ideally, these predictions use as much as possible information filled in elsewhere, regardless of exactly which cells are already filled in and which ones are not. So, at prediction time, robustness with regard to missing input information is crucial.

In both cases, at prediction time we need a model $\mathcal{M}$ that can predict some output variable(s) $\boldsymbol{Y}$ from input variable(s) $\boldsymbol{X}$, so we can regard $\mathcal{M}$ as a function from $\boldsymbol{X}$ to $\boldsymbol{Y}$. However, the actual input that is available for a particular prediction often consists of values for a strict subset $\boldsymbol{X}' \subset \boldsymbol{X}$. In the first example, this is caused by malfunctioning sensors, in the second one by empty cells in the spreadsheet. Thus, handling missing values at prediction time boils down to the task of deriving from $\mathcal{M} : \boldsymbol{X} \rightarrow \boldsymbol{Y}$ another function $\mathcal{M}' : \boldsymbol{X}' \rightarrow \boldsymbol{Y}$ with $\boldsymbol{X}' \subset \boldsymbol{X}$ which still makes maximally accurate predictions.

In a nutshell, we propose to solve this problem as follows: use a tree-based approach such as `MissForest` [17], but avoid its multiple training iterations. Given some robust prediction strategies, so-called `MERCS` models [20] could do just that. So, our proposal decomposes into two research questions:

Q1 Can a `MERCS` model be made robust to missing values at prediction time?
Q2 How does `MERCS` compare against `MissForest`, a well-established tree-based technique to deal with missing data?

First, Sect. 2 pinpoints the current gap in knowledge, and thus provides further context and motivation for our solution strategy (i.e. **Q1** and **Q2**). On one hand, we find `MissForest` [17]: a powerful tree-based approach for missing data, which is iterative and thus ill-suited for missing values at prediction time. On the other hand, we find `MERCS` [20]: a somewhat similar, but non-iterative, tree-based model, which currently lacks prediction strategies to deal with missing data effectively.

Section 3 outlines our proposal to answer **Q1** (how to extend the `MERCS` framework with robust prediction strategies) which constitutes our algorithmic contribution. Two key ideas matter here. First, *attribute importance*: this quantifies the relevance of trees for a given prediction task. Second: *chaining*: inspired by `MissForest`, `MERCS` can be made to use outputs of some trees as inputs for others.

Lastly, Sects. 4 and 5 contain experimental evaluations of **Q1** and **Q2** respectively. Ultimately, the answer to both **Q1** and **Q2** is positive: when dealing with missing input values at prediction time, `MERCS` models are a viable alternative to `MissForest`.

2 Related Work and Background

We focus on missing value handling at prediction time: given a model $\mathcal{M}$ that represents a function $\mathcal{M} : \boldsymbol{X} \rightarrow \boldsymbol{Y}$ and a *query-instance* x_q which has only values for a strict subset $\boldsymbol{X}' \subset \boldsymbol{X}$, how can we still use $\mathcal{M}$ to predict the value of $\boldsymbol{Y}$?

The discussion of related work is organized into four parts, each covering a specific approach for missing value handling. First, we consider naive approaches for handling missing values. Second, we discuss probabilistic graphical models, which handle unobserved values so naturally that the term "missing values handling" is typically not even used in that context. Third, we discuss iterative approaches in general, and `MissForest` in particular. `MissForest` is a popular tree-based technique for missing value imputation at training time. Lastly, we discuss `MERCS`, another tree-based framework in some ways similar to `MissForest`, but which could be more suitable in the specific case of missing values at prediction time.

2.1 Naive Methods

A generally applicable approach is what we call naive methods: guess the missing values. Concretely, this comes down to a one-size-fits-all strategy: simply fill in the mean, median or mode of the variable. We call this "naive", as it just fills in the same value for all instances.

The **advantage** of this technique is its low cost, both in time and memory. The obvious **disadvantage** is the limited accuracy of a naive approach. In our context, i.e. where predictive accuracy matters, naive methods can still serve as a baseline to compare other methods to, but nothing more.

2.2 Probabilistic Methods

Probabilistic graphical models (PGMs), such as Bayesian networks or Markov random fields [8,12,13], model the probability distribution $P_{\boldsymbol{X}}$ over all variables.

From this distribution, any marginal distribution $P_{\boldsymbol{X}'}$ with $\boldsymbol{X}' \subset \boldsymbol{X}$ can be computed, as well as any conditional distribution $P_{\boldsymbol{Y}|\boldsymbol{X}'}$ with $\boldsymbol{X}', \boldsymbol{Y} \subseteq \boldsymbol{X}$. As the joint distribution uniquely defines all marginal and conditional distributions, the target variable $\boldsymbol{Y}$ can be predicted from any subset $\boldsymbol{X}'$ that is equal to the set of all known variables.

Their versatility is the main **advantage** of probabilistic methods. In a sense, the "problem of missing values" simply does not even exist in this context: the optimal way of handling them follows naturally from the probabilistic model itself.

The **disadvantage** are the computational costs involved. Explicitly deriving the marginal/conditional probabilities in a PGM is NP-hard in the general case. Performing probabilistic inference in the original PGM is NP-hard too. In practice, approximate inference in the original PGM is used at prediction time, but even that can be costly. Another issue are data-types: in practice, PGMs work best on nominal data. Numeric data or mixtures of nominal/numeric data can be challenging for probabilistic approaches.

2.3 Iterative Approaches

Iterative approaches [5,18] gradually refine their imputations by means of a simple loop. First, for each variable in your dataset, you learn a predictive model, using the other variables of the dataset as inputs. Second, you use that model to fill in any missing values of that variable. This too, is repeated for each variable in the dataset. Third, you repeat this entire process (both training and prediction) until you reach a stopping criterion which indicates when no more progress is being made.

`MissForest` [17] is a specific implementation of the aforementioned idea. In `MissForest`, the underlying predictive models are Random Forests. The stopping criterion is dual; the loop is stopped when the resulting change from iteration i to iteration $i + 1$ is less then a user-defined parameter γ or when a certain maximum number of iterations n is exceeded.

The **advantages** of `MissForest` are twofold. First, Random Forests are non-parametric and make relatively few assumptions about the underlying data distributions. Second, they work well on both numeric and nominal data, or on mixtures of the two. This versatility with regard to data-types is often highlighted [21] as the "killer-feature" which makes `MissForest` such an attractive option in real-world scenarios.

The **disadvantage** of `MissForest` (or any iterative approach for that matter) is that, essentially by definition, it is geared towards missing data at training time: this is not the problem we set out to solve. Indeed, the training phase at each step of the iteration involves significant costs in time (i.e. you need several training rounds) and memory (i.e. you always need to have data to train on). When an incomplete dataset is given, iterative approaches are perfectly equipped to fill in the gaps introduced by the missing data. But at prediction time, when unseen, incomplete query-instances x_q come in one by one and need a prediction for $\boldsymbol{Y}$ right away, this double cost of keeping a training set in memory and retraining your model each time a new query-instance pops up, quickly becomes a bottleneck.

2.4 MERCS

MERCS [20] is a method for learning *multi-directional* ensembles of decision trees. This as opposed to classical ensembles of decision trees, which are *uni-directional*: a single function $f_{\boldsymbol{X} \to \boldsymbol{Y}}$ is learned that predicts some output variable(s) $\boldsymbol{Y}$ from input variable(s) $\boldsymbol{X}$, and it is known at training time what $\boldsymbol{X}$ and $\boldsymbol{Y}$ are. Bagging [3], Random Forests [4] and Gradient Boosted Trees [7] are all examples of methods that learn such uni-directional ensembles. In a multi-directional ensemble, a single tree may have multiple target variables (so-called multi-target trees), and different trees may have different sets of target variables. Such ensembles can be learned using a method that is quasi identical to the learning algorithm for Random Forests: the only difference is that for each new tree $T^i_{\boldsymbol{X}^i \to \boldsymbol{Y}^i}$ that is learned, a new set of target variables $\boldsymbol{Y}^i$ is chosen. Learning methods for MERCS models differ mostly in terms of how they choose $\boldsymbol{Y}^i$ for each tree. For instance, using one target variable per tree often gives slightly higher accuracy for individual trees, but having many target variables in one tree can reduce the size of the ensemble without reducing the number of trees available for predicting a given variable. Cf. Van Wolputte et al. [20] for more details.

What is interesting here is that MERCS is somewhat similar to MissForest: both are multi-directional ensembles of decision trees, where any variable of the dataset can be predicted by at least one tree of the ensemble. But whereas MissForest was originally conceived to do missing value imputation in a given dataset, MERCS was not. MERCS originated as a fast, tree-based alternative to PGMs: learn a *model* $\mathcal{M}$ from dataset $\mathcal{D}$.

This begs the question: could MERCS, like MissForest, become a powerful tool for missing value imputation? We believe it does. The **advantages** of MERCS in this context are twofold. First, like probabilistic approaches, MERCS learns a model $\mathcal{M}$ from training data $\mathcal{D}$. Afterwards, there is no need to keep this training data around, all the necessary knowledge is encoded in the model itself. As a consequence, MERCS would be particularly interesting for missing value imputation at prediction time, a regime where iterative approaches struggle. Second, like MissForest, MERCS is a tree-based approach, which means a.o. that MERCS can also deal with (mixtures of) nominal and numeric variables.

At this point, the main **disadvantage** is that it remains unclear whether MERCS can handle missing values effectively. In order to be proficient in such a regime, MERCS needs a prediction strategy which is *robust to missing data*: given an unseen instance x_q, MERCS should still able to do a high quality prediction for the value of $\boldsymbol{Y}$, even if x_q has some missing values. How to achieve this will be the topic of Sect. 3.

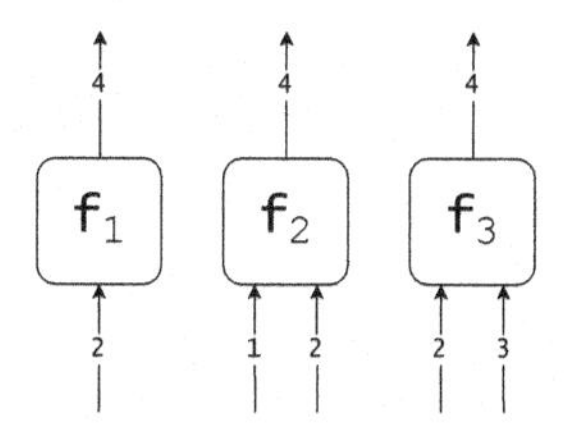

(a) A Random Forest (Eq. 1)[4]. The input attributes $\boldsymbol{X}^i$ of the component trees $T^i_{\boldsymbol{X}^i \to \boldsymbol{Y}}$ are random subsets of $\boldsymbol{A} \setminus \boldsymbol{Y}$.

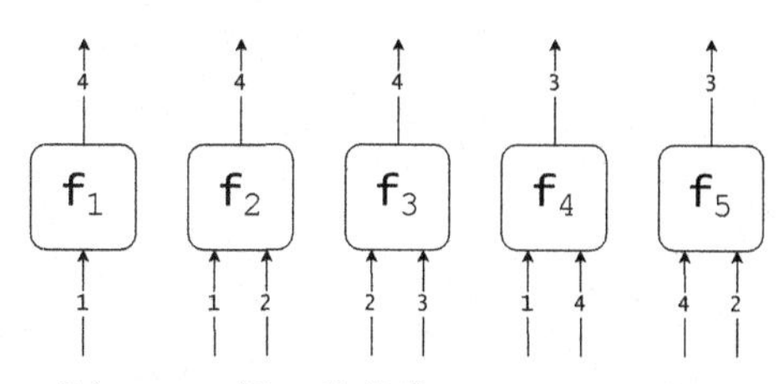

(b) MERCS (Eq. 2) [20]. MERCS generalizes Random Forests and also selects output attributes $\boldsymbol{Y}^i$ at random.

Fig. 1. Random Forests and MERCS. Attributes $A_j \in \boldsymbol{A}$ are depicted as lines annotated with their respective indices j. A decision tree, $T^i_{\boldsymbol{X}^i \to \boldsymbol{Y}^i}$, is depicted as a box connecting its input ($\boldsymbol{X}^i$) to its output ($\boldsymbol{Y}^i$) attributes.

3 Robust Prediction Strategies for MERCS

This section outlines our answer to research question **Q1**: how to make the MERCS framework robust to missing values at prediction time. The motivation behind this approach is explained in Sect. 2, whereas the experimental evaluation happens in Sect. 4.

In the following, we use T^i ($i = 1 \dots k$) to denote the different trees in the model $\mathcal{M}$. $\boldsymbol{X}^i$ refers to the set of input attributes used by tree T^i, and $\boldsymbol{Y}^i$ to the set of output (or target) attributes of T^i. Similarly, we use $q_{\boldsymbol{I} \to \boldsymbol{O}}$ to denote a particular prediction task or *query*, where $\boldsymbol{I}$ denotes the set of attributes whose value is given (i.e. input attributes of $q_{\boldsymbol{I} \to \boldsymbol{O}}$) and $\boldsymbol{O}$ the set of attributes to be predicted (i.e. output attributes of $q_{\boldsymbol{I} \to \boldsymbol{O}}$). Furthermore, $\boldsymbol{A}$ simply refers to the set of all the attributes of a given dataset $\mathcal{D}$.

Take a Random Forest (Fig. 1a),

$$RF(\boldsymbol{X}, \boldsymbol{Y}) = \{T^i_{\boldsymbol{X}^i \to \boldsymbol{Y}} | \boldsymbol{X}^i \subset \boldsymbol{A} \setminus \boldsymbol{Y}\}, \tag{1}$$

and introduce randomness in the target attributes. In this way, $RF(\boldsymbol{X}, \boldsymbol{Y})$ generalizes to a multi-directional ensemble of decision trees, or a MERCS model (Fig. 1b),

$$M(\boldsymbol{A}) = \{T^i_{\boldsymbol{X}^i \to \boldsymbol{Y}^i} | \boldsymbol{X}^i, \boldsymbol{Y}^i \subset \boldsymbol{A}, \, \boldsymbol{X}^i \cap \boldsymbol{Y}^i = \emptyset\}. \tag{2}$$

Now, to answer an arbitrary query $q_{\boldsymbol{I} \to \boldsymbol{O}}$ a MERCS model needs a *prediction strategy*. This has two reasons. First, note that $q_{\boldsymbol{I} \to \boldsymbol{O}}$ is not known at training time, and second, learning a dedicated decision tree for every possible $q_{\boldsymbol{I} \to \boldsymbol{O}}$ is simply not feasible. Thus, this prediction strategy decides how to optimally use the available T^i, present in the MERCS model (Eq. 2), to answer any incoming $q_{\boldsymbol{I} \to \boldsymbol{O}}$ as accurately as possible.

Rather than a single prediction strategy, we define a naive baseline and three, increasingly complex, strategies. These subdivide into two groups. First, *single-layer* strategies, where *attribute importance* quantifies the relevance of

T^i for a given prediction task $q_{I \to O}$. Second, *multi-layer* strategies that, like MissForest, use *chaining*: take the prediction of one tree T^i as the input for another tree T^j.

3.1 Attribute Importance and Single-Layer Prediction Strategies

Assume that for a given tree T^i, only some of its input attributes $\boldsymbol{X}^i$ are known. The more inputs are missing, the less accurate we expect the predictions of T^i to be. But not all attributes in $\boldsymbol{X}^i$ are equally important for the prediction. One way to measure this is **attribute importance** [10]:

$$I(A_j, T^i) \propto \sum_{\{a(\tau)=A_j\}} p(\tau) \Delta i(\tau) \tag{3}$$

where τ ranges over all nodes of the tree, $p(\tau)$ is the proportion of instances sorted into τ, $a(\tau)$ is the attribute tested at τ, and $\Delta i(\tau)$ is the expected reduction of impurity achieved by that node. So, the attribute importance $I(A_j, T^i)$ is essentially the normalized sum of the impurity decreases achieved by splitting on attribute A_j.

Consider a query $q_{I \to O}$, meaning that attributes $\boldsymbol{I}$ are given. The less important the missing input attributes ($\boldsymbol{X}^i \setminus \boldsymbol{I}$) of T^i are, the more accurate T^i likely is. Therefore, we use the sum of importances of the known attributes ($\boldsymbol{I}$) to quantify the relevance of T^i to make predictions in this context.

We call this sum the **input relevance** of T^i for a set of given attributes $\boldsymbol{I}$:

$$IR(T^i, \boldsymbol{I}) = \sum_{A_j \in \boldsymbol{X}^i \cap \boldsymbol{I}} I(A_j, T^i). \tag{4}$$

Now that we have established the notion of input relevance, we define our two single-layer strategies. We distinguish between a naive Random Forest baseline (*RF-prediction*) and *MRAI-prediction* which does exploit input relevance.

Random Forest (Baseline). The most basic strategy is as follows: each T^i that predicts some attributes in $\boldsymbol{O}$, that is, $\boldsymbol{Y}_i \cap \boldsymbol{O} \neq \emptyset$, is regarded as equally relevant. $\mathcal{M}$'s prediction of an individual target attribute is obtained by aggregating the predictions of all T^i in $\mathcal{M}$ that predict that attribute. A standard aggregation (majority vote, mean, ...) is used, without taking input relevance into account (Fig. 2a).

MRAI-Prediction. A second strategy, MRAI-prediction[1], does take input relevance into account. T^i is considered relevant if $\boldsymbol{Y}_i \cap \boldsymbol{O} \neq \emptyset$ and $IR(T^i, \boldsymbol{I}) \geq \theta$, for some threshold θ. That is, trees that rely too strongly on attributes whose values are missing are not included in the set of predictors (Fig. 2b).

[1] MRAI stands for most relevant attribute importance.

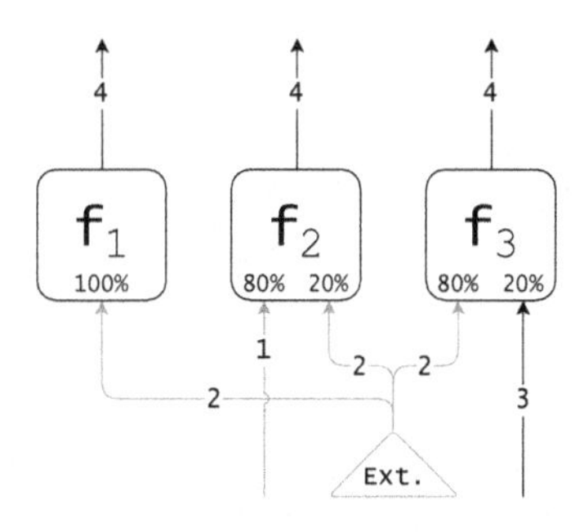

(a) **RF-prediction**. Selects all $T^i_{\boldsymbol{X}^i \to \boldsymbol{Y}^i}$ that predict an attribute in $\boldsymbol{O}$ regardless of missing input attributes ($\boldsymbol{X}^i \setminus \boldsymbol{I}$).

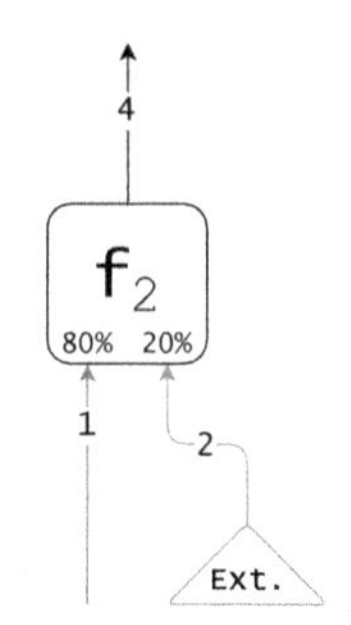

(b) **MRAI-prediction**. Selects only the most relevant trees, based on their *input relevance* (Eq. 4), which takes into account *attribute importance* (Eq. 3).

Fig. 2. RF-prediction (baseline) and MRAI-prediction build an ad hoc ensemble of relevant trees. A naive fallback procedure (depicted as red triangle) takes care of missing inputs if necessary. Attribute importances are indicated in gray. (Color figure online)

MRAI-prediction can be understood as a refinement of the MA-prediction strategy introduced in Van Wolputte et al. [20]. This MA-prediction essentially is MRAI-prediction minus attribute importance: each attribute is deemed equally important. Preliminary experiments revealed MRAI-prediction to consistently outperform MA-prediction. Therefore, the old MA-prediction strategy is omitted from subsequent experiments in favor of its superior cousin: the novel MRAI-prediction strategy.

3.2 Chaining and Multi-layer Prediction Strategies

Assume $\boldsymbol{Y}^i \cap \boldsymbol{O} \neq \emptyset$ for some tree T^i, but $IR(T^i, \boldsymbol{I}) < \theta$. Now, if more input attributes of T^i had been known, $IR(T^i, \boldsymbol{I})$ might have met the threshold θ. In fact, this can readily be achieved. After all, a `MERCS` model is multi-directional and thus contains at least one predictor for each attribute. Concretely, we can make T^i meet this threshold θ by predicting some of its missing input attributes ($\boldsymbol{X}^i \setminus \boldsymbol{I}$), using other trees T^j with $\boldsymbol{Y}^j \cap (\boldsymbol{X}^i \setminus \boldsymbol{I})$. Afterwards, we treat these predictions of T^j as known values. To decide which T^j to use, we can use exactly the same criterion as we did before: T^j is a suitable predictor if it predicts some of the missing input attributes of T^i and if $IR(T^j, \boldsymbol{I}) > \theta$. If some of T^j's input attributes are missing, the same procedure can be repeated.

This principle is known as *chaining* [14]. In our multi-layer algorithms, chaining is exploited in two different manners: bottom-up and top-down. Consequently, we distinguish between *BU-prediction* and *TD-prediction* respectively.

BU-Prediction. The BU-prediction strategy is a recursive application of MRAI-prediction. It works in a bottom-up fashion: we keep a set of known

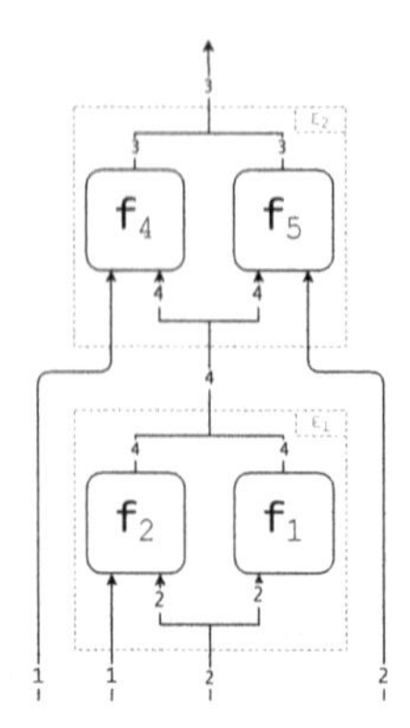

(a) **BU-prediction**. In each layer, an ensemble of relevant T^i is selected according to $IR(T^i, \boldsymbol{K})$. This makes more attributes available to the next layer, which means other T^i become relevant. Eventually, we obtain a chain of ensembles to predict $\boldsymbol{O}$.

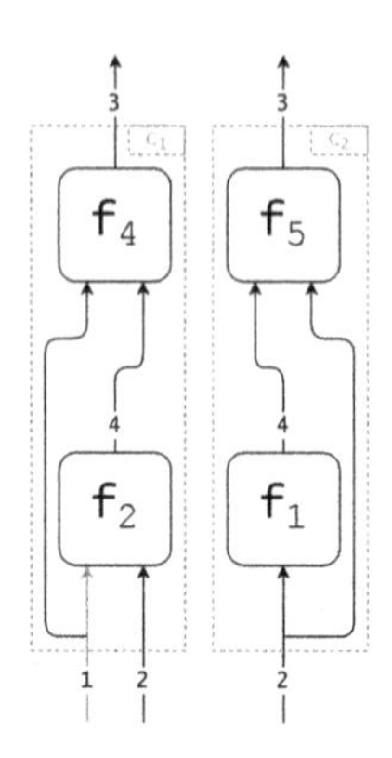

(b) **TD-prediction**. Builds a chain of relevant T^i, the probability of including a model (Eq. 6) is proportional to $IR(T^i, \boldsymbol{I})$. Repeated application of this idea yields an ensemble of chains to predict $\boldsymbol{O}$.

Fig. 3. BU-prediction and TD-prediction use *chaining*.

attributes $\boldsymbol{K}$, whose initial value is $\boldsymbol{I}$. For each T^i with $IR(T^i, \boldsymbol{K}) > \theta$, add the variables in $\boldsymbol{Y}^i$ to $\boldsymbol{K}$. That concludes one step. Repeat this until $\boldsymbol{O} \subseteq \boldsymbol{K}$ (Fig. 3a).

If, at a given step, the threshold θ is set too high, there may not be any trees with a sufficiently high input relevance. This means no progress is made and the procedure ends with $\boldsymbol{O} \nsubseteq \boldsymbol{K}$. In order to make progress, simply repeat that step with a lower value for θ, and proceed.

TD-Prediction. TD-prediction exploits the MRAI-principle in a top-down manner. Rather than extending a set of known attributes $\boldsymbol{K}$ until it covers $\boldsymbol{O}$ (as BD-prediction does), TD-prediction starts from the output attributes $\boldsymbol{O}$ instead.

First, we define a set of unknown attributes of interest $\boldsymbol{U}$, whose initial value is $\boldsymbol{O}$. Then, take the subset of trees which predict at least one attribute in $\boldsymbol{U}$,

$$C = \{T^i_{\boldsymbol{X}^i \to \boldsymbol{Y}^i} | \boldsymbol{Y}^i \cap \boldsymbol{U} \neq \emptyset\} \tag{5}$$

and continue by defining a probability distribution,

$$p(i) = \frac{IR(T^i, \boldsymbol{I})}{\sum_{T^i \in C} IR(T^i, \boldsymbol{I})}, \tag{6}$$

which assigns to each $T^i \in C$ a probability proportional to $IR(T^i, \boldsymbol{I})$. Using p, we can randomly choose a tree T^j such that more suitable trees are more likely to get chosen. This concludes one step.

For the next step, first adjust $\boldsymbol{U}$ accordingly, i.e. $\boldsymbol{U} = \boldsymbol{X}^j \setminus \boldsymbol{I}$. That means that $\boldsymbol{U}$ now contains the missing input attributes of the tree T^j selected in the previous step. Repeat the procedure from the definition of C on (Fig. 3b).

Essentially, this procedure does a random walk through the random forest. Starting from $\boldsymbol{U} = \boldsymbol{O}$, it randomly chooses a tree T^i that predicts (part of) $\boldsymbol{O}$; more suitable trees are more likely to get chosen. If that tree has missing inputs, choose a tree that predicts some of those inputs. Keep repeating this up to some maximum depth or until there are no missing inputs left. As the TD-procedure is randomized, it can be repeated multiple times. Each time, a different path through the random forest is followed.

It is instructive to compare BU-prediction and TD-prediction by viewing them as searches through a graph. Let G be a bipartite graph with nodes being attributes and trees; trees have incoming edges from their input attributes and outgoing edges to their output attributes. BU constructs a subgraph of G that connects $\boldsymbol{I}$ to $\boldsymbol{O}$ using only tree nodes whose IR is above some threshold. TD is a randomized search for paths that end in $\boldsymbol{O}$ but may begin at any point, and tends to contain tree nodes with high IR. Neither BU nor TD entirely avoid the use of external procedures for missing value imputation. BU only avoids them when $\theta = 1$ leads to a solution. TD only avoids them on paths where each tree happens to predict all the missing inputs of the tree that comes behind it in the path.

4 Comparison of Prediction Strategies in MERCS

This experiment is set up to answer our first research question **Q1**: how to make the MERCS framework robust to missing values at prediction time? Here, we compare all the prediction strategies for MERCS we introduced in Sect. 3, across different degrees of missing data. This allows us to see which prediction strategies are actually robust to missing data at prediction time. As an external baseline, we also add a PGM.

Datasets. Our experiments use a standard benchmark suite[2] of 28 real-world datasets. Our focus on multi-directionality requires adequate datasets in the sense that it should be possible to think of several potentially interesting prediction tasks. Prior appearance in studies on structure learning [9], make this benchmark a natural fit for our current setting. Lastly, PGMs are less flexible with regard to data-types (cf. Sect. 2), but in this benchmark, that will not be an issue, since all variables in these datasets are binary.

Methodology. For each dataset, we train both a MERCS model and a PGM. For PGMs, we rely on the SMILE-engine[3] for structure learning and inference.

[2] Cf. `github.com/UCLA-StarAI/Density-Estimation-Datasets` and [1,11,19].

[3] The SMILE-engine is a part of the powerful and widely used BayesFusion system, cf. `bayesfusion.com/publications`.

For structure learning, we use the *greedy thick thinning* algorithm. For inference, we use the approximate *EPIS-sampling* algorithm. In MERCS, trees are randomly assigned 60% of attributes as inputs, 2 output attributes and are limited to a maximum depth of 16. We ensure each attribute occurs 4 times as an output attribute, meaning we have $2m$ trees in total, m being the amount of attributes in the dataset. The Random Forest baseline, essentially MERCS with a trivial prediction strategy, uses the exact same trees to ensure consistency.

To see the effect of missing input attributes on performance, we consider an extensive set of queries $q_{I \to O}$. For each dataset, we randomly pick 10 output attributes. For each of those, we build a series of 10 increasingly difficult queries; the first one has no missing input attributes, and in each consecutive query of the series, we omit (at random) an additional 10% of its input attributes. In the end, this amounts to 2800 distinct prediction tasks.

Evaluation Criteria. For predictive performance, we look at F1-score [6,15] on a test set. The random selection of output attributes ($\boldsymbol{O}$) in our queries $q_{I \to O}$ means we cannot exclude very unbalanced targets. For these, high predictive accuracy is meaningless. F1-score is not susceptible to this kind of effect [16] and therefore more suitable for our needs. For runtime, we report prediction times, relative to the PGM-baseline.

Results. BU-prediction is the most robust prediction strategy in MERCS (Fig. 4a). When less than half of the inputs is missing, PGMs exhibit lower predictive performance than MERCS. MRAI-prediction outperforms the naive Random Forest baseline (RF), indicating that *input relevance* (Eq. 4) works, and consequently that *attribute importance* (Eq. 3) is a useful heuristic. In its turn, BU-prediction improves upon MRAI, showing that *chaining* indeed improves robustness. However, TD-prediction does not, and additionally is much slower than BU-prediction (Fig. 4b), which indicates that bottom-up chaining is recommended.

In terms of runtime, note that roughly speaking, all prediction strategies in MERCS do offer order(s) of magnitude of speedup over PGMs (Fig. 4b) across the board. PGMs rely on probabilistic inference. This makes them very robust to missing values, but also comes at a significant overhead in prediction time.

5 MERCS vs. MissForest

This experiment is set up to answer the second research question **Q2**: how does MERCS compare to MissForest? Concretely, we try to evaluate whether MERCS can succeed where MissForest struggles, namely when missing values are only introduced at prediction time. We expect MissForest to experience a bottleneck, since it its iterative nature requires retraining for each new query-instance q_k. The question is whether MERCS, can offer similar predictive performance, without the need to retrain.

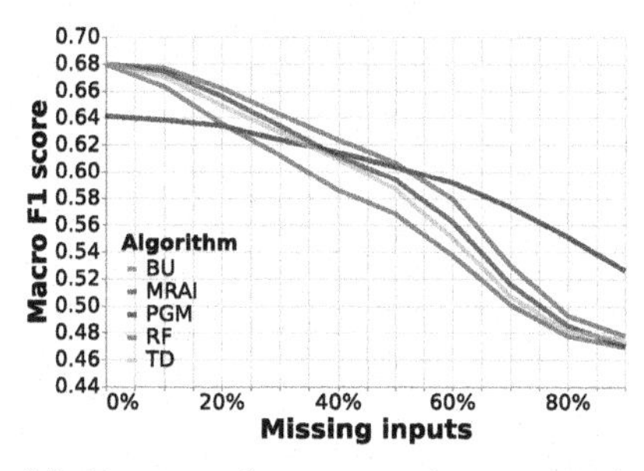

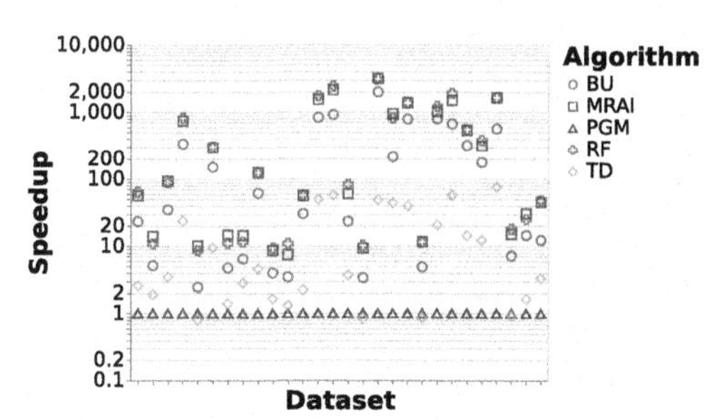

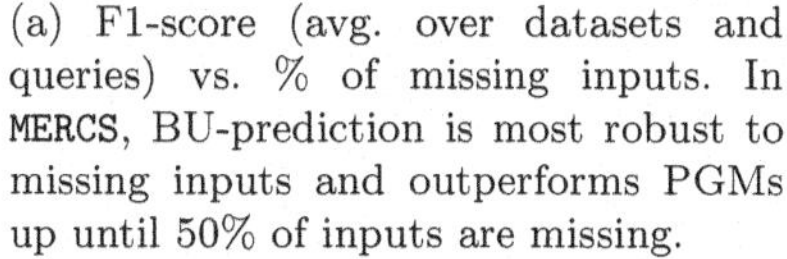
(a) F1-score (avg. over datasets and queries) vs. % of missing inputs. In MERCS, BU-prediction is most robust to missing inputs and outperforms PGMs up until 50% of inputs are missing.

(b) At prediction time, MERCS offers order(s) of magnitude speedup over PGMs (which rely on probabilistic inference). In MERCS, TD-prediction is the most costly.

Fig. 4. Avg. F1-scores, and relative prediction times of all prediction strategies. When at least half the inputs are given and prediction time matters, MERCS (and in particular, BU-prediction) works.

Datasets. This experiment uses a curated benchmark suite of classification problems, known as OpenML-CC18 [2]. Here, we are interested not so much in the multi-directional aspect (as in Sect. 4), but really on our core problem: handling missing values at prediction time. Therefore, this benchmark, with well-defined (categorical) target variables, is ideally suited.

Methodology. Each dataset is divided into a train set and a test set. This division is already defined in the OpenML-CC18 itself, which enhances reproducibility. Now, since we are interested in how MERCS and MissForest handle missing values *at prediction time*, we use the test set to generate query-instances x_q. This happens as follows: from the test set, take an instance x. From this instance x, omit a fixed number input variables at random (i.e. make those missing). This defines a query-instance x_q. This is repeated 100 times, yielding 100 query-instances per dataset. The pattern of which attributes are missing can vary from instance to instance.

Both for MERCS and MissForest the goal is, given a query-instance x_q, predict the value of its output attribute $\boldsymbol{Y}$.

In the case of MissForest, we add the query-instance x_q to the entire training set, and run the MissForest algorithm on all these instances. Since the target variable of the query-instance is unknown and therefore missing, it will also be imputed. For each query-instance, this loop has to be repeated in full.

In the case of MERCS, we can clearly distinguish between a training phase and a testing phase. First, we train a MERCS model $\mathcal{M}$ on the training set. Second, given a query-instance x_q, we can ask $\mathcal{M}$ to predict the target variable $\boldsymbol{Y}$, from the non-missing input variables. We can repeat this for all 100 query-instances, without the need of retraining. We use the BU-prediction strategy, since it is the most robust to missing values (cf. Sect. 3 and Fig. 4).

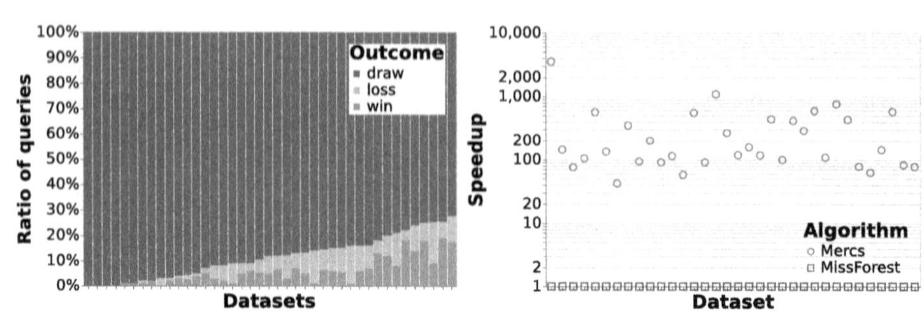

(a) MERCS vs. MissForest, win-draw-loss comparison. The overwhelming majority of draws shows that MERCS and MissForest perform at par under the conditions examined here.

(b) MERCS consistently offers multiple orders of magnitude speedup over MissForest. The iterative nature of MissForest means that it needs to retrain, whereas the MERCS model only trains once.

Fig. 5. MERCS vs. MissForest.

Evaluation Criteria. Our primary interest here is to determine of either MERCS or MissForest is clearly superior to its competitor, and if so, at which cost. Since we are dealing with classification problems, a prediction for a single query-instance is either correct or incorrect. Thus, if approach A is correct and approach B is incorrect, that constitutes a *win* for approach A on that query-instance (and vice-versa a *loss* for approach B). If both approaches are (in)correct, that constitutes a *draw*. In terms of cost, we simply measure prediction times, averaged across queries.

Results. In terms of predictive performance, it is clear from the amount of draws (Fig. 5a) that in the overwhelming majority of queries and datasets, it really does not matter whether you choose MERCS or MissForest. Both predict the same value in the large majority of cases, and when they differ, each is about equally likely to win, taken over all datasets.

In terms of runtime, although the robust BU-prediction strategy in MERCS is slower than the naive Random Forest baseline (Fig. 4b), it still entails a significant speedup (up to 3 orders of magnitude in some cases) over MissForest (Fig. 5b), across all datasets.

6 Conclusions

To conclude, let us simply answer our original two research questions, **Q1** and **Q2**.

6.1 Q1: How to Make MERCS Robust to Missing Values at Prediction Time?

In Sect. 3, we extend the original MERCS framework with three new prediction strategies: MRAI, BU and TD. All of these rely on *attribute importance* (Eq. 3) to select the most relevant trees for the task at hand. Additionally, BU and TD make use of *chaining*: the outputs of one decision tree can serve as inputs for another one. In Sect. 4, these proposed prediction strategies are compared experimentally.

The answer to **Q1** (and consequently, our contribution to the original MERCS-framework [20]) is that both attribute importance and chaining can improve robustness, and BU-prediction is found to be the best strategy (Fig. 4a) for MERCS. Additionally, the computational costs associated with the BU-prediction strategy are acceptable (Fig. 4b).

6.2 Q2: How Does MERCS Compare Against MissForest?

In Sect. 2, we made the argument that MERCS would make an interesting replacement for MissForest when dealing with missing values at prediction time. The reason being that an iterative approach such as MissForest is really geared towards dealing with the missing value problem at training time, since the iterative procedure requires multiple training rounds. Of course, this first required MERCS itself to be somewhat robust against missing values, which was dealt with in research question **Q1**. What remains is to see whether MERCS can actually *improve* upon MissForest.

The answer to **Q2** is that, when query-instances x_q come in one by one, MERCS improves upon MissForest. In terms of predictive performance, both approaches yield similar results (Fig. 5a). But, in terms of runtime, MERCS is orders of magnitude faster than MissForest (Fig. 5b). The iterative nature of MissForest makes it particularly ill-suited to tackle missing data at prediction time: it needs to retrain for each query-instance. MERCS, which never needs to retrain, is thus orders of magnitude faster when these query-instances come in one by one (Fig. 5b).

Acknowledgments. This research received funding from the Flemish Government under the '*Onderzoeksprogramma Artificiële Intelligentie (AI) Vlaanderen*" programme. Elia Van Wolputte is supported by the European Research Council (ERC) under the European Union's Horizon 2020 research and innovation programme, grant agreement No. 694980 "SYNTH: Synthesising Inductive Data Models". Lastly, we wish to acknowledge Simon Slangen, who in his Msc. thesis was the first to investigate advanced prediction strategies in MERCS.

References

1. Bekker, J., Davis, J., Choi, A., Darwiche, A., Van den Broeck, G.: Tractable learning for complex probability queries. In: Advances in Neural Information Processing Systems 28 (NIPS), December 2015
2. Bischl, B., et al.: Openml benchmarking suites. arXiv preprint arXiv:1708.03731 (2017)
3. Breiman, L.: Bagging predictors. Mach. Learn. **24**(2), 123–140 (1996)
4. Breiman, L.: Random forests. Mach. Learn. **45**(1), 5–32 (2001)
5. Buuren, S.V., Groothuis-Oudshoorn, K.: mice: multivariate imputation by chained equations in R. J. Stat. Softw. **45**(3), 1–67 (2011). https://doi.org/10.18637/jss.v045.i03
6. Chinchor, N.: MUC-4 evaluation metrics. In: Proceedings of the 4th Conference on Message Understanding, pp. 22–29 (1992)
7. Friedman, J.H.: Greedy function approximation: a gradient boosting machine. Ann. Stat. **29**(5), 1189–1232 (2001). http://www.jstor.org/stable/2699986
8. Koller, D., Friedman, N., Getoor, L., Taskar, B.: Graphical models in a nutshell. In: Introduction to Statistical Relational Learning, vol. 43 (2007)
9. Liang, Y., Bekker, J., Van den Broeck, G.: Learning the Structure of Probabilistic Sentential Decision Diagrams. In: Proceedings of the 33rd Conference on Uncertainty in Artificial Intelligence (UAI), August 2017
10. Louppe, G., Wehenkel, L., Sutera, A., Geurts, P.: Understanding variable importances in forests of randomized trees. In: Advances in Neural Information Processing Systems, pp. 431–439 (2013)
11. Lowd, D., Davis, J.: Learning Markov network structure with decision trees. In: 2010 IEEE International Conference on Data Mining, pp. 334–343 (2010)
12. MacKay, D.J.: Information Theory, Inference and Learning Algorithms. Cambridge University Press, Cambridge (2003)
13. Neapolitan, R.E.: Learning Bayesian Networks, vol. 38. Pearson Prentice Hall, Upper Saddle River (2004)
14. Read, J., Pfahringer, B., Holmes, G., Frank, E.: Classifier chains for multi-label classification. Mach. Learn. **85**(3), 333–359 (2011)
15. Rijsbergen, C.J.V.: Information Retrieval, 2nd edn. Butterworth-Heinemann, Newton (1979)
16. Sasaki, Y.: The truth of the F-measure. Teach Tutor mater, p. 5 (2007)
17. Stekhoven, D.J., Bühlmann, P.: Missforest–non-parametric missing value imputation for mixed-type data. Bioinformatics **28**(1), 112–118 (2012)
18. Van Buuren, S.: Flexible Imputation of Missing Data. CRC Press, Boca Raton (2018)
19. Van Haaren, J., Davis, J.: Markov network structure learning: a randomized feature generation approach. In: Proceedings of the Twenty-Sixth AAAI Conference on Artificial Intelligence. AAAI Publications (2012)
20. Van Wolputte, E., Korneva, E., Blockeel, H.: MERCS: multi-directional ensembles of regression and classification trees. In: Proceedings of the Thirty-Second AAAI Conference on Artificial Intelligence, New Orleans, Louisiana, USA, pp. 4276–4283. AAAI Publications (2018)
21. Waljee, A.K., et al.: Comparison of imputation methods for missing laboratory data in medicine. BMJ Open **3**(8), e002847 (2013)

Multi-directional Rule Set Learning

Jonas Schouterden[1,2](✉), Jesse Davis[1,2], and Hendrik Blockeel[1,2]

[1] Department of Computer Science, KU Leuven,
Celestijnenlaan 200A, box 2402, 3001 Leuven, Belgium
{jonas.schouterden,jesse.davis,hendrik.blockeel}@cs.kuleuven.be
[2] Leuven.AI - KU Leuven Institute for AI, Leuven, Belgium

Abstract. A rule set is a type of classifier that, given attributes X, predicts a target Y. Its main advantage over other types of classifiers is its simplicity and interpretability. A practical challenge is that the end user of a rule set does not always know in advance which target will need to be predicted. One way to deal with this is to learn a multi-directional rule set, which can predict any attribute from all others. An individual rule in such a multi-directional rule set can have multiple targets in its head, and thus be used to predict any one of these. Compared to the naive approach of learning one rule set for each possible target and merging them, a multi-directional rule set containing multi-target rules is potentially smaller and more interpretable. Training a multi-directional rule set involves two key steps: generating candidate rules and selecting rules. However, the best way to tackle these steps remains an open question. In this paper, we investigate the effect of using Random Forests as candidate rule generators and propose two new approaches for selecting rules with multi-target heads: MIDS, a generalization of the recent single-target IDS approach, and RR, a new simple algorithm focusing only on predictive performance. Our experiments indicate that (1) using multi-target rules leads to smaller rule sets with a similar predictive performance, (2) using Forest-derived rules instead of association rules leads to rule sets of similar quality, and (3) RR outperforms MIDS, underlining the usefulness of simple selection objectives.

Keywords: Rule learning · Multi-directional models · Association rule mining · Decision trees

1 Introduction

Rule sets are classifiers predicting one target Y given attributes X. Their popularity stems from their simplicity and interpretability. A problem in practice is that a rule set's user might not know during training which attribute needs to be predicted. Examples of such cases are missing value imputation, where there are gaps in the data, or anomaly detection, where a value of a suspicious instance might be compared with a value representative of the training data. In such cases, the user would need to learn a separate rule set for each attribute.

A. Appice et al. (Eds.): DS 2020, LNAI 12323, pp. 517–532, 2020.
https://doi.org/10.1007/978-3-030-61527-7_34

Learning one rule set per attribute negatively impacts the collective interpretability, as the bodies of rules predicting correlated targets cannot be shared. If rules could predict multiple targets, the rule sets (1) might be more interpretable by using fewer rules (as a single rule can predict multiple targets at once (Sect. 3.2)), and (2) might explicate correlations between different targets.

While using multi-target rules might help, current rule set algorithms selecting a subset of rules $\mathcal{R}_{sel}$ out of a candidate rule set $\mathcal{R}_{cand}$ only work with single-target rules. To work with multi-target rules, they would need to simultaneously optimize the predictive performance for multiple targets.

Another problem is that as the candidate rule set $\mathcal{R}_{cand}$ typically consists of association rules, the user must set a confidence and support threshold in advance without knowing what the size or quality of the resulting rule set will be. Too low thresholds cause $\mathcal{R}_{cand}$ to become too large, potentially making both the rule set generation and rule set selection intractable. Too high thresholds may result in a small $\mathcal{R}_{cand}$ limiting the number of rules that can be selected, which might result in an selected subset $\mathcal{R}_{sel}$ of lesser quality. As association rule mining is often very sensitive to these thresholds, a small change in value might lead to candidate sets of widely varying sizes.

In summary, current rule set methods based on selecting a subset of candidate rules have the following problems: (1) they require the user to specify the target in advance, (2) they cannot select multi-target rules, and (3) they often use association rules, which are difficult to control in number and quality.

To address these problems, this paper investigates how to learn a *multi-directional rule set* able to predict any attribute given all other attributes, thus no longer requiring the user to specify the target in advance. We propose two multi-target rule selection approaches: a generalization of *Interpretable Decision Sets (IDS)* [11], and *RR*, a new algorithm focusing only on selecting a rule set with a high predictive performance for all targets. Finally, we propose to derive rule sets from Random Forests, as the number and size of trees in a Random Forest is easy to control, and they are learned to do prediction. Our experiments indicate that (1) using multi-target rules leads to smaller rule sets with a similar predictive performance, (2) using tree rules instead of association rules leads to rule sets of similar quality, and (3) RR outperforms MIDS, underlining the usefulness of simple selection objectives.

The rest of this paper is structured as follows. After Sect. 2 provides references to related work, Sects. 3 and 4 introduce the predictive settings, rule (set) representations and rule generation approaches used in this paper. Sections 5 and 6 describe RR and MIDS. An experimental evaluation is provided in Sect. 7, after which Sect. 8 gives a conclusion.

2 Related Work

Rule learning [5] can be divided into (1) predictive approaches for building classifiers, and (2) descriptive approaches for discover interesting patterns in data in the form of rules. These two groups are bridged by the LeGo framework [6], of which associative classifiers [2,12] are a prototypical instantiation. Associative classifiers are typically learned in three stages [6]. First, a set of candidate rules $\mathcal{R}_{cand}$ is mined from data [1]. Second, a subset of those rules $\mathcal{R}_{sel} \subseteq \mathcal{R}_{cand}$ is selected which optimizes some rule set objective. Third, the selected rules are combined to form a classifier. Different candidate rule generation and rule selection approaches can be combined, as they are often independent. CBA [12] is one of the oldest and best-known associative classifiers, selecting association rules based on their confidence. In this paper, we propose two multi-directional associative classifiers: MIDS and RR. MIDS generalizes the recent IDS [11] to support multi-target rules. RR is a a new algorithm. However, other multi-target classifiers exist [15]. Predictive clustering rules [16] is a coverage-based multi-target rule learning approach keeping a clear separation between descriptive and target attributes. Other examples are PGMs [10], which use a graph structure instead of logical rules, and MERCS [14] models, which use decision trees.

3 Preliminaries

In this section, we first introduce the single-target and multi-directional prediction settings used in this paper. Second, we define the representation of rules and rule sets. Third, we point out the necessity of tie-breaking strategies and default predictions in associative classification.

3.1 Predictive Settings

In the single-target setting, a learned model predicts a designated target attribute Y from m descriptive attributes $X_j \in \boldsymbol{X}$. Here, the training data $\mathcal{D} = \{(\boldsymbol{x}_i, y_i)\}_{i=1}^{N}$ contains N attribute-value examples. In a multi-directional setting, the target is not known in advance: the learned model must be able to predict any attribute given all other attributes. Here, the training set $\mathcal{D} = \{\boldsymbol{x}_i\}_{i=1}^{N}$ has m attributes X_j and no distinction is made between descriptive and target attributes. The value of attribute X_j for datapoint $\boldsymbol{x}$ is $\boldsymbol{x}[X_j]$.

3.2 Rule Set Representations

This paper considers rules of the form:

$$r = \mathit{body}(r) \rightarrow \mathit{head}(r) = b_1 \wedge \cdots \wedge b_{r_b} \rightarrow h_1 \wedge \cdots \wedge h_{r_h}$$

where each h_i and b_i is of the form $(X_j, \mathit{operator}, \mathit{value})$. Abusing notation, $\mathit{head}(r)$ and $\mathit{body}(r)$ denote both the set and conjunction of those literals, and $\mathit{length}(r) = |\mathit{head}(r)| + |\mathit{body}(r)|$ denotes the length of rule r. Using attr to denote the attributes in a head or body, $\mathit{attr}(\mathit{head}(r)) \cap \mathit{attr}(\mathit{body}(r)) = \emptyset$, and both $\mathit{head}(r)$ and $\mathit{body}(r)$ are not empty. A rule is single-target if $|\mathit{head}(r)| = 1$ and otherwise it is multi-target. All literals in the *head* use equality as the *operator*.

A single-target rule set consist of only single-target rules, but a multi-directional rule set may consist of either single-target rules (with different rules predicting different targets) or multi-target rules (where a single rule can predict multiple targets at once).

3.3 Tie Breaking Functions and Default Predictions

As the rules of an associative classifier might overlap, an instance might be covered by multiple rules. As a result, a tie-breaking strategy is necessary to get a single prediction. Different strategies exist, such as (weighted) voting, or only using the rule with the highest F1-score. Also, as the rules might not cover the whole instance space, a default prediction is necessary when no rule applies. A common choice is the attribute's mode in the training data.

4 Rule Generation

In this paper, a candidate rule set is generated with either association rule mining or decision tree ensembles. To mine association rules, each example $\boldsymbol{x} \in \mathcal{D}$ is transformed into a transaction containing m items of the form '$X_j = v_j$' with v_j in the domain of X_j ($v_j \in dom(X_j)$), on which frequent itemset mining can be used. As this requires categorical attributes, numerical attributes are discretized.

To derive a rule set from a tree ensemble, each tree is converted into its corresponding rule set [13]. Each rule corresponds to a path in a decision tree from the root to a leaf node. The rule's body consists of the tests in the inner nodes, while its head consists of the predictions in the leaf node.

5 RR: A Simple Multi-target Rule Selection Approach

This section proposes *RR*, a new algorithm that greedily selects multi-target rules from a candidate rule set. *RR* is purely based on maximizing the predictive performance of the resulting classifier.

Algorithm 1 outlines *RR*, which iteratively selects a rule increasing the F1-score of one target while limiting a possible score decrease on the other targets. Its input is a multi-target candidate rule set $\mathcal{R}_{cand}$. It starts with an empty set initial classifier $\mathcal{R}_{sel}$. *RR* adds rules to $\mathcal{R}_{sel}$ by selecting one rule at a time from $\mathcal{R}_{cand} \setminus \mathcal{R}_{sel}$. To select rules, the algorithm loops over the target attributes in a round-robin fashion (thus the name *RR*), focusing on each target in turn. When focusing on a target attribute X_j, *RR* must select a rule r_{sel} that increases the F1-score of the complete rule set $\mathcal{R}_{sel}$ for X_j. However, selecting a multi-target rule changes the F1-scores for all targets in the head of that rule. That is, adding a rule increasing the F1-score of the current target X_j might decrease the F1-scores of other targets $X_o \neq X_j$. To deal with this, *RR* first finds the rule r_{best} that if added to $\mathcal{R}_{sel}$ results in the largest F1-score increase for the current target X_j. Second, it finds all rules $\mathcal{R}_{X_j,\delta}$ that, when individually added to $\mathcal{R}_{sel}$, result in a F1-score that differs by at most δ from the F1-score for $\mathcal{R}_{sel} \cup \{r_{best}\}$ when predicting the current target X_j. Third, *RR* selects the rule from $\mathcal{R}_{X_j,\delta}$ that decreases the F1-scores on the other targets the least. That is, δ allows trading off selecting the better rule for the current target with the 'damage' done to other targets. *RR* stops if no rule can be found that increases the F1-score of $\mathcal{R}_{sel}$ on any target X_t by at least ϵ, to prevent overfitting. At the end, $\mathcal{R}_{sel}$ contains the rules to be used as classifier.

Algorithm 1. Round-Robin (RR). Note: $\mathcal{R} + r$ is short for $\mathcal{R} \cup \{r\}$.

Require:
$\mathcal{R}_{cand}$, the candidate rule set.
ϵ, the minimally required increase in F1-score when adding a rule.
δ, the maximum distance a rule can be to the best rule to be considered.
$score_i(\mathcal{R})$, the F1-score of rule set $\mathcal{R}$ for attribute X_i on the training data.

```
1:  procedure ROUND_ROBIN
2:      R_sel ← ∅
3:      while ∃ target X_t : select_rule_for(X_t, R_sel) ≠ None do
4:          X_j ← the next target in a round-robin fashion.
5:          r_sel ← select_rule_for(X_j, R_sel)
6:          if r_sel ≠ None then
7:              R_sel ← R_sel + r_sel
8:      return R_sel
9:  procedure SELECT_RULE_FOR(target X_j, R_sel)
10:     R_{X_j} ← {r ∈ R_cand \ R_sel | X_j ∈ head(r) ∧
                     score_j(R_sel + r) − score_j(R_sel) > ε}
11:     if R_{X_j} == ∅ then
12:         return None
13:     else
14:         r_best ← argmax_{r ∈ R_{X_j}} score_j(R_sel + r)
15:         R_{X_j,δ} ← {r ∈ R_{X_j} | score_j(R_sel + r_best) − score_j(R_sel + r) < δ}
16:         r_sel ← argmax_{r ∈ R_{X_j,δ}} ( min_{X_o ∈ head(r)\X_j} score_o(R_sel + r) − score_o(R_sel)))
17:         return r_sel
```

6 MIDS: Multi-target IDS

As a second multi-target rule selection approach, we propose *Multi-target Interpretable Decision Sets (MIDS)*, a generalization of *Interpretable Decision Sets (IDS)* [11]. We choose IDS as it is a recent single-target approach offering a high predictive performance and interpretability with a small rule set size. Section 6.1 introduces IDS on a high level. In Sect. 6.2, we generalize the IDS objective function to support multi-target rules.

6.1 IDS: (Single-target) Interpretable Decision Sets

First, IDS specifies to'ssociation rule set $\mathcal{R}_{cand}$ using Apriori, which we substitute for the more efficient FP-growth [7]. Second, IDS selects a subset $\mathcal{R}_{sel} \subseteq \mathcal{R}_{cand}$ that (locally) maximizes an objective function $f(\mathcal{R})$. The objective function is a weighted sum of several heuristics f_i indicating the rule set quality, such as the predictive performance, the size and the interpretability of the rule set:

$$\mathcal{R}_{sel} = \underset{\mathcal{R} \subseteq \mathcal{R}_{cand}}{\arg\max}\, f(\mathcal{R}) = \underset{\mathcal{R} \subseteq \mathcal{R}_{cand}}{\arg\max} \sum_{i=1}^{7} \lambda_i f_i(\mathcal{R}) \quad (1)$$

Section 6.2 explains the different sub-objectives f_i and how we generalize them to support multi-target rules.

The final subset $\mathcal{R}_{sel}$ is used for classification. IDS suggests using the rule with the highest F1-score as a tie-breaking strategy and to predict the majority class label in the training data as default prediction. However, the user is free to choose other tie-breaking and default prediction strategies.

Unconstrained Submodular Maximization. Finding the best subset $\mathcal{R}_{sel} \subseteq \mathcal{R}_{cand}$ that maximizes some objective function f (Eq. 1) corresponds to a combinatorial optimization problem. By formulating f as a non-negative non-normal unconstrained submodular maximization problem, a general algorithm for this problem type can be used for IDS. However, as maximizing an unconstrained submodular function is NP-hard [4], polynomial algorithms only guarantee to find a local optimum. Originally, IDS [11] specified to use the Smooth Local Search (SLS) algorithm [4]. However, as using SLS with IDS can be prohibitively slow [8], we choose to use the more recent Randomized Double Greedy Search algorithm [3], which is considerably faster and has better theoretical guarantees.

6.2 From IDS to MIDS: Adding Support for Multi-target Rules

To select a rule set $\mathcal{R}_{sel} \subseteq \mathcal{R}_{cand}$, IDS maximizes an objective function composed of 7 sub-objectives (see Eq. 1). The sub-objectives can be loosely divided into four groups, quantifying different aspects of a rule set $\mathcal{R}$. The first group focuses on rule set conciseness, the second on non-overlapping decision boundaries, the third on explaining as many attribute-values as possible, while the fourth group focuses on making accurate predictions. To trade off the importance of each of these aspects, the original work suggests that the weights λ_i can either be set by the user or be found using coordinate ascent.

Next, we modify the IDS sub-objectives in two ways[1]: we add (1) normalization and (2) support for multi-target rules. First, we normalize each of the sub-objectives to be in the interval $[0, 1]$. While the sub-objectives of the original IDS are non-negative, they do not have a clear upper bound. This might result in some sub-objectives dominating over others, but it also makes it difficult for the user to choose weights λ_i. When compared to the original specification, our normalization corresponds to multiplying the weight λ_i with a constant dependent on the candidate rule set $\mathcal{R}_{cand}$.

Second, we modify the IDS sub-objectives to support multi-target rules. Our generalization collapses to the original formulation when using single-target rules. To stay close to the original IDS specification, we do not further modify the sub-objectives, but indicate possible improvements as footnotes.

[1] Note that while IDS specifies it uses association rules, no modifications are necessary to support rules derived from decision trees. Any rule type for which the coverage and overlap can be calculated is supported.

Rule Set Conciseness. The first two sub-objectives f_1 and f_2 directly correspond to those of IDS, apart from the normalization. The first minimizes the number of rules selected from the candidate rule set, while the second minimizes the total number of literals in the rule set[2]:

$$f_1(\mathcal{R}) = 1 - \frac{|\mathcal{R}|}{|\mathcal{R}_{cand}|}$$

$$f_2(\mathcal{R}) = 1 - \frac{1}{L_{max} \cdot |\mathcal{R}_{cand}|} \sum_{r \in \mathcal{R}} length(r)$$

$$L_{max} = \max_{r \in \mathcal{R}_{cand}} length(r)$$

Non-overlapping Decision Boundaries. Two IDS sub-objectives f_3 and f_4 minimize the overlap of rules predicting a value for its target attribute Y. IDS assumes that a rule set with lower overlap is easier interpret, as fewer rules make predictions for a given example. While rule overlap in IDS is implicitly relative to its single target, we generalize the definition of rule overlap to be relative to a given target attribute X_j.

We define two rules r_1, r_2 to overlap relative to an attribute X_j if they share a covered example and both predict a value for attribute X_j:

$$cover(r) = \{\boldsymbol{x} \in \mathcal{D} \mid \boldsymbol{x} \models body(r)\}$$

$$overlap_j(r_1, r_2) = \begin{cases} cover(r_1) \cap cover(r_2) & \text{if } X_j \in attr(head(r_1)) \cap attr(head(r_2)) \\ \emptyset & \text{if } X_j \notin attr(head(r_1)) \cap attr(head(r_2)) \end{cases}$$

The goal of f_3 is to minimize the overlap of rules predicting the same value for a given target attribute. To generalize this to the multi-target case, we average over the m different targets, normalizing each contribution. Following the original IDS, $N \cdot |R_{cand,X_j}|^2$ is used as a simple upper bound for the maximal overlap relative to an attribute X_j given a training set of N instances[3]:

$$f_3(\mathcal{R}) = \frac{1}{m} \sum_{j=1}^{m} \Bigg[1 - \frac{1}{N \cdot |R_{cand,X_j}|^2} \sum_{\substack{r_k, r_l \in \mathcal{R} \\ k<l \\ (X_j = c_k) \in head(r_k) \\ (X_j = c_l) \in head(r_l) \\ c_k = c_l}} |overlap_j(r_k, r_l)|\Bigg]$$

$$R_{cand,X_j} = \{r \in \mathcal{R}_{cand} \mid X_j \in attr(head(r))\}$$

Sub-objective f_4 minimizes the overlap of rules predicting a *different* value for a given target attribute, which corresponds to substituting $c_k = c_l$ by $c_k \neq c_l$ when filtering the sum in the formulation of f_3 above.

[2] A better denominator is to use $\sum_{r \in \mathcal{R}_{cand}} length(r)$ instead of $L_{max} \cdot |\mathcal{R}_{cand}|$.

[3] A stricter upper bound is $\frac{N}{2} \cdot |R_{cand,X_j}| \cdot (|R_{cand,X_j}| - 1)$.

Predicting All Attribute-Values. IDS sub-objective f_5 formulates the assumption that a user wants for each value c in a target's domain $dom(Y)$ at least one rule that explains it. We generalize this for multi-target rules by averaging the normalized contributions for each different target:

$$f_5(\mathcal{R}) = \frac{1}{m}\sum_{j=1}^{m}\frac{1}{|dom(X_j)|}\sum_{c' \in dom(X_j)} \mathbb{1}\left[\exists r \in \mathcal{R} \mid (X_j = c') \in head(r)\right]$$

Predictive Performance. Two sub-objectives focus on the predictive performance of the rule set. To generalize these objectives to a multi-directional setting, we first define the (in)correct coverage of a rule as the set of (in)correctly classified examples relative to a given target attribute:

$$\begin{aligned} correct\text{-}cover_j(r) &= \{\boldsymbol{x} \in cover(r) \mid (X_j = c_j) \in head(r) \text{ and } \boldsymbol{x}[X_j] = c_j\} \\ incorrect\text{-}cover_j(r) &= \{\boldsymbol{x} \in cover(r) \mid (X_j = c_j) \in head(r) \text{ and } \boldsymbol{x}[X_j] \neq c_j\} \end{aligned}$$

Sub-objective f_6 prefers rules predicting few examples incorrectly. Its generalization to the multi-directional setting averages the number of mistakes each rule makes over that rule's target attributes:

$$f_6(\mathcal{R}) = 1 - \frac{1}{N \cdot |\mathcal{R}_{cand}|}\sum_{r \in \mathcal{R}} avg\text{-}incorrect\text{-}cover\text{-}size(r)$$

$$avg\text{-}incorrect\text{-}cover\text{-}size(r) = \frac{1}{|attr(head(r))|}\sum_{X_j \in attr(head(r))} |incorrect\text{-}cover_j(r)|$$

Sub-objective f_7 focuses on each attribute-value of each instance being correctly predicted by at least one rule:

$$f_7(\mathcal{R}) = \frac{1}{N \cdot m}\sum_{\boldsymbol{x} \in \mathcal{D}}\sum_{j=1}^{m} \mathbb{1}\left[|\{r \in \mathcal{R} \mid \boldsymbol{x} \in correct\text{-}cover_j(r)\}| \geq 1\right]$$

7 Experimental Evaluation

In this section, we use two questions to experimentally investigate whether our proposed rule generation and rule selection approaches can lead to better rule models than using single-target association rules.

First, **(Q1)** "do tree rules lead to better models than association rules?" To answer this, we compare association rules with tree rules in a single-target prediction setting using IDS as the single-target rule selection approach.

Second, **(Q2)** "does learning a multi-directional model from multi-target rules have advantages over using a collection of single-target rule models?" To answer this, we compare single-target and multi-target tree-derived rules in a multi-directional prediction setting using three rule selection methods: IDS for the single-target rules, RR and MIDS for the multi-target rules.

Our Python code is available on GitHub.[4]

[4] https://github.com/joschout/Multi-Directional-Rule-Set-Learning.

7.1 General Methodology

We use 7 UCI datasets, provided in discretized form by the arcBench benchmarking suite [9]: iris, diabetes, glass, segment, breast-w and vehicle. The discretization is required for association rule mining. We learn and evaluate all models using 10-fold cross validation. When comparing two models, we use a Wilcoxon signed-rank test with a significance level $\alpha = 0.05$.

We use the same tie-breaking and default prediction strategies for all models. As the tie-breaking strategy, we use weighted voting, where each rule gets a vote weighted by the rule's confidence in the training set. As a default prediction for each target attribute, we use its majority value in the training set.

For (M)IDS, we use as optimization algorithm 'Double Greedy Local Search' (unlike the original IDS; see Sect. 6.1). Since the optimization uses randomization to find a *locally* optimal rule set $\mathcal{R}_{sel}$, we run each (M)IDS configuration 10 times and pick the rule set with the highest objective function value. For both IDS and MIDS, we use the same implementation including normalization of the subobjectives with all weights set to $\lambda_i = 1$.

Compared Metrics. We investigate the predictive performance, model induction time, model size and interpretability of the selected rule models $\mathcal{R}_{sel}$.

The predictive performance of a rule set is measured with the micro-averaged F1-score. In the single-target setting, we measure the rule set's micro-averaged F1-score on the given target. In the multi-directional setting, we separately measure the micro-averaged F1-score on each target attribute and report the average.

To compare run time, we measure both the rule generation time and the rule selection time, the sum of which we call the total run time.

The model size of $\mathcal{R}_{sel}$ is indicated by three different metrics: (1) the number of literals in $\mathcal{R}_{sel}$ as the sum of its rule lengths, (2) its average rule length and (3) the number of rules in $\mathcal{R}_{sel}$.

Although the interpretability of a rule set is related to its model size, we also use three interpretability metrics as proposed for IDS [11]. First, we use $f_5(\mathcal{R})$ to measure the fraction of values occurring in the test data that are predicted by at least one rule (Sect. 6.2). Second, we consider the fraction of test set examples not covered by any rule. Third, we use the fraction of bodily overlap, which indicates how much the bodies of a rule set $\mathcal{R}$ overlap with respect to a test dataset of M instances, independent of the targets predicted by the rules:

$$\textit{fraction-bodily-overlap}(\mathcal{R}) = \frac{2}{|\mathcal{R}| \cdot (|\mathcal{R}| - 1)} \sum_{\substack{r_k, r_l \in \mathcal{R} \\ k<l}} \frac{|\textit{overlap}(r_k, r_l)|}{M}$$

7.2 Single-Target Tree Rules Vs. Association Rules

Methodology. To investigate **(Q1)** , we generate for both rule types a candidate single-target rule set of the same size. For both rule sets, we then use IDS to select a single-target model. We call IDS using association rules *AR-IDS*, and using tree rules *T-IDS*. After rule selection, we compare the resulting models.

The two candidate rule sets are generated in two steps to ensure they have the same size. First, we generate the association rules using FP-growth [7] for a given support and confidence (instead of Apriori; see Sect. 6.1). Second, we learn a Scikit-learn Random Forest by increasing its number of trees until it corresponds to a rule set with the same number of rules or more. If it generates more rules, we sample without replacement as many tree rules as there are association rules. For both approaches, we use a minimum support of 0.1 and a maximum rule length of 7. For the tree rules, this corresponds to setting a minimum fraction of examples per leaf node of 0.1 and a maximum tree depth of 7.

For each dataset, we use two different candidate rule set sizes. We obtain these rule set sizes by using two different minimum confidence levels for the association rules: 0.75 and 0.95. Using a higher confidence results in a smaller candidate rule set (Fig. 1). Limiting the number of candidate rules is important because the computational cost of the rule selection step increases with the candidate rule set size. Note that other metrics than the confidence can be used to limit the number of association rules [17].

Results.

Model size. For a given candidate set size, the rule models selected by AR-IDS and T-IDS do not significantly differ in their number of rules or their number of literals (Fig. 1). Which of the two approaches selects more rules or contains more literals differs over the datasets. However, the average rule length is significantly shorter for tree rules than for association rules. For high confidence levels, this is to be expected, as association rules with a higher confidence are typically longer. But the tree rules are also shorter for confidence 0.75.

Run time. The total run time seems independent of the rule type (as shown in Fig. 1). The rule generation time is negligible compared to the rule selection time, i.e. the time inducing an IDS model dominates over the rule generation time. Surprisingly, the time to generate association rules is not significantly different from the time to create tree rules. The rule selection time seems independent of the rule type, but increases with the candidate set size: selecting rules is much faster for confidence level 0.95 than for confidence level 0.75.

Predictive performance. While we expected tree rules to lead to more accurate models than association rules, AR-IDS does not differ in micro-averaged F1-score from T-IDS in a statistically significant way. Our experiments also suggest there is no difference between both candidate set sizes.

Interpretability. First, we see that rules selected by T-IDS have a significantly higher overlap than the rules selected by AR-IDS. Thus, tree rules require more tie-breaking. Second, while almost all examples in a test set are covered by the

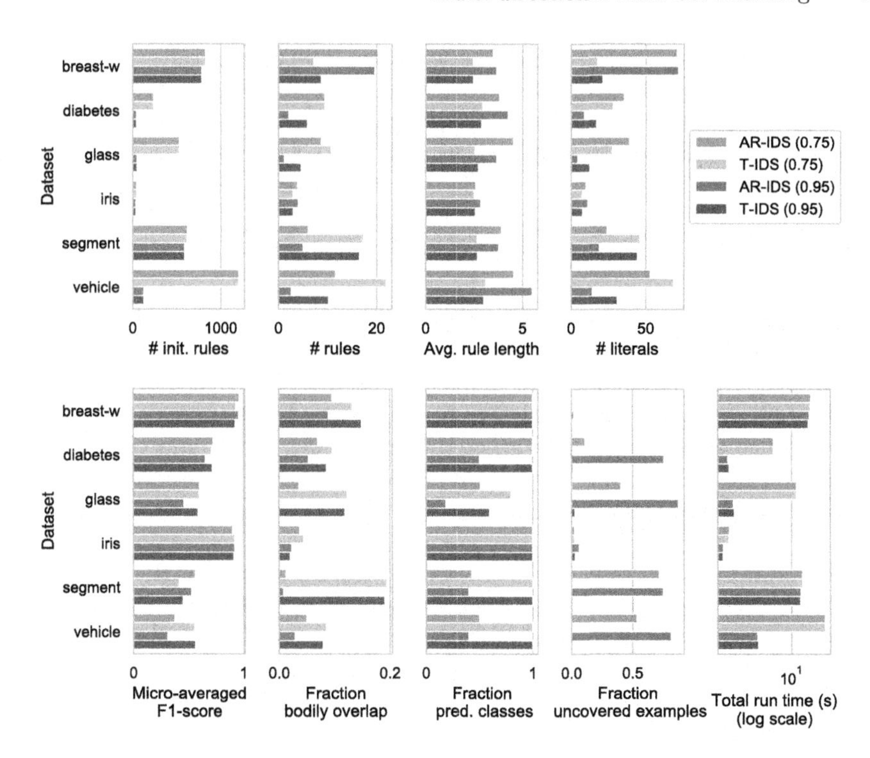

Fig. 1. Metrics quantifying the rule sets $\mathcal{R}_{sel}$ selected by IDS from single-target association rules and tree rules. For each dataset and rule type, two candidate rule set sizes were used by filtering the association rules on confidence 0.75 and 0.95.

T-IDS model, a large fraction is not covered by any rule in the AR-IDS model, thus requiring a default prediction without an explanation. Third, we see that T-IDS predicts more values in the target's domain than AR-IDS.

Discussion. Answering **(Q1)**, comparing T-IDS and AR-IDS indicates that tree and association rules lead to rule sets that do not significantly differ in predictive performance, model size and run time. However, they differ in interpretability. First, the T-IDS models explain more predictions than AR-IDS models, as they cover more instances; AR-models fall back on unexplained default predictions more frequently. Second, the explained predictions are less clear for the T-IDS models than for the AR-IDS models, since the larger overlap indicates more rules have to be interpreted for a prediction. Third, the T-IDS models predict more values than AR-IDS. A possible explanation for the difference in coverage and overlap of the selected rule sets can be found in the similar difference in the candidate rule sets. The candidate tree rules also have a high overlap and coverage, as every point in the instance space is covered by as many rules as there are trees in the corresponding ensemble. In contrast, the candidate association rules do not have to cover the whole instance space, even though they can overlap.

Thus, our results suggest that for interpretability, tree rules are preferred for explaining predictions for as many instances as possible, or for having more class values explained by at least one rule. But if it is acceptable that a rule model cannot always make a prediction and might have to use a default value, association rules can give clearer predictions for the instances that are covered.

7.3 Multi-target vs. Single-target Rules

To investigate **(Q2)**, we compare single-target and multi-target rules in a multi-directional setting. We use tree rules, as their number is easy to control and our previous experiment indicates that tree rules and association rules lead to models similar in size and predictive performance. For the single-target rules, we use IDS to select one single-target IDS model per target and combine these models in a multi-directional ensemble called *eIDS*. For the multi-target rules, we use RR and MIDS as rule selectors.

Rule generation. Both rules types are derived from Scikit-learn Random Forests using a minimum support of 0.1 and maximum rule length of 7.

For each attribute, a single-target candidate set is generated by (1) learning a Random Forest that predicts it, and (2) converting that Forest to rules. Each Random Forest contains 50 trees and has a maximum tree depth of 7.

One multi-target candidate set is constructed per dataset as follows. First, all attributes are randomly partitioned in groups of 2. For each group, a Random Forest of 10 trees is learned predicting those 2 attributes simultaneously. The attribute partitioning and Random Forest construction is repeated 5 times. As a result, each attribute is predicted by 5 Random Forests of 10 trees, or 50 trees in total. Then, one multi-target candidate set is generated for all target attributes by converting the trees of all Forest to rules. To ensure the rules have at most 7 literals, we use a maximum tree depth of 5 (as each tree predicts 2 targets).

Note that although each attribute is initially predicted by 50 trees in both the single-target and multi-target case, the number of rules predicting an attribute differs between the single-target and multi-target candidate rule sets. This results from each multi-target tree predicting 2 attributes. Thus, when combining the single-target rule sets, there are more candidate single-target rules than multi-target rules (*# init. rules* in Fig. 2).

Rule selection. From each single-target candidate set, we use IDS to select a model. These single-target models are combined in one multi-directional ensemble model per dataset, called *eIDS*.

For each multi-target rule set, we use two rule selectors: RR and MIDS. For both RR and MIDS, one model is learned per dataset. We use RR with the same tie-breaking function and default predictions as used for (M)IDS, i.e. weighted voting and the majority class label. We set $\epsilon = 0.1$ and $\delta = 0.01$. (Sect. 5)

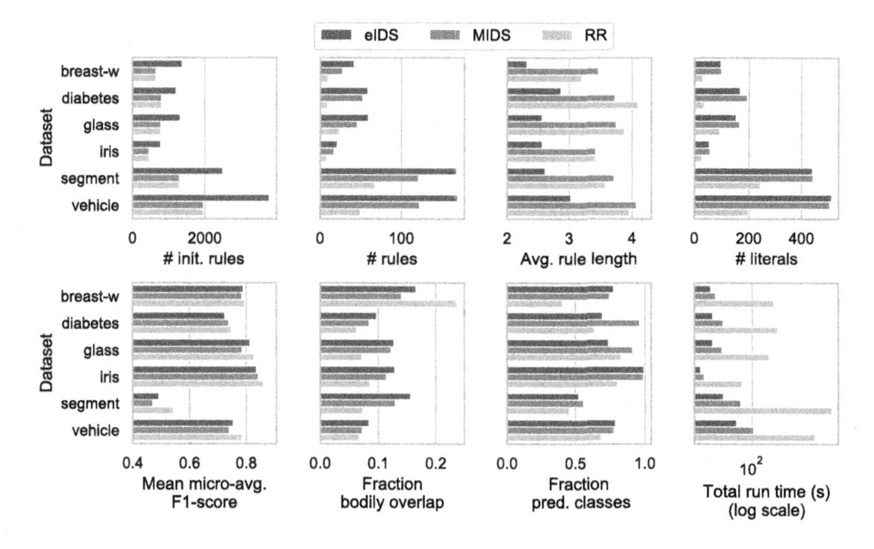

Fig. 2. Metrics quantifying the rule sets found using eIDS, MIDS and RR.

Results.

Run time. For all three approaches, the rule generation time is negligible compared to the rule selection time. When comparing rule selection time, RR is orders of magnitude slower than eIDS and MIDS. Learning one MIDS model takes more time than learning the eIDS model, which can be explained by IDS selecting from smaller candidate sets and having a simpler objective function.

Model size. RR results in a smaller model than both eIDS and MIDS (Fig. 2). The RR models contain significantly fewer literals than eIDS and MIDS, while the number of literals in the eIDS and MIDS models are similar.

When comparing the number of rules, the multi-target selection approaches result in the smallest rule sets. RR selects the smallest number of rules, while MIDS also selects significantly fewer rules than eIDS.

However, the multi-target selection approaches select significantly longer rules than eIDS. The average rule lengths of MIDS and RR are comparable, which can be expected, as they are built from the same candidate rule sets. The average rule lengths are longer for RR and MIDS than for eIDS, since the former use multi-target rules, whereas eIDS uses single-target rules.

Predictive performance. While RR outperforms both eIDS and MIDS in micro-averaged F1-score, the micro-averaged F1-scores of eIDS and MIDS do not differ in a statistically significant manner.

Interpretability. RR has a lower overlap than both MIDS and eIDS, while MIDS has a lower overlap than eIDS.

RR predicts fewer values occurring in the training data than eIDS and MIDS, between which there is no statistically significant difference.

As all three approaches cover almost all test instances with at least one rule, the fraction of uncovered instances is excluded from Fig. 2.

Discussion. Answering **(Q2)**, our results for MIDS and eIDS indicate that in a multi-directional setting, learning a single model using multi-target rules instead of naively learning multiple single-target models can lead to fewer rules and less overlap between rules, but a similar predictive performance. A possible explanation is that the selected multi-target rules explicate correlations between different targets, which cannot occur in an ensemble of single-target rule models.

Our results also indicate it is better to use RR than MIDS or eIDS in a multi-target prediction setting. Unsurprisingly, RR outperforms (M)IDS in micro-averaged F1-score, as this is the only focus of RR, while the composite (M)IDS objective function also focuses on model size and interpretability. However, RR also outperforms (M)IDS on model size and interpretability. Not only does RR select rule sets with fewer rules and literals than the (M)IDS rule sets, RR also has the lowest rule overlap of the three approaches. The only benefits of using eIDS or MIDS over RR is that they are faster and their resulting rule sets provide explanations for a larger variety of values. This highlights it is often better to use a simple rule selection objective. Although it might be possible to find parameters λ_i for (M)IDS resulting in a similar model size and predictive performance as the RR models, this would require a potentially expensive hyperparameter optimization.

8 Conclusion

In this paper, we proposed how to train a multi-directional rule set based on multi-target tree rules, as a user might not know in advance which target will need to be predicted or which support and confidence thresholds to use with association rule mining. We proposed two new methods able to select multi-target rules: the greedy RR, focused on providing a high predictive performance on all targets, and MIDS, a generalization of IDS. Our experiments indicate that tree and association rules lead to models of similar size and predictive performance, although with different interpretability characteristics. Tree rules lead to models with a higher coverage, but association rules lead to clearer decision boundaries. We also showed that, compared to naively merging a collection of single-target rule models, using a multi-directional model built using multi-target rules results in fewer rules with lower overlap but with a similar predictive performance. Lastly, the usefulness of simple objective functions was demonstrated, as our RR models were not only more accurate than IDS and MIDS, they were also smaller with a lower overlap.

Future work. While we compared single-target association and tree rules in the context of IDS, a similar comparison using other rule selection methods would be useful. Similarly, comparing RR and MIDS with other single-target rule selectors can help position these methods more clearly. Also, it would be interesting to generalize other rule selectors than IDS to handle multi-target rules.

Acknowledgments. This research received funding from the KU Leuven Research Fund (C14/17/070, "SIRV") and the Flemish Government under the "Onderzoeksprogramma Artificiële Intelligentie (AI) Vlaanderen" programme.

References

1. Borgelt, C.: Frequent item set mining. Wiley Interdisc. Rev. Data Min. Knowl. Discov. **2**(6), 437–456 (2012)
2. Bringmann, B., Nijssen, S., Zimmermann, A.: Pattern-Based Classification: A Unifying Perspective (2011)
3. Buchbinder, N., Feldman, M., Naor, J.S., Schwartz, R.: A tight linear time (1/2)-approximation for unconstrained submodular maximization. SIAM J. Comput. **44**(5), 1384–1402 (2015)
4. Feige, U., Mirrokni, V.S., Vondrák, J.: Maximizing non-monotone submodular functions. SIAM J. Comput. **40**(4), 1133–1153 (2011)
5. Fürnkranz, J., Gamberger, D., Lavrač, N.: Foundations of Rule Learning. Cognitive Technologies, p. XVIII, 334. Springer, Berlin, Heidelberg (2014). https://doi.org/10.1007/978-3-540-75197-7
6. Fürnkranz, J., Knobbe, A.: Guest editorial: global modeling using local patterns. Data Min. Knowl. Discov. **21**(1), 1–8 (2010)
7. Han, J., Pei, J., Yin, Y.: Mining frequent patterns without candidate generation. SIGMOD Rec. **29**(2), 1–12 (2000)
8. Ignatiev, A., Pereira, F., Narodytska, N., Marques-Silva, J.: A SAT-based approach to learn explainable decision sets. In: Galmiche, D., Schulz, S., Sebastiani, R. (eds.) IJCAR 2018, vol. 10900, pp. 627–645. Springer, Cham (2018). https://doi.org/10.1007/978-3-319-94205-6_41
9. Kliegr, T.: Quantitative CBA: Small and Comprehensible Association Rule Classification Models, pp. 1–24 (2017)
10. Koller, D., Friedman, N.: Probabilistic Graphical Models: Principles and Techniques. Adaptive Computation and Machine Learning. The MIT Press, Cambridge (2009)
11. Lakkaraju, H., Bach, S.H., Leskovec, J.: Interpretable decision sets: a joint framework for description and prediction. In: 22nd International Conference on Knowledge Discovery and Data Mining. KDD'16, pp. 1675–1684. ACM (2016)
12. Liu, B., Hsu, W., Ma, Y.: Integrating classification and association rule mining. In: Agrawal, R., Stolorz, P.E., Piatetsky-Shapiro, G. (eds.) 4th International Conference on Knowledge Discovery and Data Mining. KDD'98, pp. 80–86. AAAI Press, New York (1998)
13. Quinlan, J.R.: Generating production rules from decision trees. In: McDermott, J.P. (ed.) 10th International Joint Conference on Artificial Intelligence, pp. 304–307. Morgan Kaufmann, Los Altos (1987)
14. Van Wolputte, E., Korneva, E., Blockeel, H.: MERCS: multi-directional ensembles of regression and classification trees. In: 32nd AAAI Conference on Artificial Intelligence, pp. 4276–4283 (2018)

15. Waegeman, W., Dembczyński, K., Hüllermeier, E.: Multi-target prediction: a unifying view on problems and methods. Data Min. Knowl. Discov. **33**(2), 293–324 (2018). https://doi.org/10.1007/s10618-018-0595-5
16. Ženko, B., Džeroski, S.: Learning classification rules for multiple target attributes. In: Washio, T., Suzuki, E., Ting, K.M., Inokuchi, A. (eds.) PAKDD 2008. LNCS, vol. 5012, pp. 454–465. Springer, Berlin, Heidelberg (2008). https://doi.org/10.1007/978-3-540-68125-0_40
17. Zimmermann, A., De Raedt, L.: CorClass: correlated association rule mining for classification. In: Suzuki, E., Arikawa, S. (eds.) DS 2004. Lecture Notes in Computer Science, vol. 3245, pp. 60–72. Springer, Heidelberg (2004). https://doi.org/10.1007/978-3-540-30214-8_5

On Aggregation in Ensembles of Multilabel Classifiers

Vu-Linh Nguyen[1(✉)], Eyke Hüllermeier[1], Michael Rapp[2], Eneldo Loza Mencía[2], and Johannes Fürnkranz[3]

[1] Heinz Nixdorf Institute and Department of Computer Science, Paderborn University, Paderborn, Germany
vu.linh.nguyen@uni-paderborn.de, eyke@upb.de
[2] Knowledge Engineering Group, TU Darmstadt, Darmstadt, Germany
{mrapp,eneldo}@ke.tu-darmstadt.de
[3] Computational Data Analytics Group, JKU Linz, Linz, Austria
juffi@faw.jku.at

Abstract. While a variety of ensemble methods for multilabel classification have been proposed in the literature, the question of how to aggregate the predictions of the individual members of the ensemble has received little attention so far. In this paper, we introduce a formal framework of ensemble multilabel classification, in which we distinguish two principal approaches: "predict then combine" (PTC), where the ensemble members first make loss minimizing predictions which are subsequently combined, and "combine then predict" (CTP), which first aggregates information such as marginal label probabilities from the individual ensemble members, and then derives a prediction from this aggregation. While both approaches generalize voting techniques commonly used for multilabel ensembles, they allow to explicitly take the target performance measure into account. Therefore, concrete instantiations of CTP and PTC can be tailored to concrete loss functions. Experimentally, we show that standard voting techniques are indeed outperformed by suitable instantiations of CTP and PTC, and provide some evidence that CTP performs well for decomposable loss functions, whereas PTC is the better choice for non-decomposable losses.

Keywords: Ensembles of multilabel classifiers · Predict then combine · Combine then predict · Hamming loss · F-measure · Subset 0/1 loss

1 Introduction

The setting of *multilabel classification* (MLC), which generalizes standard multi-class classification by relaxing the assumption of mutual exclusiveness of classes, has received a lot of attention in the recent machine learning literature—we refer to [21] and [23] for comprehensive survey articles on this topic.

Like for other types of classification problems, the idea of *ensemble learning* [5] has also been applied for MLC (cf. Sect. 3). However, somewhat surprisingly, the question of how to aggregate the predictions of the individual members of an

A. Appice et al. (Eds.): DS 2020, LNAI 12323, pp. 533–547, 2020.
https://doi.org/10.1007/978-3-030-61527-7_35

ensemble has so far received little attention in MLC. Instead, most approaches are based on simple voting techniques, which are typically applied in a label-wise manner: For each label, the predictions—either binary predictions of relevance or, more generally, label probabilities—of all ensemble members are collected, averaged, and thresholded to obtain a final prediction for this label.

An obvious disadvantage of this simple approach is that the aggregation is independent of the underlying performance measure, i.e., the aggregation procedure is not tailored to a specific loss function. This, however, would supposedly be important: In contrast to standard classification, where a loss function compares a predicted class label with a ground truth, an MLC loss compares a *subset* of labels predicted to be relevant with a ground-truth subset. As there are various ways in which subsets can be compared with each other, a wide spectrum of loss functions is commonly used in MLC, and it is well known that different losses may call for different (Bayes-optimal) predictions [3,4]. Naturally, the idea of customizing an MLC predictor to a specific loss function should not only be considered at the level of individual predictors, but also at the level of the ensemble as a whole, and hence also concern the way in which the predictions are combined.

In this paper, we study the problem of aggregation in ensembles of multi-label classifiers (EMLC) in a systematic way. To this end, we introduce a formal framework, in which we distinguish two principal approaches: "predict then combine" (PTC), where the ensemble members first make loss minimizing predictions which are then combined, and "combine then predict" (CTP), which first aggregates information such as marginal label probabilities from the individual ensemble members, and then derives a prediction from this aggregation. While both approaches generalize common voting techniques as mentioned above, they also include more general variants and, moreover, allow one to explicitly take the target loss into account. In other words, concrete instantiations of CTP and PTC can be tailored to concrete loss functions. In an extensive experimental study, we demonstrate that such loss-based aggregation functions do indeed outperform simple voting techniques, and also investigate the question which type of aggregation is more suitable for which loss functions.

2 Multilabel Classification

Let $\mathcal{X}$ denote an instance space, and let $\mathcal{L} = \{\lambda_1, \ldots, \lambda_K\}$ be a finite set of class labels. We assume that an instance $\boldsymbol{x} \in \mathcal{X}$ is (probabilistically) associated with a subset of labels $\Lambda = \Lambda(\boldsymbol{x}) \in 2^{\mathcal{L}}$; this subset is often called the set of relevant labels, while the complement $\mathcal{L} \setminus \Lambda$ is considered as irrelevant for $\boldsymbol{x}$. We identify a set Λ of relevant labels with a binary vector $\boldsymbol{y} = (y_1, \ldots, y_K)$, where $y_k = [\![\lambda_k \in \Lambda]\!]$.[1] By $\mathcal{Y} = \{0,1\}^K$ we denote the set of possible labelings.

We assume observations to be realizations of random variables generated independently and identically according to a probability distribution $\boldsymbol{p}$ on $\mathcal{X} \times \mathcal{Y}$, i.e., an observation $\boldsymbol{y} = (y_1, \ldots, y_K)$ is the realization of a corresponding random vector $\mathbf{Y} = (Y_1, \ldots, Y_K)$. We denote by $\boldsymbol{p}(\mathbf{Y} \mid \boldsymbol{x})$ the conditional distribution of

[1] $[\![\cdot]\!]$ is the indicator function, i.e., $[\![A]\!] = 1$ if the predicate A is true and $= 0$ otherwise.

$\mathbf{Y}$ given $\mathbf{X} = \boldsymbol{x}$, and by $\boldsymbol{p}_k(Y_k \,|\, \boldsymbol{x})$ the corresponding marginal distribution of Y_k:

$$\boldsymbol{p}_k(b \,|\, \boldsymbol{x}) = \sum_{\boldsymbol{y} \in \mathcal{Y}: y_k = b} \boldsymbol{p}(\boldsymbol{y} \mid \boldsymbol{x}) \,. \tag{1}$$

Moreover, we denote by $p_k = p_k(\boldsymbol{x}) = \boldsymbol{p}_k(1 \,|\, \boldsymbol{x})$ the probability of relevance of the label λ_k.

Given training data in the form of a finite set of observations

$$\mathcal{D} = \left\{(\boldsymbol{x}_n, \boldsymbol{y}_n)\right\}_{n=1}^{N} \subset \mathcal{X} \times \mathcal{Y} \,, \tag{2}$$

drawn independently from $\boldsymbol{p}(\mathbf{X}, \mathbf{Y})$, the goal in MLC is to learn a predictive model in the form of a multilabel classifier $\boldsymbol{h}$, which is a mapping $\mathcal{X} \longrightarrow \mathcal{Y}$ that assigns a (predicted) label subset to each instance $\boldsymbol{x} \in \mathcal{X}$. Thus, the output of a classifier $\boldsymbol{h}$ is a vector of predictions

$$\boldsymbol{h}(\boldsymbol{x}) = (h_1(\boldsymbol{x}), \ldots, h_K(\boldsymbol{x})) \in \{0,1\}^K \,, \tag{3}$$

also denoted as $\hat{\boldsymbol{y}} = (\hat{y}_1, \ldots, \hat{y}_K)$.

2.1 MLC Loss Functions

The main goal in MLC is to induce predictions (3) that generalize well beyond the training data (2), i.e., predictions

$$\hat{\boldsymbol{y}} = \operatorname*{argmin}_{\bar{\boldsymbol{y}}} \sum_{\boldsymbol{y} \in \mathcal{Y}} \ell(\boldsymbol{y}, \bar{\boldsymbol{y}}) \, \boldsymbol{p}(\boldsymbol{y} \mid \boldsymbol{x}) \,, \tag{4}$$

that minimize the expected loss with respect to a specific MLC loss function $\ell : \mathcal{Y}^2 \longrightarrow \mathbb{R}$. Two important loss functions, both generalizing the standard 0/1 loss commonly used in classification, are the *Hamming loss* and the *subset 0/1 loss*:

$$\ell_H(\boldsymbol{y}, \hat{\boldsymbol{y}}) := \frac{1}{K} \sum_{k=1}^{K} [\![y_k \neq \hat{y}_k]\!] \,, \tag{5}$$

$$\ell_S(\boldsymbol{y}, \hat{\boldsymbol{y}}) := [\![\boldsymbol{y} \neq \hat{\boldsymbol{y}}]\!] \,. \tag{6}$$

The (*instance-wise*) *F-measure* compares a set of predicted labels to a corresponding set of ground-truth labels via the harmonic mean of precision and recall:

$$F(\boldsymbol{y}, \hat{\boldsymbol{y}}) = \frac{2 \sum_{k=1}^{K} \hat{y}_k \, y_k}{\sum_{k=1}^{K} \hat{y}_k + \sum_{k=1}^{K} y_k} \,. \tag{7}$$

The goal of classification algorithms in general is to capture dependencies between input features and the target variable. In MLC, dependencies may not only exist between the features and each target, but also between the targets $Y_1, \ldots, Y_K$ themselves. The idea to improve predictive accuracy by capturing such dependencies is a driving force in research on multilabel classification.

Not all loss functions capture label dependencies to the same extent: A *decomposable loss* can be reduced to loss functions for the individual labels, i.e., it can be expressed in the form

$$\ell(\boldsymbol{y}, \hat{\boldsymbol{y}}) = \sum_{k=1}^{K} \ell_k(y_k, \hat{y}_k), \tag{8}$$

with suitable binary loss functions $\ell_k : \{0,1\}^2 \longrightarrow \mathbb{R}$. A *non-decomposable loss* does not permit such a representation. It can be shown that, for making optimal predictions $\hat{\boldsymbol{y}} = \boldsymbol{h}(\boldsymbol{x})$ which minimize the expected loss, knowledge about the marginals (1) is sufficient in the case of a decomposable loss (such as Hamming), but not in the case of a non-decomposable loss [3]. Instead, if a loss is non-decomposable, higher-order probabilities are needed, and in the extreme case even the entire distribution $\boldsymbol{p}(\mathbf{Y} \mid \boldsymbol{x})$ (like in the case of the subset 0/1 loss).

On an algorithmic level, this means that MLC with a decomposable loss can be tackled by what is commonly called binary relevance (BR) learning, i.e., by learning one binary classifier for each individual label, whereas non-decomposable losses call for more sophisticated learning methods that are able to take label dependencies into account.

2.2 Risk Minimization

In the most general case, the problem of finding a risk-minimizing (Bayes-optimal) prediction is tackled by producing a prediction $\boldsymbol{p}(\cdot \mid \boldsymbol{x})$ of the conditional joint distribution of labelings, and explicitly solving (4) as a combinatorial optimization problem. Obviously, this approach is infeasible unless the number of class labels is very low. Fortunately, the problem can be solved more efficiently for specific loss functions, including those considered in this paper.

In the case of the *Hamming loss*, the Bayes-optimal prediction can be obtained by thresholding the marginal probabilities, regardless of whether the labels are independent or not:

$$\hat{y}_k = [\![p_k(\boldsymbol{x}) > 1/2]\!]. \tag{9}$$

Thus, it is sufficient to have good estimates for the marginal probabilities, which can be accomplished by simple techniques such as binary relevance [3].

For *subset 0/1 loss*, the Bayes-optimal prediction is not the marginal but the *joint* mode of the distribution $\boldsymbol{p}(\cdot \mid \boldsymbol{x})$:

$$\hat{\boldsymbol{y}} \in \operatorname*{argmax}_{\bar{\boldsymbol{y}} \in \mathcal{Y}} \boldsymbol{p}(\bar{\boldsymbol{y}} \mid \boldsymbol{x}).$$

Thus, label dependence needs to be taken into account for optimal performance.

The *F-measure* is in a sense in-between these two extremes. It can be shown that, while the entire distribution $\boldsymbol{p}(\cdot \mid \boldsymbol{x})$ is not needed to find a Bayes-optimal prediction for this measure, marginal probabilities (1) do not suffice either. Instead, probabilities on pairwise label combinations are required in the general case, whereas under the assumption of conditional label independence, marginal probabilities again provide sufficient information [22].

3 Ensembles of MLC

In general, an ensemble approach to multilabel classification (EMLC) learns a set of M multilabel classifiers, each of which predicts a binary label vector $\hat{\boldsymbol{y}}_j$. Given a query instance $\boldsymbol{x} \in \mathcal{X}$, these are then aggregated into a final prediction $\hat{\boldsymbol{y}} = \mathrm{agg}(\hat{\boldsymbol{y}}_1, \ldots, \hat{\boldsymbol{y}}_M)$. For this aggregation, variants of label-wise *majority voting (MV)* are typically used:

- **Binary majority voting (BMV)** assigns to each label $\lambda_k \in \mathcal{L}$ the prediction given by the majority of the classifiers:

$$\hat{y}_k := \operatorname*{argmax}_{y_k \in \{0,1\}} \sum_{j=1}^{M} [\![y_k = \hat{y}_{j,k}]\!] . \tag{10}$$

- **Graded majority voting (GMV)**, also known as *weighted voting*, adds up confidence scores $\boldsymbol{p}_j = (p_{j,1}, p_{j,2}, \ldots, p_{j,K})$ for each label $\lambda_k \in \mathcal{L}$:

$$\hat{y}_k := \operatorname*{argmax}_{y_k \in \{0,1\}} \sum_{j=1}^{M} p_{j,k}^{y_k} (1 - p_{j,k})^{1-y_k} . \tag{11}$$

Several ensemble-based multilabel classifiers have been tried in the literature, which typically use the above-mentioned voting techniques for combining the predictions of the ensemble members [6,7,10,18,19]. While we aim at optimizing the predictions for a particular loss function, a different line of work—orthogonal to our approach—aims at simultaneously optimizing for multiple loss functions [16,17]. In the following, we briefly recall some commonly used EMLC methods, which will serve as baselines in our experimental evaluation. We refer to [11] for an extensive discussion on ensembles of MLC classifiers.

- **Ensembles of Binary Relevance Classifiers (EBR)** use bagging [1] to construct K independent ensembles of binary classifiers, one for each label $\lambda_k \in \mathcal{L}$ [20]. At prediction time, the predictions of these classifiers are combined for each label using majority voting, as is commonly used in bagged ensembles. Obviously, like all BR methods, EBR ignores any relationships between the labels and implicitly assumes them to be independent. Moreover, EBR is computationally expensive, since $K \cdot M$ classifiers are required in order to have an "actual ensemble" of cardinality M.
- **Ensembles of Classifier Chains (ECC).** The classifier chains (CC) method [14] also trains K binary classifiers h_k, $k \in [K] := \{1, \ldots, K\}$, one for each label. Yet, to capture label dependencies, h_k is trained on an augmented input space $\mathcal{X} \times \{0,1\}^{k-1}$, taking the (binary) values of the $k-1$ previous labels as additional attributes. More specifically, h_k predicts $\hat{y}_{\sigma(k)} \in \{0,1\}$ using

$$\left(\boldsymbol{x}, \hat{y}_{\sigma(1)}, \hat{y}_{\sigma(2)}, \ldots, \hat{y}_{\sigma(k-1)}\right) \in \mathcal{X} \times \{0,1\}^{k-1}$$

as input, where σ is some permutation of $[K]$.

Practically, it turns out that the order of labels on the chain, defined by σ, has an impact on predictive performance [2,15]. As finding an optimal order appears to be difficult, [15] suggest to use an ensemble of CCs over a (randomly chosen) set of permutations and combine their predictions. In the original CC, the final prediction is derived in a label-wise manner using BMV. In a probabilistic variant of CC, we allow each classifier h_k, $k \in [K]$, to produce a score in $[0,1]$, namely an estimation of the conditional probability

$$\boldsymbol{p}\left(y_{\sigma(k)} = 1 \mid \boldsymbol{x}, y_{\sigma(1)}, \ldots, y_{\sigma(k-1)}\right) . \tag{12}$$

(12) can be seen as a *dependent* marginal probability, i.e., a marginal probability which to some extent takes label dependence into account.

- **Ensembles of Multi-Objective Decision Trees (EMODT)** are a computationally efficient EMLC method [8]. Similar to conventional decision trees (DT) [12,13], a multi-objective decision tree (MODT) partitions the instance space $\mathcal{X}$ into (axis-parallel) regions $R_1, \ldots, R_L$ (i.e., $\bigcup_{i=1}^{L} R_i = \mathcal{X}$ and $R_i \cap R_j = \emptyset$ for $i \neq j$), corresponding to individual leaves of the tree. In a probabilistic setting, each leaf of the MODT is associated with a complete marginal probability vector, where the marginal probability corresponding to a particular label is simply estimated as the proportion of the training instances in the leaf for which the label is relevant. The binary label vector predicted by an EMODT can be derived with GMV on the probability vectors provided by the individual MODTs in a label-wise manner. Due to the label-wise voting, EMODT is also tailored to decomposable performance measures.

4 A Formal Framework

In the following, we define a formal framework for EMLC.

4.1 Intermediate Relevance Information

Most MLC methods are two-step approaches in the sense that, prior to making a final prediction $\hat{\boldsymbol{y}} \in \mathcal{Y}$, intermediate results about the relevance of labels, their interdependencies, or similar information is compiled. We refer to such results as *relevance information*, which we distinguish from the final prediction. Important examples include the following:

- Estimates of *marginal probabilities* (1), which provide important information for the minimization of decomposable loss functions, or loss minimization in the case of label independence.
- The entire *joint distribution* $\boldsymbol{p}(\cdot \mid \boldsymbol{x})$, which might be needed for the minimization of non-decomposable losses in cases where the labels are not independent.
- *Probability estimates of a more general kind.* For example, [22] require the probabilities $\boldsymbol{p}(y_k = 1, s_{\boldsymbol{y}} = s)\,, k, s \in [K]$, for loss minimization in the case of the F-measure.

In general, of course, the relevance information does not need to be probabilistic, but might be of a more general nature.

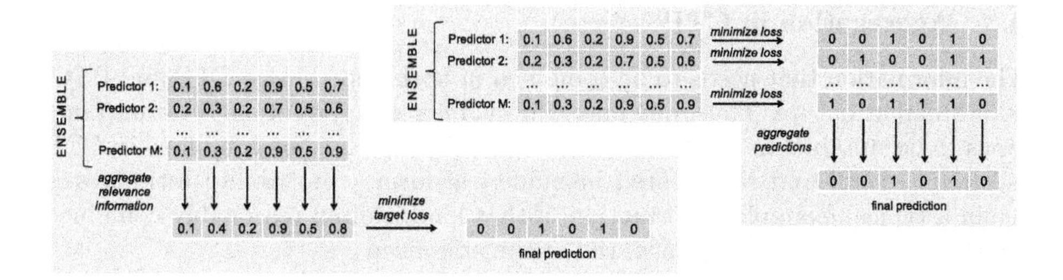

Fig. 1. Illustration of the "combine then predict" (left) and "predict then combine" (right) approaches for the case where relevance information consists of marginal probabilities.

4.2 CTP Versus PTC

In the context of ensemble learning, an important distinction between methods can be made depending on whether the relevance information provided by the different ensemble members is combined first, and a prediction is obtained afterwards, or whether individual predictions are produced first and then combined into an overall prediction (see Fig. 1 for an illustration). We refer to the former as "combine then predict" (CTP) and the latter as "predict then combine" (PTC).

In CTP, the relevance information $\mathcal{R} = \{R_1, \ldots, R_M\}$ provided by the individual ensemble members is first combined into a single condensed representation

$$R = \text{CTP-agg}\left(R_1, \ldots, R_M\right). \tag{13}$$

Then, a final prediction $\hat{\boldsymbol{y}}$ is produced on the basis of this representation, typically (though not necessarily) taking the underlying target loss ℓ into account, i.e., minimizing expected loss with regard to ℓ (cf. Sect. 2.2). Denoting the prediction step by Pred_ℓ, this can be written compactly as follows:

$$\hat{\boldsymbol{y}} = \text{Pred}_\ell\Big(\text{CTP-agg}\left(R_1, \ldots, R_M\right)\Big). \tag{14}$$

In PTC, each member of the ensemble first predicts a (loss minimizing) label combination $\hat{\boldsymbol{y}}_j = \text{Pred}_\ell(R_j)$. Then, in a second step, these predictions $\hat{\boldsymbol{y}}_m$, $m \in [M]$, are combined into an overall prediction $\hat{\boldsymbol{y}}$ using a suitable aggregation function:

$$\hat{\boldsymbol{y}} = \text{PTC-agg}\Big(\text{Pred}_\ell(R_1), \ldots, \text{Pred}_\ell(R_M)\Big).$$

Note that the commonly used techniques of weighted and binary voting as described in Sect. 3 can be seen as specific instantiations of CTP and PTC: Binary majority voting (BMV) first maps vectors of marginal label probabilities into label predictions, which are then combined via majority voting, and is thus an instance of PTC. Graded majority voting (GMV) first adds up the label probabilities into a single vector of marginal label probabilities, which are then thresholded for a final prediction, and is thus a special case of CTP. However, both voting methods are oblivious to specific loss functions.

4.3 Aggregation in CTP

The information that needs to be combined in both approaches, CTP and PTC, is of different nature. Thus, one may expect different types of aggregation functions to be suitable. In particular, relevance information to be combined in CTP is often *gradual* and represented in numerical form—probability estimates is again a typical example. Information of that kind is often reasonably combined through *averaging*. For instance, the arithmetic mean

$$\bar{p}_i = \frac{1}{M} \sum_{j=1}^{M} p_{i,j} \,, \tag{15}$$

produced by the ensemble members for the label λ_i, will be an improved estimate of the true marginal probability of that label. Of course, aggregation functions other than the arithmetic mean are also conceivable; for example, the median is known to be more robust toward outliers.

Moreover, aggregation does not necessarily need to be label-wise as in (15). Instead, it depends on what kind of relevance information is produced in the first place. Imagine, for example, that each ensemble member yields an estimate $\hat{\boldsymbol{p}}_j(\cdot \,|\, \boldsymbol{x})$ of the joint label distribution on $\mathcal{Y}$. Aggregation should then be done at the same level, and averaging is again an obvious way for doing so:

$$\hat{\boldsymbol{p}}(\boldsymbol{y} \,|\, \boldsymbol{x}) = \frac{1}{M} \sum_{j=1}^{M} \hat{\boldsymbol{p}}_j(\boldsymbol{y} \,|\, \boldsymbol{x}) \,, \forall \boldsymbol{y} \in \mathcal{Y} \,.$$

As already said, an approach of that kind might be advantageous in the case of non-decomposable losses, although it will not be tractable in general.

4.4 Aggregation in PTC

In PTC, the problem is to combine (binary) predictions. More specifically, recalling the goal to minimize a given target loss ℓ, the problem can be stated as follows: Given predictions $\hat{\boldsymbol{y}}_1, \ldots, \hat{\boldsymbol{y}}_M$, which are all supposed to minimize ℓ in expectation, what is a Bayes-optimal overall prediction $\hat{\boldsymbol{y}}$? The answer to this question is far from obvious and, to the best of our knowledge, has not been studied systematically in the literature so far. In fact, a formal analysis of this problem probably presupposes additional assumptions about how the predictions (15) may differ from the true Bayes-optimal prediction (obviously, they cannot all be Bayes-optimal at the same time, unless they all coincide).

In any case, it should be clear that averaging will be less suitable. First of all, binary predictions $\hat{\boldsymbol{y}}$ are *discrete* entities, and by averaging them one does not again end up with a discrete entity. This is to some extent comparable to the difference between ensemble *regression* (numerical case) and ensemble *classification* (categorical case): While arithmetic averaging is often used in the former, counting or "voting" techniques are more commonly applied in the latter. Second, even when solving this technical issue by turning an average into a discrete entity, for example by thresholding, undesirable effects might be produced, as shown by a simple example, in which the (conditional) ground-truth distribution $\boldsymbol{p}(\cdot \mid \boldsymbol{x})$ on the label space $\mathcal{Y} = \{0,1\}^3$ are given as follows:

$\boldsymbol{y}$	$(0,0,0)$	$(1,1,1)$	$(0,1,1)$	$(1,0,1)$	$(1,1,0)$
$\boldsymbol{p}(\boldsymbol{y} \mid \boldsymbol{x})$	$1/4$	$3/16$	$3/16$	$3/16$	$3/16$

Obviously, the Bayes-optimal prediction for the subset 0/1 loss is $(0,0,0)$, and ideally, this prediction is produced by each classifier in the ensemble. Now, since these classifiers are not perfect, suppose that the different label combinations are predicted in proportion to their conditional probabilities, i.e., $(0,0,0)$ is predicted with probability $1/4$, $(1,1,1)$ with probability $3/16$, etc. One easily verifies that, for each of the three labels, the probability of it being predicted as relevant ($9/16$) exceeds the probability for irrelevant ($7/16$). Therefore, by taking the arithmetic average over the ensemble members' predictions, and then thresholding at $1/2$, one will likely end up with the suboptimal prediction $(1,1,1)$.

The reader may have noticed that this example is actually less problematic for the Hamming loss, for which the prediction $(1,1,1)$ is indeed Bayes-optimal, and would be produced by the label-wise aggregation sketched above. More generally, it is plausible that a label-wise combination of predictions is indeed suitable for decomposable losses like Hamming, but suboptimal for non-decomposable losses.

Based on the discussion so far, we propose two aggregation functions for PTC, which can be seen as implementations of different types of voting, and will be used in our experimental study below:

- **Label-wise voting (PTC-lw)**: For each individual label λ_i, the number of positive (relevant) and negative (irrelevant) votes in the predictions $\hat{\boldsymbol{y}}_1, \ldots, \hat{\boldsymbol{y}}_M$ is counted, and the majority is adopted.
- **Mode (PTC-mode)**: Counting is done at the level of the entire predictions, i.e., we predict the label combination $\hat{\boldsymbol{y}}$ that occurs most frequently:

$$\hat{\boldsymbol{y}} = \underset{\bar{\boldsymbol{y}}}{\operatorname{argmax}} \sum_{j=1}^{M} [\![\hat{\boldsymbol{y}}_i = \bar{\boldsymbol{y}}]\!] \, . \tag{16}$$

In case the maximum is not unique, ties are broken by choosing the maximal prediction with the highest score

$$s(\hat{\boldsymbol{y}}) = \sum_{k=1}^{K} \sum_{j=1}^{M} [\![\hat{y}_k = \hat{y}_{k,j}]\!] \, . \tag{17}$$

Table 1. Datasets used in the experiments

#	Name	# Inst.	# Nom. Feat.	# Num. Feat.	# Lab.
1	Cal500	502	0	68	174
2	Emotions	593	0	72	6
3	Scene	2407	0	294	6
4	Yeast	2417	0	103	14
5	Mediamill	43907	0	120	101
6	Flags	194	9	10	7
7	Medical	978	1449	0	45
8	Bibtex	7395	1836	0	159

5 Experimental Evaluation

We perform experiments on eight standard benchmark datasets (cf. Table 1) from the MULAN repository[2], following a 10-fold cross-validation procedure. Our primary goal is to confirm that the loss-based aggregation methods PTC and CTP outperform the commonly used voting techniques. Moreover, we conjecture that CTP performs better than PTC for decomposable losses, and PTC better than CTP for non-decomposable losses. This is because accurate marginal probabilities are of utmost importance for decomposable losses — which is exactly what CTP accomplishes through averaging label-wise predictions. Likewise, PTC is more apt at capturing label dependencies, which is important for non-decomposable losses, because it aggregates over several predictions tailored to the target loss (instead of producing only a single one, as CTP).

We conducted three series of experiments using the ensemble methods EMODT, EBR, and ECC with their cardinality set to 50. We employed logistic regression as the base classifiers for EBR and ECC and let them produce probabilistic predictions. Thus, each ensemble member provides a complete marginal probability vector.

The detailed results are shown in Table 2. The best way for getting an insight into the respective performances is to consider the averages of these ranks in the final column of the table. In particular, each column shows the results of one dataset, each line shows the results of a combination of ensemble technique, loss function, and aggregation technique. For each combination of ensemble, loss function, and dataset, we also report the respective ranks for the obtained losses over the aggregation approaches. The bold value indicates the best performance on each data set. According to the Friedman/Nemenyi test, differences are statistically significant for a critical distance between the average ranks of 1.10/1.25 for $\alpha = 0.1/0.05$ for EMODT, and similarly 1.94/2.16 and 2.08/2.31 for ECC and EBR, respectively. The Friedman test fails for all Hamming loss comparisons.

[2] http://mulan.sourceforge.net/datasets.html. The source code will be available at https://github.com/nvlml/DS2020-EMLC.

Table 2. Predictive performance (in percent) and rank (small number) of aggregation methods with respect to the Hamming loss, the subset 0/1 loss and the F1-measure.

		Cal500	Emotions	Scene	Yeast	Flags	Medical	Bibtex	Mediamill	Avg. ranks
						EMODT				
Hamming loss ↓	GMV	**13.65** 1	18.72 2	9.45 3	19.57 2	**23.48** 1	1.55 2	1.27 3	2.67 3	2.13
	BMV	13.75 2	**18.52** 1	9.11 2	**19.38** 1	25.09 3	1.59 3	**1.25** 1.5	**2.65** 1.5	**1.88**
	CTP					equivalent to GMV				
	PTC-lw					equivalent to BMV				
	PTC-mode	14.41 3	19.66 3	**8.14** 1	19.77 3	24.58 2	**1.52** 1	**1.25** 1.5	**2.65** 1.5	2.00
Subset 0/1 loss ↓	GMV	**100** 2	69.82 3	49.48 3	85.48 3	79.34 2	55.70 2	87.59 3	84.79 3	2.63
	BMV	**100** 2	67.81 2	46.16 2	83.62 2	83.00 3	57.26 3	86.61 2	84.57 2	2.25
	CTP					equivalent to GMV				
	PTC-lw					equivalent to BMV				
	PTC-mode	**100** 2	**64.45** 1	**27.21** 1	**74.43** 1	**78.82** 1	**50.96** 1	**85.22** 1	**79.21** 1	**1.13**
F1-measure ↑	GMV	33.31 5	58.18 5	53.45 5	58.53 5	74.91 4	51.17 4	25.33 5	59.52 5	4.75
	BMV	37.33 4	61.79 4	57.02 4	60.99 4	74.33 5	50.72 5	27.69 4	60.74 4	4.25
	CTP	**48.31** 1	**68.30** 1	74.90 2	**66.61** 1	75.89 3	**76.17** 1	**48.77** 1	**63.88** 1	**1.38**
	PTC-lw	46.45 2	68.29 2	71.78 3	64.87 3	76.21 2	71.81 3	37.79 3	62.71 2	2.50
	PTC-mode	42.30 3	68.25 3	**77.55** 1	65.38 2	**76.48** 1	75.38 2	45.59 2	62.09 3	2.13
						ECC				
Hamming loss ↓	GMV	14.08 2	**20.28** 1	**8.63** 1	20.24 2	23.26 2	0.89 3	**1.28** 1	—	**1.71**
	BMV	**14.06** 1	20.31 2	8.76 2	**20.11** 1	23.48 3	**0.88** 1.5	1.29 2.5	—	1.86
	CTP					equivalent to GMV				
	PTC-lw					equivalent to BMV				
	PTC-mode	14.24 3	20.48 3	8.77 3	20.39 3	**22.52** 1	**0.88** 1.5	1.29 2.5	—	2.43
Subset 0/1 loss ↓	GMV	**100** 2	69.46 2	32.82 2	79.35 3	72.32 2	29.49 3	81.61 2	—	2.29
	BMV	**100** 2	70.47 3	33.07 3	78.98 2	74.87 3	28.45 2	81.64 3	—	2.57
	CTP					equivalent to GMV				
	PTC-lw					equivalent to BMV				
	PTC-mode	**100** 2	**68.45** 1	**29.04** 1	**77.28** 1	**66.61** 1	**28.35** 1	**81.37** 1	—	**1.14**
F1-measure ↑	GMV	32.13 5	63.20 5	72.84 4	62.63 5	72.80 4	81.74 4	40.08 5	—	4.57
	BMV	32.50 4	64.00 4	71.75 5	62.81 4	72.56 5	81.10 5	40.10 4	—	4.43
	CTP	**45.28** 1	**67.65** 1	**77.05** 1	**64.85** 1	74.21 2	**85.17** 1	**49.30** 1	—	**1.14**
	PTC-lw	42.05 2	67.58 2	75.52 3	64.69 2	**74.53** 1	84.85 2	48.95 2	—	2.00
	PTC-mode	41.98 3	66.81 3	75.57 2	64.06 3	74.13 3	84.58 3	48.77 3	—	2.86
						EBR				
Hamming loss ↓	GMV	**13.95** 1	**20.15** 1	9.82 2	19.89 3	24.05 2	**0.92** 2	**1.22** 1.5	—	**1.79**
	BMV	14.02 2	20.32 3	9.83 3	**19.87** 1	**23.62** 1	**0.92** 2	**1.22** 1.5	—	1.93
	CTP					equivalent to GMV				
	PTC-lw					equivalent to BMV				
	PTC-mode	14.10 3	20.21 2	**9.76** 1	19.88 2	24.49 3	**0.92** 2	1.23 3	—	2.29
Subset 0/1 loss ↓	GMV	**100** 2	**72.86** 1	46.11 3	84.78 3	82.45 3	30.71 3	82.12 3	—	2.57
	BMV	**100** 2	73.53 3	46.03 2	84.49 2	81.92 2	30.39 2	81.88 2	—	2.14
	CTP					equivalent to GMV				
	PTC-lw					equivalent to BMV				
	PTC-mode	**100** 2	73.37 2	**45.53** 1	**84.32** 1	**80.87** 1	**30.09** 1	**81.72** 1	—	**1.29**
F1-measure ↑	GMV	34.17 5	58.38 4	61.99 5	61.37 5	72.60 4	79.18 5	39.75 5	—	4.71
	BMV	34.40 4	58.11 5	62.07 4	61.49 4	73.42 3	79.76 4	40.17 4	—	4.00
	CTP	**47.62** 1	66.67 3	76.14 3	**65.07** 1	**75.18** 1	**85.27** 1	**50.78** 1	—	**1.57**
	PTC-lw	47.46 2	66.96 2	**76.29** 1	65.01 2	74.80 2	84.63 2	49.02 3	—	2.00
	PTC-mode	47.33 3	**67.07** 1	76.18 2	64.99 3	71.41 5	84.58 3	49.43 2	—	2.71

- **Loss-based aggregation vs. voting.** Especially for F1 and subset 0/1 loss, there are (statistically significant) large differences between the voting-based decompositions on the one side, and PTC/CTP on the other side. This confirms our expectation that GMV and BMV are poorly suited for the case of non-decomposable performance measures. Only for Hamming loss, the voting-based techniques are in the same range, and, in fact, sometimes even better (yet, no significant difference).
  This result is also expected because Hamming loss is decomposable, so that its performance primarily depends on accurate marginal probabilities. In fact, in this case, label-wise PTC and CTP are equivalent to binary and graded voting, respectively. For subset 0/1 loss, assuming label independence and marginal probability as the relevance information, our loss-based instantiations of CTP and PTC-lw are equivalent to GMV and BMV, respectively. As can be seen from the results, however, this assumption is most likely invalid for the investigated datasets, because PTC-mode, which addresses the problem of finding the mode of the joint label distribution, typically outperforms the alternatives.
- **PTC vs. CTP.** With respect to the two different approaches, the mode-based PTC decomposition performs significantly better for subset 0/1 loss, whereas CTP (or, in this case, equivalently GMV) seems to perform better for Hamming loss. These results provide clear evidence in favor of our conjecture. The results for F1 are a bit more difficult to interpret but also consistent. Given marginal probabilities, we derive the loss minimizer for F1 under the assumption of label independence, and in this case, accurate marginal probabilities are again crucial. This is probably the reason for why CTP has an advantage over PTC.

We also conducted a series of experiments using EMODT with the number of ensemble members varying from 1 to 100 ($M \in \{1, 5, 10, 20, 30, \ldots, 100\}$). Here, our interest was to study the influence of the ensemble size on the performance of the aggregation methods. For each value of the ensemble cardinality, we have run a 10 times 10-fold cross-validation, for which we report the average scores.

As expected, the results shown in Fig. 2 confirm that the MLC scores typically improve with an increasing size of the ensembles. This is in agreement with the observation on the performance of ECC reported in [9]. More importantly, we also see differences between the different aggregation methods, and that suitable instantiations of CTP and PTC can indeed reach better performance than standard voting techniques. In particular, the visible gaps for the subset 0/1 loss re-confirm the superiority of PTC-mode for non-decomposable losses. Finally, we note that the performances change rapidly in the beginning and tend to converge when the number of ensemble members reaches moderate values (i.e., 30 or 40), except for the subset 0/1 loss and PTC-mode. This is again in agreement with our expectations, because PTC-mode does voting at the level of the entire predictions, and the number of possible predictions increases exponentially with the number of labels, so that more iterations are necessary for convergence. A similar effect can be observed for PTC-lw/BMV, whose label-wise votings converge

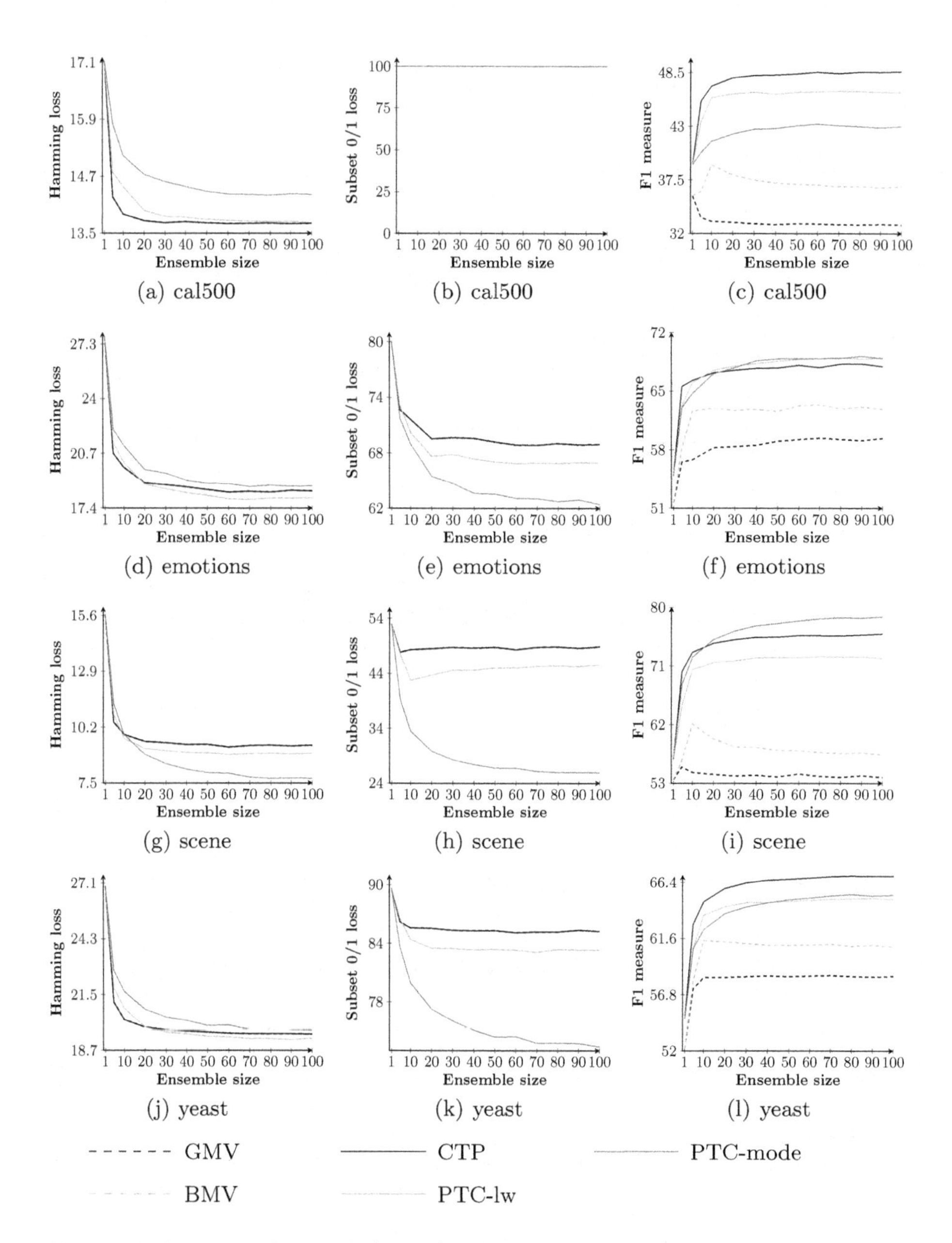

Fig. 2. Predictive performance (y-axis) of aggregation methods as a function of the cardinality of ensembles (x-axis) in terms of Hamming loss (left column), subset 0/1 loss (middle), and F1 (right column) for four datasets.

less rapidly to accurate marginal probability estimates than CTP/GMV, but are able to catch up with increasing number of votes. For EMODT, there seems to be even an advantage in the end for using the vote distributions, possibly due to less accurate probability estimates of the trees.

Results similar to those shown in Fig. 2 have been obtained for EBR and ECC and omitted due to space limitations.

6 Conclusion

This paper studied the question of how to aggregate the predictions of individual members of an ensemble of multilabel classifiers in a systematic way. We introduced a formal framework of ensemble multilabel classification, in which we distinguish two principal approaches, referred to as "predict then combine" (PTC) and "combine then predict" (CTP). Both approaches generalize voting techniques commonly used for EMLC, while allowing one to explicitly take the target performance measure into account. Our framework supports the analysis of existing EMLC methods as well as the systematic development of new ones. Besides, it suggests a number of interesting theoretical problems, like the question of how to combine predictions in PTC in a provably optimal way. Experimentally, we showed that standard voting techniques are indeed outperformed by suitable instantiations of CTP and PTC. Moreover, our results suggest that CTP performs well for decomposable loss functions, whereas PTC is the better choice for non-decomposable losses.

Acknowledgements. This work was supported by the German Research Foundation (DFG) under grant number 400845550.

References

1. Breiman, L.: Bagging predictors. Mach. Learn. **24**(2), 123–140 (1996)
2. Cheng, W., Hüllermeier, E., Dembczyński, K.J.: Bayes optimal multilabel classification via probabilistic classifier chains. In: Proceedings of the 27th International Conference on Machine Learning (ICML), pp. 279–286 (2010)
3. Dembczyński, K., Waegeman, W., Cheng, W., Hüllermeier, E.: On label dependence and loss minimization in multi-label classification. Mach. Learn. **88**, 5–45 (2012). https://doi.org/10.1007/s10994-012-5285-8
4. Dembczyński, K., Waegeman, W., Hüllermeier, E.: An analysis of chaining in multi-label classification. In: Proceedings of the 20th European Conference on Artificial Intelligence (ECAI), pp. 294–299. IOS Press (2012)
5. Dietterich, T.G.: Ensemble methods in machine learning. In: Kittler, J., Roli, F. (eds.) MCS 2000. LNCS, vol. 1857, pp. 1–15. Springer, Heidelberg (2000). https://doi.org/10.1007/3-540-45014-9_1
6. Gharroudi, O.: Ensemble multi-label learning in supervised and semi-supervised settings. Ph.D. Thesis, Université de Lyon (2017)
7. Gharroudi, O., Elghazel, H., Aussem, A.: Ensemble multi-label classification: a comparative study on threshold selection and voting methods. In: Proceedings of the 27th IEEE International Conference on Tools with Artificial Intelligence (ICTAI), pp. 377–384. IEEE Computer Society (2015)

8. Kocev, D., Vens, C., Struyf, J., Džeroski, S.: Ensembles of multi-objective decision trees. In: Kok, J.N., Koronacki, J., Mantaras, R.L., Matwin, S., Mladenič, D., Skowron, A. (eds.) ECML 2007. LNCS (LNAI), vol. 4701, pp. 624–631. Springer, Heidelberg (2007). https://doi.org/10.1007/978-3-540-74958-5_61
9. Li, N., Zhou, Z.-H.: Selective ensemble of classifier chains. In: Zhou, Z.-H., Roli, F., Kittler, J. (eds.) MCS 2013. LNCS, vol. 7872, pp. 146–156. Springer, Heidelberg (2013). https://doi.org/10.1007/978-3-642-38067-9_13
10. Madjarov, G., Kocev, D., Gjorgjevikj, D., Džeroski, S.: An extensive experimental comparison of methods for multi-label learning. Pattern Recogn. **45**(9), 3084–3104 (2012)
11. Moyano, J.M., Gibaja, E.L., Cios, K.J., Ventura, S.: Review of ensembles of multi-label classifiers: models, experimental study and prospects. Inf. Fusion **44**, 33–45 (2018)
12. Murthy, S.K.: Automatic construction of decision trees from data: A multi-disciplinary survey. Data Min. Knowl. Disc. **2**(4), 345–389 (1998). https://doi.org/10.1023/A:1009744630224
13. Quinlan, J.R.: Induction of decision trees. Mach. Learn. **1**(1), 81–106 (1986). https://doi.org/10.1007/BF00116251
14. Read, J., Pfahringer, B., Holmes, G., Frank, E.: Classifier chains for multi-label classification. In: Buntine, W., Grobelnik, M., Mladenić, D., Shawe-Taylor, J. (eds.) ECML PKDD 2009. LNCS (LNAI), vol. 5782, pp. 254–269. Springer, Heidelberg (2009). https://doi.org/10.1007/978-3-642-04174-7_17
15. Read, J., Pfahringer, B., Holmes, G., Frank, E.: Classifier chains for multi-label classification. Mach. Learn. **85**(3), 333 (2011)
16. Saha, S., Sarkar, D., Kramer, S.: Exploring multi-objective optimization for multi-label classifier ensembles. In: Proceedings of the IEEE Congress on Evolutionary Computation (CEC), pp. 2753–2760. IEEE, Wellington (2019)
17. Shi, C., Kong, X., Fu, D., Yu, P.S., Wu, B.: Multi-label classification based on multi-objective optimization. ACM Trans. Intell. Syst. Technol. **5**(2), 1–22 (2014)
18. Shi, C., Kong, X., Yu, P.S., Wang, B.: Multi-label ensemble learning. In: Gunopulos, D., Hofmann, T., Malerba, D., Vazirgiannis, M. (eds.) ECML PKDD 2011. LNCS (LNAI), vol. 6913, pp. 223–239. Springer, Heidelberg (2011). https://doi.org/10.1007/978-3-642-23808-6_15
19. Tsoumakas, G., Vlahavas, I.: Random *k*-labelsets: an ensemble method for multi-label classification. In: Kok, J.N., Koronacki, J., Mantaras, R.L., Matwin, S., Mladenič, D., Skowron, A. (eds.) ECML 2007. LNCS (LNAI), vol. 4701, pp. 406–417. Springer, Heidelberg (2007). https://doi.org/10.1007/978-3-540-74958-5_38
20. Tsoumakas, G., Katakis, I., Vlahavas, I.: Mining multi-label data. In: Maimon, O., Rokach, L. (eds.) Data Mining and Knowledge Discovery Handbook, pp. 667–685. Springer, Heidelberg (2009). https://doi.org/10.1007/978-0-387-09823-4_34
21. Tsoumakas, G., Katakis, I., Vlahavas, I.: Random k-labelsets for multilabel classification. IEEE Trans. Knowl. Data Eng. **23**(7), 1079–1089 (2010)
22. Waegeman, W., Dembczyński, K., Jachnik, A., Cheng, W., Hüllermeier, E.: On the Bayes-optimality of F-measure maximizers. J. Mach. Learn. Res. **15**(1), 3333–3388 (2014)
23. Zhang, M.L., Zhou, Z.H.: A review on multi-label learning algorithms. IEEE Trans. Knowl. Data Eng. **26**(8), 1819–1837 (2014)

Neural Networks and Deep Learning

Attention in Recurrent Neural Networks for Energy Disaggregation

Nikolaos Virtsionis Gkalinikis(✉), Christoforos Nalmpantis, and Dimitris Vrakas

School of Informatics, Aristotle University of Thessaloniki, 54124 Thessaloniki, Greece
{virtsion,christofn,dvrakas}@csd.auth.gr
https://www.csd.auth.gr/en/

Abstract. Energy disaggregation refers to the separation of appliance-level data from an aggregate energy signal originated from a single-meter, without the use of any other device-specific sensors. Due to the fact that deep learning caught great attention in the last decade, numerous techniques using Artificial Neural Networks (ANN) have been developed to accomplish this task. Whereas most of the current research focuses on achieving better performance, the goal of this paper is to design a computationally light deep neural network based on attention mechanism. A thorough analysis shows how the proposed model is implemented and compares the performance of two different attention layers in the problem of energy disaggregation. The novel architecture achieves fast training and inference with minor performance trade-off when compared against other computationally expensive state-of-the-art models.

Keywords: Energy disaggregation · Non-intrusive load monitoring · Artificial neural networks · Attention

1 Introduction

Energy disaggregation provides the ability to estimate the electrical energy consumption of an appliance, using only the total power consumption of a house. It is also known as non-intrusive load monitoring (NILM). Further analysis can identify inefficiencies of the various appliances, in order to reduce their energy usage. Additionally, with the use of NILM, the electrical energy management may be improved towards a direction of nullifying the unnecessary waste of energy usage, one of the crucial factors of climate change and global warming.

Smart houses integrate home energy management systems (HEMS) in order to monitor and manage electrical appliances, reducing energy cost for consumers. In HEMS appliance load monitoring (ALM) can be achieved with either intrusive or non-intrusive monitoring methods [15]. The main advantage of NILM against intrusive-loading monitoring (ILM) is that it requires measurements from a single mains meter instead of multiple meters. Load monitoring is cheaper and more straightforward, although ILM offers higher accuracy.

This paper contributes to the research of NILM in two major points. Firstly, with the design of a lightweight model using artificial neural networks. Thus,

A. Appice et al. (Eds.): DS 2020, LNAI 12323, pp. 551–565, 2020.
https://doi.org/10.1007/978-3-030-61527-7_36

faster training and inference times were achieved with a minor performance decrement in comparison to a state-of-the-art architecture. Secondly, with the introduction of Attention in the task of energy disaggregation alongside with promising results.

The structure of this article is as described bellow. To begin with, the related work about NILM and energy disaggregation is presented. Secondly, the method of Attention is described. Section 3 includes a short explanation of the calculations inside the attention mechanism. In Sect. 4, there is an in depth analysis of the novel architecture and a presentation of its purpose and benefits. Next, the methodology of experiments is described. In Sect. 6, there is a presentation of the most important results. Finally there are conclusions and proposals for future work.

2 Related Work

The problem of energy disaggregation states back to mid 1980s when it was firstly introduced by Hart. Hart in [5] proposed a combinatorial optimization method in order to extract the optimal states of the target appliances so that the sum of power consumption would be the same as the meter reading. This method is applicable only on devices that have finite number of states, thus it cannot be used on appliances with variable consumption.

NILM research interest has raised a lot with the rise of internet of things. For a long time one of the most popular methods solving the energy disaggregation problem was Factorial Hidden Markov Models (FHMM), an extension to Hidden Markov Models (HMMs). In FHMM the architecture consists of multiple independent HMMs in parallel, where the observed output is a combination of all the hidden states. Kolter and Jaakkola used additive FHMMs, where the output was the sum of all the independent HMMs outputs [10]. The rise of machine learning and deep learning pushed researchers to use techniques from the sectors of Natural language processing (NLP), Computer Vision and Time Series Analysis. In 2015, Kelly and Knottenbelt [7] described three novel architectures using three different kinds of artificial neural networks (ANNs), an LSTM recurrent neural network (RNN), a denoising autoencoder architecture and a network to regress start/end time and power. These models outperformed Hart's algorithm and FHMM on experiments executed on the UK-DALE [6] data set. Mauch and Yang [14] investigated another method using a recurrent network with LSTM neurons on low frequency real power data. The experiments were executed on REDD [11] dataset alongside with synthetic data. This approach showed good performance for appliances with recurring patterns.

In 2017 Zhang et al. [22] implemented an architecture called Sequence-to-Point using convolutional neural network (CNN) layers, outperforming the results of Kelly and Knottenbelt [7]. A key point of difference of this model in respect to the recurrent architectures in [7] and [14] is that a window of aggregate data is considered in order to predict the appliance consumption on a single time step, thus the name Sequence-to-Point. On the other hand, in [7] and [14] a single time step of the aggregate signal is used to predict the device power consumption at a the same time step. Krystalakos et al. [12] used Gated Recurrent Units (GRUs) instead of LSTMs alongside with dropout layers in order to improve the efficiency of previous RNN architectures.

Due to the lack of a benchmark method, the comparison of various methods and models most of the time is questionable. In an effort to efficiently tackle this challenge, Symeonidis et al. [21] proposed a set of experiments as a benchmark basis. In addition, the Stacking method of five popular architectures is explored resulting in promising results on 2-state devices. In a nutshell, regarding the matters of reproducibility and comparability of NILM frameworks, it is suggested to be a standardization of the assessment procedures [9,16].

NILM research mostly focuses on designing one model per device, resulting that a complete NILM system should integrate as many models as the number of devices the target environment contains. Thus, these type of architectures are not directly applicable in real time situations, where energy measurements provide huge quantities of data even at low sampling frequencies. The creation of lightweight architectures is a first and important step in order to achieve successful deployment of NILM on embedded systems. The next step is to consider multi-label machine learning models, where one model is trained in order to identify more than one appliances. Basu et al. [2,3] were the first to introduce the multi-label classification in NILM tasks, with the use of known machine learning algorithms such as decision trees and boosting. The most recent work considering multi label classification in energy disaggregation was published by Nalmpantis and Vrakas [18], where a novel framework called multi-NILM is proposed. In multi-NILM approach, a dimensionality reduction technique called Signal2vec [17] is combined with a lightweight disaggregation model, achieving very promising results.

3 Attention Mechanism

One of the most common tasks in machine learning is to extract input-output relations such as in machine translation and image captioning. In Deep Learning, the most popular way of dealing with this format of tasks is with sequence to sequence models (seq2seq). The original seq2seq architecture (Sutskever et al. [20]) consists of two essential RNNs; the encoder and the decoder. The encoder's role is to compress the sequential input into a context vector of fixed length, which contains a summary of the source sequence. On the other side, given the context vector, the decoder's purpose is to construct the target sequence.

The purpose of attention, as introduced by Bahdanau et al. [1], is to assist the decoder to focus on the most important parts of the input. It provides information between the entire input sequence and the decoder output at each time step. The idea is that at every time step of the decoder an alignment vector is computed containing the score between the input's sequence and the decoder's output at the corresponding moment. As a result, the context vector is a combination of the alignment vector and the encoder's output. The model successfully focuses on the relevant parts of the input sequence.

There are different types of attention, depending on how scores and alignments are computed. The most common ones are the Additive [1] and the Multiplicative/Dot [13]. In addition, Cheng et al. [4] proposed a different attention mechanism called Self-Attention which is also referred as intra-attention. The benefit is that different positions of the same inputs are related. Self-Attention can adopt both Bahdanau's and Luong's scoring functions. The neural network that is proposed in this research incorporates Additive and Dot attention mechanisms.

In general an Attention layer receives three kinds of vectors; query, key and value. Depending on the query, attention computes an output based on the key and value. The steps to calculate the output are described below and depicted in Fig. 1.

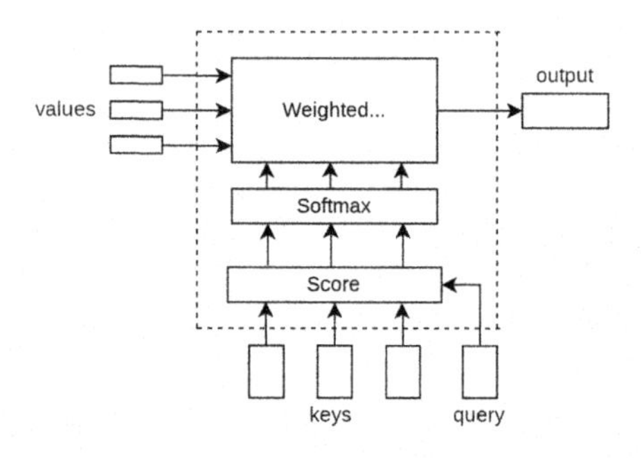

Fig. 1. Inside Attention mechanism.

Firstly, a score function is used to measure the similarity between a query (q) and a key (k_i) and for each query-key pair, scores (a_i) are computed.

$$a_i = score(q, k_i) \tag{1}$$

Secondly, these scores are normalized to add to one, using a softmax.

$$b_i = \frac{exp(a_i)}{\sum_j \exp(a_j)} \tag{2}$$

The last step is to combine the values (v) and the attention weights (b) as a weighted sum.

$$output = \sum_{i=1}^{n} b_i v_i \tag{3}$$

The main difference between Additive and Dot attention mechanisms is the scoring function. Dot mechanism scores between keys and queries are computed by calculating the dot product, whereas Additive attention computes scores as a non-linear sum. For the purpose of using an attention layer between the CNN and GRU layers the idea of Self-Attention was used. In Self-Attention the aim is to learn the dependencies between all the parts of the same input sequence. In this set up as query, key and value inputs the Attention layer receives the output of the CNN layer.

4 Neural Network Architecture

The goal of this paper is to design a computationally light neural network. Being inspired by Window GRU (WGRU) [12], a lightweight architecture has been developed. The novel model is called Self-Attentive-Energy-Disaggregation (SAED) and is up to 7.5 times faster in training and up to 6.5 times faster in inference, while there is trivial trade-off in performance.

WGRU [12] consists of the following layers: a convolutional layer, two Bidirectional GRU layers and one Dense layer before the output. The method of Dropout [19] between layers is used to prevent overfitting. To predict the power consumption of a device at a single time point, WGRU utilizes a sliding window of past aggregate data points as opposed to other similar recurrent neural network architectures [7,20] that use a single point of the aggregate time series. Therefore, the model receives more information about the target time series and is more capable to recognize useful patterns. The sliding window method is also used in the proposed architecture.

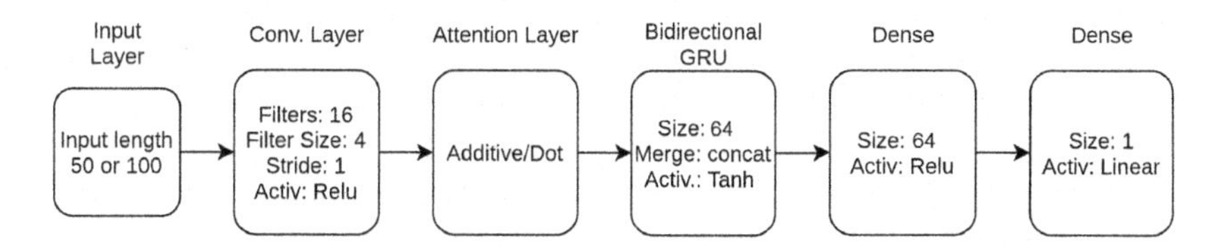

Fig. 2. Architecture of the Attention model.

In the proposed architecture, the first GRU is replaced by the Attention Layer. SAED combines the benefits of three different types of layers. Firstly a 1D convolution layer extracts new features. 1D convolution layers can recognize local patterns in a sequence at certain positions of the sequence, which can later be recognized at different positions. As a consequence, 1D convnets are time invariant. Next, the attention mechanism learns to focus on the most important features. Following a recurrent neural network is capable of extracting sequential patterns. Lastly, the dense layer acts as a regressor, giving the final result. The architecture is shown in Fig. 2. It is important to point out that in the proposed model the Attention layer functions as a Self-Attention mechanism receiving as input only the output of the CNN layer. In this paper the model comes with either Additive or Dot attention mechanism, mentioned as SAED-add and SAED-dot correspondingly.

5 Methodology of Experiments

For the experiments only real data was used with sampling period 6 seconds and batch size 1024. The devices that were chosen are dish washer, fridge, kettle, microwave and washing machine. The optimal size of the input vector is device-dependent [12]. The sliding windows were 50 samples for all devices except for washing machine in which case it was 100. All the models were trained for 5 epochs following the benchmark methodology described in [21], where the experiments are divided in four categories; Single Building NILM, Single Building learning and generalization on same dataset, Multi building learning and generalization on same dataset and Generalization to different dataset.

The first category is about experiments where training and testing are applied on the same house at different time periods, in order to evaluate the model in the same environment where training took place. If a model doesn't perform well in this category of experiments it is probably weak [21]. The second category of experiments refers to training and inference on different buildings of the same data set. The purpose of these experiments is to inspect the generalization potential of the model on unseen buildings. Different buildings mean that different appliances are used, the residents have different habits, resulting in divergent energy patterns. Within the same data set though, similarities of the energy footprint of each building are also expected. These similarities are mainly attributed to properties of the electricity grid, common seasonality or weather conditions and regionality. In the third category, training data is collected from different houses of the same data set and inference is executed on an unseen building. In the last category of experiments training data is also collected from different houses but the model is tested on houses of a different data set.

The experiments where the training data is composed from different houses, evaluate the sufficiency of the model in learning from multiple/different sources. In addition, the challenge for the model is higher in the last category, because it has to successfully learn from high variety data and infer on unseen data from a different data set. These two categories (especially the last) are considered as tough tasks and if a model excels in them then it is considered very strong [21].

All the models were trained and tested using the UK-DALE [6] data for the first three categories, while as test data for the fourth category of experiments we used the REDD [11] data set. These data sets are considered dissimilar, because they are originated from different countries; UK-DALE contains measurements of house-hold devices in UK and REDD measurements of house-hold devices in USA. The experiments are summed up in Table 1. For Kettle the fourth category of experiments was not executed due to the lack of kettle device in the REDD data. For categories 1 and 2 the training on house 1 from UK-DALE is during the first 9 months of 2013 while the inference contains the last 3 months of the same year. For categories 3 and 4 the ratio of test versus training data depends on each device. In addition, for some devices the REDD data contains very few measurements which resulted in bad results even for the state of the art model.

In order to evaluate and compare the models with Attention versus the state of the art WGRU model, three metrics are used; F1 score, Relative Error in

Table 1. Buildings used for train and test. In Categories 1–3, UK-DALE was used for training and testing. In Category 4, UK-DALE was used for training and REDD was used for testing

Device	Category1		Category2		Category3		Category4	
	Train	Test	Train	Test	Train	Test	Train	Test
Dish Washer	1	1	1	2,5	1, 2	5	1, 2	1, 2, 3, 4, 6
Fridge	1	1	1	2, 4, 5	1, 2, 4	5	1, 2, 4	1, 2, 3, 5, 6
Kettle	1	1	1	2, 3, 4, 5	1, 2, 3, 4	5	–	–
Microwave	1	1	1	2, 3, 5	1, 2	5	1, 2	1, 2, 3, 5
Washing M.	1	1	1	2, 4, 5	1, 5	2	1, 5	1, 2, 3, 4, 5, 6

Total Energy (RETE) and Mean Absolute Error (MAE). The purpose of the F1 score is to evaluate the ability of model to detect on/off energy states. MAE (measured in Watts) and RETE (dimensionless) are used in order to measure how capable is the model in predicting the actual electrical power consumed by the device. Considering as E' the predicted total energy, E the true value of total energy, T the length of the predicted sequence, yt' the inferred power consumption and yt the true value of power consumption at time point t, the metrics are calculated as:

$$F1 = 2\frac{Precision * Recall}{Precision + Recall} \tag{4}$$

$$RETE = \frac{|E' - E|}{max(E', E)} \tag{5}$$

$$MAE = \frac{1}{T}\sum |y'_t - y_t| \tag{6}$$

$$Precision = \frac{TP}{TP + FN} \tag{7}$$

$$Recall = \frac{TP}{TP + FP} \tag{8}$$

For the investigation of the generalization capabilities of the proposed architecture on a deeper level, the necessity of more metrics arose. Klemenjak et al. [8] proposed a novel set of metrics which consider the number of seen and unseen buildings where the experiments were conducted. The basic idea behind this concept is to calculate the generalization loss (G-loss) of a metric between a seen and an unseen building. Depending on whether the metric under consideration evaluates the event detection or the energy estimation ability of the model, the G-loss is given from Eq. 9 or Eq. 10 accordingly. The u indicator stands for unseen buildings, whereas the s for seen buildings. A G-loss of 10% on F1 score denotes that the measured F1 score on the unseen building data is 10% lower than on the seen building, where the training was executed. On the other hand,

10% G-loss on MAE designates that the error measured on the unseen building is 10% higher than the error measured on the seen building data.

$$G\text{-}loss = 100(1 - \frac{F1_u}{F1_s}) \tag{9}$$

$$G\text{-}loss = 100(\frac{MAE_u}{MAE_s} - 1) \tag{10}$$

$$MGL = \frac{1}{N}\sum_{i}^{N} G\text{-}loss_i \tag{11}$$

$$AUH = \frac{1}{N}\sum_{i}^{N} F1_{ui} \tag{12}$$

$$EUH = \frac{1}{N}\sum_{i}^{N} MAE_{ui} \tag{13}$$

The overall performance loss is represented by the mean generalization loss (MGL), the mean value of the G-loss of all the unseen buildings. In addition, the generalization ability of a model can be measured with accuracy on unseen houses (AUH) and the error on unseen houses (EUH). In Eq. 11–13 the N represents the number of the unseen houses and the i points to each house.

6 Results and Comparisons

The comparison of the models is threefold; firstly on performance, where the performance of the models in different categories of experiments is compared. Secondly, on the generalization capability, where the generalization loss of the metrics on unseen data is computed. Lastly, on the scalability, where the train and inference times of the models are measured for different sizes of data.

6.1 Performance Comparison

Due to the size of the results, the most important of them are presented in Tables 2, 3, 4, 5, 6, 7, 8, 9, 10, 11, 12, 13, 14, 15, 16, 17, 18, 19 and 20, where the best are highlighted. Also, the average duration of a training epoch, measured in seconds, is mentioned as time(s). The complete set of results alongside with the supplementary code are provided in the following github repository: https://github.com/Virtsionis/SelfAttentiveEnergyDisaggregator.

As shown in Table 2, in Category 1 of Dish Washer the SAED models perform on par with WGRU in up to 7.1 times faster training time per epoch. In Category 2, SAED-dot is the clear winner with similar metric values as the SAED-add model, but with almost half the training time per epoch. In Category 3 of the same device, Table 4, the SAED models show better performance. Specifically the SAED-add performs better in respect of F1 and RETE, whereas in terms of

Table 2. Dish Washer, Category 1, Train and Test house 1 UK-DALE

Model	F1	RETE	MAE	Time(s)
WGRU	**0.33**	**0.17**	13.22	550
SAED-dot	0.28	0.31	13.03	77
SAED-add	0.25	**0.17**	**12.03**	141

Table 3. Dish Washer, Category 2, Train house 1, Test house 2 UK-DALE

Model	F1	RETE	MAE	Time(s)
WGRU	0.26	0.77	37.47	550
SAED-dot	**0.63**	**0.62**	**33.48**	77
SAED-add	0.6	0.63	34.31	141

Table 4. Dish Washer, Category 3, Train on houses 1, 2 and Test on house 5 of UK-DALE.

Model	F1	RETE	MAE	Time(s)
WGRU	0.23	0.42	43.33	575
SAED-dot	0.25	0.46	**43.3**	74
SAED-add	**0.52**	**0.37**	44.48	138

Table 5. Dish Washer, Category 4, Train on houses 1, 2 UK-DALE, Test on house 1 REDD.

Model	F1	RETE	MAE	Time(s)
WGRU	0.39	0.6	34.35	575
SAED-dot	**0.41**	0.19	**27**	74
SAED-add	0.18	**0.13**	36.16	138

MAE all the models perform the same. As presented in Table 5, in Category 4 the SAED-dot achieves better F1 score and MAE, while SAED-add has lower METE. The general conclusion is that SAED shows promising results on Dish Washer in comparison to the WGRU, with faster training and better performance in Categories 2–4.

In similar manner, Tables 6, 7, 8 and 9 present the results on the Washing Machine for Categories 1–4 accordingly. In Category 1, SAED-dot is 7.5 times faster than WGRU trading of maximum 10% performance regarding the metrics F1 and MAE. As presented in Table 7, in Category 2 SAED-dot performs on par with WGRU but with 7.5 times faster training time per epoch. Results for Category 3 are shown in Table 8, where the SAED-add has best F1 score and RETE. In terms of METE in this category of experiments, the SAED models are better than the WGRU. Results of the fourth category of experiments can be found in Table 9. In this category, the SAED models are trained 7.2 times faster and with lower RETE and METE values than the WGRU.

It is notable that disaggregating Dish Washer and Washing Machine, the SAED models have comparable or better performance with the WGRU while training time per epoch was up to 7.5 times faster.

Results for the Fridge are summed in Tables 10, 11, 12 and 13. As presented in Table 10, in Category 1 WGRU achieves greater F1 score, while SAED-add shows promising results with the smallest RETE and MAE, reaching up to 4 times faster training times. In Category 2, WGRU is a clear winner, whereas in Categories 3 and 4 the SAED models perform the same as the WGRU showing good generalization capabilities.

In Categories 1 and 2 of the Kettle, shown in Tables 14 and 15, the three models have comparable RETE and MAE values, but the WGRU achieves the best F1 score in 7.7 slower training time. In the third category of experiments

Table 6. Washing M., Category 1, Train and Test on house 1 of UK-DALE.

Model	F1	RETE	MAE	Time(s)
WGRU	**0.54**	**0.12**	**16.55**	1097
SAED-dot	0.51	0.26	18.51	147
SAED-add	0.45	0.29	28.55	416

Table 7. Washing M., Category 2, Train on house 1 and Test on house 2 of UK-DALE.

Model	F1	RETE	MAE	Time(s)
WGRU	**0.34**	0.43	**10.45**	1097
SAED-dot	0.3	**0.34**	13.1	147
SAED-add	0.3	0.53	22.01	416

Table 8. Washing M., Category 3, Train on houses 1, 5 and Test on house 2 of UK-DALE.

Model	F1	RETE	MAE	Time(s)
WGRU	0.12	0.36	22.74	1097
SAED-dot	0.19	0.36	**14.66**	147
SAED-add	**0.2**	**0.21**	15.18	416

Table 9. Washing M., Category 4, Train on houses 1, 5 UK-DALE, Test on house 1 REDD.

Model	F1	RETE	MAE	Time(s)
WGRU	**0.26**	0.66	43.65	585
SAED-dot	0.18	**0.39**	50.65	81
SAED-add	0.18	0.7	**41.93**	81

Table 10. Fridge., Category 1, Train and Test on house 1 of UK-DALE.

Model	F1	RETE	MAE	Time(s)
WGRU	**0.63**	0.27	33.29	562
SAED-dot	0.27	0.23	12.11	73
SAED-add	0.27	**0.13**	**10.94**	145

Table 11. Fridge, Category 2, Train house 1, Test house 2 UK-DALE.

Model	F1	RETE	MAE	Time(s)
WGRU	**0.82**	**0.13**	**28.46**	562
SAED-dot	0.62	0.6	35.25	73
SAED-add	0.66	0.65	32.31	145

Table 12. Fridge, Category 3, Train houses 1,2,4, Test house 5 UK-DALE.

Model	F1	RETE	MAE	Time(s)
WGRU	**0.52**	**0.18**	**51.18**	519
SAED-dot	**0.52**	0.29	51.35	69
SAED-add	**0.52**	0.22	50.52	70

Table 13. Fridge, Category 4, Train houses 1, 2, 4 UK,Test house 1 REDD

Model	F1	RETE	MAE	Time(s)
WGRU	**0.53**	0.32	**52.57**	519
SAED-dot	0.49	**0.29**	50.89	69
SAED-add	0.5	0.33	51.39	70

presented in Table 16, the WGRU is the winner in terms of F1 and RETE, whereas in MAE all the models perform the same. The above results reveal that, comparing to the WGRU, the SAED models show difficulties in disaggregating devices with simple behavior, such as the Fridge and the Kettle. The Kettle is a two-state device, while the Fridge has a finite number of states and repetitive time series. Especially, in Categories 1–2 of the Fridge and the Kettle the SAED has low values on F1 score, but it achieves good results in Categories 3–4 of

Table 14. Kettle, Category 1, Train and Test house 1 UK-DALE

Model	F1	RETE	MAE	Time(s)
WGRU	**0.65**	**0.09**	**7.35**	563
SAED-dot	0.44	0.14	8.57	73
SAED-add	0.34	0.26	9.46	143

Table 15. Kettle, Category 2, Train house 1, Test house 2 UK-DALE

Model	F1	RETE	MAE	Time(s)
WGRU	**0.9**	0.31	**14.04**	563
SAED-dot	0.62	0.3	19.03	73
SAED-add	0.49	**0.28**	17.35	143

the Fridge. The low values of F1 score indicate the difficulty of the models to identify the On/Off states of the test devices.

Table 16. Kettle, Category 3, Train houses 1, 2, 3, 4, Test house 5 UK-DALE

Model	F1	RETE	MAE	Time(s)
WGRU	**0.41**	**0.05**	**9.92**	1096
SAED-dot	0.27	0.27	12.24	141
SAED-add	0.31	0.18	10.95	271

The results of the experiments on the Microwave are displayed in Tables 17, 18, 19 and 20. As presented in Tables 17 and 18, in Categories 1–2 the WGRU performs better than the SAED models in terms of F1. In the same categories, the SAED performs on par with the WGRU regarding the RETE and MAE metrics. In the third category of experiments SAED models outperform the WGRU, where in Category 4 the WGRU achieves 17% better F1 score in 10 times slower training time. Considering that the Microwave is a multi-state device with variable power consumption and on-state duration, the SAED models show descent performance comparing with the WGRU.

Overall, the SAED models achieve good performance in disaggregating multi-state devices instead of simpler devices. Furthermore, the SAED performs good in experiments of Categories 3–4, a fact that reveals the great generalization capability of the proposed models.

Table 17. Microwave, Category 1, Train and Test house 1 UK-DALE

Model	F1	RETE	MAE	Time(s)
WGRU	**0.32**	**0.09**	**6.29**	560
SAED-dot	0.16	0.14	7.51	74
SAED-add	0.18	0.16	7.61	144

Table 18. Microwave, Category 2, Train house 1, Test house 2 UK-DALE

Model	F1	RETE	MAE	Time(s)
WGRU	**0.44**	0.25	**4.36**	560
SAED-dot	0.25	0.19	5.97	74
SAED-add	0.26	**0.17**	5.98	144

Table 19. Microwave, Cat.3, Train houses 1, 2, Test house 5 UK

Model	F1	RETE	MAE	Time(s)
WGRU	0.08	0.59	60.53	440
SAED-dot	0.21	0.58	56.93	41
SAED-add	**0.22**	**0.51**	**59.36**	41

Table 20. Microwave, Cat.4, Train houses 1, 2 UK, Test house 1 REDD

Model	F1	RETE	MAE	Time(s)
WGRU	**0.41**	**0.19**	**23.53**	440
SAED-dot	0.34	0.2	25.67	41
SAED-add	0.34	0.15	25.13	41

6.2 Generalization Evaluation

To explore on a deeper level the generalization ability of the SAED, in comparison to the WGRU, a computation of more metrics took place. Table 21 presents the values of AUH, EUH alongside with the corresponding MGL calculations. These metrics are calculated using the F1 scores and MAE measured in the Category 1 of experiments. Because of the size of experiments only some of the measurements are used. To compare the models the interest concentrates on MGL values, where lower means better.

In terms of MGL and Classification Accuracy, the SAED models achieve lower values than the WGRU on all the test devices. Thus, SAED shows great generalization ability when detecting on/off events. Also, the negative values of MGL indicate that the SAED models perform better on the unseen houses than on the seen house. Regarding the MGL and Estimation Accuracy, mixed results are observed with the SAED showing finer values than the WGRU on Dish Washer, Washing Machine and Kettle. As a result, on these test devices, SAED generalizes better than the WGRU in terms of power estimation levels. The above results strongly highlight the generalization power of the SAED approach in the task of NILM.

6.3 Scalability Comparison

An important and frequently neglected parameter when comparing models is the inference time. In a large scale application a deployed disaggregation model will be fed with batches of data from many houses. The cost of this application is critical and depends heavily on the scalability of inference time of the model. Since we don't have access to data from many houses, the scalability is simulated by increasing the time period of disaggregation from one day to 3 months. Next, the three models under investigation are compared by measuring the inference time for the various sizes of test data. The results are summarized in Fig. 3, where 1 day of data is equal to 14351 samples. Given 1 day of test data, inference time of WGRU was 5.77 s, while SAED-add and SAED-dot achieve 1.56 s and 2.27 s respectively. As a consequence, SAED-add model is 3.7 times faster than the WGRU and almost 1.5 times faster than the SAED-dot. For 1 week of test data, SAED models are more than 5.2 times faster than the WGRU completing inference in almost 6.8 s instead of 35.6 s. Given 1 month of test data, the SAED-dot is 5.7 times faster than WGRU with similar test time as the SAED-add.

Table 21. Classification and Estimation Accuracy of the SAED in comparison to the WGRU.

Device	Seen	Unseen	Model	Classification accuracy			Estimation accuracy		
				F1s	AUH	MGL[%]	MAEs	EUH[W]	MGL[%]
Dish Washer	1	2,5	WGRU	0.33	0.26	19.3	13.22	31.29	136.7
			SAED-dot	0.28	0.48	−72.5	13.03	30.42	**133.6**
			SAED-add	0.25	0.45	**−82**	12.03	31.72	163.6
Washing M.	1	2,5	WGRU	0.54	0.29	46.9	16.55	25.02	51.2
			SAED-dot	0.51	0.27	48.1	18.51	23.56	27.3
			SAED-add	0.45	0.26	**43.4**	28.55	35.1	**22.9**
Fridge	1	2,5	WGRU	0.63	0.69	**−9.8**	33.3	34.08	**2.3**
			SAED-dot	0.27	0.59	**−119**	12.11	37.05	205.9
			SAED-add	0.27	0.62	**−129**	10.93	35.17	221.7
Kettle	1	2,5	WGRU	0.66	0.59	9.9	7.35	24.44	232.5
			SAED-dot	0.44	0.45	**−2**	8.57	23.49	174.1
			SAED-add	0.33	0.37	**−10.4**	9.46	21.05	**122.5**
Microwave	1	2,5	WGRU	0.32	0.33	**−1.7**	6.29	12.79	**103.5**
			SAED-dot	0.16	0.26	**−68.6**	7.5	18.07	140.9
			SAED-add	0.18	0.28	**−53.9**	7.61	17.59	131.2

Finally, with test data size of 3 months, the SAED-dot is almost 6.5 times faster than the WGRU and 1.2 times faster than the SAED-add. In this case WGRU inference time was 468.87 s versus 72.56 s of SAED-dot and 87.89 s of SAED-add.

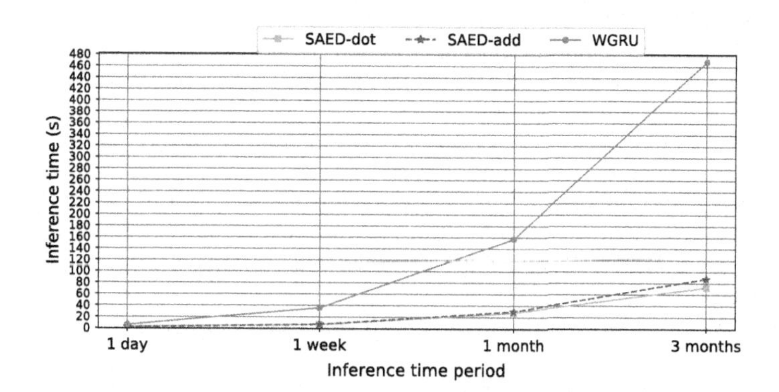

Fig. 3. Inference time versus inference time period for kettle.

7 Conclusions and Proposals for Future Work

In general, the proposed lightweight SAED models showed good performance and in some cases better results in comparison to one of the best and more lightweight models, named WGRU. Interestingly, the SAED seems to perform better on multi-state devices than on two-state devices or devices with simple time signatures. In order to extract more insight on this matter, experiments on different devices should be executed. Furthermore, achieving good performance on the Categories 3 and 4 of experiments, points out the generalization power of the novel architecture. In terms of speed, the SAED models was up to 7.5 and 6.5 faster than the WGRU in training and inference accordingly, resulting that the SAED is more eligible for deployment on embedded systems. Between the SAED-dot and SAED-add, there is not a clear winner, although the SAED-dot has faster training. Additionally, training and testing on different devices and data sets should be executed in order to evaluate and compare these mechanisms in detail.

To summarize, the use of Attention mechanism granted great generalization ability to a simple and light model, making it possible to achieve good performance in short amount of training and inference times. Therefore, Attention may be used in other architectures in order to improve them in the task of NILM.

Acknowledgments. This research has been co-financed by the European Regional Development Fund of the European Union and Greek national funds through the Operational Program Competitiveness, Entrepreneurship and Innovation, under the call RESEARCH-CREATE-INNOVATE (project code: 95699 - Energy Controlling Voice Enabled Intelligent Smart Home Ecosystem).

References

1. Bahdanau, D., Cho, K., Bengio, Y.: Neural machine translation by jointly learning to align and translate. In: ICLR (2015)
2. Basu, K., Debusschere, V., Bacha, S.: Load identification from power recordings at meter panel in residential households. In: 2012 XXth International Conference on Electrical Machines, pp. 2098–2104. IEEE (2012)
3. Basu, K., Debusschere, V., Bacha, S.: Residential appliance identification and future usage prediction from smart meter. In: 39th Annual Conference of the IEEE Industrial Electronics Society, IECON 2013, pp. 4994–4999. IEEE (2013)
4. Cheng, J., Dong, L., Lapata, M.: Long short-term memory-networks for machine reading. In: Proceedings of the 2016 Conference on Empirical Methods in Natural Language Processing, pp. 551–561 (2016)
5. Hart, G.W.: Nonintrusive appliance load monitoring. Proc. IEEE **80**(12), 1870–1891 (1992)
6. Jack, K., William, K.: The UK-DALE dataset domestic appliance-level electricity demand and whole-house demand from five UK homes. Sci. Data **2**, 150,007 (2015)
7. Kelly, J., Knottenbelt, W.: Neural NILM: deep neural networks applied to energy disaggregation. In: Proceedings of the 2nd ACM International Conference on Embedded Systems for Energy-Efficient Built Environments, pp. 55–64 (2015)
8. Klemenjak, C., Faustine, A., Makonin, S., Elmenreich, W.: On metrics to assess the transferability of machine learning models in non-intrusive load monitoring. arXiv preprint arXiv:1912.06200 (2019)

9. Klemenjak, C., Makonin, S., Elmenreich, W.: Towards comparability in non-intrusive load monitoring: on data and performance evaluation. In: 2020 IEEE Power & Energy Society Innovative Smart Grid Technologies Conference (ISGT), pp. 1–5. IEEE (2020)
10. Kolter, J.Z., Jaakkola, T.: Approximate inference in additive factorial HMMs with application to energy disaggregation. In: Artificial Intelligence and Statistics, pp. 1472–1482 (2012)
11. Kolter, J.Z., Johnson, M.J.: REDD: a public data set for energy disaggregation research. In: Workshop on Data Mining Applications in Sustainability (SIGKDD), San Diego, CA, vol. 25, pp. 59–62 (2011)
12. Krystalakos, O., Nalmpantis, C., Vrakas, D.: Sliding window approach for online energy disaggregation using artificial neural networks. In: Proceedings of the 10th Hellenic Conference on Artificial Intelligence, pp. 1–6 (2018)
13. Luong, M.T., Pham, H., Manning, C.D.: Effective approaches to attention-based neural machine translation. In: Proceedings of the 2015 Conference on Empirical Methods in Natural Language Processing, pp. 1412–1421 (2015)
14. Mauch, L., Yang, B.: A new approach for supervised power disaggregation by using a deep recurrent LSTM network. In: 2015 IEEE Global Conference on Signal and Information Processing (GlobalSIP), pp. 63–67. IEEE (2015)
15. Naghibi, B., Deilami, S.: Non-intrusive load monitoring and supplementary techniques for home energy management. In: 2014 Australasian Universities Power Engineering Conference (AUPEC), pp. 1–5. IEEE (2014)
16. Nalmpantis, C., Vrakas, D.: Machine learning approaches for non-intrusive load monitoring: from qualitative to quantitative comparation. Artif. Intell. Rev. **52**(1), 217–243 (2019)
17. Nalmpantis, C., Vrakas, D.: Signal2Vec: time series embedding representation. In: Macintyre, J., Iliadis, L., Maglogiannis, I., Jayne, C. (eds.) EANN 2019. CCIS, vol. 1000, pp. 80–90. Springer, Cham (2019). https://doi.org/10.1007/978-3-030-20257-6_7
18. Nalmpantis, C., Vrakas, D.: On time series representations for multi-label NILM. Neural Comput. Appl. (2020, early access)
19. Srivastava, N., Hinton, G., Krizhevsky, A., Sutskever, I., Salakhutdinov, R.: Dropout: a simple way to prevent neural networks from overfitting. J. Mach. Learn. Res. **15**(1), 1929–1958 (2014)
20. Sutskever, I., Vinyals, O., Le, Q.V.: Sequence to sequence learning with neural networks. In: Advances in Neural Information Processing Systems, pp. 3104–3112 (2014)
21. Symeonidis, N., Nalmpantis, C., Vrakas, D.: A benchmark framework to evaluate energy disaggregation solutions. In: Macintyre, J., Iliadis, L., Maglogiannis, I., Jayne, C. (eds.) EANN 2019. CCIS, vol. 1000, pp. 19–30. Springer, Cham (2019). https://doi.org/10.1007/978-3-030-20257-6_2
22. Zhang, C., Zhong, M., Wang, Z., Goddard, N., Sutton, C.: Sequence-to-point learning with neural networks for nonintrusive load monitoring. In: AAAI (2018)

Enhanced Food Safety Through Deep Learning for Food Recalls Prediction

Georgios Makridis[1](✉), Philip Mavrepis[1], Dimosthenis Kyriazis[1], Ioanna Polychronou[2], and Stathis Kaloudis[2]

[1] University of Piraeus, 185 34 Pireas, Greece
{gmakridis,pmav,dimos}@unipi.gr
[2] Agroknow, 151 26 Marousi, Greece
{ioanna.polyxronou,stathis.kaloudis}@agroknow.com

Abstract. Several application domains/sectors such as logistics, healthcare, industry and transportation, are exploiting the added value of deployed sensors to obtain information relevant to the domain and exploit it in different contexts (e.g. for processes optimization, for actions adaptation, for decision support, etc.). The same applies to the agriculture sector, through the deployment of smart devices and sensors that provide a wealth of datasets for irrigation tuning, crops assessment, food supply chain operations monitoring, etc. Furthermore, emerging machine and deep learning data analytics techniques are utilized as a means to obtain insights and optimize the aforementioned processes. In this context, one significant challenge refers to the enhancement of the food safety across the food supply chain given that goods and products can become unsafe for plenty of reasons, such as mislabeling allergens, contamination etc. To address this challenge, in this paper we introduce a set of deep and machine learning techniques employing time series forecasting to provide insights regarding the risk associated with each product category concerning potential food recalls. Additionally, we propose an approach based on reinforcement learning which utilizes historical recall announcements for predicting future recalls (by their type) that leads to timely recalls and contributes to enhanced food safety across the supply chain. We also evaluate and demonstrate the effectiveness and added-value of the proposed approaches through a real-world scenario that yields promising results.

Keywords: Deep learning · Time series forecasting · Reinforcement learning · Surrogate data · Food safety

1 Introduction

Food safety has attracted considerable public concern and press attention in recent years. Food safety and certification throughout all the supply chain pose very strict requirements, so that global associations such as the FDA ("Food and drug administration") or the WHO ("World Health Organization") are continuously in activity to develop and promote more selective methods of monitoring

A. Appice et al. (Eds.): DS 2020, LNAI 12323, pp. 566–580, 2020.
https://doi.org/10.1007/978-3-030-61527-7_37

and control (FAO/WHO, 2007). Furthermore, many products found in the food-marketplace may be harmful not only for individuals but also for the general economy and health system [5,11].

Given the advances in data analytics, it is of high interest to be able to predict emerging trends in regards with products that have a high probability of been recalled. For example, the most recalled products in the USA are bakery products [5,8]. Utility across different domains, denotes that the extraction of useful knowledge via temporal data [7] is an active research area that could also be exploited in the food domain, with an emphasis on time series analysis, which is motivated by the challenge of reducing future uncertainty. The methods for time series prediction rely on historical data since they include intrinsic patterns that convey useful information for the future description of the phenomenon under investigation.

In this context, the paper proposes:

(a) Deep learning techniques to provide information regarding the risk associated with food products and their potential recalls.
(b) The usage of synthetically produced surrogate data as a way of enriching the original dataset to improve performance of deep learning models.

The techniques mainly focus on time series forecasting and reinforcement learning in order to predict the number of food incidents based on specific industry scenarios. The data are real-world data that have been provided by Agroknow [1].

The remaining of the paper is structured as follows. Section 2 presents related work in the area of study of this paper, while Sect. 3 delivers an overview of the proposed approach, introduces the overall architecture and details regarding the collection of the data streams and how these are utilized within the models. Section 4 presents experiments that have been conducted to demonstrate and evaluate the operation of the implemented algorithms. The performance of the proposed mechanisms is depicted in the results and the evaluation section. Finally, Sect. 5 concludes with a discussion on future research and potentials for the current study.

2 Related Work

Regarding the food industry one promising research took place at Boston University School of Medicine, where a BERT-based AI algorithm that can detect unsafe food based on Amazon customer reviews was developed. The BERT-based AI program identified thousands of recalled products with an accuracy rate of 74% [5]. Also, in [22] drug recalls ordered by FDA were predicted using attributes that quantify the change in query volume at state level. Their results show that future drug recalls can indeed be identified with an AUC of 0.791 and a lift of 5% or approximately 6% when predicting a recall occurring one day ahead. This performance degrades as prediction is made for longer periods ahead, proposing that aggregated Internet search engine data can be used in early warning of medicine's faulty batches.

To the best of the authors knowledge, there is no existing prediction based on pure historical time series data in the pertinent bibliography considering food recalls in time domain representation. On the other hand, there are plenty of applications in other sectors of industry where ML models are leveraged to forecast future events. Nevertheless, as mentioned in [13], where an inspection of the 117 papers discovered by the systematic review of which Fig. 1 illustrates some interesting statistical results.

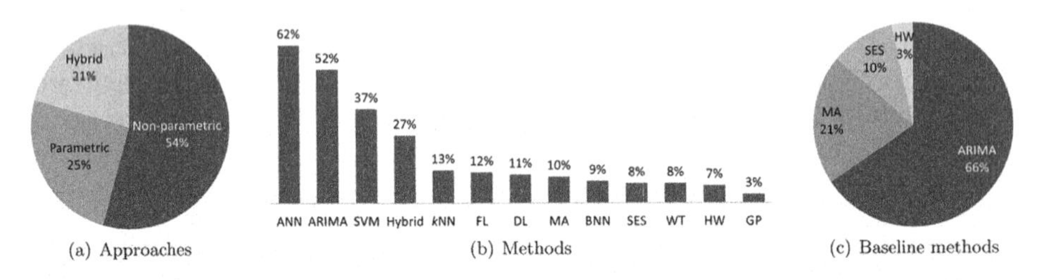

(a) Approaches (b) Methods (c) Baseline methods

Fig. 1. (a) shows the percentage of use of each prediction approach in 68 papers. (b) portrays the frequency in which the most popular methods appeared in the 117 publications. (c) graphically summarizes the algorithms most used as baselines in 29 empirical studies involving both statistical and machine learning methods

Subsequently, a comparison between some of the prevalent algorithms showed that SARIMA is the only statistical method able to outperform, but without a statistical difference, the following machine learning algorithms: ANN, SVM, and kNN-TSPI. However, such forecasting accuracy comes at the expense of a larger number of parameters [13]. Taking into account this research we didn't use the SARIMA method. Instead, we used the GluonTS framework [2] as a deep learning library that bundles components, models and tools for time series applications such as forecasting. Moreover, it should be noted that for time series forecasting, the task of using observed time series in the past to predict values in a look-ahead horizon gets proportionally harder as this horizon widens [10].

Currently, multi-step ahead prediction consists of predicting the next values of the time series. This task is achieved by two different ways. The first one, called independent value prediction, consists of training a direct model to predict the exact steps ahead value. The second strategy, called iterative method, consists of repeating one-step ahead predictions to the desired horizon. The iterative prediction only uses one model to forecast all the horizons needed; the objective is to analyze a short sequence of data and try to predict the rest of the data sequence until a predefined time step is reached. The main drawback of this approach is the accumulative nature of the error.

Beyond the time series domain, a research domain of interest for the given analysis refers to Reinforcement Learning (RL) [20], which refers to algorithms that are "goal-oriented." They can learn how to attain a complex objective, i.e. a goal by maximizing along a specific dimension over a number of iterations. These algorithms are penalized when they make wrong decisions and rewarded when they make the correct ones, this is how they reflect the concept of reinforcement. The power of RL models has been demonstrated in various field, for example in games against world champions and human experts [18]. The field in which RL-based models on time series data are thriving is Finance and Trading. In [3] the authors explore deep reinforcement learning algorithms to automatically generate consistently profitable, robust, uncorrelated trading signals in any general financial market implementing a novel Markov decision process model to capture the financial trading markets.

Presently, RL literature in time series forecasting can be broken down to three main categories: critic-only, actor-only, actor-critic [23].

In the critic-only approach the agent tries to learn a state-action value function Q or an approximation of Q in order to create a mapping $S, A \rightarrow v$ representing the appropriateness of a particular action given the state, where S and A are the state and action spaces accordingly. While there are many implementations falling into this category the most prominent is the deep Q-learning (DQN) with many improvements such as fixed Q-targets, double DQN's, dueling DQN (DDQN) and prioritized experience replay (PER) [9]. For this approach to work it is necessary to either have a discrete action space or discretise a continuous one with methods such as tile-coding [19], coarse coding, function approximation etc.

Regarding the actor-only approach the agent's goal is to optimize the objective function, creating an approximation of an optimal policy, without taking into consideration the value of each action in a state. This allows actor-only methods to be generalized to continuous action spaces. In most of actor-only literature the distribution of a policy is studied using Monte Carlo methods or Policy Gradient Theorem where the model is updated during each episode, while the are some methods (gradient-ascent) used in recent work [9,23] to optimize the objective function. All these methods require lots of samples to correctly approximate the optimal policy as some bad actions can be mistreated as good ones due to the overall reward being positive.

The actor-critic approach comes as an improvement to actor-only approach as it tries to solve its learning problem. Here the policy is updated in real time by alternating between an actor model, controlling the actions of the agent given the specific state, and a critic model assessing these actions.

The main difference of the proposed work compared to the aforementioned ones, is that we introduce an innovative actor-only approach to map the problem as a time series case. More specifically, the approach includes the description of a suitable preprocessing technique for the acquired dataset, and the realization of a RL agent that can estimate the four month ahead time-based prediction of 30 categories. The main contribution of this paper is a framework leveraging a custom environment for food recalls, and a customized and optimized reward function, for food recalls forecasting.

3 Proposed Approach

3.1 Context Formulation

In order to incorporate RL in the prediction of recalls in the food domain, the latter needs to be defined and framed as a RL problem. To this, we define a custom environment to express the 'context' which our agent can interact with, a custom set of actions (A), which is specific to the time series at hand and a reward $Ra(s, s')$ function (1). The environment consists of a set of states (S) where each state is set to be an array of the last twelve data points of the time series. A shallow Neural Network was employed as the mechanism to select the appropriate action for the agent. The agent presented in this paper, is trained to predict the next value in a products' time series in terms of percentage change. To facilitate this, the actions available are a discrete set of numbers (20), drawn from statistical features of each time series (distribution, mean, median, min and max) which the agent can predict. A uniform experience replay mechanism is also employed to help the agent remember its past experiences and their corresponding rewards due to the nature of time series forecasting problems. The design of the reward function depends on the actions we want the agent to favor. In this work, the reward Rt at time t is:

$$R_t = 1/log(MSE(Y_t, \hat{Y_t})) + k \tag{1}$$

where Y_t is the actual number of recalls for the given time step, $\hat{Y}_t$ is the predicted one and k is a small positive integer.

3.2 Dataset

The utilized dataset comprises of 30 time series, representing the daily number of food recalls of 30 categories since 2000. Figure 2 depicts some indicative categories of interest such as "cereal and baked products", "crustaceans", "fish", "fruits and vegetables", "herbs and spices", "meat", "nuts and seeds" and "poultry". Figure 3 shows the seasonality component of the examined time series, highlighting that the majority of food recalls occurs on Fridays (considering days of week as time frame).

Additionally, it is observed that during the summer months we have less food recalls than in other months. Furthermore, the aforementioned streams of data are represented and handled as time series data. Thus, the stationarity of the data needs to be checked by using augmented Dickey-Fuller testing [4] to prove whether they are stationary. This action is of major importance for any predictive method that exploits historical data since these methods are usually based on the assumption that the data generation mechanism does not change over time.

To effectively predict future behavior of a time series the need arises to firstly reject the "white-noise" null hypothesis and then, if nonlinear methods such as machine learning, deep learning and reinforcement learning are to be used, reject

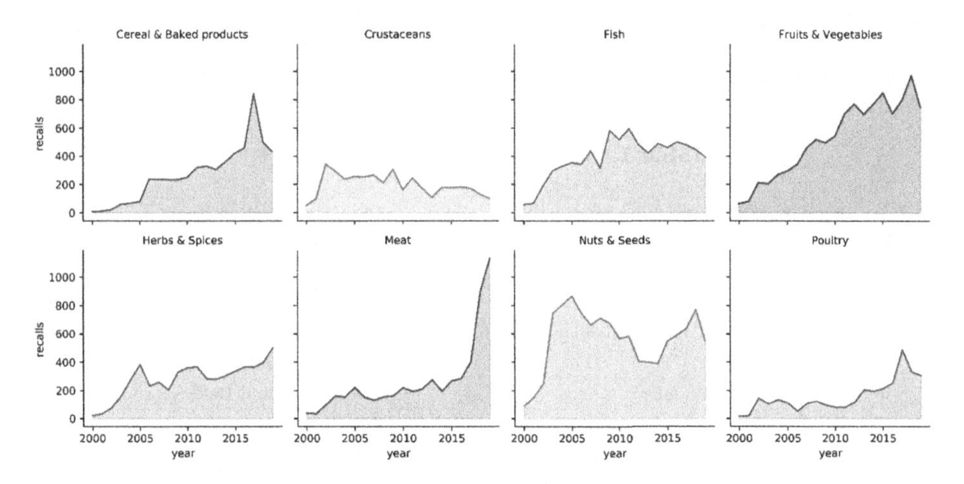

Fig. 2. Food recalls of specific categories in time domain representation

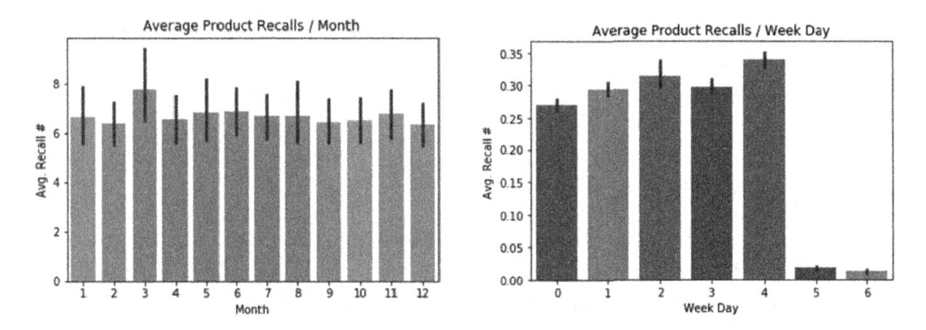

Fig. 3. Average product recalls aggregated by month and day of the week

the null hypothesis regarding the existence of solely temporal linear correlations in the time series.

A statistically sound framework for the aforesaid test is that of surrogate data. It refers to time series data that reproduce various statistical properties like the autocorrelation structure of a measured data set. The null hypothesis is represented by the surrogate data which are compared with the original data under a nonlinear discriminating statistic to reject or approve the null hypothesis. In this work the usage of surrogate data was twofold; Test the nonlinearity hypothesis and enrich existing dataset with the newly generated data. The latter is proposed due to the nature of most deep learning models which can benefit from the increase both in diversity and volume of training data. It is expected that the usage of analogous data will help the generalization of some if not all deep learning models in the case of time series forecasting.

Even though there are multiple methods to generate such data the IAFFT method [6] was chosen as described in the next subsection. The original dataset size after preprocessing is 30 * 147 and produced surrogate data matches its size, providing a combined dataset with of 30 * 294 data points. During our analysis, we refer to the corresponding data based on their ID number as cited in Table 1.

Table 1. Categories of time series dataset

0	Food Additives and Flavorings	15	Other food products
1	Fruits and Vegetables	16	Food contact materials
2	Poultry meat and products	17	Non-alcoholic beverages
3	Prepared dishes and snacks	18	Fats and Oils
4	Dietetic foods, Food supplements, fortified foods	19	Sugars and Syrups
5	Feed materials	20	Nuts, Seeds and products
6	Honey and Royal Jelly	21	Cereals and Bakery products
7	Bivalve mollusks and products	22	Crustaceans and products
8	Meat and products	23	Eggs and products
9	Feed additives	24	Alcoholic beverages
10	Milk and Milk products	25	Cocoa, Coffee and Tea
11	Herbs and Spices	26	Gastropods
12	Confectionery	27	Cephalopods and products
13	Soups, Broths, Sauces and Condiments	28	Pet Feed
14	Fish and products	29	Ices and Desserts

3.3 Surrogate Data

The most commonly used techniques for generating surrogate data for statistical analysis of nonlinear processes include random shuffling of the original time series, Fourier-transformed surrogates, amplitude adjusted Fourier-transformed (AAFT surrogates), and iterated AAFT surrogates (IAAFT) [6]. In our work we incorporated the IAAFT method to addresses the issue of power spectrum whitening, as the main drawback of AAFT method, by performing a series of iterations in which the power spectrum of an AAFT surrogate is adjusted back to that of the original data before the distribution is rescaled back to that of the original data.

Through this proposed approach, with each iteration the change to the distribution that occurs when the Fourier amplitudes are adjusted will be smaller than in the previous iteration, and thus the alteration of the power spectrum when the rescaling is performed will also be smaller than in the previous iteration. In fact, Schreiber [17] showed that, for a nonlinearly transformed autoregression process, the iteration procedure will converge towards the power spectrum of the original data until a saturation point is reached, where the Fourier amplitude adjustment is so small that the rescaling puts the data into the exact order it had before the amplitude adjustment [16].

3.4 Preprocessing

Based on the real industry requirements, we have identified the main parameters that need to be predicted. These refer to the number of recalls and the rate of change for every product category. It should be noted that in terms of predictions, the timeframes of the recalls may be 4 months, 6 months, or 12 months. Since the predictions will assist the quality assurance and food safety professionals to ensure the continuity of their supply chain, minimize future risks and financial losses, we chose a 4-month prediction window. Based on the latter, three options stand out:

i) Use the dataset as is, with its daily frequency resulting in a 120-time step window of prediction. This is not recommended as most of the time series produced are sparse leaving no obvious pattern to be learnt from.
ii) Resample the data in weeks resulting in a 16-time step window of prediction. While some patterns begin to emerge, the data in some cases are still very sparse and the cumulative error from the 16-step prediction is theoretically relatively big.
iii) Resample the data in months providing stationary time series with visible patterns and a lower theoretical accumulated error of prediction, since the window has been reduced to 4-time steps.

After performing evaluations on the above three scenarios, we confirmed our initial hypothesis regarding the proposed resampling, thus the third case. The forecasting task was addressed as a univariate and multivariate problem utilizing all the given time series assuming that complex non-linear feature interactions are present in our data as different categories of food recalls are concerned. Consequently, taking also into consideration that 23 out of 30 in total time series were stationary according to the Augmented Dickey–Fuller test that was conducted in the first place, we used the percent- age change of the given time series as the greatly different statistical features between products' time series (min, max, variance etc.) would have a negative effect on some model's performance while favor some others. This enables the model to generalize better (arithmetic rate) while it also transforms non-stationary time series to stationary. Although normalization is a common preprocessing step, we chose not to apply it since we wanted both, our prediction and data, to be readable and understandable. Specifically, regarding the given data, to apply percentage change $((Xnew - Xold))/Xold$, we had to deal with zero values. In most of the equivalent tasks handling those cases requires domain expertise as there is no "right" methodology. We added one recall across all time- series, which is insignificant and does not convey real change, to avoid having a zero in the percentage change denominator. Furthermore, the possibility of applying the logarithmic transformation of the percentage change was examined, as it both rescales and normalizes the data.

3.5 Data Driven Models

Several models have been utilized through the GluonTS framework. These models are briefly introduced below:

- A **DeepAR Estimator**, which implements an (Recurrent Neural Network) RNN-based model, close to the one described in [15]. More specifically it applies a methodology for producing accurate probabilistic forecasts, based on training an autoregressive recurrent neural network model on many related time series. Recurrent Neural Network is a feed-forward neural network that has an internal memory. For making a decision regarding every output of every layer, it considers the current input and the output that it has learned from the previous input.
- A **Simple Feed Forward Estimator** which implements a simple Multi-layer Perceptrons (MLP) model predicting the next target time steps given the previous ones. MLP is a supervised learning algorithm that learns a function by training on a dataset, where n is the number of dimensions for input and o is the number of dimensions for output. Given a set of features $X = x1, x2, ..., xn$ and a target y, it can learn a non-linear function of approximation for either classification or regression. Implemented as described and depicted in [14].
- **One Deep Factor Estimator**, an implementation of the 2019 ICML paper [21]. It uses a global RNN model to learn patterns across multiple related time series and an arbitrary local model to model the time series on a per time series basis. In the current implementation, the local model is an RNN.
- **Seasonal Naive Estimator**, based on data seasonality. This model predicts $Y(T + k) = y(T + k - h)$ where T is the forecast time, $k \epsilon [0, predictionlength - 1]$ and h = season length. If a time series is shorter than season length, then the mean observed value is used as prediction.
- A **WaveNet Estimator** based on WaveNet architecture [12] were used. WaveNet, is a deep neural network created for generating raw audio waveform. The model is fully probabilistic and auto-regressive, with the predictive distribution for each audio sample conditioned on all previous ones, yielding state of the art results in tasks such as text-to speech conversions, source disaggregation etc.

Besides implementing the aforementioned models, a reinforcement learning approach was developed, including both a custom environment and agent and is shortly presented below.

An RL agent interacts with its environment in discrete time steps. At each time t, the agent receives an observation O_t, which typically includes the reward R_t. It then chooses an action from the set of available actions At, which is subsequently sent to the environment. The environment moves to a new state and the reward associated with the transition (S_t, A_t, S_{t+1}) is determined. As mentioned above with Q-learning the agents seeks to learn a policy that maximizes the total reward for the selected set of actions in given environment. In practice, a Q-table is a [state, action] table where the values of each action are stored. In this paper a DQN architecture with experience replay was implemented as seen below:

Algorithm 1: Deep Q-learning with experience replay

```
1  Initialize replay memory D to capacity N
2  Initialize action function Q with random weights θ
3  Initialize target action function Q̂ with weights θ̄
4  for episode = 1,M do
5      Initialize the environment and get the state s_1
6      for t = 1,T do
7          with probability ε select a random action α_t
8          otherwise select α_t = argmax_α Q(s_t, α; θ)
9          execute action α_t in the environment and observe the reward r_t and
           next state s_{t+1}
10         store transition (s_t, α_t, r_t, s_{t+1}) in D
11         use a random batch of transitions from D
12         if episode terminates at step j + 1 then
13             set y_j = r_j
14         end
15         else
16             set y_j = r_j + γ max_α Q(s_{j+1}, α; θ̄)
17         end
18         perform gradient descent on (y_j − Q(s_j, α_j; θ))^2 with respect to
           parameters θ
19         every C steps reset Q̂ = Q
20     end
21 end
```

Where γ is the discount factor, used to balance importance of future and immediate reward, α (learning rate) defines the rate in which the newly calculated value of Q affects the old one. While this works for problems with small action and state spaces, some have a prohibitive size of S_t, A_t. In these problems a Neural Network is used to approximate and compress the Q-table, where updating the weights w corresponds to updating the Q-values. One improvement of Deep Q-learning algorithm is using an additional NN with the same architecture but with fixed weights $\hat{w}$ which are updated every n iterations to break the correlation between updated values w and Δ_w.

4 Evaluation Results

In the scope of evaluation, the total training data consists of a table N of dimensions (30, 147), where "30" corresponds to the time series of products and "147" to the time steps of each, corresponding to the last 147 months. In order to obtain more reliable results, we trained every model 47 times on a rolling window of 100 time steps forming 47 tables with dimension size (30, 100) (ex. [30, 1–100], [30, 2–101] etc.). While all models share some hyper-parameters such as epoch size (200), the prediction and context length of four (4) and eight (12) months accordingly, some others are specific only to some models.

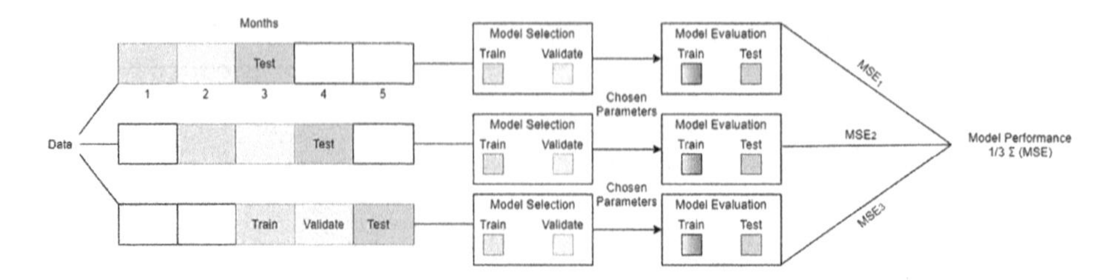

Fig. 4. Proposed one step forward validation scheme

These specific parameters were fine-tuned using grid-search techniques and intuition, as a fundamental step of ML pipeline.

In common ML usage, cross-validation methods, such as the k-fold cross validation provide an elegant way to evaluate the results. It should be mentioned that this method requires shuffle and randomly spilt the data in k different folds. A typical extension of k-fold cross validation is the leave one out validation, which creates an even more extreme peeking into the future considering the shuffle and random split. These issues reflect a pitfall in a time series forecasting framework, as they result in a significant overlap between train and test data. Thus, the better approach is to simulate models in a "walk-forward" sequence, periodically re-training the model to incorporate specific chunk of data available at that point in time. This procedure is depicted in Fig. 4. Regarding the validations scheme followed in the case of the RL model, the Algorithm 2 has been applied.

Algorithm 2: DEEP Q-LEARNING with experience replay

1 **for** $validation_fold$ = *1*, $Validation_folds$ **do**
2   Initialize the custom time series environment and get the state s regarding $validation_fold$
3   Load the pre-trained DQN agent as *agent* regarding $validation_fold$
4   initialize list $predictionsgets[]$
5   **for** *step in lengthofstepsahead* **do**
6         $prediction = agent.predict(s)$
7         $state = state[1:].append(prediction)$
8         $predictions.append(prediction)$
9   **end**
10    $evaluate(y_true, predictions)$
11 **end**

Based on the evaluation, we observed the following results in terms of the Mean Squared Error (MSE) with the usage of the original data, the data after applying the preprocessing mentioned in the corresponding section and with the enriched dataset using surrogate data in Table 2. Furthermore the relevant plots are depicted in Fig. 5, showing the estimators on specific time series.

Table 2 is divided into three parts where the comparison of the models in different cases is shown. The first case uses univariate data (predictions are based solely on one time series at a time) while the second uses multivariate

Table 2. Model comparison on Uni/Multivariate analysis & RL approach

	Univariate				Multivariate					RL
	Deep AR	Simple FF	Seasonal	WaveNet	Deep AR	Simple FF	Deep Factor	Seasonal	WaveNet	
0	0.922 + 0.818	1.895 - 0.045	3.268 - 0.001	2.175 - 0.531	0.154	0.198	1.777	0.244	0.187	1.752
1	0.196 - 0.123	0.442 - 0.374	0.787 - 0.681	0.697 - 0.627	1.919	1.851	3.266	3.267	1.736	0.104
2	0.037 + 0.181	0.081 + 0.136	0.144 + 0.319	0.131 + 0.122	5.04	6.678	7.181	23.423	5.681	0.171
3	0.453 + 0.902	1.307 + 0.273	2.765 + 0.021	1.178 + 0.236	0.817	0.959	2.406	1.567	0.862	0.238
4	0.367 +3.644	0.969 + 2.939	1.688 + 6.353	1.432 + 2.269	1.55	1.755	2.806	2.287	1.534	1.256
5	0.076 + 0.991	0.207 + 0.966	0.244 + 0.958	0.234 + 0.849	0.233	0.213	1.805	0.394	0.270	1.011
6	6.021 - 4.700	8.377 - 7.013	8.39 - 5.961	8.093 - 6.912	1.514	1.582	3.155	2.786	1.636	0.309
7	0.539 + 0.480	1.841 - 0.746	3.046 - 0.825	1.511 - 0.556	0.951	1.196	2.748	1.202	1.195	0.282
8	0.228 + 5.020	0.782 + 4.695	1.374 + 4.867	1.235 + 4.018	0.318	0.380	1.943	0.557	0.314	2.52
9	0.284 +0.660	0.93 +0.031	1.811 - 0.244	0.892 - 0.116	4.530	6.278	7.375	8.786	5.777	0.055
10	0.138 + 0.420	0.376 + 0.027	0.557 + 0.230	0.509 - 0.084	1.406	1.484	2.901	3.045	1.428	1.144
11	0.03 + 0.030	0.085 - 0.003	0.106 + 0.038	0.112 - 0.029	0.083	0.07	1.619	0.106	0.083	0.149
12	0.058 + 1.080	0.212 + 0.929	0.236 + 2.527	0.215 + 0.856	0.413	0.420	2.052	0.787	0.428	0.683
13	1.374 - 0.737	3.603 - 2.959	5.399 - 3.707	3.044 - 2.435	2.111	2.792	4.280	5.396	2.654	0.105
14	0.108 + 0.06	0.22 - 0.021	0.394 - 0.15	0.329 - 0.146	3.950	3.893	5.377	8.041	3.619	0.648
15	2.305 + 5.780	7.129 + 0.996	8.789 - 0.403	6.067 + 1.865	0.643	0.828	2.353	1.811	0.800	14.329
16	0.459 + 0.970	0.985 + 0.484	1.609 + 1.436	0.953 + 0.316	1.165	1.180	2.593	2.763	1.089	0.586
17	0.13 + 0.550	0.531 + 0.149	1.026 + 0.348	0.524 + 0.092	0.921	0.784	2.424	1.181	0.954	0.357
18	0.108 + 0.640	0.228 + 0.596	0.625 + 1.186	0.317 + 0.466	0.621	0.681	2.131	1.692	0.597	0.874
19	0.375 - 0.010	0.836 - 0.465	1.179 - 0.622	1.221 - 0.866	0.812	0.981	2.558	2.221	0.825	0.213
20	0.513 - 0.380	1.76 - 1.617	2.286 - 2.050	2.644 - 2.500	0.085	0.079	1.751	0.144	0.081	0.083
21	2.285 + 0.290	11.554 - 8.806	23.353 - 17.957	9.327 - 6.721	0.842	0.828	2.381	1.609	0.790	3.068
22	0.086 + 0.170	0.277 - 0.066	0.462 - 0.068	0.315 - 0.090	7.311	8.071	9.351	8.387	8.030	0.474
23	0.581 + 5.500	1.889 + 4.299	2.7866 + 5.999	1.858 + 4.023	4.916	5.493	6.682	6.241	5.167	N/A
24	1.575 - 0.600	4.22 - 3.368	8.045 - 6.436	3.744 - 3.002	0.555	0.689	2.191	1.027	0.492	0.593
25	0.348 + 0.120	1.389 - 0.910	1.203 - 0.176	2.595 - 2.135	0.267	0.247	1.764	0.625	0.310	0.084
26	0.615 - 0.360	1.404 - 1.148	2.429 - 1.804	2.345 - 2.068	0.244	0.22	1.74	0.463	0.232	0.232
27	0.314 + 0.490	1.087 - 0.273	2.22 - 1.039	1.036 - 0.274	0.173	0.142	1.777	0.236	0.160	0.053
28	2.129 - 0.530	5.206 - 3.471	6.244 - 3.957	4.68 - 3.126	1.139	1.316	2.657	2.429	1.315	0.622
29	0.313 + 4.500	1.049 + 5.337	1.568 + 21.855	1.02 + 4.731	0.672	0.783	2.394	1.374	0.647	9.747

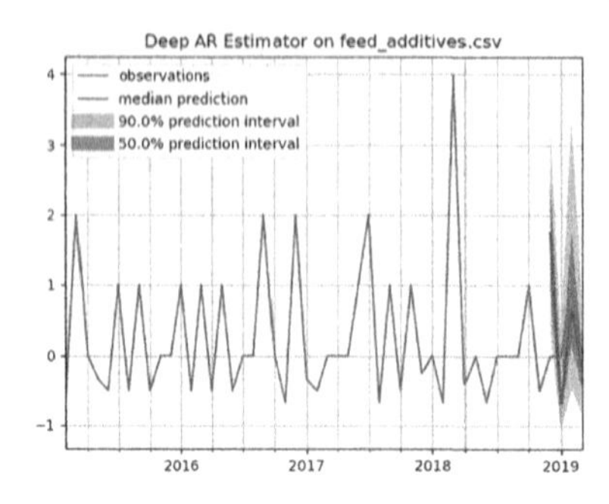

DeepAR Estimator on feed additives time series

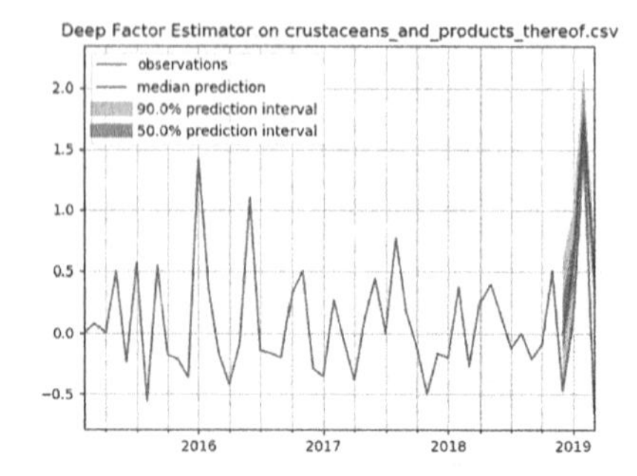

Deep Factor Estimator on crustaceans and products thereof

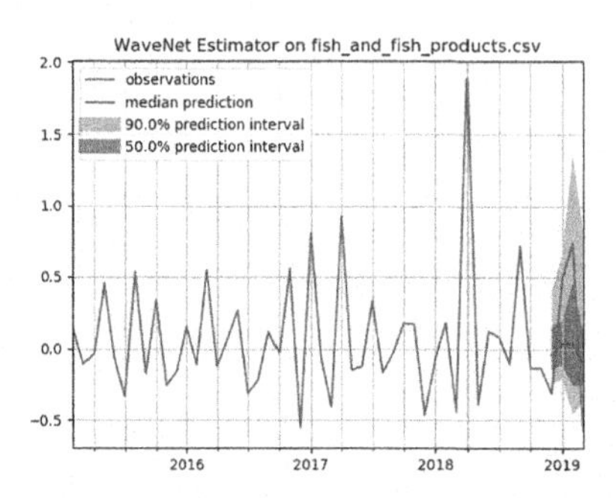

Wavenet Estimator on fish and fish products

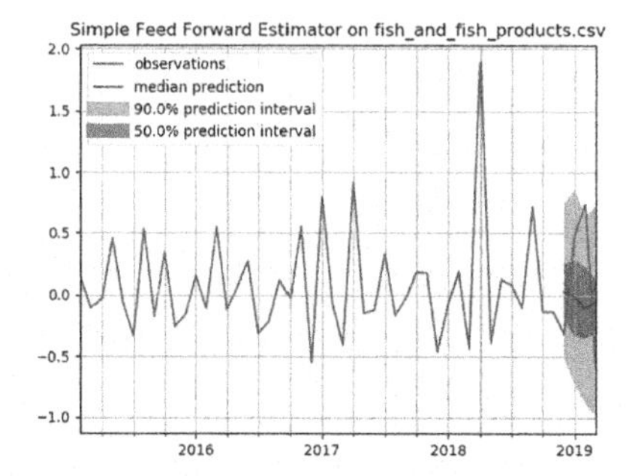

Simple Feed Forward on fish and fish products

Fig. 5. Indicative examples of estimator on specific time series

data using all thirty time series to predict each one. The last part of the table, provides a comparison with the proposed RL approach.

The Deep Factor model is omitted in the univariate case as it is not applicable due to the existence of a global RNN model as described in [21], used to learn patterns across multiple related time series.

The results presented in Table 2, highlight that the model with the best performance in most of the time series is the Deep AR model, while the WaveNet as proposed by Google Deep Mind follows up due to its capability of capture long term dependencies like LSTMs but with less training. The Simple Feed Forward network wins in 4/30 time series in the multivariate setup.

As expected, even though the seasonal model achieves mostly low errors due to the way of predicting values it never outperforms all deep models. Another interesting result is that utilizing univariate model seems to be more accurate. This, contradicts our initial hypothesis that by using multivariate data streams, complex non-linear feature interactions will emerge, facilitating the optimization of the models. As mentioned above the data were normalized in case of different scales. The inferiority of the multivariate dataset arises from the fact that a wide but not deep set of data is exploited rendering harder to distinguish between signal and noise as well as from the lack of correlation between time series.

We can express guarded optimism for the usage of analogous data as an enrichment technique, which can be further explored, since results presented in Table 2, marked in light blue, demonstrate a successful trial of the proposed approach in some cases.

Finally, the RL model that utilized our custom environment yields promising results. According to Table 2, it outperforms all other models in 9/30 datasets. Taking under consideration the fact that the RL model was trained on univariate time series without surrogate data we can reinforce the belief that using univariate data is appropriate for this task.

5 Conclusion

This paper presents and defines an important and growing challenge in which food safety can benefit greatly from modern techniques of time series forecasting. We presented specific approaches in the field of time series forecasting while addressing key challenges that include interpretation, scale, accuracy and complexity (which are inherent in many cases of time series manipulation). Though the experimentation and evaluation, we compared a variety of approaches based on deep neural networks and statistical terms. The complementary model, which consists of an RL (DQN) model, provides promising results in terms of food recalls prediction. Future research work will be focused around applying continuous action space RL model, and utilizing a multivariate and multi-step actions environment. Another interesting and informative task would be to test the statistical characteristics of each time- series in order to evaluate the difference in performance of the models and propose the corresponding adaptations and enhancements.

Acknowledgement. The research leading to the results presented in this paper has received funding from the European Union's Project CYBELE under grant agreement no 825355.

References

1. https://www.agroknow.com/
2. Alexandrov, A., et al.: GluonTS: probabilistic time series models in Python. arXiv preprint arXiv:1906.05264 (2019)
3. Calabuig, J., Falciani, H., Sánchez-Pérez, E.: Dreaming machine learning: lipschitz extensions for reinforcement learning on financial markets. Neurocomputing (2020)
4. Cheung, Y.W., Lai, K.S.: Lag order and critical values of the augmented dickey-fuller test. J. Bus. Econ. Stat. **13**(3), 277–280 (1995)
5. Devlin, J., Chang, M.W., Lee, K., Toutanova, K.: BERT: pre-training of deep bidirectional transformers for language understanding. arXiv preprint arXiv:1810.04805 (2018)

6. Dolan, K.T., Spano, M.L.: Surrogate for nonlinear time series analysis. Phys. Rev. E **64**(4), 046128 (2001)
7. Fu, T.C.: A review on time series data mining. Eng. Appl. Artif. Intell. **24**(1), 164–181 (2011)
8. Gendel, S.M., Zhu, J.: Analysis of US food and drug administration food allergen recalls after implementation of the food allergen labeling and consumer protection act. J. Food Prot. **76**(11), 1933–1938 (2013)
9. Hessel, M., et al.: Rainbow: combining improvements in deep reinforcement learning. In: Thirty-Second AAAI Conference on Artificial Intelligence (2018)
10. Lai, G., Chang, W.C., Yang, Y., Liu, H.: Modeling long-and short-term temporal patterns with deep neural networks. In: The 41st International ACM SIGIR Conference on Research & Development in Information Retrieval, pp. 95–104 (2018)
11. Nyachuba, D.G.: Foodborne illness: is it on the rise? Nutr. Rev. **68**(5), 257–269 (2010)
12. Oord, A.V.D., et al.: WaveNet: a generative model for raw audio. arXiv preprint arXiv:1609.03499 (2016)
13. Parmezan, A.R.S., Souza, V.M., Batista, G.E.: Evaluation of statistical and machine learning models for time series prediction: identifying the state-of-the-art and the best conditions for the use of each model. Inf. Sci. **484**, 302–337 (2019)
14. Pedregosa, F., et al.: Scikit-learn: machine learning in Python. J. Mach. Learn. Res. **12**, 2825–2830 (2011)
15. Salinas, D., Flunkert, V., Gasthaus, J., Januschowski, T.: DeepAR: probabilistic forecasting with autoregressive recurrent networks. Int. J. Forecast. **36**, 1181–1191 (2019)
16. Schreiber, T., Schmitz, A.: Improved surrogate data for nonlinearity tests. Phys. Rev. Lett. **77**(4), 635 (1996)
17. Schreiber, T., Schmitz, A.: Surrogate time series. Phys. D **142**(3–4), 346–382 (2000)
18. Schrittwieser, J., et al.: Mastering atari, go, chess and shogi by planning with a learned model. arXiv preprint arXiv:1911.08265 (2019)
19. Sherstov, A.A., Stone, P.: Function approximation via tile coding: automating parameter choice. In: Zucker, J.-D., Saitta, L. (eds.) SARA 2005. LNCS (LNAI), vol. 3607, pp. 194–205. Springer, Heidelberg (2005). https://doi.org/10.1007/11527862_14
20. Sutton, R.S., Barto, A.G., et al.: Introduction to Reinforcement Learning, vol. 135. MIT Press, Cambridge (1998)
21. Wang, Y., Smola, A., Maddix, D.C., Gasthaus, J., Foster, D., Januschowski, T.: Deep factors for forecasting. arXiv preprint arXiv:1905.12417 (2019)
22. Yom-Tov, E.: Predicting drug recalls from internet search engine queries. IEEE J. Transl. Eng. Health Med. **5**, 1–6 (2017)
23. Zhang, Z., Zohren, S., Roberts, S.: Deep reinforcement learning for trading. J. Financ. Data Sci. **2**(2), 25–40 (2020)

FairNN - Conjoint Learning of Fair Representations for Fair Decisions

Tongxin Hu[1], Vasileios Iosifidis[2], Wentong Liao[1](✉), Hang Zhang[1], Michael Ying Yang[3], Eirini Ntoutsi[2], and Bodo Rosenhahn[1]

[1] Institut für Informationsverarbeitung, Leibniz University of Hanover, Hanover, Germany
{hu,liao,zhang,rosenhahn}@tnt.uni-hannover.de

[2] L3S Research Center, Leibniz University Hanover, Hanover, Germany
{iosifidis,ntoutsi}@l3s.de

[3] Scene Understanding Group, University of Twente, Enschede, The Netherlands
michael.yang@utwente.nl

Abstract. In this paper, we propose *FairNN* a neural network that performs joint feature representation and classification for fairness-aware learning. Our approach optimizes a multi-objective loss function which (a) learns a fair representation by suppressing protected attributes (b) maintains the information content by minimizing the reconstruction loss and (c) allows for solving a classification task in a fair manner by minimizing the classification error and respecting the equalized odds-based fairness regularizer. Our experiments on a variety of datasets demonstrate that such a joint approach is superior to separate treatment of unfairness in representation learning or supervised learning. Additionally, our regularizers can be adaptively weighted to balance the different components of the loss function, thus allowing for a very general framework for conjoint fair representation learning and decision making.

Keywords: Fairness · Bias · Neural networks · Auto-encoders

1 Introduction

The wide usage of AI-based systems, mostly powered nowadays by data and machine learning algorithms, in areas of high societal impact raises a lot of concerns regarding accountability, fairness, and transparency [25] of their decisions. Such systems can become discriminatory towards groups of people or individuals based on *protected attributes* like gender, race, religious beliefs etc., as it has been already showcased in a variety of cases [3,5,9]. For example, [3] shows that Google's ad-targeting system was displaying more highly paid jobs to men than to women, thus making discriminatory decisions based on gender. Such incidents call for methods that explicitly target bias and discrimination in AI-systems, while maintaining their predictive power. The ever increased interest in this area is already reflected in the large, given the recency of the field, body

A. Appice et al. (Eds.): DS 2020, LNAI 12323, pp. 581–595, 2020.
https://doi.org/10.1007/978-3-030-61527-7_38

of literature on fairness-aware learning and responsible AI, in general (see [23] for a recent survey).

However, despite the large number of methods and approaches for fairness-aware machine learning proposed thus far, most of these approaches refer to supervised learning upon a given feature representation. Some approaches that target fair representation learning also exist, e.g., [28] but they focus on learning a fair lower dimensional representation of the data which can be used either as a standalone result (e.g., for visualization purposes), or as an input to some other learning task (e.g., for learning a classifier upon the reduced representation) [26]. Only a few approaches exist that jointly target fairness in both representation learning *and* supervised learning, e.g., [6,19,22,24].

In this work we argue that a joint tackling of fairness in the machine learning pipeline (data → algorithm → model) is superior to the separate treatment of unfairness in representation- or supervised-learning. This is because bias-related corrections in representation learning do not guarantee that a model derived from the corrected data will be fair. Instead, the learning algorithm might still pick up certain data peculiarities that lead to discriminatory outcomes. Therefore, a joint *goal-oriented* consideration in the pipeline is much more effective, as also demonstrated in our experimental results. To this end, we aim for a fair representation learning that preserves as much as possible the original data while obfuscating information on the protected attribute so decisions based on the protected attribute in the latent space are not possible. Additionally, the learned representation should structure itself in such a fashion, that a task-goal, such as a classification task, can still be appropriately solved.

Our contributions can be summarized as follows:

- We propose a neural network that learns a fair representation and a fair classifier *jointly* in an end-to-end manner.
- The contribution of the different components during training can be adjusted, leading to a very flexible and competitive framework.
- Our experiments demonstrate that *FairNN* with a goal-oriented fair representation is superior to a plain fair classifier without explicit representation constraints as well as to a standard fair representation learner without an explicit classification goal.
- The source code is available[1].

The rest of the paper is organized as follows: Related work is summarized in Sect. 2. Necessary background is provided in Sect. 3. Our joint goal-oriented approach to fairness-aware learning is introduced in Sect. 4. Experimental results are presented in Sect. 5. Finally, Sect. 6 concludes our work and identifies interesting directions for future research.

2 Related Work

The domain of fairness-aware machine learning can be categorized into pre-processing, in-processing and post-processing approaches to fairness depending

[1] git@github.com:wtliao/FairNN.git.

on whether they focus on mitigating discrimination at the data, algorithms or model output, respectively.

Mitigating Fairness in Supervised Learning: *Pre-processing approaches to fairness* assume that there exist encoded (e.g., societal) biases in the data which they try to eliminate before "feeding" the data to some learning algorithm. For example, [14] proposes instance re-weighting, label swapping, and data augmentation to eliminate discrimination in the input data. Similar ideas, but for the online scenario, were proposed by [13]. Data augmentation has also been used in [11] in order to force the model so as to learn efficiently all the population segments. In [10] a bagging schema is proposed to equalize the data distributions for the different population segments. In [2] a probabilistic framework for discrimination-preventing preprocessing in supervised learning is introduced with the goal to preserve the utility of the data for the learning task while controlling the correlation between the protected attributes and class and minimizing instance distortion. *In-processing approaches to fairness* aim to explicitly consider fairness into the learning algorithm by constraining or regularizing the model during the training phase. It comprises the most popular category to fairness mitigation, which however depends on the algorithm per se. For example, in [29] the authors tweak the objective function of the linear SVM and Logistic Regression models by inserting convex-concave fairness-related constraints (they use equalized odds as fairness measure). In [15], a fairness-aware splitting criterion for decision trees is proposed that evaluates not only the splitting quality w.r.t. the class but also the discrimination effect of a potential split. The work is extended in [30] for online learning, using Hoeffding Trees as the underlying model. In [12] the authors aim to eliminate discrimination in sequential learning scenarios (in particular, boosting) by dynamically adapting the data distributions over the training rounds using a cumulative version of equalized odds. In [17] it is assumed that there exist latent fair class labels (non-observable) which are estimated via an iterative process. Finally, *post-processing approaches to fairness* work directly at the output of a model and change its outcomes until a chosen fairness notion is satisfied. For example, [7] shifts the decision boundary of AdaBoost w.r.t a protected attribute until statistical parity is achieved. In [8] different thresholds are introduced for different population segments to enforce equal error rates. In [16] the predictions of probabilistic classifiers and ensemble models for instances close to the decision boundary are altered until statistical parity is fulfilled. Our *FairNN* belongs to the category of in-processing approaches as the objective function of the NN is altered to account for fairness. In contrast to the majority of the previous approaches however, our method comprises a *joint* approach for fair-feature representation- and classifier-learning.

Fair Representation Learning Approaches: Fair representation learning aims to learn a transformation to a lower dimensional space where the protected and non-protected groups are indistinguishable. In [28] the authors propose Fair-PCA, an extension of PCA, that forces similar reconstruction errors between protected and non-protected groups. In [18], the Variational Fair Auto Encoder is proposed that is able to also learn fair non-linear functions, which can be used after as input to other learning models. Our *FairNN* also derives non-linear transformations via autoencoders, however on the contrary to [18], we dont only focus on fair-representation learning but rather on joint representation-and classifier-learning. In [27] an approach for learning individually fair representations is proposed using an end-to-end model with autoencoders. On the contrary, our *FairNN* aims at learning representations that are fair for each group (i.e., protected and non-protected).

Closer to our work are the joint approaches [6,19] that aim at both fair representation- and classifier-learning. In [6,19] instead of using some constraining to reduce the dependencies on the sensitive attribute in the latent space (e.g., by minimizing KL-divergence as in our *FairNN*), they train an adversary classifier to discriminate between the protected and non-protected groups. In particular, in [6] they optimize for statistical parity, whereas [19] extends the idea for more fairness measures. It is not clear in what circumstances a constraint-based approach or an adversary one should be preferred [6], but we include [19] in our experimental analysis.

3 Basic Concepts and Definitions

Let $A = \{A_1, ..., A_d\}$ be a d-dimensional feature space of mixed attribute types. We assume the existence of a protected attribute $S \in A$, e.g., $S = gender$. We assume S is binary: $S = \{s, \bar{s}\}$, with s denoting the protected group (e.g., $s = female$), and $\bar{s}$ the non-protected group e.g.., $\bar{s} = male$. An instance $X \in A_1 \times A_2 \cdots \times A_n$ is a d-dimensional feature vector representing an object in the vector space A. Each instance is assigned a label $c \in C$ by some unknown target function $g : A \rightarrow C$. For simplicity, we assume the class attribute is also binary, i.e., $C = \{+, -\}$. We use the notation s_+ (s_-), $\bar{s}_+$ ($\bar{s}_-$) to denote the protected and non-protected group for the positive (negative, respectively) class.

The target function $g()$ is unknown, instead a training set $D = \{(X_i, c)\}$ of i.i.d. instances drawn from the joint attribute-class space $A \times C$ is available and can be used for approximating $g()$. The goal of fairness-aware supervised learning is to approximate $g()$ via a mapping function $f()$ that does not only map correctly future unseen instances of the population from A into C, but also mitigates discriminatory outcomes. The former aspect corresponds to the typical objective of supervised learning achieved through empirical risk minimization. The latter aspect is evaluated in terms of some fairness measure (c.f. Sect. 2).

3.1 Formalizing Fairness

In this work, we employ Equalized Odds [8] (shortly *Eq.Odds*) as our fairness measure. *Eq.Odds* accounts for the percentage difference among protected and non-protected groups in the model's outcomes. In particular, let δFPR (δFNR) be the difference in false positive rates (false negative rates, respectively) between the protected and non-protected groups, defined as follows:

$$\begin{aligned} \delta FPR &= P(c \neq \dot{c}|\bar{s}_-) - P(c \neq \dot{c}|s_-) \\ \delta FNR &= P(c \neq \dot{c}|\bar{s}_+) - P(c \neq \dot{c}|s_+) \end{aligned} \tag{1}$$

where $\dot{c}$ are the predicted labels. The goal of *Eq.Odds* is to minimize both differences:

$$Eq.Odds = |\delta FPR| + |\delta FNR| \tag{2}$$

where $Eq.Odds \in [0, 2]$, with 0 indicating no discrimination and 2 indicating maximum discrimination.

Eq.Odds has become quite popular among recent state-of-the-art fairness-aware methods [8,12,17,19,29]. In contrast to the well-known statistical parity [14], which uses only the positive predicted outcomes without the aid of true labels, or equal opportunity [8], which accounts only for the false negative difference among s and $\bar{s}$, *Eq.Odds* is able to locate discriminatory outcomes for both classes. Furthermore, statistical parity is prone to favor groups by discriminating on specific individuals [4].

3.2 Auto-encoders

An auto-encoder (AE) is an unsupervised neural network that learns an approximation of the identity function such that the output of the network is similar to its input. A reduced/compressed representation is learned by placing constraints in the structure of the network, e.g. by using a bottleneck layer.

In this work, we consider mixed attribute type data of numerical and nominal attributes. Reconstructing the numerical attributes could be considered as a regression task, so we use the *Mean Square Error* as the loss function for numerical attributes. Since for the nominal attributes there is no order among their values, reconstructing their values could be considered as a classification task, so we use the *Cross Entropy* as the loss function for nominal attributes. We assume there exist K numerical and N nominal features, such that $K + N = d$. We combine the feature-type specific loss functions in the overall objective function of the auto-encoder as follows (we compute the loss per batch of B instances):

$$L\left(X, \hat{X}\right) = \frac{1}{B}\sum_{b=1}^{B}\left(\sum_{k=1}^{K}\left(X_{b,k} - \hat{X}_{b,k}\right)^2 - \sum_{j=1}^{N}\sum_{l_j=1}^{M_j} X_{b,l_j}\log\left(p_{b,l_j}\right)\right) \tag{3}$$

where X is the original instance, $\hat{X}$ is the reconstructed instance and X_{ji} is the value of instance j in dimension i. The first term of the above equation refers to the loss of numerical attributes: $X_{b,k}$, $\hat{X}_{b,k}$ denotes the original and reconstructed data of numerical attributes, respectively. The second term of the above equation refers to the loss of nominal attributes. For each nominal attribute j, l_j represents the class label and M_j the number of values of the feature. For the j-th nominal attribute in instance b, X_{b,l_j} has the binary value (positive or negative) which indicates if the class label l_j is the correct classification, p_{b,l_j} represents the predicted probability of class l_j.

4 *FairNN*

In this section, we introduce our proposed method, namely *FairNN* that jointly learns a fair representation and a fair mapping function for classification.

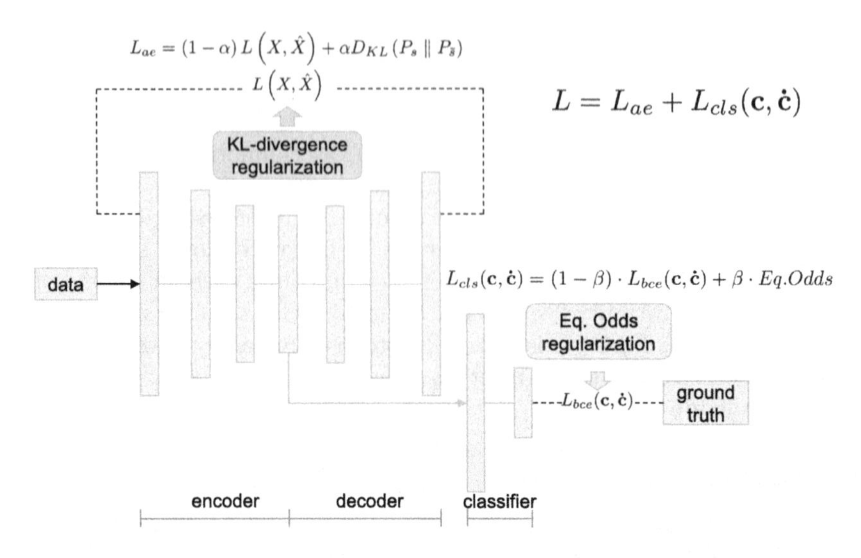

Fig. 1. An overview of *FairNN* that jointly learns a fair representation and a fair mapping function for classification. The auto-encoder (left part) is responsible for representation learning; the KL-divergence constraint forces the representation to be fair. The loss function of the classifier (right part) is tweaked towards fairness through the Eq.Odds regularization. Both aspects are reflected in the joint objective

An overview of our approach is depicted in Fig. 1. The architecture consists of two parts, an auto-encoder block aiming at learning a fair latent representation of the data (left) and a classification block aiming at learning a fair classifier (right). We explicitly consider fairness in the representation learning by adding an additional constraint to the latent space of the auto-encoder in order to obfuscate the information on the protected attribute (Sect. 4.1). Likewise, we explicitly consider fairness in the classification part by adding an additional constraint to the loss function based on the Equalized Odds fairness notion (Eq. 2) (Sect. 4.2). We consider these aspects *jointly* and optimize a multi-loss objective function that balances the importance of the different components in training (Sect. 4.3).

4.1 Fair Representation Learning via KL-Divergence Regularization

In order to learn fair feature transformations for the protected and non-protected groups, *KL divergence* is added to the loss function to train the auto-encoder, which constrains the learned features of different groups to have similar distribution properties. With this constraint, the auto-encoder is trained to mix up the protected attribute information and meanwhile to maintain good reconstruction ability. In practice, we use the *KL divergence* as an additional regularization in the objective function. Based on the values of protected attributes, we divide the data points into protected group s and non-protected group $\bar{s}$. Without loss of

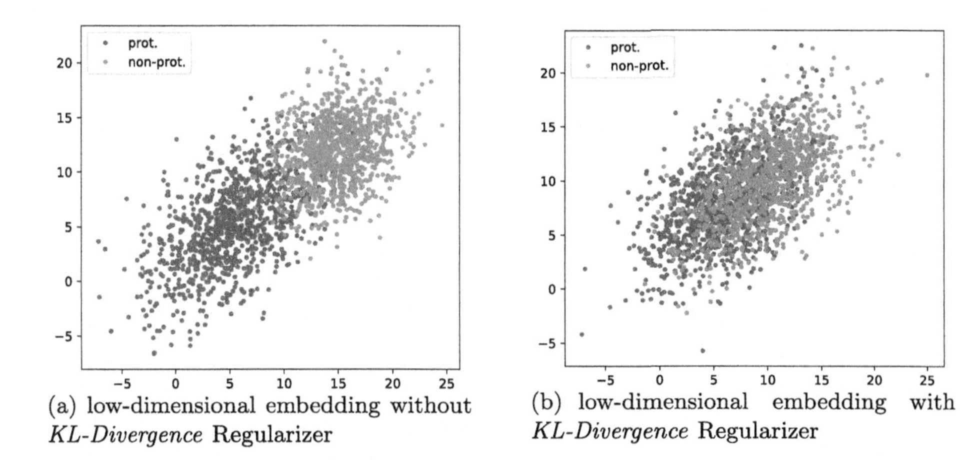

(a) low-dimensional embedding without *KL-Divergence* Regularizer

(b) low-dimensional embedding with *KL-Divergence* Regularizer

Fig. 2. Effect of the *KL-Divergence* Regularizer in (fair) representation learning

generality, we assume their distribution in the latent space as *d-dimensional normal distributions* with means μ_s, $\mu_{\bar{s}}$ and covariance matrices Σ_s, $\Sigma_{\bar{s}}$ respectively. Then, the *KL divergence* between the their distributions is given as:

$$D_{KL}\left(P_s \parallel P_{\bar{s}}\right) = \frac{1}{2}\left(\log\frac{det\left(\Sigma_{\bar{s}}\right)}{det\left(\Sigma_s\right)} - d + tr\left(\Sigma_{\bar{s}}^{-1}\Sigma_s\right) + \left(\mu_{\bar{s}} - \mu_s\right)^T \Sigma_{\bar{s}}^{-1}\left(\mu_{\bar{s}} - \mu_s\right)\right) \tag{4}$$

where, $det(\Sigma)$ is the determinant of the covariance matrix Σ, and $tr(\cdot)$ is the trace of the matrix, which is the sum of elements on the main diagonal of the matrix. With the *KL-Divergence* Regularization, the original reconstruction loss function of the auto-encoder (c.f., Eq. 3) is rewritten as:

$$L_{ae} = (1-\alpha)\, L\left(X, \hat{X}\right) + \alpha D_{KL}\left(P_s \parallel P_{\bar{s}}\right) \tag{5}$$

where $\alpha \in [0, 1)$, is a coefficient for balancing the two terms.

Figure 2 demonstrates the impact of our KL-divergence regularizer, as distribution of data points in a low-dimensional feature space, in contrast to a transformation that has been learned without KL-Divergence regularization. The protected and non-protected groups are denoted in blue and orange respectively. Figure 2(a) shows that the data points belonging to different groups are easy to be separated in the latent space with direct implications to fairness. The regularizer mixes-up the distributions of the two groups making it hard to predict the protected attribute, c.f., Fig. 2(b).

4.2 Fair Classifier Learning via Equalized Odds Regularization

The classifier is an MLP with two FC layers followed by *Relu* activation. The output is a scalar that is squashed by the sigmoid function between 0 and 1 for our binary classification task. The *Binary Cross Entropy* is used as loss function to train the classifier as follows:

$$L_{bce}(\mathbf{c}, \dot{\mathbf{c}}) = -\frac{1}{B} \sum_{n=1}^{b} ((c_b \log(\dot{c}_b) + (1 - c_b) \log(1 - \dot{c}_b))) \tag{6}$$

where c_b is the true label and $\dot{c}_b$ is the predicted probability of the data point b having the label c_b.

Our goal is to improve the fairness performance without losing the classification performance. This motivates us to add an additional fairness measurement as a regularization term in the objective function. As we mentioned before, among different fairness measurements, *Equalized Odds* does not only consider the predicted outcome but also compares it to the actual outcome recorded in the dataset. It considers both the samples with actual positive labels and also those with negative labels. Therefore, *Equalized Odds* (Eq.Odds) is used as the constraint term and added to the classification loss Eq. (6):

$$L_{cls}(\mathbf{c}, \dot{\mathbf{c}}) = (1 - \beta) \cdot L_{bce}(\mathbf{c}, \dot{\mathbf{c}}) + \beta \cdot Eq.Odds \tag{7}$$

where $\beta \in [0, 1)$, is a balancing coefficient between the classification loss L_{bce} and the $Eq.Odds$ fairness regularization.

4.3 Fair Representation and Classifier-Learning via Joint Optimization

By combining the two parts of our network, which are the auto-encoder (Eq. 5) and classifier loss (Eq. 7), the acquired multi-loss function can be expressed as:

$$L = L_{ae} + L_{cls}(\mathbf{c}, \dot{\mathbf{c}}). \tag{8}$$

It is known that neural networks can easily be over-parameterized and tend to overfit, given limited training data. The additional constraints in our architecture, together with the auto-encoder component enforces better generalization, as demonstrated in our experiments (Sect. 5). We implemented *FairNN* in the Python framework using PyTorch.

5 Experiments

We evaluate the predictive and fairness performance of *FairNN* and compare the results with recent state-of-the-art methods. Additionally, we perform several ablation studies to demonstrate the importance of each component in our proposed framework. Accuracy and balanced accuracy are reported for evaluating the *predictive* performance and *Equalized Odds* for *fairness* performance. Since *Equalized Odds* reports the difference between two groups and we also want to maintain the predictive performance for both groups, we also report the actual *TPR* and *TNR* of both groups.

Table 1. An overview of the datasets.

	#Instances	#Attributes	Protected attribute	Protected group	Class ratio (+:−)	Positive class
Adult census	45,175	14	*Gender*	*Female*	1:3.03	>50K
Bank marketing	40,004	16	*Marital status*	*Married*	1:7.57	*Yes*

5.1 Experimental Setup

5.1.1 Datasets

We evaluate our method on two real-world datasets, summarized in Table 1:

- **Adult Census Income Dataset** [1] is extracted from the 1994 American Census Database. The task is to predict whether a person's income is over 50K a year. People with label >*50K* belong to the positive class. $S = gender$ is considered as the protected attribute, $s = female$ the protected group and $\bar{s} = male$ the non-protected group.
- **Bank Marketing Dataset** [21] is collected from a Portuguese bank that focuses on selling long-term deposits over the phone. The task is to predict whether a client will make a deposit subscription. We take $S =$ *marital status* as the protected attribute, $s = married$ the protected group and $\bar{s} =$ *single/divorced* as the non-protected group.

5.1.2 Experimental Settings

The nominal attributes are encoded to one-hot vector and max-normalization is applied to the numerical attributes to ensure the values are in $[0, 1]$. In the auto-encoder block, both the encoder and decoder have three fully-connected linear layers and each is followed by a *ReLU* activation. Following the evaluation setup in [12,17,29], 50% of the data is used for training in which 20% of them are used for validation, and the other 50% is for testing. All experiments are evaluated using 10 random splits. We train the auto-encoder and classifier simultaneously by minimizing the objective function Eq. 8. For training, we use the Adam optimization method, with batch size $B = 512$ and a learning rate 0.002. In order to get the best $\alpha - \beta$ combination (see Eq. (5) and 7), grid search is operated within $\alpha \in [0.4, 0.5, 0.6, 0.7, 0.8, 0.9]$ and $\beta \in [0.1, 0.2, 0.3, 0.4, 0.5]$. Finally, $\alpha = 0.9, \beta = 0.2$ for the Adult Census Income Dataset and $\alpha = 0.8, \beta = 0.4$ for the Bank Marketing Dataset are selected.

To further improve the performance, the well known preferential sampling [14] is applied. The samples are ranked according to their classification scores ascendingly. Centered on classification score 0.5, K samples whose scores >0.5 are duplicated while K samples whose scores <0.5 are skipped. K is computed based on the size of sensitive attributes and labels.

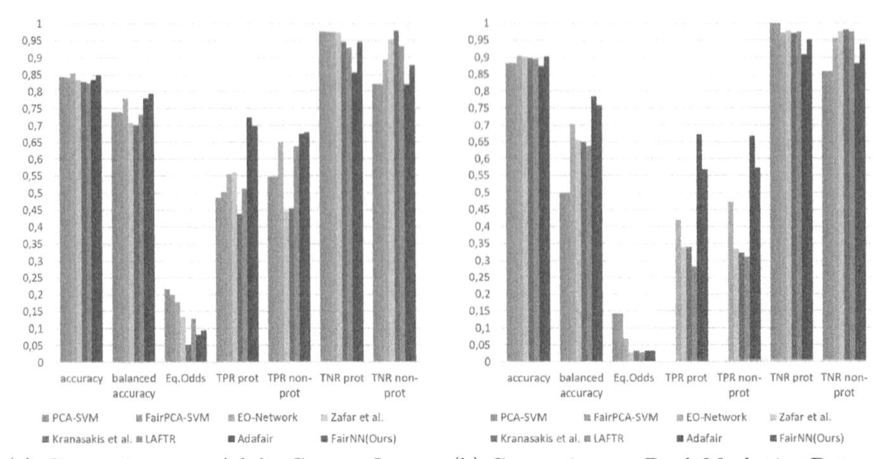

(a) Comparison on Adult Census Income Dataset. (b) Comparison on Bank Marketing Dataset.

Fig. 3. Comparison with the state-of-the-art methods on Adult Census Income dataset and Bank Marketing dataset. For fairness measurement *Eq.Odds*, lower values are better; For others, higher are better.

5.2 Comparison with Other Methods

We compared our approach with the recently proposed state-of-the-art in-processing approaches which mainly aim to minimize *Eq.Odds*.

AdaFair [12]: a boosting model which assigns fairness related weights in each boosting round by observing the cumulative fairness behavior of the ensemble.

LAFTR [19]: a holistic approach that learns a latent fair representation using an encoder/decoder and an adversary (where the encoder/decoder seek to minimize the adversary's objective), and at the same time trains a fair classifier on the latent space.

FairPCA-SVM [28]: aims to find a low dimensional representation of the original data while maintaining similar fidelity for two groups. We project the data to the Fair PCA space and use SVM for binary classification.

PCA-SVM: Similar to FairPCA-SVM, we project the data to the PCA space and use an SVM classifier. This is only a naive baseline method for comparison.

EO-Network [20]: A two-layer neural network, with *Eq.Odds* as a constraint in the loss function. This can be seen as our model without the auto-encoder part.

Krasanakis et al. [17]: In this work, the authors assume the existence of a latent fair class distribution, which they approximate through the CULEP model by re-estimating the instance weights iteratively.

Zafar et al. [29]: In this work, the authors formulate fairness as a set of convex-concave constrains which are embedded in the objective function of a logistic regression model.

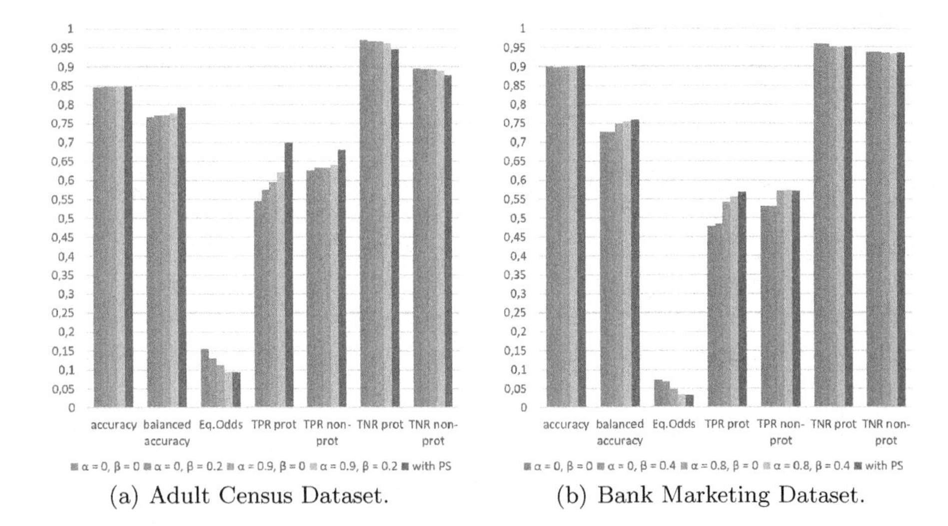

(a) Adult Census Dataset.

(b) Bank Marketing Dataset.

Fig. 4. Ablation Study (for *Eq.Odds* lower values are better - for the rest, higher values are better.)

The experimental results from different methods on two datasets are depicted in Fig. 3, detailed discussion on each dataset follows hereafter. The results of [12,17,29] are taken from [12].

5.2.1 Adult Census Income

Figure 3(a) displays the baselines, state-of-the-art and our final experimental results on the Adult Census Dataset. Our method achieves the highest accuracy and balanced accuracy rates. The lowest *Eq.Odds* is achieved by Krasanakis et al. However, its TPRs for both *protected* (TPR prot) and *non-protected* (TPR non-prot) groups are much lower than the other methods (the lowest TPR prot and the second-lowest TPR non-prot). Fair-PCA aims to learn a fair feature representation in the low-dimensional space. But the learned representation may be unsuited for the binary classification task. It achieves fairer decision-making (lower *Eq.Odds*) comparing to PCA yet performs worse compared to our method. The comparison of our method with EO-Network demonstrates an 8% decrease in *Eq.Odds* and 14% improvement in TPR prot, revealing the effectiveness of generating low-dimensional features. Similar to our approach, LAFTR also leverages the joint-learning thought, but ours is more effective comparing to theirs: balanced accuracy is 5% higher and *Eq.Odds* 3% lower. Our method also brings a significant increase in TPR prot (18% higher). The superior performance from our method indicates that our method is able to learn the fair representation. It balances the balanced accuracy and *Eq. Odds* well.

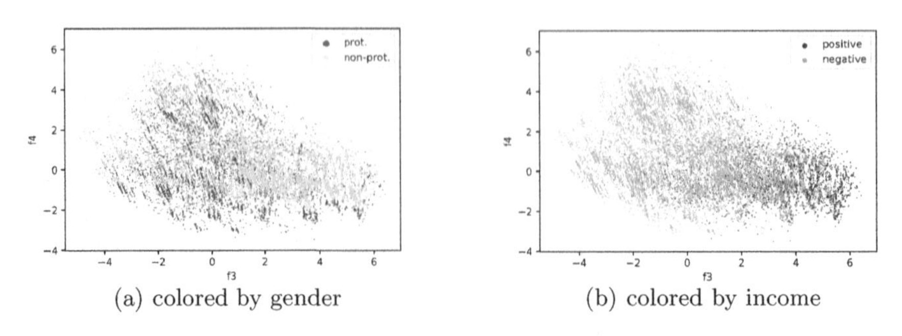

(a) colored by gender (b) colored by income

Fig. 5. Visualization of learned features colored by (a) gender and (b) income.

5.2.2 Bank

In Fig. 3(b), we report experimental results on the Bank Marketing Dataset. Due to the class imbalance problem, both PCA-SVM and FairPCA-SVM perform poorly on this dataset. They output all zeros for the binary classification task which result in balanced accuracy 0.5, TPR prot and TPR non-prot are 0. In EO-Network, the weight parameter of the *Eq.Odds* constraint is the same as used in our method i.e., 0.4. LAFTR reaches the lowest *Eq. Odds* result but its TPRs for both groups are also the lowest. There is a minor difference in *Eq.Odds* between LAFTR and our method, yet ours achieves the much higher balanced accuracy rate, TPR prot, and TPR non-prot. Compared to Zafar et al. and Krasanakis et al., our method reports a higher balanced accuracy rate, higher TPRs for both groups and also the comparable *Eq.Odds*. It proves that our method maintains classification performance while achieving fairness.

5.3 Ablation Study

We perform ablation studies to evaluate how different parts influence the predictive and fairness performance of our method. In Fig. 4, $\alpha = 0$ represents the outcome without *KL-Divergence* regularization and $\beta = 0$ without *Eq.Odds* regularization respectively. Figure 4(a) demonstrates the ablation study on the Adult Census Income Dataset and Fig. 4(b)the Bank Marketing Dataset. We can see that, integrating only the *KL-Divergence* regularization is more effective than integrating *Eq.Odds* regularization only (comparing the second and third bars in Fig. 4(a)). Applying both regularizations further improves the performance (the fourth bars in Fig. 4(a)). Preferential sampling further improves the TPR prot and TPR non-prot while almost not affecting *Eq.Odds*. The ablation study results on the Bank Dataset (as shown in Fig. 4(b)) display a similar tendency yet Preferential sampling does not bring much improvement on TPRs.

5.4 Feature Visualization

To better understand what kind of features are learned from the auto-encoder part, we visualize the extracted features by randomly selecting 2 dimensions

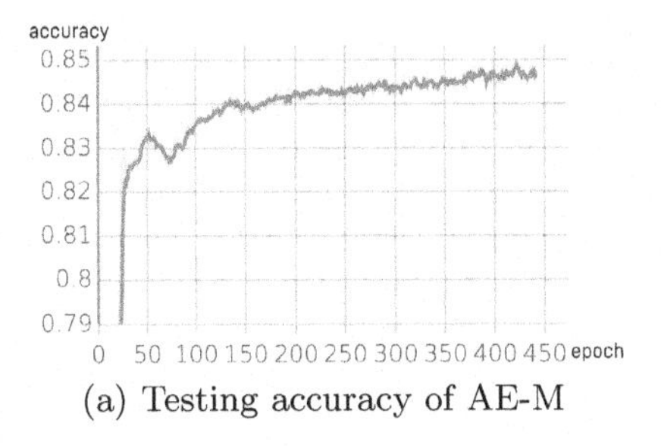

(a) Testing accuracy of AE-M

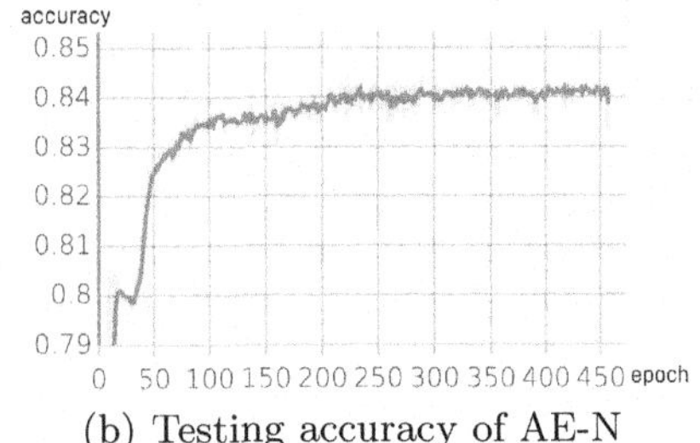

(b) Testing accuracy of AE-N

Fig. 6. Comparison of auto-encoder with MSE+Cross-Entropy loss (left) and with normal MSE loss function (right).

of the 10 dimensional latent space and color them according to the protected attribute (Fig. 5(a)) and by the label (Fig. 5(b)) respectively. Figure 5(a) illustrates that the protected attribute information is mixed up in the latent space, which indicates that the fair representation is learned. Figure 5(b) shows that the label information is distinguishable. The learned representation is not only fair but also suitable for the binary classification task which follows afterwards.

5.5 The Effect of the Multi-loss Function on the Accuracy

In this experiment we evaluate the effect of the multi-loss function on the accuracy and compare the auto-encoder with MSE loss + Cross Entropy loss (we call the network AE-M), to an auto-encoder with normal MSE loss (AE-N). We set $\alpha = 0, \beta = 0$, which means to ignore the Eqs. 5 and 7.

By observing the testing accuracy shown in Fig. 6(a) and Fig. 6(b), we can conclude that AE-M does not perform worse but even achieves a slightly better predictive performance (testing accuracy is 0.76% higher than AE-N).

6 Conclusion

We proposed *FairNN*, a neural network approach for fairness-aware learning that jointly learns a feature representation and classification model. The neural network consists of two parts, an autoencoder component for fair representation learning and a classification component for fair decision making. Our approach optimizes a multi-objective loss function which (a) learns fair representation by suppressing protected attributes (b) maintains the information content by minimizing the reconstruction loss and (c) allows for fair classification by minimizing the classification error and respecting the equalized odds-based fairness regularizer. Our experiments demonstrate that such a joint approach is superior to a separate treatment of unfairness in representation learning or classifier learning. Our method achieves the highest accuracy and balanced accuracy rates. All components are important as demonstrated by the ablation study.

Note that our architecture contains a branch of an auto-encoder which allows unsupervised learning. Thus, our framework is suited for semi-supervised learning with sparsely labeled data. We will elaborate on this aspect in future works.

Acknowledgements. The work is supported by BIAS (*Bias and Discrimination in Big Data and Algorithmic Processing. Philosophical Assessments, Legal Dimensions, and Technical Solutions*) a project funded by the Volkswagen Foundation within the initiative *AI and the Society of the Future* for which the last authors are Principal Investigators.

References

1. Bache, K., Lichman, M.: UCI machine learning repository (2013)
2. Calmon, F., Wei, D., Vinzamuri, B., Ramamurthy, K.N., Varshney, K.R.: Optimized pre-processing for discrimination prevention. In: NeurIPS, pp. 3992–4001 (2017)
3. Datta, A., Tschantz, M.C., Datta, A.: Automated experiments on ad privacy settings. Priv. Enhanc. Technol. **2015**(1), 92–112 (2015)
4. Dwork, C., Hardt, M., Pitassi, T., Reingold, O., Zemel, R.S.: Fairness through awareness. In: Innovations in Theoretical Computer Science 2012, Cambridge, MA, USA, 8–10 January 2012, pp. 214–226 (2012)
5. Edelman, B.G., Luca, M.: Digital discrimination: The case of airbnb. com (2014)
6. Edwards, H., Storkey, A.J.: Censoring representations with an adversary. In: Bengio, Y., LeCun, Y. (eds.) 4th International Conference on Learning Representations, ICLR 2016, San Juan, Puerto Rico, 2–4 May 2016, Conference Track Proceedings (2016)
7. Fish, B., Kun, J., Lelkes, Á.D.: A confidence-based approach for balancing fairness and accuracy. In: Venkatasubramanian, S.C., Jr, W.M. (eds.) Proceedings of the 2016 SIAM International Conference on Data Mining, Miami, Florida, USA, 5–7 May 2016, pp. 144–152. SIAM (2016)
8. Hardt, M., Price, E., Srebro, N.: Equality of opportunity in supervised learning. In: Advances in Neural Information Processing Systems 29: Annual Conference on Neural Information Processing Systems 2016, Barcelona, Spain, 5–10 December 2016, pp. 3315–3323 (2016)
9. Ingold, D., Soper, S.: Amazon doesn't consider the race of its customers. should it. Bloomberg, April 2016
10. Iosifidis, V., Fetahu, B., Ntoutsi, E.: FAE: a fairness-aware ensemble framework. In: 2019 IEEE International Conference on Big Data (Big Data), pp. 1375–1380. IEEE (2019)
11. Iosifidis, V., Ntoutsi, E.: Dealing with bias via data augmentation in supervised learning scenarios. Jo Bates Paul D. Clough Robert Jäschke, p. 24 (2018)
12. Iosifidis, V., Ntoutsi, E.: AdaFair: cumulative fairness adaptive boosting. In: Proceedings of the 28th ACM International Conference on Information and Knowledge Management, CIKM 2019, Beijing, China, 3–7 November 2019, pp. 781–790 (2019)
13. Iosifidis, V., Tran, T.N.H., Ntoutsi, E.: Fairness-enhancing interventions in stream classification. In: Database and Expert Systems Applications - 30th International Conference, DEXA 2019, Linz, Austria, August 26–29 2019, Proceedings, Part I, pp. 261–276 (2019)

14. Kamiran, F., Calders, T.: Data preprocessing techniques for classification without discrimination. Knowl. Inf. Syst. **33**(1), 1–33 (2011)
15. Kamiran, F., Calders, T., Pechenizkiy, M.: Discrimination aware decision tree learning. In: ICDM, pp. 869–874. IEEE Computer Society (2010)
16. Kamiran, F., Mansha, S., Karim, A., Zhang, X.: Exploiting reject option in classification for social discrimination control. Inf. Sci. **425**, 18–33 (2018)
17. Krasanakis, E., Xioufis, E.S., Papadopoulos, S., Kompatsiaris, Y.: Adaptive sensitive reweighting to mitigate bias in fairness-aware classification. In: Proceedings of the 2018 World Wide Web Conference on World Wide Web, WWW 2018, Lyon, France, 23–27 April 2018, pp. 853–862 (2018)
18. Louizos, C., Swersky, K., Li, Y., Welling, M., Zemel, R.S.: The variational fair autoencoder. In: Bengio, Y., LeCun, Y. (eds.) 4th International Conference on Learning Representations, ICLR 2016, San Juan, Puerto Rico, 2–4 May 2016, Conference Track Proceedings (2016)
19. Madras, D., Creager, E., Pitassi, T., Zemel, R.: Learning adversarially fair and transferable representations. arXiv preprint arXiv:1802.06309 (2018)
20. Manisha, P., Gujar, S.: A neural network framework for fair classifier. arXiv preprint arXiv:1811.00247 (2018)
21. Moro, S., Cortez, P., Rita, P.: A data-driven approach to predict the success of bank telemarketing. Decis. Support Syst. **62**, 22–31 (2014)
22. Navarin, N., Oneto, L., Donini, M.: Learning deep fair graph neural networks (2020)
23. Ntoutsi, E., et al.: Bias in data-driven artificial intelligence systems - an introductory survey. WIREs Data Mining and Knowledge Discovery (2020)
24. Oneto, L., Donini, M., Maurer, A., Pontil, M.: Learning fair and transferable representations. arXiv preprint arXiv:1906.10673 (2019)
25. U.S.E.O. of the President, Podesta, J.: Big data: Seizing opportunities, preserving values. White House, Executive Office of the President (2014)
26. Rudolph, M., Wandt, B., Rosenhahn, B.: Structuring autoencoders. In: Third International Workshop on "Robust Subspace Learning and Applications in Computer Vision" (ICCV), August 2019
27. Ruoss, A., Balunovic, M., Fischer, M., Vechev, M.T.: Learning certified individually fair representations. CoRR abs/2002.10312 (2020)
28. Samadi, S., Tantipongpipat, U.T., Morgenstern, J.H., Singh, M., Vempala, S.S.: The price of fair PCA: one extra dimension. In: Advances in Neural Information Processing Systems 31: Annual Conference on Neural Information Processing Systems 2018, NeurIPS 2018, Montréal, Canada, December 3–8 2018, pp. 10999–11010 (2018)
29. Zafar, M.B., Valera, I., Gomez-Rodriguez, M., Gummadi, K.P.: Fairness beyond disparate treatment & disparate impact: Learning classification without disparate mistreatment. In: Proceedings of the 26th International Conference on World Wide Web, WWW 2017, Perth, Australia, 3–7 April 2017, pp. 1171–1180 (2017)
30. Zhang, W., Ntoutsi, E.: FAHT: an adaptive fairness-aware decision tree classifier. In: Proceedings of the Twenty-Eighth International Joint Conference on Artificial Intelligence, IJCAI 2019, Macao, China, 10–16 August 2019, pp. 1480–1486 (2019)

Improving Deep Unsupervised Anomaly Detection by Exploiting VAE Latent Space Distribution

Fabrizio Angiulli(✉), Fabio Fassetti, and Luca Ferragina

DIMES, University of Calabria, 87036 Rende, CS, Italy
{f.angiulli,f.fassetti,l.ferragina}@dimes.unical.it

Abstract. Anomaly detection methods exploiting autoencoders (AE) have shown good performances. Unfortunately, deep non-linear architectures are able to perform high dimensionality reduction while keeping reconstruction error low, thus worsening outlier detecting performances of AEs. To alleviate the above problem, recently some authors have proposed to exploit Variational autoencoders (VAE), which arise as a variant of standard AEs designed for generative purposes. The key idea of VAEs is take into account a regularization term constraining the organization of the latent space. However, VAEs share with standard AEs the problem that they generalize so well that they can also well reconstruct anomalies. In this work we argue that the approach of selecting the worst reconstructed examples as anomalies is too simplistic if a VAE architecture is employed. We show that outliers tend to lie in the sparsest regions of the combined latent/error space and propose a novel unsupervised anomaly detection algorithm, called *VAEOut*, that identifies outliers by performing density estimation in this augmented feature space. The proposed approach shows sensible improvements in terms of detection performances over the standard approach based on the reconstruction error.

Keywords: Anomaly detection · Variational autoencoder · Nearest neighbor density estimation

1 Introduction

Outlier detection is a fundamental and widely applicable discovery problem. Outliers can arise due to many reasons like mechanical faults, fraudulent behavior, human errors, instrument error or simply through natural deviations in populations. Generally speaking, the problem of outlier detection consists in isolating samples suspected of not being generated by the same mechanisms as the rest of the data. Approaches to outlier detection can be classified in supervised, semi-supervised, and unsupervised [1,13]. Supervised methods take in input data labeled as normal and abnormal and build a classifier. The challenge there is posed by the fact that abnormal data form a rare class. Semi-supervised methods,

A. Appice et al. (Eds.): DS 2020, LNAI 12323, pp. 596–611, 2020.
https://doi.org/10.1007/978-3-030-61527-7_39

also called one-class classifiers or domain description techniques, take in input only normal examples and use them to identify anomalies. Unsupervised methods detect outliers in an input dataset by assigning a score or anomaly degree to each object. Several statistical, data mining and machine learning approaches have been proposed to detect outliers, namely, statistical-based [10,14], distance-based [6–9,23], density-based [11,20], reverse nearest neighbor-based [3–5,16,27], isolation-based [26], angle-based [25], SVM-based [28,30], deep learning-based [12,15], and many others [1,13].

Deep learning anomaly detection approaches exploiting autoencoders (AE) have shown good performances [2,12,17]. Autoencoder-based anomaly detection consists in training an autoencoder to reconstruct a set of examples and then to detect as anomalies those inputs that show a sufficiently large reconstruction error. This approach is justified by the observation that, since the reconstruction process includes a dimensionality reduction step (the *encoder*) followed by a step mapping back representations in the compressed space (also called the *latent space*) to examples in the original space (the *decoder*), regularities should be better compressed and, hopefully, better reconstructed [17].

Unfortunately, deep non-linear architectures are able to perform high dimensionality reduction while keeping reconstruction error low. Ideally, an expressive enough architecture could reduce arbitrarily large dimensional data to one dimensional data while performing the reverse transformation with negligible loss. This problem is in part due to the lack of regularity in the latent space. *Variational autoencoders* (VAE) arise as a variant of standard autoencoders designed for *generative* purposes [22]. The key idea of variational autoencoders is to regularize the standard loss function consisting in the reconstruction error by including a regularization term constraining the organization of the latent space. Basically, variational autoencoders encode each example as a normal distribution over the latent space, instead of encoding them as single points, and regularize the loss by maximizing similarity of these distributions with the standard normal distribution. This encoding is conducive to obtain a *continuous* latent space, namely a latent space for which close points will lead to close decoded representation, thus avoiding the severe overfitting problem affecting standard autoencoders, for which some points of the latent space will give meaningless content once decoded.

As already pointed out, variational autoencoders were initially proposed as a tool for generating novel realistic examples by sampling and then decoding points of the latent space. Due to similarities to standard autoencoders some authors also proposed their use to detect anomalies. However, it has been noticed that variational autoencoders share with standard autoencoders the problem that they generalize so well that they can also well reconstruct anomalies [2,12,21,29].

The main contribution of this work can be summarized as follows: we argue that the approach of selecting the worst reconstructed examples as anomalies is too simplistic if a variational autoencoder architecture is employed and, specifically, we show that the anomaly detection process can greatly benefit from taking into account the VAE latent space distribution together with the associated reconstruction error. Indeed, *we show that outliers tend to lie in the sparsest regions of the combined latent and reconstruction error space and propose a novel unsupervised anomaly detection algorithm, called VAEOut, that identifies outliers by performing density estimation in this augmented feature space.* The proposed approach shows sensible improvements in terms of detection performances over the standard approach based on the reconstruction error.

The rest of the paper is organized as follows. Section 2 presents preliminary definitions and discusses related work. Section 3 introduces the *VAEOut* unsupervised anomaly detection algorithm. Section 4 illustrates experimental results. Finally, Section 5 concludes the work.

2 Preliminaries and Related Work

An *autoencoder* (AE) is a deep neural network trained with the aim of outputting a reconstruction $\hat{x}$ of an input sample x as close as possible to x [15,18,24]. An autoencoder consists in two parts, an encoder f_ϕ and a decoder g_θ. An *enconder* f_ϕ is a mapping of a sample from the input feature space to a hidden representation in a *latent space*, and is univocally determined by parameters ϕ. A *decoder* g_θ is a mapping of a hidden representation from the latent space to a reconstruction in the input feature space, and is univocally determined by parameters θ.

Given an autoencoder $\langle f_\phi, g_\theta \rangle$, let x be a sample and let $z = f_\phi(x)$ be the latent variable where the sample x is mapped by the encoder, the *reconstruction* $\hat{x}$ of x is given by $\hat{x} = g_\theta(z) = g_\theta(f_\phi(x))$ and the *reconstruction error* $E(x)$ of the autoencoder is a measure of dissimilarity of x with respect to $\hat{x}$. A common reconstruction error is the *mean squared error* (MSE), defined as

$$E(x) = \|x - g_\theta(f_\phi(x))\|_2^2.$$

The autoencoder tries to minimize the reconstruction error.

A *variational autoencoder* (VAE) is a stochastic generative model aimed at outputting a reconstruction $\hat{x}$ of a given input sample x [22]. To this aim, VAE are composed by an encoder f_ϕ which outputs parameters of $q_\phi(z|x)$, that is the posterior distribution of observing the latent variable z given x, and a decoder g_θ computing parameters of $p_\theta(x|z)$, that is the likelihood of x given the latent variable z. The prior distribution of the latent variable z is denoted by $p_\theta(z)$. Thus, the actual values of z are sampled from $q_\phi(z|x)$. Given the latent variable z, the reconstruction $\hat{x}$ is obtained as a realization of $p_\theta(x|z)$.

As for the distributions associated with the latent variable z, that are $p_\theta(z)$ and $q_\phi(z|x)$, the common choice is the isotropic normal. The distribution of the likelihood $p_\theta(x|z)$ depends on the nature of the data: Bernoulli for binary data or multivariate Gaussian for continuous data. In these cases, $g_\theta(z)$ outputs the mean of the distribution and usually the reconstruction $\hat{x}$ is given by $g_\theta(z)$.

Given a variational autoencoder $\langle f_\phi, g_\theta \rangle$ and a sample x, the *reconstruction error* is represented by the *cross entropy* of the distribution $q_\phi(z|x)$ relative to the distribution $p_\theta(x|z)$:

$$E(x) = -\mathbf{E}_{q_\phi(z|x)} \left[\log p_\theta \left(x|z \right) \right].$$

For example, given x and its reconstruction $\hat{x}$, the corresponding contribution $e(x, \hat{x})$ to the above error is given by $e(x, \hat{x}) = -\log \hat{x}^x (1 - \hat{x})^{(1-x)} = -x \log \hat{x} - (1-x) \log(1-\hat{x})$ for Bernoulli data and $e(x, \hat{x}) \propto -\log \exp -\|x - \hat{x}\|_2^2 = \|x - \hat{x}\|_2^2$ for continuous data.

The reconstruction error can be computed through a Monte Carlo estimation. Thus, by letting L be the number of samples $z^{(1)}, z^{(2)}, \ldots, z^{(L)}$ from $q_\phi(z|x)$,

$$E(x) = -\frac{1}{L} \sum_{l=1}^{L} \log p_\theta \left(x|z^{(l)} \right).$$

The loss of the variational autoencoder is given by

$$L_{\phi,\theta}(x) = -\mathbf{E}_{q_\phi(z|x)} \left[\log p_\theta \left(x|z \right) \right] + \beta \cdot D_{KL}\left(q_\phi(z|x) \parallel p_\theta(z) \right),$$

where the second term represents the KL divergence between the distribution $q_\phi(z|x)$, modelled as a multivariate normal distribution with independent components, and the prior $p_\theta(z)$, modelled as a multivariate normal standard distribution, and plays the role of a regularization term forcing the posterior distribution to be similar to the prior distribution. The hyper-parameter β can be used to balance the two terms of the loss [19]. In such a case, the variational autoencoder is also called a β-VAE.

The classic use of standard AE for anomaly detection is based on the idea that, after the training, these networks are able to better reproduce in output the inlier data than the outlier and, hence, the loss or the reconstruction error of the network is used as an anomaly score [17]. In [2] this idea is applied to VAEs, by using as anomaly score the *reconstruction probability*, corresponding to the negative cross entropy

$$score(x) = recprob(x) = \mathbf{E}_{q_\phi(z|x)}[\log p_\theta(x|z)] = \frac{1}{L} \sum_{l=1}^{L} \log p_\theta(x|z^{(l)}).$$

The experimental results obtained in [2] show that VAE outperforms, in terms of AUC, standard AE and PCA for a semi-supervised anomaly detection setting.

A slightly different approach is pursued in [31], where it is considered the whole negative loss function

$$score(x) = -L_{\phi,\theta}(x)$$

as anomaly score instead of the reconstruction probability, which is only a term of it. The authors justify this choice with the slightly better results they obtain in their experiments compared to reconstruction probability.

It has been observed that sometimes VAEs share with standard AE the problem that they generalize so well that they can also reconstruct anomalies, which leads to view some anomalies as normal data. Thus, in [21] the authors try to overcome this problem by modifying the structure of VAEs in order to make them able to support supervised learning and to be trained with both anomalies and normal data. In particular it is adopted an a priori distribution in the latent space that encourages the separation between normal and anomalous data which leads to non-standard loss function and anomaly score.

3 Method

Let $\mathcal{I}$ denote the *input space* (usually $\mathcal{I} \subseteq \mathbb{R}^d$), let $\mathcal{L}$ denote the *latent space* (usually $\mathcal{L} \subseteq \mathbb{R}^k$ with $k \ll d$), and let $\mathcal{E}$ denote the *reconstruction error space* (usually $\mathcal{E} \subseteq \mathbb{R}$). As above pointed out, the traditional approach pursued to detect anomalies using (variational) autoencoders is to compare the input to its reconstruction by means of the reconstruction error, thus it is based on exploiting only the input and reconstruction error spaces. We argue that the approach of selecting the worst reconstructed examples as anomalies is too simplistic if a variational autoencoder architecture is employed. Specifically, we show that the anomaly detection process can greatly benefit of taking into account the latent space distribution together with the associated reconstruction error.

To illustrate this, we considered the MNIST dataset of handwritten digits and created a training-set consisting of the 6000 digits from the class 0 (the inliers) plus 90 randomly picked digits from the classes 1–9 (the outliers). Figure 1(a) reports the two-dimensional latent space of a variational autoencoder trained on the above set of examples (details on the architecture are provided in Sect. 4). In particular, we reported the means of the distributions associated with training examples (standard deviations are not shown for the ease of visualization): inliers are the (blue) dots and outliers are the (red) asterisks.

First of all we note that, since regular examples (the inliers) form the majority of the data, they will be encoded as distributions better complying with the standard normal one. In other words, the associated latent distributions will tend to distribute around the origin of the latent space and, more importantly, means tend to be closer and supports will overlap more.

Nonetheless, not all the normal data complies with the above behavior and, thus, a non-negligible fraction of inliers spreads also over more peripheral regions. As for the abnormal examples, typically they spread over a wide portion of the latent space, including both boundary regions and the central region of the space, their location depending on the similarities they share with normal examples. This means that *neither the location of the distributions in the latent space nor their degree of overlapping alone are sufficient to separate inliers from outliers.* Indeed, in Fig. 1(a) the sparsest regions of the latent space contain both normal and abnormal examples.

Consider now Fig. 1(b) where the reconstruction error is associated with each latent distribution. It can be seen that even in this case *the reconstruction error*

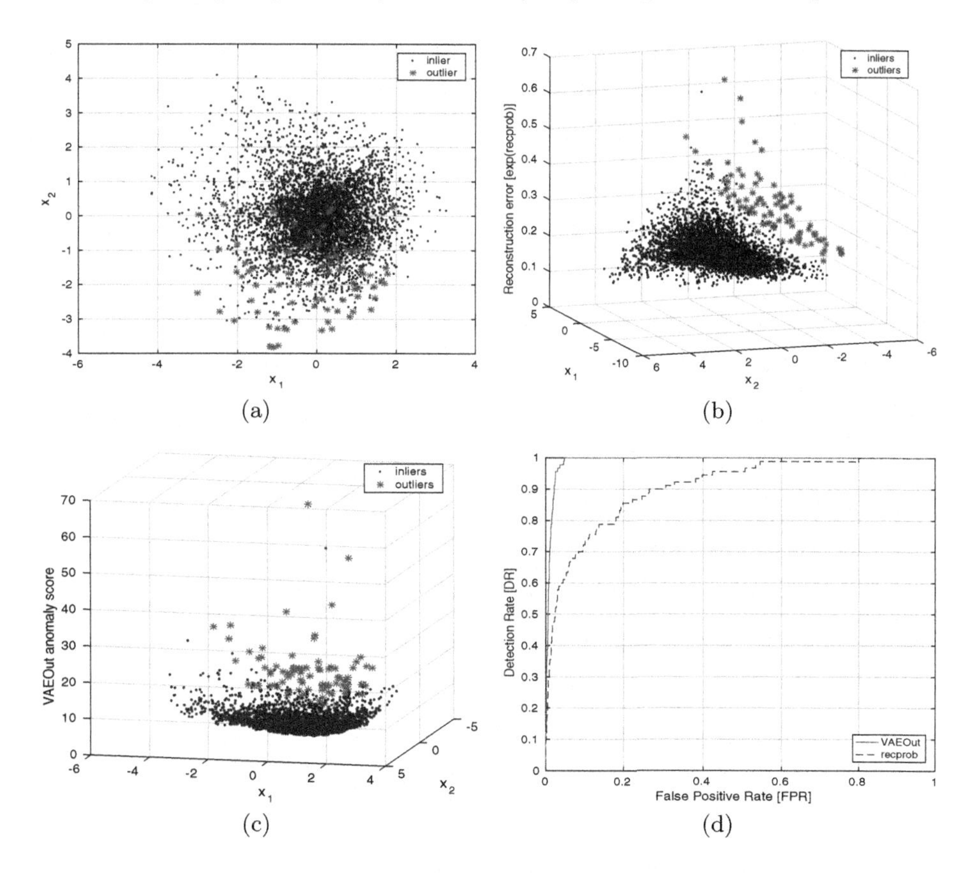

Fig. 1. Comparing *VAEOut* and *recprob* anomaly scores. (Color figure online)

alone is not sufficient to guarantee a good separation between inliers and outliers. Indeed, though some clear anomalies can be recognized by means of a very high reconstruction error, most of the outliers have relatively low reconstruction errors. However, Fig. 1(b) also suggests that *outliers tend to lie in the sparsest regions of the latent/reconstruction error feature space.* This can be understood since outliers have two properties: (1) *they are few,* and (2) *their reconstruction error, even when it is not exceptionally large, is still significantly larger than that of their most similar inliers.* All this tends to move away in the augmented feature space the outliers from the other points.

In light of these observations, the key idea of the proposed approach, called *VAEOut*, is to simultaneously exploit the two above highlighted informations, namely the latent space distribution and the reconstruction error distribution, by *constructing the novel feature space* $\mathcal{F} = \mathcal{L} \times \mathcal{E}$, *consisting of the juxtaposition of the latent space and of the reconstruction error space*, and then by measuring the degree of overlapping of the examples in this novel feature space $\mathcal{F}$, namely the *density* of the distribution of examples. *Outliers will be the points lying in the sparsest regions of the feature space* $\mathcal{F}$.

Specifically, given a dataset $S = \{x_1, x_2, \dots, x_n\}$ our goal is to detect the outliers contained in S. With this aim we first train a variational autoencoder $\langle f_\phi, g_\theta \rangle$ to reconstruct examples in S. Given an example x_i, let z_{x_i} denote the point

$$z_{x_i} = (z_i, \hat{e}(x_i, \hat{x}_i)) \in \mathcal{F}$$

where $z_i \sim q_\phi(z|x_i)$ is a latent space point sampled from the posterior distribution $q_\phi(z|x_i)$ and $\hat{e}(x_i, \hat{x}_i)$ is a measure related to the reconstruction error $e(x_i, \hat{x}_i)$ associated with the reconstruction $\hat{x}_i$ of x_i obtained by means of z_i. Specifically, if $e(x_i, \hat{x}_i)$ is a log-likelihood we can take the exponential $\hat{e}(x_i, \hat{x}_i) = \exp e(x_i, \hat{x}_i)$ since all the other features are on a non-log scale, otherwise $\hat{e}(x_i, \hat{x}_i)$ could be equal to $e(x_i, \hat{x}_i)$.

Given a dataset $S = \{x_1, \dots, x_n\}$, by z_S we denote the transformed dataset $z_S = \{z_{x_1}, \dots, z_{x_n}\}$ and by $\overline{z}_S = \{\overline{z}_{x_1}, \dots, \overline{z}_{x_n}\}$ we denote the standardized versions of z_S, that is the dataset obtained by normalizing each feature according to its mean and standard deviation. Standardization is needed here to handle non-homogeneous features.

To measure the *density* of a point x_i in a set of points S we use *nearest neighbor density estimation* and specifically the *average k-nearest neighbor distance* of point x_i from points in S, denoted as $k\text{-NN}_S(x_i)$. However, instead of employing the distance defined in the original feature space, we employ as distance $\text{dist}(x_i, x_i)$ between x_i and x_j the distance separating their images $\overline{z}_{x_i}$ and $\overline{z}_{x_j}$ in the transformed dataset.

Thus, the *VAEOut anomaly score* of x_i in the dataset S consists of a k-nearest neighbor estimate of the *density of* $\overline{z}_{x_i}$ *in the dataset* $\overline{z}_S$. To take into account Monte Carlo estimation, L samples $z_{x_i}^{(l)}$ ($l \in \{1, \dots, L\}$) can be used for each example x_i and the distance $\text{dist}(x_i, x_j)$ is obtained as the average distance between pair of samples $\overline{z}_{x_i}^{(l)}$ and $\overline{z}_{x_i}^{(l)}$.

Figure 1(c) shows the latent samples and their associated anomaly score. It can be seen that now there is a marked separation between inliers and outliers in terms of the anomaly score. Inliers tend to have low scores, while almost all the outliers are associated with the largest anomaly scores of the population as a consequence of their inherent sparsity. Figure 1(d) compares the ROC curves obtained by our method (*VAEOut*, the solid red line), with the ROC curve obtained by exploiting the reconstruction error of a variational autoencoder (*recprob* [2], the dashed blue line). Note that the AUC = 0.9063 of the standard VAE increases to the value AUC = 0.9908 if *VAEOut* is employed.

Algorithm 1 details the steps of the proposed technique. First of all, a variational autoencoder VAE is trained by exploiting input examples in S. This allows the encoder f_ϕ and the decoder g_θ to output parameters of q_ϕ and p_θ. Next, each example $x_i \in S$ can be mapped to the novel feature space $\mathcal{F} = \mathcal{L} \times \mathcal{E}$. In particular, L mappings of x_i to $\mathcal{F}$ are built. The mappings $z_i^{(l)}$ of x_i to $\mathcal{L}$, with $l \in \{1, \dots, L\}$, are obtained by sampling values from $q_\phi(z|x_i)$ while the mapping of x_i to $\mathcal{E}$ are obtained by considering the reconstruction $\hat{x}_i^{(l)} = g_\theta(z_i^{(l)})$ of x_i

Algorithm 1: VAEOut

Input: Dataset S, number N of outliers (expected contamination), number of k nearest neighbors, number ℓ of runs, parameter β of the β-VAE
Output: The top N outliers

```
// VAE training
1  train a β-VAE ⟨f_φ, g_θ⟩ by using examples in S;
   // map examples x_i to points z_{x_i}
2  foreach run l = 1, ..., L do
3      foreach example x_i ∈ S do
4          sample z_i^(l) ~ q_φ(z|x_i);
5          obtain the reconstruction x̂_i^(l) = g_θ(z_i^(l));
6          build the transformed point z_{x_i}^(l) = (z_i^(l), ê(x_i, x̂_i^(l)));
   // map points z_{x_i} to points z̄_{x_i}
7  foreach feature h = 1, ..., k+1 do
8      μ_h = (1/(nL)) Σ_{i=1}^{n} Σ_{l=1}^{L} z_{x_i,h}^(l);
9      σ_h^2 = (1/(nL)) Σ_{i=1}^{n} Σ_{l=1}^{L} (z_{x_i,h}^(l) − μ_h)^2;
10     foreach run l = 1, ..., L do
11         foreach example x_i ∈ S do
12             z̄_{x_i,h}^(l) = (z_{x_i,h}^(l) − μ_h) / σ_h;
   // compute anomaly scores
13 foreach example x_i ∈ S do
14     foreach example x_j ∈ S do
15         compute the distance dist(x_i, x_j) = (1/L) Σ_{l=1}^{L} ||z̄_{x_i}^(l) − z̄_{x_j}^(l)||_2^2;
16     compute score(x_i) = (1/k) Σ_{t=1}^{k} d_i^(t), where d_i^(t) denotes the t-th smallest
       distance in {dist(x_i, x_j) | j = 1, ..., n};
17 return the N examples x_i scoring the highest values of score(x_i)
```

provided by the decoder, and, then, by computing the measure $\hat{e}(x_i, \hat{x}_i^{(l)})$ related to reconstruction error.

Once the L mappings $z_{x_i}^{(l)}$ of x_i to $\mathcal{F}$ have been generated, they are normalized by standardizing each feature with respect to its mean and standard deviation. Next, the distance between all pairs of examples x_i and x_j can be computed by averaging the Euclidean distances between mappings of x_i and x_j to $\mathcal{F}$. Finally, the k nearest neighbors of x_i according to the above illustrated distance are detected and the outlier score is computed as the mean distance between x_i and such neighbors.

4 Experimental Results

We start by describing the experimental settings. In order to generate an unsupervised setup, we considered some labelled dataset and, for each class label, we created a novel dataset having as inliers all the examples of the considered class and as outliers some randomly picked examples from the other classes. Precisely, we selected s examples ($s \in \{10, 100\}$) from each different dataset class label, so that the total number of outliers is $s \times (m-1)$, where m denotes the number of classes.

In the following we consider the *MNIST*[1] and *Fashion-MNIST*[2] datasets. Both datasets consist of 60000 grayscale 28×28 pixels images partitioned in 10 classes: *MNIST* contains handwritten digits, while *Fashion-MNIST* contains Zalando's article images. The number of outliers within each dataset is also called its (absolute) *contamination* c. Since both the above datasets consist of 10 classes, their contamination corresponds to $c = 9s$.

If not otherwise stated, during experiments the parameter k is held fixed to $0.25c$, thus $k = 15$ for $s = 10$ and $k = 150$ for $s = 100$. Later, we will study the effect of the parameter k on the accuracy. According to the literature [19], we employ large values for the parameter β in order to allow the variational autoencoder to properly organize the latent space, and specifically $\beta = 10^4$. As for the parameter L, we verified that for large β values it has a small impact on the accuracy and, hence, in the following we report results for $L = 1$. All the experimental results are obtained by averaging over ten runs, thus we report both the mean and the standard deviation of performance measures.

As for the autoencoder architecture, the encoding part is composed by an initial sequence of convolutional layers that reduce the size of the data to 14×14, a flattening layer that transforms the data into a vectorial form and two dense layers that brings the data to the latent space having dimension d. The decoder consist in a layer that reshapes the data into a bi-dimensional form and a sequence of convolutional layers that transform the data back into the original 28×28 shape.

***VAEOut* versus *recprob*.** First of all, we investigated the impact of the proposed strategy on the accuracy of the variational autoencoder-based outlier detection approach, by comparing the Area Under the ROC Curve (AUC) of *VAEOut* with that of *recprob*, that is the standard strategy based on exploiting the VAE reconstruction error. Comparisons are conducted by considering the influence of the latent space dimension on the quality of the detection. Figure 2 reports the AUCs of *VAEOut* (red circle-marked lines) and *recprob* (blue square-marked lines) for the latent space dimension d ranging in the interval $[2, 32]$ and $s = 10$. Due to the lack of space, results for $s = 100$ are summarized in Table 1.

The results highlight that the proposed strategy is able to improve accuracy of VAE-based outlier detection. Indeed, in many runs *VAEOut* improves over *recprob*, and for almost all the digits the achieved improvement is sensible.

[1] http://yann.lecun.com/exdb/mnist/.
[2] https://github.com/zalandoresearch/fashion-mnist.

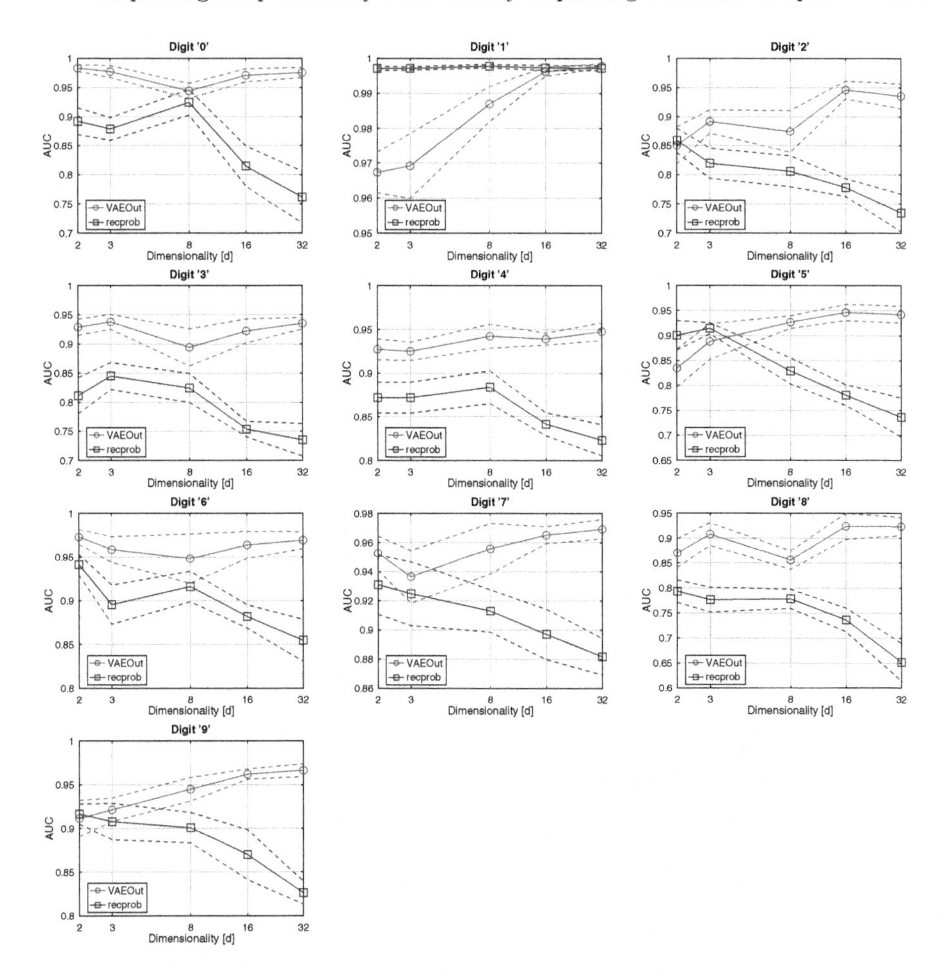

Fig. 2. MNIST dataset ($s = 10$): AUCs of *VAEOut* and *recprob*. (Color figure online)

The experiments also show that accuracy of *VAEOut* is positively affected by the latent space dimension, while this do not seem to be the case for the standard VAE. We explain this behavior since lower dimensions constrain distributions within the latent space to overlap more, thus worsening the separation induced by the density associated with latent points. From these experiments, we conclude that a good choice for the latent space dimension d is in the order of a few tens, namely $d \in [16, 32]$.

Note that intervals of AUC values reported on the vertical axis of the plots are not identical. As for digit 1, it must be pointed out that the variational autoencoder is very able to reconstruct it, probably since it is the easiest digit in the set, and this explains why the *recprob* AUC is very close to 1. *VAEOut* shows a slightly smaller AUC for low latent dimensions, but reaches a similar AUC for sufficiently large dimensions.

Table 1. AUC for the MNIST datasets ($s = 100$).

c	$d = 8$		$d = 16$		$d = 32$	
	VAEOut	*recprob*	*VAEOut*	*recprob*	*VAEOut*	*recprob*
0	*0.928 ± 0.016*	0.767 ± 0.015	*0.945 ± 0.017*	0.743 ± 0.033	**0.954 ± 0.010**	0.603 ± 0.044
1	0.990 ± 0.003	*0.995 ± 0.001*	0.993 ± 0.001	*0.995 ± 0.001*	**0.995 ± 0.001**	0.995 ± 0.001
2	*0.808 ± 0.037*	0.690 ± 0.016	**0.863 ± 0.021**	0.691 ± 0.015	*0.826 ± 0.035*	0.609 ± 0.100
3	*0.866 ± 0.017*	0.726 ± 0.012	**0.898 ± 0.024**	0.708 ± 0.021	*0.887 ± 0.024*	0.663 ± 0.073
4	*0.905 ± 0.013*	0.832 ± 0.008	*0.910 ± 0.015*	0.820 ± 0.011	**0.910 ± 0.010**	0.779 ± 0.023
5	*0.896 ± 0.019*	0.722 ± 0.020	**0.906 ± 0.043**	0.717 ± 0.025	*0.895 ± 0.016*	0.654 ± 0.070
6	*0.934 ± 0.018*	0.830 ± 0.011	**0.944 ± 0.013**	0.817 ± 0.021	*0.941 ± 0.009*	0.735 ± 0.054
7	*0.926 ± 0.021*	0.883 ± 0.007	**0.934 ± 0.014**	0.878 ± 0.004	*0.933 ± 0.006*	0.863 ± 0.008
8	*0.864 ± 0.017*	0.679 ± 0.012	*0.888 ± 0.018*	0.660 ± 0.020	**0.889 ± 0.020**	0.600 ± 0.086
9	*0.921 ± 0.012*	0.850 ± 0.010	**0.940 ± 0.010**	0.841 ± 0.016	*0.936 ± 0.008*	0.747 ± 0.056

Table 2. MNIST dataset *Prec@n* for n set to the contamination $c = 9s$.

c	$s = 10$		$s = 100$	
	VAEOut	*recprob*	*VAEOut*	*recprob*
0	**0.462 ± 0.044**	0.227 ± 0.026	**0.654 ± 0.033**	0.295 ± 0.048
1	**0.762 ± 0.045**	0.744 ± 0.028	**0.904 ± 0.005**	0.898 ± 0.008
2	**0.388 ± 0.042**	0.204 ± 0.039	**0.429 ± 0.050**	0.239 ± 0.083
3	**0.377 ± 0.036**	0.160 ± 0.036	**0.519 ± 0.037**	0.272 ± 0.087
4	**0.497 ± 0.071**	0.453 ± 0.041	**0.598 ± 0.019**	0.516 ± 0.040
5	**0.371 ± 0.032**	0.210 ± 0.044	**0.524 ± 0.027**	0.281 ± 0.080
6	**0.490 ± 0.051**	0.357 ± 0.046	**0.632 ± 0.024**	0.393 ± 0.074
7	**0.528 ± 0.048**	0.493 ± 0.042	**0.647 ± 0.026**	0.624 ± 0.021
8	**0.276 ± 0.049**	0.109 ± 0.045	**0.494 ± 0.051**	0.230 ± 0.080
9	**0.470 ± 0.040**	0.333 ± 0.048	**0.628 ± 0.022**	0.390 ± 0.145

Precision. Another measure employed to evaluate outlier detection approaches is the *Precision*. Specifically, since to goal is to isolate the most deviating dataset examples, we used the *Prec@n* measure, representing the percentage of true outliers among the examples associated with the top n anomaly scores. We set n to the absolute contamination $n = c$. Table 2 compares the *Prec@n* achieved by *VAEOut* and *recprob* on *MNIST* ($d = 32$). The results point out that *VAEOut* is able to significantly increase the percentage of true anomalies among the examples ranked in the very first positions. Moreover, in different cases the precision is doubled.

Note that despite the case $s = 10$ shows slightly larger AUCs, the *Prec@n* is higher for the case $s = 100$. We explain this behavior by noticing that while the inliers of the two datasets are the same, the outliers for the case $s = 100$

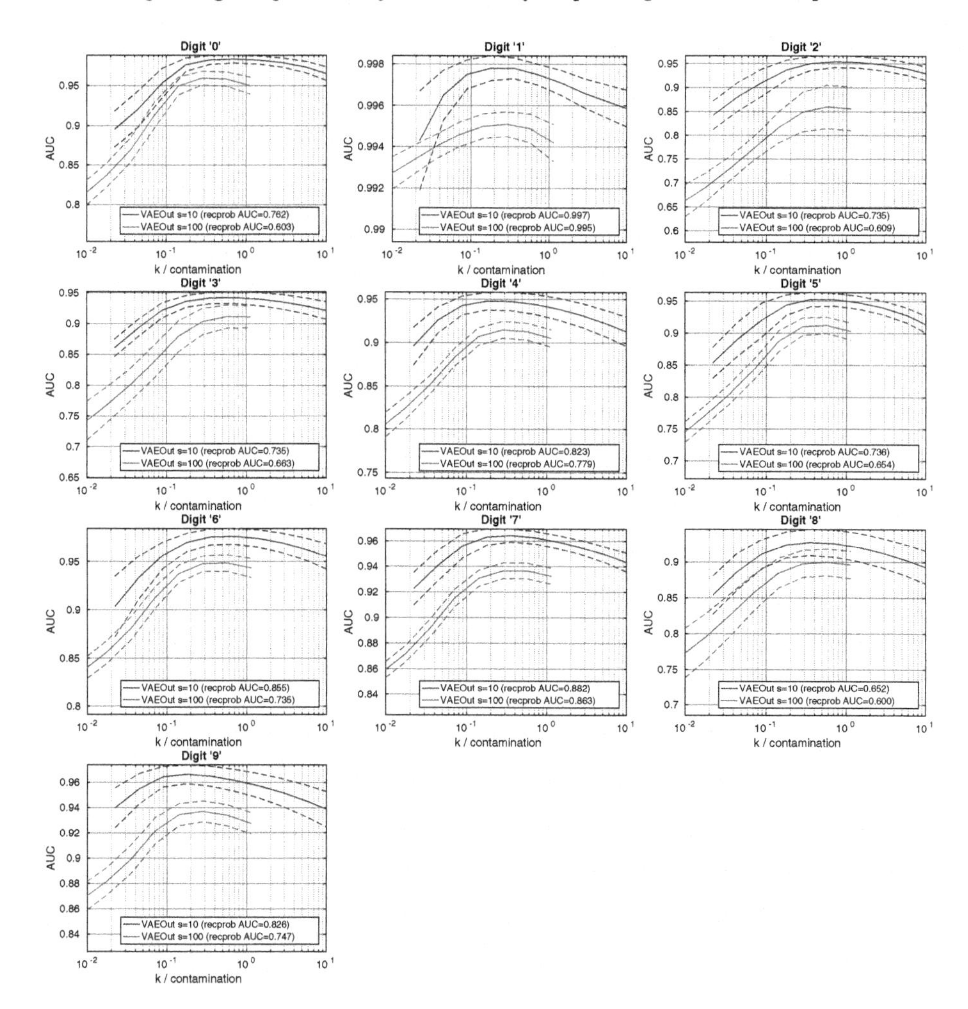

Fig. 3. MNIST dataset: AUC of *VAEOut* for varying k values.

have increased tenfold and this means that the probability that largest scores are assigned to outliers is increased, although overall the outliers are ranked slightly worse according to the AUC.

Sensitivity Analysis for the Parameter k. Experiments reported in Fig. 3 are aimed at determining the optimal value for the parameter k, by performing a sensitivity analysis with respect to this parameter. With this aim, we took into account log-spaced values k in the interval $[2, 1024]$ and determined the AUC of *VAEOut* on the MNIST dataset for $s = 10$ and $s = 100$. In these experiments, the latent space dimension d is held fixed to $d = 32$.

To help understand the effect of k on the accuracy, on the horizontal axis we reported the value $k/c = k/(9s)$ of k normalized on the absolute contamination

Table 3. MNIST ($s = 10$) AUC for $d = 32$ ($k = 30$).

Class	*AE*	*VAE*	*VAEOut*
0	0.7053 ± 0.0525	*0.8147± 0.0443*	**0.9825 ± 0.0056**
1	0.9913 ± 0.0031	*0.9973± 0.0007*	**0.9978 ± 0.0005**
2	0.6407 ± 0.0534	*0.7780± 0.0152*	**0.9504 ± 0.0143**
3	0.6844 ± 0.0354	*0.7535± 0.0133*	**0.9415 ± 0.0092**
4	0.7743 ± 0.0278	*0.8415± 0.0133*	**0.9477 ± 0.0106**
5	0.6776 ± 0.0323	*0.7811± 0.0208*	**0.9523 ± 0.0111**
6	0.7651 ± 0.0282	*0.8819± 0.0133*	**0.9758 ± 0.0083**
7	0.8635 ± 0.0120	*0.8970± 0.0172*	**0.9645 ± 0.0052**
8	0.5993 ± 0.0328	*0.7363± 0.0237*	**0.9277 ± 0.0185**
9	0.7781 ± 0.0449	*0.8698± 0.0287*	**0.9649 ± 0.0079**

Table 4. *Fashion-MNIST* ($s = 10$) AUC for $d = 32$ ($k = 30$).

Class	*AE*	*VAE*	*VAEOut*
T-shirt/top	*0.8388± 0.0146*	0.4701 ± 0.0369	**0.8946 ± 0.0117**
Trouser	**0.9792 ± 0.0048**	0.9520 ± 0.0102	*0.9599± 0.0111*
Pullover	*0.8288± 0.0240*	0.3472 ± 0.0278	**0.8757 ± 0.0138**
Dress	0.6857 ± 0.0101	*0.7867± 0.0242*	**0.8883 ± 0.0132**
Coat	*0.8420± 0.0232*	0.4805 ± 0.0452	**0.8752 ± 0.0153**
Sandal	0.7740 ± 0.0210	*0.8738± 0.0152*	**0.9094 ± 0.0165**
Shirt	*0.7490± 0.0270*	0.3208 ± 0.0190	**0.8419 ± 0.0153**
Sneaker	*0.9587± 0.0129*	0.9322 ± 0.0178	**0.9729 ± 0.0125**
Bag	*0.6763± 0.0503*	0.4269 ± 0.0308	**0.8866 ± 0.0288**
Ankle boot	*0.8905± 0.0189*	0.6860 ± 0.0363	**0.9260 ± 0.0183**

c of the dataset, also called *normalized neighborhood.* Each plots reports also the AUC achieved by *recprob*. It can be seen that for a wide range of values of the parameter k the AUC of *VAEOut* is sensibly larger than that of *recprob*. In most cases the above property is valid for all the reported values of k.

This experiment witnesses that, although *VAEOut* requires an additional parameter with respect to a standard VAE, the selection of the right value for this parameter is not critical, being almost always guaranteed an improvement. Moreover, the optimal value for the normalized neighborhood appears to be located within the interval $[10^{-1}, 10^{0}]$. Thus, the normalized neighborhood provides a tool for selecting a reasonable value for k. As a rule of thumb, we recommend to use $k \approx N/3$, where N is the user-specified expected absolute contamination or, vice versa, to return $N \in [3k, 5k]$ anomalies when k is user-specified.

Impact on the Neural Architecture. In this experiment we compare the detection performances of Auto-Encoder based anomaly detection (*AE*), Variational Auto-Encoder based anomaly detection (*VAE*), and *VAEOut* based anomaly detection. The aim of this experiment is not to determine the best configuration for each approach, but instead to compare the performances of these three autoencoder based approaches when the architecture is held fixed. Thus, all the results are relative to the equivalent network architectures and for the same common hyper-parameters. Specifically, the *AE* has the same structure of the *VAE*, except for employing a deterministic latent space and for the loss consisting only of the reconstruction error, while *VAEOut* builds on the same *VAE* architecture described at the beginning of this section.

Tables 3 and 4 report the AUC of the three methods on the *MNIST* and *Fashion-MNIST* datasets with $s = 10$, respectively, for $d = 32$ and k set to 30, that is to one third of the dataset contamination. While on the *MNIST* dataset *VAE* performs better than *AE*, on the *Fashion-MNIST* dataset with the same loss hyper-parameter β, *VAE* perform worse than the corresponding deterministic architecture.

Importantly, *VAEOut* always shows clear improvements over the corresponding *VAE* architecture. On *MNIST*, for some critical classes, see for example digit 8 of *MNIST*, the performance are resolutely winning. On *Fashion-MNIST*, despite the sometimes poor performances of the *VAE* reconstruction error, by exploiting the latent space information *VAEOut* is able to achieve excellent detecting performances, almost always filling the gap between the *AE* and *VAE* results and going even further.

5 Conclusions

The main goal of this work is to show that, within the context of autoencoder neural networks architectures, the outlier detection process can greatly benefit of taking into account the latent space distribution together with the associated reconstruction error. Specifically, we observed that outliers tend to lie in the sparsest regions of the combined latent/error space and proposed a novel unsupervised anomaly detection algorithm, called *VAEOut*, that exploits this property to identify outliers. The novel approach showed sensible improvements in terms of detection performances over the basic autoencoder architecture to which it is applied. As far as the future work is concerned, we are currently enlarging the experimental campaing and investiganting other measures of density and other rules for combining latent space and reconstruction error.

References

1. Aggarwal, C.C.: Outlier Analysis. Springer, New York (2017). https://doi.org/10.1007/978-1-4614-6396-2
2. An, J., Cho, S.: Variational autoencoder based anomaly detection using reconstruction probability. Technical report, 3, SNU Data Mining Center (2015)

3. Angiulli, F.: Concentration free outlier detection. In: Ceci, M., Hollmén, J., Todorovski, L., Vens, C., Džeroski, S. (eds.) ECML PKDD 2017. LNCS (LNAI), vol. 10534, pp. 3–19. Springer, Cham (2017). https://doi.org/10.1007/978-3-319-71249-9_1
4. Angiulli, F.: On the behavior of intrinsically high-dimensional spaces: distances, direct and reverse nearest neighbors, and hubness. J. Mach. Learn. Res. **18**, 170:1–170:60 (2018)
5. Angiulli, F.: CFOF: a concentration free measure for anomaly detection. ACM Trans. Knowl. Disc. Data (TKDD) **14**(1), 4:1–4:53 (2020)
6. Angiulli, F., Basta, S., Pizzuti, C.: Distance-based detection and prediction of outliers. IEEE Trans. Knowl. Data Eng. **2**(18), 145–160 (2006)
7. Angiulli, F., Fassetti, F.: DOLPHIN: an efficient algorithm for mining distance-based outliers in very large datasets. ACM Trans. Knowl. Disc. Data (TKDD) **3**(1), Article 4 (2009)
8. Angiulli, F., Pizzuti, C.: Fast outlier detection in large high-dimensional data sets. In: Proceedings of International Conference on Principles of Data Mining and Knowledge Discovery (PKDD), pp. 15–26 (2002)
9. Angiulli, F., Pizzuti, C.: Outlier mining in large high-dimensional data sets. IEEE Trans. Knowl. Data Eng. **2**(17), 203–215 (2005)
10. Barnett, V., Lewis, T.: Outliers in Statistical Data. Wiley, Hoboken (1994)
11. Breunig, M.M., Kriegel, H., Ng, R., Sander, J.: LOF: identifying density-based local outliers. In: Proceedings of International Conference on Management of Data (SIGMOD) (2000)
12. Chalapathy, R., Chawla, S.: Deep learning for anomaly detection: a survey (2019)
13. Chandola, V., Banerjee, A., Kumar, V.: Anomaly detection: a survey. ACM Comput. Surv. **41**(3), 1–58 (2009)
14. Davies, L., Gather, U.: The identification of multiple outliers. J. Am. Stat. Assoc. **88**, 782–792 (1993)
15. Goodfellow, I., Bengio, Y., Courville, A.: Deep Learning. MIT Press, Cambridge (2016)
16. Hautamäki, V., Kärkkäinen, I., Fränti, P.: Outlier detection using k-nearest neighbour graph. In: International Conference on Pattern Recognition (ICPR), Cambridge, UK, 23–26 August 2004, pp. 430–433 (2004)
17. Hawkins, S., He, H., Williams, G., Baxter, R.: Outlier detection using replicator neural networks. In: Kambayashi, Y., Winiwarter, W., Arikawa, M. (eds.) DaWaK 2002. LNCS, vol. 2454, pp. 170–180. Springer, Heidelberg (2002). https://doi.org/10.1007/3-540-46145-0_17
18. Hecht-Nielsen, R.: Replicator neural networks for universal optimal source coding. Science **269**(5232), 1860–1863 (1995)
19. Higgins, I., et al.: β-vae: learning basic visual concepts with a constrained variational framework. In: International Conference on Learning Representations (ICLR) (2017)
20. Jin, W., Tung, A., Han, J.: Mining top-n local outliers in large databases. In: Proceedings of ACM SIGKDD International Conference on Knowledge Discovery and Data Mining (KDD) (2001)
21. Kawachi, Y., Koizumi, Y., Harada, N.: Complementary set variational autoencoder for supervised anomaly detection. In: IEEE ICASSP, pp. 2366–2370 (2018)
22. Kingma, D.P., Welling, M.: Auto-encoding variational bayes (2013)
23. Knorr, E., Ng, R., Tucakov, V.: Distance-based outlier: algorithms and applications. VLDB J. **8**(3–4), 237–253 (2000)

24. Kramer, M.A.: Nonlinear principal component analysis using autoassociative neural networks. AIChE J. **37**(2), 233–243 (1991)
25. Kriegel, H.P., Schubert, M., Zimek, A.: Angle-based outlier detection in high-dimensional data. In: Proceedings of International Conference on Knowledge Discovery and Data Mining (KDD), pp. 444–452 (2008)
26. Liu, F., Ting, K., Zhou, Z.H.: Isolation-based anomaly detection. ACM Trans. Knowl. Disc. Data (TKDD) **6**(1), 1–39 (2012)
27. Radovanović, M., Nanopoulos, A., Ivanović, M.: Reverse nearest neighbors in unsupervised distance-based outlier detection. IEEE Trans. Knowl. Data Eng. **27**(5), 1369–1382 (2015)
28. Schölkopf, B., Platt, J.C., Shawe-Taylor, J., Smola, A.J., Williamson, R.C.: Estimating the support of a high-dimensional distribution. Neural Comput. **13**(7), 1443–1471 (2001)
29. Sun, J., Wang, X., Xiong, N., Shao, J.: Learning sparse representation with variational auto-encoder for anomaly detection. IEEE Access **6**, 33353–33361 (2018)
30. Tax, D.M.J., Duin, R.P.W.: Support vector data description. Mach. Learn. **54**(1), 45–66 (2004)
31. Wiewel, F., Yang, B.: Continual learning for anomaly detection with variational autoencoder. In: IEEE ICASSP, pp. 3837–3841 (2019)

Spatial, Temporal and Spatiotemporal Data

Detecting Temporal Anomalies in Business Processes Using Distance-Based Methods

Ioannis Mavroudopoulos and Anastasios Gounaris(✉)

Aristotle University of Thessaloniki, Thessaloniki, Greece
{mavroudo,gounaria}@csd.auth.gr

Abstract. Outlier detection in process mining refers to either infrequent behavior in relation to the underlying business process models or to anomalous latencies of task execution (temporal anomalies). In this work, we focus on the latter form of anomalies and we propose distance-based methods. Compared to solutions relying on probability distribution analysis and based on the experimental evaluation presented, our proposal is shown to be capable of covering both trace and event outliers, and being more efficient and effective. More specifically, running times of our technique are lower by up to an order of magnitude, while we achieve significantly higher precision and recall.

1 Introduction

Nowadays a lot of businesses turn to Business Process Management (BPM) in order to improve their processes and become more efficient. BPM is the art and science of overseeing how work is performed in an organization to ensure consistent outcomes and to take advantage of improvement opportunities [16]. Thus the main focus of BPM is to improve the processes in a business. A business process is represented as a set of tasks and their flows, which are orchestrated to achieve a common business goal [14]. Since BPM execution is supported by a breadth of software tools, automated log collection is not just feasible but easy. These data can be processed using data mining techniques to provide more knowledge about the business than just monitoring [1].

In this work, we focus on finding anomalies (or equivalently, outliers) in the monitored data. An outlier is *"an observation which deviates so much from other observations as to arouse suspicions that it was generated by a different mechanism"* [9]. The outlying data often contains useful information about the abnormal behavior, which is the reason that is used in applications such as intrusion detection, credit card fraud, and so on [2]. In business processes, an abnormal behavior can be related to a variety of issues, such as delay in a task execution, or wrong sequences of tasks in a trace.

Most studies in the field focus on finding outlier patterns in the sequence of events in a trace, e.g., [7,20] are proposals that detect such structural anomalies. A complementary approach is to identify anomalies in the logged behavior of the task execution in terms of non-typical runtime. These anomalies are termed here

A. Appice et al. (Eds.): DS 2020, LNAI 12323, pp. 615–629, 2020.
https://doi.org/10.1007/978-3-030-61527-7_40

as *temporal* ones, and as discussed in [18], it is important to check the behavior of a complete interdependent set of tasks to derive more insightful conclusions, e.g., whether it is (probably) a measurement error rather than actual abnormal task behavior.

The contribution of this work is to propose a distance-based outlier detection approach [2,12,13] to dealing with temporal anomalies in business processes. We present how this can be achieved, we discuss implementation issues and we thoroughly evaluate the proposed solution. According to our results, we can claim that our proposal is superior in terms of coverage, effectiveness and efficiency compared to the proposal in [18], which is based on probability distribution analysis. The running times of our technique are lower by up to an order of magnitude for practical settings, while we achieve significantly higher precision and recall.

The remainder of this work is structured as follows. In Sect. 2, we give the details of the problem and our solution. We present the experimental evaluation next. Section 4 discusses the related work and we conclude in Sect. 5.

2 A Distance-Based Temporal Outlier Detection Approach

In this section, first we formalize event logs. Then we describe the methods for discovering outlying traces and events. Following the terminology in [1,16], a *task* is an atomic *activity*, and in this work, without loss of generality, when we refer to activities, we imply atomic ones; therefore the terms tasks and activities are used interchangeably.

2.1 Preliminaries

We follow the same terminology and event log definition as in many other works in business process mining, e.g., [7,18]. More specifically, we assume that businesses have a mechanism to record the corresponding event logs in place in order to analyze their processes. An *event log* is composed of a set of traces. Each *trace* corresponds to a specific process instance execution and is identified by a unique *case* identifier. The instance execution is manifested as a recorded sequence of events. Each *event* records the execution of an *activity* in a particular trace(case).

Definition 1. *Event Log: let $A = \{a_1, a_2, \ldots, a_m\}$ be a finite set of activities (tasks) of size m. A log L is defined as $L = (E, C, \gamma, \delta, ts, \preceq)$ where E is the finite set of events, C is the finite set of Cases (Traces), $\gamma : E \to C$ is a surjective function assigning events to Cases, $\delta : E \to A$ is a surjective function assigning events to activities, ts records the timestamp denoting the finish of task execution, and $\preceq$ is a strict total ordering over events, normally based on execution timestamps.*

From the definition above, it is straightforward to calculate the latency of each task based on the difference between its timestamp and the timestamp of its immediate predecessor; process instance start and termination events are included in the logs. The latency includes both the task duration and any waiting time.[1]

2.2 Temporal Trace Outliers

To the best of our knowledge there is no method that finds temporal trace outliers. Here, we propose a method that finds outlier traces based on the number of executions and the mean time spend in every activity in a specific process instance. We do not pose any restriction on (i) the order the activities may be executed, (ii) whether they all appear in the same instance and (iii) on the times they are executed in a single instance, since all these factors may differ between process instances, e.g., [7,19].

We group L entries by trace, i.e., case identifier. In order to define the distance between two traces, we first convert every trace to a vector of size $2m$. Each position to this vector corresponds to a distinct attribute.

Definition 2. *Trace Vector: given a trace t, the Trace Vector of this trace, is a vector* $[n_0, ..., n_m, t_0, ..., t_m]$ *where m is the number of different activities in the process to which L refers,* n_i, $i = 1 \ldots m$ *is the number of executions of the same task in the same trace (e.g., due to loops) and* t_i *is the mean execution time of* a_i.

In the definition above, execution time should be interpreted as equivalent to task latency. In the next step, we normalize the data, so that no attribute dominates the distance calculation during the outlier detection process. We create a trace matrix $[t; t'; t''; ...]$, where each row is a trace vector and there are $2m$ columns. We then apply the z-score normalization Eq. (1) to every attribute, so that its mean value becomes equal to 0 and its standard deviation is equal to 1 [2].

$$Z = \frac{X - mean(X)}{std(X)}, \; X \; is \; a \; column \; of \; the \; trace \; matrix \tag{1}$$

Definition 3. *Trace Vectors Distance: given two trace vectors* t_1, t_2*, where* $t_1 = [t_1^1, \ldots, t_{2m}^1]$ *and* $t_2 = [t_1^2, \ldots, t_{2m}^2]$*, we define the distance between them as*

$$dist = \sqrt{\sum_{i=1}^{2m} \frac{(t_i^1 - t_i^2)^2}{2m}} \tag{2}$$

The Eq. (2) is similar to the traditional RMSE (Root Mean Square Error). We take the second power of their difference, so that the distance between two traces is dominated by the attributes for which they have the biggest difference.

[1] If the logs contain the start and end finish time of each task explicitly, then our approach to detecting latency anomalies can be applied to detecting anomalous task durations in a straightforward manner.

Algorithm 1 Find Outlier Traces

```
    traceVectors ← [ ]
    traces ← extract traces from L
    for all trace ∈ traces do
4:     traceVectors append preprocess(trace)
    end for
    normalizedTraces ← normalize(traceVectors)
    mtree ← constructMTree(normalizedVectors)
8:  outlierFactors ← [ ]
    for all trace ∈ normalizedTraces do
       kneighbors ← mtree.kneirestneighbors(trace,k)
       outlierFactor ← 0
12:    for all neighbor ∈ kneighbors do
          outlierFactor ← outlierFactor + dist(trace,neighbor)
       end for
       outlierFactors append outlierFactor
16: end for
    sort outlierFactors
    return the top ζ traces
```

The pairwise distance between traces is used in a straightforward manner to detect outliers. Each trace is assigned an outlier factor, which is equal to the sum of distances from the k-nearest neighbors. As such, the traces that, when depicted in a euclidean space, are located in areas with low density, will be reported as outliers. Once we have calculated the outlier factor for every trace, we can report the top ζ outliers. Formally, the temporal trace outliers are defined as follows.

Definition 4. *Temporal trace outlier: given a normalized trace matrix, a (normalized) trace vector t is an outlier if the sum of the distances $dist(t, t')$ with the closest k other trace vectors t' is in the top ζ values, where k and ζ are parameters defined by the user.*

In real scenarios, there are thousands of traces, so we cannot afford to calculate the distance between every pair of traces due to the quadratic complexity of such a process. That is why we need a data structure called M-Tree to mitigate the performance impact of range queries. M-Tree is a tree data structure that is constructed using a metric and relies on the triangle equality for efficient range and k-nearest neighbors queries [6]. The pseudo-code is in Algorithm 1.

In the last step of the algorithm, instead of reporting the top ζ outliers, we can report the traces for which their outlier factor deviates more than x times the standard deviation from the mean value. Or we can resort to the outlier definition used in the subsequent section. However, our approach is more suitable for a high-dimensionality space, since $2m$ can grow large. This is because a key characteristic of the outlier definition above is that it does not rely on estimating the underlying probability distribution, which is notoriously difficult in multiple dimensions.

2.3 Temporal Event Outliers

The previous method of outlier detection refereed to traces as a whole. Event outliers, examine each event type separately, i.e., log entries in L are grouped by the activity they refer to through the function δ. The aim of temporal event outlier detection is to identify the event log entries, for which the corresponding execution latency is highly dissimilar compared to the rest of the executions of the same activity type.

This problem is first addressed in [18], where Rogge-Solti *et al.* presented a method to find temporal anomalies, i.e. anomalies concerning the running time of an activity in a process, using probability distribution fitting. The key point is that, most commonly in real cases, the distribution of the task execution latencies does not follow a normal distribution; so the proposal in [18] leverages a more robust curve fitting method, which relies on the work in [23].

In this work we propose a method that uses the pairwise event log distances to determine the outliers. The distance between two task entries is defined as the absolute value of the difference between their execution latency. Then, the definition employed for trace outlier detection is transferred to events in a straightforward manner.

Definition 5. *Temporal event outlier: given a set of task execution latencies referring to the same activity $a_i \in A$, a logged task latency is an outlier if the sum of the distances with the closest k other task latencies is in the top ζ values, where k and ζ are parameters defined by the user.*

Alternatively, the following definition can be used, which is on the one hand the same as in the seminal work of [12,13] but is more sensitive to the user-defined input parameters.

Definition 6. *Temporal event outlier (alternative definition): given a set of task execution latencies referring to the same activity $a_i \in A$, a logged task latency is an outlier if it has less than k neighbors, where another task latency is neighbor if its distance is less than or equal to R, and k and R are parameters defined by the user.*

Both definitions, for efficient implementation, require a data-structure, such as R-tree or M-tree to perform the range queries involved.[2] Another similarity is that they do not require data normalization, e.g., through Eq. (1). Their main difference is that Definition 5, gives a list of ζ values and implies that the end user will post-process this list to check whether all the reported events are outliers indeed, or more outliers exist, e.g., through examining the relative differences of the sums. In Definition 6, there is no post-processing, but the correct setting of the R value, which is scenario-dependent, rests with the user.

---

[2] It is out of our scope in this work to compare R-tree vs M-tree.

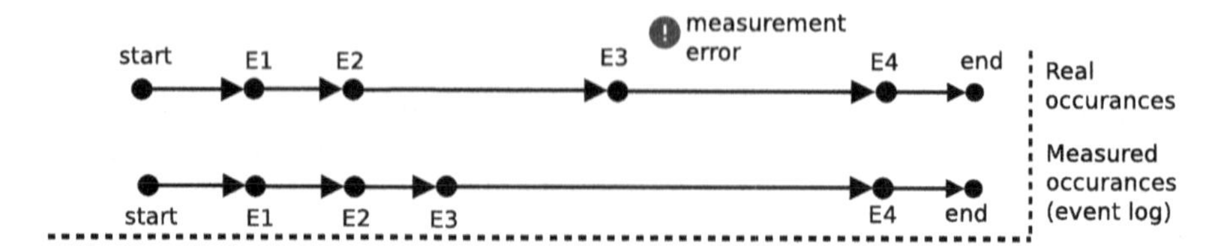

Fig. 1. Example of a measurement error from [18]. In the first row, the real execution times of each event are shown. After a measurement error at event E2, which also effected the execution time of E3, the second row shows the entries in the log file.

2.4 Temporal Outliers of Event Pairs

Reporting individual temporal event outliers is hard to interpret because such temporal anomalies can be either true anomalies or measurement errors. In the latter case, the timestamp of an event has been falsely logged, causing an event to appear as outlier. This problem was first defined in [18], where the key observation is that measurements errors typically affect two consecutive tasks in the trace in a negatively correlated manner. More specifically, if there is no abnormal behavior in the real execution, but due to delayed (resp. early) recording, a task instance has a long (resp. short) latency in the logs, it is expected that the subsequent event will have a recorded short (resp. long) latency. This is (most probably) a measurement error and has to be distinguished from real outliers (see Fig. 1).

The approach to identifying measurement errors in [18] consists of two components: (i) estimating the probability distribution function of the task latencies, as mentioned in the previous section, and (ii) employing a Bayesian Network to define the dependencies between activities; where the nodes in the Bayesian Network are the activities and the edges represent the succeeding-preceding relations. Here, we suggest to replace the first part with distance-based outlier detection techniques, extending the proposal in Sect. 2.3. More specifically, we introduce two techniques for detecting measurement errors (or better, probable measurement errors). The first one transfers the problem of outlying detection of event pairs to the problem of distance-based outlier detection of 2-dimensional points, so that Definitions 5 and 6 apply with simple extensions. The second one uses exactly the same setting as in Sect. 2.3.

Mapping Consecutive Events in a 2-Dimensional Space. First, we normalize the task latencies similarly to the approach in Sect. 2.2. Second, we take all the ordered pairs of consecutive events as they appear in the log file. For each such ordered pair, we keep only the ordered pair of their normalized latencies. Then, each ordered pair of latency values can be treated as a 2-dimensional point. Regardless of any probability distribution of the task latencies, for every activity, the mean latency is 0 and its standard deviation equals to 1. Overall, the whole log set is transformed to a set of points. The size of this set is smaller

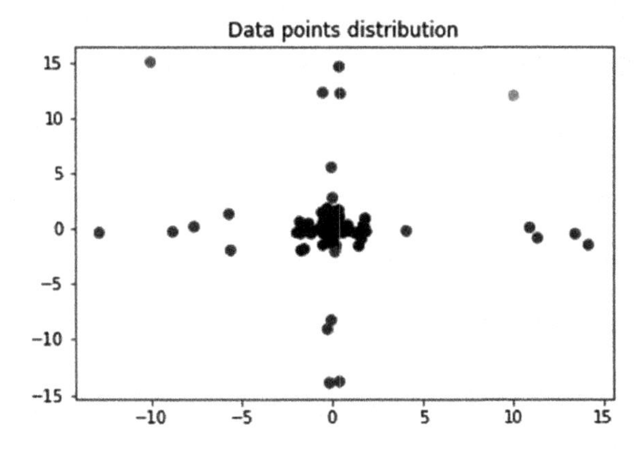

Fig. 2. Example data point distribution after normalization. The blue data points are true outliers. The green data point at the top-right is also a true outlier because its coordinates have the same sign. However, the red point at the top-left part is (probably) a measurement error. (Color figure online)

than the number of log entries in L because we produce a point for each event in a trace apart from the start one.

Figure 2 shows an example of a set of log entries transformed to a set of points. We notice that most of the data points are near the center (0, 0), but there are several points that are far from it. The data points that deviate significantly from the center, are reported as outliers using our distance-based techniques (either definitions), because they have a few points in their neighborhood. In the example, most of the outliers deviate only with regards to a single dimension; these are considered as true outliers. True outliers are also the outliers in the top-right and left-bottom part of such an illustration. However, any points in the top-left or right-bottom part, like the red point in the figure, are reported as (probable) measurement errors because the sum of their execution times is normal and there is a suspicion that it is simply due to delayed or premature recording of the finish timestamp of the preceding event.

We employ Definition 5 for outlier detection, using the euclidean distance. Instead of a M-tree, we can also use a R-tree (see also Algorithm 2). As previously, we report the top ζ points.

We use again as an outlying factor, the sum of distance for the k-nearest neighbors for every data point. In order to make fast knn queries we implement a tree data structure, named R-Tree, which is suitable for range queries and knn queries in the space. As a distance metric we are using the Euclidean distance.

The m data points with the highest outlying factor, will be reported as outliers. In Algorithm 2 we show the procedure of detecting outlying pairs of events.

Algorithm 2 Temporal outliers of event pairs

```
   Input traces
   Output outliers
   eventPairs ← [ ]
   for all trace ∈ traces do
     for all e1, e2 ∈ trace do
       eventPairs append [normalized(e1.latency), normalized(e2.latency)]
5:   end for
   end for
   rtree ← constructRTree(eventPairs)
   outlierFactors ← [ ]
   for all pair ∈ eventPairs do
10:  kneighbors ← rtree.kneirestneighbors(pair)
     outlierFactor ← 0
     for all neighbor ∈ kneighbors do
       outlierFactor ← outlierFactor + euclidean-distance(pair,neighbor)
     end for
15:  outlierFactors append outlierFactor
   end for
   sort outlierFactors
   outliers ← top ζ pairs
```

The technique so far has produced the list of the consecutive logged events that are temporal outliers. The next step is the assessment of the type of outlierness. As already discussed, if these points have their coordinates with opposite signs, i.e., one positive and another negative, there are reported as measurement errors; otherwise they are reported as normal outliers. However, the normalized latencies may create a problem if the absolute mean latency of one task in the pair is much larger than the absolute mean latency of the other; in such a case, deviating the same amount of standard deviations in both dimensions does not counterbalance each other. Therefore, we denote as (d_1, d_2) the pair of the non-normalized latencies for each result item in Algorithm 2, we adopt the following definition and we run Algorithm 3.

Definition 7. *Measurement error: given a set of pairs of outliers as reported by Algorithm 2, the constituent outliers are reported as (probable) measurement errors if*

$$|(mean(d_1) - d_1) - (mean(d_2) - d_2)| \leq \tau \tag{3}$$

Outlier Detection of Pairs: A More Efficient Approach. In the previous technique, for each pair, we calculate the sum of distances from its k nearest neighbor. Even though with the use of R-Tree, we manage to reduce the running time, most of the computations were unnecessary, because they involve normal latencies. Based on this, we keep the Algorithm 3 the same, where outliers differentiate between *normal outliers* and *measurement errors*, but the input is not

Algorithm 3 OutliersOrMeasurementErrors

```
   Input outliers from Algorithm 2, threshold τ
   Output normalOutliers, measurementErrors
   normalOutliers ← o ∈ outliers and (o.x same sign as o.y)
2: measurementErrors ← [ ]
   for all pair ∈ outliers-normalOutliers do
4:     if pair is measurement error based on Equation (3) then
           measurementErrors append pair
6:     else
           normalOutliers append pair
8:     end if
   end for
10: return normalOutliers and measurementErrors
```

provided by Algorithm 2. Instead, we use the techniques in Sect. 2.3 to find isolated outliers. For each such outlier, we check whether its succeeding event is an outlier as well, and if this is the case, the pair is considered in Algorithm 3. As will be reported in the evaluation, avoiding to map the task pairs to a 2d space yields performance improvements while producing exactly the same results.

3 Evaluation

We used both real-world and synthetic datasets to evaluate the performance of the proposed methods. We start by presenting the datasets, followed by the evaluation of the trace outlier detection method. Then we compare the two pair-based methods that use distance and at the end we compare the best of these methods with the one proposed in [18]. All tests were conducted in a machine with 16 GB of RAM and 3.2 GHz CPU with 8 cores. The source code for all of the proposed distance-based outlier detection methods is publicly available on GitHub[3].

The real-world datasets are taken from the Business Process Intelligence (BPI) Challenges, and more specifically from the 2012 and 2017 ones. BPI12[4] is an event log of a loan application process. It consists of 13087 traces that contain a total number of 262200 events. The mean amount of events per trace is 20.03 and the minimum and maximum amount is 3 and 175, respectively. BPI17[5] is an event log, which also corresponds to a loan application of an Dutch financial institute. It includes 31509 traces, which contain over 1M (1202267) events in total. The mean, max and min number of events per trace for this dataset are 38.15, 10 and 180, respectively. We have also created a synthetic dataset (details will be discussed later).

[3] https://github.com/mavroudo/BPM-outlierDetection-distance-based.
[4] https://data.4tu.nl/repository/uuid:3926db30-f712-4394-aebc-75976070e91f.
[5] https://data.4tu.nl/repository/uuid:5f3067df-f10b-45da-b98b-86ae4c7a310b.

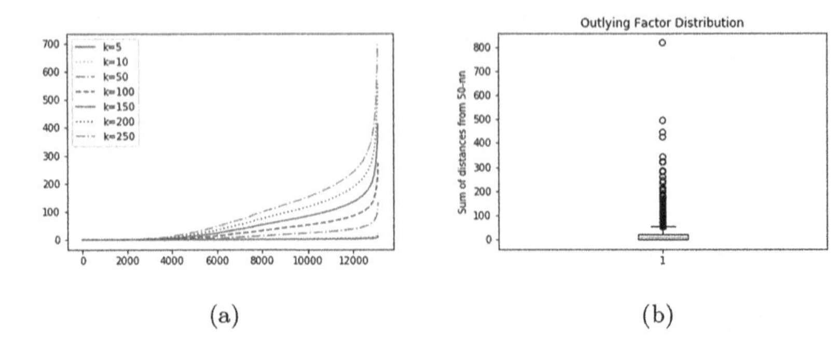

(a) (b)

Fig. 3. Sum of the distances from the k-nearest neighbor for different values of k (a) and sum value distributions for $k = 50$ (b)

3.1 Evaluation of Trace Outlier Detection

Our first experiment aims at verifying the effectiveness of our approach. For this, we use the read-world dataset BPI12. Figure 3(a) shows how the sum of distances from k-nearest neighbors is changing with different values of k. It also helps us identifying the most suitable value of k, which corresponds to the plot that is initially as parallel as possible with the horizontal axis, and then, after a sharp change, becomes parallel to the vertical one. Such a behavior allows a clearer distinction of outliers, and in our experiment is the plot for $k = 50$.

In Fig. 3(b), the box-plot shows how the sums of distances of the 50 nearest neighbors are distributed for all the traces. As shown, the mean value is very low (actual value is 14.4) and there are only a few traces that have sum of distances more than 400. After executing the trace outlier detection method for $k = 50$ and a relatively large ζ value, we report as real outliers the traces with outlying factor greater than the mean value of all the sums plus 4 times the standard deviation; in this manner, we keep 52 outlying traces. Two examples are as follows:

1. $Trace_2$, which contains the activity "W_Wijzigen contractgegevens". This activity was only executed in 12 out of 13082 traces.
2. $Trace_{856}$ because its mean execution time of the activity "W_Afhandelen leads" is 2327510.471 s. This activity has mean execution time 586 s in the complete log.

The previous experiments referred to the effectiveness of the approach. Regarding the efficiency, even though that we use an M-Tree to reduce the time for k-nearest neighbors queries, as the dimensions increase, the running time for these queries increases in a quadratic manner, as shown in Fig. 4.

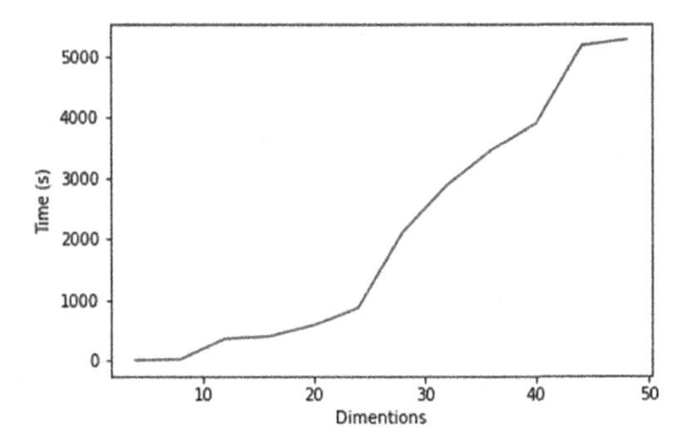

Fig. 4. Execution time for varying number of dimensions

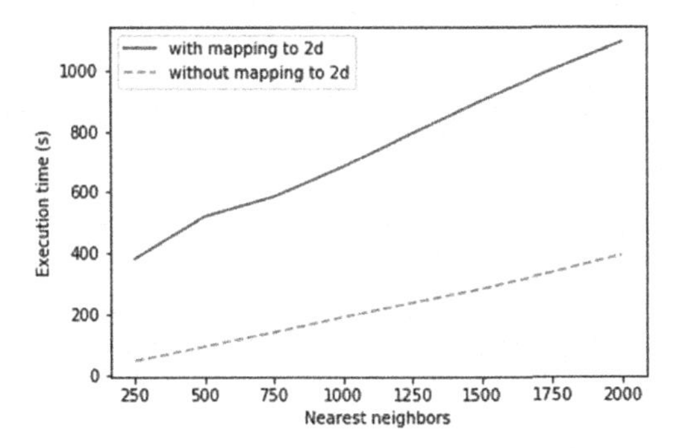

Fig. 5. Execution times of distance-based outlier detection with and without mapping event pairs to a 2-dimensional space

3.2 Evaluation of Event Outlier Detection

In Sect. 2.4 we proposed two different methods that use the distance to find anomalous pairs of events and then classifying them as true outliers or measurement errors. We conduct an experiment using the BPI12 dataset to compare the execution times between these two methods and we show the running times in Fig. 5. From the figure, we can conclude that not mapping to a 2-dimensional space but directly relying on individual event outlier detection is more efficient by up to an order of magnitude for small k values. Therefore, next, we employ solely the second method from Sect. 2.4.

Comparison Against [18]. To compare our distance-based outlier detection against the probability distribution-based in [18], we employ both BPI12 and BPI17 and also a synthetic data set. The synthetic dataset contains 4 activities with different distributions, namely (i) a combination of two normal distributions (see Fig. 6), (ii) a combination of two alpha distributions, (iii) a combination of two exponential distributions and (iv) a power lognormal distribution. It includes 526 traces of 8K events overall, where every trace contains between 5 to 25 events. The synthetic dataset is provided along with the source code.

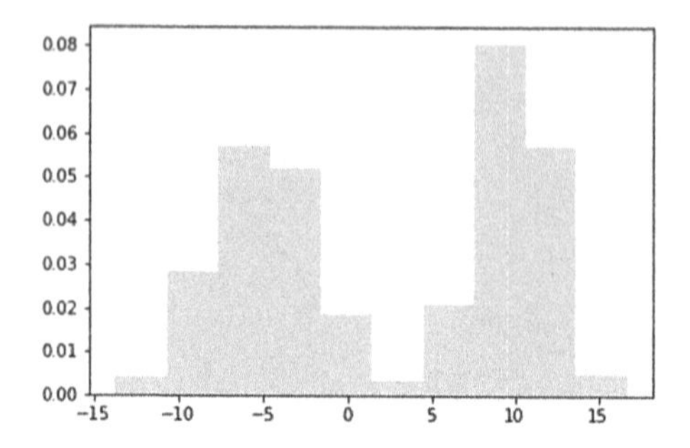

Fig. 6. Distribution of task execution latencies for the first activity in the synthetic dataset.

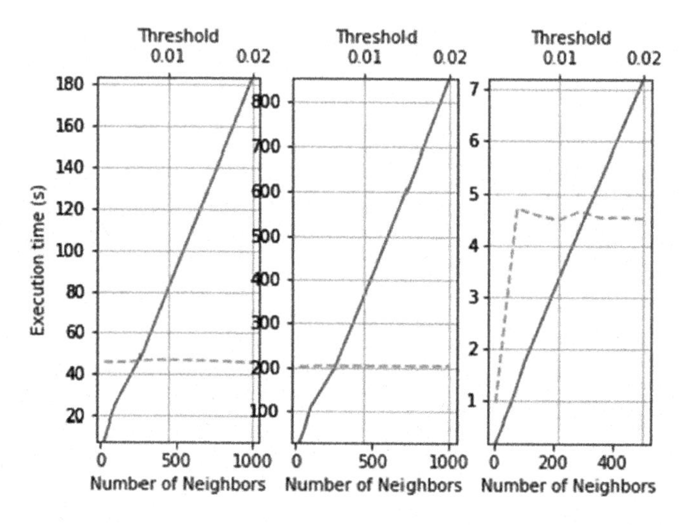

Fig. 7. Running times for the BPI12 (left), BPI17 (middle) and synthetic (right) datasets, respectively. The yellow (doted) line represents execution time based on threshold and the blue line represents execution time based on the number of neighbors (Color figure online)

The execution times for both methods are presented in Fig. 7. We confirm that the time is linear dependent on k, i.e., the number of nearest neighbors that we take into account in the distance-based method. The execution time for the distribution fitting method does not depend on the input parameters (i.e., probability fitting threshold ranging from 0.001 to 0.2). For a value of k close to 50 that we have shown that make more sense, our method runs faster by an order of magnitude.

We test the effectiveness of each method in the synthetic dataset. Each event is classified as outlier or normal based on whether the probability density function is below 0.01. The precision and recall results appear in Fig. 8. Our method has significantly higher precision and recall, whereas the distribution

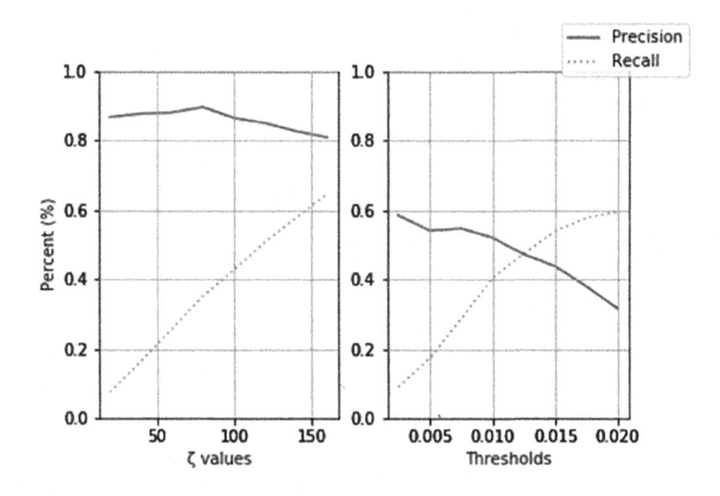

Fig. 8. Precision and recall as a function of ζ for the distance-based method with $k = 50$ (left) and the probability fitting threshold for the technique in [18] (right).

fitting method cannot approximate correctly the underlying distributions. E.g., for the distribution in Fig. 6 (first activity in the synthetic dataset), it fails to correctly report outliers with task latency between 1 and 4 time units. Overall, Fig. 8 shows that no matter how the technique in [18] is configured, there is better configuration of our technique in terms of precision and recall.

4 Related Work

There are several proposals about outlier detection in business processes. However, almost all of them focus on detecting outliers regarding the structure of the underlying process model. In this direction, outliers can be used in order to predict the failure of an ongoing process. In [5,11], different approaches to predicting the next tasks of an active trace, and based on this prediction to determine if a trace will fail to execute properly, are presented. In addition, as a business process log trace is typically in the form of a sequence of tasks, outliers can be found in these sequences, e.g., [7,15]. Anomaly detection methods that take into account both the structure of the model and the data attributes are known as multi-perspective, have been developed, such as the ones in [3,4,17]. However, none of these approaches deal with temporal anomalies in the way we do, and thus are not directly comparable to our approach. Temporal outlier detection, in the context of BPM, have been addressed in [18], and we have directly compared our solution against it in the previous section. The approach in [10], even though it calculates the distance between traces based on task duration, it only considers those traces that contain identical tasks. Hence, it cannot find global temporal outliers and does not deal with measurement errors.

In addition, a typical application is to clear the log dataset removing infrequent behavior in order to facilitate process discovery; process discovery aims to derive the underlying process model out of event logs. Another example of dealing with variations in the process model structure is to allow configurable models, as thoroughly covered in [19]. All the proposals above are orthogonal to our work, which focuses on temporal outliers.

From the outlier detection point of view, there exist several textbooks, e.g., [2]. The techniques based on statistics were the first to be proposed, e.g., through determining data values at the tails of a univariate distribution and the corresponding level of statistical significance. Distance-based outlier detection, which is leveraged in this work, is a representative of proximity-based anomaly detection. Proximity-based methods define a data point as an outlier if its neighborhood is sparsely populated. In distance-based techniques, a data point is considered as an outlier, if its distance from its k-th closest neighbor is longer than a predefined radius. Some of the advantages of these techniques are the linear scalability in the size of the dataset [8,12], the ability to interpret the results and operate in a streaming and/or massively parallel environment, e.g., [22], and their wide applicability as reported in [21].

5 Summary

In this work, we advocate the usage of distance-based outlier detection methods for identifying anomalous behavior in event logs in terms of the running time of tasks. We explain the implementation details, and compared to the existing methods that rely on probability distribution approximation, our proposal is broader, in the sense that it applies to complete traces, more efficient, in the sense that runs faster, and more effective, in the sense that achieves higher precision and recall. Our implementation is provided as open-source.

Acknowledgment. The research work was supported by the Hellenic Foundation for Research and Innovation (H.F.R.I.) under the "First Call for H.F.R.I. Research Projects to support Faculty members and Researchers and the procurement of high-cost research equipment grant" (Project Number: 1052).

References

1. van der Aalst, W.M.P.: Process Mining - Data Science in Action, 2nd edn. Springer, Heidelberg (2016). https://doi.org/10.1007/978-3-662-49851-4
2. Aggarwal, C.C.: Outlier Analysis. Springer, Cham (2017). https://doi.org/10.1007/978-3-319-47578-3
3. Böhmer, K., Rinderle-Ma, S.: Multi-perspective anomaly detection in business process execution events. In: International Conference on Cooperative Information Systems (CoopIS) 2016, October 2016. http://eprints.cs.univie.ac.at/4785/
4. Böhmer, K., Rinderle-Ma, S.: Mining association rules for anomaly detection in dynamic process runtime behavior and explaining the root cause to users. Inf. Syst. **90**, 101–438 (2020)
5. Borkowski, M., Fdhila, W., Nardelli, M., Rinderle-Ma, S., Schulte, S.: Event-based failure prediction in distributed business processes. Inf. Syst. **81**, 220–235 (2019)

6. Ciaccia, P., Patella, M., Zezula, P.: M-tree: an efficient access method for similarity search in metric spaces. In: International Conference on Very Large Data Bases (VLDB) (2001)
7. Conforti, R., Rosa, M.L., ter Hofstede, A.H.M.: Filtering out infrequent behavior from business process event logs. IEEE Trans. Knowl. Data Eng. **29**(2), 300–314 (2017)
8. Dai, Q.Z., Xiong, Z.Y., Xie, J., Wang, X.X., Zhang, Y.F., Shang, J.X.: A novel clustering algorithm based on the natural reverse nearest neighbor structure. Inf. Syst. **84**, 1–16 (2019)
9. Hawkins, D.: Identification of Outliers. Springer, Netherlands (1980). https://doi.org/10.1007/978-94-015-3994-4
10. Hsu, P.Y., Chuang, Y.C., Lo, Y.C., He, S.C.: Using contextualized activity-level duration to discover irregular process instances in business operations. Inf. Sci. **391–392**, 80–98 (2017)
11. Kang, B., Kim, D., Kang, S.H.: Real-time business process monitoring method for prediction of abnormal termination using KNNI-based LOF prediction. Expert Syst. Appl. **39**(5), 6061–6068 (2012)
12. Knorr, E.M., Ng, R.T.: Algorithms for mining distance-based outliers in large datasets. In: Proceedings of the 24rd International Conference on Very Large Data Bases, pp. 392–403 (1998)
13. Knorr, E.M., Ng, R.T.: Finding intensional knowledge of distance-based outliers (1999)
14. Kueng, P., Kawalek, P.: Goal-based business process models: creation and evaluation. Bus. Process Manag. J. **3** (1996)
15. de Lima Bezerra, F., Wainer, J.: Algorithms for anomaly detection of traces in logs of process aware information systems. Inf. Syst. **38**, 33–44 (2013)
16. Dumas, M., La Rosa, M., Mendling, J., Reijers, H.: Fundamentals of Business Process Management. Springer, Heidelberg (2019). https://doi.org/10.1007/978-3-662-56509-4
17. Nolle, T., Seeliger, A., Thoma, N., Mühlhäuser, M.: DeepAlign: alignment-based process anomaly correction using recurrent neural networks. In: Dustdar, S., Yu, E., Salinesi, C., Rieu, D., Pant, V. (eds.) CAiSE 2020. LNCS, vol. 12127, pp. 319–333. Springer, Cham (2020). https://doi.org/10.1007/978-3-030-49435-3_20
18. Rogge-Solti, A., Kasneci, G.: Temporal anomaly detection in business processes. In: Sadiq, S., Soffer, P., Völzer, H. (eds.) BPM 2014. LNCS, vol. 8659, pp. 234–249. Springer, Cham (2014). https://doi.org/10.1007/978-3-319-10172-9_15
19. Rosa, M.L., van der Aalst, W.M.P., Dumas, M., Milani, F.: Business process variability modeling: a survey. ACM Comput. Surv. **50**(1), 2:1–2:45 (2017)
20. Satyal, S., Weber, I., Paik, H.Y., Ciccio, C.D., Mendling, J.: Business process improvement with the AB-BPM methodology. Inf. Syst. **84**, 283–298 (2019)
21. Subramaniam, S., Palpanas, T., Papadopoulos, D., Kalogeraki, V., Gunopulos, D.: Online outlier detection in sensor data using non-parametric models. In: VLDB, pp. 187–198 (2006)
22. Toliopoulos, T., Gounaris, A., Tsichlas, K., Papadopoulos, A., Sampaio, S.: Parallel continuous outlier mining in streaming data. In: 5th IEEE International Conference on Data Science and Advanced Analytics (DSAA) (2018)
23. Yeung, D.Y., Chow, C.: Parzen-window network intrusion detectors. In: Object Recognition Supported by User Interaction for Service Robots, vol. 4, pp. 385–388 (2002)

Mining Constrained Regions of Interest: An Optimization Approach

Alexandre Dubray(✉), Guillaume Derval, Siegfried Nijssen, and Pierre Schaus

UCLouvain - ICTEAM/INGI, Louvain-la-Neuve, Belgium
{alexandre.dubray,guillaume.derval, siegfried.nijssen,pierre.schaus}@uclouvain.be

Abstract. The amount and diversity of mobile and IoT location and trajectory data are increasing rapidly. As a consequence, there is an emerging need for flexible and scalable tools for analyzing this data. In this work we focus on an important building block for analyzing location data, that is, the problem of partitioning a space into regions of interest (ROIs) that are densely visited. The extraction of ROIs is of great importance as it constitutes the first step of many types of data analysis on mobility data, such as the extraction of trajectory patterns expressed in terms of sequences of ROIs. However, in this paper we argue that unconstrained ROIs are not meaningful and useful in all applications. To address this weakness, we propose the problem of constraint-based ROI mining, and identify two types of constraints: intra- and inter-ROI constraints. Subsequently, we propose an integer linear programming formulation of the task of discovering a fixed number of constrained ROIs from a binary density matrix. We extend the approach to discover automatically the number of ROIs by relying on the Minimum Description Length Principle. Our experiments on real data show that the approach is both flexible, scalable and able to retrieve constrained ROIs of higher quality than those extracted with existing approaches, even when no constraints are imposed.

Keywords: Data mining · Constrained optimization · Integer linear programming · Regions of interest · Constrained clustering

1 Introduction

The number and diversity of tracking devices are constantly increasing and so does the volume of recorded location data. Innovative applications exploiting these data can be imagined if some form of meaningful aggregated information can be discovered. An important building block for summarizing trajectory data is the extraction of regions of interest (ROIs). Informally, a ROI is a densely visited space. The discovery of ROIs is of practical importance as it can be instrumental for other tasks. Examples of such tasks related to trajectory mining are:

- In [8], the authors propose to discover trajectory patterns expressed in terms of ROIs. They first rewrite the trajectories as a sequence of the extracted

A. Appice et al. (Eds.): DS 2020, LNAI 12323, pp. 630–644, 2020.
https://doi.org/10.1007/978-3-030-61527-7_41

ROIs. A frequent sequence mining algorithm [1,4] can then be applied on the sequence database to extract sequential patterns with a minimum support.
- Another possible use of ROIs is location prediction. This task consists in, given a database of trajectories and the start of a new trajectory of a moving object, predicting what will be the next location of the moving object [12,15].
- In the area of urban management [18], the authors proposed a system relying on ROIs to help taxis to wait in a region likely to contain their next trip request.

However, not all ROIs are equally useful and meaningful in all applications. For example, in the case of tourist spot recommendation, it may be desirable that the extracted ROIs are close to public transport access; in an application suggesting visiting a city by bicycle it is useful to impose a constraint that extracted ROIs are close enough to bike paths, and are within reasonable distance from each other. No existing approaches for identifying ROIs take such constraints into account. For this reason, in this paper we introduce the problem of *constraint-based ROI mining*. We categorize these constraints into two types: *intra-ROI constraints*, which impose requirements on the individual ROIs, and *inter-ROI constraints*, which impose requirements on the relationships between ROIs. In this work, we propose an Integer Linear Program, that can directly incorporate the two types of constraints, to solve this problem.

2 Preliminary Concepts

A well-known algorithm for identifying ROIs is the *PopularRegion* algorithm [8] that is both easy to implement and scalable. This algorithm extracts non-overlapping rectangular ROIs from a 2D grid of density values $\mathcal{G}$ of size $N \times M$ (N rows and M columns). This grid-based approach enables application dependent density definitions. For analyzing trajectory data, the density of a cell can be the number crossing trajectories with or without interpolation between consecutive points. If one is rather interested to detect geographic regions where users stay for a significant amount of time (Stay Points) [11], one can define the density as the relative fraction of time spent in the cell by a trajectory.

The *PopularRegion* algorithm works as follows. Starting from a small ROI, it greedily expands the rectangle ROI in one of the four directions as long as the average density of the rectangle remains above a given threshold. Using the same notation as in [8], c_{ij} is the cell at row i and column j ($1 \leq i \leq N$, $1 \leq j \leq M$), θ is a user defined minimum density threshold and $\mathcal{G}^* = \{c \in \mathcal{G} \mid \text{density}(c) \geq \theta\}$ is the set of all dense cells. The algorithm works as follows:

1. Take the cell in $\mathcal{G}^*$ with the highest density that is not already in a ROI. If there is none, return the set of ROIs.
2. Create a ROI with this single cell.
3. While there is a direction in which we can extend the ROI, extend it in the direction that gives the highest average density.
4. Add the ROI to the set of ROIs and go to 1.

The main advantage of the *PopularRegion* algorithm is its scalability and ease of implementation. However, it clearly also has a number of weaknesses. First its output is ill-defined; there is no clear characterization of an objective function that is minimized. Furthermore, as explained in [9], it is easy to create examples where the greedy algorithm ends up finding very large ROIs that may hinder the creation of interesting subregions. This is illustrated in Fig. 2 and 3, which show the initial dense cells as well as the regions discovered by *PopularRegion*, for two different data sets. As can be observed, for both data sets *PopularRegion* identifies regions that cover large part of the city, which is not satisfying. Finally, *PopularRegion* does not allow constraining the discovered ROIs and it can only generate rectangular ROIs.

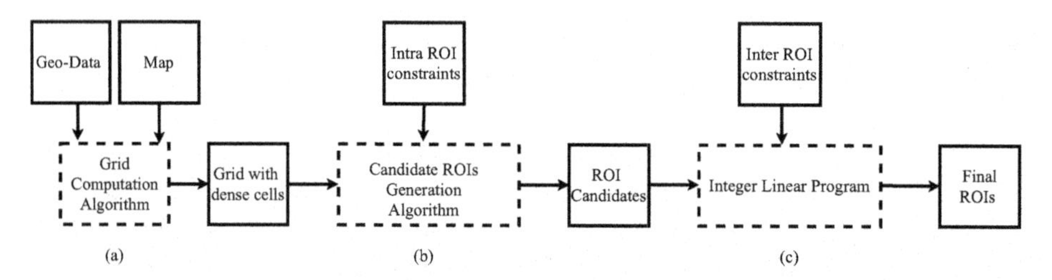

Fig. 1. Our approach is decomposed into consecutive steps: (a) The grid $\mathcal{G}$ is created from the geolocalized data; (b) A set of candidate ROIs is generated satisfying intra-ROIs constraints; (c) The set of ROIs is selected among the candidate ROIs solving the Integer Linear Program taking into account inter-ROI constraints.

Our contribution is the process for discovering constraint-based ROIs given in Fig. 1:

- The grid and the dense cells are computed based on the map and the geo-localized data. Many alternatives are possible depending on the application.
- A set of candidate ROIs are computed. The final ROIs will be selected from this set. These ROIs must satisfy the intra-ROI constraints such as the minimum distance to public transportation, the shape constraints, etc.
- An Integer Linear Program (ILP) selects the final ROIs. It consists in finding the most parsimonious representation of all the dense cells. Two variants are proposed: one with a fixed number K of regions and one in which this number is chosen automatically by relying on the *minimum description length* (MDL) principle [14]. The ILP can easily accommodate inter-cluster constraints such as the minimum distance between any two selected ROIs.

This paper focuses mainly on the generation of candidate ROIs and the ROI selection algorithm (the ILP). The grid and dense cells generation is left to the user: it is an orthogonal task which must be adapted to the task at hand.

This approach for detecting ROIs addresses a number of weaknesses of the *PopularRegion* algorithm. In particular, it can easily accommodate constraints on ROIs and the optimization problem for discovering the ROIs is well-defined.

We evaluate the new approach qualitatively and compare it with the *PopularRegion* and OPTICS [3] algorithms on real-life data. As alternative approaches do not support constraints, we also evaluate our approach without constraints.

An example of regions discovered by our method is illustrated in Figs. 2c and 3c. As can be seen, our method finds more fine-grained ROIs and avoids selecting all the isolated cells.

Related work is discussed in Sect. 3. Our optimization model is introduced in Sect. 4. The candidate ROIs generation is discussed in Sect. 5, as it is dependent on the optimization model, and the addition of constraints is describe in Sect. 6. The experiments are described in Sect. 7. We conclude in Sect. 8.

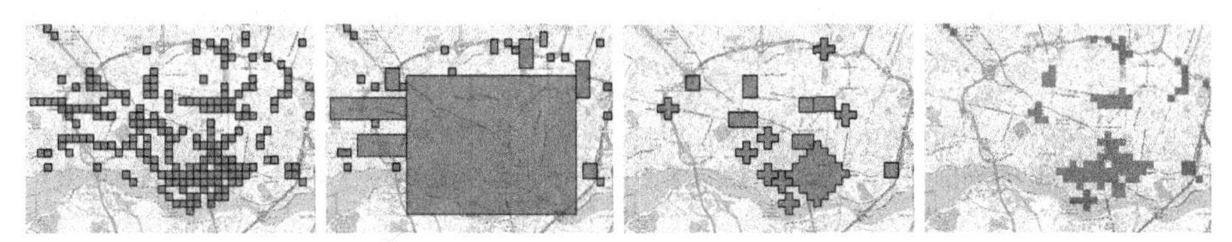

(a) Dense cells on a 100 × 100 grid (b) *PopularRegions* (c) Our method with ratio constraints on the rectangles (d) OPTICS, each color represents a cluster

Fig. 2. Visualization of the output of the different methods for the Kaggle data set.

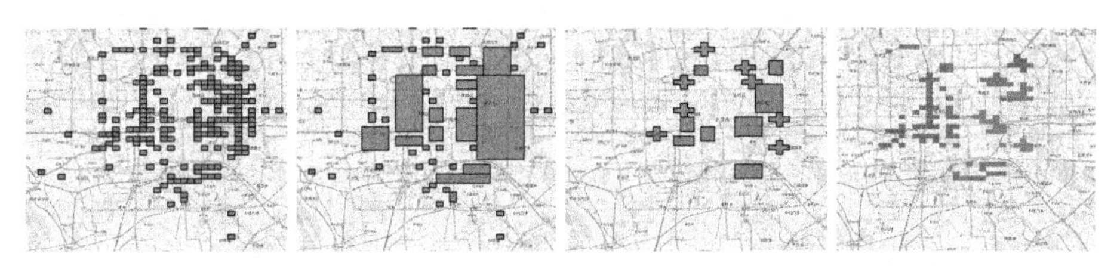

(a) Initial set of dense cells on a 200× 200 grid (b) *PopularRegions* (c) Our method with ratio constraints on the rectangles (d) OPTICS, each color represents a cluster

Fig. 3. Visualization of the output of the different methods for the T-Drive data set.

3 State of the Art and Related Work

Like *PopularRegions*, Gorawski and Jureczek [9] proposed a grid-based approach to identify ROIs. The algorithm is essentially *PopularRegions* with a limit on the size of the rectangles during the extension process. It requires additional parameters and does not permit constraining the ROIs.

The approach of [5] is not grid-based. Starting from geo-tagged locations on the map, it discovers dense convex polygons around predefined points-of-interest

(PoI). The fixed PoI setting limits the use cases of the approach and the fact that it is not grid-based also limits the possible applications. Furthermore, shape constraints on the ROIs are not possible.

The task of finding ROIs on a grid is similar to clustering. Starting from a grid of dense cells, any clustering method can be used to group dense cells close to each other. However, the problem is not exactly the same. Clusters of dense cells are not necessarily connected regions. DBSCAN [7] is one of the most popular density-based clustering algorithms. It does not require to specify the number of clusters and is also able to identify outlier points. OPTICS [3] is another well-known method to perform density-based clustering that is able to deal with clusters of varying density. Examples of output of OPTICS are shown in Figs. 2d and 3d. OPTICS identifies clusters of various forms since they are not constrained by the algorithm.

In [6], the authors propose a clustering method computing connected component sets of dense cells starting from the rectangular regions found by *Popular-Region*. This method is not able to filter outlier cells like DBSCAN or OPTICS and does not accept constraints on the ROIs.

4 An Optimization Model for ROIs

This section describes the optimization model used in step (c) of Fig. 1. The model is in charge of selecting the final ROIs from a set of precomputed candidate ROIs denoted $\mathcal{S}$ (shapes). We formalize the problem as an *integer linear program* (ILP). For simplicity we first assume that $\mathcal{S}$ is composed of rectangles and that the desired number of ROIs to select is fixed to K. Subsequently, we will extend the approach to discover automatically the number of regions K, by using the Minimum Description Length Principle [14]. We will first introduce our approach when no constraints are given; how to deal with constraints is discussed in Sect. 6.

4.1 Selection of K ROIs

Assuming that the set of candidate rectangles is composed of all the possible rectangles, our approach aims to find K non-overlapping rectangles that cover the dense cells well and avoid covering the non-dense ones. For a $N \times M$ grid, there are less than $N^2M^2 = |\mathcal{G}| \times |\mathcal{G}|$ such rectangles, that is, the total number of possible pairs of coordinates.

The approach can be interpreted as discovering a classification model for predicting the dense-non-dense status of a cell solely based on its coordinates. The prediction function to discover is chosen from a hypothesis space composed of the power-set of non-overlapping shapes from $\mathcal{S}$. Of course, such a prediction model will make a number of errors: the non-dense cells contained in some selected rectangle and the dense cells not covered by any selected rectangle. In the example of Fig. 4, the model has selected two rectangles and makes four prediction errors: the cells (4, 3) and (6, 7) are non-dense cells covered by a rectangle, and the cells (6, 2) and (7, 8) are dense cells not covered by a rectangle.

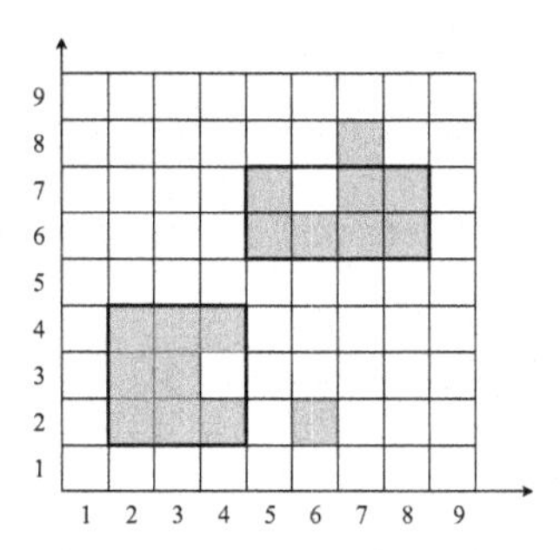

Fig. 4. Model example

The Integer Linear Program. The selection status of every candidate $R_i \in \mathcal{S}$ is modeled with one binary variable $x_i \in \{0, 1\}$. The set of selected shapes is noted $\mathcal{R} = \{R_i \in \mathcal{S} \mid x_i = 1\}$. By abuse of notation, we also use $\mathcal{R}$ to denote the set of covered cells $\bigcup_{x_i=1} R_i$. (Un)covered cells are captured in the model using binary variables $cov_c \in \{0, 1\}$, where $cov_c = 1 \iff c \in \mathcal{R}$.

Given that $\mathcal{G}$ ($\mathcal{G}^*$) is the set of (dense) cells, the dense cells not covered by any rectangle are denoted as $error^+ = \{c \in \mathcal{G}^* \mid cov_c = 0\}$ and the non-dense cells covered by some rectangles are denoted as $error^- = \{c \in \mathcal{G} \setminus \mathcal{G}^* \mid cov_c = 1\}$. We hence wish to discover the set of rectangles that minimizes the error: $\arg\min_{\mathcal{R}} |error^+| + |error^-|$. The complete model is given next.

$$\text{minimize} \sum_{c \in \mathcal{G}^*} (1 - cov_c) + \sum_{c \in (\mathcal{G} \setminus \mathcal{G}^*)} cov_c \tag{1a}$$

subject to

$$\textstyle\sum_{R_i \in \mathcal{S}} x_i \leq K \tag{1b}$$

$$\textstyle\sum_{R_i \in \mathcal{S} | c \in R_i} x_i \leq 1 \qquad \forall c \in \mathcal{G} \tag{1c}$$

$$x_i \leq cov_c \qquad \forall R_i \in \mathcal{S}, \forall c \in R_i \tag{1d}$$

$$cov_c \leq \textstyle\sum_{R_i \in \mathcal{S} | c \in R_i} x_i \qquad \forall c \in \mathcal{G} \tag{1e}$$

$$x_i \in \{0, 1\} \qquad \forall R_i \in \mathcal{S} \tag{1f}$$

$$cov_c \in \{0, 1\} \qquad \forall c \in \mathcal{G} \tag{1g}$$

The constraint (1b) limits the number of selected rectangles to K. The constraints (1c) prevent selecting overlapping rectangles. The constraints (1d) and (1e) ensure $cov_c = 1 \Leftrightarrow \exists x_i = 1 : c \in R_i$.

We further improve this model to get rid of the $|G| \times |\mathcal{S}|$ constraints (1d) and (1e), and the binary variables cov_c. This new model relies on the next theorem stating that the value $|error^+| + |error^-|$ can be inferred solely based on the number of dense and non-dense cells covered by the rectangles.

Theorem 1. *By denoting d_i (resp. u_i) the number of dense (resp. non-dense) cells covered by the rectangle R_i, it follows that*

$$\arg\min_{\mathcal{R}} |error^+| + |error^-| \Leftrightarrow \arg\min_{\mathcal{R}} \sum_{R_i \in \mathcal{R}} (u_i - d_i).$$

Proof. The term $|error^+|$ can be written as $|\mathcal{G}^*| - \sum_{R_i \in \mathcal{R}} d_i$. The term $|error^-|$ is $\sum_{R_i \in \mathcal{R}} u_i$. It follows that

$$\begin{aligned} \arg\min_{\mathcal{R}} |error^+| + |error^-| &= \arg\min_{\mathcal{R}} |\mathcal{G}^*| - (\sum_{R_i \in \mathcal{R}} d_i) + (\sum_{R_i \in \mathcal{R}} u_i) \\ &= \arg\min_{\mathcal{R}} \sum_{R_i \in \mathcal{R}} (u_i - d_i) \end{aligned}$$

□

The linear program to solve is then the following:

$$\text{minimize} \sum_{R_i \in \mathcal{S}} x_i \cdot (u_i - d_i) \tag{2a}$$

subject to

$$\sum_{R_i \in \mathcal{S}} x_i \leq K \tag{2b}$$

$$\sum_{R_i \in \mathcal{S} | c \in R_i} x_i \leq 1 \quad \forall c \in \mathcal{G} \tag{2c}$$

$$x_i \in \{0, 1\} \quad \forall R_i \in \mathcal{S} \tag{2d}$$

with Eq. (2b) limiting the number of regions and Eq. (2c) enforcing non-overlap between the regions. The problem of ROI selection is thus reduced to an instance of the Maximum Weighted Independent Set problem with an additional cardinality constraint [10] that is generally solved with integer programming solvers.

ROIs of Arbitrary Shape. The integer linear model (2) does not require that candidate regions are rectangular. Any shape that covers a set of cells can be included in the candidate set $\mathcal{S}$. In particular, a circular region $Circ = (row, col, radius)$ defined by its center and radius is a natural ROI candidate. The circular region covers the cells $Circ = \{c_{ij} \in \mathcal{G} \mid |i - row| + |j - col| \leq radius\}$ (assuming Manhattan distance). We further discuss the generation of candidate regions in Sect. 5.

4.2 A Parameter-Free Approach

Fixing the limit K for the maximum number of ROIs can in some cases be an arbitrary decision. We can use the *Minimum Description Length* (MDL) principle [14] to determine the size of a model in a principled manner. MDL trades off the description length of the data given the model, and the description length of the model itself. More precisely, let us assume that we have a set of models (hypothesis) $\mathcal{H}$. The description *length* of the model $L(H)$ is the number of bits needed to encode the model; the description length of the data $L(D \mid H)$ is the number of bits needed to encode the data given the model H. The MDL principle tells us to prefer the model that minimizes $L(D, H) = L(H)+L(D \mid H)$.

The model described in the previous sections is composed of a choice of multiple ROIs, indicating where the cells must be dense, along with errors of the model, that is, coordinates of cells which are included in a selected ROI but are non-dense, and dense cells outside the selected ROIs. Each of these can be encoded using a different number of integers:

- 4 integers per rectangle (top-left and bottom-right corners' coordinates)
- 3 integers per circle (center coordinate and radius)
- 2 integers per wrongly classified cells (coordinates of the cell)

A fixed number of bits are used for every integer.

The length to encode the prediction model $\mathcal{S}$ (selected ROIs), and the input cells in this model is:

$$L(\mathcal{S}) = \sum_{R_i \in \mathcal{S}} size(R_i) \qquad L(\mathcal{G} \mid \mathcal{S}) = 2 \cdot (|\mathcal{G}^*| + \sum_{R_i \in \mathcal{S}} (u_i - d_i)),$$

where $size(R_i)$ is the number of integers required to encode R_i (3 if R_i is a circle and 4 if it is a rectangle, for example). The integer $|\mathcal{G}^*| + \sum_{R_i \in \mathcal{S}}(u_i - d_i)$ counts the number of errors made by the model and the factor 2 accounts for encoding the two coordinates of each exception cell.

We can use the MDL criterion to discover the ROIs without fixing their number in advance. To find regions that minimize the MDL criterion, we solve the following ILP model:

$$\text{minimize} \sum_{R_i \in \mathcal{S}} x_i \cdot (2(u_i - d_i) + size(R_i)) \tag{3a}$$

subject to

$$\textstyle\sum_{R_i \in \mathcal{S} | c \in R_i} x_i \leq 1 \qquad \forall c \in \mathcal{G} \tag{3b}$$

$$x_i \in \{0, 1\} \quad \forall R_i \in \mathcal{S} \tag{3c}$$

Note that we do not exactly minimize $L(\mathcal{G}, \mathcal{S})$. Indeed we removed the constant $2 \cdot D$ as it does not impact the optimization. This problem is an instance of the Maximum Weighted Independent Set problem.

5 Generation of Candidate ROIs

In this work we assume a generic generate and filter approach based on a set of predicates for the candidate regions. The time needed to solve the ILP grows with the number of candidate shapes as each one requires the introduction of one binary decision variable. We show how to reduce the number of candidates while ensuring that the solution found is still optimal.

Table 1. Impact of the filtering on the set of possible candidates before and after the filtering.

Min. density threshold	Grid Size	# candidates	# remaining candidates
2	100	25 502 500	17 218
	150	128 255 625	7 703
	200	404 010 000	3 330
5	100	25 502 500	2 523
	150	128 255 625	1 255
	200	404 010 000	448

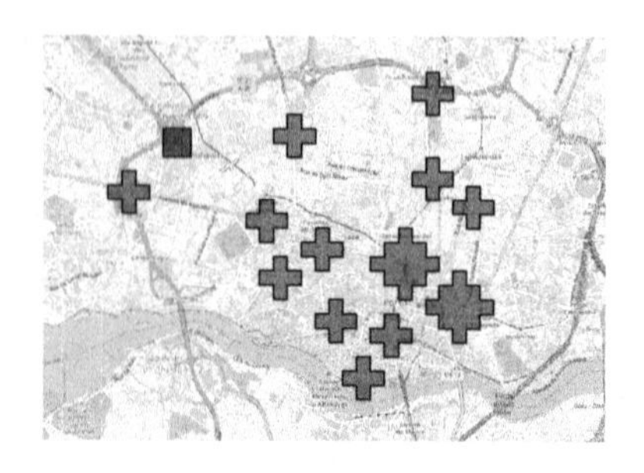

Fig. 5. Output of our method with a minimum distance constraint of 2 and a maximum diameter constraint of 5.

In the worst case, without any filtering, the number of possible rectangles is still polynomial in the size of the grid; more precisely, for a $N \times M$ grid, there are less than N^2M^2 possible rectangles (all the coordinates $(x^1, y^1), (x^2, y^2)$). There is also a polynomial number of circles. Fortunately, one can avoid generating all the candidates. Obviously, we can directly filter out all the candidates R_i for which $2(u_i - d_i) + size(R_i) > 0$. Indeed, in that case the cost of taking the candidate $(2u_i + size(R_i))$ is higher than the cost of not selecting it $(2d_i)$. Moreover if a rectangle contains a set of contiguous rows (or columns) that cover u non-dense cells, d dense cells and the inequality $u > d + 2$ holds, then this rectangle is not part of the optimal solution. The intuition behind this property is that by removing the rows (or columns) from the rectangle, we create two rectangles that yield a smaller description length. Indeed, the gain in description length $(2u)$ is higher than its increase $(2d + 4)$.

As an example of the effectiveness of the filtering, Table 1 shows, for multiple configurations on a Kaggle data set, the total number of distinct rectangles before and after the filtering.

6 Finding Constrained ROIs

As explained in Fig. 1 the ROIs can be constrained in two different ways: with intra- or with inter-ROI constraints.

Intra-ROI Constraints are the ones that must be satisfied independently by each ROI such as "a ROI contains at least one bus stop" or "is at a distance less than 100 m from a train station". These constraints define predicates that must be satisfied by each region. These constraints are taken into account by the algorithm that generates the set of candidates; in its most basic form, this algorithm generates candidates which are filtered using the constraints.

Inter-ROI Constraints are the ones that involve more than one ROI. For instance, "two ROIs must be separated by a minimum distance to ensure diversity in a tourist recommendation system". Such constraints can be modeled in the integer linear program within the non-overlapping constraint (3b) by also including in the sum range all the candidate ROIs within a given radius distance from the cell. The ILP model can also be extended to accommodate constraints that concern only a subset of the regions that cover a cell; in principle, any constraint that can be modeled using linear equations can be added to the model.

Figure 5 shows the output of our method with a minimum distance constraint of 2 between the ROIs and a maximum diameter of 5, and illustrates how the introduction of constraints allows the ROIs to be more diverse.

7 Results and Comparison

Our experiments compare the PopularRegion algorithm with our new approach on both real and synthetic data. Clustering techniques are not producing ROIs with predefined shapes and thus explore an incomparable hypothesis space. We nevertheless include the OPTICS clustering algorithm as an optimistic baseline in our comparisons, assuming that the clusters discovered constitute the prediction function for the density status of cells.

We did not include in this experiments the works of [2,13] since they use application dependent semantic information. The method proposed in [6] also is not evaluated as it finds an exact cover of all the dense cells without generalizing with regions excluding outlier cells like OPTICS.

For the rest of this section, we denote by *ILP* our full model (i.e. Eqs. (3a)–(3b) that includes rectangular and circular ROIs while *ILP-rectangles* denote a restricted model containing only rectangular ROIs[1]. In both models we impose a ratio constraint on the width and height of the rectangles (one can not be more than two times the other) to avoid pathological solutions.

In this section we will address the following questions: i) How well does our method perform compared to *PopularRegion* and OPTICS? ii) How efficient is our approach and what is the computation bottleneck ? iii) Is our method robust to noise in data?

7.1 Performances with Respect to the MDL Criterion

We first describe an experiment performed on two real-world data sets. The first one comes from the taxi destination prediction challenge that was organized by the 2015 ECML/PKDD conference and proposed as a Kaggle competition. This data set contains more than 1.6 million trajectories from taxis of the city of Porto[2]. The second data set is the T-Drive data set from Microsoft and contains

[1] The Python code of our model is accessible here https://github.com/AlexandreDubray/mining-ROI.

[2] The data set can be downloaded at this link https://www.kaggle.com/crailtap/taxi-trajectory/home. We filtered out incomplete trajectories and the few trajectories that went too far away from Porto.

GPS traces from taxis of Beijing [16,17]. For the Kaggle data set, the density threshold will be expressed as a percentage of the total number of trajectories and we used a 100×100 grid. For the T-Drive data set, we used a 200×200 grid and, since we do not have separate trajectories, the density threshold is a percentage of the maximum density in the grid.

OPTICS requires two parameters: $minPts$, a threshold to be a core point, and ξ, a distance ratio to separate the clusters. Details about the parameters can be found in [3]. In our experiments, we set $minPts = 3$ since it is the threshold at which our method considers a candidate interesting. We set $\xi = 0.05$, but this parameter has almost no effect on the results in our experiments.

Figures 6b and 6d show the number of integers needed to encode the errors made by the models (2 per cell wrongly classified), in function of the minimum density threshold. As explained before and illustrated in Figs. 2b and 3b, for low-density thresholds, *PopularRegion* tends to create large regions, which results in a high number of errors since it covers many non-dense cells.

OPTICS selects in its clusters all the cells not considered noise; it is thus expected that it will make few errors, at the expense of a larger model length. Recall that OPTICS does not explore the same hypothesis space. It can thus only be interpreted as a baseline when comparing the errors. Our method discovers regions that generalize well the initial distribution of the dense cells, and allows some non-dense cells in the ROIs. The number of errors is generally between the ones of *PopularRegion* and OPTICS. When the minimum density threshold increases, OPTICS and our approach perform slightly worse than *PopularRegion*. The reason is that *PopularRegion* will overfit perfectly the isolated dense cells by creating one region for each, which is obviously not the expected behavior of an algorithm for detecting ROIs. As expected, the addition of circular shapes permits decreasing slightly the number of errors over the rectangle model since it augments the capacity of the prediction function.

Figures 6a and 6c show the number of integers needed to encode the ROIs (i.e. the first part of the MDL criterion, excluding the values needed to encode errors). Our method always gives a smaller value, with and without circular regions. It can be seen that for a high density threshold, the number of ROIs tends to zero as it is more advantageous to store the exceptions directly rather than using ROIs (the number of dense cells decreasing). When the minimum density threshold becomes larger, the dense cells become sparse over the map and OPTICS considers them as noise without identifying any cluster.

Figure 6 shows that our method outperforms *PopularRegion* on low threshold values by having less errors and nonetheless using fewer ROIs. For a higher value, our methods maintain a similar number of errors as *PopularRegion* while using at least four times less ROIs. Compared to OPTICS, we have a more errors due to the inclusion of non-dense cells in the ROIs, but our ROIs require fewer integers for their encoding. This is only valid due to the balance imposed by the usage of MDL: in general, for a fixed number of ROIs our method will have a smaller error than PopularRegion, and for a fixed error it will have a smaller number of ROIs, by design.

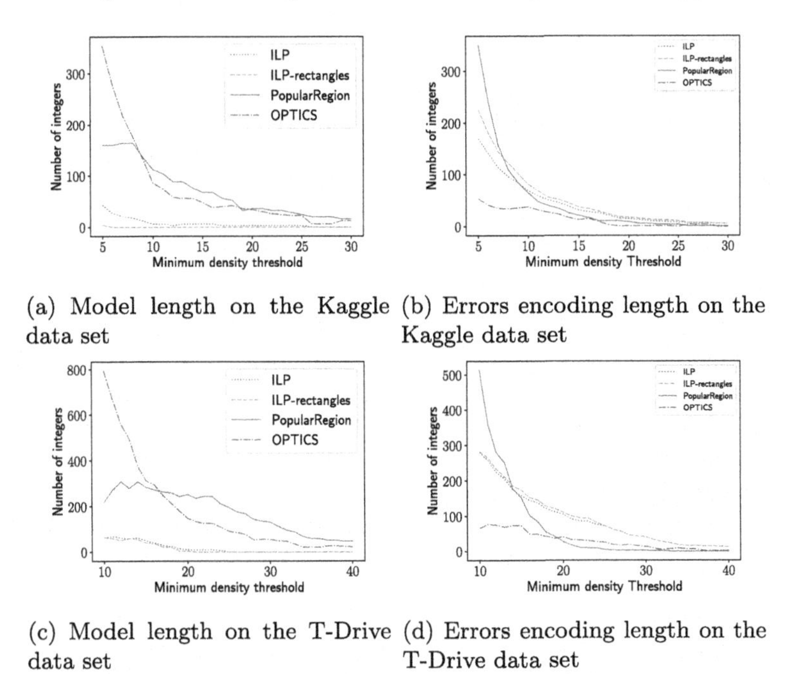

(a) Model length on the Kaggle data set

(b) Errors encoding length on the Kaggle data set

(c) Model length on the T-Drive data set

(d) Errors encoding length on the T-Drive data set

Fig. 6. Error percentage and length of the models in function of the minimum density threshold on the Kaggle data set (a)–(b) and the T-Drive data set (c)–(d).

7.2 Execution Times

Table 2 shows the run time of the methods for two minimum density thresholds and three grid sizes for the Kaggle data set. We limit the size of the grid to 200×200, which corresponds a cell size of 50×50 m. Working beyond this limit seems unreasonable given the accuracy of GPS data. For the ILP model, we show the time needed to solve the optimization problem defined in Eqs. (3a)–(3b). The table also shows the total number of dense cells in the grid as well as the number of candidate shapes.

Table 2. Run time of the methods for different grid sizes and minimum density thresholds for the Kaggle data set.

Minimum density threshold	2%			5%		
Grid side size	100	150	200	100	150	200
Number of dense cells ($\|\mathcal{G}^*\|$)	571	597	537	230	178	137
Number of ILP candidates	23 814	7 779	3 399	2 880	1 232	434
ILP optimization time (s)	4.328	0.464	0.109	0.113	0.044	0.029
PopularRegion run time (s)	0.003	0.005	0.006	0.002	0.003	0.004
OPTICS run time (s)	0.209	0.222	0.200	0.084	0.065	0.051

With its greedy behavior, *PopularRegions* obtains the best run time for all configurations. The run time of our method is mostly determined by the number of candidate shapes as these correspond to the number of variables in the model. We see that when the number of candidates becomes low enough our method has a run time that is similar to OPTICS. In a more constrained application, the set of candidates is expected to be smaller and the constraints stronger, which makes our method practical for identifying constrained regions of interest.

7.3 Robustness to Noise

To evaluate the robustness of the approaches to noise, we start from the Kaggle dataset, which consists of trajectories (i.e. series of points in space and time), and generate the grid by dividing the space in 100×100 cells of uniform size. The dense cells are chosen as being the ones with a minimum density threshold of 0.05 (i.e. at least 5% of the trajectories visit these cells). By running the methods, we obtain for each of them a set of selected ROIs $\mathcal{R}$. We then introduce noise by modifying the trajectory data points: for a level of noise p, each element of a trajectory has a probability p to be moved; if it is moved, its new position is chosen randomly in the square of 10×10 cells around the initial point. By running the methods again, we obtain new sets of selected ROIs under noise $\mathcal{R}'$.

We compute the recall $|\mathcal{R} \cap \mathcal{R}'|/|\mathcal{R}|$, the precision $|\mathcal{R} \cap \mathcal{R}'|/|\mathcal{R}'|$ and the F1-measure $(2 \cdot precision \cdot recall)/(precision + recall)$. Figure 7 shows how these metrics evolve with the level of noise.

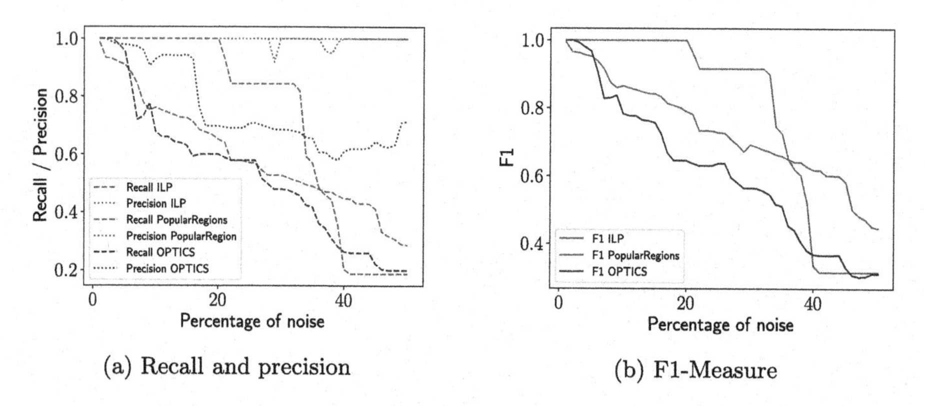

(a) Recall and precision

(b) F1-Measure

Fig. 7. Recall, precision an F1-measure w.r.t the original data in function of the percentage of noise, on the Kaggle data set.

Both *PopularRegions* and our method obtain almost always a precision of 1.0. This means that these methods do not cover areas that were not covered before. However, their recalls decrease, meaning that the found regions will tend to shrink as the amount of noise increases.

While for *PopularRegions*, the recall decreases smoothly with the level of noise, it decreases stepwise for our method. The reason is that our method uses a threshold to define the binary density status of the cells. It thus requires enough noise in order to flip the status of a cell. For OPTICS it can be seen that its precision and recall are lower than for the other two methods, for most of the noise levels. In the beginning, as for the ILP-based method, it still produces the same solution since it only considers the state of the cells. But unlike our method, it is not able to generalize well as it can only cover dense cells. As a consequence, its recall drops faster. For the same reason, it will never return a non-dense cell that was initially dense, thus causing a drop in recall. However, it will return dense cells that were non-dense (and thus not in the initial solution), decreasing its precision. At the light of the F1-Measure, these combined effects are in favor of our method. On any of the metrics, the ILP provides better results as long as the noise remains reasonable. When the noise level becomes significant ($\simeq 40\%$), the dense cells become very sparse and the results are much less relevant to interpret.

8 Conclusion and Future Work

Mining approaches for discovering regions of interest (ROIs) are an important building block for any application wishing to extract knowledge from location data. In order to be useful, the extracted ROIs generally need to satisfy application dependent constraints. This last requirement was missing in existing approaches. Inspired by the approach introduced in [8], we introduced an alternative approach for discovering constrained ROIs. It relies on an efficient Integer Linear Program (ILP) to extract the ROIs from a set of predefined ROIs candidates. The model can be used in a setting where the number of ROIs is fixed, or it can work in a parameter free setting by relying on the *minimum description length principle*. Our approach is flexible as it can discover ROIs satisfying various types of constraints that can be enforced either at the step of the candidate ROI generation, or directly in the integer linear programming model. We have reported various experiments showing the flexibility of the proposed approach on both real and synthetic data sets. The results have shown that it was able to retrieve constrained ROIs of higher quality than those extracted with existing approaches such as the PopularRegion algorithm [8] and clustering techniques. Despite the larger computation time, we showed that the approach is able to scale on real-world data sets using fine-grained grids.

As future work one could solve the candidate generation problem using a custom constraint-based search algorithm rather than with a generate and filter one. Although less generic, this could be more efficient if many regions need to be filtered out. The ILP does not require the shapes to be defined on the grid. As future work, it could be interesting to extend our work with ROIs defined in the continuous space. Finally, it would be interesting to extend the approach to work with continuous density values rather than binary ones that require a threshold parameter.

References

1. Agrawal, R., Srikant, R.: Mining sequential patterns. In: IEEE International Conference on Data Engineering (ICDE), vol. 95, pp. 3–14 (1995)
2. Alvares, L.O., Bogorny, V., Kuijpers, B., de Macedo, J.A.F., Moelans, B., Vaisman, A.: A model for enriching trajectories with semantic geographical information. In: GIS (2007)
3. Ankerst, M., Breunig, M.M., Kriegel, H.P., Sander, J.: OPTICS: ordering points to identify the clustering structure. ACM SIGMOD Rec. **28**(2), 49–60 (1999)
4. Aoga, J.O.R., Guns, T., Schaus, P.: An efficient algorithm for mining frequent sequence with constraint programming. In: Frasconi, P., Landwehr, N., Manco, G., Vreeken, J. (eds.) ECML PKDD 2016. LNCS (LNAI), vol. 9852, pp. 315–330. Springer, Cham (2016). https://doi.org/10.1007/978-3-319-46227-1_20
5. Belcastro, L., Marozzo, F., Talia, D., Trunfio, P.: G-ROI: automatic region-of-interest detection driven by geotagged social media data. TKDD **12**(3), 1–22 (2018)
6. Cai, G., Hio, C., Bermingham, L., Lee, K., Lee, I.: Sequential pattern mining of geo-tagged photos with an arbitrary regions-of-interest detection method. Expert Syst. Appl. **41**(7), 3514–3526 (2014)
7. Ester, M., Kriegel, H.P., Sander, J., Xu, X., et al.: A density-based algorithm for discovering clusters in large spatial databases with noise. In: KDD (1996)
8. Giannotti, F., Nanni, M., Pinelli, F., Pedreschi, D.: Trajectory pattern mining. In: SIGKDD (2007)
9. Gorawski, M., Jureczek, P.: Regions of interest in trajectory data warehouse. In: Nguyen, N.T., Le, M.T., Świątek, J. (eds.) ACIIDS 2010. LNCS (LNAI), vol. 5990, pp. 74–81. Springer, Heidelberg (2010). https://doi.org/10.1007/978-3-642-12145-6_8
10. Kalra, T., Mathew, R., Pal, S.P., Pandey, V.: Maximum weighted independent sets with a budget. In: Gaur, D., Narayanaswamy, N.S. (eds.) CALDAM 2017. LNCS, vol. 10156, pp. 254–266. Springer, Cham (2017). https://doi.org/10.1007/978-3-319-53007-9_23
11. Li, Q., Zheng, Y., Xie, X., Chen, Y., Liu, W., Ma, W.Y.: Mining user similarity based on location history. In: SIGSPATIAL. ACM (2008)
12. Monreale, A., Pinelli, F., Trasarti, R., Giannotti, F.: Wherenext: a location predictor on trajectory pattern mining. In: SIGKDD. ACM (2009)
13. Palma, A.T., Bogorny, V., Kuijpers, B., Alvares, L.O.: A clustering-based approach for discovering interesting places in trajectories. In: SAC (2008)
14. Rissanen, J.: Modeling by shortest data description. Automatica **14**(5), 465–471 (1978)
15. Ying, J.J.C., Lee, W.C., Weng, T.C., Tseng, V.S.: Semantic trajectory mining for location prediction. In: SIGSPATIAL. ACM (2011)
16. Yuan, J., Zheng, Y., Xie, X., Sun, G.: Driving with knowledge from the physical world. In: SIGKDD. ACM (2011)
17. Yuan, J., et al.: T-drive: driving directions based on taxi trajectories. In: SIGSPATIAL. ACM (2010)
18. Yuan, N.J., Zheng, Y., Zhang, L., Xie, X.: T-finder: a recommender system for finding passengers and vacant taxis. IEEE TKDE **25**(10), 2390–2403 (2013)

Mining Disjoint Sequential Pattern Pairs from Tourist Trajectory Data

Siqi Peng(✉) and Akihiro Yamamoto

Kyoto University, Kyoto, Japan
peng.siqi.37r@st.kyoto-u.ac.jp, yamamoto.akihiro.5m@kyoto-u.ac.jp

Abstract. Route mining from trajectory databases, or trajectory data mining, has become an important and valuable task since the popularization of GPS devices. Sequential pattern mining based approaches are well applied to trajectory data mining, while they often suffer the problem of high redundancy and low comprehensibility. In this paper, we solve this problem by proposing a novel approach of disjoint sequential pattern pair mining, which takes on a new perspective to this problem by focusing on extracting extra valuable information, *i.e.*, hyper patterns from the "redundant" patterns instead of just removing them. We conduct experiments on a real tourist trajectory database as well as an artificial one. We show the practical applicability of our approach and the effectiveness and efficiency of our mining algorithm by analyzing the mining results.

Keywords: Route mining · Trajectory data mining · Sequential pattern mining · Disjoint sequential pattern pair mining

1 Introduction

Trajectory databases, the databases keeping records of the moving trajectories of vehicles and/or pedestrians, have recently become more available and cost effective due to the popularization of GPS technology. Thus, extracting information or knowledge from such databases, *i.e.*, route mining, has become a valuable and challenging task. One of the most preferred approaches for route mining is to convert each trajectory into a sequence and then apply *sequential pattern mining* (SPM) [1,6] for mining the frequent segments of these trajectories [3,5,8,12]. In Fig. 1, we give an example for the basic procedure of SPM based route mining.

These SPM based approaches are relatively easy to implement, but may suffer the problem of producing an output with much redundancy [2,3], and therefore in the previously proposed SPM based route mining systems, various techniques have been implemented to improve the quality of the outputs [3,4,9,11,12,14,15]. In this research, we propose a new approach to solve this problem from another perspective. The goal of this approach is to mine

Supported by CREST and North Grid Co., Ltd.

A. Appice et al. (Eds.): DS 2020, LNAI 12323, pp. 645–658, 2020.
https://doi.org/10.1007/978-3-030-61527-7_42

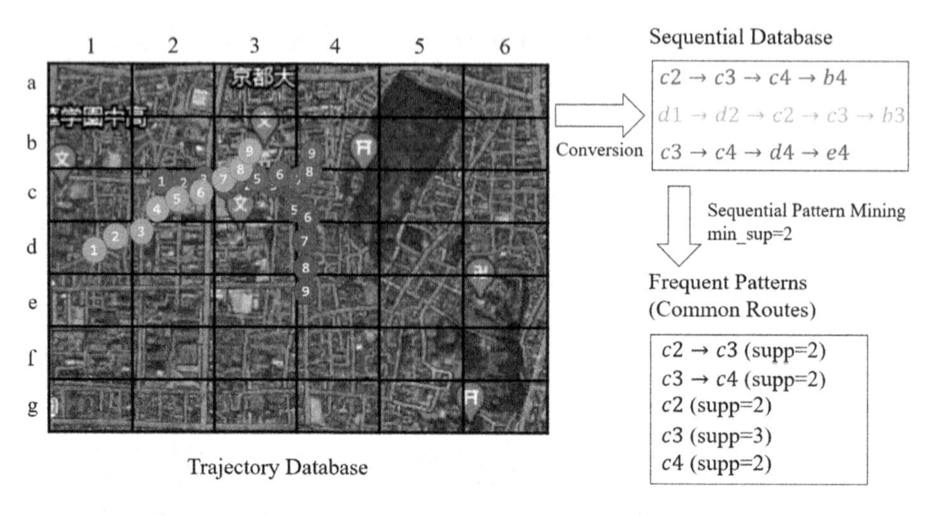

Fig. 1. An example for SPM based route mining. In the example, three trajectories are shown on the map provided by Google Earth. The map is divided into several mesh grids and all grids are marked with chessboard coordinates so that we can represent each trajectory with a sequence of grids. For instance, the red trajectory successively passes through grid $c2, c3, c4, b4$, so it can also be represented with sequence $c2 \rightarrow c3 \rightarrow c4 \rightarrow b4$. After conversion, the trajectory database turns into a sequential database so we can use SPM to mine frequent patterns, *i.e.*, frequent segments of trajectories. (Color figure online)

hyper-patterns called *disjoint sequential pattern pairs*, *i.e.*, associated pattern pairs with a *disjoint* relationship. These hyper-patterns are proved to provide more concise and valuable information compared with the original sequential patterns. That is, by applying disjoint sequential pattern pair mining, we can get more concise outputs with little redundancy and much valuable knowledge which is difficult to be extracted with original SPM based methods.

In the latter part of this paper, we first summarize the background knowledge of SPM based route mining and previously proposed approaches for output optimization. Then, we introduce our newly proposed approach of disjoint sequential pattern pair mining. The introduction includes our definitions of frequent, closed and maximal disjoint sequential patterns and our algorithms for mining them. Next, we describe our procedures for processing a real tourist trajectory database and generating an artificial database. At last, we conduct comparative experiments on the two databases and discuss the results.

2 Preliminary Knowledge

2.1 Sequential Pattern Mining (SPM) and Route Mining

Sequential Pattern Mining (SPM) is the task of extracting ordered patterns from a sequential database [1,6]. The task has lots of variations for applications in different fields, while in this paper, we only introduce the variation mainly applied in route mining [3,8,12]: Given a *sequential database*

$DB = \{S^{(1)}, S^{(2)}, \ldots, S^{(\|DB\|)}\}$ consisting of *sequences*, the goal of the task is to list all short sequences with a support value larger than a threshold. Here, a sequence $A = A_1 \to A_2 \to \cdots \to A_m$ is an ordered list of *items*. The support value of a short sequence S is defined as below:

Definition 1. *The support value of a sequential pattern S in database DB, denoted by $supp_{DB}(S)$, is defined as the number of sequences in the database which support S. Here, we say sequence P supports sequence Q (denoted as $P \sqsupseteq Q$ or $Q \sqsubseteq P$) if Q is a subsequence of P, i.e., $\exists\{x_1, x_2, \ldots, x_{\|Q\|}\}$ such that $1 \leq x_1 \leq x_2 \leq \ldots \leq x_{\|Q\|} \leq \|P\|$ and $1 \leq \forall i \leq \|Q\|$ $Q_i = P_{x_i}$. Here $\|P\|$ means the length of sequence P.*

The task of route mining can be reduced to SPM with techniques converting the trajectories into a sequence of road segments [3,12] or zones [5,8], as is previously introduced in Fig. 1. However, such basic form of SPM based route mining may generate large quantities of redundant patterns [3,14,15]. To get a practical output, researchers have proposed methods and techniques for reducing the size and increasing the quality of the mined patterns.

3 Related Work on Improving the Output Quality

A common solution for reducing the redundancy of the output is to limit the output size by adding constraints to the patterns to be mined. These methods are categorized as *constraint-based sequential pattern mining*. To the furthest of our preliminary survey, the most preferred constraint for SPM based route mining is the *contiguous constraint* [3,14,15], *i.e.*, the constraint that only patterns matching with contiguous road sections are to be mined. These methods are effective in reducing the redundancy, but they rely on foreknown features of the database to choose the proper type of constraint to be applied. For example, the well-preferred contiguous constraint is only applicable to vehicle trajectory data mining because other types of trajectories (like bicycles) may not match perfectly with contiguous road sections [5]. That is, for databases with unknown features, a more general method for reducing the redundancy is needed.

Another common solution is to extend the mined sequential patterns to more informative and valuable forms with the help of external data sources so that the extended outputs will contain less redundancy. For example, the *POI-Visit pattern mining module* [4] utilizes the internal traffic time database to extend the sequential patterns into *POI-Visit patterns* which is proved to have much value and little redundancy. These methods can work on any trajectory databases, but become unavailable when we have no access to additional data sources.

Inspired by these ideas, we propose our new solution of *disjoint sequential pattern pair mining*. As is being proved in the following sections, by applying the solution, we can not only reduce the sizes of the outputs, but also extend the patterns in the outputs into associated pattern pairs from which we can extract more valuable knowledge without access to external databases.

4 Disjoint Sequential Pattern Pair Mining

We first give the basic definition of *disjoint sequential pattern pairs* as follows. Note that in this paper, a right arrow is used for denoting not only the order of items but also the concatenation of sequences and elements (*e.g.* $a \rightarrow F, B \rightarrow C$), where sequences are denoted with uppercase letters and items are denoted with lowercase letters like p or uppercase letters with subscripts like P_1.

Definition 2. *Sequential pattern* $P = P_1 \rightarrow P_2 \rightarrow \cdots \rightarrow P_{\|P\|}$ *and* $Q = Q_1 \rightarrow Q_2 \rightarrow \cdots \rightarrow Q_{\|Q\|}$ *form a disjoint sequential pattern pair (DSPP) if all conditions below are satisfied.*

- $P_1 = Q_1$ *and* $P_{\|P\|} = Q_{\|Q\|}$.
- $\{P_2, P_3, ..., P_{\|P\|-1}\} \cap \{Q_2, Q_3, ..., Q_{\|Q\|-1}\} = \emptyset$.
- *For any integer pair* (i, j) *satisfying* $1 < i < \|P\|$ *and* $1 < j < \|Q\|$, *there must be* $P_i \notin \{P_1, P_{\|P\|}\}$ *and* $Q_j \notin \{Q_1, Q_{\|Q\|}\}$ *and* $supp_{DB}(P_0 \rightarrow P_i \rightarrow Q_j \rightarrow P_{\|P\|}) = 0$ *and* $supp_{DB}(Q_0 \rightarrow Q_j \rightarrow P_i \rightarrow Q_{\|Q\|}) = 0$.

From the definition, we know that the disjoint sequential pattern pair is a hyper-pattern representing the disjoint relation of two routes from the starting place to the destination place. Here "disjoint" means a tourist moving from the starting place to the destination place may only choose either one of the route, and once they choose the route, he or she should never switch to another route before he or she arrives at the destination. These disjoint route pairs widely exist in our real life. For example, say we have six spots located in zone a, b, c, d, e, f and we find that pattern $a \rightarrow c \rightarrow e \rightarrow f$ with $a \rightarrow b \rightarrow d \rightarrow f$ associates a disjoint pattern pair. Figure 2 depicts two possible cases where the disjoint relationship may exist. In the first case, a lake is located between walking path $a \rightarrow b \rightarrow d \rightarrow f$ and $a \rightarrow c \rightarrow e \rightarrow f$, suggesting that the direct access from b or d to c or e is impossible. In the second case, zone b and d are located on a different bus route with zone c and e. Since tourists are unlikely to switch from a bus route to another one, it is also unlikely for a tourist to visit c or e directly after visiting b or d. Such information of disjoint routes can be quite important for tourists to make their plan of where to visit and where not to visit, while the information could be difficult to extract by human inspection from a large SPM output. Therefore, it is beneficial for us to develop a method for mining disjoint sequential pattern pairs.

Now that we have defined disjoint sequential pattern pairs, we still need an indicator measuring the quality or the importance of a pattern pair. In SPM, we have a support threshold functioning as the indicator of importance. Following this idea, we define the *support value* of disjoint sequential pattern pairs and the task of *disjoint sequential pattern pair mining (DSPPM)* as follows:

Definition 3. *Given a sequential database DB and a support threshold min_sup, the task of DSPPM is to find all frequent disjoint sequential pattern pairs (FDSPP), i.e., all disjoint sequential pattern pairs* (P, Q) *satisfying* $supp_{DB}(P) \geq min\_sup$ *and* $supp_{DB}(Q) \geq min\_sup$.

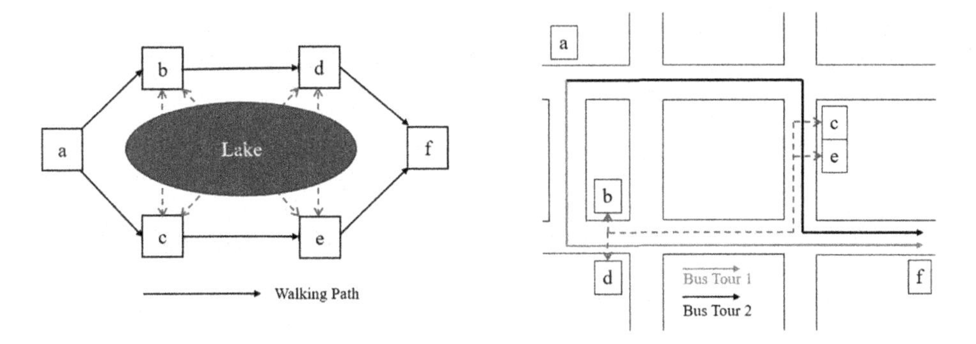

Fig. 2. Two examples for real-life disjoint route pairs. Note that while we only mark out 6 zones, every point on the map should be assigned into a zone.

The most important feature of the support value is the property of anti-monotonicity, as is presented in Lemma 1. Note that here $P \sqsupset Q$ means $Q \sqsubseteq P$ and $\|Q\| < \|P\|$.

Lemma 1. *Suppose* $(s \to P \to t, s \to Q \to t)$ *to be a frequent disjoint sequential pattern pair, then for all* $P' \sqsubseteq P$ *and* $Q' \sqsubseteq Q$, $(s \to P' \to t, s \to Q' \to t)$ *is also a frequent disjoint sequential pattern pair.*

Proof. We use $\{P\}$ for denoting the set $\{P_1, P_2, \ldots, P_{\|P\|}\}$. Then, it is sure that for all $P' \sqsubseteq P$, there must be $\{P'\} \subseteq \{P\}$. Thus, for any $p' \in \{P'\}$ and $q' \in \{Q'\}$, we have $p' \in \{P\}$ and $q' \in \{Q\}$. Since $s \to P \to t$ and $s \to Q \to t$ are disjoint, we have $supp_{DB}(s \to p' \to q' \to t) = supp_{DB}(s \to q' \to p' \to t) = 0$, which leads to the conclusion that $s \to P' \to t$ and $s \to Q' \to t$ are disjoint. Hence, for any $S \in DB$ such that $s \to P \to t \sqsubseteq S$, there must be $s \to P' \to t \sqsubseteq S$. This suggests that $supp_{DB}(s \to P \to t) \leq supp_{DB}(s \to P' \to t)$. Similarly, we get $supp_{DB}(s \to Q \to t) \leq supp_{DB}(s \to Q' \to t)$. In summary, $(s \to P' \to t, s \to Q' \to t)$ is a frequent disjoint sequential pattern pair.

The lemma could be useful for enumerating FDSPPs, while it also indicates that the set of FDSPPs may contain some redundancy. Thus, we propose the definition of *closed-disjoint sequential pattern pairs* and *maximal-disjoint sequential pattern pairs* following the idea of *closed sequential pattern mining* [6,13] and *maximal sequential pattern mining* [6,10].

Definition 4. *Sequential pattern* $s \to P \to t$ *and* $s \to Q \to t$ *form a closed-disjoint sequential pattern pair (CDSPP) with respect to support threshold* min_sup *if all following conditions are satisfied.*

- $(s \to P \to t, s \to Q \to t)$ *is a frequent disjoint sequential pattern pair.*
- *For all* $P' \sqsupset P$, *if* $s \to P' \to t$ *and* $s \to Q \to t$ *are disjoint, there must be* $supp_{DB}(s \to P' \to t) < supp_{DB}(s \to P \to t)$.
- *For all* $Q' \sqsupset Q$, *if* $s \to Q' \to t$ *and* $s \to P \to t$ *are disjoint, there must be* $supp_{DB}(s \to Q' \to t) < supp_{DB}(s \to Q \to t)$.

Definition 5. *Sequential pattern* $s \to P \to t$ *and* $s \to Q \to t$ *form a maximal-disjoint sequential pattern pair (MDSPP) with respect to support threshold* min_sup *if all following conditions are satisfied.*

- $(s \to P \to t, s \to Q \to t)$ *is a frequent disjoint sequential pattern pair.*
- *For all* $P' \sqsupset P$, *if* $s \to P' \to t$ *and* $s \to Q \to t$ *are disjoint, there must be* $supp_{DB}(s \to P' \to t) < min\_sup$.
- *For all* $Q' \sqsupset Q$, *if* $s \to Q' \to t$ *and* $s \to P \to t$ *are disjoint, there must be* $supp_{DB}(s \to Q' \to t) < min\_sup$.

It is clear that all CDSPPs are FDSPPs, and all MDSPPs are CDSPPs. The conclusion could be useful for developing an enumerate-and-check algorithm for listing CDSPPs and MDSPPs.

5 Algorithms

We first propose an algorithm for basic DSPPM. Although we address it as a generate-and-test algorithm, it never means that we are to generate all frequent sequential pattern pairs and check them one by one because it is very inefficient hence impossible. See the following case for example:

Example 1. Consider the following database $DB = \{a \to b \to c, a \to d \to c, a \to b \to d \to c\}$ with our min_sup set to 2. If we use the strategy that lists all frequent sequential patterns and check the disjoint relation of every frequent pattern pair with the information from the output, we will not even get a correct result. The reason is that both $a \to b \to c$ and $a \to d \to c$ will be included in the set of frequent sequential patterns, while $a \to b \to d \to c$ will not, which means that we will fail to recognize that $a \to b \to c$ and $a \to d \to c$ are actually not disjoint unless we access back into the database.

In all, a well-designed algorithm is required. In this research, we discover that most candidate pattern pairs can be pruned out during the generating period exploiting the features of *a projected database* [7]. The term is firstly proposed in *PrefixSpan algorithm* [7], while in this research, we present a modified and generalized version of projected databases as follows:

Definition 6. *Given a sequence* P *and two short sequences* S *and* T. *Let* n *be the smallest integer satisfying* $S \sqsubseteq P_1 \to P_2 \to \cdots \to P_n$ *and* m *be the largest integer satisfying* $T \sqsubseteq P_m \to P_{m+1} \to \cdots \to P_{\|P\|}$. *The projected sequence of* P *with prefix* S *and suffix* T, *denoted with* $P|_S^T$, *is defined to be* $P_{n+1} \to P_{n+2} \to \cdots \to P_{m-1}$.

Definition 7. *Given a database* $DB = \left\{S^{(1)}, S^{(2)}, \ldots, S^{(\|DB\|)}\right\}$, *the projected database of* DB *under prefix* P *and suffix* Q, *denoted with* $DB|_P^Q$, *is defined to be* $\left\{S^{(i)}|_P^Q \,\middle|\, \|S^{(i)}|_P^Q\| > 0\right\}$. *Hence, the ID set of the projected database* $DB|_P^Q$, *denoted with* $IDset(DB|_P^Q)$, *is defined as* $\left\{i \,\middle|\, \|S^{(i)}|_P^Q\| > 0\right\}$.

From this definition, we can derive the following theorem allowing us to enumerate all FDSPPs in a divide-and-conquer order, which is also quite similar to the basic principle of the PrefixSpan algorithm [7].

Theorem 1. *For any sequential pattern $P \rightarrow a \rightarrow Q$, $supp_{DB}(P \rightarrow a \rightarrow Q) = supp_{DB|_P^Q}(a)$.*

Proof. For any sequence $S \in DB$ satisfying $P \rightarrow a \rightarrow Q \sqsubseteq S$, there must be $S|_P^Q \in DB|_P^Q$ for that $\|S|_P^Q\| > 0$. Now we support that $a \not\sqsubseteq S|_P^Q$ and let $S|_P^Q = S_{n+1} \rightarrow S_{n+2} \cdots \rightarrow S_{m-1}$. Because $a \not\sqsubseteq S|_P^Q$, we should have either $a \sqsubseteq S_1 \rightarrow S_2 \rightarrow \cdots \rightarrow S_n$ or $a \sqsubseteq S_m \rightarrow S_{m+1} \cdots \rightarrow S_{\|S\|}$. However, because $P \rightarrow a \rightarrow Q \sqsubseteq S$, if we have $a = S_i$, we must have $P \sqsubseteq S_1 \rightarrow S_2 \rightarrow \cdots \rightarrow S_{i-1}$ if $i \leq n$ or $Q \sqsubseteq S_{i+1} \rightarrow S_{i+2} \rightarrow \cdots \rightarrow S_{\|S\|}$ if $i \geq m$, which breaks the rule of projected sequence. Therefore, $supp_{DB}(P \rightarrow a \rightarrow Q) \leq supp_{DB|_P^Q}(a)$. Hence, for any sequence $T \in DB|_P^Q$ such that $a \sqsubseteq T$, there must be $T' \in DB$ such that $T'|_P^Q = T$. According to the definition of projected sequence, we are sure to have $P \rightarrow a \rightarrow Q \sqsubseteq T'$, which suggests $supp_{DB}(P \rightarrow a \rightarrow Q) \geq supp_{DB|_P^Q}(a)$. In all, $supp_{DB}(P \rightarrow a \rightarrow Q) = supp_{DB|_P^Q}(a)$.

Also, we can carry out another important property of DSPPs:

Theorem 2. *For any sequential pattern $P = s \rightarrow P' \rightarrow t$ and $Q = s \rightarrow Q' \rightarrow t$ with $s, t \notin \{P'\} \cup \{Q'\}$ and $\{P'\} \cap \{Q'\} = \emptyset$, P and Q can form a disjoint sequential pattern pair if and only if $1 \leq \forall i \leq \|P\|$ $1 \leq \forall j \leq \|Q\|$ $IDset(DB|_{s \rightarrow P_i}^t) \cap IDset(DB|_{s \rightarrow Q_j}^t) = \emptyset$.*

Proof. We first give the proof of necessity. Let $DB = \{D^{(1)}, D^{(2)}, \ldots, D^{(\|DB\|)}\}$ and $R = D^{(r)}$. Suppose that $r \in IDset(DB|_{s \rightarrow P_i}^t) \cap IDset(DB|_{s \rightarrow Q_j}^t)$. Then, according to Definition 7, we have $\|R|_{s \rightarrow P_i}^t\| > 0$ and $\|R|_{s \rightarrow Q_j}^t\| > 0$. Because $P_i \neq Q_j$, we must have either $s \rightarrow P_i \rightarrow Q_j \rightarrow t \sqsubseteq R$ or $s \rightarrow Q_j \rightarrow P_i \rightarrow t \sqsubseteq R$, which suggests that P and Q cannot form a disjoint sequential pattern pair. To give the proof of the adequacy, suppose that P and Q are not disjoint, then there must be $1 \leq \exists i \leq \|P\|$ $1 \leq \exists j \leq \|Q\|$ $supp_{DB}(s \rightarrow P_i \rightarrow Q_j \rightarrow t) > 0$ or $supp_{DB}(s \rightarrow Q_j \rightarrow P_i \rightarrow t) > 0$. Because $P_i \neq Q_j$, there must exist pattern $R' = D^{(r')}$ such that $Q_j \sqsubseteq R'|_{s \rightarrow P_i}^t$ or $P_i \sqsubseteq R'|_{s \rightarrow Q_j}^t$. This suggests that $r \in IDset(DB|_{s \rightarrow P_i}^t) \cap IDset(DB|_{s \rightarrow Q_j}^t)$, which contradicts with the condition that $IDset(DB|_{s \rightarrow P_i}^t) \cap IDset(DB|_{s \rightarrow Q_j}^t) = \emptyset$.

With Theorem 2, we can develop the basic flow of the algorithm for DSPPM as follows: First we enumerate item pair (s, t) satisfying $\|DB|_s^t\| \geq min\_sup$. Then, for each pair (s, t), we generate item pair (p, q) satisfying $IDset(DB|_{s \rightarrow p}^t) \cap IDset(DB|_{s \rightarrow q}^t) = \emptyset$ and add all these pairs into the set $DisjointPairs$. Hence, for each pair $(p, q) \in DisjointPairs$ we also add p and q into another set $CandidateItems$. Next, we are to generate all frequent sequential patterns $s \rightarrow P \rightarrow t$ with only the items in the set of $CandidateItems$ for that according to Theorem 2, it is impossible for a pattern containing items not belonging to this set to form a disjoint sequential pattern pair. After that we get the set of candidate frequent sequential patterns $FreqPatterns$. Finally, we enumerate all pattern pairs $(P, Q) \in FreqPatterns$, and check the disjoint relation between them individually with $DisjointPairs$ referring to Theorem 2, and we will get the set of FDSPPs. The whole procedure is described with pseudocode in Algorithm 1.

Algorithm 1. Basic Flow for DSPPM

```
for each item pair (s, t) such that ||DB|_s^t|| > min_sup do
    DisjointPairs ← ∅, CandidateItems ← ∅.
    for each item pair (p, q) do
        if p ≠ q ≠ s ≠ t and IDset(DB|^t_{s→p}) ∩ IDset(DB|^t_{s→q}) = ∅ then
            Add (p, q) into DisjointPairs
            Add p and q into CandidateItems
        end if
    end for
    FreqPatterns ← all frequent sequential patterns s → P → t such that ∀p ∈
{P} p ∈ CandidateItems.
    for each pair (s → P → t, s → Q → t) ∈ FreqPatterns do
        if ∀p ∈ {P} ∀q ∈ {Q} (p, q) ∈ DisjointPairs then
            Output pair (s → P → t, s → Q → t)
        end if
    end for
end for
```

Now, the only unsolved problem is how to list all frequent sequential patterns $s \rightarrow P \rightarrow t$ with items from the set of $CandidateItems$. Here we choose to apply the PrefixSpan algorithm. The pseudocode is listed in Algorithm 2. With Theorem 1, we can confirm that this algorithm will produce a correct result. A more detailed proof of correctness and completeness of PrefixSpan algorithm can be found in [7].

Algorithm 2. Listing all $s \rightarrow P \rightarrow t$ with PrefixSpan

```
function SEARCH(CurDB, CurPrefix)
    for each p ∈ CandidateItems do
        if ||CurDB|_p|| ≥ min_sup then
            Add CurPrefix → p → t into FreqPatterns.
            SEARCH(CurDB|_p, CurPrefix → p)
        end if
    end for
end function
SEARCH(DB|_s^t, s)
Output FreqPatterns.
```

Now that we have listed all FDSPPs, it would not be a hard job to list all CDSPPs and MDSPPs because we only need to traverse the set of FDSPPs and check if each pattern satisfies the conditions of a CDSPP or an MDSPP. The pseudocode is presented in Algorithm 3. Although this idea seems to be simple and direct, in the next section, we will show that the algorithm will not result in a low efficiency when we conduct experiments on a real trajectory database.

Algorithm 3. Listing closed-disjoint and maximal-disjoint sequential pattern pairs

```
for each disjoint sequential pattern pair (s → P → t, s → Q → t) do
    checkClosed ← True
    checkMaximal ← True
    for each item p ∈ CandidateItems do
        for 1 ≤ i ≤ |P| + 1 do
            Insert p between P_{i-1} and P_i and get P'.
            if (s → P' → t, s → Q → t) is in the output of DSPPM then
                checkMaximal ← False
            end if
            if supp_DB(s → P' → t) = supp_DB(s → P → t) then
                checkClosed ← False
            end if
        end for
        for 1 ≤ i ≤ |Q| + 1 do
            Insert p between Q_{i-1} and Q_i and get Q'.
            if (s → P → t, s → Q' → t) is in the output of DSPPM then
                checkMaximal ← False
            end if
            if supp_DB(s → Q' → t) = supp_DB(s → Q → t) then
                checkClosed ← False
            end if
        end for
    end for
    if checkClosed then
        Output (s → P → t, s → Q → t) as a closed-disjoint sequential pattern pair.
    end if
    if checkMaximal then
        Output (s → P → t, s → Q → t) as a maximal-disjoint sequential pattern
pair.
    end if
end for
```

At last, we are to analyze the complexity of the algorithms. We denote $|FreqPatterns|$, the number of all items, the number of all mined FDSPPs with F, I, R respectively, and regard other factors as constants. For Algorithm 1, the complexity of enumerating all frequent pattern pairs and (p, q) pairs is $O(F^2)$ and $O(I^2)$. For Algorithm 2, the complexity should be $O(IF)$ because each time we call the *Search* function we enumerate all items and output at most one pattern into *FreqPattern*. For Algorithm 3, the complexity should be $O(IR)$. From these theoretical derivations, we know that generally, the overall complexity of the algorithms is linear related with the size of the final output. That is, the fewer FDSPPs are there in the output, the shorter time our algorithms will spend in mining them.

6 Experiment

All experiments in this research are conducted on an Intel Core i7 CPU with 20 GB of RAM. The databases we use include a real trajectory database provided by North Grid Co., Ltd. and an artificial database generated with respect to the features of the real database. The real trajectory database contains 1,412 tourist trajectories formed with 605,772 tracking points. To convert it into a sequential database, we replace each tracking point in the database with its Japanese standard address (see Fig. 3 for an example). We use this strategy instead of the uniform mesh introduced in Fig. 1 mainly because sequences of addresses are more comprehensible than sequences of grid numbers, so it allows us to inspect and analyze the mined patterns in an easier way.

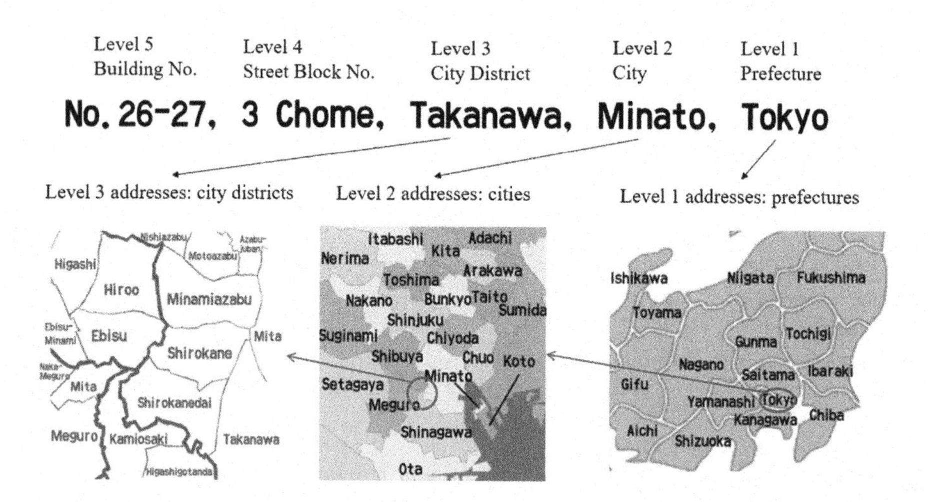

Fig. 3. The structure of Japanese standard addresses. As is shown in the figure, the addressing system has a hierarchy structure, which is equivalent to a multi-level non-uniform mesh. In this research, we finally use the addresses exacted to Level 3 for we have not found reliable data of higher-level addresses.

After the pre-processing, the database is converted into a sequential database with 17,344 different items *i.e.*, addresses. The processed data can be accessed at GitHub[1]. Note that in the public data, each address is encrypted with an integer ranged from 0 to 17,343.

Now that we get the sequential database DB_R, we are to create an artificial database DB_A. It contains the same number of sequences as DB_R, where each sequence is generated with the following steps.

1. Calculate the length distribution of sequences with:

$$P_L(X=i)=\frac{|\{L\,\|L|=i\}\,|}{\|DB_R\|}$$

2. Calculate the generative probability of each item by regarding them as independent variables with:

$$P_G(X=x_i)=\frac{\sum_{P^{(i)}\in DB_R}\sum_{1\leq j\leq \|P^{(i)}\|}\mathbf{1}_{P_i=x_i}}{\sum_{P^{(i)}\in DB_R}\|P^{(i)}\|}$$

   where $\mathbf{1}_{cond}$ is the indicative function which takes the value of 1 if *cond* is true and 0 if it is false.
3. When generating a sequence in DB_A, we first generate a random length L from the length distribution P_L, and then generate L items as a sequence from the generative probability distribution P_G. Repeat generating sequences until $\|DB_A\|=\|DB_R\|$.

The artificial database can also be accessed in the same repository[2]. From the description above, we know that database DB_A copies the superficial features of DB_R but discards the mutual relationships of items in DB_R. Here, the mutual relationship refers to the probability of a tourist to move from one place (address) to another. It is clear that the probability also has a strong connection with the formation of disjoint sequential pattern pairs. Thus, by conducting experiments on these two databases, we can study how these mutual relationships of a real database will influence the output size.

We apply our algorithm with several different parameter settings on both DB_R and DB_A. The experiments can be reproduced with the Python scripts available in the repository mentioned before[3]. The running time of our algorithm on the real database is plotted in Fig. 4 (a). Here, the whole procedure of our algorithm is divided into 3 different stages, where Stage 1 is for candidate generation (presented in Algorithm 1); Stage 2 is for disjoint checking (presented in Algorithm 2); and Stage 3 is for maximal-disjoint and closed-disjoint checking (presented in Algorithm 3). The running time of each stage is plotted in separate polylines. From the figure, we discover that generally, our estimated complexity for the three stages are correct. That is, as we tune *min_sup* to a lower value, the

[1] https://github.com/DSPPM/Disjoint-Sequential-Pattern-Pair-Mining/blob/master/all_trans_lv3.csv.

[2] https://github.com/DSPPM/Disjoint-Sequential-Pattern-Pair-Mining/blob/master/dummy_trans_lv3.csv.

[3] https://github.com/DSPPM/Disjoint-Sequential-Pattern-Pair-Mining.

running time as well as the output size increases in the similar rate. However, there is an exception that the running time of Stage 3 is increasing at a higher rate than Stage 1 despite the fact that they should have the similar complexity. This is probably because that we are using a hash table for indexing a pattern in the set of FDSPPs. The operation is $O(1)$, but its constant increases as the hash table, *i.e.*, the output size becomes larger. The increasing of the constant finally causes Stage 3 to be the slowest stage when *min_sup* is set to a low value.

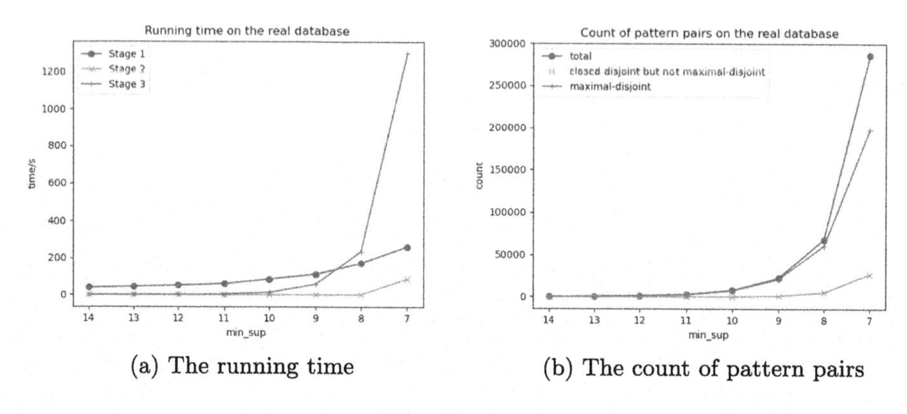

(a) The running time (b) The count of pattern pairs

Fig. 4. The results on the real database.

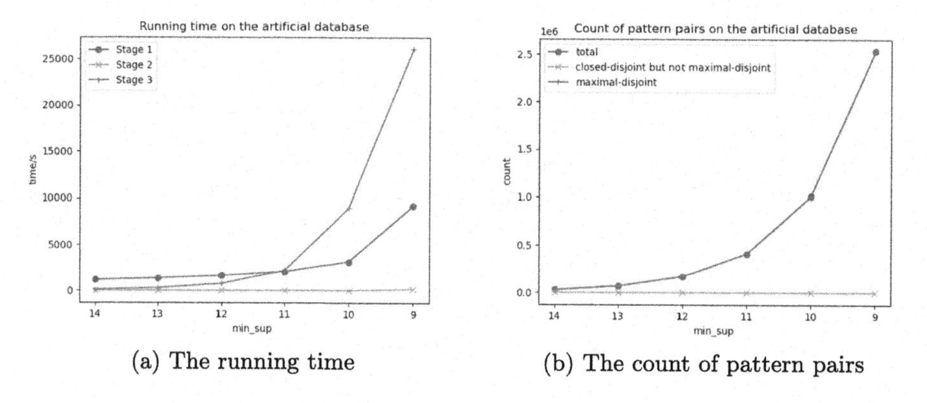

(a) The running time (b) The count of pattern pairs

Fig. 5. The results on the real database.

Also, we find from Fig. 5 (a) that the running time of our algorithm on the artificial database is a lot slower than that of the real database. Hence, from Fig. 5 (b) we find that the output size is also a lot larger than that of the real database. It is probably the mutual relationship between items that makes this to happen. As is analyzed before, in the real world, the most possible reason for the absence of pattern $a \rightarrow b$ in the database is the inaccessibility of the two locations. The fewer inaccessible location pairs are, the fewer disjoint sequential pattern pairs there will be. Intuitively, there should not be many inaccessible locations for that the transportation system today is well-developed. Hence the number of disjoint route pairs should also remain at a reasonable level. This perfectly matches with our experimental results. Also, the results give another empirical proof of the conclusion that the complexity of our algorithm is approximately linear related with the output size.

Now that we have confirmed our algorithms can produce an output with a reasonable size, we are to describe the practical values of the output with a real mined example. As is shown in Fig. 6, by conducting SPM, we can find many popular travel routes starting from Nishi-shinjuku, Tokyo and ending at Jingu-mae, Tokyo. Some of the routes take a detour around the famous Mount. Fuji, while some other routes pass through central Tokyo where many famous sightseeing spots are located. However, by applying DSPPM, we discover that the route via Mount. Fuji is disjoint with any other popular route which passes through the sightseeing spots in Tokyo. The most possible reason for the case is that none of the bus tour from Nishi-shinjuku to Jingu-mae via Mount. Fuji include sightseeing spots in Tokyo into the tour plan, and thus, we should never plan to visit these spots when we decide to take these tours. In all, DSPPM is effective in extracting such valuable information which is unavailable when using vanilla SPM. Therefore, we say DSPPM is capable of producing an output with a reasonable size and high values.

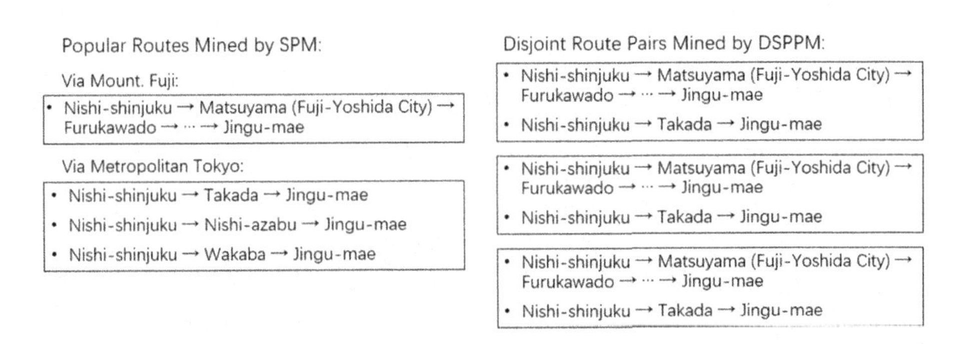

Fig. 6. An example showing the values of FDSPPs, which demonstrate that the preferred travel route from Nishi-shinjuku to Jingumae detouring around Mount. Fuji is disjoint with any other routes via metropolitan Tokyo.

7 Conclusion

In this paper, we propose the task of disjoint sequential pattern pair mining aiming at finding a reasonable volume of hyper-patterns, *i.e.*, disjoint sequential pattern pairs, which can also provide valuable information for tourists. We developed algorithms for mining frequent disjoint, maximal-disjoint and closed-disjoint sequential pattern pairs. We conduct experiments on a real trajectory database and an artificial database, and the results indicate that the volume of pattern pairs mined from the real database is reasonable and the practical values of the mined pattern are high.

However, it also needs to be noted that our algorithm still has the deficit that the candidate generation is unavoidable. If we could propose an algorithm without candidate generation, *i.e.*, enumerate the disjoint sequential pattern pairs only, we may reduce the overall complexity of the algorithm to $O(R)$. Also, we plan to study more hyper patterns other than disjoint sequential pattern pairs so that more valuable information can be extracted from the database.

References

1. Agrawal, R., Srikant, R.: Mining sequential patterns. In: Proceedings of the Eleventh International Conference on Data Engineering, pp. 3–14. IEEE (1995)
2. Atev, S., Miller, G., Papanikolopoulos, N.P.: Clustering of vehicle trajectories. IEEE Trans. Intell. Transp. Syst. **11**(3), 647–657 (2010)
3. Bermingham, L., Lee, I.: Mining distinct and contiguous sequential patterns from large vehicle trajectories. Knowl.-Based Syst. **189**, 105076 (2020)
4. Bin, C., Gu, T., Sun, Y., Chang, L.: A personalized poi route recommendation system based on heterogeneous tourism data and sequential pattern mining. Multimedia Tools Appl. **78**(24), 35135–35156 (2019)
5. Chen, L., Lv, M., Ye, Q., Chen, G., Woodward, J.: A personal route prediction system based on trajectory data mining. Inf. Sci. **181**(7), 1264–1284 (2011)
6. Fournier-Viger, P., Lin, J.C.W., Kiran, R.U., Koh, Y.S., Thomas, R.: A survey of sequential pattern mining. Data Sci. Pattern Recogn. **1**(1), 54–77 (2017)
7. Han, J., et al.: Prefixspan: mining sequential patterns efficiently by prefix-projected pattern growth. In: Proceedings of the 17th International Conference on Data Engineering, pp. 215–224. Citeseer (2001)
8. Ibrahim, R., Shafiq, M.O.: Detecting taxi movements using random swap clustering and sequential pattern mining. J. Big Data **6**(1), 39 (2019)
9. Parent, C., et al.: Semantic trajectories modeling and analysis. ACM Comput. Surv. (CSUR) **45**(4), 1–32 (2013)
10. Srikant, R., Agrawal, R.: Mining sequential patterns: generalizations and performance improvements. In: Apers, P., Bouzeghoub, M., Gardarin, G. (eds.) EDBT 1996. LNCS, vol. 1057, pp. 1–17. Springer, Heidelberg (1996). https://doi.org/10.1007/BFb0014140
11. Takimoto, Y., Sugiura, K., Ishikawa, Y.: Extraction of frequent patterns based on users' interests from semantic trajectories with photographs. In: Proceedings of the 21st International Database Engineering & Applications Symposium, pp. 219–227 (2017)
12. Wang, Y., Zheng, Y., Xue, Y.: Travel time estimation of a path using sparse trajectories. In: Proceedings of the 20th ACM SIGKDD International Conference on Knowledge Discovery and Data Mining, pp. 25–34 (2014)
13. Yan, X., Han, J., Afshar, R.: Clospan: mining: closed sequential patterns in large datasets. In: Proceedings of the 2003 SIAM International Conference on Data Mining, pp. 166–177. SIAM (2003)
14. Yang, C., Gidófalvi, G.: Mining and visual exploration of closed contiguous sequential patterns in trajectories. Int. J. Geogr. Inf. Sci. **32**(7), 1282–1304 (2018)
15. Zhang, J., Wang, Y., Yang, D.: CCspan: mining closed contiguous sequential patterns. Knowl.-Based Syst. **89**, 1–13 (2015)

Predicting the Health Condition of mHealth App Users with Large Differences in the Number of Recorded Observations - Where to Learn from?

Vishnu Unnikrishnan[1(✉)], Yash Shah[1], Miro Schleicher[1], Mirela Strandzheva[2], Plamen Dimitrov[2], Doroteya Velikova[2], Ruediger Pryss[3], Johannes Schobel[4], Winfried Schlee[5], and Myra Spiliopoulou[1]

[1] Otto-von-Guericke University Magdeburg, Magdeburg, Germany
{vishnu.unnikrishnan,yash.shah,miro.schleicher,myra}@ovgu.de
[2] National Center for Public Health and Analyses, Sofia, Bulgaria
{m.strandzheva,p.dimitrov,d.velikova}@npcha.government.bg
[3] University of Würzburg, Würzburg, Germany
ruediger.pryss@uni-wuerzburg.de
[4] Ulm University, Ulm, Germany
johannes.schobel@uni-ulm.de
[5] University of Regensburg, Regensburg, Germany
winfried.schlee@tinnitusresearch.org

Abstract. Some mHealth apps record user activity continuously and unobtrusively, while other apps rely by nature on user engagement and self-discipline: users are asked to enter data that cannot be assessed otherwise, e.g., on how they feel and what non-measurable symptoms they have. Over time, this leads to substantial differences in the length of the time series of recordings for the different users. In this study, we propose two algorithms for wellbeing-prediction from such time series, and we compare their performance on the users of a pilot study on diabetic patients - with time series length varying between 8 and 87 recordings.

Our first approach learns a model from the few users, on which many recordings are available, and applies this model to predict the 2nd, 3rd, and so forth recording of users newly joining the mHealth platform. Our second approach rather exploits the similarity among the first few recordings of newly arriving users. Our results for the first approach indicate that the target variable for users who use the app for long are not predictive for users who use the app only for a short time. Our results for the second approach indicate that few initial recordings suffice to inform the predictive model and improve performance considerably.

1 Introduction

Recent trends in consumer electronics towards affordable and relatively powerful devices capable of sensing health-related attributes have been matched by an increase in research interest in exploiting this data to assist the healthcare

A. Appice et al. (Eds.): DS 2020, LNAI 12323, pp. 659–673, 2020.
https://doi.org/10.1007/978-3-030-61527-7_43

practitioner. Not only do these devices help in diagnostics, by recording values of attributes related to health and subjective well-being; they also allow that the disease may be monitored with only asynchronous involvement of the practitioner. Self-monitoring of the disease contributes thus to patient empowerment, and also delivers precious data that can be used for personalization, i.e. for treatments tailored to the individual needs and characteristics. This potential requires adequate data to build upon.

A major challenge of mobile health platforms that collect user inputs is that the amount of data users contribute can vary substantially. As we reported in [15] when analysing user recordings on the mobile health platform "TrackYourTinnitus" [9], a minority of users interact intensively with the system and contribute a disproportionately large amount of data, while the majority of users contribute very few inputs. In this study, we investigate whether predictions can be made for this majority of users by learning a model on the few users who provide many data to the system. Also, differently from our work in [15], we focus here on a one-step-ahead forecast instead of classification.

We propose an approach that learns from users who contribute long sequences of inputs to predict the subjective perception of wellbeing for users who contribute only short sequences of input data, including users that have very recently joined the platform. Each user in the system is required to fill in a "End of Day Questionnaire", where he reports among other things the overall "feeling in control", the variable of prediction interest. These user-level timestamped observations therefore constitute one user-centric time series, the length of which varies depending on how long the user has been in the system, and how the doctor's recommendation of filling in the questionnaire at the end of every day has been followed. We denote the set of users with long sequences of recordings as U_{long} and the users with few recordings as U_{short}. Our approach deals with the following three questions:

- RQ1: How well can we predict the behaviour of users in U_{short} given the data from the users in U_{long}?
- RQ2: Can we predict the entire sequence of observations of a user in U_{short} with a model trained only on data from users in U_{long}? (i.e, does a model learned on data from users with long sequences transfer to those with short ones?)
- RQ3: How can we incorporate early recordings of users in U_{short} incrementally into the model to improve predictive performance?

The paper is organised as follows: Sect. 2 introduces related literature, followed by Sect. 3, which introduces the m-Health application on which the this work is based. Section 4 discusses our proposed solution, followed by a discussion in Sect. 5, and closing remarks in Sect. 6.

2 Related Work

In our work, we concentrate on time series in applications of health and wellbeing. The early study [5] by Madan et al. reported on the potential of mobile technologies to capture epidemiological behaviour change, including physiological symptoms like running nose, and mental health conditions like stress. For example, they found that total communication of the affected persons decreased for the response "sad-lonely-depressed" (cf. [5] for the definition of this response). While a change in communication intensity can be captured by Bluetooth connection activity or absence thereof, the information on how a person feels demands user inputs. Ecological Momentary Assessments (EMA) are a widespread tool for this purpose [3,14].

EMA is an instrument for assessing "behavioral and cognitive processes in their natural settings" [14]. From the technical perspective, EMA recording is feasible and well-supported. For example, in their survey on sleep self-management apps [2], Choi et al. list the recording of user-entered data as an important functionality, and stress that all investigated apps do support this functionality. However, next to the technical modalities, EMA relies also on self-discipline and adherence.

As Mohr et al. stress in [6], "although a number of small studies have demonstrated the technical feasibility of sensing mood, these findings do not appear to generalize". In the meanwhile, there are large studies involving EMA recordings of more participants for longer time periods. However, the emphasis seems still to be on users who interact intensively with the mobile health application. In their insightful comparison of the results of EMA recordings with the TrackYourTinnitus mHealth app versus retrospective ratings of the users, only users with at least 10 days of interaction were considered [11]. For findings with the TrackYourStress platform that records EMA geolocation, only users with at least 10 recordings per day were considered [10].

This provokes the question of whether users with few recordings belong to the same population as users with many recordings. In [8], Probst et al. considered both users with few days of recordings and users with many days of recordings for their Multi-Level Analysis (median number of days: 11, with range from 1 to 415 days), but demanded at least 3 EMA per day, each of them containing answers for the three EMA items under study [8]. In this work, we do not attempt to win insights that pertain to a specific group of users, but rather to assess whether the EMA of users with few recordings can be predicted by models learned on users with many recordings.

The EMA of mobile health app users constitute multivariate time series. The challenge posed by short time series is discussed by Palivonaite and Ragulskis in their work on short-term forecasting [7], where they associate the length of the time series to the reliability of longer-term forecasts.

Dynamic Time Warping (DTW) or one of its numerous enhancements can be used to compare time series of different lengths and exploit their similarity for learning. DTW is a very old method, cf. [16], for an early citation to DTW by authors Yfantis and Elison who proposed a faster alternative. Such methods can be used to enhance algorithms like [1,15], which do predictions by building a model for each time series, but can also exploit information from similar time series. Despite this potential, the amount of data per user in some mHealth applications is very small, so that we opt for similarity-based methods that capitalize more on the similarity of values than on the ordering of the values – albeit both are taken into account.

3 EMA with the TrackYourDiabetes Mobile Health App

As part of two pilot studies on empowerment of diabetes patients, a mobile crowdsensing framework was adjusted to implement the TrackYourDiabetes mHealth platform [4,9]. Figure 1 summarizes the entire procedure of the app from the patient's point of view. The pilot studies were conducted in regions of Spain and Bulgaria, and involved patient recruitment and exposition to two variants of the app, while under remote supervision by a practitioner.

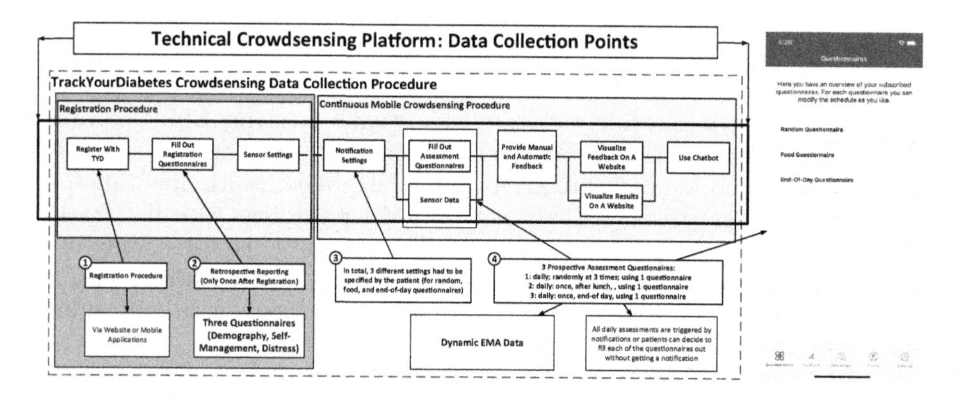

Fig. 1. Mobile crowdsensing collection procedure of TrackYourDiabetes

The platform comprises two mobile applications (i.e., native Android and iOS apps), a relational database, a RESTful API [12], and a web application. The mobile applications are only used by the patients, while the web application was used by the patients as well as their related healthcare professionals. The latter were enabled by the web application to monitor the data of the patients as well as to provide individual feedback if wanted or required.

Before starting interaction with the app, study participants registered with the platform by using the mobile apps or a web application ①. After that, they have to fill out three registration questionnaires once ②: one registration questionnaire collects demographic data, one collects information on the self-management of the patients with his/her diabetes, and one captures the extend to which diabetes causes distress to the patient.

There were EMA recordings more than once a day, concerning physical activity and food intake, and EMA recordings at the end of each day, using the End-of-Day questionnaire items depicted in Table 1. Furthermore, individualised messages based on given answers of daily assessments were provided with the goal

Table 1. Variables in the dataset: questions in the end-of-day questionnaire

#	Question	Set of answers/Data type
01	How often do you have measured your sugar level today?	Numeric
02	For how many minutes have you performed physical activity or sports today?	Numeric
03	How many bread units have you eaten today?	Numeric
04	Did you have signs of hyper- or hypoglycemia today?	"Don't know", "No", "Both", "Hypoglycemia", or "Hyperglycemia"
05	Did you feel to be in control of your diabetes today?	Numeric, [0–100]

to better motivate the patients in using the platform. The healthcare professional(s) responsible for the participants could also provide individualised feedback. Finally, a chatbot was integrated, which could be used by the patients to discuss questions on their diabetes. For the analysis of the proposed approach, we concentrated on the Bulgarian pilot study and investigated solely the user inputs to the End-of-Day questionnaire; no further features were considered.

4 Our Method

We investigate a prediction problem on timestamped data, transferring a predictor learned on the data of one set of users, U_{long}, to another set of users U_{short}. In all cases, our goal is to predict many observations of a user, not just the next one, as is typical in many time series prediction problems.

4.1 Core Concepts and Core Elements

This section offers a brief overview of the terms used in this work and their exact definitions, which is followed by a broader description of our workflow in Sect. 4.2.

User Sequences: Each user p who uses the mHealth app generates a time-ordered sequence of observations $x_{p,t}$, where p is the user, and t denotes time.

We distinguish between users with *short sequences* of observations, constituting a set U_{short}, and users with *long sequences* of observations, constituting a set U_{long}. For the partitioning of users into these two strata, we consider a threshold τ_{length}.

In our experiments, we set τ_{length} on the basis of the user-observations distribution, which has shown a gap. In distributions that follow a power law, τ_{length} serves to separate between the short head and the long tail. More generally, we may decide to place into U_{short} those users who have very recently started their interaction with the app and thus have contributed only few initial observations.

Observations: An observation is a multi-dimensional vector of values from a feature space F. In our application scenario, an observation is an EMA recording comprised of answers to questions from a questionnaire. Accordingly, an observation is a mix of numerical and categorical variables.

Handling Categorical Data: A term Frequency-Inverse Document Frequency (TF-IDF) Inspired Approach: Before training the models, it is important to consider the exact way in which categorical attributes in the input data are used. Of the various questions in the questionnaire answered by the users, the questions that generate categorical data (chosen from a drop-down list) need to be treated to accommodate for the fact that not all answers are equally likely. Compared to simply using a standard method like one-hot encoding, this step brings the answer closer to the user's history, for e.g., by more accurately capturing the information that a user who commonly answers a question with "no" has said "yes", even if "yes" frequently appears in the dataset.

We treat the answers to this categorical data as 'words', and each session where the questionnaire is answered as a 'document'. During preprocessing, given the exact answer chosen by a user during a particular day, we replace the binary flag marking the presence of that word with a new value that is adjusted to reflect the amount of "surprise" in seeing that data point given the user through the use of the TF-IDF (see [13]) inspired formula: $preprocessed\_value = f_{term} \cdot -log\frac{n_{term}}{N}$, where f_{term} can only be binary, since the categorical answer only picks one term from a list of several. The inverse document frequency component measures how often the term has appeared in the user history.

Core Learning Method: Given data P_p for all users $p \in U_{long}$ for time points $1 \ldots t$, we have $P_p = \{x_{p,1} \ldots x_{p,t}\}$. Using this data, we create a linear regression model that for each possible $i \in 1 \ldots t-1$, learns to predict the target variable for time point $i+1$. Naturally, since there is no known label for the last time point t, each user p with a sequence length of t only provides $t-1$ time points of training data. This model is only used for predicting the labels for the observations $\{x_{p,1} \ldots x_{p,t-1}\}$ for all users $p \in U_{short}$.

Augmented Method: We augment the above method by creating predictions specific to the users U_{short}: In addition to the above model which only learns on the users of U_{long}, we add an additional K-NN Regression model that is trained only on the user's own history of observations. This means that given an observation $x_{p,t}$, we can generate predictions for $x_{p,t+1}$ from two models, the model trained on all users in U_{long}, and additionally the K-NN Regression model that has only been trained on the observations of the user p seen so far, i.e. $x_1 \ldots x_{t-2}$ (Note: The training data for $p \in U_{short}$ ends at $t-2$, because x_{t-1} is used as the label for the training point x_{t-2}, and the true label for x_{t-1} has not been observed yet).

4.2 Prediction Workflows

Proof-of-Concept Step: The basic workflow we propose has a preliminary step and two components. The preliminary step is designed to check that the task is indeed *learnable*, and success at this stage can ensure that the further steps in the workflow are applicable. For this, instead of training a model on only data from users in U_{long}, a model is trained on 75% of all data, and the performance is analysed to confirm that the model can learn given the data. By framing the problem as a regressor and not as a time series forecast, we avoid the problem of having insufficient data to train a time series forecasting model. This model can unfortunately not be used as a baseline to compare against since it does not learn on the same amount of data as the model learning only on U_{long}, and also because the number of data points available for testing over users in U_{short} is very small (often only a couple of observations). However, the performance of this model can still be considered a benchmark for the upper limit of performance for the transfer learning model.

Basic Workflow: In this workflow, we find a subset of the dataset D comprising of only the data $x_{p,t}$, where $p \in U_{long}$. This creates a model trained only on the data from users with long sequences, the performance of which is tested on users of U_{short}. It is important to remember that the model has the challenging task of making predictions for users that have never been seen by the system, and predictions for them are made based only on what has been learned over the user. This is arguably more challenging than predicting unseen observations for users who have already contributed observations to the training set. Additionally, since these users have not adhered to instructions of the physician to use the application for the prescribed period of 2 months, it is possible that these users differ somehow in the expression or the perception of the disease in some way. However, it is still possible that a model learned on those data points from long users bring a modest predictability to the disease development of users in U_{short}. Similarly to the model introduced above, we learn to predict the numeric value of the target variable for the next observation given the questionnaire answers of the current observation (including the current value of the target variable). A graphical overview of the workflow is shown under 'Basic Workflow' of Fig. 2.

K-Nearest Neighbour (K-NN) Augmented Workflow: If the users in U_{short} are indeed different from the users in U_{long}, then using a model that transfers the parameters learned on U_{long} is not expected to bring reliable predictions to the users in U_{short}. However, since the users in U_{short} do not bring enough data to train complex models, only simple techniques can be used to try and incrementally improve predictive performance over users in U_{short} by capturing the idiosyncratic patterns in the user's disease development/answering style. This design aims to balance the tradeoff between keeping as much data as we can use to learn about how the disease develops, while also staying close to the idiosyncratic ways in which the user may answer questions. In this work,

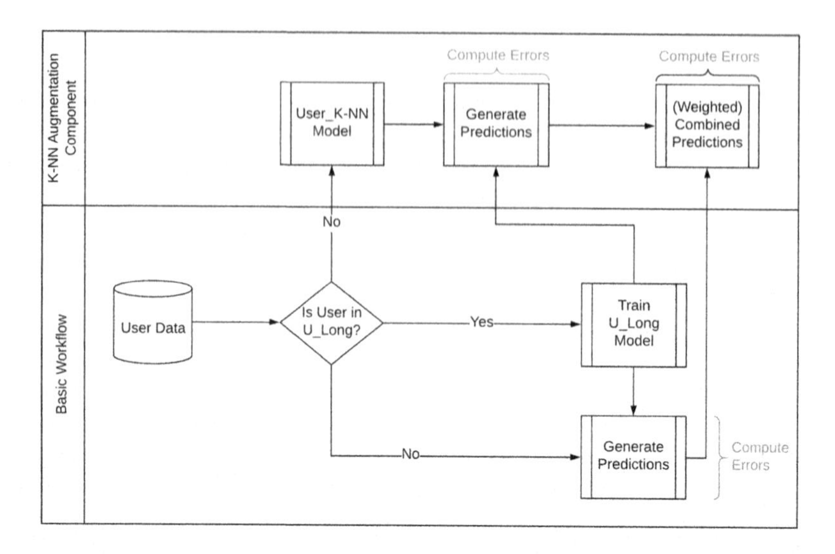

Fig. 2. Prediction workflows

we propose the use of a K-Nearest Neighbours regressor trained over the user's own history, the predictions of which are used to augment the predictions from the U_{long} model weighted on their past errors (similarly to [1]). Restricting the K-NN regressor to the user's own sequence also has the unintended consequence of out-of-the-box support for data-privacy, something that is especially relevant in the medical domain.

During use, the K-NN regressor is incrementally trained on the user sequence as more of it becomes available, and the errors are recorded for comparison to the standard U_{long} model. Figure 2 shows an overview of the model training process with the K-NN augmentation component.

5 Results

We describe the dataset of our evaluation in Subsect. 5.1, and then explain in Subsect. 5.2 how the number of users with short and long sequences affect the prediction tasks and the settings of K-NN in the augmented workflow. We evaluate using Mean Absolute Error (MAE). The results of the *proof-of-concept* experiment are in Subsect. 5.3, while the results for the basic workflow and the KNN-augmented workflows are in Subsect. 5.4.

5.1 The Dataset

For our evaluation, we used the dataset of the Bulgarian pilot study. This dataset contains observations from 11 study participants. While the inclusion of the users from the pilot study in Spain is desirable, a model that learns on the combined data of the two pilots is not done for two reasons: (a) The two countries are different in the dominant diabetes type that the users have, and (b) Many users in the Spain pilot use continuous blood sugar measuring devices, strongly influencing the accuracy of the "self-assessed" blood sugar estimations, and therefore, the "feeling in control". We set $\tau_{length} = 30$ days, whereupon 6 of the users belong to U_{long} (30+ days) and 5 users are in U_{short} (8–13 days) after eliminating users with 5 users with less than 3 days of data. We denote this dataset as L6+S5_dataset hereafter, to stress the number of users per length-stratum.

Figure 3 depicts the number of days of interaction for all users. It can be observed that there is a clear separation between users in U_{long} compared to the rest of the users.

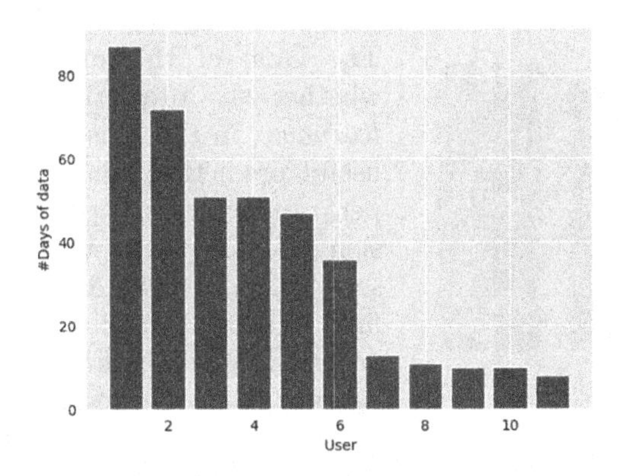

Fig. 3. Number of days with EOD observations per user; user #1 is the user with the largest number of EOD observations, user #11 has the smallest number

Of the 5 variables of the EOD questionnaire filled by the pilot study participants (cf. Sect. 3), the target variable is the 5th one on Table 1, i.e. each user's self-reported 'feeling in control', on a scale of 0 to 100. We denote this variable as 'EOD_feel' hereafter.

5.2 Prediction Tasks and Imposed Restrictions on Training

For the proof-of-concept step in Subsect. 4.2, we train a predictor on the first 75% of the observations of the users in *U_long* of L6+S5_dataset and predict the subsequent 25% observations. As can be seen on Fig. 3, the 6 users in U_{long} contribute unequally to learning: user #1 contributes more than 60 (out of ca. 85) observations to the training dataset, while user #6 contributes less than 30 (ie half as many). Similarly, we predict the EOD_feel value of more than 20 observations of user #1 and ca. 10 of user #6.

For the basic workflow of Subsect. 4.2, the prediction task is to predict *all* observations of the 5 users in U_{short} of the L6+S5_dataset, without having seen any observations on them during training. This amounts to 47 predictions.

For the K-NN augmented workflow, some observations of each user in U_{short} of the L6+S5_dataset are disclosed and used for augmentation of the model learned on all of the U_{long} observations in the L6+S5_dataset. User #7 has less than 15 observations, user #11 has 8 (cf. Fig. 3). This imposes an upper limit to K: if we set $K = 8$, we cannot do any predictions on user #11. On the other hand, K-NN based regression needs at least 2 observations per user to learn.

Larger values of K allow for a more robust regression model and make the prediction task easier, since less EOD_feel values are predicted. To investigate whether the very few first observations on a user can inform a model learned on U_{long}, we have set $K = 2$. This amounts to 37 predictions.

5.3 Learning a Predictor on U_{long}: Proof-of-Concept Experiment

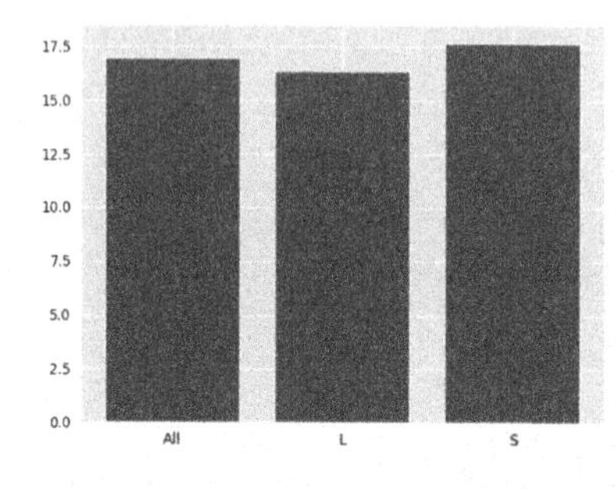

Fig. 4. MAE in $U_{long} \cup U_{short}$ (All), in U_{long} (S) and in U_{short} (S)

The Goal of this experiment is to check whether the prediction problem is indeed *learnable*, in the sense that we can derive a useful prediction model. Figure 4 shows the performance of the proof-of-concept regression model for the first prediction task of Subsect. 5.2 on L6+S5_dataset, learning on the first 75% of all user observations (All), and accordingly on the first 75% of the observations in U_{long} (L), resp. U_{short} (S).

For "All" (leftmost part), MAE remains around 17%, decreasing slightly within "L" (U_{long}) and increasing slightly within U_{short}. However, MAE within "S" (U_{short}) is rather unreliable, since there are less than two observations per user in the testset (more precisely: 1.4). Hence, these MAE values serve only as lower limit for the errors of the transferred models.

5.4 Learning on Users in U_{long} to Make Predictions for Users in U_{short}: Transfer Learning Experiments

Since we have a baseline (albeit weak, since errors for U_{short} are not reliable) for the performance of a model on the data from all users, we can now investigate the transfer learning case where the model is only learned on the users of U_{long}. As already described above, there are two workflows that use such a model, a more basic workflow that uses a model learned over U_{long} only, and another model that augments the basic workflow with a user-specific K-Nearest observations regressor. The models are all evaluated against the absolute errors they make in their predictions. The 'mean' in the Mean Absolute Error may either be computed over all predictions that a model has made, or may be restricted to the predictions for particular users. Given the rather short sequence lengths of the users in U_{short}, it is necessary to not rely on point-estimates like means, but consider the 'spread' in the errors as well. We therefore present box plots over all the prediction errors for the basic and K-NN Augmented workflow. The entire test set contains 47 observations for which predictions are required.

Basic Workflow. In this workflow, instead of learning a model over data from all users, as described in Sect. 5.3, we will learn a model only on those users who have contributed more than 30 days of data, the necessary criterion for their addition to the set U_{long}. Figure 5 shows a box plot of the absolute prediction errors for the transferred model. Compared to the basic model in Sect. 4.2, the MAE over all predictions for all users has increased from 17.5 to 24.76 (indicated by the green triangle). However, since not all users in U_{short} have the same sequence lengths, the MAE is biased towards the users with longer sequences. The blue dots inside the box plot shows the MAE for each user separately, and we can see that the users who are best predicted have errors as low as 21, with the worst-predicted users showing MAE in excess of 35. The mean being closer to 21 indicates that the well-predicted users are indeed the ones with longer sequences. This indicates that they are more similar to the users in U_{long} than other shorter members of U_{short}.

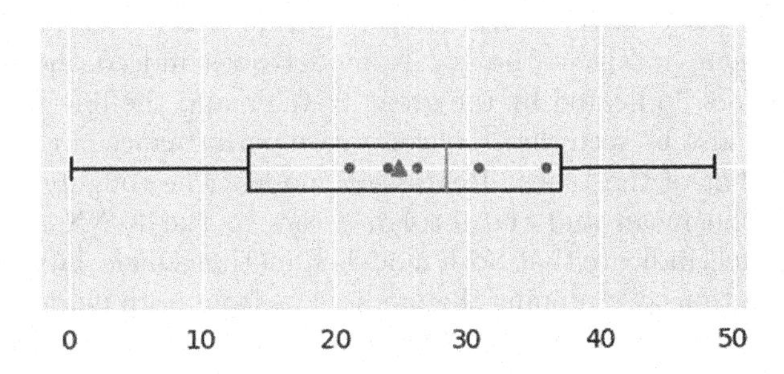

Fig. 5. Basic workflow: box plot of errors for predicted next-observation EOD_feel, along with mean (Green Triangle), median (Yellow Line), and user-level MAEs (blue dots) for all users in U_{short}. (Color figure online)

K-NN Augmented Workflow: This section discusses the results for the more advanced "K-NN Augmented Workflow" detailed in Sect. 4, where a user-level model learned on data from U_{short} augments the predictions of the model discussed above. Figure 6 shows a box plot of the absolute prediction errors for the K-NN Regressor, along with comparisons against the U_{long} model's errors. The box plot on the far right shows the errors in the case where the predictions of each method are combined as a weighted average on their cumulative errors for the user to form a final prediction. Since the users in U_{short} can have as few as 8 observations, our choice of K is quite strongly limited to very low numbers, as the K-NN Regressor cannot create predictions until it sees at least K observations. In these cases, the K-NN is assumed to make the same prediction as the Linear Regression model over U_{long}, since it is necessary to compare the errors of the two models for the same number of predictions.

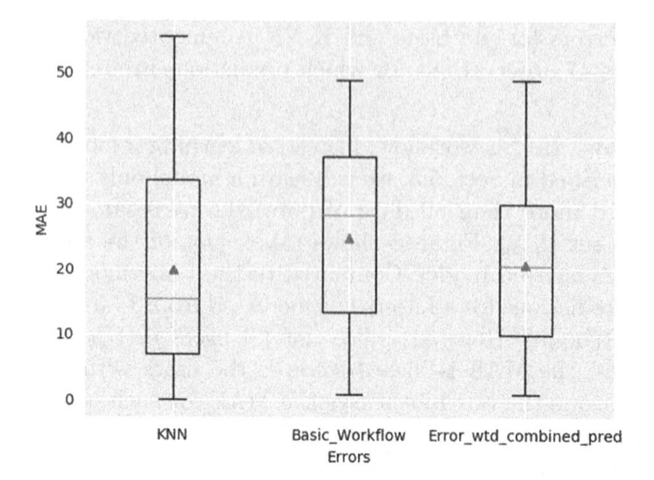

Fig. 6. K-NN Augmented workflow: bloxplots of absolute errors for KNN model, basic workflow, and the combined workflow. Means are denoted by the green triangle and medians by the yellow horizontal line (Color figure online)

It can be seen in Fig. 6, the K-NN model does indeed show lower mean and median errors, indicated by the green triangle and the line in the box plot. However, it can also be seen that the worst-case performance of the K-NN model is worse than that of the Linear Regression model. The roughly similarly sized gaps between the mean and the median errors in the K-NN and the Linear Regression models indicate that both models sometimes make large errors, albeit in different directions. Combining the predictions from both models does seem to mitigate these worst-case errors, since the mean and the median absolute errors are observed to be very close, at around 20.

In addition to the boxplots of the error itself, Fig. 7 shows how the error develops over time for users in U_{short} as they stay longer in the system. The X-Axis shows the observation number, with the MAE on the Y-axis. The MAE at each time point is averaged over the individual prediction errors over all users at that time point. Until the Kth observation, the K-NN predictor does not generate predictions, but we have used the linear regression model prediction errors in order to not unfairly favour any algorithm. From the 3rd observation, however, we see that the user-level K-NN predictor almost always outperforms the linear regression model (and therefore the weighted average model). It is also noteworthy that until the 7th time point, the error-weighted combination of both models is very close to the K-NN model. This shows that augmenting the predictions of the basic workflow with the K-NN regressor does improve performance. The results beyond the 8th observation get progressively less reliable since all users in U_{long} have at least 8 observations, but the number of users contributing to the averages after time point 8 get unreliable, though it is possible that users in U_{short} are more and more predictable given the history of their own observations with time.

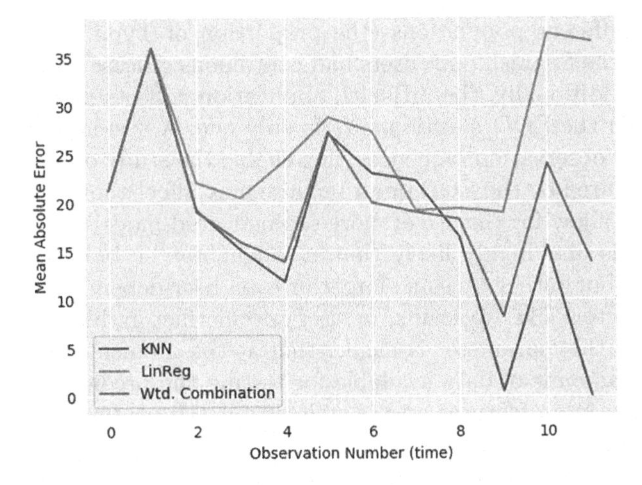

Fig. 7. K-NN augmented workflow: development in error with time

6 Conclusions and Closing Remarks

In this work, we studied if the data from users of a diabetes self-management app with more than 30 days of data (U_{long}) can be used to infer something about the future of less intensive users with less data. Since neither the number of patients ($N = 11$) nor the number of observations for the longest-sequence user (87) is very long, we investigate simpler models like linear regression. The model is trained to predict the next observation for user-reported "feeling in control", the last question of the *End Of Day* questionnaire, given the answers to all questions of End Of Day questionnaire for the current observation. The categorical information in the dataset is handled using a method inspired by TF-IDF to capture the 'surprise element' in an answer given a user. i.e., when a user answers a question like (s)he usually does, that answer gets a smaller weight than if the answer is unexpected.

Further, we investigate whether transfer learning can be used to learn a model on the users of U_{long} in order to make predictions for the observations of users in U_{short}. We saw that the transferred model predictably shows a higher error, which can be mitigated by combining the predictions of the U_{long} model with a K-Nearest Neighbours Regressor over the patient's own past data. The short sequences necessitate that the K is limited to quite low values, but the predictor that combines the predictions of both models does eliminate some extreme errors, bringing the mean and the median errors closer.

The primary threat to validity of this work is the size of the dataset from which the conclusions have been drawn. The large disparity between the lengths in U_{long} and U_{short} make further analysis of the K-NN Augmented predictor less reliable, making the findings more qualitative than quantitative. Although two pilots exist from which data can be analysed, this study focused the investigation only on data from Bulgaria because the users for the two studies are drawn from different populations (the proportion of Type 2 diabetics is very different, and the Spanish pilot users had continuous glucose monitoring devices implanted). Additionally, the mHealth application collects more data from the users, of which the EOD questionnaire is only one. A system with either more users or longer observation sequences may enable the study of how other dimensions not measured by the EOD questionnaire may affect the subjective "Feeling in control", or allow for the use of more sophisticated models than simple linear regression. It is also highly likely that x_t might not be best predicted by the value of x_{t-1}, but rather by some larger or even user-dependent lag, depending on external factors like weekends, or user-specific factors like exercise routine. The testing of this parameter is challenging at the moment because it further decreases the amount of data available for testing the predictions over users in U_{short}, or adds more features and complexity in the context of already scarce data. If such a large disparity did not exist between the lengths of users in U_{long} and U_{short}, it would also be possible to investigate the aspects that characterise users who transition from U_{short} to U_{long}.

Acknowledgements. This work was funded by the the CHRODIS PLUS Joint Action, which has received funding from the European Union, in the framework of the Health Programme (2014–2020).

References

1. Beyer, C., Unnikrishnan, V., Niemann, U., Matuszyk, P., Ntoutsi, E., Spiliopoulou, M.: Exploiting entity information for stream classification over a stream of reviews. In: Proceedings of the 34th ACM/SIGAPP Symposium on Applied Computing, pp. 564–573 (2019)
2. Choi, Y.K., et al.: Smartphone applications to support sleep self-management: review and evaluation. J. Clin. Sleep Med. **14**(10), 1783–1790 (2018)
3. Csikszentmihalyi, M., Larson, R.: Validity and reliability of the experience-sampling method. In: Csikszentmihalyi, M. (ed.) Flow and the Foundations of Positive Psychology, pp. 35–54. Springer, Dordrecht (2014). https://doi.org/10.1007/978-94-017-9088-8_3
4. Kraft, R., et al.: Combining mobile crowdsensing and ecological momentary assessments in the healthcare domain. Front. Neurosci. **14**, 164 (2020)
5. Madan, A., Cebrian, M., Lazer, D., Pentland, A.: Social sensing for epidemiological behavior change. In: Proceedings of the 12th ACM international conference on Ubiquitous computing, pp. 291–300 (2010)
6. Mohr, D.C., Zhang, M., Schueller, S.M.: Personal sensing: understanding mental health using ubiquitous sensors and machine learning. Annu. Rev. Clin. Psychol. **13**, 23–47 (2017)
7. Palivonaite, R., Ragulskis, M.: Short-term time series algebraic forecasting with internal smoothing. Neurocomputing **127**, 161–171 (2014)
8. Probst, T., et al.: Does tinnitus depend on time-of-day? An ecological momentary assessment study with the "trackyourtinnitus" application. Front. Aging Neurosci. **9**, 253 (2017)

9. Pryss, R.: Mobile crowdsensing in healthcare scenarios: taxonomy, conceptual pillars, smart mobile crowdsensing services. In: Baumeister, H., Montag, C. (eds.) Digital Phenotyping and Mobile Sensing. SNPBE, pp. 221–234. Springer, Cham (2019). https://doi.org/10.1007/978-3-030-31620-4_14
10. Pryss, R., et al.: Machine learning findings on geospatial data of users from the trackyourstress mhealth crowdsensing platform. In: 2019 IEEE 20th International Conference on Information Reuse and Integration for Data Science (IRI), pp. 350–355. IEEE (2019)
11. Pryss, R., et al.: Prospective crowdsensing versus retrospective ratings of tinnitus variability and tinnitus-stress associations based on the TrackYourTinnitus mobile platform. Int. J. Data Sci. Anal. **8**(4), 327–338 (2019)
12. Pryss, R., Schobel, J., Reichert, M.: Requirements for a flexible and generic API enabling mobile crowdsensing mhealth applications. In: 2018 4th International Workshop on Requirements Engineering for Self-Adaptive, Collaborative, and Cyber Physical Systems (RESACS), pp. 24–31. IEEE (2018)
13. Rajaraman, A., Ullman, J.D.: Data Mining, pp. 1–17. Cambridge University Press (2011). https://doi.org/10.1017/CBO9781139058452.002
14. Stone, A.A., Shiffman, S.: Ecological momentary assessment (EMA) in behavorial medicine. Ann. Behav. Med. **16**, 199–202 (1994)
15. Unnikrishnan, V., et al.: Entity-level stream classification: exploiting entity similarity to label the future observations referring to an entity. Int. J. Data Sci. Anal. **9**(1), 1–15 (2019). https://doi.org/10.1007/s41060-019-00177-1
16. Yfantis, E., Elison, J.: Vector interpolation for time alignment in speech recognition. WIT Trans. Modell. Simul. **23**, 6 p. (1970). https://doi.org/10.2495/BT990391

Spatiotemporal Traffic Anomaly Detection on Urban Road Network Using Tensor Decomposition Method

Leo Tišljarić[1(✉)], Sofia Fernandes[2], Tonči Carić[1], and João Gama[2]

[1] Department of Intelligent Transport Systems, Faculty of Transport and Traffic Sciences, University of Zagreb, Zagreb, Croatia
ltisljaric@fpz.unizg.hr

[2] LIAAD - INESC TEC, University of Porto, Porto, Portugal

Abstract. Tensor-based models emerged only recently in modeling and analysis of the spatiotemporal road traffic data. They outperform other data models regarding the property of simultaneously capturing both spatial and temporal components of the observed traffic dataset. In this paper, the nonnegative tensor decomposition method is used to extract traffic patterns in the form of Speed Transition Matrix (STM). The STM is presented as the approach for modeling the large sparse Floating Car Data (FCD). The anomaly of the traffic pattern is estimated using Kullback–Leibler divergence between the observed traffic pattern and the average traffic pattern. Experiments were conducted on the large sparse FCD dataset for the most relevant road segments in the City of Zagreb, which is the capital and largest city in Croatia. Results show that the method was able to detect the most anomalous traffic road segments, and with analysis of the extracted spatial and temporal components, conclusions could be drawn about the causes of the anomalies. Results are validated by using the domain knowledge from the Highway Capacity Manual and achieved a precision score value of more than 90%. Therefore, such valuable traffic information can be used in routing applications and urban traffic planning.

Keywords: Road traffic anomaly detection · Tensor decomposition methods · Speed probability distribution · Intelligent transport systems · Traffic state estimation

1 Introduction

Anomaly detection on the urban road network is one of the most attractive research topics for the researches in the field of Intelligent Transportation Systems (ITS). Anomaly detection, in general, can be defined as a process of finding unexpected behavior of some instances in the set of data that is observed. The importance of anomaly detection and analysis lies in the potentially useful, actionable information for the domain of the ITS. The traffic patterns that indicate anomaly on the urban road networks could identify the severe traffic accident, traffic congestion, or a violation of the regulations.

Regarding the type of traffic anomaly, there are two most common types: (i) non-recurrent and (ii) recurrent traffic anomalies. Non-recurrent traffic anomalies are usually caused by some unexpected events such as traffic accidents or special social events,

A. Appice et al. (Eds.): DS 2020, LNAI 12323, pp. 674–688, 2020.
https://doi.org/10.1007/978-3-030-61527-7_44

and recurrent traffic anomalies are occurring daily, mostly caused by commuters. Chow et al. [5] report that 85% of all congestions occurring on the urban road network are of recurrent type. The recent review that covers methods for the urban road traffic anomaly detection [6], states that this research topic is in early stages and challenges like new detection algorithms, optimizations, and high-performance computing need to be further addressed.

This paper presents a tensor decomposition-based method for the recurrent anomaly detection on the urban road network, modeled using FCD. The proposed framework consists of the three main steps: (i) data preprocessing, (ii) STMs generation, and (iii) anomaly detection using CANDECOMP/PARAFAC (CP) method. The validation is conducted by using the domain knowledge from the Highway Capacity Manual (HCM), which reports the relations between traffic speed values and the level of service on the road segments. The method achieved the precision score of 91.68% in the detection of the extreme traffic conditions (recurrent traffic anomalies).

Contributions of this paper are as follows: (i) framework for the anomaly detection using Nonnegative Tensor Decomposition (NTD) method and Kullback–Leibler Divergence (KLD) values, (ii) the usage of the tensor composed of STMs to model the traffic patterns, and (iii) results of the anomaly detection are analyzed on the urban road network segments in the City of Zagreb, Croatia.

The rest of the paper is organized as follows. In Sect. 3, related work on recent developments on tensor-based models for traffic data modeling is presented. Section 4 presents the methodology used in this paper. Section 5 presents the conducted experiment, including data processing, validation, and the analysis of the anomalous spatiotemporal patterns. The conclusion and future work suggestions are given in Sect. 6.

2 Related Work

One of the essential topics related to the ITS domain is detecting the spatiotemporal traffic anomalies that influence traffic flow on the urban road networks. A significant number of studies were carried out investigating traffic dynamics focused on extracting and describing the spatiotemporal traffic patterns [12, 18].

The conventional way of representing the traffic data is using time series represented with vectors [7]. The vectors mostly represent a change of the typical traffic parameter through a defined time intervals. Largest limitation of the techniques based on the data represented by vectors is that it can not take into the account spatial component, such as the spatial correlation between road traffic segments. In contrast, matrix models can model more complex behavior of the traffic data [12, 13]. Commonly, matrix dimensions are $m \times n$ where m represents the number of links and n number of the observed time intervals. Matrix models can be used for the extraction of spatial and temporal dependencies between the observed traffic parameters, but only if a matrix can represent spatial and temporal components. In the case of common traffic data represented with origin-destination matrices, to extract the temporal component, one more dimension must be added. One of the most used matrix decomposition methods is Principal Component Analysis (PCA). PCA can interpret the data in terms of a smaller number of components, which can be used to detect some anomalous behavior in the data.

Wang et al. [25] report that PCA is not applicable in the traffic analysis because traffic data consists of many outliers. In [8], authors also claim that PCA can not capture spatiotemporal patterns from the traffic data because it relaxes three-dimensional data into a bidimensional form.

Tensor models emerged as a technique that does not suffer from the limitations mentioned in the previous approaches. Tensor decomposition presents as a powerful tool for the spatiotemporal data analysis. It is used in a wide range of research areas such as signal processing [17], pattern recognition [26], recommendation systems [27], etc. In the traffic-related application, tensor decomposition methods are used for missing data imputation [4], traffic prediction [14, 19], spatiotemporal correlation analysis [20], travel time estimation [21], anomaly detection [24], and modeling spatiotemporal structure of time-evolving traffic networks [23].

When using large sparse Global Navigation Satellite System (GNSS) datasets, vectors that are representing speed profiles could have large deviations because a large amount of speed data is aggregated into a single value that represents the speed on the observed road segment in a single time interval. If a large sparse dataset is used, data represented with origin-destination matrices could have a lot of missing values in some time intervals. Also, if datasets consist of delivery vehicles, origin-destination matrices could present wrong spatiotemporal patterns due to the predefined delivery locations.

In this paper, the problem of the large sparse GNSS data analysis by using NTD is addressed in order to detect the traffic anomalies. Tensor is constructed using novel data representation called STM. It can be used regardless of the delivery vehicles and does not suffer from the large deviations as data is not aggregated in such way. The method is proposed to overcome the mentioned limitations regarding the analysis of sparse GNSS datasets.

3 Background

Tensors are multidimensional arrays, or more formally, products of N vector spaces. A first-order tensor is a vector, second-order represents a matrix, and three or more order tensors are called higher-order tensors [10]. For spatiotemporal traffic analysis, authors mostly use a third-order tensor composed using $origin \times destination \times time$ and $profile \times roadsegments \times time$ where profile represents the speed or volume time series on the observed road network segment.

The well-known tensor decomposition method, CP in its nonnegative form, is used to extract the spatiotemporal traffic patterns. The CP decomposition factorizes a tensor into a sum of component rank-one tensors. For tensor $\mathcal{T}$ CP is the following:

$$\mathcal{T} \approx \sum_{r=1}^{R} a_r \circ b_r \circ c_r \tag{1}$$

where R is a positive integer that represents the decomposition rank. Then, rank one components can be expressed as factor matrices $\mathbf{A} \in \left(\mathbf{a}^{(1)}\ \mathbf{a}^{(2)}\ ...\ \mathbf{a}^{(R)}\right)$, $\mathbf{B} \in \left(\mathbf{b}^{(1)}\ \mathbf{b}^{(2)}\ ...\ \mathbf{b}^{(R)}\right)$, and $\mathbf{C} \in \left(\mathbf{c}^{(1)}\ \mathbf{c}^{(2)}\ ...\ \mathbf{c}^{(R)}\right)$.

In the majority of papers, there is an assumption that tensor rank can be determined in advance with specific knowledge of the phenomena that are under observation [23]. In this paper, Core Consistency Diagnostic (CORCONDIA) [2] method for the tensor rank estimation is applied to determine the best match for the experiment.

4 Methodology

Given a set of FCD data, the goal is to obtain a method for the spatiotemporal congestions pattern extraction on the urban road network to detect recurrent traffic flow anomalies. Figure 1 illustrates the proposed framework, and this section briefly describes: (i) data preprocessing, (ii) STMs generation, and (iii) anomaly detection using CANDECOMP/PARAFAC (CP) method and KLD values. After preprocessing the GNSS dataset, the STMs were generated using the large, real-life FCD data. The next step was the anomaly detection based on the nonnegative CP tensor decomposition method. As input to the decomposition method, the spatiotemporal tensor is proposed, which is composed of flattened STMs, transitions (spatial component), and time intervals (temporal component). The validation is conducted by using the domain knowledge from the Highway Capacity Manual (HCM), which reports the relations between traffic speed values and the level of service on the road segments. The method achieved the precision score of 91.68% in detection of the extreme traffic conditions (recurrent traffic anomalies). Notations and abbreviations are adopted from Kolda and Bader [10].

4.1 Speed Transition Matrix

Most of the authors represent traffic data as a time series vector $\mathbf{v} \in \mathbb{R}^{1 \times n}$ [7] or a two-dimensional matrix $\mathbf{M} \in \mathbb{R}^{m \times n}$ [11]. Dimensions m and n refer to the numbers of the road network segments (the spatial component) and the number of time intervals (the temporal component) of the observed road network. Values in the cells of the vector or matrix are the values of the traffic parameter under observation, most commonly speed, volume, or density. For this research, the STM as form to represent traffic data is used.

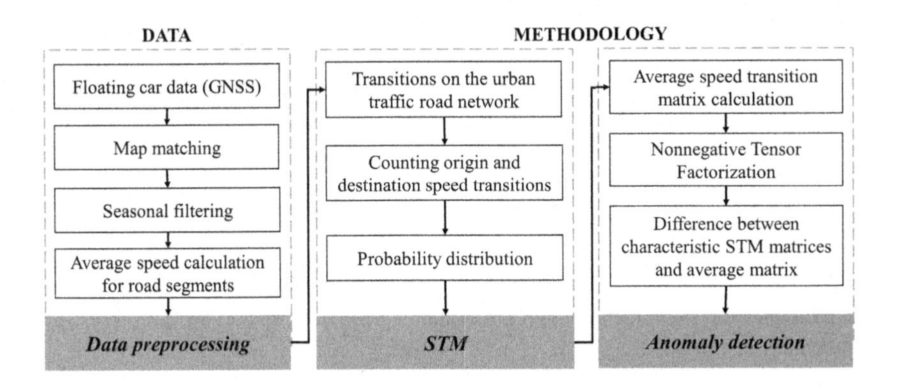

Fig. 1. Proposed methodology for the anomaly detection

The concept is proposed based on Markov chain theory, where the transition matrix shows the probability of transition from one state to another. The STM is used to represent the probability of changing the speed value when a vehicle travels between two consecutive network segments. In this paper, road network is represented as a directed graph $G = (V, E)$ where V is a set of vertices representing the intersections, and E is a set of edges representing road segments which connect two adjacent intersections. The transition is defined as a spatial change in vehicle trajectory when traveling from edge e_i to edge e_j in time interval t. As a traffic parameter under observation, the average speed is used. Average speed is calculated on e_i and labeled as the origin speed s_o and average speed on the e_j segment is labeled as destination speed s_d. Two examples of the transition are visually represented in Fig. 2 (a) with red and blue colors. The transition describes the vehicle that is traveling between edge bounded with vertices 7 and 4. Origin speed s_o is an average speed on the edge h and destination speed s_d is an average speed on edge f. Then, the STM matrix $\mathbf{X}$ is constructed as follows. First, all the changes from s_o to s_d between e_i and e_j are discretized and then counted for the particular time interval t. Each obtained value will represent the count of transitions between s_o and s_d. Figure 2 (b) and (c) show two examples of the transition counts: the anomalous traffic flow that needs to be detected and normal traffic flow, respectfully.

The speed counts are further transformed into the speed transition probability distribution to get the probabilities for every transition. Values are put into the matrix $\mathbf{X}$, and its dimensions depend on the chosen resolutions (sensitivity) of the speed change and the maximal speed that can be captured. In this paper, 5 km/h is chosen as the discretization period and 100 km/h for the maximal possible speed, which resulted in matrix dimensions 20×20. The specific maximal speed value is chosen because experiments are conducted on the road segments with a speed limit between 50 and 80 km/h. Equation (2) presents the STM where every value p_{ij} represents the probability that vehicle had origin speed s_o and destination speed s_d in the observed transition at time interval t.

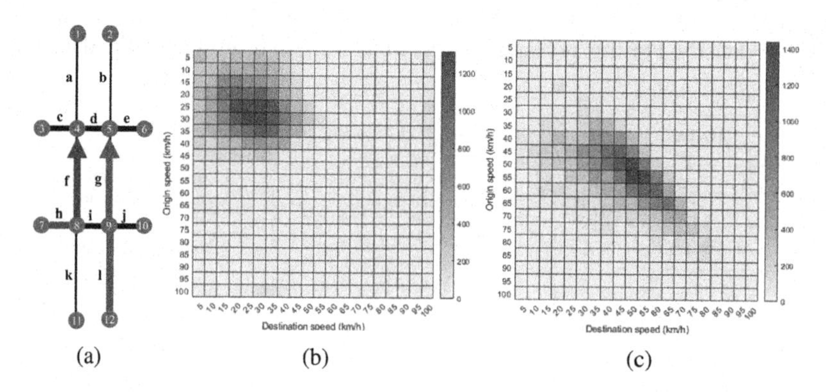

Fig. 2. Examples for: (a) transition, (b) STM containing anomaly, and (c) STM representing normal traffic

$$\mathbf{X} = \begin{pmatrix} p_{11} & p_{12} & \cdots & p_{1n} \\ p_{21} & \ddots & & \vdots \\ \vdots & & \ddots & \vdots \\ p_{m1} & \cdots & \cdots & p_{mn} \end{pmatrix} \tag{2}$$

4.2 Tensor Construction

In this paper, as a method for modeling the traffic data, a spatiotemporal tensor is proposed. The tensor is composed of flattened STMs, transitions (spatial component), and time intervals (temporal component) presented in Fig. 3. Tensor $\mathcal{T} \in \mathbb{R}^{m \times n \times t}$ is constructed, where m represents the flattened size of the STM, n represents the number of observed transitions in road network (pairs of the adjacent road segments), and t represents number of the time intervals. Frontal slices of tensor $\mathcal{T}$ can be represented with matrix $\mathbf{T}_{::t} \in \mathbb{R}^{m \times n}$, where every STM matrix $\mathbf{X}$ is flattened into a vector $\mathbf{x} \in \mathbb{R}^{m \times 1}$ and placed into the matrix $\mathbf{T}_{::t}$ as column. Dimension m had the value of 400 as STM size is 20×20. Instead of using one tensor with all the data, data is divided into several smaller tensors $\mathcal{T}^{(1)}, \mathcal{T}^{(2)}, ..., \mathcal{T}^{(N)}$ where $\mathcal{T}^{(i)} \in \mathbb{R}^{400 \times 100 \times 8}$. Regarding smaller spatial dimension, this approach allows to capture the anomalies from many different parts of the road traffic network, and more diverse traffic patterns can be captured in the anomaly detection process.

Tensor Rank Estimation. In this paper, CORCONDIA is applied to as the tensor rank estimation method. It is essential to mention that tensor rank estimation methods are used to get recommendations more than the exact actual value of the rank. Algorithm $AutoTen$ is applied that extends CORCONDIA adaptation to KLD [15]. The algorithm was run five times on randomly chosen tensor $\mathcal{T}^{(i)}$ and the average estimated rank resulted with a value of $R = 10$, which is the rank used for the experiments.

Factor Matrix Discussion. The result of the tensor decomposition can be represented with the three factor matrices $\mathbf{A} \in \mathbb{R}^{400 \times 10}$, $\mathbf{B} \in \mathbb{R}^{100 \times 10}$, and $\mathbf{C} \in \mathbb{R}^{8 \times 10}$ as presented

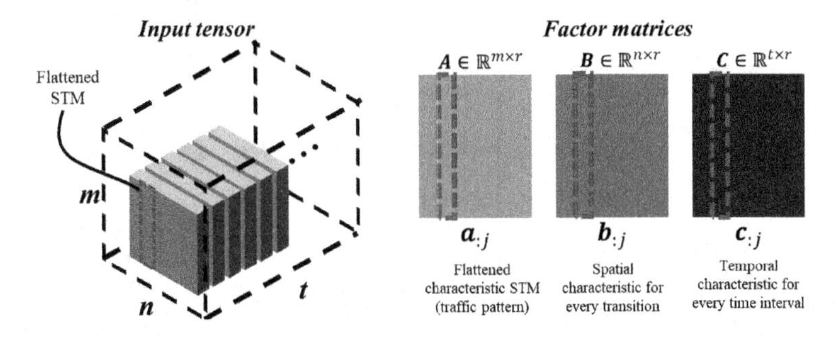

Fig. 3. Constructed tensor with corresponding factor matrices

in Fig. 3. Factor matrix **A** consists of extracted characteristic traffic patterns on the road network that is under observation. If the columns $a_{:j} \in \mathbb{R}^{400\times 1}$ of the factor matrix **A** are reshaped into the matrix 20×20 it represent the characteristic STM (traffic patterns). The goal of anomaly detection is to find the anomalous traffic patterns and link them to the corresponding values in spatial and temporal factor matrices. The matrix **B** represents the spatial factor matrix, and the values in the rows $b_{i:}$ represent how well each of the characteristic STM represent the traffic flow on the corresponding transition on index i. The values in the columns $b_{:j}$ show how well each characteristic matrix describes each of the transitions (spatial components) in the observed road network. The matrix **C** represents the temporal factor matrix. The values in the rows $c_{i:}$ represent how well each of the characteristic STM represents the corresponding time interval on index i, and the values in the columns $c_{:j}$ show how well each characteristic matrix describes each of the time interval (temporal components). The larger values in the factor matrices **B**, and **C** suggest the grater impact of the spatial or temporal components on the corresponding factor [16].

4.3 Anomaly Detection

Anomaly detection is divided into five steps, as shown in Fig. 4. The first step is defining the input tensor size. For input tensors $\mathcal{T}^{(i)} \in \mathbb{R}^{400\times 100\times 8}$ number of transitions is limited to 100. Using the smaller spatial dimension resulted in more diverse spatiotemporal traffic patterns identified because smaller parts of the network are being analyzed. Then, in the second step, nonnegative CP decomposition is applied on every input tensor $\mathcal{T}^{(i)}$. Decomposition resulted in three matrices $\mathbf{A}_i$, $\mathbf{B}_i$, and $\mathbf{C}_i$ that represent characteristic STM, spatial, and temporal components, respectively. The third step is calculating the difference between the characteristic and the average STM labeled as **M**. Every column of the factor matrix $a_{:j}$ is reshaped to matrix 20×20 to represent calculated STM

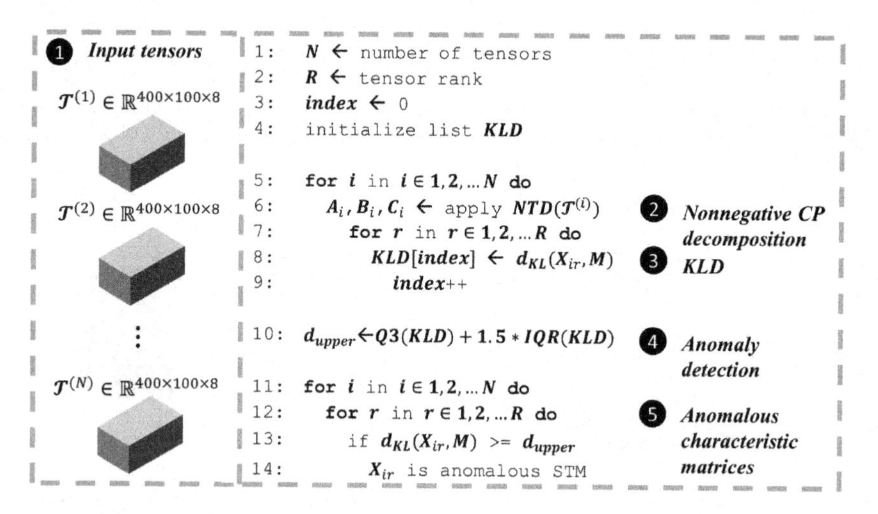

Fig. 4. Methodology for the anomaly detection using tensor CP method and KLD

characteristic matrix $\mathbf{X}_{ir}$ where r is a rank number. KLD is calculated to measure the difference between $\mathbf{X}_{ir}$ and $\mathbf{M}$:

$$d_{KL}(\mathbf{X}_{ir}, \mathbf{M}) = \sum_{x_i \in X} \mathbf{X}_{ir}(x_i) \log \left(\frac{\mathbf{X}_{ir}(x_i)}{\mathbf{M}(x_i)} \right) \quad (3)$$

As shown in Sect. 4.1, the STM represents the speed transition probability distribution, which is the reason why KLD is used as a difference measure as it calculates the difference between two probability distributions. The standard anomaly classification method for the KLD values is used. The characteristic matrix that had d_{KL} value larger than upper bound calculated as $Q3 + 1.5 \cdot IQR$, and labeled as d_{upper}, out of all calculated KLD values, is declared as an anomaly, which concludes step four. The last step is related to detecting the anomalies for every characteristic STM matrix $\mathbf{X}_{ir}$. For each input tensor $\mathcal{T}^{(i)}$, for every calculated X_{ir}, if d_{KL} is larger or equal than d_{upper}, then X_{ir} was declared as characteristic STM representing the anomaly.

5 Experiments

This section presents the data and the results of a conducted experiment on detecting the traffic flow anomalies using NTD and KLD values. For the data preprocessing, Python NumPy package was used [22], and the NTD and anomaly detection was carried out on MATLAB using Tensor Toolbox [1].

5.1 Data

The FCD used for the experiments is based on the raw GNSS data acquired from the vehicles equipped with the tracking devices. The data summary is given in Table 1. Each record contains a time-stamp, geographical longitude and latitude, speed, and heading. Due to the storage limitation, most of the data is sampled in the following way: sampling rate of 100 m for vehicles in driving mode and every 5 min for turned off vehicles. Raw data is map matched to the road segments in a digital map. GNSS data for the road network of Croatia were recorded for five years between August 2009 and October 2014 by approximately 4200 by tracked vehicles. The tracked vehicle fleet is versatile and mostly consists of delivery vehicles (vans and caddies) and taxi cars. The historical tracked data, which consists of $6, 55$ billion records, were provided by the company Mireo Inc. as a part of the SORDITO project [3,7]. Data is analyzed and anomalies are detected using the proposed method for large road segments in the City of Zagreb, which is the capital and largest city in Croatia. In the European Union context, Zagreb represents a medium-size city with approximately 800000 citizens. The seasonality of the traffic flow is considered to lower the deviation. Summer months, July and August, are not considered in the experiment. They significantly influence the results due to the different, and lower traffic flows caused by vacations [28]. Data is further divided into two groups: working days and weekend days. Working days data, Monday to Friday, are different from the weekend data for Saturday and Sunday, mostly due to the daily commuters. This filtering is used to extract only the most relevant and extreme congestion

Table 1. Data summary

Number of GNSS traces	6,55 billion
Sampling rate	100m/5 min
Time-span	August 2008–October 2014
Number of vehicles	4200
Number of road segments (Croatia)	2000000
Number of road segments (Zagreb)	86900

conditions on the urban road network. Therefore, the dataset used for the experiments includes only working days. For the construction of input tensors, eight time intervals are defined based on [3]. Intervals are defined as follows: (i) $05:30-06:45$ as morning interval with very small traffic volume, (ii) $06:45-07:25$ as interval before the morning rush hour, (iii) $07:25-08:20$ as morning rush hour, (iv) $08:20-15:30$ as interval between morning and evening rush hour, (v) $15:30-17:05$ as evening rush hour, (vi) $17:05-19:00$ as interval after evening rush hour, (vii) $19:00-22:00$ as late evening interval, and (viii) $22:00-05:30$ as night interval.

5.2 Validation

The Highway Capacity Manual (HCM) contains the guidelines, concepts, and the procedures for computing the road traffic parameters for calculating the capacity and quality of service for different road infrastructure facilities [9]. It reports the six level of service defined for the urban road segments depending on the traffic flow speed relative to the free flow speed. The speed limit on the road or the night speed are mostly used values for estimating the free flow speed. Levels of service are labeled with letters from A to F, where A represents the best traffic conditions with vehicle speeds larger than 80% of the free flow speed, and F represents the most extreme congestion where vehicle speeds are less than 30% of the free flow speed. To evaluate the process of the anomaly detection, cross-validation was adopted. The 2000 STMs were labeled by using the HCM data for the level of service. STMs are labeled as anomalous only if the traffic conditions can be classified by the HCM as extreme traffic conditions (level of service values E and F). In other cases, STMs are labeled as not anomalous. Firstly, 500 anomalous, and 500 STMs without the anomaly were selected randomly from the labeled data as a training dataset. Then the results of our approach were compared to the HCM classification as the ground truth. We report the precision calculated as $TP/(TP+FP)$, recall $TP/(TP+FN)$, and $F-1$ score in Table 2.

Table 2. Validation results of the proposed method by using the domain knowledge data

Anomalous STMs	Normal STMs	Precision	Recall	F-1
500	500	91.68%	84.30%	87.64%

5.3 Spatiotemporal Traffic Patterns

The experiment includes all road traffic segments in the City of Zagreb that have a speed limit between 50 and 80 km/h. This type of filtering is chosen because the aim of the research is anomaly detection on the large urban road network segments. The conducted experiment resulted in 140 extracted traffic patterns in the form of the characteristic STMs. After the anomaly detection method is applied, 34 anomalous characteristic matrices were extracted with different temporal characteristics. For convenience, the top five anomalous characteristic matrices and five unusual patterns are described and explained. The STM is labeled as unusual if temporal characteristic showed that anomaly is not occurring in the rush hours, where most anomalies occur. In Table 3 and 4 five characteristic STMs are shown with its corresponding temporal component extracted from columns $\mathbf{c}_{:j}$ of the temporal factor matrix. The rush hour intervals are highlighted with the vertical red dashed lines. All corresponding spatial components are located and shown on the map (Fig. 6). Column values ID of the tables correspond to the ID values on the map. In contrast to anomalous characteristic STM, Fig. 5 (a) shows the examples of the characteristic STMs that represent the normal traffic flow. Normal traffic flow is characterized by speeds that do not have a strong deviation when compared to free-flow speed [9]. As shown in Fig. 5 (b), higher values of the temporal characteristics suggest that normal characteristic matrices are best for describing traffic conditions on road segments primarily before and after, and to some extent in between rush hours.

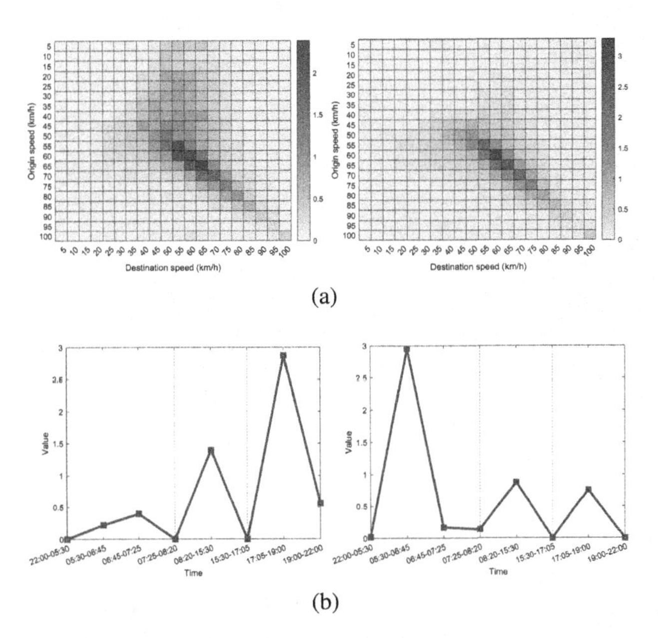

Fig. 5. Examples of spatiotemporal patterns representing the normal traffic flow (a) with corresponding temporal components (b)

Table 3. Ranking of most anomalous characteristic matrices

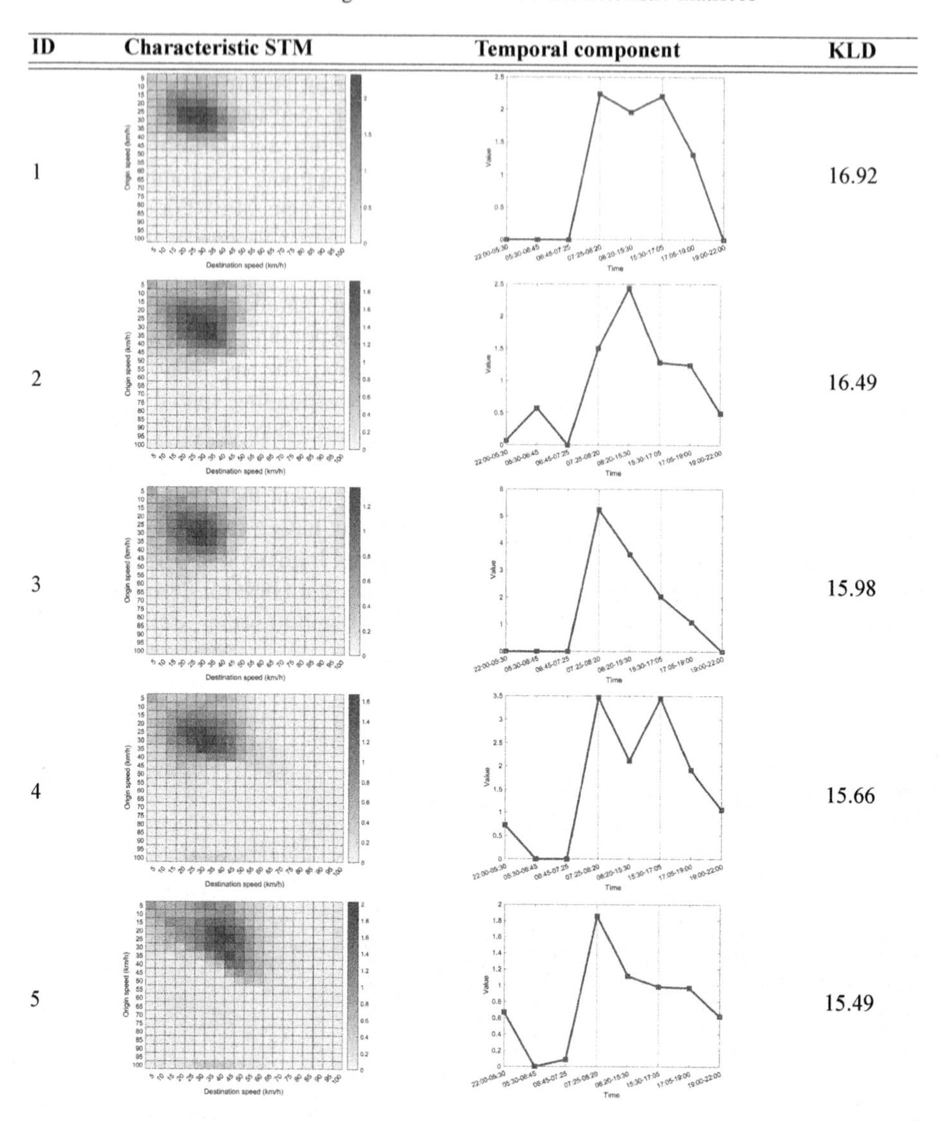

ID	Characteristic STM	Temporal component	KLD
1			16.92
2			16.49
3			15.98
4			15.66
5			15.49

Locations in Fig. 6 are extracted using spatial component matrix **B**. Where the column values $\mathbf{b}_{(:,j)}$ represent spatial coefficients. The highest value of the spatial coefficient indicates the transition that the anomalous characteristic matrix represents had the most impact. Every anomaly location is labeled with the index from 1 to 10 that matches the ID values from Tables 3 and 4. Indexes 1 and 2 point to Zagreb's business district, where the most traffic congestion occurs due to daily commuters. Index 4 points to the most congested large roundabout in the City of Zagreb and index 3 to one of the roads leading to the roundabout. This part of the city is vital because the bridge represents

Table 4. Unusual anomalous characteristic matrices

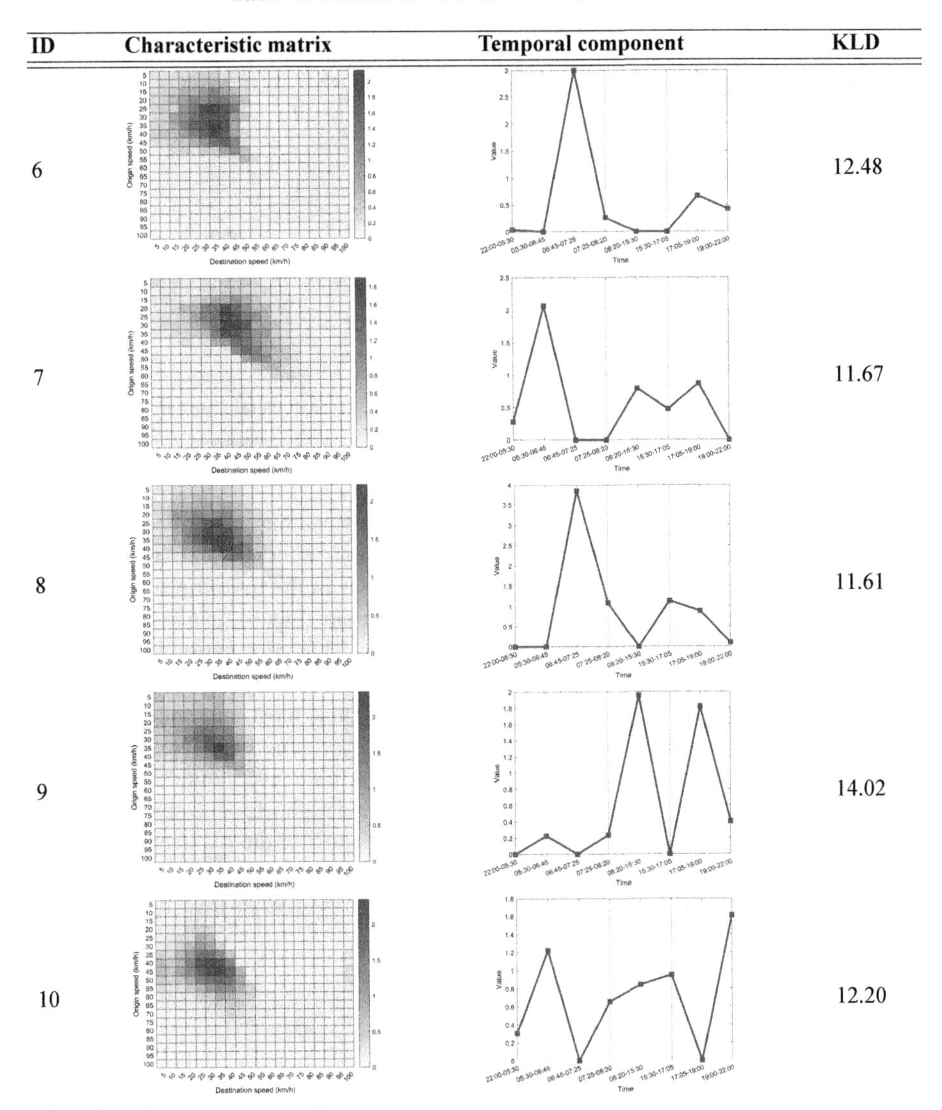

ID	Characteristic matrix	Temporal component	KLD
6			12.48
7			11.67
8			11.61
9			14.02
10			12.20

the connection between the northern and the southern part of the city. With this fact, a bridge can be identified as most congested among all the bridges that connect these two parts of the city. This information could be valuable to authorities because it can indicate the need for the building of another bridge that connects the northern and the southern part of the city. Index 5 points to street which is the entry point to the downtown area in the city center and incorporates many traffic modes, including car, tram, bicyclists, and pedestrians.

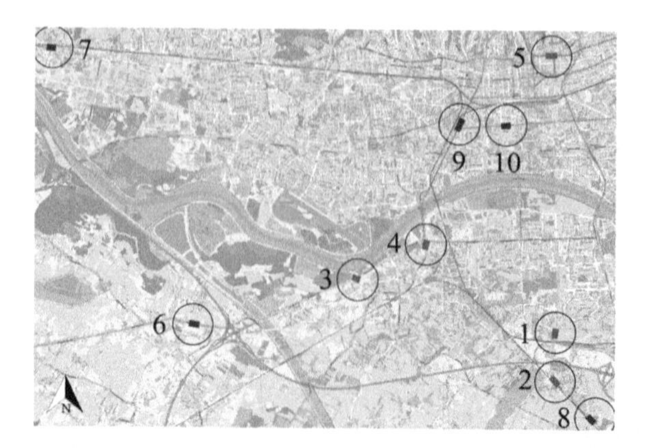

Fig. 6. Map with top five most anomalous transitions (red) and five unusual anomalous transitions (green) (Color figure online)

Table 4 presents five anomalous characteristic matrices that describe anomaly in the traffic flow in the intervals that are not the rush hour intervals and are labeled as unusual. Matrices with id $6-8$ represent the anomalies in the morning time intervals before the morning rush hour. When located on the map (Fig. 6), it can be seen that these road segments are located at the edges of the city. This information indicates that anomalous behavior occurs due to commuters that are traveling to work from more distant locations. Matrices with id 9 and 10 indicate that most anomalous behavior can be detected in the interval between rush hours and evening time intervals. Traffic demand is increased because road segments are located in the city center that attracts a lot off traffic during the evening due to city attractions and nightlife. The detection of the unusual anomalies is a valuable traffic insight because most of the recurrent anomalies are expected to occur in the rush hours. Results point out the road segments that showed different behavior of traffic flow patterns and needed to be further investigated by the traffic experts.

6 Conclusions

In this paper, a method for anomaly detection on the urban road network is presented. The conducted experiment indicates that our method can be used to detect spatiotemporal anomalies and to detect the most anomalous road traffic segments. Also, the unusual recurrent anomalies can be detected and analyzed by its spatial or temporal components.

There are some drawbacks that need to be addressed in the future work: (i) flattening of STM could cause some loss of information and sensitivity analysis must be performed, (ii) other types of f-divergence should be explored to confirm the edge cases that could be affected by selection of the distance measure, and (iii) calculation of the STM can be time consuming process, and the running time requirements of the technique must be addressed.

The result presents the valuable traffic insights that are useful for the routing application especially in non-rush hour periods, responsible urban planners, or to the road infrastructure maintenance authorities. It can be used as valuable traffic information about the need for infrastructure expansion, additional improvement strategies, or to analyze the traffic influence of the new road infrastructure. Future work should include training the neural network to detect the anomalies, based on the characteristic STMs that will be used to label the extracted spatiotemporal traffic patterns.

Acknowledgment. This research has been supported by the European Regional Development Fund under the grant KK.01.1.1.01.0009 (DATACROSS). Data used for this research is collected during the SORDITO project (RC.2.2.08-0022). Authors are also very grateful to industrial partner MIREO Inc. Sofia Fernandes acknowledges the support of FCT (Fundação para a Ciência e a Tecnologia) via the PhD scholarship PD/BD/114189/2016.

References

1. Bader, B.W., et al.: Matlab tensor toolbox version 3.1 (2019). https://www.tensortoolbox.org
2. Bro, R., Kiers, H.A.L.: A new efficient method for determining the number of components in parafac models. J. Chemometr. **17**(5), 274–286 (2003). https://doi.org/10.1002/cem.801
3. Carić, T., Fosin, J.: Using congestion zones for solving the time dependent vehicle routing problem. Promet-Traffic Transp. **32**(1), 25–38 (2020). https://doi.org/10.7307/ptt.v32i1.3296
4. Chen, X., He, Z., Chen, Y., Lu, Y., Wang, J.: Missing traffic data imputation and pattern discovery with a Bayesian augmented tensor factorization model. Transp. Res. Part C: Emerg. Technol. **104**(2018), 66–77 (2019). https://doi.org/10.1016/j.trc.2019.03.003
5. Chow, A.H., Santacreu, A., Tsapakis, I., Tanasaranond, G., Cheng, T.: Empirical assessment of urban traffic congestion. J. Adv. Transp. **48**(8), 1000–1016 (2014). https://doi.org/10.1002/atr.1241
6. Djenouri, Y., Belhadi, A., Lin, J.C., Djenouri, D., Cano, A.: A survey on urban traffic anomalies detection algorithms. IEEE Access **7**, 12192–12205 (2019). https://doi.org/10.1109/ACCESS.2019.2893124
7. Erdelić, T., Ravlić, M., Carić, T.: Travel time prediction using speed profiles for road network of Croatia. In: 2016 International Symposium ELMAR, pp. 97–100 (2016). https://doi.org/10.1109/ELMAR.2016.7731763
8. Fanaee Tork, H., Gama, J.: Event detection from traffic tensors: a hybrid model. Neurocomputing **203**, 22–33 (2016). https://doi.org/10.1016/j.neucom.2016.04.006
9. HCM2010: Highway capacity manual, transportation Research Board, National Research Council (2010)
10. Kolda, T.G., Bader, B.W.: Tensor decompositions and applications. SIAM Rev. **51**(3), 455–500 (2009). https://doi.org/10.1137/07070111X
11. Liu, X., Liu, X., Wang, Y., Pu, J., Zhang, X.: Detecting anomaly in traffic flow from road similarity analysis. In: Cui, B., Zhang, N., Xu, J., Lian, X., Liu, D. (eds.) WAIM 2016. LNCS, vol. 9659, pp. 92–104. Springer, Cham (2016). https://doi.org/10.1007/978-3-319-39958-4_8
12. Ma, X., Dai, Z., He, Z., Ma, J., Wang, Y., Wang, Y.: Learning traffic as images: a deep convolutional neural network for large-scale transportation network speed prediction. Sensors (Switz.) **17**(4), 1–16 (2017). https://doi.org/10.3390/s17040818
13. Nguyen, H., Liu, W., Chen, F.: Discovering congestion propagation patterns in spatio-temporal traffic data. IEEE Trans. Big Data **3**(2), 169–180 (2017)
14. Pan, P., Wang, H., Li, L., Wang, Y., Jin, Y.: Peak-hour subway passenger flow forecasting: a tensor based approach. In: 21st International Conference on Intelligent Transportation Systems, pp. 3730–3735 (2018). https://doi.org/10.1109/ITSC.2018.8569577

15. Papalexakis, E.E.: Automatic unsupervised tensor mining with quality assessment. In: Proceedings of the International Conference on Data Mining, pp. 711–719 (2016). https://doi.org/10.1137/1.9781611974348.80
16. Qi, G., Huang, A., Guan, W., Fan, L.: Analysis and prediction of regional mobility patterns of bus travellers using smart card data and points of interest data. IEEE Trans. Intell. Transp. Syst. **20**(4), 1197–1214 (2019)
17. Qi, N., Shi, Y., Sun, X., Wang, J., Yin, B., Gao, J.: Multi-dimensional sparse models. IEEE Trans. Pattern Anal. Mach. Intell. **40**(1), 163–178 (2018)
18. Shi, Y., Deng, M., Yang, X., Gong, J.: Detecting anomalies in spatio-temporal flow data by constructing dynamic neighbourhoods. Comput. Environ. Urban Syst. **67**, 80–96 (2018). https://doi.org/10.1016/j.compenvurbsys.2017.08.010
19. Tan, H., Wu, Y., Shen, B., Jin, P.J., Ran, B.: Short-term traffic prediction based on dynamic tensor completion. IEEE Trans. Intell. Transp. Syst. **17**(8), 2123–2133 (2016). https://doi.org/10.1109/TITS.2015.2513411
20. Tan, H., Yang, Z., Feng, G., Wang, W., Ran, B.: Correlation analysis for tensor-based traffic data imputation method. Procedia - Soc. Behav. Sci. **96**, 2611–2620 (2013). https://doi.org/10.1016/j.sbspro.2013.08.292
21. Tang, K., Chen, S., Liu, Z.: Citywide spatial-temporal travel time estimation using big and sparse trajectories. IEEE Trans. Intell. Transp. Syst. **19**(12), 4023–4034 (2018). https://doi.org/10.1109/TITS.2018.2803085
22. Walt, S., Colbert, C., Varoquaux, G.: The numpy array: a structure for efficient numerical computation. Comput. Sci. Eng. **13**(2), 22–30 (2011)
23. Wang, J., Gao, F., Cui, P., Li, C., Xiong, Z.: Discovering urban spatio-temporal structure from time-evolving traffic networks. In: Chen, L., Jia, Y., Sellis, T., Liu, G. (eds.) Web Technologies and Applications, pp. 93–104. Springer International Publishing, Cham (2014)
24. Wang, X., Fagette, A., Sartelet, P., Sun, L.: A probabilistic tensor factorization approach to detect anomalies in spatiotemporal traffic activities. In: IEEE Intelligent Transportation Systems Conference, pp. 1658–1663 (2019). https://doi.org/10.1109/ITSC.2019.8917169
25. Wang, Z., Hu, K., Xu, K., Yin, B., Dong, X.: Structural analysis of network traffic matrix via relaxed principal component pursuit. Comput. Networks **56**(7), 2049–2067 (2012)
26. Xie, Q., Zhao, Q., Meng, D., Xu, Z.: Kronecker-basis-representation based tensor sparsity and its applications to tensor recovery. IEEE Trans. Pattern Anal. Mach. Intell. **40**(8), 1888–1902 (2018). https://doi.org/10.1109/TPAMI.2017.2734888
27. Yu, L., Huang, J., Zhou, G., Liu, C., Zhang, Z.: Tiirec: a tensor approach for tag-driven item recommendation with sparse user generated content. Inf. Sci. **411**, 122–135 (2017). https://doi.org/10.1016/j.ins.2017.05.025
28. Żochowska, R., Karoń, G.: ITS Services Packages as a Tool for Managing Traffic Congestion in Cities, pp. 81–103. Springer International Publishing, Cham (2016). https://doi.org/10.1007/978-3-319-19150-8_3

Time Series Regression in Professional Road Cycling

Arie-Willem de Leeuw[1(✉)], Mathieu Heijboer[2], Mathijs Hofmijster[3,4], Stephan van der Zwaard[1], and Arno Knobbe[1]

[1] Leiden University, Niels Bohrweg 1, 2333 CA Leiden, The Netherlands
a.de.leeuw@liacs.leidenuniv.nl
[2] Team Jumbo-Visma, Rietveldenweg 47e, 5222 AP Den Bosch, The Netherlands
[3] VU Amsterdam, De Boelelaan 1105, 1081 HV Amsterdam, The Netherlands
[4] University of Applied Sciences Amsterdam, Spui 21, 1012 WX Amsterdam, The Netherlands

Abstract. With the recent explosive developments in sensoring capabilities and ubiquitous computing in road cycling, large quantities of detailed data about performance are becoming available. In this paper, we will demonstrate that this rich data in cycling offers several non-trivial data science challenges. The primary task that we focus on is a regression task: given a collection of results in previous races of a specific rider, predict the performance in a future race solely based on the characteristics of said rider and the stage profile. To make these predictions, we have developed a predictive pipeline that consists of three consecutive rider-specific models. First, we transform the distance-altitude profile into a time profile, by using a climb-descent model that describes the relationship between the speed of the cyclist and the slope of the terrain. Second, we introduce an *effective profile* that includes the rider-specific physiological capabilities. Third, we predict the performance based on the characteristics of the effective profile, by using a model constructed from the historical records of our cyclist. To demonstrate the relevance of this work, we show that for a professional cycling team, important information for making tactical decisions can be obtained from our modeling approach.

Keywords: Temporal data mining · Time series regression · Predictive modeling

1 Introduction

Road cycling is a complex sport where the final result in competition is affected by many aspects [1,7,8], such as the physiological capacities of a rider [14], weather conditions and course details. In this sport, the differences between success and failure are small and therefore even slight improvements in one of the many performance indicators might have large consequences. Since most professional cyclists nowadays use so-called *cycling computers* to frequently monitor important characteristics during their rides, e.g., heart rate and produced power, and data science flourishes in the presence of this abundance of data, the application of these

A. Appice et al. (Eds.): DS 2020, LNAI 12323, pp. 689–703, 2020.
https://doi.org/10.1007/978-3-030-61527-7_45

novel techniques in the context of road cycling has the potential to reveal valuable insights for improving the performance of professional cyclists.

In this article, we consider the following data science challenge in road cycling: given a future stage profile and a database of historical races of a rider, can we predict the loss or gain of time for this rider, relative to his competitors? Notice that the particular task we are focusing on here is one of personalization, that is, we are using rider-specific data in order to learn his specific strengths and weaknesses compared to the competition. This regression problem, more specifically a *time series regression* problem, can be seen as the regression-counterpart of the *time series classification* problem [2]. Thus, instead of attaching a label to future time series, we want to predict a numeric value for an unseen time series. To the best of our knowledge, we are the first to tackle this type of problem. Although it is tempting to think that the task at hand is an instance of one of the various temporal prediction settings, such as forecasting or time series classification, we stress here that this is not the case.

To address the aforementioned challenge, we have developed a three-stage pipeline. First, we construct a climb-descent model to transform a given distance profile into a time profile. This is necessary for including the physiological response of a rider as this is measured on a time scale, whereas the stage profile is provided on a distance scale. In the second step, we incorporate the rider-specific physiology by transforming the time profile into an *effective profile*. We calculate the effective profile by performing a convolution of the time series with an exponential kernel. This form of the kernel is inspired by the response to exercise of the human body. In the third step of the pipeline, we apply feature extraction to the effective profiles and use two types of regression modeling for constructing the final model.

The work described here is the result of a collaboration with one of the world-leading cycling teams, Team Jumbo-Visma. The team has kindly made available several years of detailed rider data, which was enriched with public data about race results and competitors. Based on this data, rider-specific models were developed and subsequently applied to stage profiles of one of three then upcoming Grand Tours[1] in 2019, which we report upon.

2 Related Work

In this article, we are dealing with time series analysis. A large branch of research in this field deals with the specific task of *forecasting*, where extrapolation is used to make predictions for future values of the time series. There are several methods for making these predictions, such as ARIMA models [4]. In this work, we do *not* consider forecasting. Instead, we will focus on determining a target value for a yet to be predicted time series based on a collection of historical time series and target values, i.e., a *time series regression* challenge. Where in forecasting, one is interested in predicting the next value in a sequence, in time series regression we are primarily interested in learning how to aggregate a time series into a single number (in our case win or loss of time).

[1] The most important stage races: *Tour de France*, *Giro d'Italia* and *Vuelta a España*.

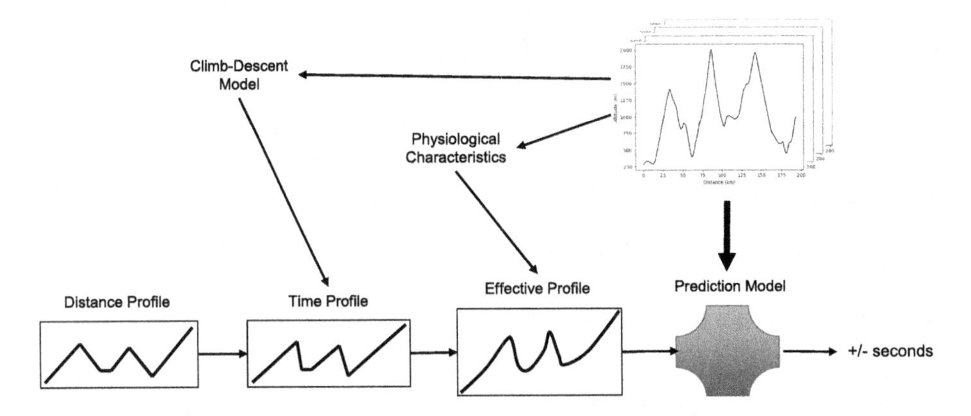

Fig. 1. Schematic overview of the different parts of the pipeline for predicting the performance of our general classification rider. The historical races are used for constructing a climb-descent model, obtaining the rider-specific characteristics and determining a prediction model. By using these three components, the distance profile of a stage can be translated into a prediction for the time difference with the ten best riders in the general classification at the end of the stage race.

There are many possibilities for approaching this regression problem, but for a number of application-specific reasons, we here opt for a *feature extraction* approach. There are several approaches for feature construction with time series data [5,17,18,21,22], where most recent developments focus on deep learning methods [3,19,23]. However, traditional approaches are still commonly used as they can compete or even outperform these methods in accuracy [15] and have the advantage that the results are interpretable. To avoid a dependence on domain knowledge to find the appropriate features, a method has been developed that applies automated feature selection [9].

Usually, the behavior of the time series prior to a certain time is implicitly taken into account in the feature construction process. Contrary to this, we have a pre-processing step, where we explicitly focus on how the previous values in the time series should be taken into account to optimize the performance of the regression model. Besides improving the accuracy of the results, the advantage of this approach is that it makes explicit which timescales are important and with what weight the previous values should be taken into account. In many practical cases, and also here, this is useful information for the domain experts.

3 Modeling Approach

We consider the following time series regression challenge in road cycling.

Given: A collection of time series data with corresponding race results of a specific cyclist and a distance-altitude profile for a stage in a future multiple-day race that is of the form $\{(d_0, h_0), \ldots, (d_f, h_f)\}$, where f indicates the integer of the final element of the series, d_j is the covered distance and h_j is the corresponding altitude at a specific point in the race.

Goal: Predict the performance of the rider in the stage solely based on the characteristics of the said rider and the stage profile.

To tackle this problem, we have developed a three-stage pipeline, i.e, see Fig. 1. Below, we will discuss the different parts of this pipeline separately, starting with the last step, and working to the front along the pipeline.

3.1 Time Series Regression

The most important (and last) part of our pipeline is the step in which we apply regression. In a general time series regression setting, we have a collection of $N \in \mathbb{N}$ different pairs v. Each pair is of the form (d^v, y^v), with d^v a time series and y^v the target value that belongs to this time series. Hence,

$$v = (d^v, y^v) = (\{(t_0^v, x_0^v), \ldots, (t_{f^v}^v, x_{f^v}^v)\}, y^v), \tag{1}$$

where t_i^v is the temporal component of the data, x_i^v the corresponding variable that is monitored as a function of time and $f^v \in \mathbb{N}$ denotes the last element of the time series. Note that the superscript v in the latter denotes that the length of the time series can be different for each v. In the remainder of this section, we will omit the superscript v for convenience. The main regression task is as follows

Problem 1: Given a collection of time series data with corresponding numerical target as defined in (1), find the relationship between the time series and target variable.

This problem is challenging for several reasons. First of all, there are no constraints on the length of the time series, and different examples will have different lengths. Next, the predictive information necessary for making the actual regression may be contained in the entire time series, but it may also just depend on segments of the data, say only the last 15 min of the race. Additionally, what aspects of the time series are of specific relevance to the prediction may differ from task to task. For example, in our case the target may depend on continuous, aggregated features such as the total amount of climb meters, but it might also require more discrete information like the number of climbs.

There are several options for addressing this problem. In principle, most methods that are developed for time-series classification problems, including distance-based, shapelets, and feature-extraction approaches [2], can also be applied to our time-series regression setting. For our application, we have used an aggregation-based feature construction approach. Moreover, we used LASSO regression [20] to select the most relevant features and to reduce overfitting. We opted for this method to provide useful feedback to the cycling team. Additionally, to demonstrate the versatility and advantages of our approach, we will also apply a non-linear method that focuses on finding local patterns, i.e., Subgroup Discovery [10].

3.2 Effective Time Series

In the second step of our pipeline, we explicitly incorporate the integration over time introduced by the physiology of the human body, by transforming the data into an *effective* time series. More specifically, for a time series as defined in (1), the effective time series is defined as $\{(t_0, z_0), \ldots, (t_f, z_f)\}$, where for a given integer $m \in [0, f]$, z_m is defined as

$$z_m = \sum_{i=0}^{m} h(t_i) \cdot x_{m-i}, \tag{2}$$

Note that this procedure is similar to taking the convolution of the time series with a yet to be mentioned kernel $h(t)$. Therefore, we also have the following optimization task.

Problem 2: Given a collection of time series data with corresponding numerical target as defined in (1), transform the learning time series data into an effective time series as in (2) and find the function $h(t)$ that maximizes the accuracy of the regression model.

To address this problem, we consider a class of continuous functions $h_\tau(t)$ on the interval $[a, b]$ for $a, b \in \mathbb{R}$ that satisfy

$$\lim_{\tau \to a} h_\tau(t_k) = \delta_{k,0}, \tag{3}$$

and

$$\lim_{\tau \to b} h_\tau(t_k) = 1/(f+1). \tag{4}$$

Here, $\delta_{k,0}$ is a Kronecker delta that is defined as

$$\delta_{k,0} = \begin{cases} 1 & \text{if } k = 0, \\ 0 & \text{otherwise,} \end{cases} \tag{5}$$

and recall that $f + 1$ denotes the length of the time series.

For these functions $h_\tau(t)$, z_m only contains the current values of the time series x_m if τ tends to a. On the other hand, in the limit $\tau \to b$, all previous elements of the time series are included in a sum with equal weights for each element. Because the kernel is continuous in τ between the two above-mentioned extreme cases, we can capture the most important distinct kind of temporal effects by only changing the value of a single parameter τ between a and b. In our application for professional road cycling, we set

$$h_\tau(t) = (\delta_{t,0} + \Theta(t) \cdot e^{-t/\tau})/(1+\tau), \tag{6}$$

where $\delta_{t,0}$ is the Kronecker delta as defined in (5) and the Heaviside step function $\Theta(t)$ is given by

$$\Theta(t) = \begin{cases} 1 & \text{if } t > 0, \\ 0 & \text{otherwise.} \end{cases} \tag{7}$$

This kernel consists of two different parts. The first term is included to make sure that

$$z_m = \lim_{\tau \to 0} \sum_{i=0}^{m} h_\tau(t_i) x_{m-i} = \sum_{i=0}^{\infty} \delta_{i,0} x_{m-i} = x_m. \tag{8}$$

The second term in (6) is incorporated to model the recovery abilities of the cyclist, where τ specifies how fast an athlete recovers from previous efforts within a race. Note that this procedure introduces a parameter τ that needs to be optimized.

3.3 Climb-Descent Model

For our application, we start with a distance profile that specifies the altitude of the terrain and the covered distance since the start of the race. To include the rider-specific recovery abilities, we need to transform the distance profile into a time profile as the recovery abilities depend on the time rather than the distance between subsequent efforts. Note that this first step in the pipeline is specific for the application in road cycling, while the aforementioned two steps are also relevant in other time series regression settings.

To achieve this transformation, we use the historical races of our rider to construct a climb-descent model that relates the speed of the cyclist to the slope of the terrain. For constructing this model, we collect the speed and corresponding slope of the terrain in all historical races, including both mass-start stages and time trials, which could amount to millions of data points per rider. Subsequently, we apply the method that is introduced in [13] to find the relationship between slope and speed. In this method, we consider polynomial regression and use cross-validation to select the degree of the polynomial that gives the most accurate model.

4 Materials

4.1 Data

We used data of a single rider of one of the world-leading professional road cycling teams. During a period of several years, a large variety of important characteristics, such as physical measurements, temperature, altitude and speed, were collected once every second during all his rides by using a cycling computer. For this work, the relevant information is the speed, distance and altitude in competition. Moreover, we have used an online cycling results database to obtain the results of the races in which the rider competed [11].

For our application, there are two complications that need to be addressed. First, the stages can be divided into two different categories. On the one hand, there are stages with a mass start. On the other hand, there are time trials in which each cyclist rides alone or only with his teammates. In our data collection, there are only few time trials and most stages start with a mass start. Therefore,

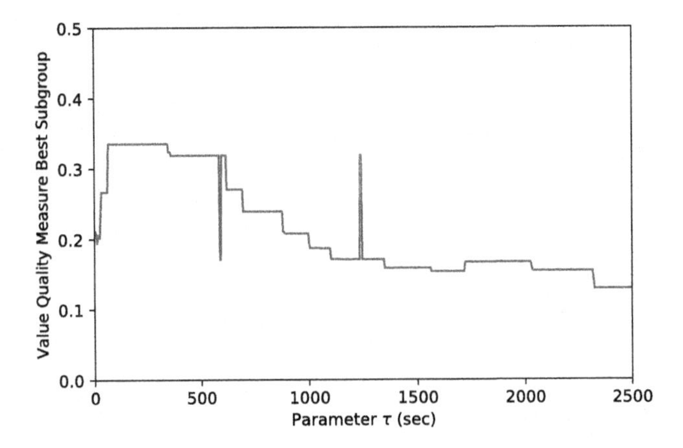

Fig. 2. The results for the parameter optimization with Subgroup Discovery. For every τ, we display the value of the quality measure *explained variance* for the best subgroup at search depth 1. The subgroups with the highest quality is obtained for $\tau \in [65, 345]$.

from now on we will focus on the mass start stages. The second complication is the purpose of the race within the training-competition regime. Depending on the importance of a race, the aim can be optimal performance, or race-specific training for a more important race to come. Since we want to predict the performance, we need to make sure to only include the former. Therefore, we consulted the coach of the team to select the stages where the rider was aiming for an optimal performance. This resulted in a selection of 122 mass-start stages that are used for further analyzes. In addition, we have a validation set comprised of the 21 stages of one of the Grand Tours in 2019 where the rider participated, with great success.

4.2 Target

The first task in our analyzes is defining a target variable, i.e., a measure that characterizes the performance of the cyclist in competition. For a proper definition, we need to consider some contextual information. First, our cyclist is a general classification rider. This implies that during stage races, i.e., races that last for multiple days, in every stage the goal is to gain time or at least limit time losses compared to the competitors. Thus, a good performance is not necessarily a high classification in individual stages, but rather, faster finishing times compared to the other general classification riders. Therefore, for every stage, we compare the finish time of our cyclist with the average finish time of the ten best riders in the general classification at the end of the multi-day race as in the cycling community this is considered a benchmark for a good performance[2]. Moreover, we consider the general classification at the end of the multi-day race

[2] This specific information is only available at the end of the race, which makes our analysis a post-hoc one.

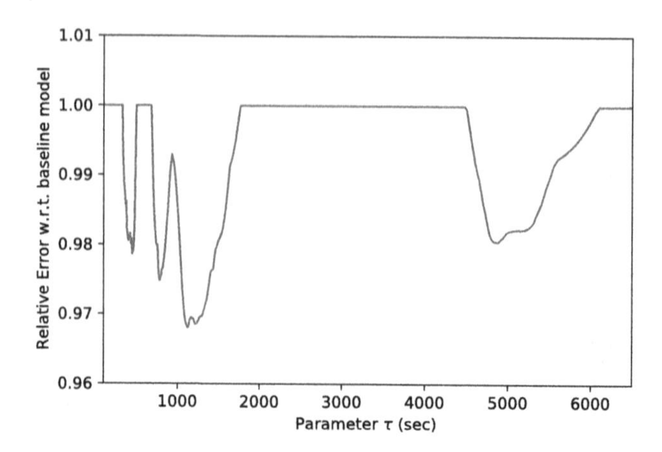

Fig. 3. The results for the parameter optimization with LASSO regression. For every value of the parameter τ that models the recovery abilities of our cyclist, we choose the value of the LASSO regularization parameter α that gives the smallest error. The error is relative to the baseline model, which is the linear model that only includes the intercept and for which all other coefficients are equal to zero. The best model is obtained for $\tau = 1124$ and $\alpha = 0.27$.

as it is quite common that throughout the race there are riders on high positions in the general standings that are not aiming for a good position in the final general classification.

4.3 Feature Engineering

Our data collection contains distance profiles that specify the altitude of the terrain at different points in the race. First, we transform these distance profiles into new profiles that characterize the change in altitude from the start of the race as this is a more appropriate measure for the heaviness. Hereafter, we calculate the time profiles by using the climb-descent model. Subsequently, for each τ, we consider the kernel as defined in (6) and we determine the effective profile. Finally, the calculated effective profile is used to determine features. Each feature is an aggregate of either the *effective altitudes* or the *changes in effective altitudes* in a specific window.

In cycling, the last part of the stage is the part where most time differences between the general classification riders occur. Therefore, to predict the performance of our cyclist, besides considering the distribution of the entire stage, we also consider the last 30 min, the last 20 min, the last 10 min, the last 30 to 10 min, the last 30 to 20 min and the 20 to 10 min, separately. For each window, we determine the mean, standard deviation, interquartile range, skewness, kurtosis and difference between maximum and minimum. Moreover, we consider the nth-percentiles for n equal to 0, 25, 50, 75, 80, 85, 90, 95 or 100. Note that we focused on the high percentiles, as we expect that the largest values of the effective altitude and changes herein are most decisive.

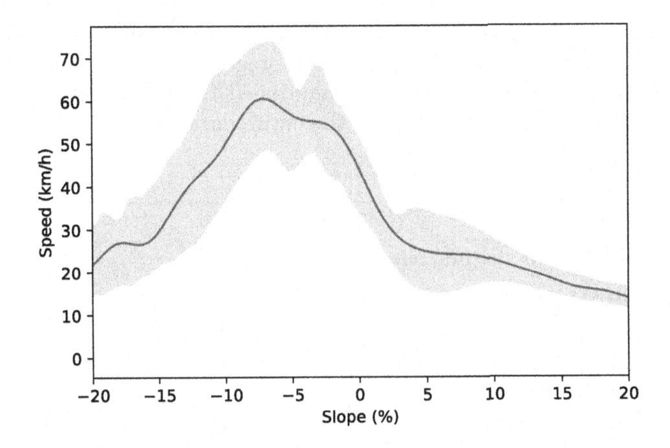

Fig. 4. Speed as a function of the slope at the cycling stage course. The solid line is the polynomial model of degree 43. The shaded area denotes a confidence interval that we calculate by binning the slope into bins of a percent, where subsequent bins overlap for 50%. The upper (lower) part is the average speed plus (minus) one standard deviation.

In total, we consider the effective altitude and changes in effective altitude, take into account seven different windows and use fifteen different aggregates. Hence, for every stage we construct 210 features.

5 Results

5.1 Climb-Descent Model

In the climb-descent model, we determine the relationship between the speed of our cyclist and the slope of the terrain. The speed of the cyclist is defined as the average speed at a time t and one second later. To determine the corresponding slope $S(t)$, we need the covered distance D and the change in elevation. The covered distance D is calculated by using the average speed. To determine the difference in elevation, we first bidirectionally filtered the altitude data using an exponential weighted moving average with smoothing factor 1/2 to introduce a smoothed altitude $H(t)$. Now, the slope S(t) is defined as

$$S(t) = 100 \cdot \frac{H(t) - H(t-1)}{\sqrt{D^2 - (H(t) - H(t-1))^2}}. \tag{9}$$

Due to failures of the GPS tracking, there is a small percentage of the data collection for which we obtain extreme and unreliable results. Since 38% is the steepest slope on earth, we remove the data points with larger slopes. In the end, we are left with almost 2 million instances.

To find the climb-descent model, we apply the method that is introduced in [13]. We performed the polynomial regression one hundred times, of which an optimal degree of 43 was found in 98 instances. The final model can be seen in Fig. 4. As expected, starting from large ascents, the speed of our cyclist increases

Table 1. Overview of the features that are included in the final model to predict the performance in road cycling based on the stage profile. To compare the values of coefficients of the different features, we multiplied the coefficients by the standard deviation of the feature that is present in the entire data collection.

Feature	Coefficient
Intercept	−8.56
Total effective altitude change minimum	−7.43
Effective altitude change Kurtosis last 10–20 min	−4.40
Effective altitude change difference maximum and minimum last 10–20 min	−4.11
Effective altitude change skewness last 0–20 min	−1.30
Total effective altitude change skewness	−1.11
Effective altitude change Kurtosis last 10–30 min	−1.01
Effective altitude change Kurtosis last 20–30 min	−0.96
Effective altitude change minimum last 10–20 min	1.17
Effective altitude change minimum last 0–20 min	4.36
Effective altitude change skewness last 0–10 min	5.02
Effective altitude interquartile range last 0–10 min	6.92
Effective altitude change skewness last 20–30 min	8.40

if the slope decreases. Indeed, the steeper a climb becomes, the slower the speed is, but to begin with, this progression is already nonlinear. The interesting part is when you descend. Initially, at moderate slopes, the speed is correlated with the slope angle. However, at greater angles, the speed drops to almost the level of an ascent, mostly since the road will involve more hairpins, and the riders will have to frequently break to get down safely. Note that we derive a climb-descent model from the data that, again, is rider-specific. Although we expect that the overall behavior is similar for most cyclists, there can be some differences. For example, the fact that some riders are known to be better descenders than others will influence the behavior of the climb-descent model at negative slopes.

To investigate the quality of the regression, we determine the explained variance R^2, defined as

$$R^2 = 1 - \frac{\mathrm{SS_{res}}}{\mathrm{SS_{tot}}}, \tag{10}$$

where $\mathrm{SS_{res}}$ is the sum of squares of the residuals and $\mathrm{SS_{tot}}$ denotes the total sum of squares. For the climb-descent model, we obtain an explained variance of 0.657.

For interpreting this value, we determine the statistical significance by using an F-test. The F-statistic F^* for testing the null hypothesis that all coefficients of the model are zero, i.e., there is no relationship between slope and speed, is equal to

$$F^* = \frac{R^2}{1-R^2} \frac{n-k-1}{k}, \tag{11}$$

where n is the number of data points and k the number of parameters in the model. In this case $k = 44, n = 1934056$ and $R^2 = 0.657$. This gives a p-value that is very close to zero and therefore we can reject the null hypothesis at a significance level of 0.05. Hence, our climb-descent model is statistically significant.

5.2 Regression

First, we have applied the exploratory data analysis technique Subgroup Discovery to address the regression challenge. Specifically, we use the Cortana tool [16], to find the subgroup at search depth 1, i.e., subgroups that are characterized by a condition on a single feature. In our experiments, we have used a beam search and the *explained variance* R^2 as the quality measure [12]. Moreover, to obtain subgroups of substantial size, we restricted our search to subgroups between 5% and 95% of the entire data collection. As displayed in Fig. 2, the best subgroups are obtained for $\tau \in [65, 345]$. Note that there is not a unique value for τ, as the effective profiles only slightly change as a function of τ. The subgroup with the largest value for the quality measure is characterized by a condition on the minimum value of the change in effective altitude in the last 30 min of the stage. If this minimum value is above a certain positive-valued threshold, where the precise value for this threshold depends on the value for parameter $\tau \in [65, 345]$ that is considered, our rider had some bad performances. The value for the quality measure for the best subgroup is equal to 0.335. To interpret this value, we use the *distribution of false discoveries* [6]. In this case, the threshold for finding statistically significant results at confidence level 0.95 is equal to 0.187. Hence, our results are statistically significant.

Second, we have applied LASSO regression. Again, we first need to find the optimal values for the kernel parameter τ and the regularization parameter α. From Fig. 3, we observe that for LASSO regression the best model is obtained for $\tau = 1124$ and $\alpha = 0.27$. As for Subgroup Discovery, we find an optimal value for τ that is different from 0. This implies that including the physiological properties of our cyclist by introducing an effective profile is a crucial step in our pipeline. Moreover, note that this value for τ is different from the one that we obtained in the previous experiments. This is a consequence of the different nature of the two experiments. In LASSO regression, we only consider linear effects and we are optimizing a global model. On the other hand, Subgroup Discovery focuses on finding the interesting local parts of the data and is a non-linear technique.

From our LASSO regression procedure, we find a model that includes twelve parameters and an intercept. In Table 1, we display the features in our model and the corresponding values of the coefficients. A negative value for the coefficient implies that the cyclist improves his performance if the value of this feature is increased. On the other hand, a positive value for the coefficient indicates that the performance of our rider is reduced if the value of this feature is increased.

We obtained a model with $R^2 = 0.255$. To interpret this value for the explained variance, we test the null hypothesis that all coefficients in the model are zero. By using (11), we find $F^* = 2.84$ and obtain a p-value of 0.002. Therefore, we find that the model is statistically significant at confidence level 0.95.

5.3 Validation

After we developed the model in the beginning of 2019, the model was applied to the stage profiles of one of the Grand Tours in 2019, and the predictions were communicated to the team in the weeks prior to the race. Before we discuss the lessons that are learned from the implementation of our approach, we need to mention that the opportunity for validation provided by a single Grand Tour is rather limited for several reasons, but we specifically emphasize two here. First, there are only few data points available, as of the 21 stages, not all can be considered. Namely, each Grand Tour contains (team) time trials and stages can be altered or even be neutralized due to extreme weather conditions. In our case, only 17 stages could be considered. Second, the models are based on historic data while the subject might have improved and be in exceptionally good shape at the validation event.

Despite these limitations, we still evaluated the usefulness of our approach in this event. The stage duration predictions produced by the climb-descent model are quite accurate. The predictions had a $R^2 = 0.79$, with a RMSE of 1 298 s, or less than 8% of the average stage duration. On the whole, the predictions of stage performance were more disappointing. In this Grand Tour, there were six mass-start stages where the race dynamics required an optimal performance. For these stages, we predicted on average a loss of 22.5 s. All six predictions were losses, whereas in reality, three of the stage proved a gain. However, as mentioned earlier, this might be a consequence of the rider being in exceptionally good shape at the validation event. Indeed, our rider performed above expectation and fared quite well overall. This is reflected in the six stages, where an average gain of -4.9 s was made, compared to the predicted 22.5 s loss.

It is good to remember that the aim of our work was not to provide accurate predictions of time differences, but rather assess to what extent the various stages provided an opportunity for time gain (or loss) to the rider. As pointed out earlier, whether that opportunity blossomed out depends on the race dynamics. This effect is clearly present in this case. Specifically, the difference between an opportunity and actual gain is apparent in 11 of the 17 stages available. In these stages, we predicted gains/losses ranging from -13.2 to 32.3 s but none panned out, since the stage ended in a bunch sprint or breakaway (both resulting in the majority of the peloton achieving zero gain). In other words, there was an intrinsic (dis)advantage predicted for our rider, but no gain or loss was obtained.

6 Implications for Road Cycling

In this section, we will put some emphasis on the implications of the results from the point of view of professional road cycling.

6.1 Climb-Descent Model

The climb-descent model can successfully be used to predict the duration of the entire stage and also estimate the time at which the rider arrives at a specific point in the race. This information can be used for composing an optimal nutrition scheme. For example, sport directors can advice the rider on the nutrition intake during the race, as the model estimates the time that is still needed to finish the race at every point in the race.

6.2 Effective Profile

The effective profile describes the heaviness that is experienced by the rider. This is essential information for deciding upon the team strategy during the stage. At the toughest points in the race, a team wants to give maximal support to the general classification rider. Therefore, the team directors can instruct the team such that at these points the rider is well-surrounded by fellow team members. This can for example be achieved by sending teammates forward in an earlier part of the race.

6.3 Prediction Model

With the prediction model, we can distill the characteristics of stage profiles that have most influences on the performance of our rider. In this case, we observe that the performance is mainly influenced by *changes in* the effective altitude, and the effective altitude itself is less relevant. Moreover, the rider performs well in stages where the effective altitude in the entire stage increases or if the decrease is minimized.

The performance predictions can be used to finalize the tactics during a stage. For example, road cycling is a team sport and therefore the teammates can put more emphasis on protecting the team leader if the cyclist is expecting to lose time on his opponents. On the other hand, the team could apply a more offensive strategy in stages where the general classification rider is expected to gain time on his rivals. Second, by applying our method to all general classification riders in the same team and comparing the predictions for these riders in the different Grand Tours, for each Grand Tour the team can select the general classification rider that has the highest chance of being successful.

6.4 Future Work

Although the small size of the validation set prevents us from properly evaluating our approach, several lessons are learned that can be addressed in future work. As the predictions for the stage duration were quite accurate, most of the improvements can be made in the performance prediction part. The complexity of road cycling results in an interaction of several components that all have their influence on performance of a rider. Therefore, most importantly, it would be worthwhile to take a multi-dimensional approach. Thus, instead of the approach that is discussed here, where predictions are based on one particular factor, i.e., the altitude-distance profile, it would be interesting to include also other factors, such as the present physiological shape.

7 Conclusions

In this paper, we have considered time series regression challenges in professional road cycling. We have introduced a three-stage pipeline to determine rider-specific predictions based on the characteristics of the race route, with the final two steps of the developed pipeline also being applicable to any other time series regression problem. By applying LASSO regression and Subgroup Discovery, we have demonstrated that explicitly taking into account temporal effects increases the accuracy of the results.

For our road cycling example, we have shown that several interesting results can be gleaned from our approach. First, from our climb-descent model, we have demonstrated that the relationship between the slope of the terrain and speed of the cyclist is nontrivial and highly nonlinear. Above a certain threshold, the instances with negative slopes are dominated by hairpin corners in which the riders can only ride slowly. Second, we have calculated an effective profile that specifies the actual exertion that is experienced by the rider throughout the stage. Third, we have used a feature-aggregation based approach to construct a model for predicting the performance by calculating the time that is gained or lost with respect to the direct rivals of the rider. Therefore, this model can be used to identify stages with opportunities to gain time or potential hindrances. Moreover, our model describes the most important performance indicators, which in this case are several descriptions of changes in the effective profile in the last part of the race. Finally, we discussed the lessons that are learned from the deployment of the models in one of the Grand Tours in 2019 and we explained how the results of our approach can be relevant for a professional cycling team.

References

1. Atkinson, G., Davison, R., Jeukendrup, A., Passfield, L.: Science and cycling: current knowledge and future directions for research. J. Sports Sci. **21**, 767–787 (2003)
2. Bagnall, A., Lines, J., Bostrom, A., Large, J., Keogh, E.: The great time series classification bake off: a review and experimental evaluation of recent algorithmic advances. Data Min. Knowl. Disc. **31**, 606–660 (2017)
3. Bengio, Y., Courville, A.C., Vincent, P.: Unsupervised feature learning and deep learning: review and new perspectives (2012). CoRR abs/1206.5538
4. Box, G., Jenkins, G.M.: Time Series Analysis: Forecasting and Control. Holden Day, San Francisco (1976)
5. Deng, H., Runger, G., Tuv, E., Vladimir, M.: A time series forest for classification and feature extraction. Inf. Sci. **239**, 142–153 (2013)
6. Duivesteijn, W., Knobbe, A.J.: Exploiting false discoveries-statistical validation of patterns and quality measures in subgroup discovery. In: Proceedings of the 2011 IEEE 11th International Conference on Data Mining, ICDM'2011, USA, pp. 151–160. IEEE Computer Society (2011)

7. Faria, E., Parker, D., Faria, I.: The science of cycling: Factors affecting performance-part 2. Sports Med. (Auckland, N.Z.) **35**, 313–337 (2005)
8. Faria, E., Parker, D., Faria, I.: The science of cycling: physiology and training-part 1. Sports Med. (Auckland, N.Z) **35**, 285–312 (2005)
9. Fulcher, B.D., Jones, N.S.: Highly comparative feature-based time-series classification. IEEE Trans. Knowl. Data Eng. **26**, 3026–3037 (2014)
10. Herrera, F., Carmona, C.J., González, P., del Jesus, M.J.: An overview on subgroup discovery: foundations and applications. Knowl. Inf. Syst. **29**(3), 495–525 (2011)
11. https://www.procyclingstats.com (2019)
12. Knobbe, A.J., Orie, J., Hofman, N., van der Burgh, B., Cachucho, R.E.: Sports analytics for professional speed skating. Data Min. Knowl. Disc. **31**, 1872–1902 (2017)
13. de Leeuw, A.W., Meerhoff, R., Knobbe, A.J.: Effects of pacing properties on performance in long-distance running. Big Data **6**(4), 248–261 (2018)
14. Lucia, A., Hoyos, J.J., Chicharro, L.: Physiology of professional road cycling. Sports Med. (Auckland, N.Z.) **31**, 325–337 (2001)
15. Makridakis, S., Spiliotis, E., Assimakopoulos, V.: Statistical and machine learning forecasting methods: concerns and ways forward. PLoS One **13**(3), 1–26 (2018)
16. Meeng, M., Knobbe, A.J.: Flexible enrichment with cortana-software demo. In: Proceedings of BeneLearn, the annual Belgian-Dutch conference on machine learning, pp. 117–119 (2011)
17. Mörchen, F.: Time series feature extraction for data mining using DWT and DFT (2003)
18. Nanopoulos, A., Alcock, R., Manolopoulos, Y.: Feature-based classification of time-series data. Int. J. Comput. Res. **10**, 49–61 (2001)
19. Song, H.A., Lee, S.Y.: Hierarchical representation using NMF. In: Lee, M., Hirose, A., Hou, Z.G., Kil, R.M. (eds.) Neural Information Processing, pp. 466–473. Springer, Berlin Heidelberg (2013)
20. Tibshirani, R.: Regression shrinkage and selection via the lasso. J. Royal Stat. Soc. Ser. B (Methodol.) **58**(1), 267–288 (1996)
21. Wang, X., Smith-Miles, K., Hyndman, R.: Characteristic-based clustering for time series data. Data Min. Knowl. Disc. **13**, 335–364 (2006)
22. Wang, X., Wirth, A., Wang, L.: Structure-based statistical features and multivariate time series clustering. In: Seventh IEEE International Conference on Data Mining (ICDM 2007), pp. 351–360, October 2007
23. Xiong, M., Chen, J., Wang, Z., Liang, C., Zheng, Q., Han, Z., Sun, K.: Deep feature representation via multiple stack auto-encoders. In: Advances in Multimedia Information Processing-PCM 2015, pp. 275–284. Springer International Publishing, Cham, September 2015

Author Index

Printed in the United States
By Bookmasters